Lecture Notes in Mathematics

Edited by A. Dold and B. Eckmann

1007

Geometric Dynamics

Proceedings of the International Symposium
held at the Instituto de Matématica Pura e Aplicada
Rio de Janeiro, Brasil, July – August 1981

Edited by J. Palis Jr.

Springer-Verlag
Berlin Heidelberg New York Tokyo 1983

Editor

J. Palis Jr.
IMPA – Instituto de Matématica Pura e Aplicada
Rua Luiz de Camoes 68, Rio de Janeiro, Brasil

AMS Subject Classifications (1980): 58

ISBN 978-3-540-12336-1 Springer-Verlag Berlin Heidelberg New York Tokyo
ISBN 978-0-387-12336-3 Springer-Verlag New York Heidelberg Berlin Tokyo

Library of Congress Cataloging in Publication Data. Main entry under title: Geometric dynamics. (Lecture notes in mathematics; 1007) English and French. 1. Global analysis (Mathematics)--Congresses. I. Palis Júnior, Jacob. II. International Symposium on Dynamical Systems (1981: Rio de Janeiro, Brazil). III. Instituto de Matématica Pura e Aplicada (Brazil). IV. Series: Lecture notes in mathematics (Springer-Verlag); 1007.
QA3.L28 no. 1007 [QA614] 510s [514'.74] 84-14723
ISBN 978-0-387-12336-3 (U.S.)

2146/3140-543210

DEDICATED TO MAURICIO MATOS PEIXOTO

ON HIS SIXTIETH BIRTHDAY

F O R E W O R D

These are the Proceedings of an International Symposium on Dynamical Systems, that took place at the Instituto de Matemática Pura e Aplicada (IMPA), Rio de Janeiro, in July-August, 1981.

The Meeting was to celebrate the opening of IMPA's new building "Edifício Lélio Gama" , a dream of more than 10 years that came true.

Many friends (mathematicians) contributed to the success of the Meeting. Financial support came from many sources especially CNPq and FAPESP (Brasil), Brasilian Universities, CNRS (France), GMD (Germany), British Council, NSF (U.S.A), Internationa Mathematical Union (IMU) and Research Councils of Argentina, Chile, Mexico and Venezuela.

We are thankful to all and, in particular, to Lindolpho de Carvalho Dias, the Director of IMPA.

C O N T E N T S

AN INDEX FOR THE GLOBAL CONTINUATION OF RELATIVELY ISOLATED SETS OF PERIODIC ORBITS

Kathleen T. Alligood
John Mallet-Paret
James A. Yorke

§1. Introduction

Since Poincaré the techniques of fixed point theory have been used to study the continuation of periodic orbits of parametrized differential equations, (i.e., systems

$$\frac{dx}{dt} = f(x,\lambda) \quad , \tag{1.1}$$

where M^n is a smooth n-manifold (usually R^n) and $f:M^n \times R \to M^n$ is C^1). We generally write R^n but the results are the same for M^n. The first fixed-point type index for (periodic) orbits was developed by F. B. Fuller [10] in 1967. With it, he obtained local continuation results for orbits of non-zero index. This index (the Fuller index) is invariant under orbit bifurcations as λ is varied. A periodic orbit can be thought of as being in (x,λ)-space or alternatively, as in Fuller's approach, as being in (x,λ,τ)-space, where τ is a multiple of the minimum period. His orbit index could take on different values for each multiple of the period of the orbit in (x,λ)-space. (An example in Sec. 2 will indicate more clearly some limitations of this approach.) C. Conley [9] has an index for isolated invariant sets (the "generalized Morse" index); however, he addresses a different type of problem since there is no way of determining whether the invariant sets specifically contain periodic orbits. A new orbit index (the ϕ-index) for orbits in (x,λ)-space was developed in [12]. Like the Fuller index, this orbit index was invariant under orbit bifurcations as λ was varied. In addition, it was coupled with a center index for stationary points so that the sum of orbit indices plus the center indices of stationary points was invariant even in the presence of Hopf bifurcations. As originally conceived in [12] the orbit index (and center index) is designed only for generic situations. By introducing the concept of the virtual periods, it was found in [4] that global results concerning components of orbits in (x,λ)-space could be obtained for a general C^1 system by considering such a system as the limit of generic ones. Because of the difficulties in this approximation approach, that paper concentrated on global results that could be obtained without reference to any kind of orbit index. The purpose of this paper is to apply this approach of approximating f by generic vector fields to the problem of defining the orbit index for general f and then to demonstrate the use of this index. We note that while Fuller defined his index using homology theory, it also can be defined and shown to be an invariant by using generic approximations (see [7]). In some cases, the orbit index gives more information about continuation than the Fuller index. As an example, we investigate vector fields on S^3, a setting which presented Fuller with his primary application. The Fuller index also has been used for studying global Hopf bifurcation, and it will be shown in [5] how sharper results can be obtained using the orbit index. When n=3 additional global continuation results can be obtained (see [2]) because of the manner of inter-linking of orbits and unstable manifolds that must occur.

In Section 4, we demonstrate that for dimension n=3 or 4, global continuability results can be stated in terms of periods rather than virtual periods. This result was found in collaboration with S. N. Chow.

§2. Background

A periodic orbit γ is viewed as a subset of $R^n \times R$ (or $M^n \times R$). For a point (x_o,λ_0) on a periodic orbit γ of (1.1) at $\lambda=\lambda_0$, let T be the Poincaré return map, defined on an n-1 disk in $R^n \times \{\lambda_0\}$ transverse to γ at (x_o,λ_0). A <u>multiplier</u> of γ is an eigenvalue of $D_x T(x_o,\lambda_0)$. In the following, orbit will always refer to a periodic orbit whose minimum period is non-zero. "Period" will mean minimum period. First, we describe the continuability of orbits of generic systems. The generic class considered here is discussed in detail in [4], and for non-Hamiltonian systems permits only three types of orbits, briefly described as follows:

(0) orbits having no multipliers which are roots of unity;

(I) bifurcation orbits having a simple multiplier equal to +1 and no other multipliers that are roots of unity, and from which precisely two arcs of orbits emanate; and

(II) bifurcation orbits having a simple multiplier equal to -1 and no other multipliers that are roots of unity, and from which three arcs of orbits emanate -- one arc with orbits whose periods are approximately twice as long as those on the other two arcs.

The bifurcations at type I orbits are called "jug handle" bifurcations; those at type II orbits are called period-doubling bifurcations. We let K be the set of those C^3 functions $f: R^n \times R \to R^n$ such that every periodic orbit of $\dot{x}=f(x,\lambda)$ is of type 0, I, or II. The set K is residual in the C^3 topology. (See the Appendix to this paper for justification of this result, which is based on the local results of Brunovský [6] coupled with the global methods of Peixoto [13]. See also Sotomayor [16].) In the following, the term "generic" is applied not only to the properties of vector fields in K, but also more loosely to the vector fields themselves and to families (i.e., components) of orbits of these vector fields.

Let γ be an orbit of (1.1), where $f \in K$. Suppose that γ has no multipliers that are roots of unity, has k^+ multipliers in $(1,\infty)$ and k^- multipliers in $(-\infty,-1)$, counted with multiplicities. The orbit index (or ϕ-index) $\phi(\gamma)$ is defined to be

$$\phi(\gamma) = \begin{cases} (-1)^{k^+} & \text{if } k^- \text{ is even} \\ 0 & \text{if } k^- \text{ is odd.} \end{cases}$$

An orbit with ϕ-index 0 is called a <u>Möbius orbit</u>. (For such an orbit, the unstable manifold is necessarily non-orientable.)

The definitions of global continuability for an orbit of (1.1) are based on the natural ways in which components of orbits can terminate, including the case where a family of orbits converges to a "center" of (1.1). A center is a stationary point (x_o,λ_o) for which $D_x f(x_o,\lambda_o)$ has some non-zero imaginary eigenvalues.

<u>Definition 2.1.</u> For a periodic orbit γ of (1.1), let Γ be the component of orbits of (1.1) containing γ. We say γ is <u>P-globally continuable</u> if either of the following conditions hold:

(a) $\Gamma-\gamma$ is connected; or

(b) each of two components Γ_i $(i=1,2)$ of $\Gamma-\gamma$ satisfies one of the following:

(1) Γ_i is unbounded in (x,λ)-space,

(2) $\bar{\Gamma}_i$ contains a center, or

(3) the periods of orbits in Γ_i are unbounded.

In the case of fixed points, local continuability in fact implies global continuability; see [1]. The same result, however, does not hold for orbits, even in the generic case. An example presented in [3] demonstrates the existence of orbits which are locally continuable, (i.e., for which the fixed point index of the Poincaré map and the Fuller index are non-zero), but which do not satisfy any of the conditions of Defn. 2.1. The construction of the example can be outlined as follows (see Figure 1):

Let $g:R^4 \to R^4$ be a C^1 function such that $\dot{x}=g(x)$ has a Möbius orbit solution γ. We define a homotopy $f(\cdot,\lambda)$ of g such that:

(1) $f(\cdot,\lambda_0) = g$;

(2) γ is contained in a family Γ of Möbius orbits for λ near λ_0;

(3) a second family Γ_2 of orbits (with approximately twice the period) bifurcates from Γ at $\lambda = \lambda_1$; and

(4) the family Γ_1 (the low-period continuation of Γ for $\lambda > \lambda_1$) and the family Γ_2 coalesce and annihilate each other at $\lambda = \lambda_2$.

The only orbits contained wholly within some ε-neighborhood of this component C of orbits are those in the families described. In addition, this example persists under small C^1 perturbations and can be made real analytic.

An analysis of this example via the Fuller index demonstrates the advantage of a minimum-period approach to orbit continuation theory. The Fuller index of the set of orbits contained within a given bounded set W in (x,τ)-space (where τ is the period variable) is invariant under homotopies of the underlying vector field -- provided no orbits cross the boundary of W throughout the homotopy. For an appropriate set W in $R^4 \times R$, the initial Möbius orbit γ of the example above has non-zero index. It appears on the surface that W could be chosen so that throughout the homotopy described, no orbits would cross the boundary of W. Of course, this would imply that for $\lambda > \lambda_2$, W must continue to contain orbits, which is not the case. The apparent contradiction here results from the fact that when we do not deal specifically with the minimum periods of orbits, the component C containing γ is unbounded in τ-space. Refer again to Fig. 1 and consider a path of orbits beginning with γ which follows the low-period branch through the period doubling bifurcation at λ_1, goes around the loop of orbits to λ_1 again (at which point the periods of orbits have doubled), as in Fig. 2 top, and continues back into the loop -- this time with double the minimum period of those orbits, as in Fig. 2 bottom. Continuing in this way, each time the path goes around the loop it contains orbits with twice the

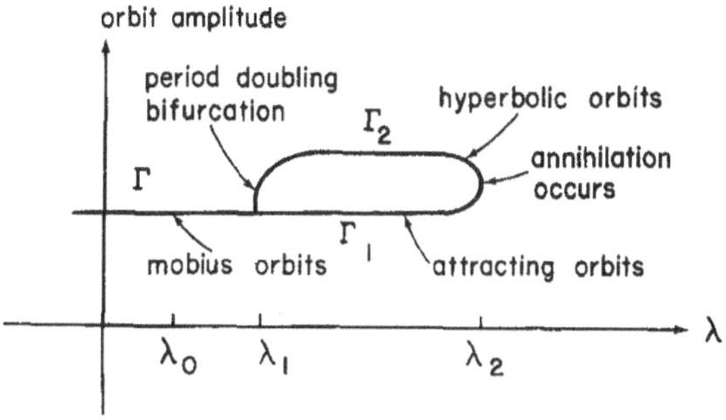

Figure 1

A schematic diagram of the Mobius example is shown, in which each point on the 1-dimensional branched curve represents a periodic orbit.

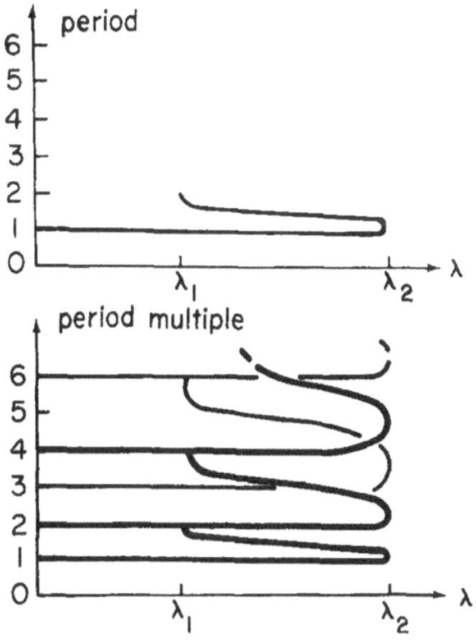

Figure 2

Two representations of the Mobius example are shown. In the top figure, each point represents a periodic orbit in $\lambda \times$ (minimum) period space; in the bottom figure, the same orbits are represented in $\lambda \times$ period multiple space.

periods of the same orbits on the previous loop. The component in Fig. 2 bottom is unbounded. Thus there is a path of orbits in C along which the periods are unbounded.

The P-global continuability of non-Möbius orbits in generic systems is proved as a special case of the index theory developed in [12]. A proof of this theorem which does not use index theory is given in [4].

Theorem 2.2. [12] Let γ be a periodic orbit of (1.1) where $f \in K$. If $\phi(\gamma) \neq 0$, then γ is P-globally continuable.

The global continuability of orbits in the general case, dropping genericity assumptions, is addressed in [4]. In order to obtain continuation results for orbits of the general system (1.1), we consider a sequence of functions $\{f_i\}_{i \in N}$, such that $f_i \in K$ for each i, and $\lim_{i \to \infty} f_i = f$; and we relate the behavior of families of orbits of (1.1) to that of the nearby generic families.

In this case, however, a different phenomenon occurs in the limit case. Consider a bounded sequence $\{\gamma_i\}$ of individual orbits, where γ_i is an orbit of $\dot{x} = f_i(x,\lambda)$ with period τ_i, for each i. If $\{\tau_i\}$ is bounded, then $\lim_{i \to \infty} \gamma_i$ is an orbit or stationary point of (1.1). (This limit is the set of all limit points of sequences where the ith point of the sequence lies on γ_i). In the former case, some subsequence $\{\tau_{i_k}\}$ of periods converges to an integer multiple of τ, the period of the limit orbit γ. The following definition is central to the analysis of the behavior of the set of limit orbits:

Definition 2.3. Let A be $D_x T(x_0)$, where T is the Poincaré map associated with an orbit γ of period τ at x_0. If there exists a point $y \in R^{n-1}$ with $y, Ay, \ldots, A^{m-1}$ distinct, but $A^m y = y$, for some $m \geq 1$, then we say $\bar{\tau} = m\tau$ is a virtual period of γ, and m is the order of the virtual period $\bar{\tau}$.

For $m = 1$, the definition is satisfied by $y = 0$; thus the period is a virtual period of the orbit. For $m > 1$, $y \neq 0$. For type 0 and I orbits, the period is the only virtual period. Notice that each orbit can have only a finite number of virtual periods. It was shown in [8], in the situation described above where $\gamma_i \to \gamma$, that if $\lim \tau_i$ exists, the limit is a virtual period of γ. We incorporate the idea of virtual period into the definition of global continuability for an orbit γ of (1.1), by replacing condition (b3) of Defn. 2.1 (P-global continuability) with the following condition:

(b3)' the virtual periods of orbits in Γ_i are unbounded.

A generalization of Thm. 2.2 for general C^1 systems is proved in [4]:

<u>Theorem 2.4.</u> Let γ be a periodic orbit of (1.1), and let T be the Poincaré map of γ at (x_o,λ_o). Assume that $(D_xT(x_o,\lambda_o))^j = I$ is non-singular for $j \geq 1$, and that γ is not a Möbius orbit. Then γ is globally continuable.

For generic systems, the only virtual periods are the actual periods plus (in the case of type II orbits only) twice the actual period. It is easily seen that if the virtual periods of orbits of a generic family go to infinity, so do the actual periods. Thus, in the generic case, Thm. 2.4 reduces to Thm. 2.2.

If a type 0 non-Möbius orbit γ satisfies <u>only</u> condition (b3)' for global continuability, then additional information as to the behavior of the virtual periods of orbits on the component containing γ is known, (see Prop. 4.5 in [4]):

<u>Proposition 2.5.</u> [4] For a type 0 non-Möbius orbit γ, assume that one of the two components of Γ-γ (as in Defn. 2.1), say Γ_1, has the property that the virtual periods of orbits on Γ_1 are unbounded and that no other conditions of Defn. 2.1 are satisfied by Γ_1. Then if α_1 and $\alpha_2 > \alpha_1$ are sufficiently large, there will be a compact, connected set S Γ_1 such that each orbit in S has virtual period in $[\alpha_1, 2\alpha_2]$. Furthermore, for each $\alpha \varepsilon [\alpha_1,\alpha_2]$ there is an orbit in S with virtual period in $[\alpha, 2\alpha)$.

§3. The Orbit Index: General Case

In this section we extend the ϕ-index for orbits of generic systems to an index for certain sets of orbits of general C^1 systems. We define global continuability for these sets and prove a theorem analogous to Theorem 2.4: if the index of a set is non-zero, then the set is globally continuable.

Let Q be a set of orbits of

$$\dot{x} = f(x) \ , \tag{3.1}$$

where $f:R^n \to R^n$ is C^1. We say Q is relatively isolated if Q is compact, if the set of virtual periods of orbits in Q is bounded, and if Q is isolated from other low-period orbits in the following sense: if there exists a sequence $\{\beta_i\}_{i\in N}$ of orbits of (3.1) not in Q which converge to an orbit β in Q, then the periods of the orbits β_i go to infinity as $\{\beta_i\}$ converges to β.

Proposition 3.2. Let Q be a relatively isolated set of orbits of (3.1), with $p_o = \max\{\tau : \tau$ is a virtual period of an orbit in Q$\}$. Let p_1 and z be real numbers such that $p_o < p_1 < 2p_o$ and $z > p_1$. Then there exists a neighborhood N of Q in R^n such that no orbit in N has period in $[p_1, z]$.

Proof. Let $q = \min\{\tau : \tau$ is the period of an orbit in Q$\}$. If $\phi(t,x)$ is the solution of (3.1) through x, we define

$$F(x) = \min_{q \le t \le z} |\phi(t,x) - x| .$$

The set of zeroes of F are points on periodic orbits of (3.1). Loosely speaking, F measures how close ϕ is to being periodic, for periods is $[q,z]$. Notice that F is constant along solution curves of (3.1). Let

$$M_\varepsilon = \{x \in R^n : |F(x)| < \varepsilon\} ,$$

and let N_ε be the union of components of M_ε containing orbits of Q. Finally, let $\tau_\varepsilon = \min\{\tau : \tau$ is the period of an orbit in $N_\varepsilon - Q\}$. Since Q is relatively isolated, $\lim_{\varepsilon \to 0} \tau_\varepsilon = \infty$. Otherwise, there exists a sequence $\{\varepsilon_i\}_{i\in N}$, where $\lim_{i\to\infty} \varepsilon_i = 0$, and and a bound $b > 0$ for which $N_{\varepsilon_i} - Q$ contains an orbit γ_i of (3.1) of period less than b. Since the periods of $\{\gamma_i\}$ are bounded, some subsequence of $\{\gamma_i\}$ converges to an orbit in Q, contradicting the fact that Q is relatively isolated. Thus there exists an $\varepsilon^* > 0$ such that $\tau_{\varepsilon^*} > z$. Since all orbits in Q have periods less than p_1, by letting $N = N_{\varepsilon^*}$, the proposition is proved. □

In the following, we define an index for relatively isolated sets of orbits.

Let Q be a relatively isolated set of orbits of (3.1), with p_o and p_1 as in Prop. 3.2, and $z > 4p_o$; and let $g : R^n \times R \to R^n$ be a C^1 function such that $g(\cdot, 0) = f$ and

$g\varepsilon K$ for $\lambda > 0$. For a neighborhood $N \subset R^n$ (as in Prop. 3.2) and a real number $\lambda_0 > 0$, let

$$J_{\lambda_0} = \Sigma\phi(\gamma) \quad ,$$

summing over the orbits γ of $x = g(x,\lambda_0)$ such that $\gamma \subset N \times \{\lambda_0\}$, and the period of γ is less than p_1. (This set of orbits is necessarily finite.)

We define the index of Q, i(Q), as follows:

$$i(Q) = \lim_{i\to\infty} J_{\lambda_i} \quad ,$$

where $\{\lambda_i\}_{i\varepsilon N}$ is a sequence of real numbers such that $\lim_{i\to\infty} \lambda_i = 0$.

Proposition 3.3. The limit in the above definition exists and is independent of the choice of g.

Proof. We assume that z is sufficiently large that there are no stationary points of

$$\dot{x} = g(x,\lambda) \tag{3.4}$$

in $N \times \{0\}$. Then there exists a $\lambda_1 > 0$ such that all orbits of (3.4) within $N \times (0,\lambda_1]$ have periods less than p_1 or greater than z. Otherwise, for some $\lambda_i \to 0$, there exists a sequence $\{\gamma_i\}_{i\varepsilon N}$ of orbits (where γ_i is an orbit of $x = g(x,\lambda_i)$) converging to an orbit γ of $x = g(x,0)$ in $N \times \{0\}$. The limit orbit γ necessarily has a virtual period in $[p_1,z]$. Since p_0 is the maximum of virtual periods of orbits in Q, γ must be in $N - Q$; however, then (by Prop. 3.2) the period of γ, and hence all virtual periods of γ, are greater than z, a contradiction.

Since all orbits of $\dot{x} = g(x,0)$ which lie in $\partial N \times \{0\}$ have periods in (p_1,∞), there exists a $\lambda_2 > 0$ such that all orbits of (3.4) for $\lambda \le \lambda_2$ which intersect $\partial N \times [0,\lambda_2]$ and lie in $N \times [0,\lambda_2]$ have periods greater than p_1. Let $\lambda_* = \min\{\lambda_1,\lambda_2\}$; and let $\{\Gamma_k\}_{1\le k\le m}$ be the set of components of orbits in $N \times (0,\lambda_*]$ containing orbits of period less than p_1.

We argue that all orbits in $\bigcup_{k=1}^{m} \Gamma_k$ have period less than p_1. For generic components -- which are in fact branched paths of orbits -- the periods of orbits along a path change continuously or by jumps of a factor of two (or one-half) at period doubling (or halving) bifurcations. In $N\times(0,\lambda_*]$, the periods along paths cannot continuously cross p_1. Since $p_1 < 2p_1 < 4p_0 < z$ and there are no orbits in $N\times(0,\lambda_*]$ with periods in $[p_1,z]$, no branches containing orbits of periods greater than p_1 can result from period-doubling bifurcations at orbits of period less than p_1. Thus all orbits in $\bigcup_{k=1}^{m}\Gamma_k$ have periods less than p_1.

Mallet-Paret and Yorke have shown in [12] that the sum of the orbit indices of orbits in $N\times(0,\lambda_*)$ of period less than p_1 is constant (as a function of λ), provided no orbits cross ∂N and no periods along components of these orbits cross p_1. Thus for i sufficiently large so that $\lambda_i<\lambda_*$, $J\lambda_i$ is constant. Hence $\lim_{i\to\infty} J\lambda_i$ exists.

In order to show that for large i, $\lim J\lambda_i$ is independent of the choice of generic perturbation, we let h_1 and h_2 be two homotopies of f (as above), such that $k_1\varepsilon K$ and $h_2\varepsilon K$, for $\lambda>0$. Suppose that J_λ is constant for orbits of $\dot{x}=h_1(x,\lambda)$ for $0<\lambda\le\lambda_1$, and for orbits of $\dot{x}=h_2(x,\lambda)$ for $0<\lambda\le\lambda_2$. For ease of conceptualization, let $h_2^*(x,\lambda)=h_2(x,-\lambda)$. Then $h=h_1\cup h_2^*$ is a C^0-map defined on $R_n\times[-\lambda_2,\lambda_1]$. We assume that all orbits of $\dot{x}=h(x,\lambda)$ contained in $(N\times\{-\lambda_2\})\cup(N\times\{\lambda_1\})$ of period less than p_1 are of type 0. (There are necessarily at most a finite number of such orbits.) Since the set K of maps is also dense in the C^0-topology, there exist $g\varepsilon K$ such that $\|g-h\|<\varepsilon$ for artitrarily small $\varepsilon>0$. Let $I=[-\lambda_2,\lambda_1]$. Any nearby perturbation of the vector field h will have the same number of orbits in $(N\times\{-\lambda_2\})\cup(N\times\{\lambda_1\})$ of period less than p_1 (as h), (with the orbits perhaps slightly shifted in position); furthermore, the ϕ-indices of these orbits will be the same as for h. Since p_0 is the maximum of virtual periods of orbits of (3.1) in $N\times\{0\}$ with period less than p_1, for small λ (here we assume for $\lambda\varepsilon I$) the corresponding orbits of $\dot{x}=h(x,\lambda)$ will have virtual periods less than p_1. Thus for g sufficiently close to h, no orbits of $\dot{x}=g(x,\lambda)$ contained $N\times I$ have periods in $[p_1,z]$, and none which also intersect $\partial N\times I$ have periods less than p_1 (since this was the case for h). For such a $g\varepsilon K$, let $\{\Gamma_i\}_{1\le i\le k}$ be the set of components of orbits of $\dot{x}=g(x,\lambda)$ in $N\times I$ containing the orbits of period less than p_1. As in the argument given above, all orbits in $\bigcup_{i=1}^{k}\Gamma_i$ have period less than p_1. Since then none of the Γ_i intersect $\partial N\times I$, the sum of the ϕ-indices of orbits on these components is constant (as a function of λ) Thus $J\lambda_1 = J_{-\lambda_2}$. □

Definition 3.5. Let Q be a set of orbits of (1.1) at $\lambda = \lambda_0$. We say Q is globally continuable if some component B of orbits of (1.1) containing orbits of Q satisfies one of the following:

(1) $\overline{B}$ contains a center;

(2) B is unbounded in (x,λ)-space; or

(3) the virtual periods of orbits in B are unbounded.

Theorem 3.7. Let Q be the set of all orbits of (1.1) at $\lambda = 0$. Assume that Q is compact and that the virtual periods of orbits in Q are bounded. If $i(Q) \neq 0$, then Q is globally continuable.

Since the proof of this theorem is similar to that of the global continuability theorem for type 0 non-Möbius orbits of (1.1), we present an outline of the proof here and refer the reader to [4] for details of the constructions and limit arguments.

Proof. Suppose that all components $\{B_j\}_{1 \leq j \leq n}$ of orbits of (1.1) containing orbits of Q are bounded in (x,λ)-space. Assume further that, for each j, $\overline{B}_j$ does not contain a center, and that the virtual periods of orbits on each B_j are bounded. Let $p_o = \max\{\tau : \tau$ is a virtual period of an orbit in $\bigcup_{j=1}^{n} B_j\}$ and let $z > 4p_o$ and $p_o < p_1 < 2p_o$. As in [4], we let W be a neighborhood of $\bigcup_{j=1}^{n} B_j$ such that

(1) W is bounded;

(2) ∂W contains no orbits of (1.1) with virtual periods less than p_1;

(3) W contains no stationary points of (1.1);

(4) W contains no orbits of (1.1) with virtual periods in $[p_1, z]$; and

(5) there exists a $\rho > 0$ such that orbits and stationary points of $\dot{x} = g(x,\lambda)$ satisfy conditions (2)-(4) above, provided $\|f-g\| < \rho$ in the C^1 topology.

We choose $g \in K$ such that $\|f-g\| < \rho$ and $\Sigma\phi(\gamma) \neq 0$, summing over all orbits γ of $\dot{x} = g(x,0)$ of period less than p_1. The analysis of generic systems in [12] implies that, since $\Sigma\phi(\gamma) \neq 0$, some component Γ of orbits of $\dot{x} = g(x,\lambda)$ containing such an orbit γ of $\dot{x} = g(x,0)$ must either intersect ∂W or contain orbits in W whose periods are unbounded. Since $g \in K$, the periods along paths of orbits in Γ change continuously or by jumps of 2 or 1/2. If the periods of Γ go to infinity in W, some orbit in W must have period in $[p_1, z]$, a contradiction. Hence Γ intersects ∂W. However, again we reach a contradiction, since, (arguing as above), Γ must then intersect ∂W in an orbit of period less than p_1. Thus Q is globally continuable.

Remarks. A stronger version of Theorem 3.7 is possible: Let Q be a relatively isolated set of orbits such that $i(Q) \neq 0$. If for some set $\{B_i\}_{1\le i\le m}$ of components of orbits of (1.1), $Q = (\bigcup_{i=1}^{m} B_i) \cap (R^n \times \{0\})$, then Q is globally continuable.

In addition, under either set of hypotheses, it follows that Q is globally continuable both to the left and right, (i.e., that each of two families of orbits -- one extending into $R^n \times (0,\infty)$ and one into $R^n \times (-\infty,0)$ from Q -- satisfies one of the conditions of Defn. 3.5).

For Prop. 3.2, the corresponding result with "period in $[p_1,z]$" replaced by "virtual period in $[p_1,z]$" also holds, however the proof is more technical.

Since the index calculation and limit techniques are local in nature, the results of this section hold for manifolds as well; i.e., where $g:M^n \times R \to M^n$ is a C^1 map on a C^∞-manifold M^n.

§4. Global Continuation in Dimensions 3 and 4.

An orbit of (1.1) has at most a finite number of virtual periods -- the maximum possible number being determined by the dimension of the x-space. In this section we examine the consequences of this fact on global continuation and obtain a stronger version of Thm. 3.7 in dimensions 3 and 4.

Lemma 4.1. A periodic orbit of $\dot{x} = f(x)$ with $x \in R^3$ or R^4 has at most one virtual period in addition to the period τ of the orbit.

Proof. Let $\mu_1, \ldots, \mu_{n-1}$ denote the n-1 characteristic multipliers of the orbit where n=3 or 4. We note the product $\mu_1 \ldots \mu_{n-1}$ is positive, since the Poincaré map is orientation preserving. Now, 2τ is a virtual period if and only if $\mu_i = -1$ for some i. Also, if $p \ge 3$ then $p\tau$ is a virtual period if and only if for some $i \neq j$

$$\mu_i = \bar{\mu}_j$$
$$\mu_i^q \neq 1 , \qquad 1 \le q < p ,$$
$$\mu_i^p = 1 .$$

This is a consequence of the fact that $n \le 4$. Finally, note that 2τ and $p\tau$ cannot both be virtual periods, otherwise $n = 4$ and there would be three characteristic multipliers, -1, $e^{i\theta}$, $e^{-i\theta}$ with a negative product $\mu_1\mu_2\mu_3 = -1$.

Theorem 4.2. Let Q be the set of orbits of (1.1) at $\lambda = 0$, as in Thm. 3.7, where $n = 3$ or 4. If $i(Q) \neq 0$, then Q is P-globally continuable; i.e., some component B of orbits of (1.1) in $R^n \times [0,\infty)$ containing orbits of Q satisfies one of the following:

(1) $\overline{B}$ contains a center;

(2) B is unbounded in (x,λ)-space; or

(3) the periods of orbits in B are unbounded.

Of course $R^n \times [0,\infty]$ can be replaced by $R^n \times (-\infty,0]$.

The proof of this theorem depends on the following lemma:

Lemma 4.3. Let $n = 3$ or 4, and let Q be the set of orbits of (1.1) at $\lambda = 0$, as in Thm. 3.7. If there exists a component B of orbits in $R^n \times [0,\infty)$ containing orbits of Q for which the virtual periods are unbounded, and no other conditions of Defn. 3.6 (global continuability) are satisfied by any component of orbits in $R^n \times [0,\infty)$ containing orbits of Q, then the periods of orbits in B are also unbounded.

Proof. Suppose all periods of orbits in B lie between $T_1 > 0$ and $T_2 > T_1$. Choose

$$\alpha_1 > T_2$$

$$\alpha_2 \ge \frac{2T_2}{T_1}\alpha_1$$

sufficiently large that Prop. 2.5 may be applied: thus there is a connected set $S \subseteq B$ of orbits each of which has a virtual period in $[\alpha_1, 2\alpha_2]$. Because $\alpha_1 > T_2$, this virtual period is different from the period, and by Lemma 4.1 is uniquely determined. Denote the periods and virtual periods by

$$\tau(x,\lambda) \quad \text{and} \quad k(x,\lambda)\,\tau(x,\lambda) \quad ,$$

where $(x,\lambda) \in S$ and $k(x,\lambda) \ge 2$. Further, by Prop. 2.5 there exist $(x_i,\lambda_i) \in S$ $i = 1,2$, with

$$k(x_i,\lambda_i)\,\tau(x_i,\lambda_i) \in [\alpha_i, 2\alpha_i)$$

and this implies $k(x,\lambda)$ is not constant on S:

$$k(x_1,\lambda_1) < \frac{2\alpha_1}{\tau(x_1,\lambda_1)} \leq \frac{2\alpha_1}{T_1} \leq \frac{\alpha_2}{T_2} ,$$

$$k(x_2,\lambda_2) \geq \frac{\alpha_2}{\tau(x_2,\lambda_2)} \geq \frac{\alpha_2}{T_2} .$$

We will obtain a contradiction by showing $k(x,\lambda)$ is constant on S.

First, we show periods are continuous on S, that is, τ restricted to S is a continuous function. If $(x_o,\lambda_o)\varepsilon S$, then $\tau(x,\lambda)$ is near either $\tau(x_o,\lambda_o)$ or $k(x_o,\lambda_o)\tau(x_o,\lambda_o)$ for $(x,\lambda)\varepsilon S$ near (x_o,λ_o). But the latter is impossible, for

$$k(x_o,\lambda_o)\,\tau\,(x_o,\lambda_o) \geq \alpha_1 > T_2$$

violating the bounds on the periods. Thus τ is continuous on S.

Now let $S_m \subseteq S$ be those points of S where $k(x,\lambda) = m$. Because τ is continuous, the characteristic multipliers $\{\mu_1, \ldots, \mu_{n-1}\}$ vary continuously, as a set, and hence S_m is closed relative to S. Also, S is a disjoint union of finitely many S_m, by the bounds on the virtual periods; hence each S_m is open relative to S. Thus $S = S_m$ for some m, since S is connected. This means $k(x,\lambda)$ is constant on S, a contradiction. □

Remarks. In the proof of Lemma 4.3, we use a slightly more general version of Prop. 2.5 than is stated in Section 2. The results of this proposition hold for any component B of orbits of (1.1) on which the virtual periods are shown to be unbounded by approximating (1.1) with generic systems which contain P-globally continuable orbits (of bounded periods) which converge to orbits of B. Thus, in some cases, particularly in dimension $n = 3$, the conclusion of Thm. 4.2 can be proved even when $i(Q) = 0$. In [2] it is shown for $f\varepsilon K$ that if there is only one orbit at $\lambda = 0$, an orbit of type 0, then even if that orbit is a Möbius orbit, it is P-globally continuable.

§5. The Seifert Example

For each $\lambda \geq 0$, let k be a C^1 vector field on S^3, continuously dependent on λ. In particular, let h_o be the vector field of the Hopf flow, whose trajectories are all (periodic) orbits with period 2π. Seifert [15] showed that for λ sufficiently small, h must have orbits. Fuller [10] gave a simpler proof of this result based on his index. Schweitzer [14] and Harrison [11] have, however, produced examples of C^1 and C^2 vector fields, respectively, on S^3 that have no orbits and no stationary points. Clearly then, C^1-homotopies h_λ of h_o exist such that for some $\lambda_1 > 0$, h_{λ_1} has no orbits. In the following proposition, we analyze the global continuability of the orbits of the Hopf flow under such a homotopy.

Proposition 5.1. For each $\lambda \geq 0$, let $h_\lambda = h(\cdot,\lambda)$ be a vector field on S^3 (as above), such that $h(\cdot,0)$ is the vector field of the Hopf flow and

$$\frac{dx}{dt} = h(x,\lambda) \tag{5.2}$$

has no orbits for $\lambda = 1$. Let Q be the set of orbits of (5.2) at $\lambda = 0$, and let Γ be the component of orbits of (5.2) containing Q. Then either (5.2) has at least one center in $\bar{\Gamma}$ or the set of periods of orbits of Γ is unbounded.

Proof. In order to calculate the index of Q, we define a homotopy f of $h(\cdot,0) = h_o$ such that $f(\cdot,0) = h_o$ and $f(\cdot,\lambda) \varepsilon K$ for $\lambda > 1$. (This construction appears in Fuller [10].) Notice that f has no stationary points for λ near 0. Let ν be the vector field on S^2 which has zeroes at the north and south poles, v_o and v_1, respectively, and moves all other points down the meridians towards v_1. Considering S^3 as a fiber bundle over S^2, lift ν to a vector field $\bar{\nu}$ on S^3.

Let

$$f(x,\lambda) = h(x,0) + \bar{\nu}(x).$$

Then for each $\lambda > 0$,

$$\dot{x} = f(x,\lambda) \tag{5.3}$$

has exactly 2 orbits -- one over each of v_o and v_1 in S^2. For a particular $\lambda_1 > 0$, let γ_o and γ_1 denote these orbits. The orbit γ_o is a repellor (with orientable unstable manifold), and the orbit γ_1 is an attractor. Thus $\phi(\gamma_o) = (-1)^2 = 1$ and $\phi(\gamma_1) = (-1)^o = 1$. Summing and taking the limit as $\lambda_1 \to 0$, we obtain $i(Q) = 2$.

By Theorem 3.7, Q is globally continuable. Since Γ is bounded and $\bar{\Gamma}$ contains no centers, the virtual periods of orbits in Γ are unbounded. By Thm. 4.2, the actual periods of orbits in Γ are unbounded. □

APPENDIX

Systems Whose Orbits are all Types 0, I, and II are Generic

Let K be the set of functions $f \in C^3(R^n \times R, R^n)$ for which each orbit of $\dot{x} = f(x,\lambda)$ is of type 0, I, or II. Our aim is to provide a justification for the statement that K is residual in $C^3(R^n \times R, R^n)$. Most of the ingredients for this result already exist: results of Brunovský [6a,6b] consider fixed points of a family of maps $x \to T(x,\lambda)$. A proof given by Peixoto [13] of the Kupka-Smale Theorem indicates how to pass from local results for maps to global results for flows, using the Poincaré map for a periodic orbit. Peixoto does not consider parameterized families, but only a single equation $\dot{x} = f(x)$; his techniques, however, carry over with minor modifications. Below we shall outline his techniques and how to modify them so that they may be applied to the Brunovský results.

We first present some of Brunovský's results in a slightly modified form adapted to our needs. The proofs of these generally can be found in [6b], although the reader may have to supply minor details and change some notation. A brief summary of much of this is given in [6a].

To begin, consider maps

$T : B \times (\alpha,\beta) \to R^n$, where

$B \subseteq R^n$ is an open ball (later we let $T:R^n \times R \to R^n$), and

$T(x,\lambda)$ is of class C^r, $3 \le r < \infty$.

A fixed point $x_o = T(x_o,\lambda_o)$ is the type 0', I' or II' according to the following definitions. It is of type 0' if and only if no eigenvalue of the Jacobian $A = D_x T(x_o,\lambda_o)$ is a kth root of unity, for any $k \ge 1$. It is of type I' provided

(Ia) 1 is a simple eigenvalue of A;

(Ib) there are no other eigenvalues of A on the unit circle; and

(Ic) finitely many generic conditions on the other derivatives $D_\lambda T(x_0,\lambda_0)$, $D_{xx}T(x_0,\lambda_0)$, ... are fulfilled.

The conditions of (Ic) are that certain polynomials in these derivatives not vanish at (x_o,λ_o). They are given precisely in Lemma 2 of [6b] as transversality conditions, namely that $T_{j_p}(x)$ meets $(TP)_o$ transversally on X_1 (in Brunovský's notation). See also Corollary 1 of [6b]. Finally, a fixed point is of type II' provided

(IIa) -1 is a simple eigenvalue of A;

(IIb) there are no other eigenvalues of A on the unit circle; and

(IIc) certain other generic conditions (as in (Ic)) hold.

The generic conditions of (IIc) appear explicitly in the proof of Theorem 3 of [6b], as $\beta^2+\gamma\neq 0$. We remark that even though Brunovský considers only $x\varepsilon R^2$ here, there is no loss of generality; for the case of R^n, equation (16) of [6b] simply becomes vector valued.

If $T(x,\lambda)$ is a Poincaré map of the differential equation $\dot{x}=f(x,\lambda)$, then fixed points x_o of type 0', I', or II' lie on periodic orbits of type 0, I, or II (although the converse is not in general true). So, for example, a point of type II' lies on a curve of fixed points of T of type 0', and has also another curve of fixed points of T^2 of type 0' emerging from it. Moreover, the generic conditions (Ic) and (IIc) imply an openness property for the fixed points of type I' and II': the sets $\Sigma_1(T)$ and $\Sigma_2(T)$ of fixed points (x_o,λ_o) of types I' and II' are discrete, and depend continuously on $T\varepsilon C^r(B\times(\alpha,\beta),R^n)$. That is, these points are preserved under perturbations of the map T. (The topology on the space C^r is the Whitney topology.) The above results essentially can be found in [6b]. The following generic result is also proved there.

<u>Theorem (Brunovský)</u> <u>If</u> $3\leq r<\infty$, <u>then there is a residual set</u> $F\subseteq C^r(B\times(\alpha,\beta),R^n)$ <u>of maps for which each fixed point has type either</u> 0', I', <u>or</u> II".

We now study generic properties of one parameter families $\dot{x}=f(x,\lambda)$ of vector fields. To simplify matters, we only deal with periodic orbits, and avoid critical points of the flow. Fix compact sets

$$W\subseteq L\subseteq R^n\times R$$

and a sufficiently small neighborhood U of $f_o\varepsilon X$, where X is the Banach space of C^r functions from $M\times[a,b]\supseteq L$ into R^n. Here M is a sufficiently large closed ball, fixed, $a < b$ is fixed, and $3 \le r < \infty$ is fixed. We study periodic orbits $x(t)$ of

(1) $\dot{x} = f(x,\lambda) \qquad f \in U$

with the property that

(2) $$\begin{aligned} &(x(0),\lambda) \in W \\ &(x(t),\lambda) \in L \qquad \text{for all } t. \end{aligned}$$

For convenience we assume

$$f \in U \Rightarrow f(x,\lambda) \neq 0 \quad \text{on} \quad W,$$

so critical points are avoided. We wish to show the following result.

<u>Theorem</u>. <u>For any</u> $T > 0$, <u>the set</u>

$U(T) = \{f \in U$: if $(x(t),\lambda)$ is a periodic orbit of (1), satisfying (2), and with period in $(0,T]$, then $(x(t),\lambda)$ is of type 0', I', or II'$\}$

<u>is residual in</u> U. <u>Hence so is</u>

$U(\infty) = \bigcap_{T=1}^{\infty} U(T)$

$= \{f \in U$: every periodic orbit of (1) satisfying (2) is of type 0', I', or II'$\}$.

An orbit is defined to be of type m' if, for its Poincare map $T(x,\lambda)$, the corresponding fixed point is of type m'.

We outline the proof of this result. First note there is a lower bound, say $\tau > 0$, uniform for $f \in U$, on the periods of all orbits of (1) satisfying (2). In particular it follows that

(3) $U(\tfrac{1}{2}\tau) = U$,

and this will allow us to begin an induction on the periods T, as in [13].

Next, we show that if U(T) is dense in U, then it is residual. To see this, set

$U_k(T) = \{f \in U$: every periodic orbit of (1) satisfying (2) is of type 0_k, I', or II'$\}$.

An orbit is defined to be of type 0_k if for every characteristic multiplier μ (i.e eigenvalue of $D_xT(x_0,\lambda_0)$) we have

$$\mu^j \neq 1 \quad \text{if} \quad 1 \le j \le k \ .$$

An orbit is of type 0' if and only if it is of type 0_k for every k, and hence

$$(4) \qquad \begin{aligned} & U_1(T) \supseteq U_2(T) \supseteq \cdots \\ & U(T) = \cap_{k=1}^{\infty} U_k(T) \ . \end{aligned}$$

We claim that $U_k(T)$ is open in U if $k > T/\tau$; by (4) this is enough to show U(T) dense implies U(T) residual. The openness of $U_k(T)$ follows from the openness results mentioned above for type I' and II' orbits, and similar openness results for type 0_k orbits. In particular, if f $U_k(T)$ is perturbed to a nearby g, and $(x(t),\lambda)$ is a type 0_k orbit for f with period in $[\tau,T]$, there is a unique nearby orbit $(y(t),\lambda)$ of type 0_k, with nearby period, for g. Because $k\tau > T$, no additional orbits with periods in $[\tau,T]$ are introduced nearby.

To complete the proof of the theorem, we follow Peixoto [13] (in particular the arguments of his Section 3), and prove density of U(T) in U by inducting on T. Equation (3) begins the induction. Assuming U(T) is dense in U, we shall prove U(3T/2) also is dense.

With $f \in U(T)$ fixed, let $\Gamma \subseteq L$ be the set of closed orbits of $\dot{x} = f(x,\lambda)$, satisfying (2), with period in [7T/8, 13T/8]. Unlike the situation in [13], Γ may not be closed owing to the possibility of period doubling bifurcations. However,

$$\bar{\Gamma} = \Gamma \cup \Gamma_*$$

where Γ_* is the set of type II' orbits, satisfying (2), with period in [7T/16, 13T/16] Moreover, points of Γ near Γ_* are all type 0' orbits. Choose a sufficiently small neighborhood W of Γ_*:

$$\Gamma_* \subseteq W \ ;$$

and cover the compact set $\bar{\Gamma}$-W with finitely many neighborhoods as follows:

$$\begin{aligned} & \bar{\Gamma} - W \subseteq \cup (V_i \times (\alpha_i,\beta_i)) \qquad \text{(finite union)} \\ & V_i \subseteq \bar{V}_i \subseteq U_i \end{aligned}$$

Here U_i and V_i are neighborhoods in the phase space (x-space) as in [13] and (α_i, β_i) are intervals in the parameter space. For appropriate choices of neighborhoods, Poincaré maps

$$x \in \Sigma_i \to T_i(x,\lambda) \in \Sigma_i'$$

are defined on cross sections of V_i and U_1, as in [13].

Now, by Brunovský's result, arbitrarily small perturbations $\tilde{T}_i$ of the maps T_i can be made so that all fixed points of $\tilde{T}_i$ are of type 0', I', or II'. As in [13] these together correspond to a perturbation g of the vector field f. Since all orbits of g of period in $[T, 3T/2] \subseteq [7T/8, 13T/8]$ appear as fixed points of the $\tilde{T}_i$, all such orbits are of type 0', I', or II'.

Furthermore, if $k > T/\tau$ is given, then the perturbation can be made small enough so that $g \in U_k(T)$. These facts together say that $g \in U_k(3T/2)$, and so (since $U(T)$ is dense in U)

$$U_k(\tfrac{3}{2}T) \text{ is dense in } U \text{ if } k > \frac{T}{\tau}, \text{ and open if } k > \frac{3T}{2\tau} .$$

So by (4), $U(3T/2)$ is dense in U. This proves our theorem. □

Since the countable intersection of residual sets is residual, we have the following corollary:

Corollary: $U(\infty)$ is residual in $C^3(R^n \times R, R^n)$.

Kathleen T. Alligood
Department of Mathematics
College of Charleston
Charleston, South Carolina 29424

James A. Yorke
Institute for Physical Science
and Technology and Department
of Mathematics
University of Maryland
College Park, Maryland 20742

John Mallet-Paret
Lefshetz Center for Dynamical Systems
Division of Applied Mathematics
Brown University
Providence, Rhode Island 02912

and

Mathematics Department
Michigan State University
East Lansing, Michigan 48824

Research Support. The research of each of the authors was supported in part by the National Science Foundation.

REFERENCES

1. Alexander, J. C., and Yorke, J. A., The implicit function theorem and global methods of cohomology, J. Functional Anal. 21 (1976), 330-339.

2. Alexander, J. C., and Yorke, J. A., Families of periodic orbits of differential equations in R^3, J. Differential Equations, to appear.

3. Alligood, K. T., Mallet-Paret, J., and Yorke, J. A., Families of periodic orbits: local continuability does not imply global continuability, J. Differential Geometry, to appear.

4. Alligood, K. T., and Yorke, J. A., Families of periodic orbits: virtual periods and global continuation, preprint.

5. Alligood, K. T., and Yorke, J. A., Virtual periods and generalized Hopf bifurcation, in preparation.

6. Brunovský, P., (a): One-parameter families of diffeomorphisms, Symposium on Differential Equations and Dynamical Systems, University of Warwick, 1969, Springer Lecture Notes 206. (b): On one-parameter families of diffeomorphisms, Commentationes Mathematicae Universitatis Carolinae, 11 (1970), 559-582.

7. Chow, S. N., and Mallet-Paret, J., The Fuller index and global Hopf bifurcation, J. Differential Equations 29 (1978), 66-85.

8. Chow, S. N., Mallet-Paret, J., and Yorke, J. A., A bifurcation invariant: degenerate periodic orbits treated as clusters of simple orbits, appearing in this proceedings.

9. Conley, C. C., Isolated invariant sets and the Morse index, CBMS Regional Conference Series, No. 38, 1978.

10. Fuller, F. B., An index of fixed point type for periodic orbits, Amer. J. Math. 89 (1967), 133-148.

11. Harrison, J., A C^2 counterexample to the Seifert conjecture, preprint.

12. Mallet-Paret, J., and Yorke, J. A., Snakes: oriented families of periodic orbits, their sources, sinks, and continuation, J. Differential Equations, to appear.

13. Peixoto, M. M., On an approximation theorem of Kupka and Smale, J. Differential Equations 3 (1966), 214-227.

14. Schweitzer, P. A., Counterexamples to the Seifert conjecture and opening closed leaves of foliations, Annals of Math. (1971), 386-400.

15. Seifert, H., Closed integral curves in 3-space and isotopic two-dimensional deformations, Proc. Amer. Math. Soc. 1 (1950), 287-302.

16. Sotomayor, J., Generic bifurcations of dynamical systems, <u>Salvador Symposium on Dynamical Systems</u>, M. Peixoto, Ed., Academic Press, 1973, 561-581.

KATHLEEN T.ALLIGOOD
Department of Mathematics
College of Charleston
Charleston,South Carolina
29424 - USA

JAMES A.YORKE
Inst. for Physical Science
and Technology
and
Department of Mathematics
Univ. of Maryland
College Park, Maryland 20742
USA

JOHN MALLET-PARET
Lefschetz Center for Dyn.Systems
Division of Applied Mathematics
Brown University
Providence-R.I. 02912 - USA
and
Mathematics Department
Michigan State University
East Lansing, Michigan
48824 - USA

Unremovable Closed Orbits

by D. Asimov and J. Franks

A classical problem of topology deals with the question of when a fixed point of a map $f:X \to X$ can be removed by deforming f through a homotopy. Or put another way, what is the minimum number of fixed points for any map homotopic to f. This question can often be answered by Nielsen theory (see [B]).

In this article we consider an analagous problem for flows obtained as the suspension (or mapping torus) of a diffeomorphism $f:M \to M$. Recall that the suspension of f is the flow on the mapping torus $M \times I/(fx,0) \sim (x,1)$ which is tangent to the I factors on $M \times I$. Closed orbits of this flow are easily seen to be in one-to-one correspondence with periodic orbits of f. Thus the question of the minimal number of closed orbits can sometimes be answered by Nielsen theory. In the case M is a surface (which is of primary interest to us) this question has been treated by Thurston [Th].

In this article we address the more delicate question of which isotopy classes of closed orbits must be represented by a closed orbit for the suspension flow of g whenever g is isotopic to f. The answer involves the following equivalence relation on periodic orbits. We denote the orbit of a point p under f by orb(p,f).

<u>Definition</u>: If $f_0, f_1 : M \to M$ are isotopic diffeomorphisms and p_0, p_1 are points of (least) period n for f_0 and f_1 respectively then we say that (p_0, f_0) and (p_1, f_1) are in the same <u>strong Nielsen class</u> provided there exists an isotopy f_t from f_0 to f_1 and a smooth arc γ joining p_0 to p_1 such that the point

$\gamma(t)$ is a point of (least) period n for the diffeomorphism f_t.

It is easy to see that this is an equivalence relation. The classical definition of the Nielsen class is defined only when $n = 1$ (i.e., for fixed points) and it requires only that $f(\gamma(t))$ be homotopic to $\gamma(t)$ rel(p_0, p_1) not that $f(\gamma(t)) = \gamma(t)$. From this the following result is immediate.

Proposition 1: *If p_0, p_1 have period n and (p_0, f_0) is in the same strong Nielsen class as (p_1, f_1) then (p_0, f_0^n) and (p_1, f_1^n) are in the same Nielsen class.*

It also is clear that $(f_0^i(p_0), f_0)$ and $(f_1^i(p_1), f_1)$ are in the same strong Nielsen class. However (p_0, f_0) and $(f(p_0), f_0)$ can be in quite different strong Nielsen classes and often (p_0, f_0^n) and $(f(p_0), f_0^n)$ are even in different (ordinary) Nielsen classes.

Proposition 2: *If (p_0, f_0) and (p_1, f_1) are in the same strong Nielsen class then the closed orbits they determine in their suspension flows are isotopic.*

In fact the map which assigns to t the closed orbit in the suspension of $f_t: M \to M$ through $\gamma(t)$ gives the isotopy. We note that the mapping torus on which these flows occur fibers over S^1, and the isotopy we have described respects the fibers of this fibration. The fact that $\gamma(t)$ has least period n implies that throughout this deformation all of the closed orbits are embedded.

Definition: If p is an isolated point of period n for f, we say that (p,f) is unremovable provided that for any g isotopic to f, there is a point q of period n for g with (p,f) and (q,g) in the same strong Nielsen class.

If p is an isolated point of period n we define the Lefschetz index $L(p,f^n)$ to be the Lefschetz number for the fixed point p of f^n. This is easily seen to be equal to $L(f(p),f^n)$, $L(f^2(p),f^n)$, etc.

Main Theorem: If $f:M \to M$ is a diffeomorphism of a compact manifold with isolated points of period n and p has period n then (p,f) is unremovable provided:

(a) The points p, f(p), $\cdots$ $f^{n-1}(p)$ are all in different (ordinary) Nielsen classes with respect to the map f^n, and

(b) $\Sigma L(q,f) \neq 0$ where the sum is over all q such that (q,f) and (p,f) are in the same strong Nielsen class.

Corollary: Suppose that $f:M \to M$ is a diffeomorphism of a compact manifold, with a point p of period n which is the only point in its Nielsen class with respect to the map f^n, and $L(p,f^n) \neq 0$. Then (p,f) is unremovable.

Any periodic orbit of an Anosov or pseudo-Anosov diffeomorphism of a compact surface satisfies the hypothesis of the corollary so all of their periodic orbits are unremovable [Aster]

Proof of Corollary: Since (p,f^n) is the only point in its Nielsen class, part (a) of the hypothesis of the main theorem is satisfied. But also by our first proposition (p,f) is the only orbit in its strong Nielsen class (with respect to f) and by hypothesis $L(p,f^n) \neq 0$ so part (b) of the hypothesis of the main theorem is also satisfied. q.e.d.

Proof of Main Theorem: Given f and p satisfying the hypothesis, and g isotopic to f, we must find q so that (p,f) and (q,g) are in the same strong Nielsen class.

We do this first under the assumption that both f and g are Kupka-Smale diffeomorphisms (see [Sm]). In particular we assume that all periodic points of f and g are hyperbolic (i.e., if $f^n(q) = q$, the derivative $Df^m: TM_q \to TM_q$ has no eigenvalues of modulus one). With this hypothesis a result of Brunovsky [Brun] (see also [Sot]) says that, given $n>0$, an isotopy from f to g can be approximated by an isotopy f_t, $f_0 = f$, $f_1 = g$ with the property that all orbits of periods $\leq 3n$ undergo only two types of bifurcation during the deformation. We will argue that neither of these bifurcations change $\Sigma L(q,f_t)$ where the sum is over orbits of f_t in the same strong Nielsen class as (p,f). Since this quantity is independent of t and non-zero it follows $\Sigma L(q,g) \neq 0$ which implies there exists a q with (q,g) in the same strong Nielsen class as (p,f).

The two types of bifurcations are the saddle node (see fig. 1) where two orbits of adjacent Morse index coalesce and cancel each other. Since $L(p,f_{t-\varepsilon})$ and $L(p',f_{t-\varepsilon})$ are +1 and -1 or -1 and +1 respectively $\Sigma L(q,f_t)$ is not changed as t goes

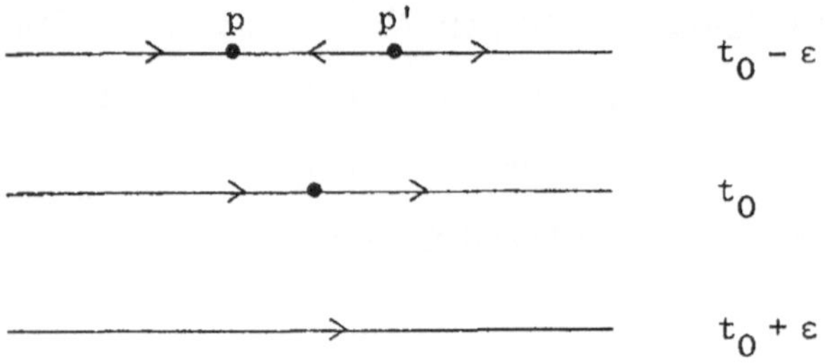

fig. 1

from $t_0 - \varepsilon$ to $t_0 + \varepsilon$.

The second bifurcation is called the period doubling bifurcation (see fig. 2). Here a point p_0 of period n has an eigenvalue (of the derivative Df_t^n) go through -1 as t passes through t_0 and the result is the creation of a new periodic orbit of period 2n. One sees easily that $L(p_0, f^n_{t_0-\varepsilon}) = L(p_0, f^n_{t_0+\varepsilon})$ so $\Sigma L(q, f_t)$ has not changed. Of course the deformation may

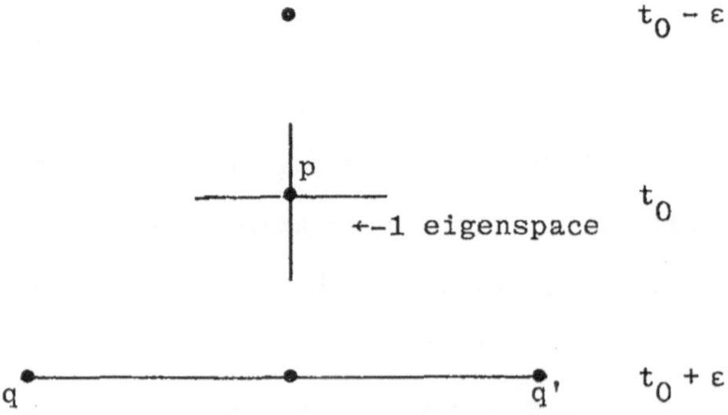

fig. 2

be going the other way and a closed orbit of period 2n definitely can disappear at t_0. However this cannot be in the class we are interested in because if q is on this orbit (for $f_{t_0+\varepsilon}$) and ε is sufficiently small, q and $q' = f^n(q)$ will be very close to p_0 and hence close together. So close together, in fact, that a short arc joining them and f^n of this arc will both lie in a contractible neighborhood of p_0 and hence $(q, f^{2n}_{t_0+\varepsilon})$ and $(q', f^{2n}_{t_0+\varepsilon})$ are in the same Nielsen class. If the strong Nielsen class of (p,f) equals that of $(q, f_{t_0+\varepsilon})$ then (p, f^{2n}) and (p', f^{2n}) where $p' = f^n(p)$ are in the same Nielsen class, which contradicts part (a) of our hypothesis. This completes the proof when f and g are both Kupka-Smale.

We now consider the case when g is not Kupka-Smale. Since Kupka-Smale diffeomorphisms are dense in Diff(M) there is a sequence g_j of such converging to g and a sequence $\{q_j\} \subset M$ such that the strong Nielsen class of (p,f) is the same as that of (q_j, g_j). Choosing a subsequence we can assume q_j converges, say to q. By continuity $g^n(q) = q$ but we need to check $g^m(q) \neq q$ for $1 \leq m < n$. If this were the case then for j sufficiently large several points on $orb(q_j, g_j)$ would lie very close together and close to q. Just as in the period doubling bifurcation this will imply there are distinct points in $orb(q_j, g_j)$ which lie in the same Nielsen class for g_j^n and hence the same would be true for orb(p,f) contradicting our hypothesis. It follows that $orb(q_j, g_j)$ converges uniformly to orb(q,g). The fact that (p,f) and (q,g) are in the same strong Nielsen class is then a consequence of the following lemma.

Lemma: If $g_j \to g$, $q_j \to q$ and n is the least period of q with respect to g and q_j with respect to g_j then for j sufficiently large (q,g) and (q_j, g_j) are in the same strong Nielsen class.

Proof: We can choose diffeomorphisms h_j isotopic to the identity such that $h_j \to id$ and $h_j(g_j^k(q_j)) = g^k(q)$. Define $\bar{g}_j = h_j \circ g_j \circ h_j^{-1}$. Then $\bar{g}_j \to g$ and $\bar{g}_j = g$ on orb(q,g). But also (q_j, g_j) and $(q, \bar{g}_j)$ are in the same strong Nielsen class with $\gamma(t) = h_j(t)(q_j)$ where $h_j(t)$ is an isotopy between h_j and id. Thus it suffices to prove that $(q, \bar{g}_j)$ and (q,g) are in the same strong Nielsen class for large j. However since $\bar{g}_j \to g$ and $\bar{g}_j = g$ on orb(q,g), for large j there is an isotopy G_t satisfying $G_0 = g$, $G_1 = \bar{g}_j$ $G_t = g$ on orb(q,g). Hence $(q, \bar{g}_j)$ and (q,g) are in the same strong Nielsen class. This completes the lemma.

Finally suppose f is not Kupka-Smale. By hypothesis we can choose a representative (p,f) of the strong Nielsen class with $L(p, f^n) \neq 0$. We choose a small neighborhood U of p containing no other points of period $\leq n$. By an arbitrarily small perturbation of f to $\hat{f}$ supported in U we can arrange that all periodic points of $\hat{f}$ in U of period $\leq n$ are hyperbolic. A Lefschetz index argument shows there must be at least one point $\hat{p} \in U$ of period n for $\hat{f}$ and $L(\hat{p}, \hat{f}^n) \neq 0$. If U and the perturbation are chosen sufficiently small it is clear from the lemma above that the strong Nielsen class of $(\hat{p}, \hat{f})$ and (p,f) agree. It is now possible to change $\hat{f}$ to make it Kupka-Smale without altering it on $\hat{p}$, $\hat{f}(\hat{p}), \cdots \hat{f}^{n-1}(\hat{p})$. In fact we don't really need Kupka-Smale, only that all points of periods up to say 3n are hyperbolic. We remark that a similar argument can be used with pseudo-Anosov diffeomorphisms (which fail to be smooth at a finite set of points) to perturb them to smooth diffeomorphisms without losing a particular strong Nielsen class we wish to show unremovable. q.e.d.

Remark: With an appropriate definition of strong Nielsen class and unremovable the proof of the main theorem would work equally well for a finite set of closed orbits instead of one.

The definition of strong Nielsen class would have a $\gamma_j(t)$ for each orbit (orbits indexed by j) and would require that for each t, $\gamma_j(t) = \gamma_k(t)$ implies $j = k$.

Part (a) of the hypothesis would require that all points in the orbits be in different Nielsen classes (with respect to some power of f which fixes the points). Part (b) would have to hold for every p in the finite set of orbits.

An application of this stronger result would be to show that many closed orbit links in suspension flows (e.g., for pseudo-Anosov maps) are unremovable.

References

[Aster] Travaux de Thurston sur les surfaces, Asterisque 66-67, 1979.

[Brun] P. Brunovsky, On One-Parameter Families of Diffeomorphisms I and II Comment. Math. Univ. Carolinae 11 (1970), 559-582; and 12 (1970), 765-784.

[Sm] S. Smale Differentiable Dynamical Systems, Bull. Amer. Math. Soc. 73, (1967), 747-817.

[Sot] J. Sotomayor, Generic One-Parameter Families of Vector Fields on Two-dimensional Manifolds, Pub. Math., I.H.E.S. 43 (1974), 5-46.

[Th] W. Thurston, On the geometry and dynamics of diffeomorphisms of surfaces, preprint.

ON LOCAL ENTROPY

by

M. Brin* and A. Katok*

The theorem proved in this paper answers the question raised by Lai-Sang Young and F.Ledrappier during the International Symposium on Dynamical Systems in Rio de Janeiro. The problem appeared in connection with their study of interrelations between the measure theoretic entropy, Lyapunov exponents and dimension-like characteristics of smooth dynamical systems. The paper was written in IMPA shortly after the Symposium. We would like to thank the hosts of the Symposium for the invitation to come and for their warm hospitality.

Let X be a compact metric space with distance function d and $f:X\to X$ be a continuous mapping preserving a Borel probability non-atomic measure m. We assume that $h_m(f)$, the entropy of f with respect to m, is finite. Denote as usual for $x,y\in X$ and a positive integer n

$$d_n^f(x,y) = \max_{0\le i\le n-1} d(f^i x, f^i y)$$

and for $r>0$ let $B_n^f(x,r)$ be the d_n^f-ball about x of radius r.

The following theorem may be considered as a local version of the charactsrization of entropy given in [1] and at the same time it is a topological version of the Macmillan-Breiman theorem [2].

Theorem. For m-almost every $x\in X$

(a) $$\lim_{\delta\to 0}\ \liminf_{n\to\infty} \frac{-\log m(B_n^f(x,\delta))}{n} =$$

*Partially supported by NSF Grant MCS-7903046

$$= \lim_{\delta \to 0} \limsup_{n \to \infty} \frac{-\log m(B_n^f(x,\delta))}{n} \overset{\text{def}}{=} h_m(f,x) ;$$

(b) $h_m(f,x)$ is f-invariant;

(c) $\int_X h_m(f,x)\, dm = h_m(f)$.

Corollary. If f is ergodic with respect to m, then for almost every x $h_m(f,x) = h_m(f)$.

There exists a proof of this statement which is easier than the argument given below for the general case.

Proof of Theorem. Fix $\delta > 0$ and consider a finite measurable partition ξ such that $\operatorname{diam} \xi \overset{\text{def}}{=} \max_{c \in \xi} \operatorname{diam}(c) < \delta$. Let $c_\xi(x)$ be the element of ξ containing x and $c_n^\xi(x)$ be the element of the partition

$$\xi_n = \xi \vee f^{-1}\xi \vee \ldots \vee f^{-n+1}\xi$$

containing x. By the Macmillan-Breiman theorem

$$\lim_{n \to \infty} \frac{-\log m(c_n(x))}{n} \overset{\text{def}}{=} m_\xi(x)$$

exists for a.e. x and

$$\int_X m_\xi(x)\, dm = h(f,\xi) \leq h_m(f).$$

Obviously, $B_n^f(x,\delta) \supset c_n^\xi(x)$ so that for every $\delta > 0$ we have

$$(1) \qquad \int_X \limsup_{n\to\infty} \frac{-\log m(B_n^f(x,\delta))}{n}\, dm \leq h_m(f) .$$

We proceed now to the estimate from below. Let ξ_e be the decomposition of X into ergodic components of f with respect to m , $p: X \to X/\xi_e = Y$ be the natural projection and $h(y)$, $y \in Y$, be the entropy of the restriction $f|p^{-1}(y)$ with respect to the conditional measure.

Fix a small positive number α , choose $M>1$ such that

$$\int_{h^{-1}([M,\infty))} h(y)\, dm_Y < \alpha$$

and let for $k = 0,1, \ldots, \left[\frac{M}{\alpha}\right] = K$

$$A_k = p^{-1}(h^{-1}([k\alpha,(k+1)\alpha))),$$

$$A_{K+1} = X - \bigcup_{k=0}^{K} A_k , \quad \text{so that} \quad m(A_{K+1}) < \alpha .$$

Denote $a = (A_0, \ldots, A_K, A_{K+1})$. Let $\xi \geq a$ be a finite measurable partition, then for $x \in A_k$, $k = 0,1, \ldots ,K$, $m_\xi(x) \leq (k+1)\alpha$. If, in addition, ξ is sufficiently fine (e.g. its elements have uniformly small diameters), then

$$(2) \qquad k\alpha - \frac{\alpha}{100} \leq m_\xi(x) \leq (k+1)\alpha , \ x \in A_k, \ k \neq K+1.$$

Fix a number $\gamma > 0$ and choose a partition $b \geq a$ such that

$$h_m(f,b) > h_m(f) - \gamma .$$

For every $x \in X$ there exists a ball centered at x of arbitrarily small radius whose boundary has measure 0. Take a finite cover of X by such balls and construct in the usual way the partition into all possible intersections of those balls. Thus, we obtain an arbitrarily fine partition ζ such that $m(\partial\zeta) = 0$, where by definition $\partial\zeta = \bigcup_{c\in\zeta} \partial c$. In particular, we can approximate b by a partition $\eta \leq \zeta$. Therefore, for any positive number q we have constructed a partition η with the following properties

(3) $$h_m(f,\eta) > h_m(f) - \gamma ;$$

for every element $c \in \eta$ there exists a set $A_{k(c)}$ such that

(4) $$m(c \cap A_{k(c)}) > (1-q)m(c) ;$$

(5) $$m(\partial\eta) = 0 .$$

Notice that the number N of elements in η depends only on α and γ but not on q.

For $\delta > 0$ let

$$U_\delta(\eta) = \{x \in X : \text{the } \delta\text{-ball about } x \text{ is not contained in } c_\eta(x)\} .$$

Since $\bigcap_{\delta>0} U_\delta(\eta) = \partial\eta$, by (5) $m(U_\delta(\eta)) \to 0$ as $\delta \to 0$, so we can choose $\delta > 0$ such that $m(U_{\delta'}(\eta)) < q$ for every $\delta' < \delta$.

Let us denote for $k = 0,1, \dots ,K$

$$A'_k = \bigcup_{c\in\eta : k(c) = k} c .$$

In other words, A'_k is the union of those elements of η which lie mostly in A_k. Furthermore, denote

$$D = \bigcup_{k=0}^{K} (A_k' - A_k) .$$

By (4), $m(D) < q$, hence, by the Birkhoff ergodic theorem and by the Chebyshev inequality applied to the function

$$\lim_{n \to \infty} \frac{1}{n} \sum_{i=0}^{n-1} \chi_D(f^i x) ,$$

we obtain that for n large enough

$$(6) \quad m(\{x \in X : \forall n' \geq n \sum_{i=0}^{n'-1} \chi_D(f^i x) < 2\sqrt{q}\, n'\}) > 1 - 2\sqrt{q} .$$

The same argument applied to $U_\delta(\eta)$ instead of D gives

$$(7) \quad m(\{x \in X: \forall n' \geq n \sum_{i=0}^{n'-1} \chi_{U_\delta(\eta)}(f^i x) < 2\sqrt{q}\, n'\}) > 1 - 2\sqrt{q} .$$

We now apply the Macmillan-Breiman theorem to the partitions η and $\xi = \eta \vee a$ and compare the functions m_ξ and m_η. Obviously

$$m_\xi(x) \geq m_\eta(x) .$$

On the other hand, by (3)

$$\int m_\eta(x)\, dm = h_m(f,\eta) \geq h_m(f) - \gamma \geq h_m(f,\xi) - \gamma = \int m_\xi(x)\, dm - \gamma .$$

Therefore, by the Chebyshev inequality

$$m(\{x \in X : m_\xi(x) - m_\eta(x) > \sqrt{\gamma}\}) < \sqrt{\gamma} .$$

Thus, for n large enough

$$(8)\quad m\left(\left\{x \in X : \frac{\log m(c_n^{\xi}(x)) - \log m(c_n^{\zeta}(x))}{n} < 2\sqrt{\gamma}\right\}\right) > 1 - 2\sqrt{\gamma}.$$

Let $F(x) = k\alpha - \frac{\alpha}{50} - 2\sqrt{\gamma}$, if $x \in A_k$, $k = 1,2,\ldots,K$, and $F(x) = 0$ if $x \in A_0$. By (2) and (8) we have

$$m\left(\left\{x \in \bigcup_{k=0}^{K} A_k : \frac{-\log m(c_n^{\xi}(x))}{n} < F(x)\right\}\right) < 3\sqrt{\gamma}.$$

Since $\frac{-\log m(c_n^{\xi}(x))}{n} \to m_{\xi}(x)$ a.e., we have

$$m\left(\left\{x \in \bigcup_{k=0}^{K} A_k : m_{\xi}(x) < F(x)\right\}\right) < 3\sqrt{\gamma},$$

so that for n large enough

$$(9)\quad m\left(\bigcup_{k=0}^{K}\left\{x \in A_k : \forall n' \geq n, \frac{-\log m(c_{n'}^{\xi}(x))}{n'} \geq F(x) - \sqrt{\gamma}\right\}\right) \geq m\left(\bigcup_{k=0}^{K} A_k\right) - 4\sqrt{\gamma}.$$

Let E be the set of points satisfying the conditions described in the left-hand parts of (6), (7) and (9) and denote

$$E_k = E \cap A_k .$$

Let $w_{\eta}^{n}(x) = (c_{\eta}(x), c_{\eta}(fx), \ldots , c_{\eta}(f^{n-1}x))$ be the (η,n)-name of x. If $y \in B_n^f(x,\delta)$, then for every $0 \leq i \leq n-1$ either $f^i x$ and $f^i y$ belong to the same element of η or $f^i x \in U_{\delta}(\eta)$. Hence, if $x \in E_k$ and $y \in B_n^f(x,\delta)$, then by (7) the Hamming distance between the (η,n)-names of x and y is less than $2\sqrt{q}$. We shall give now an upper estimate for $m(B_n^f(x,\delta))$ for most of the points in E_k. To do this we shall estimate the measure of the set of points y whose (η,n)-names are $2\sqrt{q}$-close to the (η,n)-name of x. For the number L_n of (η,n)-names which are $2\sqrt{q}$-close to the (η,n)-name of $x \in E_k$ we have (see [1], (1.3))

$$\lim_{n \to \infty} \frac{\log L_n}{n} = 2\sqrt{q}\log(N-1) - 2\sqrt{q}\log(2\sqrt{q}) - (1-2\sqrt{q})\log(1-2\sqrt{q}),$$

where N is the number of elements in η. So, if n is large enough, then

$$L_n \leq \exp(g(q,N)n), \tag{10}$$

where $g(q,N) = 2\sqrt{q}\log(N-1) - 2\sqrt{q}\log(2\sqrt{q}) - (1-2\sqrt{q})\log(1-2\sqrt{q})+q$.

Since we first choose N and then q, the number $g(q,N)$ can be made arbitrarily small, e.g.

$$g(q,N) < \frac{\alpha}{100} . \tag{11}$$

Fix k, $0 \leq k \leq K$. We want to estimate the measure of points in E_k whose (η,n)-names have an element of η_n of measure greater than $\exp(-k\alpha + \frac{\alpha}{10})n$ in their Hamming $2\sqrt{q}$-neighborhood. Obviously, the total number of such elements does not exceed

$$\exp\ (k\alpha - \frac{\alpha}{10})n ,$$

hence, by (10) and (11), the total number of elements of η_n in their Hamming $2\sqrt{q}$-neighborhood satisfies

$$(12) \quad Q_n \leq \exp(k\alpha - \frac{\alpha}{10} + g(q,N))n < \exp(k\alpha - \frac{\alpha}{10} + \frac{\alpha}{100})n \; .$$

Consider those of the Q_n elements of η_n whose intersections with E_k have positive measure. To estimate their total measure S_n we multiply their number estimated by (12) and the upper bound for their measure given by (9):

$$(13) \quad S_n \leq \exp\,(k\alpha - \frac{\alpha}{10} + \frac{\alpha}{100} - k\alpha + \frac{\alpha}{50} + 3\sqrt{\gamma})n \leq \exp\,(-\frac{\alpha}{20}n) \; ,$$

the last inequality being true if γ is chosen small compared with α.

Thus, we obtain from (13) that

$$m(\{x \in E_k : m(B_n^f(x,\delta)) > \exp\,(-k\alpha + \frac{\alpha}{10})n\}) < \exp\,(-\frac{\alpha}{20}n)$$

for $k = 1,2,\ldots,K$.

By the Borel-Cantelli lemma, for a.e. $x \in E_k$

$$(14) \qquad \liminf_{n\to\infty} \frac{-\log m(B_n^f(x,\delta))}{n} \geq k\alpha - \frac{\alpha}{10} \; .$$

We integrate (14) and get from (6), (7) and (9)

$$(15) \qquad \int_X \liminf_{n\to\infty} \frac{-\log m(B_n^f(x,\delta))}{n}\, dm \geq \sum_{k=1}^{K} (k\alpha - \frac{\alpha}{10})\; m(E_k)$$

$$\geq \sum_{k=1}^{K} k\alpha\cdot m(A_k) - \sum_{k=1}^{K} k\alpha\cdot(m(A_k) - m(E_k)) - \frac{\alpha}{10}$$

$$h_m(f) - 2\alpha - \frac{K(K+1)}{2}\alpha\,(4\sqrt{q} + 4\sqrt{\gamma}) - \frac{\alpha}{10} \; .$$

Since q and γ were chosen after K and α were, the last expression canbe made arbitrarily close to $h_m(f)$. To achieve that we may have to take δ, which depends on q and γ, very small. Statements (a) and (c) follow immediately from (1) and (15).

Since

$$B_n^f(fx,\delta) \supseteq B_{n+1}^f(x,\delta),$$

it follows from (a) that

$$h_m(f,x) \geqslant h_m(f,fx) .$$

The last inequality together with (c) gives (b).

Remark 1. We considered only the case of finite entropy which is the one interesting for applications in smooth dynamical systems. A slightly more complicated argument allows to include transformations with infinite entropy as well.

Remark 2. Our theorem easily implies Theorem 1.1 from [1] as well as its generalization to the non-ergodic case [3], which give the description of measure theoretic entropy in terms of asymptotic capacity of sets of typical orbits.

REFERENCES

1. A.Katok, Lyapunov exponents, entropy and periodic orbits for diffeomorphisms, Publ.Math.I.H.E.S., v.51(1980), 137-173.
2. P.Billingsley, Ergodic Theory and Information, New York, John Wiley, 1965.
3. A.Katok, Hyperbolicity in smooth dynamical systems, Ecole d'eté de Physique Theorique, Les Houches, 1981,Proc.,to appear.

M. Brin and A. Katok
Department of Mathematics
University of Maryland
College Park, MD 20742, USA

and

Instituto de Matemática
Pura e Aplicada
Estrada Dona Castorina, 110
Rio de Janeiro, RJ - Brasil
22460

Infinitely many moduli of strong stability in divergence free unfoldings of singularities of vector fields

by

Henk Broer* and Sebastian van Strien**

In this note we study generic unfoldings of a certain singularity of codimension one in the class of all C^∞ volume preserving (or divergence free) vector fields on $\mathbb{R}^3$, to be specified below. In [B1] it has been shown that such unfoldings, which have one parameter, are structurally stable in the weak sense, i.e:

Suppose that X^μ and Y^μ (μ being a real parameter) are two such generic arcs, unfolding the specific codimension one singularity in X^0 and Y^0 respectively. Also suppose that the arcs are sufficiently close to each other in the C^2-topology. Then there exists a one parameter family of homeomorphisms $h^\mu: \mathbb{R}^3 \to \mathbb{R}^3$ and a reparametrizing homeomorphism $\rho: (\mathbb{R},0) \to (\mathbb{R},0)$, such that for μ in a neighbourhood of 0, the homeomorphism h^μ is a local C^0-equivalence between X^μ and $Y^{\rho(\mu)}$.

Below we shall be more precise. The family h^μ, together with ρ, is called a weak equivalence between the arcs X^μ and Y^μ. Observe that in this definition we do not require the h^μ to depend continuously on the parameter μ. If the extra condition of continuity in the parameter is introduced, we speak of strong equivalence and

Part of this research was done at IMPA.

*was supported by IMPA.

**was supported by IMPA and the Netherlands Organization for the Advancement of Pure Research (Z.W.O.)

the corresponding stability-concept is called structural stability in the strong sense. Note that strong equivalence of two arcs X^{μ} and Y^{μ} has the same meaning as "ordinary" equivalence of $X = X^{\mu}(x)$ and $Y = Y^{\mu}(x)$, seen as "vertical" vector fields on R^4. For similar definitions see e.g. [PdM] or [NPT]. Also remark that in the classical bifurcation theory, see e.g. [A], the bifurcationsare classified modulo strong equivalence.

So it is a natural question to ask whether our generic, weakly stable unfoldings are also stable in the strong sense. In [B1] it was conjectured that this, in general, is not the case. Below we shall present examples, which show that there exist infinitely many moduli of strong stability. These moduli are invoked by a broken saddle connection, but they cannot be expressed in terms of some finite jet. For details see below, §1. The situation where moduli arise when the equivalences are required to depend continuously on the parameter, are not uncommon: see [NPT], [S1] and [S2]. Also compare [So] with [MP].

We like to express our gratitude to Floris Takens for stimulating and helpful discussions during the preparation of this paper.

§1. Introduction, results

Consider on $\mathbb{R}^3$ a divergence free vector field, which has the origin as a singular point, where the eigenvalues of the linear part are 0, $i\alpha$ and $-i\alpha$ for some positive α. One easily sees that such singularities have codimension one, i.e. that they may occur as a bifurcation in generic one parameter families of divergence free vector fields:
In the divergence free case the trace of the linear part in a singularity must be zero.
So let us consider such a generic one parameter family (or arc)

$X = X^{\mu}(x)$, where μ is a real parameter, which unfolds the above singularity $x = 0 \in \mathbb{R}^3$ in the vector field X^0. In [B1] a normal form theorem was proven which states that up to a μ-dependent, volume preserving change of coordinates and some reparametrization of the time, one may write $X^{\mu} = \tilde{X}^{\mu} + p^{\mu}$, where

i. Both $\tilde{X}^{\mu}$ and p^{μ} have divergence zero,

ii. $p = p^{\mu}(x)$ is flat in $(x,\mu) = (0,0) \in \mathbb{R}^3 \times \mathbb{R}$,

iii. In cylindrical coordinates r, φ and z, the vector field $\tilde{X}^{\mu}$ has the system form

$$(1.1) \qquad \begin{cases} \dot{\varphi} = 1 \\ \dot{r} = rg(r^2,z,\mu) \\ \dot{z} = h(r^2,z,\mu), \end{cases}$$

which expresses rotational symmetry. Moreover

$$\frac{\partial h}{\partial \mu}(0,0,0) = 1 \quad \text{and} \quad g(0,0,0) = h(0,0,0) = \frac{\partial h}{\partial z}(0,0,0) = 0.$$

The 2-jet of $\tilde{X} = \tilde{X}^{\mu}(x)$ in $(x,\mu) = (0,0)$ now can be written as

$$(1.2) \qquad \begin{cases} \dot{\varphi} = 1 \\ \dot{r} = (a_1 z + a_2\mu)r \\ \dot{z} = \mu + b_1 r^2 - a_1 z^2 - 2a_2\mu z + b_2\mu^2, \end{cases}$$

where a_1, a_2, b_1 and b_2 are real constants. We impose as a generic condition that $a_1 \neq 0$ and $b_1 \neq 0$. It is no essential restriction to assume that $b_1 \geq 0$, otherwise replace z by $-z$. Then, in the generic case, we obtain two situations corresponding to the sign of a_1. Here we consider the case where $a_1 > 0$ (and $b_1 > 0$), which in [B1] was labelled (3,2) II.*

If in the system (1.1) the angular component is omitted, we obtain a reduced vector field $\bar{X}^{\mu}$, defined in the (r,z)-plane. For μ in a neighbourhood of 0 in [B1] local phase portraits were obtained for the reduction $\bar{X}^{\mu}$, which turns out to be topologically-determined by its 2-jet in a weak sense. See fig. 1. Compare

[B1] Ch. 4 and [T1]. Note that the characteristic distance in these phase portraits asymptotically equals $\sqrt{|\mu|}$ as $\mu \to 0$. The corresponding phase portraits for the symmetric vector field $\tilde{X}^\mu$ in $\mathbb{R}^3$, now can be obtained by introducing again the angular component.

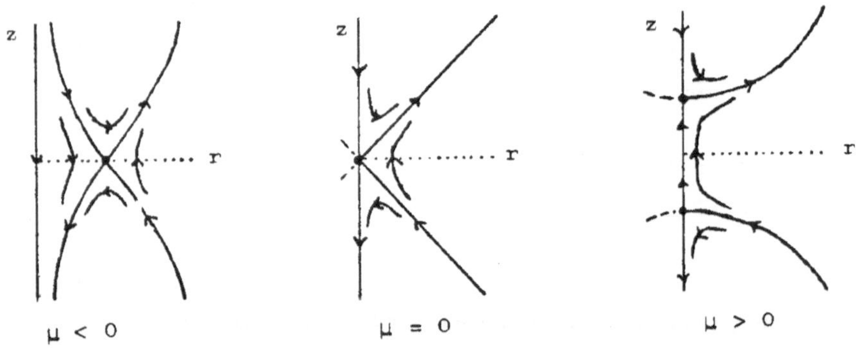

fig. 1

For $\mu > 0$ and sufficiently small, the symmetric vector field $\tilde{X}^\mu$ possesses a saddle connection, which, if one adds the flat perturbation $p = p^\mu(x)$, generically will be broken. In that case the arc $X^\mu = \tilde{X}^\mu + p^\mu$ near the point $(x,\mu) = (0,0)$ is (locally) stable in the weak sense. See [B1]. Also in [B1] it was already remarked that the unfoldings with a 2-jet of type II have a strange topology, since the symmetric specimen, with a saddle connection for all μ in some right hand (or left hand) neighbourhood of $\mu = 0$, form a dense subset as a consequence of our normal form theorem. See above. In the complement of this dense set the weakly families are open and dense. This phenomenon may already suggest that the strong stability of such arcs might be problematic. Below we shall examine this. We shall study various flat perturbation terms $p = p^\mu(x)$ and our special interest is with the consequences of the assumption that for two such terms p_1 and p_2 the corresponding arcs $X_1 = \tilde{X} + p_1$ and $X_2 = \tilde{X} + p_2$ are strongly equivalent. In fact we shall prove:

Main Theorem:

Let $\tilde{X} = \tilde{X}^{\mu}(x)$ be a symmetric, divergence free arc as above, so with a 2-jet of type II. Then for any arbitrary $N \in \mathbb{N}$ there exists a flat, divergence free perturbation p, depending C^{∞} on N extra real parameters $\alpha_1, \alpha_2, \ldots, \alpha_N$, such that the following holds:
If the arcs $X_1^{\mu}(x) = \tilde{X}^{\mu}(x) + p^{\mu}(x;\alpha_1^1,\alpha_2^1,\ldots,\alpha_N^1)$ and $X_2^{\mu}(x) = \tilde{X}^{\mu}(x) + p^{\mu}(x;\alpha_1^2,\alpha_2^2,\ldots,\alpha_N^2)$ are strongly equivalent, then for all $1 \le j \le N$ it follows that $\alpha_j^1 = \alpha_j^2$.

Remarks:

i. The arcs X_1 and X_2 are very close to each other, since their difference is flat in $(x,\mu) = (0,0)$.

ii. The fact that in the case of strong equivalence all the values of the parameters $\alpha_1, \alpha_2, \ldots, \alpha_N$ have to coincide, expresses that the α_j are moduli of strong stability, for the subclass of all generic unfoldings, as specified in the proof of the main theorem. See below. Since N may be chosen arbitrary large, we say that there are infinitely many of such moduli.

iii. Evidently there is no finite jet of these arcs in $(x,\mu) = (0,0)$ which contains information that distinguishes between their strong equivalence classes. This phenomenon is related to an example of Floris Takens concerning a class of vector fields which are not topologically determined by any finite jet. See [T2].

In the proof of the main theorem we shall use an "invariant" for strong equivalence of our arcs: the so called winding number. A related notion is used in [S2].
In §2 we give some technical details concerning this winding number and in §3 we present a proof of the main theorem.

§2. Construction of the perturbations, the notion of winding number

In this section we shall specify the class of perturbation terms, as mentioned in the previous section. For the corresponding weakly stable arcs of vector fields we shall introduce the winding number and derive some useful technical results.

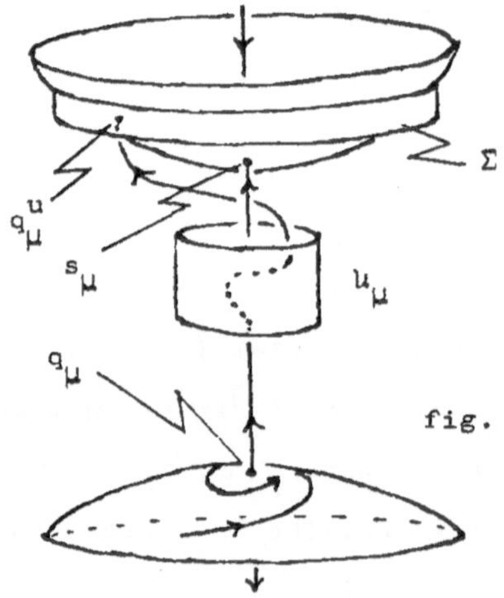

fig. 2

First consider the symmetric vector field $\tilde{X}^\mu$, for μ positive and small. Cf. fig. 1. Let q_μ and s_μ denote the saddles of $\tilde{X}^\mu$ as depicted in fig. 2. Both are lying on the z-axis and we let $q(\mu)$ and $s(\mu)$ be the z-coordinates of q_μ and s_μ respectively.

We recall that as $\mu \downarrow 0$ both $q(\mu)$ and $s(\mu)$ asymptotically have the size $\sqrt{\mu}$.

Now we choose a cylindrical box $\mathcal{U}_\mu$ as depicted in fig. 3, modulo revolution around the z-axis. The shaded and double shaded regions correspond to smaller cylinders $\mathcal{V}_\mu^1$ and $\mathcal{V}_\mu^2$ respectively.

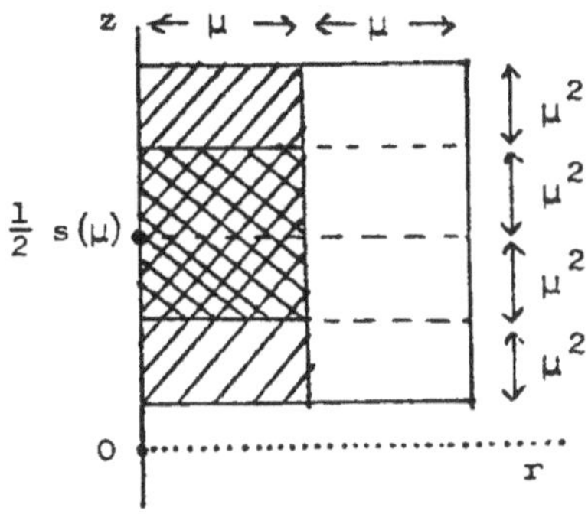

fig. 3

Observe that the size of $\mathcal{U}_\mu$ is such that for $\mu > 0$, sufficiently small, it has no intersections with the 2-dimensional invariant manifolds of q_μ and s_μ.

Our perturbations will all have their supports contained in $\mathcal{U}_\mu$ and are constructed as follows:

For $\mu > 0$ and small, we choose bumpfunctions $\gamma_\mu : \mathbb{R}^3 \to \mathbb{R}$, with $\operatorname{supp}(\gamma_\mu) \subseteq \mathcal{U}_\mu$ which have the form

$$\gamma_\mu(x_1,x_2,x_3) = \beta\left(\frac{1}{\mu}\sqrt{x_1^2+x_2^2}\right)\cdot\beta\left(\frac{1}{\mu^2}\left(x_3 - \frac{s(\mu)}{2}\right)\right),$$

where as before $x_1 = r\cos\varphi$, $x_2 = r\sin\varphi$ and $x_3 = z$ and where $\beta : \mathbb{R} \to \mathbb{R}$ is a bumpfunction with $\operatorname{supp}(\beta) = [-2,2]$ and $\beta(s) \equiv 1$ for $s \in [-1,1]$.

Analogous to [R] we consider a divergence free perturbation

$$P^\mu(x) = -\frac{\partial}{\partial x_2}(x_1\gamma_\mu(x))\cdot\frac{\partial}{\partial x_1} + \frac{\partial}{\partial x_1}(x_1\gamma_\mu(x))\cdot\frac{\partial}{\partial x_2},$$

for small, positive μ.

The perturbations to be used for our proof now will be of the form

$$p^\mu(x) = \delta(\mu)\cdot P^\mu(x),$$

where $\mu \mapsto \delta(\mu)$ is a function which is flat in $\mu = 0$. Since the derivatives of γ_μ explode only polynomially as $\mu \downarrow 0$, the scalar $\delta(\mu)$ ensures that $p = p^\mu(x)$ is a C^∞ vector field, which is flat in $(x,\mu) = (0,0)$.

In the next section we shall say more of the function δ. We there shall make it depending on the parameters $\alpha_1, \alpha_2, \ldots, \alpha_N$ as mentioned in the main theorem. For the present δ is an arbitrary flat function, with $\delta(\mu) > 0$ for $\mu > 0$.
The technical lemma 2.2, below, among other things states that under these circumstances the vector field $X^\mu = \tilde{X}^\mu + p^\mu$ is weakly stable: so there are no saddle connections for small, positive μ.
For the moment we assume this result.

Now consider a cylindrical section $\Sigma = \{r=\varepsilon\}$, see fig. 2, where ε is a positive number, to be restricted later on.
For $\mu > 0$, sufficiently small, the 1-dimensional unstable manifold $W^u(q_\mu)$ of the saddle q_μ will intersect Σ in a unique point, to be denoted by q_μ^u.
Next consider the polar angle φ, but lifted to $\mathbb{R}$ and, depending on the choice of ε, define the winding number $\varphi_X(\mu)$ by

$$\varphi_X(\mu) = \varphi(q_\mu^u).$$

It will be clear that for a weakly stable arc $X = X^\mu(x)$, as above, one has that $\varphi_X(\mu) \to \infty$ as $\mu \downarrow 0$, since the distance between s_μ and $W^u(q_\mu)$ tends to zero as $\mu \downarrow 0$. The following lemma says that the way in which this limit is "attained" is almost an invariant for strong equivalence.

2.1 <u>Lemma</u>:

> Let p_1 and p_2 be as above and assume that the arcs $X_1 = \tilde{X} + p_1$ and $X_2 = \tilde{X} + p_2$ are strongly equivalent via a reparametrization $\mu \mapsto \rho(\mu)$.
> Then $|\varphi_{X_1}(\mu) - \varphi_{X_2}(\rho(\mu))|$ is bounded for $\mu \downarrow 0$.

<u>Proof</u>: The lemma is a direct consequence of the fact that there exist C^0 equivalences $h^\mu: X_1^\mu \cong X_2^{\rho(\mu)}$ which for $\mu \downarrow 0$ depend continuously on μ.

QED

It is to be noted that lemma 2.1 does not need all the details of the above construction. The next lemma however, uses them heavily.

First we introduce some notation, which will be used from now on. For positive, real functions $c = c(\mu)$ and $d = d(\mu)$ we write

$$c(\mu) \lesssim d(\mu) \quad \text{as} \quad \mu \downarrow 0,$$

as soon as a constant $C > 0$ exists such that for all $\mu > 0$ and small one has $c(\mu) \leq C \cdot d(\mu)$.

Also

$$c(\mu) \simeq d(\mu) \quad \text{as} \quad \mu \downarrow 0$$

means that both $c(\mu) \lesssim d(\mu)$ and $d(\mu) \lesssim c(\mu)$, as $\mu \downarrow 0$. For $\mu \to \infty$ we use similar conventions.

We now formulate a result which expresses the winding number $\varphi_X(\mu)$ in terms of the flat function $\mu \mapsto \delta(\mu)$:

2.2 Lemma

For p^μ as above the arc $X^\mu = \tilde{X}^\mu + p^\mu$ is weakly stable, while $\varphi_X(\mu) \simeq \left|\frac{\log \delta(\mu)}{\sqrt{\mu}}\right|$ as $\mu \downarrow 0$.

Proof: Consider the integral curve in the unstable manifold $W^u(q_\mu)$. Let $T_1(\mu)$ be the time that this integral curve spends in the box $\mathcal{U}_\mu$ and let $T_2(\mu)$ be the time this integral curve needs to travel from $\mathcal{U}_\mu$ to Σ.

We shall show that $T_1(\mu) \simeq \mu$ and that between $\partial \mathcal{U}_\mu$ and Σ the curve $W^u(q_\mu)$ does not coincide with the z-axis; the latter implying the weak stability of our arc.

Since $T_2(\mu) \to \infty$ as $\mu \downarrow 0$ and since outside $\mathcal{U}_\mu$ the rotational velocity of X^μ equals 1, it follows that $\varphi_X(\mu) \simeq T_2(\mu)$. Below we shall estimate $T_2(\mu)$.

Observe that the behaviour of $T_2(\mu)$ is very sensitive for the exact place where the curve $W^u(q_\mu)$ leaves the box $\mathcal{U}_\mu$.

Before we start our estimations we introduce the following subsets of $\mathcal{U}_\mu$:
Let, as depicted in fig. 3, $\mathcal{V}_\mu^1 = \{x \in \mathcal{U}_\mu \mid \sqrt{x_1^2+x_2^2} \leq \mu\}$ and $\mathcal{V}_\mu^2 = \{x \in V_\mu^1 \mid |x_3 - \frac{1}{2}s(\mu)| \leq \mu^2\}$, then it follows from the defintion of γ_μ that for $x \in \mathcal{V}_\mu^1$

(2.1) $$p^\mu(x) = \delta(\mu)\gamma_\mu(x)\frac{\partial}{\partial x_2},$$

where, in fact, γ_μ only depends on x_3. For $x \in \mathcal{V}_\mu^2$ we moreover have

(2.2) $$\gamma_\mu(x) \equiv 1.$$

<u>Part 1: An estimate on</u> $T_1(\mu)$

We time-parametrize the integral curve contained in $W^u(q_\mu)$ as $(x_1(t),x_2(t),x_3(t))$, where $x_3(0) = \frac{1}{2}s(\mu) - \mu^2$: the initial value is chosen in the bottom of $\mathcal{U}_\mu$.
From (1.1) and (1.2) we recall that

$$g(r^2,z,\mu) = a_1 z + a_2\mu + \text{H.O.T.} \qquad \text{and}$$

$$h(r^2,z,\mu) = \mu + b_1 r^2 - a_1 z^2 - 2a_2\mu z + b_2\mu^2 + \text{H.O.T.}$$

which yields for $x \in \mathcal{U}_\mu$:

$$g(r^2,z,\mu) \simeq \sqrt{\mu} \quad \text{and} \quad h(r^2,z,\mu) \simeq \mu.$$

So in $\mathcal{U}_\mu$ we obtain for the $\frac{\partial}{\partial r}$ - and the $\frac{\partial}{\partial z}$ - components of X^μ respectively

(2.3) $$\dot{r} \leq C_1\sqrt{\mu}\, r + \xi(\mu) \qquad \text{and}$$

(2.4) $$\dot{z} \simeq \mu,$$

where C_1 is a positive constant and where $\mu \mapsto \xi(\mu)$ is a flat function coming from the perturbation $p^\mu(x)$.
From (2.3) we derive

(2.5) $$r(t) \leq \frac{\xi(\mu)}{C_1\sqrt{\mu}}\left(e^{C_1\sqrt{\mu}\, t} - 1\right).$$

Since the height of the cylinder $\mathcal{U}_\mu$ equals $4\mu^2$, we deduce from (2.4) and (2.5) that

(2.6) $$T_1(\mu) \simeq \mu \quad \text{and}$$

(2.7) $$r(t) \lesssim \mu\, \xi(\mu), \quad \text{for} \quad 0 \leq t \leq T_1(\mu).$$

Now the flatness of $\xi(\mu)$ yields that for sufficiently small $\mu > 0$ our integral curve remains in a μ-neighbourhood of the z-axis, during its stay in $\mathcal{U}_\mu$, i.e. in the set $\mathcal{V}^1_\mu$.
So it leaves $\mathcal{U}_\mu$ via the "top" and with help of (2.1) we replace (2.7) by

(2.8) $$r(t) \lesssim \mu\, \delta(\mu), \quad \text{for} \quad 0 \leq t \leq T_1(\mu).$$

<u>Part 2</u>: <u>A lower bound for</u> $\Delta(\mu) := r(T_1(\mu))$

In the cylinder $\mathcal{V}^1(\mu)$ the perturbation $p^\mu(x)$ has the form (2.1) and we consider the $\frac{\partial}{\partial x_2}$ - component of X^μ in $\mathcal{V}^1(\mu)$:

(2.9) $$\dot{x}_2 = x_1 + g(r^2, z, \mu)x_2 + \delta(\mu)\gamma_\mu(x).$$

Since $|x_1| \leq r$ and $|x_2| \leq r$, it follows from (2.2), (2.8) and (2.9) that during the stay of our integral curve in the smallest cylinder $\mathcal{V}^2_\mu$

(2.10) $$\dot{x}_2(t) \geq \tfrac{1}{2}\,\delta(\mu),$$

for $\mu > 0$ sufficiently small.

If $S(\mu)$ denotes the time that our integral curve $W^u(q_\mu)$ spends in $\mathcal{V}^2_\mu$, then as before we estimate

(2.11) $$S(\mu) \simeq \mu.$$

On the other hand, in $\mathcal{V}^1(\mu)$ we derive from (2.8) and (2.9):

(2.12) $$\dot{x}_2(t) \geq -C_2\, \mu\, \delta(\mu),$$

where C_2 is a positive constant and $\mu > 0$ sufficiently small. Now (2.6), (2.10), (2.11) and (2.12) together yield

$$x_2(T_1(\mu)) \geq \frac{1}{2}\delta(\mu)S(\mu) - C_2\,\mu\,\delta(\mu)[T_1(\mu)-S(\mu)] \simeq$$
$$\simeq \mu\,\delta(\mu),$$

from which together with (2.8) it follows that

$$\Delta(\mu) \simeq \mu\,\delta(\mu).$$

Part 3: Weak stability of the arc X^μ and an estimate on $T_2(\mu)$

Now we time-parametrize the integral curve contained in $W^u(q_\mu)$ as $(x_1(t),x_2(t),x_3(t))$, where $x_3(0) = \frac{1}{2}\,s(\mu) + 2\mu^2$: the initial value is chosen in the "top" of $\mathcal{U}_\mu$.
This part of the integral curve is not effected by the perturbations and so it is an integral curve of $\tilde{X}^\mu$. Since $\tilde{X}^\mu$ has divergence zero, it follows that it possesses a first integral

(2.13) $\quad H^\mu(r,z) = r^2(\mu + \frac{1}{2}\,b_1r^2 - a_1z^2 - 2a_2\mu z + b_2\mu^2 + \text{H.O.T.})$

Compare [B2].
Observe that the level surface $H^\mu(r,z) = 0$ consists precisely of the saddle manifolds of $\tilde{X}^\mu$. Cf. fig. 1.
According to part 2, the segment of $W^u(q_\mu)$ that we consider now, lies approximately in the level $(\Delta(\mu))^2\mu$ of H^μ. Since this is positive for $\mu > 0$ it follows that the arc X^μ is weakly stable: $W^u(q_\mu)$ does not coincide with the z-axis.
Note that for our estimations on $T_2(\mu)$ we may restrict to the reduction $\bar{X}^\mu$ of $\tilde{X}^\mu$ to the (r,z)-plane. Compare §1.
Of course H^μ also is a first integral for $\bar{X}^\mu$, if we consider the function defined on the (r,z)-plane.

It follows that for $\varepsilon > 0$ sufficiently small, for $0 \leq t \leq T_2(\mu)$ on our curve

(2.14) $\quad g(r^2,z,\mu) \gtrsim \sqrt{\mu}.$

This yields an upper bound for $T_2(\mu)$, since from (2.14) we deduce for the curve segment

$$\frac{\dot{r}}{r} \gtrsim \sqrt{\mu}.$$

Now since $r(0) = \Delta(\mu)$ and $r(T_2(\mu)) = \varepsilon$ it follows

$$T_2(\mu) \lesssim \left|\frac{\log \Delta(\mu)}{\sqrt{\mu}}\right| . \tag{2.15}$$

Subsequently we derive a similar lower bound for $T_2(\mu)$. Let $T_3(\mu)$ denote the time our integral curve needs to travel from ∂U_μ to the plane $\{z = 4S(\mu)\}$, and let $\tilde{r}(\mu)$ be the r-coordinate of the intersection of $W^u(q_\mu)$ with this plane. Since $s(\mu) \simeq \sqrt{\mu}$ it follows from (2.13) that

$$\tilde{r}(\mu) \simeq \sqrt{\mu}.$$

Also we have that for $\mu > 0$ sufficiently small $T_3(\mu) \le T_2(\mu)$. On the curve segment with $0 \le t \le T_3(\mu)$ we estimate

$$g(r^2, z, \mu) \lesssim \sqrt{\mu}$$

and so

$$\frac{\dot{r}}{r} \lesssim \sqrt{\mu}.$$

Then we conclude

$$T_2(\mu) \gtrsim \frac{|\log \Delta(\mu)| - |\log \sqrt{\mu}|}{\sqrt{\mu}} . \tag{2.16}$$

From Part 2 we recall that $\Delta(\mu) \simeq \mu\,\delta(\mu)$, which is flat in $\mu = 0$, such that (2.15) and (2.16) imply

$$T_2(\mu) \simeq \left|\frac{\log \delta(\mu)}{\sqrt{\mu}}\right| .$$

QED

§3. Proof of the main theorem

In this section we further specify the flat functions $\delta(\mu)$ introduced in §2. For the corresponding arcs X^{μ} we shall compute the winding numbers using lemma 2.2.
Then, for different choices of δ we consider the consequences of the assumption that the associated arcs X^{μ} are strongly equivalent. Here we shall use lemma 2.1.

Let $\sigma: \mathbb{R} \to \mathbb{R}$ be a bumpfunction with for all $s \in \mathbb{R}$ $\sigma(s) \geq 0$ and $\mathrm{supp}(\sigma) = [0,1]$. Also for $1 \leq j \leq N$ we define
$\sigma_j: [0,N] \to \mathbb{R}$ by $\sigma_j(s) = \sigma(s-j)$.
So $\mathrm{supp}(\sigma_j) = [j,j+1]$. Here $N \in \mathbb{N}$ is as in the main theorem.

Now first we extend the function σ_j periodically with period N.
Then we define for $\mu > 0$:

$$(3.1)\qquad \delta(\mu;\alpha_1,\alpha_2,\ldots,\alpha_N) = $$

$$= e^{-\mu^{-1}} + \sum_{j=1}^{N} \sigma_j(\mu^{-1})\cdot(e^{-\mu^{-\alpha_j}} - e^{-\mu^{-1}}),$$

where $\alpha_1,\alpha_2,\ldots,\alpha_N$ are real parameters with

$$(3.2)\qquad \alpha_1 > \alpha_2 > \ldots > \alpha_N > 1.$$

Clearly $\delta(\cdot;\alpha_1,\alpha_2,\ldots,\alpha_N)$ can be extended to a function which is flat in $\mu = 0$. This function is positive for $\mu > 0$ and so the associated arc X^{μ} is weakly stable.

In stead of calculating the winding number $\varphi_X(\mu)$ we rather consider

$$(3.3)\qquad \Phi_X(\mu) = \log \varphi_X(\mu),$$

since we prefer working modulo bounded additive constants rather than modulo bounded coefficients, in view of lemma 2.2. So by lemma 2.2:

$$\Phi_X(\mu) = \log \left| \frac{\log \delta(\mu;\alpha_1,\ldots,\alpha_N)}{\sqrt{\mu}} \right| + O(1) \tag{3.4}$$

as μ 0. Now we introduce

$$\mu_k = (\theta + kN)^{-1},$$

where $0 < \theta \le N$ and $k = 0,1,2,\ldots$.
Since for all k we have $\sigma_j(\mu_k^{-1}) = \sigma_j(\theta)$ we obtain the following expressions:

$$\Phi_X(\mu_k) = (\alpha_j + \tfrac{1}{2})\cdot|\log \mu_k| + O(1) \quad \text{as} \quad k \to \infty, \tag{3.5}$$

if $j < \theta < j+1$, while

$$\Phi_X(\mu_k) = \tfrac{3}{2} \cdot |\log \mu_k| + O(1) \quad \text{as} \quad k \to \infty, \tag{3.6}$$

for integer valued θ.
The graph of $\mu \mapsto \Phi_X(\mu)$ is roughly drawn in fig. 4 for $N = 2$.

fig. 4

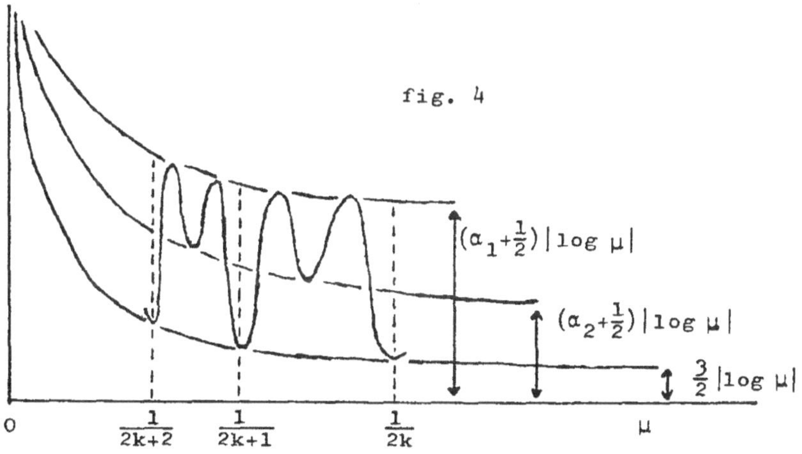

In order to prove our theorem we take two sets of parameter values $\alpha_1^1,\alpha_2^1,\ldots,\alpha_N^1$ and $\alpha_1^2,\alpha_2^2,\ldots,\alpha_N^2$, satisfying (3.2), and assume that for the by (3.1) associated functions δ_1 and δ_2 respectively, the corresponding arcs X_1^μ and X_2^μ (defined as in §2) are strongly equivalent. As before the reparametrising homeomorphism is denoted by ρ. We have to show that for all $1 \le j \le N$

$$\alpha_j^1 = \alpha_j^2 . \tag{3.7}$$

From lemma 2.1 it follows that

$$|\Phi_1(\mu) - \Phi_2(\rho(\mu))| \to 0 \quad \text{as} \quad \mu \downarrow 0, \tag{3.8}$$

where Φ_i abbreviates Φ_{X_i}, $i = 1,2$.

Firstly we show that this implies that the reparametrisation $\mu \mapsto \rho(\mu)$ is differentiable in $\mu = 0$.

3.1 <u>Lemma:</u> $\frac{\rho(\mu)}{\mu} \to 1$ as $\mu \downarrow 0$.

<u>Proof:</u> Take $\nu_k = \frac{1}{k}$, then from (3.6) we see

$$\Phi_1(\nu_k) = \frac{3}{2} |\log \nu_k| + O(1). \tag{3.9}$$

Similarly, see fig. 3, one obtains

$$\Phi_2(\rho(\nu_k)) \ge \frac{3}{2} |\log \rho(\nu_k)| + O(1). \tag{3.10}$$

Therefore from (3.8) we conclude

$$\nu_k \lesssim \rho(\nu_k).$$

Now for any $\mu \in [\nu_{k+1}, \nu_k]$ we have $\frac{1}{2}\mu \le \nu_{k+1}$ and so

$$\frac{1}{2}\mu \le \nu_{k+1} \lesssim \rho(\nu_{k+1}) \le \rho(\mu),$$

since ρ is increasing.

By reversing the rôle of Φ_1 and Φ_2 in this argument one obtains

$$\rho(\mu) \simeq \mu. \tag{3.11}$$

So in stead of (3.10) we may write

$$\Phi_2(\rho(\nu_k)) \ge \frac{3}{2} |\log \nu_k| + O(1). \tag{3.12}$$

Subsequently we shall show that

(3.13) $$\frac{1}{\rho(\nu_k)} \pmod{\mathbb{Z}} \to 0 \pmod{\mathbb{Z}} \quad \text{as} \quad k \to \infty,$$

which means that there exists a sequence $\{n_k\}_{k=1}^{\infty}$ of natural numbers, such that $\left|\frac{1}{\rho(\nu_k)} - n_k\right| \to 0$ as $k \to \infty$.
In order to see this suppose by contradiction that for some $\delta > 0$

$$\frac{1}{\rho(\nu_k)} - \left[\frac{1}{\rho(\nu_k)}\right] \in [\delta, 1-\delta],$$

for infinitely many k. Here $[\,\cdot\,]$ denotes the entier-function.
For such a sequence we have by (3.2) and (3.5)

$$\Phi_2(\rho(\nu_k)) \geq (\alpha_N + \tfrac{1}{2})\cdot|\log \nu_k| + O(1),$$

which is impossible in view of (3.8) and (3.9) and the fact that $\alpha_N > 1$.

Also we have that

(3.14) $$\frac{1}{\rho(\nu_{k+1})} - \frac{1}{\rho(\nu_k)} \to 1 \quad \text{as} \quad k \to \infty.$$

In fact, suppose that this were not true, then for an infinite number of k's either

(3.15) $$\frac{1}{\rho(\nu_{k+1})} - \frac{1}{\rho(\nu_k)} \to 0 \quad \text{as} \quad k \to \infty$$

or

(3.16) $$\frac{1}{\rho^{-1}(\nu_{k+1})} - \frac{1}{\rho^{-1}(\nu_k)} \to 0 \quad \text{as} \quad k \to \infty,$$

where ρ^{-1} is the inverse of the homeomorphism ρ.
We only deal with the case (3.15), the case (3.16) can be treated similarly.
Let $\lambda_k = \frac{1}{k+\frac{1}{2}}$. Then $\lambda_k \in (\nu_{k+1}, \nu_k)$ and from (3.15) it follows that for an infinite number of k's:

$$\Phi_2(\rho(\lambda_k)) = \tfrac{3}{2}\,|\log \lambda_k| + O(1).$$

But our choice of λ_k also implies that

$$\Phi_1(\lambda_k) \geq (\alpha_N + \tfrac{1}{2})\cdot|\log \lambda_k| + O(1).$$

As before this leads to a contradiction.

The lemma is a direct consequence of (3.13) and (3.14).

QED

The proof of the main theorem will now be completed in proving by induction on $j = 1,2,\dots,N$ that $\alpha_j^1 = \alpha_j^2$. We introduce

$$\nu_{k,j} = \frac{1}{j+kN}$$

and

$$\lambda_{k,j} = \frac{1}{\frac{1}{2}+j+kN}.$$

We formulate the <u>induction hypothesis</u> H_j:

$$(3.17)_j \qquad \frac{1}{\rho(\lambda_{k,j})} \in (j,j+1) \pmod N \quad \text{for } k \text{ large}$$

and

$$(3.18)_j \qquad \alpha_j^1 = \alpha_j^2.$$

<u>Proof of</u> H_1: Using (3.11) we write

$$\Phi_1(\lambda_{k,1}) = (\alpha_1^1 + \tfrac{1}{2}) \cdot |\log \lambda_{k,1}| + O(1)$$

and

$$\Phi_2(\rho(\lambda_{k,1})) \le (\alpha_1^2 + \tfrac{1}{2}) \cdot |\log \lambda_{k,1}| + O(1).$$

Then (3.8) gives that $\alpha_1^1 \le \alpha_1^2$.

Reversing the argument yields that $\alpha_1^1 = \alpha_1^2$, which proves $(3.18)_1$.

In order to prove $(3.17)_1$ suppose by contradiction that $\frac{1}{\rho(\lambda_{k,1})} \in (2,N) \pmod N$ for an infinite number of k's. Then

$$\Phi_2(\rho(\lambda_{k,1})) \le (\alpha_2^2 + \tfrac{1}{2}) \cdot |\log \lambda_{k,1}| + O(1),$$

so we would have that $\alpha_1^1 \le \alpha_2^2$. But this contradicts that $\alpha_1^1 = \alpha_1^2$ and that $\alpha_1^2 > \alpha_2^2$.
This proves $(3.17)_1$ and therefore the hypothesis H_1.

<u>Proof that</u> H_j <u>implies</u> H_{j+1} <u>for</u> $j < N$:

Suppose that $(3.17)_j$ and $(3.18)_j$ are true.
If one assumes that $(3.17)_{j+1}$ is not true then, using (3.13), (3.14) and the fact that ρ is monotonous easily yield that

$$\{\frac{1}{\rho(\lambda_{k,j})}\}_{k=1}^{\infty} \quad \text{or} \quad \{\frac{1}{\rho(\lambda_{k,j+1})}\}_{k=1}^{\infty}$$

have subsequences which "converge to the integers" in the sense of (3.13). As before this gives a contradiction.

Finally we prove $(3.18)_{j+1}$.
From (3.5) it follows that

$$\Phi_1(\lambda_{k,j+1}) = (\alpha_{j+1}^1 + \tfrac{1}{2})\cdot|\log \lambda_{k,j+1}| + O(1),$$

and since, by $(3.17)_{j+1}$, $\frac{1}{\rho(\lambda_{k,j+1})} \in (j+1, j+2) \pmod N$ asymptotically, we also have

$$\Phi_2(\rho(\lambda_{k,j+1})) \geq (\alpha_{j+1}^2 + \tfrac{1}{2})\cdot|\log \lambda_{k,j+1}| + O(1).$$

It follows that $\alpha_{j+1}^1 \geq \alpha_{j+1}^2$.

Reversing the argument yields that $\alpha_{j+1}^1 = \alpha_{j+1}^2$.
This finishes the proof of the induction step and therefore of the main theorem.

References:

[A] Arnol'd, V.I.: Lectures on Bifurcations and Versal Families. In: Russ. Math. Surveys 27, 54-123 (1972).

[B1] Broer, H.W.: Formal Normal Form Theorems for Vector Fields and some Consequences for Bifurcations in the Volume Preserving Case. To appear, proceedings Warwick 1980, Springer.

[B2] ----------: Bifurcations of Singularities in Volume Preserving Vector Fields. Groningen, Ph.D- thesis, 1979.

[MP] Malta, I.P., Palis, J.: Families of Vector Fields with Finite Modulus of Stability. To appear, proceedings Warwick 1980, Springer.

[NPT] Newhouse, S., Palis, J. and Takens, F.: Stable Families of Diffeomorphism. Preprint, I.M.P.A., Rio de Janeiro, 1979.

[PdM] Palis, J. and de Melo, W.C.: Introdução aos Sistemas Dinâmicos. São Paulo, E. Blücher, 1978.

[R] Robinson, R.C.: Generic Properties of Conservative Systems I, II. In: Amer. J. Math. 92, 562-603, 897-906 (1970).

[So] Sotomayor, J.: Generic One-parameter Families of Vector Fields on Two-dimensional Manifolds. In: Publ. Math. I.H.E.S. 43, 5-46 (1974).

[S1] van Strien S.J.: Saddle Connections of Arcs of Diffeomorphisms. To appear, proceedings Warwick 1980, Springer.

[S2] ---------------: Moduli of Stability for Vector Fields due to non-real Eigenvalues. Preprint, Utrecht, 1981.

[T1] Takens, F.: Singularities of Vector Fields. In: Publ. Math. I.H.E.S. 43, 48-100 (1974).

[T2] ----------: A Nonstabilisable Jet of a Singularity of a Vector Field. In: Dynamical Systems, ed. M.M. Peixoto, 583-597, Acad. Press 1973.

July 1981, Rio de Janeiro

H.W. Broer
Department of Mathematics
Ryksuniversiteit Groningen
The Netherlands.

S.J. van Strien
Mathematics Institute
Ryksuniversiteit Utrecht
The Netherlands.

AN EXTENSION OF PEIXOTO'S STRUCTURAL STABILITY THEOREM TO OPEN SURFACES WITH FINITE GENUS

by

C. Camacho, M. Krych
R. Mañe, and Z. Nitecki*

Introduction:

In a classic paper published twenty years ago, Mauricio Peixoto [Pe 1] characterized the C^1 structurally stable flows on any closed (compact, boundaryless) surface. Recently, an extension of Peixoto's conditions to flows on open (metrizable, boundaryless, noncompact) surfaces was formulated [KKN], and shown sufficient in general, and necessary on the plane, for global C^1 structural stability. In this paper, we show that the conditions of [KKN] characterize the flows on any surface of finite genus which are C^1 structurally stable in a somewhat stronger sense.

Recall that the *genus* of a surface M is the maximum number $g(M)$ of disjoint loops which can be embedded in M without disconnecting M. A surface M with $g(M) = g < \infty$ is homeomorphic to $X \setminus K$, where X is a closed surface of genus g and K is a closed, totally disconnected subset (see [Ri]). We note that finite genus is a weaker condition than finite Euler characteristic, which would require that K be a finite set.

On a closed manifold, M, a flow ϕ is C^r *structurally stable* if for any flow ψ in a (uniform) C^r neighborhood there exists an *equivalence homeomorphism* $h_\psi : M \to M$ taking ϕ-orbits to ψ-orbits. It is a consequence of Peixoto's theorem that in the case of a closed surface the equivalence homeomorphism can always be

*Supported in part by NSF grant MCS-8102122.

chosen C^0-near the identity for ψ sufficiently near ϕ.

On an open manifold, the uniform C^r topology on perturbations is replaced by the Whitney (strong) C^r topology, and it is necessary to assume some control on the equivalence homeomorphism. The notion of "global C^r structural stability" adopted in [KKN] (see §§ 1-2 there for a discussion of various notions of stability on open manifolds) requires that h_ψ can be chosen near the identity in the compact-open (C^0) topology when ψ is (Whitney C^r) near ϕ. That is, h_ψ is continuous at $\psi = \phi$, as a map from flows (with strong C^r topology) to maps (with compact-open topology). In the present paper, we require that h_ψ be continuous on a whole neighborhood of $\psi = \phi$.

Definition: A flow ϕ on the surface M is continuously C^r structurally stable if there exists a Whitney C^r neighborhood $\mathcal{U}$ of ϕ such that every flow $\psi \in \mathcal{U}$ is topologically equivalent to ϕ, via an equivalence homeomorphism h_ψ such that the map $\psi \to h_\psi$ is continuous with respect to the Whitney C^r topology on $\psi \in \mathcal{U}$ and the compact-open topology on h_ψ.

This strengthening of the notion of stability is similar in spirit, although not in content, to the "absolute stability" notions of Franks [Fr] and Guckenheimer [Gk].

The characterization of structurally stable flows involves the *limit sets* of a point

$$L^{\pm}(x,\phi) = \{y \mid \exists t_n \to \pm\infty \ .\ni.\ y_n = \phi(t_n,x) \to y\}$$

(we use the above notation in preference to $\omega(x,\phi)$ and $\alpha(x,\phi)$) and the *prolongational limit sets*

$$J^{\pm}(x,\phi) = \{y \mid \exists x_n \to x,\ t_n \to \pm\infty \ .\ni.\ y_n = \phi(t_n,x_n) \to y\}.$$

On a compact surface, this characterization deals with the *non wandering set* of ϕ

$$\Omega(\phi) = \{x \mid x \in J^+(x,\phi)\} = \{x \mid x \in J^-(x,\phi)\}$$

and the important subsets of *periodic points*

$$\mathrm{Per}(\phi) = \{x \mid \mathcal{O}(x,\phi) \text{ is compact}\}$$

and *rest points*

$$\mathrm{Fix}(\phi) = \{x \mid \mathcal{O}(x,\phi) = \{x\}\},$$

where $\mathcal{O}(x,\phi)$ denotes the orbit of x under ϕ.

Recall that a fixedpoint x is *hyperbolic* if the linearized velocity field $D\dot{\phi}(x)$ has no pure imaginary eigenvalues, while a circular orbit γ is hyperbolic if the integral of div $\dot{\phi}$ around γ is nonzero. Hyperbolic periodic orbits are sinks, sources, or saddles; of course on a surface all saddles are fixedpoints.

We will call a limit set *trivial* if it consists solely of periodic orbits. (Note that a trivial limit set automatically consists of a single orbit.) A semi-orbit

$$\mathcal{O}_{\pm}(x,\phi) = \{\phi(t,x) \mid \pm t > 0\}$$

is a *stable* (resp. *unstable*) *separatrix* if $L^{\pm}(x,\phi)$ is trivial and $J^{\pm}(x,\phi) \neq L^{\pm}(x,\phi)$. When all orbits in $\mathrm{Per}(\phi)$ are hyperbolic, the limit set of a separatrix is either empty or is a fixed saddle, σ, and in the latter case $\mathcal{O}_{\pm}(x)$ belongs to the stable (or unstable) manifold of σ. When $L_{\pm}(x,\phi) = \emptyset$, we refer to a "saddle at infinity" (note that this cannot occur on closed surfaces). The set of all stable (resp. unstable) separatrices will be denoted

$$W^{\pm}(\phi) = \{x \mid L^{\pm}(x,\phi) \text{ is trivial and } J^{\pm}(x,\phi) \setminus L^{\pm}(x,\phi) \neq \emptyset\}$$

Peixoto's theorem (whose proof in the version below requires the closing lemma [Pu, Pu R] when M is nonorientable - see [Gt]) is:

__Theorem__ *(Peixoto, 1962 [Pe 1]):*

If M is a closed surface, then a flow ϕ on M is C^1 structurally stable if and only if the following conditions hold:

(i) $\Omega(\phi) = \mathrm{Per}(\phi)$;

(ii) every orbit in $\mathrm{Per}(\phi)$ *is hyperbolic;*

(iii) $W^s(\phi) \cap W^u(\phi) = \emptyset$.

Furthermore, (i) - (iii) hold for a dense open set of flows on M.

The result of this paper is the following partial extension:

__Main Theorem:__

If M is a surface (closed or open) of finite genus, then a flow ϕ on M is continuously C^1 structurally stable if and only if:

(i) $\Omega(\phi) = \mathrm{Per}(\phi)$;

(ii) every orbit in $\mathrm{Per}(\phi)$ *is hyperbolic;*

(iii) $\mathrm{clos}\ W^s(\phi) \cap \mathrm{clos}\ W^k(\phi) \subset \mathrm{Fix}(\phi)$.

We note that condition (iii) of our theorem fails on a nonempty C^1-open set of flows on any open surface ([Pe Pu, TW, Kr]) and even the weaker condition $W^s(\phi) \cap W^u(\phi) = \emptyset$ fails on a nonempty C^1 open set of flows on certain surfaces with $g(M) < \infty$ but infinite Euler characteristic ([KKN, example 2.8]).

The main accent in this paper is on the necessity of (i) - (iii) for continuous C^1 stability. It is known [KKN, prop. 3.2] that conditions (i) - (iii) are equivalent to those obtained by replacing (i) with

(i'a) ϕ has no oscillating semi orbits

(i'b) ϕ has no nontrivial minimal sets.

Here, an *oscillating semi orbit* is one with a noncompact limit set; a *minimal set* is a compact invariant set with no compact invariant proper subsets.

It is easy to see, using the Kupka-Smale theorem in the form proved by Peixoto [Pe 2], that condition (ii) up to topological equivalence is necessary for any form of structural stability. The argument that (ii) in the differential form is necessary for stability is given in [KKN, prop. 8.8, 8.9]. A brief argument, using Schwartz's Poincaré-Bendixson theorem [S], that (i'b) is also necessary for stability, is given at the beginning of § 4 in [KKN]. Thus, it remains to show that any continuously C^1 structurally stable flow ϕ on M satisfies (i'a) and (iii). Our argument has three main steps. In §1 we recall an argument from Pugh's closing lemma [Pu] to show that nonwandering orbits outside the closure of $\mathrm{Per}(\phi)$ have trivial limit sets. In § 2, we show that the set $\mathrm{Per}(\phi)$ is closed, so that in particular oscillating orbits must themselves be wandering. Then, in § 3, we prove the necessity of

(iii') $W^s(\phi) \cap W^u(\phi) = \emptyset$.

As a consequence, we immediately obtain (for stable flows) the absence of oscillating orbits (i'a), hence the equality $\Omega(\phi) = \mathrm{Per}(\phi)$ (i), and finally using this argue that the necessity of (iii') for stability implies the necessity of (iii).

Finally, in § 4, we reconsider the proof in [KKN] that (i) - (iii) imply global C^1 structural stability to note how it actually gives continuous C^1 structural stability.

This paper is an outgrowth of Z. Nitecki's conversations with C. Camacho and R. Mañe during the International Symposium on Dynamical Systems, in which the former learned about earlier unpublished work by the latter on results similar to those in [KKN]. We thank Floris Takens, Charles Pugh and Maria Lucia Peixoto for their role in getting these conversations started. We also thank Charles Pugh and Clark Robinson for subsequent conversations concerning the closing lemma.

1. *A Closing Lemma*

The purpose of this section is to prove the following technical fact:

Proposition 1:

Suppose ϕ *is a continuously* C^1 *structurally stable flow on a surface* M, *and suppose* q *is a non-restpoint for* ϕ *such that, given any transversal* T *through* q *and any integer* N, *there exists*

$$r_N \in T \cap \Omega(\phi)$$

whose forward orbit crosses T *in at least* N *distinct points.*

Then $q \in \overline{Per(\phi)}$.

Note that the hypotheses of prop. 1 apply in particular to the limit sets of non-wandering orbits. In the next section, we will show that the periodic orbits of a continuously C^1 structurally stable flow on a surface of finite genus form a closed set, which will imply with prop. 1 that non-wandering orbits have trivial limit sets. This will prove especially useful in handling oscillatory orbits in section 3.

Proposition 1 is proved by arguments from Pugh's closing lemma [Pu, Pu R]. For completeness, we will reproduce these arguments below. For the reader familiar with technical details of the closing lemma we point out that we are handling an especially easy case. In its fullest generality, the closing lemma for an unbounded non-wandering orbit is unknown. However, we need only the version which can be obtained via perturbations in a single flowbox. In our two-dimensional setting, the linear-algebra considerations of the higher-dimension proof are unnecessary. Finally, since we are dealing exclusively with continuously C^1 structurally stable flows, we can *assume throughout this section that* ϕ *is a* C^∞ *flow,* by replacing it if necessary with a nearby C^∞ flow $\tilde{\phi}$, which is conjugate to ϕ and is itself continuously C^1 structurally stable. This avoids the technical problems in genericity arguments that can result from insufficient smoothness of flowboxes.

The basic estimates are easier to understand when the Poincaré map for a section T through q is locally linear. To make this assumption, we first establish the following technical lemma using an argument of L. Markus [Ma]:

Lemma 1:

Given $\varepsilon > 0$ *and a* C^r $(r \geq 1)$ *diffeomorphism* f *of an interval* $I = [-k,k]$ *into* $\mathbb{R}$ *with* $f(0) = 0$, *set* $f'(0) = m$. *There exists a* C^r *diffeomorphism* $g : \mathbb{R} \to \mathbb{R}$ *such that:*

(i) $|g(x) - x| < \varepsilon$ *for all* $x \in \mathbb{R}$

(ii) $|g'(x) - 1| < \varepsilon$ *for all* $x \in \mathbb{R}$

(iii) $f(g(x)) = mx$ *for* $|x| < C_1$, *some* $C_1 \in (0,k)$

(iv) $g(x) = x$ *for* $|x| > C_2$, *some* $C_2 \in (C_1,k)$.

Proof:

On some neighborhood of 0, we can define

$$h(x) = f^{-1}(mx) - x .$$

Take $\alpha: \mathbb{R} \to [0,1]$ a C^∞ function with

$$\alpha(x) = 1 \quad \text{if} \quad |x| \leq 1$$

$$\alpha(x) = 0 \quad \text{if} \quad |x| \geq 2 .$$

Now, define the one-parameter family of functions

$$g_C(x) = x + \alpha(\tfrac{x}{C})\, h(x) .$$

Note that $g_C(x) = x$ for $|x| \geq 2C$, so that g_C is well-defined for C sufficiently small, and satisfies (iii), (iv) for $C_1 = C$, $C_2 = 2C$.

To obtain (i), (ii) we need to show $g_C \to \mathrm{id}$ in the C^1 topology. First, since $h'(0) = 0$, note that

$$\lim_{x \to 0} \frac{h(x)}{x} = 0 .$$

This immediately establishes convergence in the C^0 topology, since

$$|g_C(x) - x| = |\alpha(\tfrac{x}{C})\, h(x)| \leq h(x) .$$

Furthermore, since $g_C'(x) = 1$ for $|x| \geq 2C$, we need only estimate $|g_C'(x) - 1|$ for $|x| \leq 2C$. But

$$|g_C'(x) - 1| \leq \frac{1}{C} |\alpha'(\tfrac{x}{C})|\ |h(x)| + |\alpha(\tfrac{x}{C})|\,|h'(x)| .$$

The second term on the right is bounded (for $|x| \leq 2C$) by

$$\sup_{|x| \leq 2C} |h'(x)|$$

which goes to 0 as $C \to 0$. The first term is bounded by

$$\left[\sup_{y \in \mathbb{R}} |\alpha'(y)|\right] \left[\sup_{|x| \leq 2C} \frac{|h(x)|}{C}\right]$$

The first of these factors is independent of C, while the second is bounded by

$$2 \sup_{|x| \leq 2C} \left|\frac{h(x)}{x}\right|$$

which we have already seen goes to 0 as $C \to 0$. □

Corollary 1 (Linearization lemma):

Suppose T is a transverse section for ϕ through $x \notin Per(\phi)$, and suppose the forward orbit of x under ϕ crosses T again, at y. Then there exist C^∞-reparametrizations $\rho : T \to T$ arbitrarily C^1-near the identity such that, for every point p of T on the orbit segment from x to y, $\rho(p) = p$ and the first-return map P of T to itself is linear on a neighborhood of p.

Proof:

Number the (finite set of) points $p \in T$ on the orbit segment from x to y as

$$x = p_0, p_1, \ldots, p_n = y.$$

Then P is defined from a neighborhood of p_i to a neighborhood of p_{i+1} $(i = 0, \ldots, n-1)$. Apply lemma 1 to the restriction of P to a neighborhood of p_i to obtain g_i that linearizes P near p_i and equals the identity near all p_j, $j \neq i$. Then $g = g_0 \circ g_1 \circ \ldots \circ g_{n-1}$ is as required. □

The next step in Pugh's closing lemma argument is to estimate the effect of a small push concentrated in a long, thin flow box.

Lemma 2 (Lifting lemma):

Suppose $f(t,x) \geq 0$ is a C^∞ function which is positive on $(0,1) \times (-1,1)$ and zero elsewhere. Then for every compact set $A \subset (-1,1)$ there exists a constant $k = k(A,f) > 0$ such that the general solution $\psi_t(x, \varepsilon, b)$ of

$$\dot{x} = \varepsilon\, b\, f(t, \tfrac{x}{b})$$

$$x(0) = 0$$

satisfies

$$\psi_1(ba, \varepsilon, b) - ba \geq \varepsilon\, bk$$

for all $a \in A$, $\varepsilon \in [0,1]$ and $b \in [0,1]$.

Proof:

Pick a closed interval $B \subset (-1,1)$ containing A in its interior. Pick $k > 0$ such that

$$\text{dist}(A, [-1,1] \setminus B) > k$$

$$\inf \{f | [0.2, 0.7] \times B\} \geq 2k .$$

Then for $(t, x/b) \in [0.2, 0.7] \times B$,

$$\varepsilon\, b\, f(t, x/b) \geq 2k\, \varepsilon\, b$$

and the estimate on ψ_1 follows. □

Finally, the following lemma will allow us to pick points near which to perturb in such a way as to avoid intermediate intersections of these orbits with the support of our perturbation, and to avoid passages too near the edge of our flow box.

Lemma 3 *(Fundamental lemma):*

Suppose S *is a transverse section through* $x \in \Omega(\phi)$. *Given* $\varepsilon > 0$ *and* $\tau > 0$, *there exist points* $y, z \in S$ *such that*

(i) $|x - y| < \varepsilon$, $|x - z| < \varepsilon$

(ii) $z = \phi_T(y)$ *for some* $T > \tau$

(iii) any intermediate point $w = \phi_t(y) \in S$, $\tau < t < T$, *satisfies*

$$|w - y|,\ |w - z| > \frac{1}{2}|y - z| .$$

Proof:

By shrinking S, we can assume $\phi_t(u) \notin S$ for $u \in S$ and $0 < t \leq \tau$. Since $x \in \Omega(u)$, we can pick y_0, $z_0 \in S$ with

$$|x - y_0|,\ |x - z_0| < \varepsilon_0 = \varepsilon/4$$

$$z_0 = \phi_{t_0}(y_0),\quad t_0 > \tau .$$

If y_0, z_0 fail to satisfy (iii), there exists

$$w_0 = \phi_t(y_0) \in S$$

such that either

(a) $|y_0 - w_0| \leq \frac{1}{2}|y_0 - z_0|$

or (b) $|z_0 - w_0| \leq \frac{1}{2}|y_0 - z_0|$.

Replace the pair (y_0, z_0) with a pair (y_1, z_1) consisting of w_0 and the nearer of y_0, z_0. Note that (y_1, z_1) satisfy (i) with $\varepsilon_1 = \varepsilon/4 + \varepsilon/8$ and (iii) with $t_1 \in (\tau, t_0)$.

Now, if necessary, we continue replacing each pair (y_i, z_i) with a new pair (y_{i+1}, z_{i+1}), noting that since y_i and z_i all lie on the orbit segment from y_0 to z_0, which crosses S only finitely often, this sequence of pairs eventually stops, and then we have for $y = y_n$ and $z = z_n$ (i) with $\varepsilon_n < \varepsilon$, (ii) with $\tau_n \in (\tau, \tau_0)$ and (iii) because we cannot continue. □

We immediately obtain

Corollary 2:

Given $x \in \Omega(\phi)$ and a section S through x, there exist arbitrarily short subsections $S_1 \subset S$ arbitrarily near x and arbitrarily long orbit segments joining $y \in S_1$ to $z \in S_1$ which cross S nowhere between, and such that the length of S_1 is $2|y - z|$ and the midpoint of S_1 is $(y + z)/2$.

Corollaries 1 and 2 together with lemma 2 now let us prove the following version of the closing lemma:

Lemma 4 (Closing lemma)

Suppose ϕ is a C^∞ flow and $q \notin Fix(\phi)$ has the property that for any transverse section S through q and any integer N there exists

$$r_N \in S \cap \Omega(\phi)$$

whose forward orbit meets S at least N times.

Then in any strong C^1 neighborhood $\mathcal{U}$ of ϕ there exists a flow ψ with

$$S \cap Per(\psi) \neq \emptyset .$$

Proof:

Take F a C^∞ flowbox about q with parametrization $[0,1] \times [-1,1]$ and $S = \{0\} \times [-1,1]$. Pick $\varepsilon > 0$ so that any flow ψ with $\dot\psi = \dot\phi$ off F and

$$|\dot\psi - \dot\phi| < \varepsilon , \quad |D\dot\psi - D\dot\phi| < \varepsilon \qquad \text{on } F$$

belongs to $\mathcal{U}$. Take $f : \mathbb{R}^2 \to [0,1]$ a function which is positive on $(0,1) \times (-1,1)$, 0 elsewhere, and with

$$\left|\frac{\partial f}{\partial x}\right|, \left|\frac{\partial f}{\partial t}\right| < 1 .$$

Set $A = [-1/2, 1/2]$ and let $k = k(A,f)$ be the constant given in lemma 2. Finally, pick an integer

$$N > 1/k\varepsilon .$$

Now, pick r_N as in the hypotheses of the lemma, and denote the successive intersections of the positive orbit of r_N with int S by $r_N = x_0, x_1, x_2, \ldots, x_N$. Use corallary 1 to reparametrize S (and hence F) via a perturbation of the identity on S (resp. F) whose C^1 distance from id is at most $\varepsilon/2$, in such a way that the first-return map $P : S \to S$ is linear (say $P = L_i$ near x_i) near x_i, $i = 0, 1, \ldots, N-1$. We can assume these neighborhoods of x_i are disjoint and interior to S; denote the P-image of the neighborhood of x_i by U_i, and let U_0 be the given neighborhood of $x_0 = r_N$. Any ϕ-orbit originating in U_0 crosses U_i $(i = 1, \ldots, N-1)$ in succession before (perhaps) re-entering U_0 again.

Using corollary 2, we can find points $y_0, z_0 \in U_0$ with $z_0 = \phi_T(y_0)$ for some $T > 0$, and such that the orbit segment from y_0 to z_0 does not cross the interval S_0 in S of length $2|z_0 - y_0|$ centered at $(y_0 + z_0)/2$. Now, let $y_i = L_i(y_{i-1})$, $z_i = L_i(z_{i-1})$. The linearity of L_i insures that the orbit segment from y_i to z_i does not cross the interval S_i in S of length $2(z_i - y_i)$ centered at $(y_i + z_i)/2$. In particular, $\phi_t(p_0) \in S_i$, $t > 0$ implies $t > T$.

Form the rectangle R as the union of orbit segments from S_0 to S_N; note that R is orientable even if M is not. Let R_i be the subrectangle of R formed by orbit segments in F originating at S_i, for $i = 0, \ldots, N-1$. In each R_i, pick a sign $\delta_i = \pm 1$ according to whether or not the orientation of R_i induced by that in R agrees with that induced by F.

We shall perturb ϕ only on $R \cap F = \bigcup^N R_i$. Let $\tilde{z}$ be the last point on $\mathcal{O}_-(z_0)$ in R_N. Our goal is to change ϕ on $R \cap F$ so that $\tilde{z}$ lies on the forward orbit of z_0. Without loss of generality, we can pick the orientation in R so that $\tilde{z}$ lies above z_0.

Let b_i be half the width of R_i, and scale f to R_i by defining

$$f_{b_i}(t, x) = b_i f(t, x_i/b_i)$$

where x_i is x, rescaled affinely so that x_i ranges from -1 to +1 across R_i. Then define a vectorfield $Y(t, x)$ on F with support in $R \cap F$ and

$$Y(t, x) = (0, \varepsilon\, \delta_i\, f_{b_i}(t, x)) \quad \text{on } R_i .$$

Thus, in terms of the orientation on R, Y points up. Now, define a one-parameter family of vectorfields X by

$$X_\eta = \dot{\phi} \qquad \text{off } F$$

$$X_\eta = \dot{\phi} + \eta Y \quad \text{on } F .$$

Define the flow ψ_η by

$$\dot{\psi}_\eta = X_\eta .$$

The reader can check that

$$\|X_\eta - X\|_{C^1} < \varepsilon$$

while lemma 2 guarantees that the ηY term in X_η lifts the orbit across R_i of any point beginning between y_i and z_i by at least $\eta\,\varepsilon\, b_i k$. The linearity of L_i then guarantees that this lift shows up in the U_{i+1} as $\eta\,\varepsilon\, b_{i+1}k$. Thus as long as it stays below the orbit of y_0, the orbit of z_0 in R is lifted successively at least $i\eta\,\varepsilon\, b_i k$ by the time it finishes crossing R_i. Thus, the ψ_η-orbit of z_0 either crosses the ϕ-orbit of y_0 before leaving R_N, or else it leaves R_N above

$$z_0 + N\eta\,\varepsilon\, b_N k .$$

But our estimates on N guarantees that this is above $z_0 + \eta b_N$. For $\eta = 1$, this is above y_0. Thus, for some $\eta \in (0, 1)$, the forward ψ_η-orbit of z_0 leaves R_N via $\tilde{z}$, as desired. □

Proof of Proposition 1:

First, let us assume ϕ is C^∞. Given q as in the hypotheses, we pick nested transversals S_i through q with length $\varepsilon_i < \varepsilon_{i-1}/2$. By lemma 4, each C^1-neighborhood $\mathcal{U}$ of ϕ contains a flow ψ with $\mathrm{Per}(\psi) \cap S_i \neq \emptyset$. Given ε_i, pick $\mathcal{U}_i$ using the continuous structural stability of ϕ so that ψ_i is equivalent to ϕ via a homeomorphism h_i moving points by at most ε_i on S_i. Thus, $h_i[\mathrm{Per}(\psi) \cap S_i] \subset S_{i-1}$, so that $\mathrm{Per}(\phi) \cap S_{i-1} \neq \emptyset$.

Thus, $q \in \overline{\mathrm{Per}(\phi)}$.

Now, if ϕ is not C^∞, we perturb ϕ to a C^∞ flow $\tilde{\phi}$; by stability of ϕ, there is an equivalence homeomorphism taking q to $\tilde{q}$, and the argument above shows $\tilde{q}$ $\overline{\mathrm{Per}(\tilde{\phi})}$. so that pulling back by the homeomorphism,

$$q \in \overline{\mathrm{Per}(\phi)} . \quad \square$$

2. *Accumulation of Periodic Orbits*

In this section we prove

Proposition 2: *If M is a surface of finite genus, then for any continuously C^r structurally stable flow ϕ on M, Per(ϕ) is a closed set.*

The proof of prop. 2 will be based on the following topological observation:

Proposition 3: *Suppose ϕ is a flow on the surface M and $g(M) < \infty$. Let T be a transverse section through p such that one component T_+ of $T \setminus \{p\}$ contains periodic points arbitrarily near p. Then there exists an open punctured annulus $A_+ \subset M$ containing an open subinterval of T_+ with one endpoint p.*

Proof of Prop. 3:

Since a periodic orbit crosses a transverse section in only finitely many points, we can pick $p_n \to p$ in T lying on distinct periodic orbits. We would like to show that for all m, n sufficiently large, $\mathcal{O}(p_m)$ and $\mathcal{O}(p_n)$ bound an invariant punctured annulus $A_{m,n} \subset M$.

To this end, note that if a loop γ fails to separate M, then $g(M \setminus \gamma) \leq g(M) - 1$, whereas if γ separates M into components M_1 and M_2, then $g(M) \geq g(M_1) + g(M_2)$, so that either $g(M_1) \leq g(M) - 1$ or $g(M_2) = 0$, and then M_2 is a punctured annulus.

Now, consider the nested sequence of open submanifolds

$$M_1 \supset M_2 \supset \ldots,$$

formed by taking M_{n+1} the component of $M_n \setminus \mathcal{O}(p_{n+1})$ containing p, $M_0 = M$. By the preceding paragraph, $g(M_{n+1}) \leq g(M_n) - 1$ unless $A_{n+1} = M_n \setminus [M_{n+1} \cup \mathcal{O}(p_{n+1})]$ is a punctured annulus. Since

$$\infty > g(M_0) \geq g(M_1) \geq \ldots \geq 0,$$

it follows that $g(M_{n+1}) = g(M_n)$ for all $n \geq N$, and thus A_n is a punctured annulus for $n > N$.

Since the same argument works if we skip some p_n, we see that for all m larger than some N_1, and all $n > m$,

$$A_{m-n} = A_m \cup \mathcal{O}(p_{m+1}) \cup A_{m+1} \cup \mathcal{O}(p_{m+2}) \cup \ldots \cup A_n$$

is an open punctured annulus.

But then $A_+ = \bigcup_{n>m} A_{m,n}$

is an open annulus as desired. $\square$

We note that the proof gives us as well the fact that for some subinterval I_+ of T_+ ending at p, I_+ intersects any periodic orbit at most once. We caution the reader that the annulus A_+ may be embedded in M with a complicated boundary, and when M is nonorientable it may also contain points approaching p along T from the other side; for example the flow on the projective plane obtained by identifying the edges of figure 1 has this property. Let us call an accumulation point p of periodic orbits *two-sided* if the punctured annulus A_+ can be chosen disjoint from a half-transversal through p, and *one-sided* if there exist periodic orbits which cross T arbitrarily near p on both sides (so that A_+ must contain a punctured neighborhood of p in T). Note that one-sided accumulations can occur only on nonorientable surfaces.

We shall need also the following technical lemma concerning the direction of motion of a periodic orbit under perturbation.

Lemma 5:

Let γ be a periodic sink for ϕ, A an annulus neighborhood of γ in its basin of attraction (with $\dot{\phi}$ transverse to ∂A), and $T:(-1,1) \to A$ a transversal to ϕ crossing γ only at $T(0)$. Let $\vec{i}$ be the unit vector parallel to T at $T(0)$, in the direction of increasing r.

Suppose Y is a vectorfield on A which at each point x either vanishes, is parallel to $\dot{\phi}(x)$, or has the same orientation relative to $\dot{\phi}(x)$ as $\vec{i}$ has relative to $\dot{\phi}(T(0))$. Define a local flow ψ_μ by

$$\dot{\psi}_\mu(x) = \dot{\phi}(x) + \mu\, Y(x)$$

and, for μ small, denote by $r(\mu)$ the unique r-value at which the ψ_μ-attractor crosses T.

Then, for small μ, $r(\mu)$ is a non-decreasing function of μ.

Proof of Lemma 5:

Consider the component A_+ of $A \setminus \gamma$ containing the half-transversal $r \geq 0$. We claim that clos A_+ contains a periodic orbit of ψ_μ, for $\mu > 0$ small. By hypothesis, $\dot{\psi}_\mu$ never points out of A_+ along γ, and for μ small it points in

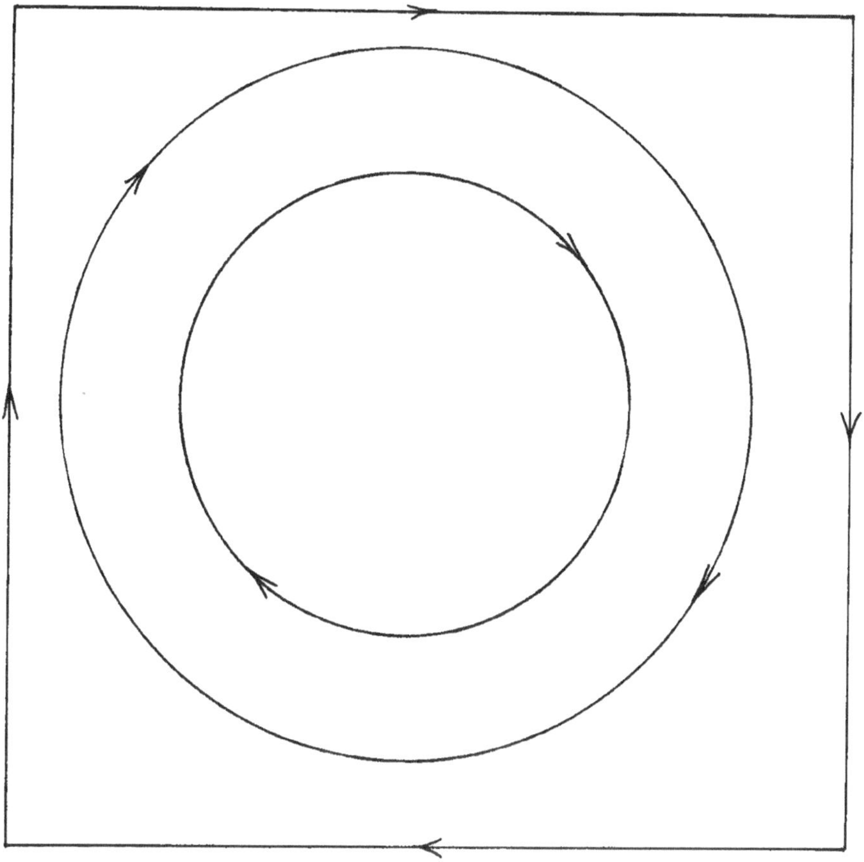

Figure 1

along the other boundary of A_+ (by transversality when $\mu = 0$). Thus, the Poincaré map of ψ_μ takes the closed transversal $r \geq 0$ into itself and has a fixedpoint, which is the required periodic orbit. Since for μ small ψ_μ has a *unique* periodic orbit in A, this shows $r(\mu) \geq 0$ for $\mu > 0$ small. But this argument could be repeated for ψ_μ near its (unique) attractor in A to show that $r(\mu)$ is non-decreasing. $\square$

Proof of Proposition 2:

Let $p_n \to p$, $p_n \in \mathrm{Per}(\phi)$, $p \notin \mathrm{Per}(\phi)$. Since $\mathrm{Per}(\phi)$ includes all fixed-points, we can assume p and p_n are nonsingular. Let F be a flowbox about p, which we can assume parametrized by $(t,y) \in (-1,1) \times (-1,1)$, with exit and entrance sets $S_\pm = \{\pm 1\} \times (-1,1)$. We can assume p is $(0,0)$ and p_n is $(0,y_n)$, with $y_n \downarrow 0$. Let $T = \{0\} \times (-1,1)$, and define $F_+ = (0,1) \times (-1,1)$, $F_- = (-1,0) \times (-1,1)$, $T_\pm = T \cap F_\pm$.

Since ϕ is Kupka-Smale, we can (reversing time if necessary and passing to a subsesequence) assume that p_n lies on a periodic sink γ_n for ϕ, which crosses T_+ only at p_n.

By prop. 3, there is a punctured annulus A_+ containing F_+ (shrinking F if necessary) and either disjoint from or containing F_-, depending on whether p is two-sided or one-sided. We separate these cases.

Case 1: p two-sided ($A_+ \cap F_- = \emptyset$)

The complement A_- of A_+ is ϕ-invariant and contains F_-. Let Y be a vectorfield with support in F which is vertical and points downward; assume it is non-vanishing in a neighborhood of p. Define flows ψ_μ on M by

$$\dot{\psi}_\mu(x) = \dot{\phi}(x) + \mu\, Y(x) .$$

Note that for $\mu > 0$ the backward orbit of $(+1,0)$ enters A_+; it crosses S_- at a point $(-1, \alpha(\mu))$ which is increasing with μ, and no periodic orbit of ψ_μ can cross the interval I_μ of S_- between $(-1,0)$ and $(-1, \alpha(\mu))$, since all orbits crossing this interval start in A_+ and end in A_-, and no ψ_μ-orbit leaves A_-.

On the other hand, by continuous structural stability there exist periodic attractors $\gamma_n(\mu)$ with $\gamma_n(0) = \gamma_n$, $n = 1, 2, \ldots$ which intersect T at $p_n(\mu)$, varying continuously with μ near 0. By the lemma, these points all move downward as μ increases, and hence $\inf p_n(\mu) \leq p$. Note that, given n, there exists $\varepsilon_n > 0$ so that $p_n(\mu) > \alpha(\mu)$ for $|\mu| < \varepsilon_n$. But since for $\mu > 0$ there are no periodic points on I_μ, there must exist n with $p_n(\mu) < p$. Continuous variation of $p_n(\mu)$ and $\alpha(\mu)$ with μ now forces $\mathcal{O}(p_n(\mu)) \cap I_\mu \neq \emptyset$ for some large n and small μ, a contradiction.

Case 2: p one-sided

By hypothesis, there exist $q_n \in T_-$ with q_n on γ_n and $q_n \uparrow p$. We see from the lemma that, for F small enough and n large enough, γ_n crosses T only at these two points. Define $p_n^{\pm}$ and $q_n^{\pm}$ as the intersection of the orbit segment of p_n (resp q_n) in F with $S_{\pm}$.

Note that the argument for prop. 3 tells us that every periodic orbit crossing F is a two-sided loop in M. Now, consider a vectorfield Y with support in F and pointing downward, and define ψ_μ on M by

$$\dot{\psi}_\mu(x) = \dot{\phi}(x) + \mu\, Y(x)$$

as before. Note that for some large n there exists a small μ such that p_n^- and q_n^+ lie on a single ψ_μ- orbit segment in F. By hypothesis, $\dot{\phi} = \dot{\psi}_\mu$ outside F, and the ϕ-orbit segment from q_n^+ to p_n^- does not intersect F. Thus q_n^+ and p_n^- lie on a periodic ψ_μ-orbit, γ_μ. But investigation of the way orbits leaving S_+ near q_n^+ first hit S_- near p_n^- shows that γ_μ is a one-sided loop in M, contradicting continuous structural stability. □

3. *Generalized Saddle Connections*

Combining propositions 1 and 2, we see the following

<u>*Corollary 3:*</u> *If ϕ is C^1 continuously structurally stable on M^2, $g(M) < \infty$, then every nonwandering orbit has trivial limit sets. In particular if $\mathcal{O}_{\pm}(x,\phi)$ is oscillating, then $\mathcal{O}(x,\phi)$ wanders, and orbits in $L^{\pm}(x,\phi)$ themselves have empty limit sets.*

This shows that the phenomenon of oscillating orbits is a special case of "saddle connections" between saddles at infinity: if x belongs to an unbounded limit set $L^{\pm}(y,\phi)$, then $L^+(x,\phi) = L^-(x,\phi) = \emptyset$ and $x \in J^+(x,\phi)$. This section will establish

Proposition 4: If $g(M) < \infty$ and ϕ is a C^1 continuously structurally stable flow on M, then

$$W^+(\phi) \cap W^-(\phi) = \emptyset .$$

We postpone proof of prop. 4 in order to consider some of its consequences. Suppose ϕ is C^1 continuously structurally stable on M, $g(M) < \infty$. First, of course, there are no oscillating orbits. But one can say more:

Corollary 4: $\Omega(\phi) = Per(\phi)$.

This is easy to see from the earlier corollary: if $x \in \Omega(\phi) \setminus Per(\phi)$ then $L_\pm(x,\phi)$ are trivial, but $x \in J^+(x,\phi)$ and $x \in J^-(x,\phi)$, so that $x \in W^+(\phi) \cap W^-(\phi)$.

An obvious consequence of this is that separatrices wander:

Corollary 5: $[W^+(\phi) \cup W^-(\phi)] \cap \Omega(\phi) = \emptyset$.

But then it becomes easy to prove the necessity of condition (iii) in our theorem from prop. 4:

Corollary 6: clos $W^+(\phi) \cap$ clos $W^-(\phi) \subset Fix(\phi)$.

Proof of Corollary 6:

If x is a non-restpoint in clos $W^+(\phi) \cap$ clos $W^-(\phi)$, then it is non-periodic, hence wandering by corollary 4 above. Thus, there exists a wandering flow-box F about x. This means that any orbit crossing F does so only once. Pick points $y_n \in S_+ \cap W^-(\phi)$ and $z_n \in S_- \cap W^+(\phi)$ tending to the orbit of x. There exist arbitrarily C^1-small perturbations ψ of ϕ in F for which y_n and z_n (for some large n) belong to a single ψ-orbit segment. But since $\mathcal{O}_-(y_n,\phi)$ and $\mathcal{O}_+(z_n,\phi)$ do not enter F, these semi-orbits are unchanged for ψ, and thus $y_n \in W^-(\psi)$, $z_n \in W^+(\psi)$. But then $W^-(\psi) \cap W^+(\psi) \neq \emptyset$, and by conjugacy the same holds for ϕ, a contradiction to prop. 4. □

Thus, it remains only to prove prop. 4. The idea of our proof is to note that any particular separatrix has trivial limit set, which means that either it tends toward a saddle fixedpoint or else it limits on a particular "point at infinity" (or technically, an "end") of the manifold M. Even though this separatrix may be an accumulation of other separatrices, a topological equivalence which is near the identity should distinguish this separatrix from nearby ones whose limit sets are far from the given one.

To make this last notion more precise, suppose $x \in W^+(\phi)$. Then $L^+(x,\phi)$ is either a single saddle point or is empty. In either case (using the standard representation of open surfaces [Ri] when $L^+(x,\phi) = \emptyset$, we can find a finite union of circles C_+ separating x from $L^+(x,\phi)$ in the sense that $M \setminus C_+$ has two components, M_0 containing x and M_+ containing $\mathcal{O}_+(x_+,\phi)$ for some $x_+ \in \mathcal{O}_+(x,\phi)$. Since $x \in W^+(\phi)$, it has some $y \in J^+(x,\phi) \setminus L^+(x,\phi)$ and by shrinking M_+ we can assume $J^+(x,\phi) \cap C \neq \emptyset$. We can also perturb C slightly so that it is in *general position relative to* ϕ, that is

(a) *Any point of tangency between* $\dot{\phi}$ *and* C *is nondegenerate — locally homeomorphic to the tangency between one of the parabolas* $y = x^2 + c$ *with a horizontal line.*

(b) *The next intersection with* C *of each* ϕ*-semiorbit of a point of tangency between* ϕ *and* C*, if any, is a transverse one.*

Now, define the sets

$$S^\pm(C) = \{y \in C \mid \mathcal{O}_\pm(y,\phi) \subset M_+\}$$

$$T^\pm(C) = \{y \in C \mid \mathcal{O}_\pm(y,\phi) \cap C \neq \emptyset, \text{ and the first intersection is transversal}\}.$$

Note that $T^+(C)$ is open in C, and that the complement of $S^+(C) \cup T^+(C)$ is the intersection of C with the set of orbit segments containing tangencies with C. The fact that $x \in W^+(\phi)$ is reflected in the fact that $x \in S^+(C) \cap \operatorname{clos} T^+(C)$ Any such point has limit set in M_+ and some prolongational limit points on C. To distinguish prolongational relations caused by behavior outside M_+ from that caused inside, we define, for any point $y \in S^+(C)$, the *prolongational limit relative to* M_+ by

$$J^+(y,\phi,M_+) = \{z \in J^+(y,\phi) \cap C \mid \text{there exist orbit segments in } M_+ \text{ with ends near } y \text{ and } z \text{, respectively}\}.$$

Now, since a non-degenerate point of tangency of $\dot{\phi}$ with C lies on the boundary of $T^\pm(C) \cup \operatorname{int} S^\pm(C)$, we can, by slightly perturbing C, move any tangency which might occur on $J^+(x,\phi,M_+)$ into an open band of orbits which cross M_+ transversally in backward time (near x) and hence make sure that $J^+(x,\phi,M_+)$ contains no tangencies. Thus, we can find a closed interval I containing x

such that

(c) $I \subset S^+(C) \cup T^+(C)$, *the endpoints of* I *are interior to* $S^+(C)$ *or* $T^+(C)$, *and for* $y \in I$, $J^+(y,\phi,M_+) \subset S^+(C) \cup T^+(C)$.

We define

$$T^+(I) = T^+(C) \cap I, \quad S^+(I) = S^+(C) \cap I.$$

We would like to show $T^+(I)$ and $S^+(I)$ are respected by equivalence homeomorphisms near the identity. Note that a homeomorphism need not preserve C_+. However, $\dot\phi$ is transverse to I, so a homeomorphism h which is near the identity on I can be retracted via the flow to an equivalence homeomorphism which takes I to C, moving points very little.

The technical base for the proof of prop. 4 is the following

Lemma 6:

Suppose C *is a finite union of circles,* $M \setminus C = M_0 \cup M_+$, *and* C *is in general position relative to* ϕ *((a), (b) above). If* $I \subset C$ *is a closed interval satisfying (c) above then there exists* $\varepsilon > 0$ *and a compact neighborhood* N *of* C *such that: for every flow* ψ *with* $\dot\phi = \dot\psi$ *on* M_+, *any equivalence homeomorphism* $h : M \to M$ *for which* $h(I) \subset C$ *and* $\| h(x) - x \| < \varepsilon$ *(whenever* $x \in N$*) must satisfy*

$$h(T^+(I)) \subset T^+(C)$$
$$h(S^+(I)) \subset S^+(C)$$

Proof:

Since the endpoints of I are interior to $S^+(C)$ or $T^+(C)$, each endpoint has a neighborhood which belongs either to $S^+(C)$ or to $T^+(C)$. We pick $\varepsilon > 0$ small enough that neither endpoint maps out of this neighborhood under h.

The fact that h is a global equivalence means in particular that it preserves the relation of belonging to the same orbit. The fact that it is near the identity on C means that, except for a small neighborhood of tangency points (say $\frac{\varepsilon}{2}$), transversal points go to points on short orbit segments that cross C transversally. Our conditions on I insure that the interaction with C of $\mathcal{O}_+(y,\phi)$ for any $y \in T^+(I)$ is outside such a neighborhood of the tangencies. Hence h maps $T^+(I)$ into $T^+(C)$. Similarly, our conditions guarantee that if $y \in S^+(I)$, then $\mathcal{O}_+(y,\phi)$ does not intersect N. Thus, $h(y) \in S^+(C)$ since otherwise some point

$z \in \mathcal{O}_+(h(y)) \cap C$ is the h-image of a point outside N, but we can guarantee that for ε small $h^{-1}(C) \subset N$. □

Corollary 7:

In the situation above, if ϕ is C^1 continuously structurally stable, then there exists a strong C^1 neighborhood $\mathcal{U}$ of ϕ such that for any one-parameter family $\psi_\mu \subset \mathcal{U}$ with $\psi_0 = \phi$ and $\dot{\phi} = \dot{\psi}_\mu$ on M_+, there is an equivalence homeomorphism which is the identity on $S^+(I) \cap \operatorname{clos} T^+(I)$.

Proof:

The set $S^+(I) \cap \operatorname{clos} T^+(I)$ is closed and totally disconnected; thus any continuous family h_μ of homeomorphisms taking this set into itself must be constant. We can find such a family for ψ_μ with $h_0 = id$, hence $h_\mu = id$ for all μ. □

Proof of Prop. 4:

Suppose $x \in W^+(\phi) \cap W^-(\phi)$. We can find circles C_+ (resp. C_-) separating x from $L_+(x)$ (resp. $L_-(x)$) but with $J^+(x) \cap C_\pm \neq \emptyset$ as in the discussion above. Since the initial and final segments of the orbit of x are disjoint from $C_\pm$ and all crossings are transversal, $\mathcal{O}(x)$ crosses $C_\pm$ only finitely many times. By some surgery, we can replace $C_\pm$ with a finite union of circles bounding connected submanifolds-with-boundary $M_\pm$ such that $\mathcal{O}(x,\phi)$ crosses $C_+ \cup C_-$ at precisely two points, $x_+ \in C_+$ and $x_- \in C_-$, such that $\mathcal{O}_+(x_+\phi) \subset M_+$, $\mathcal{O}_-(x_-,\phi) \subset M_-$, and x is a different component of $M \setminus C_\pm$ than $M_\pm$. Note that it is conceivable that M_+ and M_- overlap, for example if x tends to the same point at infinity in both directions.

Now, take a flowbox F about x, crossing $\mathcal{O}(x,\phi)$ precisely once and disjoint from $C_\pm$. Note that $x_\pm \in S^\pm(C_\pm) \cap \operatorname{clos} T\,(C_\pm)$, and by the lemma and continuous stability there exists a strong C^1 neighborhood $\mathcal{U}$ of ϕ such that any perturbation $\psi \in \mathcal{U}$ of ϕ with $\dot{\psi} = \dot{\phi}$ off F is equivalent to ϕ via a homeomorphism h that is the identity on $S^\pm(C_\pm) \cap \operatorname{clos} T^\pm(C_\pm)$ at least near $x_\pm$; we need only

$$h(x_\pm) = x_\pm .$$

We note in particular that x_+ and x_- must lie on the same ψ-orbit. However, it is easy to create arbitrarily small perturbations of ϕ, supported in F, for which x_+ and x_- lie on different orbits, and this contradiction proves prop. 4. □

4. *Sufficient Conditions for Continuous Stability*

In sections 1-3 we showed that on any surface of finite genus a continuously C^1 structurally stable flow must satisfy the conditions (i)-(iii) of our main theorem. Here, we reconsider the proof of theorem A in [KKN] to see how it gives continuous stability of a flow (on any surface) satisfying (i)-(iii):

Proposition 5: *If ϕ is a C^1 flow on any surface M and ϕ satisfies*

(i) $\Omega(\phi) = Per(\phi)$

(ii) every orbit in $Per(\phi)$ is hyperbolic

(iii) $clos\ W^s(\phi) \cap clos\ W^u(\phi) \subset Fix(\phi)$

then ϕ is continuously C^1 structurally stable.

The proof of Theorem A [KKN] gives structural stability of ϕ, that is, the existence of a strong C^1 neighborhood $\mathcal{U}$ of ϕ and an equivalence homeomorphism h_ψ between ϕ and any $\psi \in \mathcal{U}$. However, the construction of h_ψ in [KKN], using Neumann's techniques, makes it difficult to see whether h_ψ varies continuously with $\psi \in \mathcal{U}$.

The basis of the construction of h_ψ is the existence for ϕ of a structure (analogous to the "tubular families" constructed by Palis [Pa, PaS]) which we call *grand palaces*. By a positive grand palace for ϕ with base Σ_+ we mean a locally finite collection $\mathcal{P}_+$ of (generalized) flowboxes, F, each assigned an integer height $h(F) \geq 0$ such that:

(i) the flowboxes have disjoint interiors

(ii) Σ_+ is closed, and equals the union of entrance sets of flowboxes with height 0

(iii) the entrance set of each flowbox F with $h(F) = n > 0$ is interior to the exit set of a unique flowbox F' with $h(F') = n - 1$

Note that local finiteness of $\mathcal{P}_+$ together with compactness of flowboxes guarantees

(iv) the exit set of any flowbox of height n contains finitely many entrance sets for flowboxes with height $n + 1$.

Given the flow ϕ, the whole structure of P_+ is determined by the base Σ_+ and the various exit sets of the flowboxes in P_+ which we refer to as the *floors* of the palace. We will regard Σ_+ as the "ground" or 0 level, and define the level of any other floor as $n+1$ if it is the exit set of a flowbox with height n.

The primary function of a palace is to distinguish certain orbits. This is done by distinguishing towers of the palace. A *tower* of the palace P_+ is a maximal sequence $T = \{S_i\}_{i=1}^{N}$ (N finite or infinite) of successive floors S_i, at levels $i = 1, 2, \ldots, N$, such that the associated flowboxes are joined as in (iii) above. By picking the floors in T with lengths converging to 0 sufficiently quickly, it can be guaranteed that for any infinite tower T there exists a unique point $W(T,\phi) \in \Sigma_+$ whose positive semiorbit under ϕ crosses all the floors of T in succession. This property persists when ϕ is perturbed in the strong C^1 topology to ψ, and in fact this can be guaranteed simultaneously for all infinite towers of P_+ ([KKN, prop. 4.7]). Thus, if P_+ is a grand palace for ϕ, it is also a grand palace for every $\psi \in \mathcal{U}$, where $\mathcal{U}$ is some appropriate strong C^1-neighborhood of ϕ, and the set $W(P_+,\psi) \subset \Sigma_+$ defined as the union over all towers T of $W(T,\psi)$ is homeomorphic to $W(P_+,\phi)$ via a homeomorphism that preserves order on Σ_+ and takes $W(T,\psi)$ onto $W(T,\phi)$ for every tower T of P_+.

We define negative grand palaces P_-, with base Σ_- and all the related structure, as positive grand palaces for the backward flow.

When ϕ satisfies conditions (i)-(iii) of prop. 5, it is shown in [KKN] (lemma 5.4) that one can find a neighborhood Q of the fixed saddles of ϕ consisting of quadrilaterals Q_i isolating the saddles σ_i, together with two grand palaces, P_+ (positive, with base Σ_+) and P_- (negative, with base Σ_-) such that:

(i) every non-periodic orbit in clos $W^+(\phi)$ (resp. clos $W^-(\phi)$) crosses Σ_+ (Σ_-) in a unique point, which equals $W(T,\phi)$ for some infinite tower T of P_+.

(ii) the intersection of $\Sigma_{\mp}$ with any component $A_\pm$ of the complement of $\Omega(\phi) \cup$ clos $W^\pm(\phi)$ is a closed cross-section for it.

(iii) every orbit not in

$$\mathrm{Per}(\phi) \cup \operatorname{clos} W^{+}(\phi) \cup \operatorname{clos} W^{-}(\phi)$$

crosses Σ_+ and later crosses Σ_-, each in a unique point.

(iv) the edges of Q_i are transverse sections through $W^{\pm}(\sigma_i)$, and coincide with the top floors of two finite towers in each of $P_{\pm}$

(v) each infinite tower of $P_{\pm}$ is as in (i) and each finite tower is as in (iv).

A picture of this edifice is sketched in Fig. 2. Note that for each floor S of P_+, the set of points in S whose positive ϕ-semi-orbit fails to cross a floor of P_+ one level higher consists of finitely many intervals (the complement of entrance sets for flowboxes at the next height). The union of ϕ-orbit segments from one such interval until their first intersection with a floor of P_- forms a bounded set, $R(\phi)$, which is a topological disc and is a rectangle if all these segments enter P_- via a single floor. It is easy to see that the edifice described above can be constructed so that

(vi) no ϕ-orbit segment joins an endpoint of a floor of P_+ to an endpoint of a floor of P_-.

It is clear that assuming (vi) for ϕ there are local estimates on a neighborhood of $R(\phi)$ which insure that the edges of the corresponding set $R(\psi)$ vary continuously with ψ and in particular belong to the same floor.

Now, for each floor S of P_+ (resp. P_-) we let $\Sigma_+(S,\phi)$ (resp. $\Sigma_-(S,\phi)$) denote the set of points in Σ_+ (resp. Σ_-) whose positive (resp. negative) ϕ-semiorbit crosses S, and set $\Sigma_{\pm}(n,\phi)$ the union of $\Sigma_{\pm}(S,\phi)$ for all floors of level n. We note that $\Sigma_{\pm}(n,\phi)$ is a locally finite family of disjoint closed intervals in $\Sigma_{\pm}$, and $\Sigma_{\pm}(n+1,\phi)$ is interior to $\Sigma_{\pm}(n,\phi)$. Note that $\bigcap_n \Sigma_{\pm}(n,\phi)$ is the intersection of $\Sigma_{\pm}$ with the closure of the set of stable (unstable) separatrices of saddles at infinity. We also define $K_+(\phi)$ (resp. $K_-(\phi)$) as the intersection of Σ_+ (resp. Σ_-) with the ϕ-orbits of endpoints of floors of P_- (resp. P_+).

It is clear that for ψ near ϕ, the corresponding sets $\Sigma_{\pm}(S,\psi)$, $\Sigma_{\pm}(n,\psi)$, and $K_{\pm}(\psi)$ can also be defined, and that each varies continuously with ψ. Furthermore, using the boundedness of the "rectangles" $R(\psi)$ above we find that any component of $\Sigma_{\pm}(n,\phi) \setminus \Sigma_{\pm}(n+1,\phi)$ intersects only finitely many points of

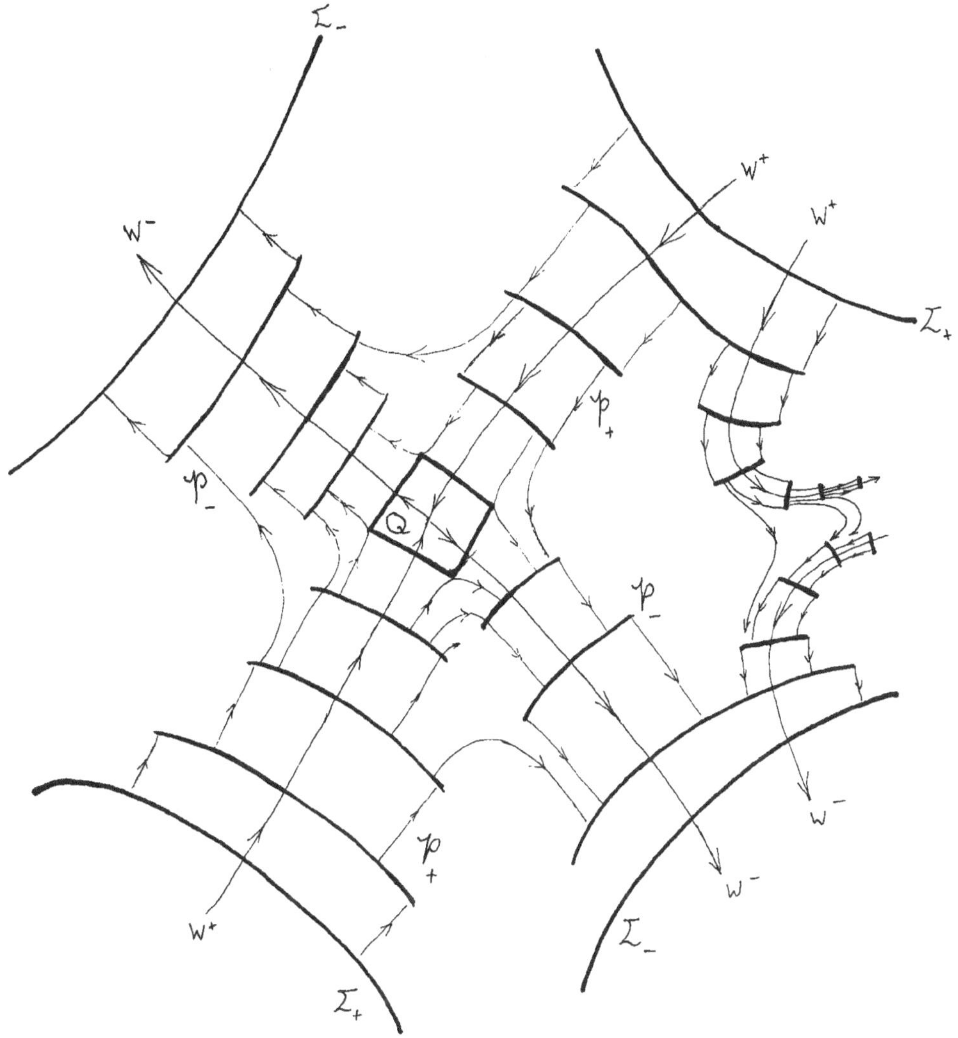

Figure 2

$K_{\pm}(\phi)$, so that $K_{\pm}(\phi)$ is locally finite and closed in the complement of $\sum_{\pm} \cap \operatorname{clos} W^{\pm}(\phi)$.

Now, given the grand palaces $P_{\pm}$ for ϕ, and ψ C^1-near ϕ, we define the homeomorphism h_{ψ} in $\sum_{+}$ by first requiring that

(i) for every floor S of P_{+}, the endpoints of $\sum_{+}(S,\phi)$ go to the endpoints of $\sum_{+}(S,\psi)$.

(ii) $K_{+}(\phi)$ goes to $K_{+}(\psi)$

(iii) h_{ψ} preserves the order along $\sum_{+}$ for the points in (i) and (ii).

These conditions define h_{ψ} uniquely on $K_{+}(\phi)$ and the endpoints of $\sum_{+}(n,\phi)$; there is then a unique continuous extension to the points of accumulation, which are precisely given by

(iv) $\bigcap_n \sum_{+}(n,\phi)$ goes to $\bigcap_n \sum_{+}(n,\psi)$.

Each finite tower of P_{+} ends in a quadrilateral Q_i which isolates a unique fixed saddle $\sigma_i(\phi)$ $(\sigma_i(\psi))$ of ϕ (resp. ψ). This tower intersects a unique stable separatrix γ for ξ_i; $\xi_i(\psi)$ varies continuously with ψ, and we can require

(v) $\xi_i(\phi) \cap \sum_{+}$ goes to $\xi_i(\psi) \cap \sum_{+}$.

Conditions (i)-(v) define h_{ψ} uniquely (given $P_{\pm}$) on a closed, nowhere-dense subset $H_{+}(\phi)$ of $\sum_{+}$, and varying continuously with ψ. Now, extend h_{ψ} to all of $\sum_{+}$ by

(vi) h_{ψ} is linear on each component of $\sum_{+} \setminus H_{+}(\phi)$.

Next, we define h_{ψ} on $\sum_{-}$. The condition that h_{ψ} be an equivalence homeomorphism forces a definition on the set $H_{-}(\phi)$ corresponding to $H_{+}(\phi)$, and here h_{ψ} varies continuously with ψ and satisfies the analogues of (i)-(v). We cannot extend linearly on the complement of $H_{-}(\phi)$ in $\sum_{-}$ as in (vi), but our definition is again forced by the condition that h_{ψ} be an equivalence homeomorphism.

We extrapolate h_{ψ} to orbit segments passing from $\sum_{+}$ to $\sum_{-}$ by making h_{ψ} piecewise linear with respect to the time parameter along each such segment. To be more precise, each ϕ-orbit segment with ends $x_{+} \in \sum_{+}$ and $x_{-} \in \sum_{-}$ crosses a finite succession of floors of P_{+}, then crosses a finite succession of floors of P_{-} with descending levels. The homeomorphism h_{ψ} defined so far on $\sum_{+} \cup \sum_{-}$ locates the ψ-orbit segment joining $h_{\psi}(x_{+}) \in \sum_{+}$ to $h_{\psi}(x_{-}) \in \sum_{-}$. The fact that

h_ψ maps $K_\pm(\phi)$ to $K_\pm(\psi)$ and $\sum_\pm (n,S,\phi)$ to $\sum_\pm n,S,\psi)$ for every floor S guarantees that this ψ-orbit segment crosses the same floors in the same succession: we define h_ψ to respect these crossings and extend linearly (in time) to their complement.

In this way, h_ψ is defined continuously, and varying continuously with ψ, on the union of orbit segments joining $\sum_+$ to $\sum_-$. We need to show that this has a unique continuous extension to the closure, which consists of the semiorbits of clos $W^\pm(\phi)$ starting from $\sum_\pm$.

We can suppose without loss of generality that the time of passage under ϕ between successive levels of $\mathcal{P}_\pm$ (when it occurs) takes unit time. One sees easily that local estimates on a neighborhood of any "rectangle" $R(\phi)$ between a part of a floor of $\mathcal{P}_+$ and the first intersections with floors of $\mathcal{P}_-$ (as discussed above) guarantee that on any component of the complement of $H_+(\phi)$, the time of passage between the last floor in $\mathcal{P}_+$ and the (unique) first floor in $\mathcal{P}_-$ is bounded, and varies continuously with ψ, when measured along orbits that correspond under h_ψ. Thus, local estimates can be used to define a strong C^1-neighborhood $\mathcal{U}$ of ϕ such that all of these (corresponding) passage times for ϕ and ψ, respectively, vary by less than 1 for $\psi \in \mathcal{U}$. In particular, if $x \in \sum_+$ and $\phi(T,x) \in \sum_-$, then $\psi(S,h_\psi(x)) \in \sum_-$ with $|S - T| \leq 1$.

Now, consider a sequence of points y_n in the domain of h_ψ (defined so far) converging to a point y_∞, say $y_\infty \in$ clos $W^+(\phi)$. We need to show that $h_\psi(y_n)$ is convergent to a point $h_\psi(y_\infty)$ clos $W^+(\psi)$. Let $x_n \in \sum_+$ be the points of intersection of the ϕ-orbit of y_n with $\sum_+$. Then $x_n \to x_\infty \in$ clos $W^+(\phi) \cap \sum_+$, and we already have $h_\psi(x_n \to h_\psi(x_\infty) \in$ clos $W^+(\psi) \cap \sum_+$, and $y_\infty \in \mathcal{O}_+(h_\psi(x_\infty), \psi)$.

Furthermore, we have $y_n = \phi(t_n, x_n)$, $h_\psi(y_n) = \psi(s_n, x_n)$, $y_\infty = \phi(t_\infty, x_\infty)$. We know that $t_n \to t_\infty$, and need to show that $s_n \to s_\infty < \infty$. Our problem is that s_n is defined by the ratio

$$s_n : S_n = t_n : T_n$$

where $\phi(T_n, x_n)$, $\psi(S_n, y_n) \in \sum_-$. But then, since $T_n \to \infty$ and $|S_n - T_n| \leq 1$, we see that

$$s_n / t_n = S_n / T_n \to 1$$

and thus in fact $s_n \to s_\infty = t_\infty$.

We have, therefore, a well-defined extension of h_ψ to an equivalence homeomorphism defined on the complement of the backward orbits of $\sum_+$ and the forward orbits of $\sum_-$, varying continuously with ψ. (Note that the previous paragraph gives us a unique extension to the backward orbits of clos $W^-(\phi)\cap\sum_-$ and hence also to the fixed saddles in Q.) Now, we can extend h_ψ forward from $\sum_-$ and backward from $\sum_+$ by making it a conjugacy $(h_\psi(\phi(t,x) = \psi(t,h_\psi(x)))$. This can present problems only at sources (resp. sinks), and that only at periodic sources (or sinks), since their periods may vary under perturbation. However, we can multiply $\dot{\psi}$ by a function supported on disjoint compact neighborhoods of the sinks and sources, and varying continuously with ψ, so that these periods remain constant. Finally, we can invoke the well-known fact that hyperbolic limit cycles have asymptotic phase (eg, [Ha], thm. 11.1) to insure that a conjugacy defined on the complement of a neighborhood of a limit cycle (with the same period for both flows) has a unique extension to the cycles themselves.

This completes the construction of h_ψ, taking ϕ-orbits to ψ-orbits and varying continuously (at ϕ) with ψ. In fact, given the floors of the palaces $P_\pm$, the definition of h_ψ is entirely determined by the flow ψ, and it is therefore clear that by fixing $P_\pm$ we have defined, for any two flows $\psi_1, \psi_2 \in \mathcal{U}$, an equivalence homeomorphism $h_{\psi_1\psi_2}$ taking ψ_1-orbits to ψ_2-orbits and varying continuously with ψ_1 and ψ_2. This proves proposition 5.

REFERENCES

Fr J. Franks, Differentiably Ω-stable diffeomorphisms, *Topology 11* (1972) 107-113.

Gk J. Guckenheimer, Absolutely Ω-stable diffeomorphisms, *Topology 11* (1972) 195-197.

Gt C. Gutierrez, Smooth nonorientable nontrivial recurrence on two-manifolds, *J. Diff. Eqns. 29* (1978) 388-395.

Ha P. Hartman, *Ordinary Differential Equations.* (N.Y.: Wiley, 1964); reissued by the author, Baltimore, 1973.

KKN J. Kotus, M. Krych, and Z. Nitecki, *Global Structural Stability of Flows on Open Surfaces.* Memoirs A.M.S., to appear.

Kr M. Krych, Two remarks on structural stability of plane dynamical systems *Astérisque 50* (1977) 197-204.

Ma L. Markus, Structurally stable differential systems, *Annals Math. 73* (1961) 1-19.

Ne D. A. Neumann, Classification of continuous flows on 2-manifolds, *Proc. A.M.S.* *48* (1975) 73-81.

Pa J. Palis, On Morse-Smale dynamical systems, *Topology* *4* (1969) 385-404.

PaS _______ and S. Smale, Structural stability theorems, *Proc. Symp. Pure Math.* *14* (Providence: A.M.S., 1970) 223-231.

Pe1 M. Peixoto, Structural stability on two-dimensional manifolds, *Topology* *1* (1962) 101-120, *2* (1963) 179-180.

Pe2 _______, On an approximation theorem of Kupka and Smale, *J. Diff. Eqns.* *3* (1966) 214-227.

PePu _______, and C. Pugh, Structurally stable systems on open manifolds are never dense, *Annals Math.* *87* (1968) 423-430.

Pu C. Pugh, The closing lemma, *Am. J. Math.* *89* (1967) 956-1009.

PuR _______ and C. Robinson, The closing lemma, including Hamiltonians. Preprin

Ri I. Richards, On the classification of noncompact surfaces, *Trans. A.M.S.* *106* (1963) 259-269.

S A. Schwartz, A generalization of a Poincaré-Bendixson theorem to closed two-dimensional manifolds, *Am. J. Math* *85* (1963) 453-458.

TW F. Takens and W. White, Vectorfields with no nonwandering points, *Am. J. Math.* *98* (1976) 415-425.

Instituto de Matemática Pura e Aplicada, Rio de Janeiro, Brazil

Instytut Matematyki Uniwersytetu Warszawskiego, Warszawa, Poland

Department of Mathematics, Tufts University, Medford, Mass., USA

SOME REMARKS ABOUT HOMOCLINIC POINTS OF SECOND ORDER DIFFERENTIAL EQUATIONS.

by Sonia de Carvalho* and Robert Roussarie.

Let $g(x)$ a $\mathscr{C}^2$ function of $x \in \mathbb{R}$ and $f(t)$ a $\mathscr{C}^2$ periodic function of t, with frequency $\nu > 0$. We look to the forced second order differential equation :

$$\ddot{x} + g(x) = - \lambda_1 \dot{x} + \lambda_2 f(t) \qquad (1)$$

where $x(t)$ is a function of t, $\dot{x}$, $\ddot{x}$ are the first and second order derivatives and $\lambda = (\lambda_1, \lambda_2)$ is a small parameter.

Examples : Duffin equation : $g(x) = - x + x^2$

Pendulum : $g(x) = \sin x$.

Many studies have been devoted to such non linear equation. (See, for instance [H_0] , [G] , [M] and [M_0]). Here, we want to investigate the appearance of homoclinic points under small perturbations ($\lambda \neq 0$). For this study, we use the methods and results of [H] . More details will appear in the thesis of Sonia de Carvalho.

We begin with the non perturbed equation ($\lambda = (\lambda_1, \lambda_2) = 0$). This equation is equivalent to the 2-dimensional Hamiltonian system :

$$(2) \quad \begin{cases} \dot{x} = y \\ \dot{y} = -g(x). \end{cases}$$

We suppose that all the zeros of g are regular : ($g(x_0) = 0 \Rightarrow g'(x_0) \neq 0$). So, the Hamiltonian function of the system : $H(x,y) = \frac{1}{2} y^2 + U(x)$ with $U(x) = \int_0^x g(\tau)d\tau$ is a Morse function. Its critical points are the points $(x_0, 0)$ where $g(x_0) = 0$. They are centers if $g'(x_0) > 0$ and saddle points, if $g'(x_0) < 0$. Now, a saddle connection is a trajectory of the system (2), in the phase space

* Research supported by CAPES - Brazil.

(x,y), $(p(t),\dot{p}(t))$, which tends to some saddle point for $t \to +\infty$ and $t \to -\infty$. The limits points p_0, p_1 may be equal (homoclinic connection) or different (heteroclinic connection). For example, the Duffin equation exhibits an homoclinic connection tending to $(0,0)$ and the Pendulum equation exhibit an heteroclinic connection from $(-\pi,0)$ to $(+\pi,0)$. A consequence of the symmetry $(x,y) \to (x,-y)$ is that heteroclinic connections appear by pair.

We give now a brief account of the perturbation theory, based on the article of Chow, Hale, Mallet-Parret [H].

So, suppose that $f(t)$ is a $\mathcal{C}^2$, $\frac{2\pi}{\nu}$-periodic function and that the non perturbed equation has a saddle connection from p_0 to p_1. The perturbed system is $\frac{2\pi}{\nu}$-periodic in t. Its Poincaré map from $\mathbb{R}^2 \times \{0\}$ to $\mathbb{R}^2 \times \{\frac{2\pi}{\nu}\}$ in the (x,y,t)-space is a diffeomorphism $A(\lambda)$. For small λ, $A(\lambda)$ has unique saddle points $p_0(\lambda)$ and $p_1(\lambda)$ near p_0 and p_1.

The perturbation theory gives a description of the intersection of $W^u(p_0(\lambda))$ the unstable manifold of $p_0(\lambda)$ and $W^s(p_1(\lambda))$, the stable manifold of $p_1(\lambda)$ for small λ. Precisely, let $\Gamma(t) = (p(t), \dot{p}(t))$ the saddle connection, with $p(0) = 0$ in the homoclinic case, $p(0)$ maximum, in the heteroclinic case. Then, there exists a $\mathcal{C}^2$ function $G(\lambda,\alpha)$, $\frac{2\pi}{\nu}$-periodic in α, giving the normal distance between $W^u(p_0(\lambda))$ and $W^s(p_1(\lambda))$, along the arc $\Gamma([0,\alpha])$, for small λ. This function G has the following expansion in λ :

$$G(\lambda,\alpha) = -\lambda_1 + \lambda_2 h(\alpha) + \tilde{G}(\lambda,\alpha)$$

where $\tilde{G}(\lambda,\alpha) = O(|\lambda|^2)$, $|\lambda| = |\lambda_1| + |\lambda_2|$

and $h(\alpha) = \frac{1}{\eta}\int_{-\infty}^{+\infty} \dot{p}(t) f(t-\alpha)\,dt$ where $\eta = \int_{-\infty}^{+\infty} \dot{p}(t)^2 dt$

(See [H]).

The formula for G is more easy to utilise in the coordinates (β,λ_2) where $\beta = \frac{\lambda_1}{\lambda_2}$ for $\lambda_2 \neq 0$. Then :

$$G(\lambda,\alpha) = \lambda_2 H(\beta,\lambda_2,\alpha) \text{ where :}$$

$$H(\beta,\lambda_2,\alpha) = -\beta + h(\alpha) + \tilde{H}(\beta,\lambda_2,\dot{\alpha})$$

and $\tilde{H} = O(\lambda_2^2)$ for bounded values of β .

The zeros of G and H are the same, with the same multiplicity in α . The study of these zeros, reduce to the study of the function $h(\alpha)$. The following proposition follows, for the essential part, from [H] :

<u>Proposition</u> : Let (α_o,β_o) such that $h(\alpha_o) = \beta_o$.

a) [H] : If $h'(\alpha_o) \neq 0$, then, there exists a neighborhood of $(\beta_o,0)$ in the (β,λ_2)-space (an angular sector in the (λ_1,λ_2)-space) , where G has a unique transversal zero near α_o (Then, there exists a transversal homoclinic point, near the point $(p(\alpha_o), p(\alpha_o))$. In the heteroclinic case, there exists heteroclinic points above and under the x-axis and also homoclinic points for $p_o(\lambda)$ and $p_1(\lambda)$).

b) [H] : If $h'(\alpha_o) = 0$ and $h''(\alpha_o) \neq 0$, there exists a differential curve in the λ-space, tangent to the line $\frac{\lambda_1}{\lambda_2} = h(\alpha_o)$ at 0, such that G has no zeros on one side and two transversal zeros on the other side, in some neighborhood of $(\beta_o,0)$ in the (β,λ)-space. (Crossing this line implies the generic birth of a pair of transversal homoclinic or heteroclinic points).

c) More generally, if h has in α_o a critical value of finite order ℓ $(h(\alpha_o) = \beta_o,\ h'(\alpha_o) = 0, \ldots h^{(\ell-1)}(\alpha_o) = 0 ,\ h^{(\ell)}(\alpha_o) \neq 0)$, then $G(\lambda,\alpha)$ has only critical points of order $\leq \ell$, near α_o, in some neighborhood of $(\beta_o,0)$ in the (β,λ_2)-space. (If ℓ is odd, then there always exists such critical point, for (β,λ_2) near $(\beta_o,0)$).

<u>Remark</u> : In [H] , the following generic condition is supposed for h : h possess only one quadratic minimum α_m and one quadratic maximum α_M. Then there exist curves C_m and C_M in the λ-space such that by crossing them homoclinic points are created.It is not completely evident for us, that there exists at least one homoclinic transversal point for each λ in an angular sector limited by C_m and C_M.

So, we look in the following to the more generic condition : h is a Morse function. Under this hypothesis, there exist transversal homoclinic points for λ in the whole angular sector limited by C_m and C_M.

Now, we want to investigate this Morse condition for h. We introduce for all $k=2,\ldots,\infty$ or $k=\omega$ the space $\mathcal{D}_\nu^k$ of $\frac{2\pi}{\nu}$ - périodic functions of class k (analytic functions for $k=\omega$).

For $f \in L^1(\mathbb{R})$, we introduce also the Fourier transform :

$$\mathcal{F}(f)(\xi)=\hat{f}(\xi)=\int_{-\infty}^{+\infty} e^{-i\xi t} f(t)dt.$$

Recall that the Fourier transform extends to the periodic functions ;

if $f(t) = \sum_{n\in\mathbb{Z}} C_n e^{i\nu nt}$, then $\hat{f} = \sum_{n\in\mathbb{Z}} C_n \delta_{\nu n}$ where δ_ξ is the Dirac mass at ξ.

Now, we note $f{*}g(\alpha) = \int_{-\infty}^{+\infty} f(t)\, g(\alpha-t)dt$, the convolution between two functions f,g.

The function $h(\alpha)=\frac{1}{\eta}\int_{-\infty}^{+\infty}\dot{p}(t)\, f(t-\alpha)dt$ is equal to : $h(\alpha)=\frac{1}{\eta}(\dot{p} * f\,(-\alpha))$ in the heteroclinic case, $h(\alpha)=-\frac{1}{\eta}(\dot{p} * f(-\alpha))$ in the homoclinic case.

So, by Fourier transform :

$$\hat{h}(\xi)=\pm\frac{1}{\eta}\,\hat{\dot{p}}(-\xi).\hat{f}(-\xi)$$

We note $\hat{\dot{p}} = P$. Let $f(t) = \sum_n C_n e^{in\nu t}$.

Then, we have the following formula for $\hat{h}$:

$$\hat{h}=\pm\frac{1}{\eta}\sum_n C_n P(-n\nu)\,\delta_{-n\nu} \qquad (3)$$

($\pm$: depending on the case).

This formula shows that the study of h depends on the distribution of the zeros of P. Let Z_p the set of zeros of P and $\tilde{Z}_p \subset Z_p$ defined by :

$\tilde{Z}_p = \{\xi\in\mathbb{R} \mid P(k\xi) = 0 \text{ for } \forall k\in\mathbb{Z}.\}$

Using the formula (3), it is easy to show the following :

Proposition : We have $\nu \in \tilde{Z}_p$ if and only if, for all k=2,...∞ ,ω ...., the set of $\dot{f} \in \mathcal{D}^k_\nu$ such that h is a Morse function is an open dense subset of $\mathcal{D}^k_\nu$. (If $\nu \in \tilde{Z}_p$, h is constant for all f).

Proof: The result follows easily from the continuity of the convolution operator $f \to \dot{p}*f$ from $\mathcal{D}^k_\nu$ to $\mathcal{D}^2_\nu$, from the result : $h = A \cos n\nu t + B \sin n\nu t$ for $P(n\nu) = A+iB$ and $f = \cos n\nu t$ and the remark that the Morse condition for h is an algebraic condition for f in each finite space of trigonometric polynomials

Remarks :

1) For the Pendulum : $P(\xi) = \dfrac{\pi}{ch\pi^2\xi}$

For the Duffin Equation : $P(\xi) = 24i\pi^2 \dfrac{\xi^2}{\cos\left(\dfrac{1}{ch2\pi^2\xi}\right)}$

In the two cases : $Z_p = \emptyset$ and the above result is true for every frequency ν .

2) In any case, under the hypothesis that g is a $\mathcal{C}^2$-function, $P(\xi)$ is analytic in the sense that P extends to an holomorphic function in a band $\{\xi+i\mu \mid |\mu| \leq \varepsilon\ \}$ for some $\varepsilon > 0$. This follows from the fact that $\dot{p}(t) \sim e^{-\Lambda_+ t}$ for some $\Lambda_+ > 0$ for $t \to +\infty$ and $\dot{p}(t) \sim e^{-\Lambda_- t}$ for some $\Lambda_- > 0$, for $t \to -\infty$ (A consequence of the hyperbolicity of $p_0(\lambda)$ and $p_1(\lambda)$). So, in any case, Z_p and $\tilde{Z}_p$ arc discret subsets of $\mathbb{R}$.

We show now that under analytic assumptions on f we always have, in général, transversal homoclinic points (And not only for generic f). We treat only the homoclinic case, but the following result is probably also true in hetéroclinic case.

Proposition : Suppose that g is $\mathcal{C}^\infty$ function with regular zeros and that system (2) has an homoclinic saddle connection. Let $f \in \mathcal{D}^\omega_\nu$.($f = \sum_n c_n e^{in\nu t}$ and $\exists R > 1$ and $M > 0$ such that : $|c_n| \leq MR^{|n|}$). Suppose that there exists $n \in \mathbb{Z}$ such that $P(-n\nu).C_n \neq 0$. Then there exists an angular sector S in the λ-space :

$\beta_1 \leq \frac{\lambda_1}{\lambda_2} \leq \beta_2$ (for some $\beta_1 < \beta_2$) such that for $\lambda \in S$, λ near 0, the perturbed system has transversal homoclinic points.

Remark : For instance if f(t) is a non-constant analytic function, the Duffin equation always exhibits transversal homoclinic points for some small values of λ.

Proof: If f is an analytic function h is also an analytic function. Now, the condition : $\exists$n such that $P(-n\nu).c_n \neq 0$ implies that h is a non-constant function (Note that n must be different from 0).

Choose $\beta_1 < \beta_2$ such that : Inf h $< \beta_1 < \beta_2 <$ Sup f.

Then, if $\beta \in [\beta_1, \beta_2]$, the function $-\beta+h(\alpha)$ has a zero of finite odd order. This implies that for small λ and λ_{1/λ_2} near β, the function $G(\lambda,\alpha)$ has also a zero of finite odd order (See the first proposition above). So, for the saddle point $p_0(\lambda)$, there exists an intersection point q of $W^u(p_0(\lambda))$ and $W^s(p_0(\lambda))$ which is an intersection point of finite odd order (This means that $W^u(p_0(\lambda))$ is given near q as graph of a function on $W^s(p_0(\lambda))$ with a zero of odd order in q). Now, the result follows from :

Lemma : Let p_0 an hyperbolic saddle point of a $\mathcal{C}^\infty$ diffeomorphism of the plane, such that the unstable manifold $W^u(p_0)$ and the stable manifold $W^s(p_0)$ have an intersection point q of finite odd order. Then, these manifolds have also some transversal intersection.

Proof of the lemma: Let $\lambda > 1$ and $\mu < 1$ the unstable and the stable eigenvalues of the given diffeomorphism G at p_0. We choose coordinates (x,y) around p_0 such that Ox is in $W^u(p_0)$, Oy is in $W^s(p_0)$ and p_0 is 0. We suppose also that for some $\varepsilon > 0$, such that $\mu + \varepsilon < 1 < \lambda - \varepsilon$, the components X(x,y) and Y(x,y) of G satisfy :

$$(\lambda-\varepsilon)x \leq X \leq (\lambda+\varepsilon)x$$

$$(\mu-\varepsilon)y \leq Y \leq (\mu+\varepsilon)y$$

(We can find the system of coordinates such that ε is as small as we desire. See below).

Now q_0, some positive iterate of q, belongs to Oy and $q_1 = (r,0)$, some negative iterate of q, belongs to Ox. Let ℓ be the contact order of $W^u(p_0)$ and $W^s(p_0)$ at q. Choose some segment J on $W^u(p_0)$ around q_0 and K some segment of $W^s(p_0)$ around q_1. The segments J and K have a contact of order ℓ with Oy and Ox respectively. The segment K is locally the graph of a function $y = \psi(x)$, $\psi(r) = 0$, with a zero of order ℓ at r.

So, for the function ψ we have the following estimates :

$$|\psi(x)| \leq N|x-r|^{\ell} \qquad (1)$$

$$\left|\frac{d\psi}{dx}\right| \geq \bar{N}|x-r|^{\ell-1} \qquad (2)$$ for some constants $N, \bar{N} > 0$.

and x near r.

Let I an interval around r, on Ox, not containing 0.

Then, for n big enough, J_n the nth iterate of J by the diffeomorphism G is, above I, the graph of a function $\varphi_n(x)$.

It is easy to show that there are constants $M_1, M_2 > 0$ independant of n, such that

$$|\varphi_n(x)| \geq M_2(\mu-\varepsilon)^n \qquad (3)$$

and $$\left|\frac{d\varphi_n}{dx}(x)\right| \leq M_1\left[\frac{\mu+\varepsilon}{(\lambda-\varepsilon)^{1/\ell}}\right]^n \qquad (4)$$

For n big enough, there exists $(x_n, y_n) \in J_n \cap K$ (Here we use the hypothesis that the contact of $W^s(p_0)$ and $W^u(p_0)$ in q is odd).

For this point (x_n, y_n), we have :

$$y_n = \varphi_n(x_n) = \psi(x_n).$$

Combining (1) and (3) we have :

$$|x_n - r|^{\ell} \geq \frac{1}{N} y_n \geq \frac{M_2}{N}(\mu-\varepsilon)^n$$

And so, with (2) , we have :

$$\left|\frac{d\psi}{d}(x_n)\right| \geqslant K \left[(\mu-\varepsilon)^{\frac{\ell-1}{\ell}}\right]^n \qquad (5)$$

for some constant $K > 0$, independant of n.

Now, by (4), we have :

$$\left|\frac{d\varphi_n}{dx}(x_n)\right| \leqslant M_1 \left[\frac{\mu+\varepsilon}{(\lambda-\varepsilon)^{1/\ell}}\right]^n$$

But $\lambda > \mu$ implies that $\mu^{1-1/\ell} > \mu/_{\lambda}1/\ell$ and for $\varepsilon > 0$ small enough :

$$(\mu-\varepsilon)^{1-1/\ell} > \frac{\mu + \varepsilon}{(\lambda-\varepsilon)^{1/\ell}}$$

It follows that, for n big enough, we have $\left|\frac{d\psi}{dx}(x_n)\right| > \left|\frac{d\varphi_n}{dx}(x_n)\right|$ and that K and J_n are transversal at the point (x_n, y_n).

[Ho] P. Holmes and J.E. Marsden : Qualitative techniques for bifurcation analysis of complex systems, in Bifurcation Theory and Application to Scientific Disciplines, New York Academy of Sciences, 1979, p.p. 608-622.

[G] R. Mc Gehee and K.R. Meyer : Homoclinic points of area preserving diffeomorphisms, Am. J. Math. 96 (1974), p.p. 409-421.

[M] V.K. Mel'nikov : One the stability of the center for time periodic solutions Transactions Moscow Math. Soc. (Trudy) 12 (1963), p.p. 3-56.

[Mo] A.D. Morozov : On the complete qualitative investigation of the equation of Duffin, Differentialniye Uravneniya, 12 (1976), p.p. 241-255.

[H] Shui-Nee CHOW, J. Mallet-Paret and J. K. Hale : An example of bifurcation to homoclinic orbits

Laboratoire de Topologie

E.R.A. 945 C.N.R.S.

Université de Dijon

ON THE LOCAL CLASSIFICATION OF HOLOMORPHIC

VECTOR FIELDS

by

Marc CHAPERON

The problem of classifying local holomorphic vector fields goes back to the second half of the nineteenth century, and has attracted many mathematicians in recent years. This is an account of the author's work on the subject, with a precise description of how a complex version of the Grobman-Hartman linearization theorem for flows is proved in [5].

The results we shall be discussing are part of my doctoral dissertation. I wish to thank Alain Chenciner, Michel Herman and René Thom, who taught me mathematics and Cesar Camacho, Freddy Dumortier, Nicolaas Kuiper, Ivan Kupka, Robert Moussu, Jacob Palis and Robert Roussarie, whose interest in the subject stimulated mine.

PART 1. DEFINITIONS AND MAIN RESULTS

1) Some equivalence relations

Let X_o and X_1 be holomorphic vector fields defined in open neighbourhoods V_o and V_1 of 0 in $\mathbb{C}^n$. For each $v \in V_i$ $(i = 0,1)$, we denote by $t \mapsto \exp_v tX_i$ the unique local holomorphic integral curve $\varphi_i : (\mathbb{C},0) \to \mathbb{C}^n$ of X_i such that $\varphi_i(0) = v$. The foliation defined by X_i is the foliation of $V_i \setminus X_i^{-1}(0)$ whose leaves are the holomorphic curves F such that $T_vF = \mathbb{C}\, X_i(v)$ for every $v \in F$ -in other words, some neighbourhood of v in F is the image of a neighbourhood of 0 by the map $t \mapsto \exp_v t X_i$.

Define X_o and X_1 to be

- holomorphically equivalent at 0 iff there exists a holomorphic local diffeomorphism $h : (\mathbb{C}^n,0) \circlearrowleft$ sending X_o onto X_1 ;
- strongly C^k-equivalent at 0 $(0 \leqslant k \leqslant \infty)$ iff there exist open neighbourhoods $U_o \subset V_o$ and $U_1 \subset V_1$ of 0 and a (real)C^k-diffeomorphism $h : U_o \to U_1$ such that the following holds : for every $v \in U_o$, the mappings $t \mapsto \exp_v tX_o$ and $t \mapsto h^{-1}(\exp_{h(v)} tX_1)$ are equal in some neighbourhood of $t = 0$. For $k \geqslant 1$, this is equivalent to saying that $h_*(X_o|_{U_o}) = X_1|_{U_1}$ and $h_*(iX_o|_{U_o}) = i\, X_1|_{U_1}$;
- weakly C^o-equivalent at 0 iff there exist U_o, U_1 as above and a homeomorphism $h : U_o \to U_1$ sending the foliation defined by $X_o|_{U_o}$ onto the foliation defined by $X_1|_{U_1}$.

As a trivial consequence of the implicit function theorem, we have

PROPOSITION 1 : If $X_o(0)$ and $X_1(0)$ are non zero, then X_o and X_1 are holomorphically equivalent at 0.

In the sequel, the letters X, X', X_k $(k \in \mathbb{N})$ will always denote holomorphic vector fields defined in some neighbourhood of 0 in $\mathbb{C}^n$, vanishing at 0, and "equivalent" will mean "equivalent at 0".

2) Invariants

Clearly, if X_o and X_1 are holomorphically or C^k- $(k \geq 1)$ equivalent, then their linear parts $L_i = dX_i(0) \in gl(n,\mathbb{C})$ $(i = 0,1)$ are conjugate, and in particular have the same eigenvalues.

This is the reason why people interested in getting structural stability results studied C^o-equivalence. Unfortunately, the situation is not as simple as in the real case : if X_o and X_1 are weakly C^o-equivalent, then [3] their linear parts L_o and L_1 must satisfy $\Delta(L_o) = \Delta(L_1)$ where the Camacho-Kuiper-Palis invariant $\Delta(L)$ of $L \in gl(n,\mathbb{C})$ is defined as follows : if $\lambda_1,\dots,\lambda_n$ denote the eigenvalues of L, counted with their multiplicities, then

$$\Delta(L) = \{c_1 \geq 0,\dots,c_n \geq 0 : \Sigma c_i = 1 \text{ and } \Sigma \frac{c_i}{\lambda_i} = 0\} .$$

Say L is in the Poincaré domain iff the convex hull $\operatorname{conv}\{\lambda_1,\dots,\lambda_n\}$ does not contain 0 in $\mathbb{C}$, in the Siegel domain otherwise (this terminology is due to Arnold [1]). Clearly, we have that $\Delta(L) = \emptyset$ in the former case, whereas the set $\{L : \Delta(L) \neq \emptyset\}$ contains the interior of the Siegel domain. Since this interior is nonvoid for $n > 2$, there is no hope for structural stability, and no reason for considering only the -very rough- weak C^o-equivalence relation.

Given any positive integer k, X_o and X_1 are said to be k-equivalent iff there exists a holomorphic local diffeomorphism $h : (\mathbb{C}^n,0) \circlearrowleft$ such that $h_* X_o$ and X_1 have the same k^{th} order Taylor expansion at 0 (of course h can be chosen polynomial of order at most k). If X_o and X_1 are k-equivalent for every k, they are formally equivalent ; if, moreover, X_o is the linear part of X_1, then X_1 is formally linearizable.

Strong C^k-equivalence $(1 \leq k < \infty)$ implies k-equivalence, whereas holomorphic or strong C^∞-equivalence imply formal equivalence.

THEOREM 0 (Poincaré) : Given $L \in gl(n,\mathbb{C})$ with eigenvalues $\lambda_1,\dots,\lambda_n$ (counted according to their multiplicities), let $P_1,\dots,P_n$ be the subsets of $\mathbb{N}^n$ defined by $P_i = \{(p_1,\dots,p_n) : \Sigma p_j > 1 \text{ and } \Sigma p_j\lambda_j = \lambda_i\}$, and let $(x_1,\dots,x_n)$ be a system of $\mathbb{C}$-linear coordinates on $\mathbb{C}^n$ such that, for some complex constants $a_{i,j}$,

$$x_i \circ L = \lambda_i x_i + \sum_{\substack{1 \leq j < i \\ \lambda_j = \lambda_i}} a_{i,j} x_j \quad \text{for every } i \in \{1,\dots,n\} .$$

Then, for every X with linear part L, there exist complex constants $b_{i,p}$ $(1 \leq i \leq n,\ p \in P_i)$ such that, for each positive integer k, X is k-equivalent to the polynomial vector field X_k given by

$$X_k . x_i = x_i \circ L + \sum_{\substack{p \in P_i \\ |p| \leq k}} b_{i,p} \, x^p \qquad (1 \leq i \leq n).$$

In particular, a necessary and sufficient condition for every such X to be formally linearizable is

$$\text{(HP)} \qquad P_1 \cup \ldots \cup P_n = \emptyset$$

See [1] or [6] for the (straight-forward) proof of this result.

Given L and $\lambda_1, \ldots, \lambda_n$ as in Theorem 0, we shall call L

- strongly hyperbolic iff $i \neq j$ implies $\lambda_i \notin \mathbb{R}\lambda_j$ (hence $\lambda_i \neq \lambda_j$);
- weakly hyperbolic iff $i \neq j$ implies $\lambda_i \notin \mathbb{R}_- \lambda_j$ (where $\mathbb{R}_- = (-\infty, 0]$).

3) Statement of our main results

THEOREM 1 [5] : Every X with strongly hyperbolic linear part L is strongly C^0-equivalent to L.

From this and the main theorem in [3], we get the following result, which was stated as a conjecture in [3] :

COROLLARY : Let X_0 and X_1 have stronly hyperbolic linear parts L_0 and L_1. Then X_0 and X_1 are weakly C^0-equivalent iff $\Delta(L_0) = \Delta(L_1)$.

In the Poincaré domain, this corollary - the only possible structural stability result in the theory- had been proved by Guckenheimer [9] (see also [3]).

THEOREM 2 [5] : If X_0 and X_1 have weakly hyperbolic linear parts and are formally equivalent, then they are strongly C^∞-equivalent.

The proof of Theorem 1 is divided into two parts : first, using basic geometric ideas and rather soft analysis, one proves Theorem 2 and

THEOREM 3 [5] : If $L \in gl(n,\mathbb{C})$ is weakly hyperbolic, then, for each positive integer k, there exists a positive integer $s(k)$ with the following property : every X with linear part L is strongly C^k-equivalent to a polynomial vector field $X_{s(k)}$ as in Theorem 0. Therefore, every X' which is $s(k)$-equivalent to X is stronly C^k-equivalent to X.

Theorem 1 is then proved using the normal form $X_{s(1)}$ and a very simple geometric idea, which will be described in part II of this paper.

NOTES : Theorem 1 remains true when L is only weakly hyperbolic, provided its eigenvalues satisfy no relation $\lambda_i = p\,\lambda_j$ with $p > 1$ an integer. Moreover, the linearizing homeomorphism we construct is Hölder-continuous with Hölder exponent α for every $\alpha < 1$.

In the Poincaré domain, a more precise version of Theorem 2 and Theorem 3 has been proved by Poincaré and Dulac : namely, if L is in the Poincaré domain, then the sets $P_1,\dots,P_n$ defined in Theorem 0 are finite, and each X with linear part L is holomorphically equivalent to X_s , where $s = \max\{1, \max\{\Sigma p_j : (p_1,\dots,p_n) \in P_1 \cup\dots\cup P_n\}\}$ - and $\max \emptyset = -\infty$. Surprisingly enough, even in this case, Theorem 1 is due to the author [4].

In the Siegel domain, Theorem 2 and Theorem 3 contain previous results by Dumortier and Roussarie [8] . Siegel has proved the existence of a (dense) subset Σ of measure zero in the Siegel domain such that every X with linear part in the complementary subset of Σ is holomorphically linearizable. It should be added that the complementary subset of Σ contains some non hyperbolic L's. See [1] for a proof of the Poincaré, Dulac and Siegel theorems.

PART II. ABOUT THE PROOFS

4) A crazy proof of the Grobman-Hartman theorem

If we replace $\mathbb{C}$ with $\mathbb{R}$ in Theorem 1, we get the Grobman-Hartman theorem for (real) flows - here, (strong or weak) hyperbolicity just means that $L \in gl(n,\mathbb{R})$ has no eigenvalue on the imaginary axis. Let E^+ and E^- denote the stable and unstable subspaces of L ; recall that $E^{\pm} = \{v \in \mathbb{R}^n : \lim_{t\to+\infty} \exp_v \pm tL = 0\}$ are L-invariant complementary vector subspaces of $\mathbb{R}^n$, and there exists a euclidean scalar product $(.\,|\,.)$ on $\mathbb{R}^n$, with associated norm $|.|$, such that $x \mapsto (Lx|x)$ is positive definite on E^- , negative definite on E^+ , and E^+ is orthogonal to E^- . Therefore, if $x_{\pm}$ denotes the orthogonal projection of $x \in \mathbb{R}^n$ onto $E^{\pm}$, each cylinder $Q_r^+ = \{x \in \mathbb{R}^n : |x_+| = r\}$ $(r > 0)$ intersects every flowline of $L|_{\mathbb{R}^n \setminus E^-}$ transversally, at exactly one point, which we express by saying that Q_r^+ is a quotient of $\mathbb{R}^n \setminus E^-$ by the $\mathbb{R}$-action ρ^1 defined by $\rho^1(t,v) = \exp_v tL$.

Given a smooth vector field X in some neighbourhood of $0 \in \mathbb{R}^n$, with linear part L , there exists (see below) a smooth action ρ of $\mathbb{R}$ over $\mathbb{R}^n$ with the following properties :

(i) Its infinitesimal generator Y is equal to X , up to a C^1 change of coordinates, in a neighbourhood of 0.

(ii) The support of $Y-L$ is compact.

(iii) There exist negative constants c_+ and c_- such that, for every $v \in \mathbb{R}^n$ and every nonnegative t , one has $|\rho(t,v)_+| \leq e^{c_+t}|v_+|$ and $|\rho(-t,v)_-| \leq e^{c_-t}|v_-|$.

In particular, E^+ and E^- are the stable and unstable manifolds of ρ and the Q_r^+'s have the same property with respect to ρ as to ρ^1.

If E^+ is nontrivial, a very natural idea is to define a homeomorphism H of $\mathbb{R}^n$, sending ρ onto ρ^1, in the following way : first decide that H and the identity have the same restriction to Q_r^+ for some positive r, which determines H on $\mathbb{R}^n \setminus E^-$, and then show that H can be nicely extended along E^-. As it stands, this "proof" of the Grobman-Hartman theorem does not always work when E^- is nontrivial, but we can make it into a true proof by introducing an "intermediate normal form" ; namely, there exists a smooth $\mathbb{R}$-action ρ', whose infinitesimal generator Y' has linear part L and satisfies (ii) and (iii) above, such that our method defines

- a homeomorphism H_1 of $\mathbb{R}^n$ sending ρ onto ρ', and
- (exchanging E^+ and E^-) a homeomorphism H_2 of $\mathbb{R}^n$ sending ρ' onto ρ^1.

The construction of ρ and ρ' goes as follows : assuming for simplicity that the eigenvalues $\lambda_1,\dots,\lambda_n$ of L are real, Theorem 3 is known [6] to be true in the real smooth case, and it is easily checked that the C^1-normal form $X_{s(1)}$ is tangent to E^+ and E^-. Therefore, by a standard extension procedure, ρ and ρ' can be so chosen that their infinitesimal generators Y and Y' satisfy $Y = X_{s(1)}$ and $Y' = X'_{s(1)}$ in some neighbourhood of O, where $X'_{s(1)}$ is obtained from $X_{s(1)}$ by cancelling the coefficients of all those x^p's ($p \in P_1 \cup \dots \cup P_n$) which vanish identically on E^-. Since every x^p with $|p| > 0$ has to vanish identically on E^+ or on E^-, it is clear that L is obtained from $X'_{s(1)}$ by cancelling the coefficients of all those x^p's which vanish identically on E^+.

Given any positive r, if $\tau : \mathbb{R}^n \setminus E^- \to \mathbb{R}$ is defined by $\rho(-\tau(v),v) \in Q_r^+$, then the diffeomorphism H_1 of $\mathbb{R}^n \setminus E^-$ sending ρ onto ρ' and equal to the identity on Q_r^+ is given by

$$H_1(v) = \rho'(\tau(v),\rho(-\tau(v),v)).$$

If $|v_+| = d(v,E^-)$ is less than r for some $v \in \mathbb{R}^n \setminus E^-$, then $\tau(v)$ is positive and, by (iii), less that $-\frac{1}{c_+} \operatorname{Log} \frac{r}{d(v,E^-)}$. Therefore, if $p \in \mathbb{N}^n$ is such that x^p vanishes identically on E^-, then each $x^p\tau^m$ with $m \in \mathbb{N}$ extends to a continuous function on $\mathbb{R}^n$, vanishing identically on E^- and locally Hölder-continuous for every Hölder exponent < 1 .

Now, by the definition of ρ and ρ', if r has been chosen small enough, there exists a neighbourhood N of O in $\mathbb{R}^n$ such that $(H_1-\mathrm{id})|_{N \setminus E^-}$ can be expressed as the sum of a convergent power series in (finitely many) variables of the form $x^p\tau^m$ with $m \in \mathbb{N}$ and $p \in P_o \cup \dots \cup P_n$
(where $P_o = \{(p_1,\dots,p_n) \in \mathbb{N}^n \setminus \{0\} : \Sigma\, p_j\lambda_j = 0\}$) such that x^p vanishes identically on E^- - if $\gamma(t;x_1,\dots,x_n)$ and $\gamma'(t;x_1,\dots,x_n)$ are the Taylor expansions of ρ and ρ' at $t = x_1 = \dots = x_n = 0$, this convergent power series is just a suitable writing

of the power series $\gamma'(t;\gamma(-t;x_1,\dots,x_n))$-id for $t=\tau$. This proves that H_1 can be (Hölder-) continuously extended by the identity on E^-. Exchanging ρ and ρ', one gets the same property for H_1^{-1}. The same reasoning applies to the diffeomorphism H_2 of $\mathbb{R}^n \setminus E^+$ sending ρ' onto ρ^1 and equal to the identity on $Q_r^- = \{v \in \mathbb{R}^n : |v_-| = r\}$ when $r > 0$ is small enough, hence the Grobman-Hartman theorem.

Of course what is crazy in this proof is its last part, which is quite elementary but somewhat technical. We refer to [1] for a description of the usual Moser-Palis-Pugh proof, and to [6] for a discussion of its geometrical content.

5) How to prove that Theorem 3 implies Theorem 1

The idea is exactly the same as in 4, except that we have to answer the following questions : in the complex case, is there anything like

a) the stable and unstable manifolds E^+ and E^- ?

b) the invariant open set $\mathbb{R}^n \setminus E^-$ and its quotient submanifolds Q_r^+ ?

The answer to question a) is the simplest possible one : let ρ^1 be the holomorphic $\mathbb{C}$-action defined by $\rho^1(t,v) = \tilde{\rho}^1(t).v = \exp_v tL$. Call a subspace of $\mathbb{C}^n$ a strongly invariant manifold (s.i.m.) of ρ^1 (or L) iff it is the unstable manifold of $\tilde{\rho}^1(t)$ for some t. If E_k denotes the eigenspace of L associated with its eigenvalue λ_k $(1 \leq k \leq n)$, the s.i.m.'s of L are all the vector subspaces $\bigoplus_{\mathcal{R}e(\lambda_k t) > 0} E_k$ with $t \in \mathbb{C}$. By hyperbolicity, the nontrivial ones are exactly the subspaces $\bigoplus_{\mathcal{R}e(\lambda_k t) \geq 0} E_k$ with $t \in \mathbb{C} \setminus \{0\}$. Thus, the s.i.m.'s of L are L-invariant, and there is but a finite number of them. Moreover, L is in the Poincaré domain iff $\mathbb{C}^n$ is one of its s.i.m.'s . We refer to [6] for further reading on s.i.m.'s.

An answer to question b) is as follows : in the real case, when L is diagonalizable, notice that it is the gradient of $F = \frac{1}{2}\sum \lambda_j x_j^2$ with respect to the euclidean metric $\sum (dx_j)^2$. Therefore, the hyperboloids $Q_b = F^{-1}(b)$ with $b > 0$ could have been used in 4 instead of the cylinders Q_r^+ (the Q_b's with $b < 0$ are quotients of $\mathbb{R}^n \setminus E^+$ by ρ^1).

In the complex case, we have [6] a similar (easy) result :

LEMMA 1 : Let $L \in gl(n,\mathbb{C})$ be strongly hyperbolic, and let $\lambda_1,\dots,\lambda_n$, $x_1,\dots,x_n$ be as in Theorem 0. For each regular value b of $F = \frac{1}{2}\sum \lambda_j |x_j|^2$, if $\mathcal{V}_b$ denotes the (L-invariant) union of those s.i.m.'s of L which do not intersect $Q_b = F^{-1}(b)$, then Q_b is a quotient of $E \setminus \mathcal{V}_b$ by ρ^1, in the sense defined in 4.

(The use of F goes back to Camacho's thesis [2]).

LEMMA 2 : Let L and F be as in Lemma 1. For every $(p_1,\dots,p_n) \in \mathbb{N}^n$, if there exist $i \neq j$ with $p_i p_j \neq 0$, then x^p vanishes identically on $\mathcal{V}_b$ for some regular value b of F .

(Take b in the angle $\{t\lambda_i + u\lambda_j : t,u \geq 0\}$ and outside $\mathbb{R}\lambda_1 \cup \ldots \cup \mathbb{R}\lambda_n$)

COROLLARY : Given X as in Theorem 1 and $X_{s(1)}$ as in Theorem 3, there exist regular values $b_1,\ldots,b_r$ of F and polynomial vector fields $Y_1,\ldots,Y_{r+1}$ on $\mathbb{C}^n$ such that $Y_1 = X_{s(1)}$, $Y_{r+1} = L$, and, for each $j \in \{1,\ldots,r\}$, Y_{j+1} is obtained from Y_j by cancelling the coefficients of those x^p's which vanish identically on $\mathcal{V}_{b_j}$.

(By strong hyperbolicity, every p in $P_1 \cup \ldots \cup P_n$ must satisfy the hypothesis of lemma 2).

In 4, we used a standard extension procedure to get global smooth $\mathbb{R}$-actions. The corresponding result in the complex case is by no means trivial :

LEMMA 3 : Let $b_1,\ldots,b_r$ and $Y_1,\ldots,Y_{r+1}$ be as in the corollary. For each $j \in \{1,\ldots,r\}$, there exist (real) C^∞ actions ρ_j and ρ_{j+1} of $\mathbb{C}(=\mathbb{R}^2)$ over $\mathbb{C}^n$ and $b \in (0,\infty)b_j$ such that the following hold for $k = j, j+1$:

(i) The infinitesimal generators of the real flows $\rho_k(t,.)$ and $\rho_k(it,.)$ are equal to Y_k and iY_k in a compact neighbourhood N_k of $0 \in \mathbb{C}^n$.

(ii) Each s.i.m. of ρ^1 is ρ_k-invariant, and $\mathcal{Q}_b$ is a quotient of $\mathbb{C}^n \setminus \mathcal{V}_b = \mathbb{C}^n \setminus \mathcal{V}_{b_j}$ by ρ_k.

(iii) If $\tau_k : \mathbb{C}^n \setminus \mathcal{V}_b \to \mathbb{C}$ is defined by $\rho_k(-\tau_k(v),v) \in \mathcal{Q}_b$, then there exist positive constants C and α such that, for every $v \in N_k \setminus \mathcal{V}_b$, one can find a continuous path $c: [0,1] \to \mathbb{C}$ from 0 to $-\tau_k(v)$ with the following property : for each $s \in [0,1]$, one has

$$\rho_k(c(s),v) \in N_k \quad \text{and} \quad |c(s)| \leq \alpha \left|\mathrm{Log}\, \frac{1}{d(v,\mathcal{V}_b)}\right| + C,$$

where $d(v,\mathcal{V}_b) = \inf \{(\Sigma\, |x_j(v-w)|^2)^{1/2} : w \in \mathcal{V}_b\}$.

A proof of Lemma 3 can be found in [5], § 7 of Chapter 1 and § 1 of Chapter 2 - a more comprehensible version of it will be contained in [7]. The idea is to define first the restriction of ρ_k to $G_1 \times \mathbb{C}^n$, where G_1 is a "generic" subgroup $\mathbb{R}\theta$ $(\theta \in \mathbb{C}\setminus\{0\})$ of $\mathbb{C}$: this can be done using standard techniques. Then, the full extension problem can be treated via a quotient space of some dense open invariant set by $\rho_k \,|\, G_1 \times \mathbb{C}^n$.

Once Lemma 3 is proved, the same reasoning as in 4 shows that, for every $j \in \{1,..,r\}$, if b is as in Lemma 3, then there exists a unique (Hölder-continuous) homeomorphism H_j of $\mathbb{C}^n$ sending ρ_j onto ρ_{j+1} and equal to the identidy on $\mathcal{Q}_b$, hence Theorem 1.

CONCLUSION

The proof of theorems 2 and 3 uses the same geometry -and nicer analysis [6] . In fact, these two results hold true for much more general smooth (germs of) abelian (or even nilpotent) Lie group actions (see [5] , [6] and [7]), whereas we have been able to prove Theorem 1 only in "most" of these cases. We hope that this paper can serve as an introduction to the general theory, in which there still exist many interesting problems to be solved. In particular, it is likely that a suitable modification of the extension procedure used in the proof of Lemma 3 will lead to an easier proof and generalization of Theorem 1 - so that section 5 above can be called a crazy proof of our linearization theorem.

REFERENCES

[1] V. I. ARNOLD : Chapitres supplémentaires de la théorie des équations différentielles ordinaires, Mir, Moscou, 1980 (translated from the russian version which appeared in 1978).

[2] C. CAMACHO : On $\mathbb{R}^k \times \mathbb{Z}$ -actions, in Differentiable dynamical systems, Peixoto ed., IMPA/Academic press, 1973.

[3] C. CAMACHO, N. H. KUIPER, J. PALIS : The topology of holomorphic flows with singularity, Publ. Math. I.H.E.S. 48 (1978), p.5-38.

[4] M. CHAPERON : Linéarisation des germes hyperboliques d'actions différentiables de $\mathbb{R}^k \times \mathbb{Z}^m$: le domaine de Poincaré, C. R. Acad. Sc. Paris, t. 289 (1979), série A, p.325-328.

[5] M. CHAPERON : Propriétés génériques des germes d'actions différentiables de groupes de Lie commutatifs élémentaires, thèse, Université Paris 7, octobre 1980.

[6] M. CHAPERON : Singularités des systèmes dynamiques, lectures given at Ecole Normale Supérieure in 1981, to appear (probably in Astérique).

[7] M. CHAPERON : To appear.

[8] F. DUMORTIER, R. ROUSSARIE : Smooth linearization of germs of $\mathbb{R}^2$ -actions and holomorphic vector fields, Ann. Inst. Fourier, Grenoble, 30, 1 (1980), 31-64.

[9] J. GUCKENHEIMER : Hartman's theorem for complex flows in the Poincaré domain, Compositio Math. 24, (1972), p.75-82 .

Centre de Mathématiques
Ecole Polytechnique
91128 Palaiseau Cedex
(France)

On Surfaces of Constant Mean Curvature in a Three-Dimensional Space of Constant Curvature

Shiing-shen Chern*

Recently W. Y. Hsiang and his collaborators found examples of immersions of the three-sphere S^3 into the four-dimensional euclidean space E^4, which have constant mean curvature but are not round [2].

By a theorem of H. Hopf such examples do not exist in one lower dimension [1]. In this note we wish to give a proof of the corresponding theorem where the ambient space is of any constant curvature.

Let M be a three-dimensional Riemannian manifold of constant curvature c. Relative to orthonormal frames $x\ell_1\ell_2\ell_3$, $x \in M$, the structure equations are

$$(1)\qquad \begin{aligned} d\theta_\alpha &= \sum \theta_\beta \wedge \theta_{\beta\alpha} , \\ d\theta_{\alpha\beta} &= \sum \theta_{\alpha\gamma} \wedge \theta_{\gamma\beta} + \Theta_{\alpha\beta} , \end{aligned}$$

where the indices have the range

$$(2)\qquad 1 \leq \alpha,\beta,\gamma \leq 3,$$

and θ_α is an orthonormal coframe, $\theta_{\alpha\beta}(= -\theta_{\beta\alpha})$ are the connection forms, and

$$(3)\qquad \Theta_{\alpha\beta} = -c\theta_\alpha \wedge \theta_\beta$$

are the curvature forms. Equations (3) express the fact that M is of constant curvature c.

Let $f : N \to M$ be an immersed two-dimensional surface. We restrict to frames $x\ell_1\ell_2\ell_3$, such that $x \in N$ and ℓ_3 is the unit normal vector to N at x, supposing N to be oriented. Then

$$(4)\qquad \theta_3 = 0 ,$$

*Work done under partial support of NSF grant MCS-8023356.

and, by (1),

(5) $$\theta_{i3} = \sum h_{ik}\theta_k ,$$

where

(6) $$h_{ik} = h_{ki}.$$

Here, and in what follows, we agree on the index range

(7) $$1 \leqq i,j,k \leqq 2.$$

The first and second fundamental forms are respectively

(8) $$I = \theta_1^2 + \theta_2^2$$
$$II = h_{11}\theta_1^2 + 2h_{12}\theta_1\theta_2 + h_{22}\theta_2^2 .$$

The invariants

(9) $$H = \frac{1}{2}(h_{11} + h_{22})$$
$$K_\ell = h_{11}h_{22} - h_{12}^2 ,$$

which are the two elementary symmetric functions of the eigenvalues of II with respect to I, are called respectively the *mean curvature* and the *total curvature* of N. The latter has an induced Riemannian metric. By (1) its *Gaussian curvature* is

(10) $$K_i = K_\ell + c.$$

N is called *totally umbilical*, (resp. *totally geodesic*) if

(11) $$II - HI = 0 \text{ (resp. } II = 0)$$

Exterior differentiation of (5) and use of (1) give, since $\Theta_{i3} = 0$,

(12) $$\sum Dh_{ik} \wedge \theta_k = 0 ,$$

where

$$(13)\qquad Dh_{ik} = dh_{ik} - \sum_j h_{ij}\theta_{kj} - \sum_j h_{kj}\theta_{ij}$$

By putting

$$(14)\qquad Dh_{ik} = \sum h_{ikj}\theta_j ,$$

we get, from (12)

$$(15)\qquad h_{ikj} = h_{ijk}.$$

Thus h_{ikj} is symmetric in any two of its indices. A symmetric tensor h_{ik}, with this symmetry property of its covariant derivatives, is called by some mathematicians a *Codazzi tensor.*

The complex structure on N is defined by

$$(16)\qquad \phi = \theta_1 + i\theta_2 .$$

By (1) its exterior derivative is given by

$$(17)\qquad d\phi = i\phi \wedge \theta_{12}$$

The form in (11),

$$(18)\qquad II - HI = \frac{1}{2}(h_{11} - h_{22})(\theta_1^2 - \theta_2^2) + 2h_{12}\theta_1\theta_2$$

has trace zero, and is the real part of the complex two-form

$$(19)\qquad \Lambda = \hat{H}\phi^2 ,$$

where

$$(20)\qquad \hat{H} = \frac{1}{2}(h_{11} - h_{22}) - h_{12}i.$$

Clearly Λ is uniquely determined by II - HI. Hence both of them are forms intrinsically associated to N, independent of choice of frames.

Theorem 1. *If* H = *const,* Λ *is a holomorphic two-form on* N.

Proof. The hypothesis implies

(21) $$h_{11i} + h_{22i} = 0.$$

By (13) and (14) we have

(22) $$\begin{aligned} dh_{11} &= 2h_{12}\theta_{12} + h_{111}\theta_1 + h_{112}\theta_2 , \\ dh_{12} &= -(h_{11} - h_{22})\theta_{12} + h_{121}\theta_1 + h_{122}\theta_2 , \\ dh_{22} &= -2h_{12}\theta_{12} + h_{221}\theta_1 + h_{222}\theta_2, \end{aligned}$$

from which it follows that

(23) $$d\hat{H} = 2i\hat{H}\theta_{12} + (h_{111} - ih_{112})\phi$$

Locally we write

(24) $$\phi = \lambda dz,$$

so that

(25) $$\Lambda = \hat{H}\lambda^2 \, dz^2$$

It suffices to show that the coefficient $\hat{H}\lambda^2$ of dz^2 in this expression is a holomorphic function of z.

By substituting (24) into (17), we get

$$d\lambda + i\lambda\theta_{12} \equiv 0, \text{ mod } dz,$$

while (23) implies

$$d\hat{H} - 2i\hat{H}\theta_{12} \equiv 0, \text{ mod } dz.$$

It follows that

$$d(\hat{H}\lambda^2) \equiv 0, \text{ mod } dz,$$

which proves the theorem.

As a corollary we have:

Theorem 2. *Let* $f : S^2 \to M$ *be an immersed two sphere with constant mean curvature, where* M *is a three-dimensional manifold of constant curvature. Then the surface is totally unimbilical.*

Proof. This follows from the fact that $\Lambda = 0$.

References

1. H. Hopf, Über Flächen mit einer Relation zwischen den Hauptkrümmungen, Math. Nachr. 4 (1950-51), 232-249.

2. Wu-yi Hsiang, Zhen-huan Teng, and Wen-ci Yu, New examples of constant mean curvature immersions of 3-sphere into euclidean 4-space to appear in Proc. Nat. Acad. Sci, USA, 1982.

Department of Mathematics
University of California
Berkeley, CA 94720
USA

A PERIODIC ORBIT INDEX
WHICH IS A BIFURCATION INVARIANT

by

Shui-Nee Chow[1], John Mallet-Paret[2], James A. Yorke[3]

§1. Introduction

The concept of the multiplicity of a solution of a system of equations (or zero of a function, or fixed point of a map) is well known. There are several ways of viewing multiplicity; perhaps the most suggestive is to think of a solution $x_0 \in R^n$ of multiplicity k as composed of a cluster of k simple, or generic, solutions, infinitesimally nearby. The number k is an algebraic count of the simple solutions, so that $k = k_+ - k_-$ where k_+ is the actual number of simple solutions counted positively and k_- the number counted negatively. More precisely, a generic perturbation of the system of equations splits x_0 into finitely many solutions, all simple; the numbers k_+ and k_- depend on the perturbation but the difference $k_+ - k_-$ does not. (Throughout this paper we assume our functions are at least C^1. A simple solution is one for which the linearized equations have only the trivial solution; in this case $k = \pm 1$ depending on the sign of a determinant.)

Multiplicity can also be approached via degree theory. If x_0 is an isolated zero of a map $F:R^n \to R^n$, then the topological degree of the map

---

1. Research supported in part under NSF Grant MCS-76-06739.

2. Research supported in part under NSF Grant MCS-79-05774-02 and Army Research Office Grant ARO-DAAG-29-79-C-0161.

3. Research supported in part under NSF Grant MCS-78-18221A02.

$$x \in S_\varepsilon^{n-1}(x_0) \longrightarrow \frac{F(x)}{|F(x)|} \in S^{n-1}$$

from a sufficiently small sphere

$$S_\varepsilon^{n-1}(x_0) = \{x \in R^n | \ |x-x_0| = \varepsilon\}$$

to the unit $(n-1)$-sphere in R^n, is called the index of x_0. It can be shown the index equals the multiplicity k above.

The situation for a fixed point x_0 of a map T is much more complicated since such a point is also fixed by each iterate T^m. Its multiplicity k_m, as a fixed point now of the map T^m, is defined to be its index as a zero of $F_m(x) = x - T^m(x)$. In general, this quantity depends on m, so we obtain a sequence of integers $\{k_1, k_2, k_3, \ldots\}$. (To do this we are assuming the fixed point x_0 is an isolated fixed point of each T^m, though the neighborhood of isolation may depend on m.)

The aim of this paper is to study the sequence $\{k_m\}$ with a view to providing a geometrical interpretation of multiplicity similar to that of the infinitesimal cluster of simple solutions mentioned above. It turns out there are rather severe restrictions on the possible sequences. Stated algebraically, these restrictions are rather technical and unenlightening; but given an appropriate infinitesimal geometrical interpretation, they are completely natural.

Our motivation for examining these questions comes originally from differential equations. In studying the flow near a periodic orbit of an autonomous ordinary differential equation, it is natural to use the Poincaré map, since fixed points of iterates of this map correspond to "orbits". In this paper the word "orbit" means periodic orbit. In this context, the index sequence $\{k_m\}$ has been used to study the continuation of orbits for parametrized systems, to follow them through successive bifurcations, and to examine their birth and death in a Hopf bifurcation.

This approach was first taken by Fuller [6], who considered the sequence $\{\frac{1}{m}k_m\}$ of so-called Fuller indices in his study of the Seifert problem. Later, Chow and Mallet-Paret [4], and Chow, Mallet-Paret and Yorke [5] examined the general question of continuation and bifurcation using the Fuller index; this followed earlier work of Alexander and Yorke [3] who studied these problems with other techniques. More recently, Mallet-Paret and Yorke [7] have employed a so-called orbit index, or φ-index, which seems to be the most natural tool here. The φ-index combines Fuller's indices into a single index, thereby overcoming some deficiencies in the statement of the global continuation theorem for periodic orbits. See [7] and [11].

The φ-index can be thought of as a bifurcation invariant. In parameterized systems, bifurcation generally occurs at multiple solutions; the φ-index algebraically counts the number of solution branches emanating on either side of the bifurcation point. In [7] only the simplest bifurcations were permitted, though here we allow much more general bifurcations and degeneracies. For examples of these see Abraham and Marsden [1] and Meyer [8]; see also [4] and [7]. In Section 5 we discuss the φ-index as a bifurcation invariant by considering maps T that depend on a parameter λ.

For the most part of this paper we consider single map T with a fixed point x_0. The φ-index is simply the average

$$\varphi(x_0) = \lim_{N\to\infty} \frac{1}{N} \sum_{m=1}^{N} k_m . \tag{1.1}$$

of the index sequence. The limit (1.1) exists since we prove the sequence $\{k_m\}$ is always periodic in m. Hence, $\varphi(x_0)$ is the rational number

$$\varphi(x_0) = \frac{1}{N} \sum_{m=1}^{N} k_m ,$$

where $k_{m+N} = k_m$ for every m.

We may interpret x_0 as a finite cluster C of simple points x^* each fixed under some iterate of T. Because the φ-index does not distinguish between fixed points of various iterates of T (since it is an average), it turns out that $\varphi(x_0)$ can be thought of as a sum

$$\varphi(x_0) = \sum_{x^* \in C_1} \varphi(x^*)$$

where $C_1 \subseteq C$ contains exactly one point from each orbit in C. As the φ-index of a simple point x^* can only equal $+1, -1$, or 0, it is not surprising (see Corollary 2.4) that

$$\varphi(x_0) \text{ is an integer };$$

this fact is not at all evident from the definition (1.1).

We stress that the cluster C is purely a conceptual device for visualizing the algebraic restriction on $\{k_m\}$. Strictly speaking, our results here concern only the relation between the Jacobian matrix $DT(x_0)$ and $\{k_m\}$: for a given Jacobian we determine which sequences $\{k_m\}$ can occur. For a family of maps $T(x,\lambda)$, as considered in Section 5, many different configurations of curves or periodic points, their periods, and their indices are possible the way in which a fixed point x_0 splits into periodic points under a perturbation cannot be predicted from the cluster C. (Indeed, C is by no means uniquely determined.) But the advantage of considering a cluster is that it captures the index sequence $\{k_m\}$ in its entirety. No single perturbation T_1 of T could be expected to do this: higher iterates of the map would require smaller perturbations. Because of this lack of uniformity, we feel a more natural interpretation might use non-standard analysis and the language of infinitesimals. The cluster C accomplishes this in a non-rigorous fashion to suggest a picture of the phenomenon .

§2. The Results and their Interpretation

We consider a map T satisfying the standing hypotheses

$$\left.\begin{array}{l} T:R^n \to R^n \quad \text{is} \quad C^1, \\ T(0) = 0; \end{array}\right\} \tag{2.1}$$

most of the time we also assume

$$\left.\begin{array}{l} 0 \text{ is an isolated fixed point of each} \\ \text{iterate } T^m\text{, though the neighborhood} \\ \text{of isolation may depend on } m. \end{array}\right\} \tag{2.2}$$

The integer k_m is defined to be the fixed point index of $0 \in R^n$ for the map T^m; certainly 0 must be an isolated fixed point of T^m before k_m is defined. The Jacobian matrix of T is denoted by

$$A = DT(0).$$

To make clear the discussion of the previous section as well as the importance of the Jacobian, we examine the relation between A and $\{k_m\}$ in several cases. We say that 0 is a <u>simple fixed point</u> of T_m if 1 is not an eigenvalue of A_m. Define the integers

$$\left.\begin{array}{l} \sigma_+ = \text{the number of eigenvalues of } A, \text{ counting} \\ \qquad \text{multiplicity, in } (1,\infty); \\ \sigma_- = \text{the corresponding number in } (-\infty, -1). \end{array}\right\} \quad (2.3)$$

Then 0 is a simple fixed point of the iterate T^m if and only if $I - A^m$ is nonsingular, in which case

$$k_m = \operatorname{sgn}\det(I-A^m) = \begin{cases} (-1)^{\sigma_+}, & m = \text{odd} \\ (-1)^{\sigma_+ + \sigma_-}, & m = \text{even}. \end{cases}$$

In the generic situation where 0 is a simple fixed point of each iterate T^m, four different sequences $\{k_m\}$ can occur; these are $k_m \equiv 1$, $k_m \equiv -1$, $k_m \equiv (-1)^m$, and $k_m \equiv -(-1)^m$, for all m, as determined by the parities (odd or even) of σ_+ and σ_-. The φ-indices for these cases are then $1, -1, 0$, and 0 respectively, by (1.1).

More generally, in the non-generic case where $I - A^m$ is singular for various m many other sequences are possible; these depend moreover on the nonlinear part of T, and not just on A. As a specific example suppose A is a 4×4 matrix with eigenvalues $\exp(\frac{\pm 2\pi i}{3})$ and $\exp(\frac{\pm 2\pi i}{5})$; we show in Theorem 2.2 that the algebraic restrictions on the indices are

$$\left.\begin{array}{ll} k_{m+15} = k_m & m \geq 1 \\ k_m = 1 & m \not\equiv 0 \pmod 3 \text{ and } m \not\equiv 0 \pmod 5 \\ k_3 = k_6 = k_9 = k_{12} \equiv 1 & \pmod 3 \\ k_5 = k_{10} \equiv 1 & \pmod 5 \\ k_{15} \equiv -1 + k_3 + k_5 & \pmod{15}. \end{array}\right\} \quad (2.4)$$

To interpret (2.4) geometrically let

$$k = (k_1, k_2, k_3, \ldots)$$

be the (infinite) vector of indices, and introduce "basis" vectors

$$\left.\begin{aligned} j_m &= (j_{m1}, j_{m2}, j_{m3}, \ldots) \\ &= (0,0,\ldots,0,m,0,\ldots,0,m,0,\ldots) \\ j_{ma} &= \begin{cases} m & \text{if } m \text{ divides } a \\ 0 & \text{otherwise.} \end{cases} \end{aligned}\right\} \qquad (2.5)$$

It is easy to see that (2.4) is equivalent to

$$\left.\begin{aligned} & k = j_1 + c_3 j_3 + c_5 j_5 + c_{15} j_{15}, \\ & c_3,\ c_5,\ c_{15} \text{ are arbitrary integers.} \end{aligned}\right\} \qquad (2.6)$$

A natural way of viewing (2.6) is to imagine the fixed point 0 as a cluster of points composed as follows:

(1) the origin itself, which is a simple fixed point of each iterate of T, and whose index sequence is $j_1 = (1,1,1,\ldots)$;

(2) $3|c_3|$ points lying in the $\exp(\frac{\pm 2\pi i}{3})$-eigenspace, infinitesimally close to, but not at the origin. These points are not fixed by T but are simple fixed points of each iterate of T^3, and so give $|c_3|$ distinct orbits. The index sequence of each point, as a fixed point of T^3, is $\pm(1,1,1,\ldots)$ where $\pm = \operatorname{sgn} c_3$. Consequently, these points contribute in all

$$c_3 j_3 = (0,0,3c_3,0,0,3c_3,\ldots)$$

to the index vector k.

(3) $5|c_5|$ points in the $\exp(\frac{\pm 2\pi i}{5})$-eigenspace, contributing $c_5 j_5$ to k, analogously to (2) above;

(4) $15|c_{15}|$ points, infinitesimally near the origin, but in neither eigenspace. These points are fixed by T^{15} but by no smaller iterate of T; they contribute $c_{15}j_{15}$ to the index vector, as above.

In the above example each orbit in the cluster C contained exactly 1,3,5, or 15 points. In general the number of points in such an orbit must belong to the set

$$M = \{m \geq 1 : \text{ there exists } y \in R^n \text{ with } y, Ay, \ldots, A^{m-1}y \text{ distinct, but } A^m y = y\}. \tag{2.7}$$

In other words, if C contains a point fixed by T^m but by no smaller iterate, then R^n must contain a point fixed by the linear transformation A^m but by no smaller power of A. Note that M is a finite set, and that $1 \in M$ since $y = 0$ can always be chosen.

Because the set M plays a central role, we make the following definition; (see also [7]).

<u>Definition 2.1</u>. Let the map T satisfy (2.1). The integers $m \in M$, in (2.7), are called the <u>virtual periods</u> for the fixed point $0 \in R^n$.

Observe also that only two of the four possible index sequences for simple fixed points occur in (2.6), namely $(1,1,\ldots)$ and $(-1,-1,-1,\ldots)$. In general, these sequences can arise from points in C with $m \in M$ points in their orbit, if and only if A^m has an even number of eigenvalues in $(-\infty,-1)$; that is, if and only if either m is even or σ_- is even. The other two kinds of sequences,

$(1,-1,1,-1,\ldots)$ and $(-1,1,-1,1,\ldots)$ can occur if and only if both m and σ_- are odd, and the resulting contribution to the vector k in this case is an integer multiple of

$$j_m - j_{2m} = (0,0,\ldots,0,m,0,\ldots,0,-m,0,\ldots).$$

The above discussion summarizes the restrictions on the sequence $\{k_m\}$, except for several conditions involving the first two iterates, to be mentioned below. We now state precisely the form of $\{k_m\}$ for a given A. Recall the definitions of $\sigma_\pm$ (2.3), j_m (2.5) and M (2.7).

Theorem 2.2 *With the standing hypotheses* (2.1) *on the map* T, *including* (2.2), *let* k_m *be the index of the fixed point* 0 *of the iterate* T^m. *Then the index vector* $k = (k_1,k_2,k_3,\ldots)$ *has the form*

$$k = \begin{cases} \sum_{m\in M} c_m j_m, & \sigma_- = \text{even}, \\ \sum_{m\in M_e} c_m j_m + \sum_{m\in M_O} c_m (j_m - j_{2m}), & \sigma_- = \text{odd}. \end{cases} \tag{2.8}$$

where $M_e = \{m \in M: m \text{ is even}\}$ *and* $M_O = M\setminus M_e$.
The coefficients c_m *are integers. Moreover,* c_1 *and* c_2 *are subject to the following restrictions:*

(1) $c_1 = (-1)^{\sigma_+}$ <u>if</u> $I - A$ <u>is nonsingular;</u>

(2) $c_1 \in \{-1,0,1\}$ <u>if</u> $I - A$ <u>has a one dimensional kernel;</u>

(3) $c_2 \in \{0,(-1)^{\sigma+1}\}$ <u>(where</u> $\sigma = \sigma_+ + \sigma_-$ <u>)</u> <u>if</u> $I - A$ <u>is nonsingular and</u> $I - A^2$ <u>has a one dimensional kernel.</u>

The proof of this theorem will be given in Section 4.

To interpret this theorem geometrically, consider the sets in R^n

$$\Lambda_m = \{y \in R^n: y, Ay, \ldots, A^{m-1}y \text{ are distinct points, and } A^m y = y\}.$$

It is clear that for $m \in M$, the Λ_m are non-empty and pairwise disjoint. Moreover, $0 \in \bar{\Lambda}_m$. We may consider the cluster C as composed of $m|c_m|$ points in each such Λ_m, infinitesimally near the origin; these comprise $|c_m|$ orbits, as each point is fixed by T^m but by no smaller iterate; and each orbit contributes either $(\operatorname{sgn} c_m)j_m$ or $(\operatorname{sgn} c_m)(j_m - j_{2m})$ to the index vector k, according to the parities of m and σ_- as in (2.8). The orbit of a point x^* is, to first order, the set $\{x^*, Ax^*, \ldots, A^{m-1}x^*\}$.

The restrictions on c_1 and c_2 arise because the dimensions of Λ_1 and Λ_2 are small in some cases, and this restricts the placement of points in the cluster. For $m \in M$, we always have $\dim \Lambda_m \geq 2$ when $m \geq 3$, since the appropriate eigenvalues of A occur in conjugate pairs. But if $I - A$ is nonsingular, as in (1), then $\Lambda_1 = \{0\}$; $C \cap \Lambda_1$ is just one point, the origin, and $c_1 = (-1)^{\sigma_+}$. Similarly, if $I - A$ has a one-dimensional kernel, then Λ_1 is precisely this kernel; the points of

$C \cap \Lambda_1$ are arranged linearly along this space and contribute alternately $\pm j_1$ (or $\pm(j_1-j_2)$) to k. Thus $c_1 \in \{-1,0,1\}$ in case (2). And in case (3), Λ_2 is a one-dimensional subspace with the origin removed; points of C occur in symmetric pairs and contribute alternately $\pm j_2$ to k.

We conjecture that this theorem is sharp: for a given Jacobian A, any k permitted by the theorem should actually occur for some nonlinearity. In low dimensions, at least, this should be relatively easy to show.

The following Corollaries are easily seen to follow from Theorem 2.2 once the definition (1.1) of the φ-index is recalled.

Corollary 2.3. The φ-index of the fixed point 0 is

$$\varphi(0) = \begin{cases} \sum_{m\in M} c_m, & \text{if } \sigma_- \text{ is even,} \\ \sum_{\substack{m\in M \\ m \text{ even}}} c_m, & \text{if } \sigma_- \text{ is odd.} \end{cases}$$

Corollary 2.4. The φ-index of the fixed point 0 is an integer.

§3. Periodic Points of Perturbations.

Shub and Sullivan [10] consider the index sequence $\{k_m\}$ in their study of periodic points of smooth maps of compact manifolds. They show this sequence is bounded, and conclude that certain homotopy classes of maps always have infinitely many periodic points. (They note also the smoothness of the map is essential for this.) Proposition 3.1 and its proof are basically rephrasings of their work, adding some details.

Proposition 3.1. Let $T: R^n \to R^n$ be C^1, with $T(0) = 0$ and $A = DT(0)$.

(1) <u>If</u> 0 <u>is an isolated fixed point of</u> T, <u>then</u>

$$x - T^a(x) = \left(\sum_{j=0}^{a-1} A^j\right)(x-T(x)) + o(|x-T(x)|);$$

(2) <u>if</u> 0 <u>is an isolated fixed point of</u> T^m,

<u>if</u> $\sum_{j=0}^{a-1} A^{jm}$ <u>is nonsingular for some</u> $a \geq 1$ <u>and if</u> $m \geq 1$, <u>then</u> 0 <u>is an isolated fixed point of</u> T^{am}; <u>moreover</u>

$$k_{am} = \operatorname{sgn} \det \left(\sum_{j=0}^{a-1} A^{jm}\right) k_m .$$

<u>Proof</u>. Let $\theta = x-T(x)$ and note $A^k = DT^k(0)$. By Taylor's theorem,

$$T^{j+1}(x) = T^j(x-\theta) = T^j(x) - DT^j(x)\theta + o(|\theta|)$$

Hence,

$$T^{j+1}(x) = T^j(x) - A^j\theta + o(|\theta|)$$

because $DT^j(x) - A^j = o(1)$ as $\theta \to 0$, since $x = 0$ is an isolated fixed point. Thus,

$$x - T^a(x) = \sum_{j=0}^{a-1} [T^j(x) - T^{j+1}(x)]$$

$$= \left(\sum_{j=|0|}^{a-1} A^j\right)\theta + o(|\theta|)$$

This proves (1). Applying (1) to T^m gives (2).

One way of obtaining further information about the indices is by keeping careful track of the sign of the determinant in (2). This shows, in particular, that $\{k_m\}$ is a periodic sequence. Indeed, it is not hard to see that if $\sum_{j=0}^{a-1} A^{jm}$ is nonsingular, then its determinant is negative if and only if m and σ_- are odd and a is even; using this and arguing as in [10] shows periodicity.

A second way of obtaining conditions on the index sequence is by making a generic perturbation of T and studying the resulting simple periodic points. In fact, this proves not only periodicity of the $\{k_m\}$, but much more. The following proposition plays a key role here in that it restricts the possible periods of points (that is, the number of points in an orbit) that can arise from a perturbation. This makes possible the interpretation of the fixed point as a finite cluster.

Proposition 3.2. (virtual period proposition) *Let the map T satisfy the standing hypotheses (2.1), and (2.2) as well. Fix an integer $m \geq 1$ and a sufficiently small disc*

$$B(\varepsilon) = \{x \in R^n: \ |x| \leq \varepsilon\}.$$

Then if $S: R^n \to R^n$ is sufficiently near T in the C^1 norm on this disc, that is, if

$$|T - S|_{C^1(B(\varepsilon))} << 1 ,$$

then a necessary condition for there to exist $x \in B(\varepsilon)$ with $x, S(x), \ldots, S^{m-1}(x) \in B(\varepsilon)$ distinct but $S^m(x) = x$, is that m is a virtual period for the fixed point 0 of T.

Proof. Suppose such an x exists. Then $\ker(I - A^m)$ has dimension greater than 0. Suppose also that m is not a virtual period. Since m is not a virtual period, there are integers a and b with $m = ab$, $a < m$ and

$$\ker(I - A^a) = \ker(I - A^m)$$

Since $a < m$, we have

$$\theta = x - S^a(x) \neq 0$$

and also

$$|\theta| = O(|x|) + o(1)$$

where $o(1) \to 0$ as $|T - S|_{C^1(B(\varepsilon))} \to 0$. We now have

$$\begin{aligned} S^{2a}(x) &= S^a(x-\theta) \\ &= S^a(x) - DS^a(x)\theta + o(|\theta|) \\ &= S^a(x) - A^a\theta + O(|x||\theta|) + o(|\theta|) \\ &= x - (I+A^a)\theta + O(|x||\theta|) + o(|\theta|). \end{aligned}$$

Continuing in this way gives

$$\begin{aligned} S^{ab}(x) &= x - \left(\sum_{j=0}^{b-1} A^{aj}\right)\theta + O(|x||\theta|) + o(|\theta|) \\ &= x. \end{aligned}$$

Because $|x|$ is small and $\theta \neq 0$, it follows that $\sum_{j=0}^{b-1} A^{aj}$ has a non-zero kernel; but then

$$\varphi \neq \ker\left(\sum_{j=0}^{b-1} A^{aj}\right) - \{0\} \subseteq \ker(I-A^m) - \ker(I-A^a),$$

a contradiction. □

§4. The Proof of Theorem 2.2

The following argument is used to prove Theorem 2.2. Fix an integer $m_0 \geq 1$, arbitrarily large, and let S be a perturbation of T small enough that Proposition 3.2 holds for all $m \leq m_0$; that is, the only periods $m \leq m_0$ that occur for S are the virtual periods of T at O. Assume moreover that S is generic in the sense that

all fixed points x of S^m in $B(\varepsilon)$ are simple, that is, $I - DS^m(x)$ is nonsingular for all m; that such a perturbation exists follows from the Kupka-Smale theorem. See Abraham and Robbin [2] and Peixoto [9].

Now the index k_m must equal the sum of the indices of the fixed points of S^m, at least for $m \leq m_0$. We shall study first the contribution to the index vector k from points fixed by S^m but by no smaller iterate; doing this will show k has the form (2.8) with integer coefficients c_m, for at least the first m_0 components, and hence for all of them since m_0 is arbitrary. Then, further arguments will show the special restrictions (1),(2) and (3) on c_1 and c_2.

To establish the form (2.8) of k, consider for some $m \leq m_0$ a point $x \in B(\varepsilon)$ with exactly m points in its orbit under S. Necessarily $m \in M$, by Proposition 3.2. Note that $I - DS^{am}(x)$ is nonsingular for $am \leq m_0$ and $DS^{am}(x)$ is near A^{am}. Three cases arise:

(1) $\det(I+A^m) > 0$; necessarily either m or σ_- is even. Each such x has an index sequence $\pm(1,1,1,\ldots)$ as a fixed point of S^m, so the whole orbit of x contributes $\pm j_m$ to k;

(2) $\det(I+A^m) < 0$; necessarily both m and σ_- are odd. The orbit now contributes $\pm(j_m - j_{2m})$ to k.

So far the contributions to k have been consistent with the parities of m and σ_- in (2.8); but the final case is somewhat critical and necessitates a slight algebraic manipulation:

(3) $\det(I+A^m) = 0$. Here the sign of $\det(I+DS^m(x))$ is unknown, so we do not know whether or not the signs of $\det(I-DS^{am}(x))$ alternate with a; the contribution to k from the orbit could be either $\pm j_m$ or $\pm(j_m - j_{2m})$, at least for iterates up to order m_0. It may be that this contribution does not respect the parities of m and σ_-; for example an index sequence j_m could occur when m and σ_- are odd. In this event a term j_{2m} must be added and subtracted so the parities are respected; this simply rearranges terms and changes the coefficient c_{2m} in (2.8). This is valid, as the form of the expression (2.8) is unchanged, since $\ker(I-A^m) \subsetneq \ker(I-A^{2m})$ and so $2m \in M$.

This completes the proof that k has the form (2.8) with integral c_m; all that remains is to prove the special restrictions on c_1 and c_2. This is easy if both $I - A$ and $I - A^2$ are nonsingular; for by the implicit function theorem S has a unique fixed point near 0; it contributes an index sequence of either $(-1)^{\sigma_+} j_1$ or $(-1)^{\sigma_+}(j_1 - j_2)$ according to whether σ_- is even or odd, as required.

If $I - A$ is nonsingular and $I - A^2$ has a kernel of two or more dimensions, then again S has a fixed point with a contribution of either $(-1)^{\sigma_+} j_1$ or $(-1)^{\sigma_+}(j_1 - j_2)$; but as before this is a critical case which may not respect the parity of σ_-; as before it may be necessary to add and subtract a term j_2; and again this can be done since $2 \in M$.

Before considering the case where $I - A$ is nonsingular and $I - A^2$ has a one dimensional kernel, let us study the case where $I - A$ has a one dimensional kernel. We must show $c_1 \in \{-1,0,1\}$; this will be so because the fixed points of S lie on a one dimensional curve near the kernel and adjacent points have indices which cancel. The Lyapunov-Schmidt procedure outlined below makes the problem of finding fixed points of S one dimensional.

Make a direct sum decomposition

$$x = (x_1,\bar{x}) \in R \oplus R^{n-1}$$

$$x_1 \in \ker(I-A)$$

and similarly decompose

$$S(x) = (S_1(x_1,\bar{x}),\bar{S}(x_1,\bar{x})) \in R \oplus R^{n-1}$$

$$\bar{S}(x) \in \text{range}(I-A).$$

By the implicit function theorem the equation

$$\bar{x} - \bar{S}(x_1,\bar{x}) = 0$$

can be solved near the origin as

$$\bar{x} = x^*(x_1,S)$$

for a unique function x^*, C^1 in both $x_1 \in R$ and $S \in C^1(B(\varepsilon))$. Therefore, for small $|x|$ solving $x - S(x) = 0$ is equivalent to solving the scalar bifurcation equation

$$x_1 - H(x_1) = 0$$

where

$$H(x_1) = S_1(x_1, x^*(x_1, S)).$$

Now we relate the indices of the fixed points of H to those of the corresponding fixed points of S. A calculation using implicit differentiation shows at such a point

$$\begin{aligned} 1 - H'(x_1) &= 1 - \frac{\partial S_1}{\partial x_1} - \frac{\partial S_1}{\partial \bar{x}} \frac{\partial x^*}{\partial x_1} \\ &= 1 - \frac{\partial S_1}{\partial x_1} - \frac{\partial S_1}{\partial \bar{x}} \left(I - \frac{\partial \bar{S}}{\partial \bar{x}}\right)^{-1} \frac{\partial \bar{S}}{\partial x_1} . \end{aligned}$$

Taking determinants in the matrix identity

$$\begin{pmatrix} 1 - \frac{\partial S_1}{\partial x_1} & - \frac{\partial S_1}{\partial \bar{x}} \\ - \frac{\partial \bar{S}}{\partial x_1} & I - \frac{\partial \bar{S}}{\partial \bar{x}} \end{pmatrix} \begin{pmatrix} 1 & 0 \\ \left(I - \frac{\partial \bar{S}}{\partial \bar{x}}\right)^{-1} \frac{\partial \bar{S}}{\partial x_1} & I \end{pmatrix}$$

$$= \begin{pmatrix} 1 - H'(x_1) & - \frac{\partial S_1}{\partial \bar{x}} \\ 0 & I - \frac{\partial \bar{S}}{\partial \bar{x}} \end{pmatrix}$$

shows that

$$\det\left(I - \frac{\partial S}{\partial x}\right) = (1 - H'(x_1)) \det\left(I - \frac{\partial \bar{S}}{\partial \bar{x}}\right)$$

and so

$$\operatorname{sgn} \det\left(I - \frac{\partial S}{\partial x}\right) = \delta \operatorname{sgn}(1 - H'(x_1)) \qquad (4.1)$$

where $\delta = \operatorname{sgn} \det\left(I - \frac{\partial \bar{S}}{\partial \bar{x}}\right)$ does not depend on the fixed point x. Because x_1 and H are scalar, the signs of $1 - H'(x_1)$ and therefore the fixed point indices of H are opposite for adjacent x_1. The sum of these indices must be ± 1 or 0; and by (4.1) the same is true for the indices of S. This proves $c_1 \in \{-1,0,1\}$.

Before continuing with the proof of Theorem 2.2, let us state a proposition which embodies the above discussion.

<u>Proposition 4.1</u>. *Let $T: R^n \to R^n$ be C^1 and have 0 as an isolated fixed point; suppose also the kernel of $I - DT(0)$ is one dimensional. Then the fixed point index of 0 for the map T is either $-1, 0,$ or 1.*

To complete the proof of the theorem we suppose $I - A$ is nonsingular and $I - A^2$ has a one-dimensional kernel; we must prove $c_1 = (-1)^{\sigma_+}$ and $c_2 \in \{0, (-1)^{\sigma_- + \sigma_+ + 1}\}$. From (2.8) we see that $k_1 = c_1$ and $k_2 = (-1)^{\sigma_-} c_1 + 2c_2$, so that in fact we must show $k_1 = (-1)^{\sigma_+}$ and $k_2 \in \{-1,1\}$. We have already noted $k_1 = (-1)^{\sigma_+}$ when $I - A$ is nonsingular. The form for k_2 is seen by applying a Lyapunov-Schmidt decomposition to $x - S^2(x)$, in the same manner as to $x - S(x)$ above. The resulting bifurcation equation $x_1 - H_2(x_1) = 0$ has an odd number of solutions: one representing the unique fixed point of S, and pairs representing points fixed by S^2 but not by S. As before, the signs of $1 - H_2'(x_1)$ at adjacent solutions of the bifurcation equation are opposite; consequently, the index k_2 equals ± 1. This completes the proof of Theorem 2.2. □

§5. A Bifurcation Invariant

In this section, we will show that the φ-index can be interpreted as a bifurcation invariant. To do this, consider the parametrized differential equation:

$$\frac{dx}{dt} = f(x,\alpha), \quad (x,\alpha) \in R^n \times R. \tag{5.1}$$

Let $p(t)$ be a nonconstant periodic solution of (5.1) for some fixed $\alpha = \alpha_0$. Then $\{(p(t),\alpha): t \geq 0\}$ is called an orbit. For simplicity, we say (p_0,α_0) is an <u>orbit point</u> of (5.1) when the solution through $p_0 \in R^n$ for $\alpha = \alpha_0$ is periodic. The minimal (or least) period of an orbit (p_0,α_0) will be denoted by $\tau(p_0,\alpha_0)$.

<u>Definition 5.1</u>. Let (p_0,α_0) be an orbit point. We will say γ is an <u>arc</u> of orbits emanating from (p_0,α_0) if

$$\gamma: [0,1] \to R^n \times R$$

is continuous and piecewise differentiable and γ satisfies the following conditions:

(1) $\gamma(\beta)$ is an orbit point for each β, with distinct β's giving points on distinct orbits; and moreover, $\gamma(0) = (p_0,\alpha_0)$;

(2) the period $\tau(\gamma(\beta))$ is continuous and bounded for $0 < \beta \leq 1$ (but is not necessarily continuous at $\beta = 0$). Note this implies

$$\lim_{\beta \to 0+} \tau(\gamma(\beta))$$

exists and is an integral multiple of $\tau(\gamma(0))$.

We will write $\Gamma(p_0,\alpha_0)$ to denote the orbit through (p_0,α_0) and we write $\Gamma(\gamma)$ for $\cup\Gamma(\gamma(\beta))$, $\beta \in [0,1]$.

Definition 5.2. We say (p_0,α_0) is a regular bifurcation orbit if at least one of its (n-1) Floquet multipliers is a root of unity and there are a finite number of arcs $\gamma_i(\beta)$, $0 \leq \beta \leq 1$, $i = 1,\ldots,K$, emanating from (p_0,α_0), and the following conditions hold:

(3) The arcs are on distinct orbits, specifically $\Gamma(\gamma_i) \cap \Gamma(\gamma_j) = \Gamma(p_0,\alpha_0)$ when $i \neq j$;

(4) if there are orbit points (p_k,α_k) converging to an orbit point $(p,\alpha_*) \in Q$ where $Q = \cup\Gamma(\gamma_i)$ and the least periods $\tau(p_k,\alpha_k)$ are bounded, then for sufficiently large k, (p_k,α_k) is in some $\Gamma(\gamma_i)$; in other words all the short period orbits near $\Gamma(p_0,\alpha_0)$ are hit by the arcs γ_i and the γ_i's do not admit secondary branches of orbits.

(5) the α coordinate, denoted by $\alpha_i(\beta)$, of each arc $\gamma_i(\beta)$ is a strictly monotone function of $\beta \in [0,1]$.

Remark 5.3. Note that condition (5) is true if for every $i = 1,\ldots,K$ and every $\beta \in (0,1]$, the orbit $\Gamma(\gamma_i(\beta))$ has no Floquet multipliers which are roots of unity; this implies also there are no secondary arcs emanating from the γ_i's.

Definition 5.4. If (p_0,α_0) is an orbit point, the φ-index of the orbit $\Gamma(p_0,\alpha_0)$ is defined to be the φ-index of the fixed point p_0 of the Poincare map of the flow (5.1) at $\alpha = \alpha_0$. (Note that in order for the φ-index of (p_0,α_0) to be well defined p_0 must be an isolated fixed point of each iterate of the Poincare map.)

We will write $\varphi(\Gamma(p_0,\alpha_0))$ and $\varphi(p_0,\alpha_0)$ interchangably.

Theorem 5.5. (Bifurcation invariant proposition) Let (p_0, α_0) be a regular bifurcation orbit with arcs $\gamma_i(\beta)$, $i = 1,\ldots,K$, $0 \leq \beta \leq 1$ emanating (p_0, α_0). Then the φ-index $\varphi(\gamma_i(\beta))$ is well defined and constant for $0 < \beta \leq 1$. Moreover, for each $0 < \beta \leq 1$

$$\sum_{\alpha_i(\beta) < \alpha_0} \varphi(\gamma_i(\beta)) = \varphi(p_0, \alpha_0) = \sum_{\alpha_i(\beta) > \alpha_0} \varphi(\gamma_i(\beta)) \tag{5.2}$$

Proof. By condition (4) and homotopy invariance of fixed point indices for $0 < \beta \leq 1$, $\varphi(\gamma_i(\beta))$ are constant along each arc, $0 < \beta \leq 1$. Let

$$n_i = \lim_{\beta \to 0+} \frac{\tau(\gamma_i(\beta))}{\tau(p_0, \alpha_0)}, \quad i = 1,\ldots,K .$$

Let $k_i(m)$ be the fixed point index of the mn_ith iterate of the Poincare map for $\gamma_i(\beta)$ for $\beta \neq 0$ where $i = 1,\ldots,K$ and $m = 1,2,\ldots,$ and $k_0(m)$ be the fixed point index of the mth iterate of the Poincare map for (p_0, α_0). It is not difficult to see that for each integer $a = 1,2,\ldots$

$$\sum_{\substack{i:\ \alpha_i(\beta) < \alpha_0 \\ \text{and } n_i | a}} n_i k(a/n_i) = k_0(a) = \sum_{\substack{i:\ \alpha_i(\beta) > \alpha_0 \\ \text{and } n_i | a}} n_i k_i(a/n_i)$$

where "$n_i | a$" says that the sums are taken only over those i for which n_i divides a. By the definition of φ-index and the periodicity of the vector of indices (Theorem 2.2), there exists an integer $N \geq 1$ such that for every i

$$\varphi(\gamma_i(\beta)) = \frac{1}{N} \sum_{\substack{a=1 \\ a:\ n_i | a}}^{N} n_i k_i(a/n_i)$$

and

$$\varphi(p_0, \alpha_0) = \frac{1}{N} \sum_{a=1}^{N} k_0(a)$$

Equation (5.2) follows from the above equalities.

REFERENCES

[1] Abraham, R. and Marsden, J., "Foundations of Mechanics", Benjamin/Cummings, Reading, Mass. 1978.

[2] Abraham, R. and Robbin, J., "Transversal Mappings and Flows", Benjamin, Amsterdam, 1967.

[3] Alexander, J.C. and Yorke, J.A., Global bifurcation of periodic orbits, Am. J. Math. 100(1978), 263-292.

[4] Chow, S.-N. and Mallet-Paret, J., The Fuller index and global Hopf bifurcation, J.Diff. Eq. 29(1978), 66-85.

[5] Chow, S.-N., Mallet-Paret, J. and Yorke, J.A., Global Hopf bifurcation from a multiple eigenvalue, Nonlin. Anal. 2(1978), 753-763.

[6] Fuller, F.B., An index of fixed point type for periodic orbits, Am. J. Math. 89(1967), 133-148.

[7] Mallet-Paret, J. and Yorke, J.A., Snakes: oriented families of periodic orbits, their sources, sinks, and continuation. Jour. Diff. Equ., to appear.

[8] Meyer, K.R., Generic bifurcation of periodic points, Trans. Amer. Math. Soc. 1970(149), 95-107.

[9] Peixoto, M.M., On an approximation theorem of Kupka and Smale, J. Diff. Eq. 3(1966), 214-227.

[10] Shub, M. and Sullivan, D., A remark on the Lefschetz fixed point formula for differentiable maps, Topology 13(1974), 189-191.

[11] Alligood, K., Mallet-Paret, J. and Yorke, J.A., Families of periodic orbit: local continuability does not imply global continuiability, Jour. Diff. Geom., to appear.

Shui-Nee Chow[1]
Mathematics Department
Michigan State University
East Lansing, Michigan 48824

John Mallet-Paret[2]
Division of Applied Mathematics
Brown University
Providence, Rhode Island 02912

James A. Yorke[3]
Molecular Physics Building
University of Maryland
College Park, Maryland 20742

AN INDEX THEORY FOR PERIODIC SOLUTIONS OF A HAMILTONIAN SYSTEM.

C. Conley and E. Zehnder

In the following we derive an estimate for the measure of the closure of the set of periodic solutions locally in a neighborhood of an elliptic equilibrium point. The result follows easily from a recent existence theorem due to J. Pöschel [10] concerning quasiperiodic solutions, which is an analytic result. In a different approach, which is of topological nature, we then use a Morse-type Index theory in order to find periodic solutions of a timedependent, asymptotically linear Hamiltonian vectorfield.

1. Periodic solutions close to an elliptic equilibrium point.

We consider a time independent Hamiltonian vectorfield $\dot{x} = Jh'(x)$, $x \in R^{2n}$ in a neighborhood of a linearly stable equilibrium point, which we assume to be $x = o$. For simplicity we assume $h \in C^{\infty}(R^{2n})$. Our aim is to estimate the set of periodic solutions close to the equilibrium point. To do so we assume the eigenvalues of the linearized Hamiltonian system at $x = o$ to be all purely imaginary and distinct from each other, such that $i\alpha_1,\dots,i\alpha_n,-i\alpha_1,\dots,-i\alpha_n$ are all the eigenvalues. For a fixed $\ell \geq 4$ we then exclude the so called resonances up to order ℓ requiring that

(1.1) $$< \alpha,k > \neq o\,, \qquad k \in Z^n, \quad 1 \leq |k| \leq \ell,$$

where $\alpha = (\alpha_1,\dots,\alpha_n)$ and $|k| = |k_1| + \dots + |k_n|$. Due to (1.1) there exist in a neighborhood of $x = o$ analytic symplectic coordinates $x = u(p,q)$ with $u(o) = o$, such that the Hamiltonian function takes, in the new coordinates, the following normal form, the so called Birkhoff-normal form:

(1.2) $$h \circ u(p,q) = h_o(J_1,\dots,J_n) + h_1(p,q),$$

where h_o is a polynomial of order ℓ in $J_1,\dots,J_n$, with $J_k(p,q) = \frac{1}{2}(p_k^2 + q_k^2)$.

and where the remainder term h_1 satisfies $|h_1|_{C^m} = O(r^{\ell+1})$ in $J_k < \frac{1}{2} r^2$ for every m. The polynomial h_o is of the form:

$$h_o(J) = < \alpha,J > + \frac{1}{2} < AJ,J > + \dots$$

with $A \in \mathcal{L}(R^n)$. The coefficients of h_o are symplectic invariants. The Hamiltonian function h_o represents an integrable system with the n involutive integrals J_k, which are linearly independend on $\{x \in R^{2n} \mid J_1 \cdot J_2 \cdots J_n(x) \neq o\}$. Therefore the system (1.2) is, near the equilibrium point, a system close to an integrable system. We now make the crucial assumption that the Hamiltonian system is nonlinear by requiring that

(1.3) $\qquad \det(A) \neq o$.

It is well known, that under this assumption the qualitative phenomena of the integrable system h_o are also present in the full system (1.2), if one looks at a small neighborhood of the equilibrium point. In fact, due to results by Kolmogorov, Arnold and Moser, every neighborhood of the equilibrium point contains an abbundance of invariant Tori of dimension n for the system with Hamilton function h. The induced flows on these tori are,moreover, linear Kronecker flows given by n rationally independent frequencies. Extending earlier results J. Pöschel [10] recently proved for the differentiable case, that in any polydisc D_r of radius r around o, the set $M \subset D_r$, which is left out by the invariant tori in D_r, can be estimated by $m(M) = O(r^{(\ell-3)/4})\, m(D_r)$, m being the Lebesgue measure. We shall use this analytically very intricate result in order to estimate the closure of the set of periodic solutions To do so we shall simply show that the invariant tori found are contained in the closure of the set of periodic solutions. This leads to the following statement, which extends an observation due to J. Moser [8].

Theorem 1.1. Let $h \in C^\infty$ satisfy (1.1) and (1.3). For $r > o$ denote by D_r a polydisc of radius r centered at o, and denote by P the set of periodic solutions in D_r. Then there is a measurable set $M \subset D_r$ containing $D_r \setminus cl(P)$ such that:

$$m(M) = O(r^{(\ell-3)/4})\, m(D_r).$$

Proof: We show that the particular invariant tori, whose existence is guaranteed by the above mentioned results are contained in cl(P). Pick such an invariant, embedded torus with frequencies $\omega \in R^n$. By its construction there is an open neighborhood in which symplectic and smooth normal coordinates $x = v(\xi,\eta)$ can be introduced, where $(\xi,\eta) \in T^n \times U$, $U \subset R^n$ being an open neighborhood of $\eta = 0$, such that $h \circ v(\xi,\eta)$ is periodic in $\xi \in T^n$ and of the form:

$$h \circ v(\xi,\eta) = c + \langle \omega,\eta \rangle + \frac{1}{2} \langle Q\eta,\eta \rangle + \hat{h}(\xi,\eta). \tag{1.4}$$

Here $c \in R$ is a constant, $Q \in \mathcal{L}(R^n)$, and $\hat{h}$ satisfies $\hat{h} = O(|\eta|^3)$. The invariant torus in question is, in these coordinates, simply $T^n \times \{0\}$. Moreover, $\det Q \neq 0$. Therefore, if ϕ^t denotes the flow of the Hamiltonian system, the symplectic map ϕ^1 satisfies the assumption of the Birkhoff-Lewis fixed point theorem, for which we refer to [9]. We conclude that ϕ^1 has a periodic point in every neighborhood of $T^n \times \{0\}$. Therefore, given the neighborhoods $V_\nu := \{\xi,\eta\} \in T^n \times U \mid |\eta| \leq \frac{1}{\nu}\}$, $\nu = 1,2,\ldots$, we find for every ν a periodic solution $p_\nu = p_\nu(t) \subset V_\nu$. Now pick a sequence of points $b_\nu \in p_\nu$, then a subsequence converges, $b_\nu \to b_\infty$. Since the η-coordinate of b_ν satisfies $|\eta(b_\nu)| \leq \frac{1}{\nu}$ we conclude that $\eta(b_\infty) = 0$ hence $b_\infty \in T^n \times \{0\}$. Also, $b_\infty \in cl(P)$. Now $T^n \times \{0\}$ and cl(P) are invariant under the flow ϕ^t, hence

$$cl\{\phi^t(b_\infty) \mid t \in R\} \subset (T^n \times \{0\} \cap cl(P)). \tag{1.5}$$

Observe that $\phi^t(b_\infty) = (\xi(b_\infty) + t\omega, 0)$ and since the frequencies $\omega = (\omega_1,\ldots,\omega_n)$ are rationally independent the left hand side of (1.5) is $T^n \times \{0\}$. Therefore $T^n \times \{0\} \subset cl(P))$ proving the claim. •

We point out that under only finitely many inequalities for the coefficients of the 4-jet of h at the equilibrium point, which formulate the required nonlinearity assumption, we conclude that in every neighborhood of the equilibrium point the closure of the set of periodic solutions is even of positive measure. The periods of these periodic solutions are, however, very large. The result depends heavily on the excessive smoothnes assumptions required for the existence of invariant tori. It is however not necessary to assume h to be C^{∞}, in fact in the case $\ell = 4$, $h \in C^{\alpha}$, $\alpha > 4n$ will be good enough. For more precise smoothnes assumptions we refer to [10].

2. Forced oscillations in timedependent Hamiltonian equations.

In contrast to above local result, which is based on an analytically intricate existence result about quasiperiodic solutions, we describe next a global result which will be proved by means of a generalized Morse theory. Here we look for T-periodic solutions of a timedependent Hamiltonian equation

$$\dot{x} = Jh'(t,x), \quad (t,x) \in R \times R^{2n}, \tag{2.1}$$

where, again $J \in \mathcal{L}(R^{2n})$ is the symplectic structure, and where $h(t+T,x) = h(t,x)$ is T-periodic in time.

In order to formulate the result, we first associate an index to a nondegenerate T-periodic solution $x_0(t) = x_0(t+T)$. Here a periodic solution is called nondegenerate, if it has no Floquet multiplier equal to 1. The linearized equation along this solution,

$$\dot{y} = Jh''(t,x_0(t))y$$

is of the form $\dot{y} = JA(t)y$, with $A(t)$ being symmetric and depending continuously and T-periodically on t. Consider the set P of all such linear equations which have the property that they do not admit Floquet multipliers equal to one. In P an equivalence relation is introduced as follows: two equations with

$A_0(t)$ and $A_1(t)$ are called equivalent, if they can be continuously deformed into each other within the set P. It turns out that the set P decomposes into countably many equivalence classes characterized by an integer j. This index is, for special equations in P defined as follows. Consider a time-independent equation $A(t) = S$, and assume JS has distinct eigenvalues. If $(\lambda, \bar{\lambda})$ is a pair of purely imaginary eigenvalues with corresponding complex eigenvectors $e, \bar{e}$ then $< \bar{e}, Je > \neq 0$ is purely imaginary, and we set $\alpha(\lambda) := \text{sign}(-i < \bar{e}, Je >) \text{ Im } \lambda$. Observe $\alpha(\lambda) = \alpha(\bar{\lambda}) = \alpha(-\lambda)$. Since, by assumption, $\exp(TJS)$ has no eigenvalue equal to 1, we conclude $\alpha(\lambda) \notin \tau Z$, $\tau = \frac{2\pi}{T}$ and hence $n\tau < \alpha(\lambda) < (n+1)\tau$ for some integer n. In this case we set $[\alpha(\lambda)] = n + \frac{1}{2}$, and define

$$j(S) = \sum_{\lambda} [\alpha(\lambda)] \in Z,$$

where the summation goes over all the purely imaginary eigenvalues. In case JS has no purely imaginary eigenvalues, we set $j(S) = 0$. The following result is proved in [11]; it is related to [12] and [13].

Theorem 2.1.

Each equivalence class of the set P *of loops contains constant loops* $A(t) = S$ *for which* $\text{ind}(A(t)) = j(S)$ *is defined as above. All such constant loops in the same equivalence class have the same index, and constant loops in different components of* P *have different indices.*

The theorem states that the index is well defined on components. It allows to associate to every nondegenerate periodic solution $x(t) = x(t+T)$ of the Hamiltonian equation the index j of the corresponding linearized equation. With this notion we can formulate the existence result:

Theorem 2.2.

Let $h = h(t,x) \in C^2(R \times R^{2n})$, $n \geq 2$ *be periodic in time of period* $T > 0$, $h(t+T,x) = h(t,x)$. *Assume (i) the Hessian of* h *is bounded:* $-\beta \leq h''(t,x) \leq \beta$ *for all*

$(t,x) \in R \times R^{2n}$ *and for some constant* $\beta > 0$. *Assume* (ii) *the Hamiltonian vector-field to be asymptotically linear*

$$Jh'(t,x) = JA_\infty(t)x + o(|x|), \text{ as } |x| \to \infty$$

uniformly in t, *where* $A_\infty(t) = A_\infty(t+T)$ *is a continuous loop of symmetric matrices. Assume* (iii) *that the trivial solution of the equation* $\dot{x} = JA_\infty(t)x$ *is nondegenerate and denote its index by* j_∞. *Then the following statements hold:*

(1) *There exists a periodic solution of period* T *for* (2.1). *If this periodic solution is nondegenerate with index* j_0, *then there is a second* T-*periodic solution, provided* $j_0 \neq j_\infty$. *Moreover if there are two nondegenerate periodic solutions there is also a third one.*

(2) *Assume all the periodic solutions are nondegenerate, then there are only finitely many of them and their number is odd. If* j_k, $1 \leq k \leq n$, *denote their indices we have the following identity:*

$$\sum_{k=1}^{n} t^{-j_k} = t^{-j_\infty} + t^{-d}(1+t)\, Q_d(t),$$

where $d > 0$ *is an integer, and where* $Q_d(t)$ *is a polynomial having nonnegative integer coefficients.*

The theorem extends earlier results in [3] and in [7]. We point out an interesting special case of the above statement, which can be viewed as a generalization to higher dimensions of the Poincaré-Birkhoff fixed point theorem for mappings in the plane. This well known theorem states that a measure preserving homeomorphism of an annulus, which twists the two boundaries in opposite directions has at least two fixed points, see G.D. Birkhoff [5] and, more recently, M. Brown and W.D. Neumann [6].

Corollary.

Let $h = h(t,x) \in C^2(R \times R^{2n})$, $n \geq 2$ *be periodic,* $h(t+T,x) = h(t,x)$ *and let the*

Hessian of h *to be bounded. Assume*

$$Jh'(t,x) = JA_\infty(t)x + o(|x|) \quad \text{as} \quad |x| \to \infty$$

$$Jh'(t,x) = JA_0(t)x + o(|x|) \quad \text{as} \quad |x| \to 0$$

uniformly in t, *for two continuous loops* $A_0(t+T) = A_0(t)$ *and* $A_\infty(t+T) = A_\infty(t)$. *Assume that the two linear systems* $\dot{x} = JA_\infty(t)x$ *and* $\dot{x} = JA_0(t)x$ *do not admit any nontrivial* T*-periodic solutions, and denote by* j_∞ *and* j_0 *the indices of these two linear systems. If* $j_\infty \neq j_0$ *then there exists a nontrivial* T*-periodic solution. Moreover, if this periodic solution is also nondegenerate then there is a second* T*-periodic solution.*

In other words, if the two linear systems with $A_0(t)$ and $A_\infty(t)$ cannot be continuously deformed into each other within the set P, then we conclude the existence of a T-periodic orbit. The corollary only claims the existence of one T-periodic solution except if the nondegeneracy condition is satisfied. This is in contrast to the Poincaré-Birkhoff fixed point theorem which always guarantees two fixed points. Birkhoff's original proof in [5] also suggests, that the integer $|j_0-j_\infty|$ is a measure for the lower bound of the number of periodic solutions of (2.1). Our proof of the above statement being based on a Morse-type index theory does not allow such a conclusion. As a sideremark we recall, however, that under additional assumptions the following result [3] has been proved by means of mini-max techniques.

Theorem 2.3.

Let h *be as in the corollary and assume, in addition,* $h(t,x) = h(t,-x)$ *for all* $(t,x) \in R \times R^{2n}$. *Moreover, let* $A_0(t) = A_0$ *and* $A_\infty(t) = A_\infty$ *be independent of* t. *Then (2.1) has at least* $|j_0-j_\infty|$ *nontrivial pairs* $(x(t), -x(t))$ *of* T*-periodic solutions.*

Proof of theorem 2.2.

The T-periodic solutions of (2.1) can be found as critical points of a functional defined on a linear space of T-periodic functions, as is well known. In order to formulate this variational problem in an abstract setting we let H be the real Hilbert space $H = L_2(o,T;R^{2n})$. We define in H the linear selfadjoint operator A: dom(A) $\subset H \to H$ by setting $\text{dom}(A) = \{u \in H^1(o,T;R^{2n}) \mid u(0) = u(T)\}$ and $Au = -J\dot{u}$, for $u \in \text{dom}(A)$. The continuous operator $F: H \to H$ is defined by $F(u)(t) = h'(t,u(t))$, $u \in H$. Its potential $\Phi(u)$ is given by

$$\Phi(u) := \int_0^T h(t,u(t))\, dt.$$

F is the gradient of Φ. Writing the equation (2.1) in the form $-J\dot{x} = h'(t,x)$ one sees immediately, that every solution $u \in \text{dom}(A)$ of the equation

$$Au = F(u) \tag{2.2}$$

defines a classical T-periodic solution of (2.1), and conversely. The equation (2.2) is the Euler equation of the variational problem extr $\{f(u) \mid u \in \text{dom}(A)\}$ where, $f(u) = \frac{1}{2} < Au,u > - \Phi(u)$. It remains to find critical points of f. As observed in [1], and already carried out for our situation in [2], the assumption (i) of theorem 2.2 allows to reduce the problem of finding critical points of f to the simpler problem of finding critical points of a function $a = a(z)$ defined on the finite dimensional space $Z = PH \subset H$, where P is the projection

$$P = \int_{-\beta}^{\beta} dE_\lambda \quad ,$$

E_λ being the spectral resolution of A. Observe the operator A has a discrete spectrum $\sigma(A) = \tau Z$, $\tau = \frac{2\pi}{T}$. The function a is of the form $a(z) := F(u(z))$, where $u: Z \to H$ is an injective differentiable map with $Pu(z) = z$ and with $u(Z) \subset \text{dom}(A)$. It can be shown [2] that in fact critical points z of $a(z)$ correspond in a one to one way t the solutions $u(z) = u$ of the equation 2.2, hence to the required T-periodic solutions of (2.1). Now critical points of a will be found by means of a Morse-type index theory for flows, which will be applied to the gradient flow $\dot{z} = a'(z)$

on Z.

In order to briefly outline this index theory for flows we consider a flow on a topological space which is not necessarily a gradient flow on a manifold. To an isolated invariant set S an index pair (N_1,N_0) can be associated, where $N_0 \subset N_1$ is roughly the "exit set" of N_1, and where $S \subset \text{int}\,(N_1 \setminus N_0)$ see [4]. The homotopy type of the pointed space N_1/N_0 then does not depend on the particular choice of index pairs for S and is called the index of S, and denoted by $h(S) := [N_1/N_0]$. We therefore can associate to an isolated invariant set S the algebraic invariant $p(t,h(S))$, which is the series in t whose coefficients are the ranks of Cech cohomology of an index pair (N_1,N_0) for S. The index theory for flows then relates the algebraic invariants of S to the algebraic invariants of a Morse decomposition of S:

Theorem 2.4.

Let S be an isolated invariant set, and let $(M_1,\ldots,M_n)$ be an ordered Morse decomposition of S, where $M_k \subset S$ are isolated and invariant. Then there is a filtration $N_0 \subset N_1 \subset \ldots \subset N_n$ for this Morse decomposition, such that (N_n,N_0) is an index pair for S and such that (N_j,N_{j-1}) is an index pair for M_j. Setting $h(M_j) = [N_j/N_{j-1}]$ and $h(S) = [N_n/N_0]$, then the following identity holds:

$$\sum_{j=1}^{n} p(t,h(M_j)) = p(t,h(S)) + (1+t)\,Q(t) ,$$

where $Q(t)$ is a series in t having only nonnegative integer coefficients. This identity can be viewed as a generalization of the Morse inequalities.

The development outlined here extends some of the results in [4]. For a complete proof of theorem 2.4 we refer to [11]. The above theorem can be viewed as a generalization of Morse theory for flows other than gradient flows on spaces other than manifolds. An index is associated not only to critical points but to any isolated invariant set of a local flow. In addition to the classical Morse theory a analogue of the

"Homotopy Axiom" of Leray-Schauder degree theory is possible in this general setting.

In order to apply this index theory to the problem at hand we first observe that due to the assumptions (ii) and (iii) of theorem 2.2, the set S of bounded solutions of the gradient flow $\dot{z} = a'(z)$ is compact, hence isolated and has, therefore, an index, see [2]. In order to compute this index, the invariance of the index under deformation is crucial. Namely using the continuation theorem in [4] together with theorem 2.1 it can be proved that the index of S is that of a hyperbolic equilibrium point in Z (defined after (2.2)), which in our situation, has been computed in [3]. Also the indices of the critical points of $a(z)$ are related to the indices of the corresponding periodic solutions of (2.1). We summarize these observations in a Lemma, which is proved in detail in [11].

<u>*Lemma*</u>

Assume that the Hamiltonian function meets the assumptions of theorem 2.2. *Then*

(i) *The set* S *of bounded solutions of* $\dot{z} = a'(z)$ *is compact, hence has an index. It is the homotopy type of a pointed sphere:* $h(S) = [\dot{S}^{m_\infty}]$ *with* $m_\infty = \frac{1}{2} \dim Z - j_\infty$. *Therefore* $p(t,h(S)) = t^{m_\infty}$.

(ii) *If* $x(t)$ *is a nondegenerate* T-*periodic solution with index* j, *then the corresponding critical point* z *of* a *is an isolated invariant set, and* $h(\{z\}) = [\dot{S}^m]$, *where* $m = \frac{1}{2} \dim Z - j$; *hence* $p(t,h(\{z\})) = t^m$.

Since by this Lemma $h(S)$ is different from the index of the empty set, we conclude $S \neq \emptyset$ and therefore there exists at least one critical point of a because we are dealing with a gradient flow. Assume that there are only finitely many critical points. They form a Morse-decomposition of S. We order them in such a way that $\{z_1, z_2, \ldots, z_n\}$ is an admissible ordering of the Morse decomposition. By the Lemma, and by theorem 2.4 we then conclude

$$\sum_{h=1}^{n} p(t,h\{z_h\}) = t^{m_\infty} + (1+t)\, Q(t). \tag{2.3}$$

Assume now, that the critical point already found represents a nondegenerate periodic orbit and call its index j. We claim that if $j \neq j_\infty$, then there is a second critical point of a. If not we conclude by (2.3) and by the above Lemma that $t^m = t^{m_\infty} + (1+t)\,Q(t)$. Setting $t = 1$, we find $Q(1) = o$ and so $Q(t) = o$ and therefore $t^m = t^{m_\infty}$ in contradiction to $m \neq m_\infty$. Similarly one shows that the existence of two nondegenerate periodic solutions implies a third one. Assume now all the T-periodic solutions of (2.1) to be nondegenerate, then there are finitely many of them, since S is compact. Let j_k, $k = 1,2,\dots,m$ be their indices. By the Lemma we have $p(t,h(\{z_k\})) = t^{m_k}$, $m_k = \frac{1}{2}\dim Z - j_k$, hence by (2.3)

$$\sum_{k=1}^{n} t^{m_k} = t^{m_\infty} + (1+t)\,Q(t).$$

We conclude that n is an odd number, and that at least one periodic solution has index j_∞. Moreover setting $d = \frac{1}{2}\dim Z$, the formula in theorem 2.2 follows multiplying the equality by t^{-d}. This finishes the proof of theorem 2.2. •

3. About a generalization of the Poincaré-Birkhoff fixed point theorem

In view of the corollary to theorem 2.2 it is tempting to interpolate a given symplectic map in R^{2n} by a timedependent Hamiltonian vectorfield in order to find a fixed point. We shall consider a very special case. Let ϕ be a symplectic diffeomorphism on R^{2n}, such that

$$\phi(x) = L_0 x, \qquad |x| \leq a$$
$$\phi(x) = L_\infty x, \qquad |x| \geq b > a,$$

for two linear symplectic maps $L_0, L_\infty \in \mathcal{L}(R^{2n})$ belonging to the set $W^* = \{L \in Sp(n,R) \mid L$ has no eigenvalue equal to $1\}$. We look for conditions on L_0, L_∞ other than $\deg(L_0) \neq \deg(L_\infty)$ guaranteeing a fixed point of ϕ in $a < |x| < b$.

We can associate to such a map ϕ an index as follows. Recall that a symplectic map M can uniquely be representet in polarform as $M = P\cdot O$, where P is symplectic and positiv definite, and where O is symplectic and orthogonal. It is of the form

$$O = \begin{pmatrix} u_1 & -u_2 \\ u_2 & u_1 \end{pmatrix}, \qquad \bar{u} = u_1 + iu_2$$

with $\bar{u}$ being unitary in $\mathcal{L}(C^n)$. To any continuous arc $\gamma : [0,1] \to Sp(n,R)$, we can therefore associate the continuous arc $\bar{u}(t)$ of unitary matrices. If then $\Delta(t)$ is a continuous function with $\det \bar{u}(t) = \exp(i\,\Delta(t))$, then $\Delta(1) - \Delta(0)$ depends only on γ. We denote this number by $\Delta(\gamma)$. If γ is a loop, then $\Delta(\gamma)$ is a multiple of 2π. It can be shown [11] that a loop is contractible in $Sp(n,R)$ if and only if $\Delta(\gamma) = 0$. Let now ϕ be as introduced above. Pick a path $x(t)$, $0 \le t \le 1$, connecting some $x(0)$ in $|x| < a$ with some $x(1)$ in $|x| > b$. Consider then the arc $\bar{\gamma} : [0,1] \to Sp(n,R)$ which connects L_0 with L_∞, and which is defined by

$$\bar{\gamma}(t) = d\phi\,(x(t)) \;,\; 0 \le t \le 1.$$

Then $\Delta(\bar{\gamma})$ is independent of the chosen path $x(t)$ between the two regions $|x| < a$ and $|x| > b$. In order to normalize, we extend $\bar{\gamma}$ by adding at the end points L_0 resp. L_∞ two arcs γ_0 resp. γ_∞ contained in W^* and connecting L_0 resp L_∞ with either $W_- = -1$ or, depending on the degrees of the maps involved, to W_+ defined by:

$$W_+ = \begin{pmatrix} 2 & & & 0 \\ & -I & & \\ & & 1/2 & \\ 0 & & & -I \end{pmatrix} .$$

Here I stands for the identity matrix in $(n-1)$ dimensions. $\Delta(\gamma_0)$ and $\Delta(\gamma_\infty)$ do not depend on the chosen arcs [11]. Take now the arc $\gamma := \gamma_0 \cup \bar{\gamma} \cup \gamma_\infty$, and define the "winding" number of ϕ as follows:

$$j(\phi) := \Delta(\gamma) \in Z.$$

This integer is dependent of the chosen extensions of $\overline{\gamma}$.

Assume now, that there is a timedependent Hamiltonian system, which satisfies the assumptions of the corollary, and which interpolates ϕ with the identity, i.e.:

$$\phi = \phi^T ,$$

where ϕ^t with $\phi^o = id$ is the flow of the timedependent vectorfield. It is easily seen that the condition $j_0 \neq j_\infty$ in the Corollary is equivalent to the condition $j(\phi) \neq o$. Therefore if $j(\phi) = o$ there is by the Corollary a nontrivial T-periodic solution, which gives rise to a fixed point $\phi(x) = x$ contained in $a < |x| < b$.

Any symplectic diffeomorphism is symplectically isotopic to the identity, as is well known. It is, however, not clear under which further conditions, if any, on the given map ϕ, such an isotopy can be chosen to satisfy, in addition, the requirements of the Corollary.

References:

[1] H. Amann: Saddle points and multiple solutions of differential equations. Math. Z. 169 (1979) 127-166.

[2] H. Amann and E. Zehnder: Nontrivial solutions for a Class of Nonresonance Problems and Applications to Nonlinear Differential Equations. Annali Scuola sup. Pisa Cl. Sc. Serie IV. Vol. VII, Nr. 4, 1980, 539-603.

[3] H. Amann and E. Zehnder: Periodic solutions of Asymptotically linear Hamiltonian systems. Manus. math. 32, (1980) 149-189.

[4] C.C. Conley: Isolated invariant sets and the Morse index, CBMS Regional Conf. Series in Math. 38 (1978) A.M.S. Providence R.I.

[5] G.D. Birkhoff: An extension of Poincaré's last geometric theorem. Acta Math. 47 (1925) 297-311.

[6] M. Brown and W.D. Neumann: Proof of the Poincaré-Birkhoff fixed point theorem. Michigan Math. Journ. 24 (1977) 21-31.

[7] K.C. Chang: Solutions of asymptotically Linear Operator equations via Morse Theory. Comm. Pure and Appl. Math. 34 (1981) 693-712.

[8] J. Moser: Quasi-periodic solutions in the three body problem, Bull. Astronomique Série 3, III(1), 1968, 53-59.

[9] J. Moser: Proof of a generalized form of a fixed point theorem due to G.D. Birkhoff. Lecture Notes in Mathematics, Vol. 597, 1976, 464-494.

[10] J. Pöschel: Integrability of Hamiltonian Systems on Cantor Sets. Preprint ETH-Zürich (1981).

[11] C. Conley and E. Zehnder: Morse-Type index theory for flows and periodic solutions of Hamiltonian Equations. Preprint RUB-Bochum (1981).

[12] I.M. Gelfand and V.B. Lidskii: On the structure of stability on linear canonical systems of differential equations with periodic coefficients. Amer. Math. Soc. Transl. (2) 8 (1958) 143-181.

[13] V.I. Arnold: On a characteristic class entering in quantization conditions. Funct. Anal. Appl. 1 (1967) 1-13.

Charles C. Conley
University of Wisconsin
Department of Mathematics
Madison, Wisconsin 53706
U.S.A.

Eduard Zehnder
Ruhr-Universität Bochum
Mathematisches Institut
Universitätsstr. 150
4630 Bochum
W.-Deutschland

Foliations That Are Not Approximable By Smoother Ones

Christopher Ennis, Morris W. Hirsch[1], and Charles Pugh[2]

§0. Introduction

We give several new examples of foliations that cannot be approximated by smoother ones. Let M be a C^∞ manifold of dimension $m > 0$ and F a foliation of M of codimension k, having smooth leaves, and defined by C^1 foliation boxes. The smoothness of F is measured by the differentiability of its tangent field

$$\tau F : M \longrightarrow G_{m-k}(M)$$

where $G_{m-k}(M)$ is the Grassmann bundle of $(m-k)$ - planes tangent to M, and τF is the section assigning to $x \in M$ the tangent plane $\tau_x F$ of the leaf F_x through x. If τF is a C^s map, we call F a C^s foliation, $0 \leq s < \infty$. This is slightly stronger than requiring F to be defined by C^s foliation boxes. This point is discussed below.

We give the space $C^s(M, G_{m-k}(M))$ the weak C^s topology. The set of C^s foliations is given the induced topology as a subset.

[1]Supported in part by NSF Grant MCS-80-02858.
[2]Supported in part by NSF Grant MCS-81-02262.

Let $0 \leqslant r \leqslant s \leqslant t$. We say that a C^s foliation F *can be* C^r *approximated by* C^t *foliations* provided every neighborhood of F in $C^r(M, G_{m-k}(M))$ contains a C^t foliation. A basic approximation question is:

Can every C^s *codimension-k foliation of every m-manifold be* C^r *approximated by* C^{s+1} *foliations* $(0 \leqslant r \leqslant s)$?

Some answers are:

(i) *Yes* if $k = m - 1$. This amounts to the Whitney approxmation theorem applied to vector fields.

(ii) *No* if $k \leqslant m - 1$ and $k, r, s \geqslant 2$, as shown by Anderson [1] using finite group actions.

(iii) *No* if $0 = r - s$, $m = 3$ and $k = 1$, proved by Rosenberg and Thurston [13] on the 3-torus T^3 .

(iv) *No* if $m \geqslant 3$ and $r = s = k = 1$. In these examples one can take $M = T^m$. See remarks below.

(v) *No* if $m \geqslant 3$ and either: $r = 0$ and $s = 1 = k$, or $r = 1 = s$ and $1 \leqslant k \leqslant m - 2$. This is the content of the following theorem, our main result. Let M^m denote a compact m-dimensional manifold without boundary.

0.1 THEOREM. Fix $m \geqslant 3$.

(a) There exists a C^1 codimension-one foliation of some compact M^m which cannot be C^0 approximated by C^2 foliations.

(b) For each k in the range $1 \leqslant k \leqslant m-1$ there is a C^1 codimension-k foliation of some M^m that cannot be C^1 approximated by C^2 foliations.

(c) The foliations in (a) and (b) can be taken to be orbit foliations of locally free actions by solvable Lie groups.

In §1 we give our basic constructions. In §5 and §6, (b) in §6, and (c) in §2 we describe the main example and prove the weaker result that it cannot be C^1 approximated by C^2 foliations. A different example of this kind is outlined in Remark 2, below. The full result in (a) is proved in §4 and §5. Parts (b) and (c) are proved in §6 and §7, respectively.

Anderson stated without proof that Rosenberg and Thurston's example (see (iii) above) can be improved to $s = 1$. Our example differs in that it has no compact leaf at which to apply Kopell's Lemma (see p. 466 of [13]).

In his paper [11], D. Pixton shows that in the space C^1 actions of $\mathbb{Z} \times \cdots \times \mathbb{Z}$ on S^1, the set of C^2 actions having at least one compact orbit is nowhere dense. By Hirsch [9] there is a nonempty open set of C^1 actions, each having a compact orbit. Therefore there exists a C^1 action $\alpha \in S$ which cannot be C^1 approximated by C^2 actions. By suspending the k generators of α, we obtain a C^1 locally free action β of $\mathbb{R}^k$ on the k-torus T^{k+1}. Let H denote the orbit foliation of β. It can be shown that any sufficiently close C^1 approximation F to H is the orbit foliation of the suspension of an action $\gamma(F)$ of $\mathbb{Z}^k$ on S^1 which is C^1 close to α; $\gamma(F) \longrightarrow \gamma$ as $F \longrightarrow H$ in the C^1 topologies; and $\gamma(F)$ is C^2 when F is C^2. It follows that H cannot be C^1 approximated by C^2 foliations.

In this way we obtain:

0.2 THEOREM. *For every integer* $k \geqslant 2$ *there is a locally free* C^1 *action of* $\mathbb{R}^k$ *on* T^{k+1} *whose orbit foliation cannot be* C^1 *approximated by* C^2 *foliations.*

This result leads to another proof of part (b) of Theorem 0.1 for the case $k = 1$. The two results differ in that all known examples of the actions described in Theorem 0.2 have compact leaves, while our main example (see §2) has no compact leaf.

The preceding theorem also gives examples of part (c) of Theorem 0.1 for the case $k = 1$. But in the proof of part (c) given in §7, we construct actions of nonabelian solvable groups.

The following conjecture is suggested by the foregoing remarks.

0.3 CONJECTURE. *Let* $\mathcal{F}$ *be a codimension-one foliation of a compact manifold which comes from a* C^1 *action of* $\mathbb{R}^n$. *If* $\mathcal{F}$ *has no compact leaves, then* $\mathcal{F}$ *can be* C^1 *approximated by* C^2 *foliations.*

In [6] D. Hart proves that if $\mathcal{F}$ is a foliation which has C^s foliation-boxes (so that $\tau\mathcal{F}$ may be only C^{s-1}) and if $s \geqslant 1$ then $\mathcal{F}$ is C^s diffeomorphic to a foliation $\mathcal{G}$ whose tangent field $\tau\mathcal{G}$ is of class C^s .

In [3] M. Cohen constructs a C^1 codimension-one foliation $\mathcal{F}$ of M^3 which cannot be C^1 approximated by C^2 foliations which are C^1 diffeomorphic to $\mathcal{F}$. Our construction in §2 is similar to his. Cohen also shows that any C^1 foliation $\mathcal{F}$ is C^1 diffeomorphic to a foliation G whose tangent field is C^∞ off some closed, nowhere dense set K in M. It remains an interesting question to determine how large K must be. For example, is there some obstruction to making K have codimension two in M ?

§1. Suspending Foliated Manifolds

Before outlining our construction we discuss some technical points.

By our definition, a foliation of class C^s has its field of tangent spaces of class C^s. Thus the set of C^s foliations of M $(s \geq 1)$ is a subset of the space of C^s maps $M \longrightarrow G_{m-k}(M)$, and we give it the subspace topology. Let us temporarily call these C^s foliations analytical.

A different, inequivalent definition of C^s foliation is in terms of foliation boxes. By a geometrical foliation of M of class C^s and codimension k, we mean a decomposition of M into disjoint subsets, called leaves, with the following property: let F_x denote the leaf through M. There is a covering of M by open sets U_i (called foliation boxes) and there are diffeomorphisms $f_i : U_i \approx \mathbb{R}^{m-k} \times \mathbb{R}^k$ such that if $f_i(x) = (y, z)$, then the component of x in $F_x \cap U_i$ is $f_i^{-1}(\mathbb{R}^{m-k} \times \{z\})$ The set of geometrical foliations has a topology derived from the C^s topology on such maps f_i ; see Hirsch [8] for details.

Every analytical foliation is geometrical, but the converse is not true: a geometrical foliation may have a tangent plane field which is only C^{s-1}, not C^s. This is a technical difficulty which has plagued the subject since its inception.

Thanks to recent work of Hart [6], these difficulties can be evaded rather simply. Let $\underline{A}$, $\underline{G}$ denote respectively the spaces of analytical and geometrical foliations of M, for some fixed codimension k and differentiability $s \geqslant 1$. Give them metrics $d_{\underline{\underline{A}}}$, $d_{\underline{\underline{G}}}$. Let $w : \underline{\underline{A}} \longrightarrow \underline{\underline{G}}$ be the natural map (weakening of structure). Hart proves (in essence) that w has <u>approximate local inverses</u> in the following sense. Fix $G \in \underline{\underline{G}}$. Then there is a neighborhood $\underline{\underline{N}} \subset \underline{\underline{G}}$ of G such that for every $\varepsilon > 0$ there are continuous maps $s_\varepsilon : \underline{\underline{N}} \longrightarrow \underline{\underline{A}}$ and $h_\varepsilon : \underline{\underline{N}} \longrightarrow \mathrm{Diff}^s(M)$, so that the following hold for all $G \in \underline{\underline{N}}$:

(i) $h_\varepsilon(G)$ maps leaves of F to leaves of $s_\varepsilon(G)$;

(ii) $d(h_\varepsilon, 1_M) < \varepsilon$ where 1_M is the identity map of M and d is the metric for $\mathrm{Diff}^s(M)$;

(iii) $d_G(ws_\varepsilon(G), G) < \varepsilon$;

(iv) if $G = w(A)$ (i.e. if τF is C^s), then

$$d_A(s_\varepsilon w(A), A) < \varepsilon .$$

It seems plausible that one can take $\underline{\underline{N}} = \underline{\underline{G}}$, but we don't need this.

It follows that for studying approximations one can pass back and forth between $\underline{\underline{A}}$ and $\underline{\underline{G}}$ (as Hart does in [6] for studying the Closing Lemma).

We shall actually use only the result of Hart that <u>every geometrical foliation is C^s diffeomorphic to some analytical foliation</u>.

In the rest of the paper we use "foliation" (without adjectives) to mean "analytical foliation."

Our basic method is that of <u>suspension</u> and <u>desuspension</u>. Let f be a C^s diffeomorphism of the m-dimensional manifold M. The <u>suspension</u>

of f is the $(m + 1)$ - dimensional manifold

$$M^f = (M \times \mathbb{R})/\mathbb{Z}$$

where the generator of $\mathbb{Z}$ acts on $M \times \mathbb{R}$ by

$$(x, t) \longmapsto (f(x), t - 1) \quad .$$

We give M^f the natural C^s structure making the orbit map

$$\pi : M \times \mathbb{R} \longrightarrow M^f$$

of class C^∞ .

Suppose now that $\mathcal{F}$ is a foliation of M of codimension k and differentiability $s \geqslant 1$, which is preserved by f ; that is,

$$f(\mathcal{F}_x) = \mathcal{F}_{f(x)}$$

for all $x \in M$. Let $\mathcal{F} \times \mathbb{R}$ denote the C^s foliation of $M \times \mathbb{R}$ whose leaves are the sets $\mathcal{F}_x \times \mathbb{R}$. Clearly the action of $\mathbb{Z}$ preserves $\mathcal{F} \times \mathbb{R}$. Therefore there is a unique foliation $\mathcal{F}^f$ of M^f such that π takes each leaf of $\mathcal{F} \times \mathbb{R}$ onto a leaf of $\mathcal{F}^f$. We call M^f and $\mathcal{F}^f$ the <u>suspensions</u> of M and $\mathcal{F}$ by f . Notice that $\mathcal{F}$ and $\mathcal{F}^f$ have the same codimension.

What is the differentiability of $\mathcal{F}^f$? The Grassmann bundles of $(m - k + 1)$ - planes are related by the commutative diagram

$$\begin{array}{ccc} M \times \mathbb{R} & \longleftarrow & G_{m-k+1}(M \times \mathbb{R}) \\ \downarrow \pi & & \downarrow \pi' \\ M^f & \longleftarrow & G_{m-k+1}(M^f) \end{array}$$

where π' is the natural map induced by π. The sections corresponding to F and F^f are related by the commutative diagram (suppressing the subscript $m - k + 1$):

$$\begin{array}{ccc} M \times \mathbb{R} & \xrightarrow{\tau F} & G(M \times \mathbb{R}) \\ {\scriptstyle\pi}\big\downarrow & & \big\downarrow{\scriptstyle\pi'} \\ M^f & \xrightarrow[\tau F^f]{} & G(M^f) \end{array}$$

Let $U \subset M \times \mathbb{R}$ be an open set which π maps diffeomorphically onto an open set $V \subset M^f$. Then $\tau F^f | V$ can be expressed as

$$\pi' \circ (\tau f) \circ (\pi | U)^{-1} : V \longrightarrow G(M^f) \quad .$$

Now π', expressed in local coordinates, involves the differential of π. Therefore π' <u>is only of class</u> C^{s-1}. In fact the natural differential structure on $G(M^f)$ is only C^{s-1}. Therefore τF^f is only of class C^{s-1}. This means that F^f is merely a C^{s-1} foliation.

On the other hand, it is easy to see that F^f is C^s as a <u>geometrical</u> foliation. Therefore, we can apply Hart's theorem to conclude that there is a C^s diffeomorphism h of M^f and a C^s analytical foliation K such that h takes leaves of F^f to leaves of K.

By construction, M^f has a C^s differential structure. By standard theorem (see e.g. [7]) M^f has a compatible C^∞ differential structure α. Then $h^*\alpha$ is <u>another</u> C^∞ differnetial structure on M^f, and

$$h : (M^f, f^*\alpha) \longrightarrow (M^f, \alpha)$$

is a C^∞ diffeomorphism. Since K is a C^s foliation of (M^f, α), it follows that F^f is a C^s foliation of $(M^f, f^*\alpha)$. We have proved:

<u>The natural</u> C^s <u>structure on</u> M^f <u>is compatible with a</u> C^∞ <u>structure in which</u> F^f <u>is a</u> C^s <u>foliation</u>.

In what follows, we shall always assume M^f has been given such a C^∞ structure. Thus, F^f <u>is of class</u> C^s .

Notice that there is a C^∞ embedding of $i : M \longrightarrow M^f$, namely the composition

$$M \xrightarrow{i_0} M \times \mathbb{R} \xrightarrow{\pi} M^f$$

where $i_0(x) = (x, 0)$. We often identify M with $i(M)$. With this identification we can recover F from F^f by

$$F = F^f \cap M \quad ;$$

that is, for each $x \in M \subset M^f$, the leaf F_x is the component of x in $(F^f)_x \cap M$.

We can also recover the C^s diffeomorphism f of M from data on M^f . The action of $\mathbb{Z}$ on $M \times \mathbb{R}$ preserves the vertical vector field $\tilde{V} = (0, \frac{\partial}{\partial t})$ on $M \times \mathbb{R}$ (where t is the coordinate in $\mathbb{R}$) . Therefore, $\tilde{V}$ is related by π to a vector field V on M^f . Let $\{\phi_t\}_{t \in \mathbb{R}}$ denote the flow of V on M^f , covered by the flow $\tilde{\phi}_t : (x, s) \longmapsto (x, s + 1)$ on $M \times \mathbb{R}$. It is easy to see that

$$\phi_1(M) = M \quad \text{and} \quad \phi_1 | M = f \quad .$$

Notice that each integral curve of V lies in a leaf of F^f , so that $\{\phi_t\}$ preserves F^f .

Now suppose G is a C^s foliation of M^f which is very C^s close to F^f . We want to C^s perturb V to a C^s vector field W

tangent to G. For this purpose, fix a C^∞ Riemannian metric on M^f (independent of G). Let $TG \subset TM^f$ be the subbundle of the tangent vector bundle of M^f which is tangent to (leaves of) G. Let $(TG)^\perp \subset TM^f$ be its orthogonal complement. Let $P : TM^f \longrightarrow TG$ be the vector bundle projection with kernel $(TG)^\perp$. Because the Grassman map $\tau G : M^f \longrightarrow G_{m-k+1}(M^f)$ is C^s, it follows that P is C^s. Therefore the vector field $W = PV$ is C^s. By construction, W is tangent to G, so its integral curves lie in leaves of G. It is easy to see that $W \longrightarrow V$ in $C^s(M^f, TM^f)$ as $G \longrightarrow F$ in the space of C^s foliations.

Set

$$M_s = \pi(M \times \{s\}) \subset M^f \quad .$$

Notice $M^0 = M$ and $M_s = M_{s+1}$. If G is sufficiently close to F, then the vector field W will be transverse to all the submanifolds M_s. From now on we assume this holds.

We now multiply the vector field W by a positive C^s function $\gamma : M^f \longrightarrow \mathbb{R}$ so that the time-one map of the flow of γW preserves M. To define γ, let $q : M^f \longrightarrow S^1 = \mathbb{R}/Z$ be the C^s map covered by the projection $M \times \mathbb{R} \longrightarrow \mathbb{R}$. Let Z be the vector field on S^1 covered by $\frac{d}{dt}$ in $\mathbb{R}$. Define $\alpha : M^f \longrightarrow \mathbb{R}$ by the equation

$$dq_p W(p) = \alpha(p) Z(q(p)) \quad , \quad p \in M^f \quad .$$

The assumption that W is transverse to the submanifolds M_t ensures that $\alpha > 0$.

Clearly α is continuous. In fact, α is C^s. To prove this, it suffices to prove that the composition

$$M \times \mathbb{R} \xrightarrow{\pi} M^f \xrightarrow{\alpha} \mathbb{R}$$

is C^s. It is easy to see that this map is the same as the composition

$$M \times \mathbb{R} \xrightarrow{\tilde{W}} T(M \times \mathbb{R}) \xrightarrow{\beta} \mathbb{R} \quad ,$$

where $\tilde{W}$ covers W, and β is the natural vector bundle map assigning to each tangent vector to $M \times \mathbb{R}$ its $\mathbb{R}$-component. (In other words, β is the derivative of the natural projection $M \times \mathbb{R} \longrightarrow \mathbb{R}$.) Since $\tilde{W}$ is C^s and β is C^∞, we find that α is C^s.

Set $\gamma = 1/\alpha$ and set $Y(x) = \gamma(x)W(x)$. Let $\{\theta_t\}$ be the flow in M^f of the vector field Y, and $\{\tilde{\theta}_t\}$ the flow in $M \times \mathbb{R}$ covering $\{\theta_t\}$. It is easy to see that for all s, $t \in \mathbb{R}$,

$$\tilde{\theta}_t(M \times \{s\}) = M \times \{s + t\} \quad .$$

Therefore there is a unique C^s map $H : M \times \mathbb{R} \longrightarrow \mathbb{R}$ such that

$$\tilde{\theta}_t(x, s) = (H(x, s), s + t) \quad .$$

It follows that $\theta_t : M_s \longrightarrow M_{s+t}$. In particular, $\theta_1(M) = M$, using the identification $M_0 = M$. It is easy to see that the C^s diffeomorphism $\theta_1 : M \longrightarrow M$ is defined by

$$\theta_1(x) = fH(x, 0) \quad .$$

Define $h(x) = fH(x, 0)$. This diffeomprhism $h : M \longrightarrow M$ is completely determined by the foliation G of M^f (once we fix the Riemannian metric on M^f). It is easy to see that $h \longrightarrow f$ in $\mathrm{Diff}^s(M)$ as $G \longrightarrow F$ in the space of C^s foliations of M^f.

Consider now the problem of C^r approximating F^f by a C^{s+1} foliation G , $0 \leq r \leq s$. If this is possible, then by desuspension we obtain a C^{s+1} foliation H which C^r approximates F , together with a C^{s+1} diffeomorphism h which preserves H and C^r approximates f . If we can show that, owing to the nature of F and f , such approximations are impossible, then we will have proved that F^f cannot be C^r approximated by C^{s+1} foliations. This is our basic method.

Now consider the C^s foliation $H = G \cap M$ of M . It is easy to see that $H \longrightarrow F$ as $G \longrightarrow F^f$, in the spaces of C^s foliations.

By construction, $\{\theta_t\}$ preserves G . Therefore, h preserves H . Thus, from the perturbation G of F^f , we obtain the perturbation H of F and the perturbation h of f , such that h preserves H . Moreover, H and h have the same differentiability as G .

It is not hard to show that there is a C^s diffeomorphism $\lambda : M^h \longrightarrow M^f$ taking H^h to G . In fact λ is covered by the map

$$M \times \mathbb{R} \longrightarrow M \times \mathbb{R} ,$$

$$(x, t) \longmapsto \tilde{\theta}_t(x, 0) .$$

Thus (H, h) is a desuspension of G .

The important element in constructing the approximating diffeomorphism $h : M \longrightarrow M$ is the vector field W tangent to G . It is not necessary to use a Riemannian metric for this purpose. An alternative method, to be followed in §4, relies instead on the existence of a foliation K of M which is transverse to F and invariant under f . Then K^f is a foliation of M^f which intersects F^f in a 1-dimensional foliation of M^f whose leaves are the integral curves of the vector field V . If G is so close to F^f that K^f is transverse to G, then $G \cap F^f$ is a 1-dimensional foliation whose leaves are integral curves of a vector field U . This vector field, suitably scaled, can be used in the same way as W . The time-one map of the flow of U will leave M invariant. Restricted to M , this map provides a C^{s+1} diffeomorphism $\tilde{f}$ of M (if K and G are C^{s+1}) which is C^r close to f (if G is C^r close to F^f) . Moreover $\tilde{f}$, by construction, has the important property of <u>preserving the foliation</u> K. In fact, $\tilde{f}$ acts on leaves of K in the same way as f .

If for some reason no such approximation $\tilde{f}$ can exist, then G cannot exist either. This method will be used in §4.

Suppose now that G is a Lie group acting on M by a locally free C^1 action, denoted by $(g, x) \longmapsto g\cdot x$, whose orbit foliation is F. Suppose that there is an automorphism Φ of G such that

$$g\cdot(f(x)) = f(\Phi(g)\cdot x)$$

for all $x \in M$, $g \in G$, where $f : M \longrightarrow M$ is the diffeomorphism which preserves F. Suppose also that Φ embeds in a 1-parameter group of automorphism $\{\Phi_t\}_{t\in\mathbb{R}}$ of G with $\Phi_1 = \Phi$. We shall extend G to a group H acting on M^f with orbit foliation F^f.

Let H be the split extension of G by $\mathbb{R}$ where $\mathbb{R}$ acts on G by: $t\cdot g = \Phi_t(g)$. By definition the underlying space of H is $G \times \mathbb{R}$ and the group operation is

$$(h, t)(g, r) = (h\Phi_t(g), t + r) .$$

Define an operation of H on $M \times \mathbb{R}$ by

$$(g, r)\cdot(x, s) = (\Phi_{s-r}(g)x, s - r)$$

for $(g, r) \in G \times \mathbb{R} = H$ and $(x, s) \in M \times \mathbb{R}$. One readily verifies that this defines a homomorphism $H \longrightarrow \mathrm{Diff}^s(M \times \mathbb{R})$, and also that the actions of H and $\mathbb{Z}$ on $M \times \mathbb{R}$ commute. (The action of $\mathbb{Z}$ is such that $(M \times \mathbb{R})/\mathbb{Z} = M^f$, i.e., $n\cdot(x, s) = (h(x), (s - n)0.$) Therefore there is induced a C^s action of H on M^f. It is evident that the orbit foliation of this action is $\mathcal{F}^f$.

Notice that H is solvable when G is solvable, since there is an exact sequence

$$0 \longrightarrow G \longrightarrow H \longrightarrow \mathbb{R} \longrightarrow 0 .$$

§2. A Suspended Solenoid

Let $T^2 = S^1 \times S^1$ be the 2-torus and $F : T^2 \longrightarrow T^2$ the DA-diffeomorphism of Smale [15]. In §3 we give a precise construction of f and prove that f leaves invariant a C^1 foliation $\mathcal{F}$ of T^2, some of whose leaves form the solenoid Σ which is the DA's attractor.

Applying the suspension construction we obtain a foliation $\mathcal{F}^f$ of $M^3 = (T^2)^f$.

<u>We claim $\mathcal{F}^f$ cannot be C^0 approximated by C^2 foliations of M^3.</u>

We first prove there is no C^1 approximation by C^2 foliations, since this proof is simpler and can be generalized in several directions.

Suppose $\mathcal{G}$ is a C^2 foliation of M^3 which is C^1 near $\mathcal{F}^f$. From the results in §1 we can desuspend $\mathcal{G}$: there is a C^2 foliation

$\mathcal{H}$ of T^2, preserved by a C^2 diffeomorphism h of T^2, such that $\mathcal{H}^h$ is C^1 diffeomorphic to $\mathcal{G}$; moreover, $\mathcal{H} \longrightarrow \mathcal{F}$ and $h \longrightarrow f$ as $\mathcal{G} \longrightarrow \mathcal{F}^f$, in the C^1 topologies.

We now use the well-known fact that <u>the DA-diffeomorphism f is structurally stable</u>. Therefore we can take $\mathcal{G}$ so close to $\mathcal{F}^f$ that h is topologically conjugate to f.

Let $g : T^2 \longrightarrow T^2$ be a homeomorphism such that $gfg^{-1} = h$. Then g carries Σ into a solenoid Σ' invariant under h. By taking h sufficiently C^1 close to f, we may assume that Σ' is a hyperbolic basic set for h, composed of unstable manifolds of h. Uniqueness of the unstable manifold foliation of a basic set implies that $\mathcal{H}_x$, for $x \in \Sigma'$, is the unstable manifold of x for h. <u>Therefore Σ' is a union of $\mathcal{H}$ leaves</u>. Now $\mathcal{H}$ is a C^2 foliation of T^2, so by Denjoy's theorem either every leaf is dense or every leaf is compact. In particular, this holds for leaves of $\mathcal{H}$ in Σ'. The same must therefore hold for leaves of $\mathcal{F}$ in Σ; but this is a contradiction. This completes the proof that $\mathcal{F}^f$ cannot be C^1 approximated by C^2 foliations of M^3.

Before proving, in §4, that there is no C^0 approximation, we study the DA in detail in §3.

§3. Foliations Invariant Under the DA-Diffeomorphism

Let f_0 be the linear Anosov diffeomorphism $\begin{bmatrix} 2 & 1 \\ 1 & 1 \end{bmatrix}$ on T^2, let $E_0^u \oplus E_0^s$ be its hyperbolic splitting, and express

$$f_0 = \begin{bmatrix} a & 0 \\ 0 & a^{-1} \end{bmatrix} \quad \text{respecting } E_0^u \oplus E_0^s$$

In fact, $a = \frac{1}{2}(3 + \sqrt{5})$, $E_0^u = \text{span}(2, -1 + \sqrt{5})$, and $E_0^s \perp E_0^u$ everywhere.

Modify f_0 according to Smale's DA construction [15] by pushing outward along E_0^s in a neighborhood of 0 . This produces a new map, the DA-diffeomorphism

$$f : T^2 \longrightarrow T^2$$

having two new saddle points on what was originally the stable manifold of 0 , while 0 itself becomes a source. The unstable manifold of 0 is open-dense in T^2 . Its complement is a <u>solenoid</u>, call it Σ , which is the closure of the unstable manifolds of the newly produced saddles. It is a hyperbolic set for f which attracts all orbits except 0 .

Notice that $a > 1$. Fix a constant α , $1 < \alpha < a$. Let $\beta : \mathbb{R} \longrightarrow \mathbb{R}$ be a C^∞ function such that $0 \leqslant \beta \leqslant 1$, $\beta(x) \equiv 1$ for x near 0 , $\beta(x) = 0$ for $|x| \geqslant 1$.

Now choose a C^∞ map $\gamma : \mathbb{R} \longrightarrow \mathbb{R}$ such that:

(1) $$0 < k = \sup \gamma'(y) \leqslant \alpha ,$$

$$\gamma(y) \equiv y/a \quad \text{for} \quad |y| \text{ near } 1 ,$$

and the graph of γ is transverse to the diagonal and crosses it 3 times: at 0 and $\pm y_0$, $0 < y_0 < 1$. Moreover, set

(2) $$c = \sup_{x,y} |\beta'(x)(\gamma(y) - y/a)|$$

and choose γ to make c so small that

$$a \geqslant \alpha\left[1 + \left(\frac{c}{1 - \alpha/a}\right)^2\right]^{\frac{1}{2}} .$$

Now define

$$f(z) = \begin{cases} f_0(z) & \text{if} \quad z \in T^2\backslash S \\ (ax,\ \beta(x)\gamma(y) + (1 - \beta(x))y/a) & \text{if} \quad z \in S \end{cases}$$

Then f is the DA-diffeomorphism, and

$$T_z f = \begin{bmatrix} a & 0 \\ c_z & k_z \end{bmatrix} \quad \text{respecting} \quad E_0^u \oplus E_0^s$$

where c_z , k_z are C^∞ functions of $z \in T^2$, and $k_z > 0$. Notice that

(3)
$$c = \sup \left|c_z\right| ,$$
$$k = \sup k_z \leqslant \alpha .$$

For if $z \in S$, then $c_z = 0$ and $k_z = a^{-1} < 1$; while if $z \in S$ then

$$c_z = \frac{\partial}{\partial x}[\beta(x)\gamma(y) + (1 - \beta(x))y/a] = \beta'(x)(\gamma(y) - y/a)$$

and

$$k_z = \frac{\partial}{\partial y}[\beta(z)\gamma(y) + (1 - \beta(x))y/a] = \beta(x)\gamma'(y) + (1 - \beta(x))/a \leqslant \alpha$$

by (1).

Let $E^u \oplus E^s$ be the hyperbolic splitting of the tangent bundle of T^2 over the solenoid Σ . The main result of this section is

(4) E^u <u>extends to a</u> C^1 <u>Tf-invariant line field</u> E^u over T^2 .

This immediately implies the result needed for §2:

<u>3.1 PROPOSITION</u>. The integral curves of E^u form an f-invariant C^1 foliation of T^2.

As remarked in §2, Denjoy showed:

(5) No C^2 1-foliation of T^2 admits a solenoid of leaves.

Thus (5) implies that E^u can only be C^1, not C^2.

To prove (4) we use the C^1 Section Theorem from [11, p. 31]. At each $z \in T^2$ consider

$$L_z = L(E^u_{0z}, E^s_{0z}) = \text{all linear maps } E^u_{0z} \longrightarrow E^s_{0z} .$$

The union of the L_z forms a bundle L over T^2, and Tf acts naturally on L as follows. Let $P \in L_z$. Then

$$P \longmapsto f_\# P = (c_z + k_z P)\, a^{-1} .$$

See [11, p. 34]. This gives a bundle map

$$\begin{array}{ccc} L & \xrightarrow{\;f_\#\;} & L \\ \downarrow & & \downarrow \\ T^2 & \xrightarrow{\;f\;} & T^2 \end{array}$$

Let $L(\ell)$ be the disk bundle of size ℓ in L : $L_z(\ell) = \{P \in L_z : |P| \leqslant \ell\}$. Observe from (3) that

$$|f_\# P| \leqslant |c_z| + k_z |P| a^{-1} \leqslant c + \alpha a^{-1} \ell \leqslant \ell$$

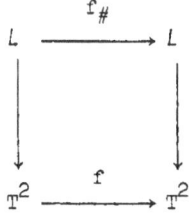

if $|P| \leq \ell$ and $\ell \geq c/(1 - \alpha/a)$. Also

$$|f_{\#}P_1 - f_{\#}P_2| \leq \alpha a^{-1}|P_1 - P_2| \qquad \text{for} \qquad P_1, P_2 \in L_z$$

Thus ([11, p. 34]), $f_{\#}$ is a fiber contraction of $L(\ell)$, $L(\ell)$ has a unique $f_{\#}$-invariant section $\sigma : T^2 \longrightarrow L(\ell)$, and σ is continuous. This produces the Tf-invariant line field as

$$E^u_z = \text{graph}(\sigma(z))$$

and proves it is continuous. Since it is an invariant continuous complement to $E^s_0 = E^s$ over Σ , it equals E^u on Σ by uniqueness of the hyperbolic splitting there.

To prove E^u is C^1 , it suffices to compare the fiber contraction rate of $f_{\#}$ versus its base contraction rate:

$$\text{fiber contraction at } z = k_z a^{-1} ,$$

$$\text{base contraction at } z = \|(T_z f)^{-1}\|^{-1} .$$

If

(6) $$\sup_z [(k_z a^{-1}) \|(T_z f)^{-1}\|] < 1$$

then E^u is C^1 [11, p. 31]. This norm $\| \ \|$ need not be the standard one -- it can arise from any convenient Finsler on T^2 . Let $\langle \ , \ \rangle_0$ enote the standard Riemann structure on T^2 as a flat torus and set

$$\langle v, w \rangle = \begin{cases} \langle v, v \rangle_0 & \text{if} \quad v = w \in E^u_z \text{ or } E^s_{0z} \\ 0 & \text{if} \quad v \in E^u_z \text{ and } w \in E^s_{0z} \end{cases} .$$

This gives a C^0 Riemann structure $\langle\ ,\ \rangle$ on T^2 agreeing with $\langle\ ,\ \rangle_0$ on E^u and E_0^s, but making E^u perpendicular to E_0^s. Now

$$T_z f = \begin{bmatrix} a_z & 0 \\ 0 & k_z \end{bmatrix} \quad \text{respecting } E^u \oplus E_0^s$$

so, in the new norm,

$$(7) \qquad \|(T_z f)^{-1}\| = \max(k_z^{-1}, a_z^{-1}) \quad .$$

Notice that a_z is the norm of $T_z f$ on E_z^u, i.e., if $\begin{bmatrix} 1 \\ w \end{bmatrix}$ spans E_z^u then

$$a_z = \frac{\left| \begin{bmatrix} a & 0 \\ c_z & k_z \end{bmatrix} \begin{bmatrix} 1 \\ w \end{bmatrix} \right|}{(1 + w^2)^{1/2}} = \left[\frac{a^2 + (c_z + k_z w)^2}{1 + w^2} \right]^{1/2} \geqslant \left[\frac{a^2}{1 + w^2} \right]^{1/2}$$

But $|w| \leqslant \ell$ since $E_z^u = \text{graph } \sigma(z)$ and $\sigma(z)$ is a linear map $E_{0z}^u \longrightarrow E_{0z}^s$ having slope $\leqslant \ell$. Besides, we were able to choose $\ell \geqslant c/(1-\alpha/a)$, so

$$a_z \geqslant \frac{a}{\left[1 + \left[\dfrac{c}{1 - \alpha/a} \right]^2 \right]^{1/2}} \geqslant \alpha$$

by (2). Thus by (3) $a_z \geqslant \alpha \geqslant k_z$, so (7) becomes

$$(8) \qquad \|(T_z f)^{-1}\| = k_z^{-1} \quad .$$

Now (6) follows from (8) since $a^{-1} < 1$. Hence E^u is C^1 and (4) is proved. Recall that (4) implies (3.1), the fact needed in our construction.

The following general result, which we shall not use, also implies (5), at least near Σ .

EXTENSION THEOREM. *Suppose* f *is a* C^2 *diffeomorphism of* M , $\Lambda \subset M$ *is a hyperbolic set for* f *with hyperbolic splitting* $E^u \oplus E^s$, E^s *is a line bundle, and* Λ *is an expanding attractor. Thus* E^u *extends to a* C^1 *Tf-invariant bundle* $\hat{E}^u$ *on some neighborhood of* Λ .

REMARKS. (a) When $\Lambda = M$, $\hat{E}^u$ must be E^u since it is known that E^u is C^1 in this case.

(b) To prove this Extension Theorem requires a new, relative version of the C^r Section Theorem [11, p. 31], so we prefer the more direct construction of $\hat{E}^u$ given above.

(c) Whether $\hat{E}^u$ can be chosen to integrate some foliation when $\dim(M) \geq 3$ is a tantalizing question.

In the next section we shall use the fact that the DA-diffeomorphism f preserves the foliation $\mathcal{G}$ tangent to E_0^s , as is evident from the formula for $f(z)$ (following (2) above). This foliation is C^2 (even real analytic) and its leaves are dense lines in T^2 .

§4. Foliations Invariant Under Perturbations of the DA

Let $f : T^2 \longrightarrow T^2$ be the DA-diffeomorphism, leaving invariant both the C^1 (but not C^2) foliation $\mathcal{F}$ and the C^2 foliation $\mathcal{K}$ by dense lines transverse to $\mathcal{F}$. Let $M^3 = (T^2)^f$. Let $\mathcal{G}$ be a C^2 foliation of M^3 which closely C^0 approximates $\mathcal{F}^f$. We seek a contradiction.

Let $\mathcal{H} = \mathcal{G} \cap T^2$ be the induced foliation of T^2. We may assume $\mathcal{H}$ and $\mathcal{K}$ are transverse to each other.

As explained at the end of §1, we can assume there is a C^2 diffeomorphism $\tilde{f}$ of T^2 which leaves both $\mathcal{K}$ and $\mathcal{H}$ invariant, and which C^0 approximates f as closely as desired. We assume $\tilde{f}$ is so close to f that <u>it carries some closed disk</u> $D \subset T^2$ <u>onto a neighborhood of itself</u>:

$$\operatorname{Int} \tilde{f}(D) \supset D \quad .$$

Here D is a neighborhood of the fixed point $(0, 0)$ (in the notation at the beginning of §3), in which the DA-diffemorphism f is constructed by deforming the linear toral automorphism f_0. Now f is an expansion in a neighborhood of 0 because $Df(0, 0) = \begin{bmatrix} a & 0 \\ 0 & k \end{bmatrix}$ with $a > 1$, $k > 1$. Therefore there is a disk D with this property for any map $\tilde{f}$ of f which is sufficiently C^0 close to f. We call such a disk D a <u>source disk for</u> $\tilde{f}$.

Our desired contradiction is a consequence of:

<u>4.1 THEOREM.</u> Let $\mathcal{H}$, $\mathcal{K}$ be 1-dimensional C^1 foliations of T^2 which are mutually transverse. Suppose neither foliaton has any exceptional minimal sets. Let g be a homeomorphism of T^2 which preserves $\mathcal{H}$ and $\mathcal{K}$ and which admits a source disk D. Then the induced endomorphism g_* of $H_1(T^2; \mathbb{R})$ has $+1$ or -1 as an eigenvalue.

To reach a contradiction, assuming 4.1, take $g = \tilde{f}$. Since $\mathcal{H}$ is C^2 it has no exceptional minimal sets by Denjoy's theorem; and since $\mathcal{K}$ has every orbit dense, it also has no exceptional minimal sets. Then $\tilde{f}_*$ has ± 1 as an eigenvalue, by 4.1. But $\tilde{f}$ is homotopic to the linear toral automorphism $f_0 = \begin{bmatrix} 2 & 1 \\ 1 & 1 \end{bmatrix}$, so the eigenvalues are $\frac{1}{2}(3 \pm \sqrt{5})$.

<u>PROOF OF 4.1</u>. Suppose H or K has a compact leaf C, representing $[C] \in H_1(T^2)$. Then $[C] \neq 0$; this follows easily from the Poincaré-Bendixson theorem applied to the foliations $\hat{H}$, $\hat{K}$ of $\mathbb{R}^2$ induced by the covering space projection $\pi : \mathbb{R}^2 \longrightarrow T^2$.

Either $g(C) = C$ or else $g(C) \cap C = \phi$, since C and $g(C)$ are leaves of a foliation. In the first case, $g_*[C] = \pm[C]$. In the second case, C and $G(C)$ bound a cylinder in T^2, and again, $g_*[C] = \pm[C]$.

From now on assume neither H nor K has a compact leaf. It is known that this implies H and K are orientable; we fix orientations for them. Since there are no exceptional minimal sets, H and K are homeomorphic to orbit foliations of irrational flows. It follows that <u>every half-leaf is dense</u>. (The positive half-leaf H_x^+ is obtained by following the positive orientation of H_x, starting at x. Similarly for K.)

Now let D be a source disk for g and put $W = \cup_{n \geqslant 0} g^n(d)$. Then W is a simply connected open set, and $g(W) = W$.

No half-leaf is contained in W. To see this, suppose $H_x^+ \subset W$. Let $\pi(y) = x$. Let $\hat{W} \subset \mathbb{R}^2$ be the component of y in $\pi^{-1}(W)$, and consider $\hat{H}_y^+$. Since every leaf of $\hat{H}$ is closed in $\mathbb{R}^2$, we know that $\hat{H}_y^+$ is closed in $\hat{W}$. Since W is simply connected, $\pi : \hat{W} \longrightarrow W$ is a homeomorphism. Therefore $H_x^+ = \pi(\hat{H}_y^+)$ is closed in W, contradicting denseness of half-leaves. Similarly for half-leaves of K.

By Brouwer's theorem D contains a fixed point p for g. Then H_p^+ meets the boundary ∂W of W. Let q be the point of $H_p^+ \cap \partial W$ which is nearest to p (in the natural ordering of H_p^+ as a half-line). Then $g(q) = q$. To see this, observe that $H_p^+ \cap \partial W$ is invariant under g. If $g(q) \neq q$ then either $g(q)$ or $g^{-1}(q)$ would be nearer to p.

Let $A \subset F_p^+$ be the interval from p to q. Since K_p^+ is dense there is a point $r \in A \cap (K_p^+ \setminus \{p\})$ which is nearest to p in the natural ordering of K_p^+. Then $g(r) = r$. Let $C_1 \subset H_p^+$ be the oriented arc from p to r and let $C_2 \subset K_p^+$ be the oriented arc from r to p. Let C be the oriented loop $C_1 \cup C_2$. Then $g(C) = C$ preserving orientation. The Poincaré-Bendixson theorem implies that $[C] \neq 0$ in $H_1(T^2)$, and evidently $g_*[C] = [C]$. QED

<u>REMARK</u>. In the case where F is C^2 and K has only dense leaves, C. Pugh has shown that the hypothesis of Theorem 4.1 is contradictory.

§5. Higher Dimensional Examples

Let E be a C^s foliation of a compact M^m which cannot be C^r approximated by C^{s+1} foliations. For example, $r = 0$, $s = 1$, $M = (T^2)^f$ and $E = F^f$, constructed in §4. Let V be any C^∞ manifold and let $E \times V$ denote the foliation of $M \times V$ whose leaves are $F_x \times V$, $x \in M$. Then $E \times V$ has the same codimension as E, and the same differentiability. Fix $v \in V$ and identify M with $M \times \{v\}$. If F_n, $n = 1, 2, \ldots$, is any sequence of C^{s+1} foliation of $M \times V$ converging C^r to $E \times V$, then for sufficiently large

j each F_j intersects M in a C^2 foliation E_j, and $\lim_{j\to\infty} E_j = E$ in the C^r topology on C^s foliations, a contradiction. This proves that $E \times V$ cannot be C^r approximated by C^s foliations. The proof of part (a) of Theorem 0.1 is complete.

§6. Higher Codimensional Examples

Let $M^3 = (T^2)^f$ be the 3-manifold from §2 on which we constructed the C^1 foliation F^f. For any compact manifold V we construct a C^1 2-dimensional foliation on $V \times M^3$ that cannot be C^1 approximated by C^2 foliations. Together with the method of 4, this gives (b) of our Main Theorem.

Let $f : T^2 \longrightarrow T^2$ be the C^∞ DA-diffeomorphism of §2, §3. Choose a point $p \in V$ and let $\phi : V \longrightarrow V$ be a C^∞ diffeomorphism isotopic to the identity having p as an attracting fixed point. The product diffeomorphism

$$\phi \times f : V \times T^2 \longrightarrow V \times T^2$$

leaves invariant the torus $p \times T^2$. We choose ϕ so that $\|(D\phi)_p\| \cdot \|(Df^{-1})_x\|^2 < 1$ for all $x \in T^2$. Then $p \times T^2$ is persistent in the sense that any C^1 approximation to $\phi \times f$ possesses a unique invariant torus $\tilde{T}$ near $p \times T^2$ and $\tilde{T}$ C^1 approximates $p \times T^2$. Besides, if the C^1 approximation to $\phi \times f$ is itself of class C^2, but does not necessarily C^2 approximate $\phi \times f$, then $\tilde{T}$ is of class C^2. See [11, pp. 39, 50-51] for the proof of these assertions about $\tilde{T}$.

Now we suspend $V \times T^2$ by $\phi \times f$. Since $\phi \times f$ is isotopic to $1_V \times f$, it follows that

$$(V \times T^2)^{\phi \times f} \approx (V \times T^2)^{1_V \times f} \approx V \times (T^2)^f = V \times M^3 \quad .$$

The foliation F of T^2, discussed in §2, extends to a foliation F' of $V \times T^2$:

$$F'_{(x, y)} = \{x\} \times F_y \quad .$$

Thus F' has 1-dimensional leaves. It is clearly preserved by $\phi \times f$.

The foliation K of T^2 extends to the foliation $V \times K$ of $V \times T^2$:

$$(V \times K)_{(x, y)} = V \times K_y \quad .$$

Then $V \times K$ has codimension-one, and is $\phi \times f$-invariant. Evidently $V \times K$ is transverse to F'.

Now consider the suspended foliation $\hat{F} = (F')^{\phi \times f}$. It is a C^1 foliation of $V \times M^3$ having 2-dimensional leaves. As remarked at the end of §1, if $\hat{F}$ can be C^1 approximated by C^2 foliations G, then $\phi \times f$ can be C^1 approximated by C^2 diffeomorphisms of $V \times T^2$ which preserve both K' and $H' = G \cap (V \times T^2)$. (We are identifying $V \times T^2$ with a submanifold of $(V \times T^2)^{\phi \times f}$ which is transverse to G. The leaves of G intersect this submanifold in leaves of a foliation $G \cap (V \times T^2)$ of $V \times T^2$). To reach a contradiction, it thus suffices to prove that there cannot exist such approxmation to $\phi \times f$. For this we will use Theorem 4.1.

As remarked above, any sufficiently C^1 close approximation $g : V \times T^2 \longrightarrow V \times T^2$ to $\phi \times f$ has an invariant torus $T \subset V \times T^2$ which is C^1 close to $\{p\} \times T^2$. Thus $g|\tilde{T}$ is a C^1 diffeomorphism of $\tilde{T}$. It preserves the foliation $\tilde{K} = \tilde{T} \cap (V \times K)$.

Next we show that <u>the leaf of H' through any point of $\tilde{T}$ lies wholly in</u> $\tilde{T}$.

If $q \in \tilde{T}$ let H_q be the tangent line to H' at x . It suffices to prove that H^q is tangent to $\tilde{T}$. These lines form a subbundle of $T_{\tilde{T}}(V \times T^2)$ which is invariant under Tg . By normal hyperbolicity of g at $\tilde{T}$, there is a Tg-invariant splitting

$$T_{\tilde{T}}(V \times T^2) = T(\tilde{T}) \oplus \nu .$$

Moreover the backward iterates of Tg drive vectors in $T_{\tilde{T}}(V \times T^2)$ which are not in $T(\tilde{T})$ toward ν (in the sense of angle in some Riemannian metric). We can assume that the angles between H_q and ν_q are bounded above zero. It follows that every H_q must be in $\tilde{T}q$.

Thus the map $g|\tilde{T}$ preserves $\tilde{K}$ and the foliation $\tilde{H} = H'|\tilde{T}$. Evidently $\tilde{K}$ (like K and $V \times K$) has dense leaves, while $\tilde{H}$ (being C^2) has no exceptional minimal sets.

The diffeomorphism $g|\tilde{T}$ of $\tilde{T}$ will, like the DA f , have a source disk if the approximation of g is sufficiently C^1 close to $\phi \times f$, which happens if G is sufficiently C^1 close to $(F')^{\phi \times f}$. (Here C^0 approximations would suffice. But we already needed C^1 approximations to get the existence and normal hyperbolicity of $\tilde{T}$.) Applying Theorem 4.1 yields a contradiction, since the induced endomorphism $(g|\tilde{T})_*$ of $H^1(\tilde{T})$ is conjugate to f_* on $H^1(T^2)$. This shows that $\hat{F}$ cannot be C^1 approximated by C^2 foliations. This completes the proof of (b) of Theorem 0.1.

§7. Group Actions

The non-approximable foliations constructed in previous sections can all be obtained, on suitable manifolds, as orbit foliatons of locally free actions of solvable groups. Consider first the foliation F of M^3 of §2.

The DA-diffeomorphism f of T^2 preserves two foliations, the C^1 foliation F and the analytic foliation K. The latter is the stable manifold foliation of the linear Anosov diffeomorphism f_0 from which f is derived.

Let X be a C^1 vector field tangent to F, such that in the hyperbolic splitting $E_0^u \oplus E_0^s$ the E^u component of X is identically 1. The DA-diffeomorphism f has the following effect on X:

$$df_p X(p) = aX(f(p)) \quad , \tag{1}$$

where $a = \frac{1}{2}(3 + \sqrt{5})$.

Let G denote the Lie group $\mathbb{R}$, acting on T^2 by the flow of X; let $s \in \mathbb{R}$ send $x \in T^2$ to $s \cdot x$. This is a free C^1 action. Define an automorphism Φ of G by $\Phi(s) = as$. Then (1) can be written

$$\frac{d}{ds}(f(s \cdot p)) = a \frac{d}{ds}(s \cdot f(p)) \quad .$$

Integrating both sides gives

$$f(s \cdot p) = (as) \cdot f(p) \quad .$$

Introduce the 1-parameter group of automorphisms $\{\Phi_t\}$ of G:

$$\Phi_t(s) = a^t s \quad ; \quad t \in \mathbb{R} \quad , \quad s \in G \quad .$$

Then $\Phi_1 = \Phi$. It follows from the construction at the end of §1 that the suspended foliation F^f of $(T^2)^f = M$ is the orbit foliation of a C^1 locally free action of the nonabelian two-dimensional solvable Lie group H whose Lie algebra is defined by $[Y, X] = aX$.

The foliation $F^f \times V$ of §5 is an orbit foliation provided V is a Lie group. There is then an obvious action of $H \times V$ on $M \times V$.

In §6 we consider the foliation F' on $V \times T^2$ whose leaf at (x, y) is $\{x\} \times F_y$. This is then suspended by $\phi \times f$ to get a foliation $\hat{F}$ of $V \times (T^2)^f$. Evidently $\hat{F}$ is the orbit foliation of H acing on $V \times (T^2)^f$ via the action of H on $(T^2)^f$ described above.

This completes the proof of part (c) of Theorem 0.1.

§8. Other Examples and a Question

In §2 we gave a proof that the foliation F^f on $(T^2)^f$ could not be C^1 approximated by C^2 foliations, which used the structural stability of the DA-diffeomorphism f of T^2 . This (unlike the later proof that F^f is not C^0 approximable) does not rely on the low dimension of F , only on its low codimension. It can therefore be imitated using DA-diffeomorphisms on other manifolds.

Let g_0 be a codimension-one Anosov diffeomorphism of a compact manifold Q , that is, the unstable manifold foliation of Q has codimension one. Let g be a C^1 DA-diffeomorphism of Q derived from g_0 . It is known that g_0 is structurally stable. Using the methods of §3 one finds a C^1 foliation F of Q , of codimension 1 , invariant under g . Imitating the first proof in §1, one shows that the C^1 foliation F^g of Q^g cannot be C^1 approximated by C^2 foliations. (One needs Sacksteder's generalization of Denjoy's theorem, [14, Theorem 5].)

Are there C^1 foliations of the n-sphere that cannot be C^1 (or C^0) approximated by C^2 foliations? None of the methods of this paper are useful for such a problem, since the manifolds they produce always have infinite fundamental group. It is amusing to note that the Reeb foliation of S^3 cannot be approximated by <u>analytic</u> foliations:

Haefliger proved there are no analytic foliations of codimension one on any manifold with finite fundamental group.

References

1. Anderson, B., Commuting diffeomorphisms and unsmoothable actions. Mimeographed, CA 1975.

2. Anosov, D., Geodesic Flows on Compact Riemannian Manifolds of Negative Curvature. Proc. Steklov Math. Inst. 90 (1967).

3. Cohen, M., Thesis, U. Calif (Berkeley), 1968.

4. Denjoy, A., Sur les courbes définies par les équations differentielle à la surface du tore. J. de Math. Pures et Appliquées 11 (1932), 333-365.

5. Harrison, J., Unsmoothable diffeomorphisms. Ann. Math. 102 (1975), 85-94.

6. Hart, D., On the smoothness of generators for flows and foliations. To appear.

7. Hirsch, M., Differential Topology. Springer-Verlag, New York, 1976.

8. ________, Stability of compact leaves of foliations. Dynamical Systems (M. Peixoto, ed.), Academic Press, New York, 1973.

9. ________ , Stability of stationary points of group actions. Ann. Math. 109 (1979), 537-544.

10. Hirsch, M. and Pugh, C., Stable manifolds and hyperbolic sets. Proc. Symp. Pure Math. 14, 133-164.

11. Hirsch, M., Pugh, C., and Shub, M., Invariant Manifolds, Lecture Notes in Math. No. 58h, Springer-Verlag, New York, 1977.

12. Pixton, D., Nonsmoothable, unstable group actions. Trans. Amer. Math. Soc. 220 (1977), 250-268.

13. Rosenberg, H. and Thurston, W., Some remarks on foliations. Dynamical Systems (M. Peixoto, ed.), Academic Press, New York, 1973.

14. Sacksteder, R., Foliations and pseudo-groups. Amer. J. Math. 88 (1o65), 79-192.

15. Smale, S., Differentiable dynamical systems. Bull. Amer. Math. Soc. 73 (1967), 747-817.

University of California
at Berkeley
29 October 1981

A PROOF OF PESIN'S STABLE MANIFOLD THEOREM

by A. FATHI
M.R. HERMAN
J.C. YOCCOZ

We present here a set of lecture notes which gives a proof of a part of Pesin's work on hyperbolicity in general dynamical systems [Pe_1] [Pe_2] . These notes correspond to the lectures we gave in the "Séminaire de théorie ergodique" of Paris VI in 1978-79 which was devoted that year to Pesin's work.

The part of Pesin's work (†), for which we provide a proof may be formulated in the following way : Given a C^2 diffeomorphism f of the manifold M , and an f-invariant probability measure μ, for μ-almost every x the stable set of x $W^s_x = \{y \in M \mid \limsup \frac{1}{n} \log d(f^n(x), f^n(y)) < 0\}$ is in fact an immersed Euclidean space. Of course, as for Anosov diffeomorphisms, it is the existence of local stable manifolds that is the most useful fact.

We give some indications on the proof, which presents only minor differences to existing proofs [Pe_1] [Ru]. We first apply Oseledec's ergodic multiplicative theorem to the tangent cocycle Tf , this shows that Tf has at μ-almost every point a well defined asymptotic behaviour. We restrict then to the set $B_{\lambda,\mu}$,

(†) Another exposition of the part of Pesin's work on the absolute continuity of the stable foliation will be given by Pugh and Shub [PS] .

where the asymptotic spectrum is disjoint from $[\exp(\lambda), \exp(\mu)]$ and then use the usual renorming technique on $TM|B_{\lambda,\mu}$ to find a norm $|||\ |||$, on which we can read the "hyperbolic behaviour". As in the case of uniform hyperbolicity, we lift the map f to a map $\mathfrak{F}$ on the space of sections of $TM|B_{\lambda,\mu}$. By doing these constructions carefully, we obtain that in fact $\mathfrak{F}$ is a small perturbation in the norm $|||\ |||$ of a linear map exhibiting a "hyperbolic" behaviour, this allows us to apply the theorem on the existence of the stable manifold of a "hyperbolic" point. As in the uniform case, we can then obtain a local stable manifold for each point of $B_{\lambda,\mu}$. The globalization of the stable manifold is then done in a straightforward way. For the convenience of the reader, we give Irwin's proof of the existence of the stable manifold of a "hyperbolic" fixed point.

As one can see from the description of the proof given above, there is nothing new in these notes. Since the informal notes in french distributed after the lectures proved useful, we decided to write this more formal account.

These notes owe much to the members of the 1978-79 "Séminaire de théorie ergodique" of Paris VI, we express to them our gratitude.

Let us fix some notations for the rest of the text. We denote the exponential function by $\exp(x)$ instead of e^x for typographical reasons. If f is a smooth map, we denote its derivative at the point x by $T_x f$, sometimes we use the notation $Tf(x)$. We call a map $C^{r,\alpha}$ $(r \in \mathbb{N}, \alpha \in]0,1])$ if it is differentiable up to r^{th} order, and its rth derivative satisfies locally a Hölder condition of order α. If a map f satisfies a Hölder condition of order α, we denote its Hölder constant of order α by $\mathrm{Lip}^{\alpha}(f)$, sometimes we use the notation $\mathrm{Lip}^{\alpha}_{\|\ \|}(f)$ to precise the choice of the norm with respect to which we are estimating the Hölder constant. When $\alpha = 1$, we use also the notation Lip(f) instead of $\mathrm{Lip}^1(f)$. When we say that a map is measurable, that <u>means</u> that the source and the target are topological spaces, and the map is <u>Borel measurable</u>.

§ 1. - SOME ESTIMATES

In the following, we obtain some estimates which follow from Oseleaec's multiplicative ergodic theorem. We fix M a compact smooth manifold without boundary of dimension m and f a C^1-diffeomorphism on M. Applying Oseledec's theorem (cf. [Os], [Ra], [Ru]) to the cocycle defined by the tangent map of f, we obtain :

THEOREM (Oseledec). There exists a Borel set B in M which has the following properties :

(i) B is invariant under f and has measure 1 for every f-invariant probability Borel measure on M.

(ii) For every x in B, there exists a splitting of the tangent space $T_xM = \bigoplus_{i=1}^{s(x)} W_i(x)$ and real numbers $\lambda_1(x) < \lambda_2(x) < \ldots < \lambda_{s(x)}(x)$ such that, for any Riemannian metric $\| \ \|$, on M :

(a) $W_i(x)$, $\lambda_i(x)$ and $s(x)$ are Borel measurable functions of x, moreover $W_i(f(x)) = T_xf(W_i(x))$ and $\lambda_i(x)$, $s(x)$ are invariant under f ;

(b) $\forall\, v \in W_i(x)$, $\lim\limits_{n\to\pm\infty} \frac{1}{n} \log \|T_xf^n(v)\| = \lambda_i(x)$;

(c) $\lim\limits_{n\to\pm\infty} \frac{1}{n} \log |\det(T_xf^n)| = \sum\limits_{i=1}^{s(x)} \lambda_i(x) \dim W_i(x)$.

Remark that $\lim\limits_{n\to\pm\infty} \frac{1}{n} \log \|T_xf^n(v)\|$ does not depend on the Riemannian metric, because if $\| \ \|'$ is another metric, there exists a constant C such that $\frac{1}{C} \| \ \| \le \| \ \|' \le C \| \ \|$.

The number $|\det T_xf^n|$ is defined in the following way : the Riemannian metric $\| \ \|$ induces a Riemannian metric on the vector bundle $\Lambda^m TM$ and $|\det T_xf^n|$ is the norm of $\Lambda^m T_xf^n$ with respect to this metric.

We fix now a C^∞ Riemannian metric $\| \ \|$ on M.

Let $[\lambda,\mu]$, $\lambda<\mu$, be a compact interval in $\mathbb{R}$. Denote by $B_{\lambda,\mu}$ the subset of B which consists of points x such that $\lambda_i(x) \notin [\lambda,\mu]$, $i=1,\dots,s(x)$, this set is invariant under f. For x in $B_{\lambda,\mu}$, we define $E_x^s = \bigoplus_{\lambda_i(x)<\lambda} W_i(x)$ and $E_x^u = \bigoplus_{\mu<\lambda_i(x)} W_i(x)$.

PROPOSITION 1. Let ϵ be a strictly positive number. There exists a real valued measurable function A_ϵ defined on $B_{\lambda,\mu}$, such that for every x in $B_{\lambda,\mu}$ we have :

(i) $\forall v \in E_x^s$, $\forall n \geq 0$, $\|T_x f^n(x)\| \leq A_\epsilon(x)\,\|v\|\exp(\lambda n)$;

$\forall v \in E_x^u$, $\forall n \geq 0$, $\|T_x f^{-n}(v)\| \leq A_\epsilon(x)\,\|v\|\exp(-\mu n)$;

(ii) $\forall n \in \mathbf{Z}$, $A_\epsilon(f^n(x)) \leq A_\epsilon(x)\exp(\epsilon|n|)$.

We need the following lemma :

LEMMA 2. Let $\langle\,,\,\rangle$ and $\langle\,,\,\rangle'$ be two scalar products on a m dimensional vector space E. Denote by $\|\;\|$ and $\|\;\|'$ the norms deduced from $\langle\,,\,\rangle$ and $\langle\,,\,\rangle'$. Suppose there exists m linearly independent vectors $u_1,\dots,u_m$ such that : $\forall i,j$, $1\leq i,j\leq m$, $\|u_i\|=1$, $\langle u_i,u_j\rangle'=\delta_{ij}$. Then for every v in E, we have :

$$\frac{1}{\sqrt{m}}\|v\| \leq \|v\|' \leq m^{\frac{m-1}{2}}\frac{\|u_1\wedge\dots\wedge u_m\|'}{\|u_1\wedge\dots\wedge u_m\|}\|v\| .$$

Proof. If $v=\sum_{i=1}^m x_i u_i$, we have :

$$\|v\|^2 \leq (\Sigma|x_i|\,\|u_i\|)^2 = (\Sigma|x_i|)^2 \leq m(\Sigma|x_i|^2) = m(\|v\|')^2 .$$

To prove the other inequality, let $v_2,\dots,v_m$ be non zero vectors in E, such that $\{v,v_2,\dots,v_m\}$ is an orthogonal family, for the scalar product $\langle\,,\,\rangle'$. We have, in $\Lambda^m E$:

$$\|v\wedge v_2\wedge\dots\wedge v_m\|' = \frac{\|u_1\wedge\dots\wedge u_m\|'}{\|u_1\wedge\dots\wedge u_m\|}\|v\wedge v_2\wedge\dots\wedge v_m\| ,$$

which implies :

$$\|v\|' \, \|v_2\|' \ldots \|v_m\|' \le \frac{\|u_1\wedge\cdots\wedge u_m\|'}{\|u_1\wedge\cdots\wedge u_m\|} \|v\| \, \|v_2\| \ldots \|v_m\| .$$

The second inequality for v is now an easy consequence of the first one applied to $v_2,\ldots,v_m$. □

Proof of proposition 1. Let x be a given point in $B_{\lambda,\mu}$. We choose a basis $w_1,\ldots,w_m$ of T_xM, with each w_i a unit vector belonging to some $W_j(x)$. For each $\ell \in \mathbb{Z}$, we define a scalar product $\langle\,,\,\rangle_\ell$ on $T_{f^\ell(x)}M$, for which $\frac{T_xf^\ell(w_i)}{\|T_xf^\ell(w_i)\|}$, $1\le i\le m$, is an orthonormal basis. By lemma 2, we have for v in $T_{f^\ell(x)}M$:

$$\frac{1}{\sqrt{m}}\|v\| \le \|v\|_\ell \le m^{\frac{m-1}{2}} \frac{\|T_xf^\ell(w_1)\wedge\cdots\wedge T_xf^\ell(w_m)\|_\ell}{\|T_xf^\ell(w_1)\wedge\cdots\wedge T_xf^\ell(w_m)\|}\|v\| .$$

Using Oseledec's theorem and the orthogonality of the $T_xf^\ell(w_i)$, $1\le i\le m$, for $\langle\,,\,\rangle_\ell$, it is easy to see that we have :

$$\lim_{\ell\to\pm\infty} \frac{1}{\ell}\log \frac{\|T_xf^\ell(w_1)\wedge\cdots\wedge T_xf^\ell(w_m)\|_\ell}{\|T_xf^\ell(w_1)\wedge\cdots\wedge T_xf^\ell(w_m)\|} = 0 .$$

It follows that there exists a constant C such that :

(1) $$\frac{1}{\sqrt{m}}\|v\| \le \|v\|_\ell \le C\|v\|\exp\left(\frac{\varepsilon}{2}|\ell|\right) .$$

Since x is in fact in some $B_{\lambda-\delta,\mu}$ where $\delta>0$, if $w_i\in W_j(x)$, there exists a constant C_i, depending on x and δ such that :

$$\forall k\in\mathbb{Z},\ \frac{1}{C_i}\exp(k\lambda_j(x)-|k|\,\delta) \le \|T_xf^kw_i\| \le C_i\exp(k\lambda_j(x)+|k|\,\delta) .$$

It follows from these inequalities that, if δ is chosen $\le\frac{\varepsilon}{4}$, there exists a constant C' such that if $w_i\in E^s_x$, we have :

(2) $$\forall n\ge 0,\ \forall k\in\mathbb{Z},\ \|T_xf^{k+n}(w_i)\| \le C'\|T_xf^k(w_i)\|\exp\left(n\lambda+\varepsilon\frac{|k|}{2}\right) .$$

Putting (1) & (2) together, we obtain, for $n\ge 0$, $k\in\mathbb{Z}$, $v=\Sigma\,\alpha_i w_i$ in E^s_x :

$$\begin{aligned}\|T_x f^{k+n}(v)\| &\le \Sigma |\alpha_i| \, \|T_x f^{k+m}(w_i)\| \\ &\le [\Sigma |\alpha_i| \, \|T_x f^k(w_i)\|\,] \, C' \exp(\lambda n + \varepsilon \tfrac{|k|}{2}) \\ &\le C' \sqrt{m} (\Sigma |\alpha_i|^2 \|T_x f^k(w_i)\|^2)^{\frac{1}{2}} \exp(\lambda n + \varepsilon \tfrac{|k|}{2}) \\ &= C' \sqrt{m} \, \|T_x f^k(v)\|_k \exp(\lambda n + \varepsilon \tfrac{|k|}{2}) \\ &\le C C' \sqrt{m} \, \|T_x f^k(v)\| \exp(\lambda n + \varepsilon |k|) \quad .\end{aligned}$$

It follows from this that the Borel function :

$$\bar{A}_\varepsilon(x) = \sup \{ \frac{\|T_x f^{k+n}(v)\|}{\|T_x f^k(v)\|} \exp(-\lambda n - \varepsilon |k|) \mid n \ge 0, \, k \in \mathbb{Z}, \, v \in E^s_x \}$$

is finite at each point of $B_{\lambda,\mu}$.

In the same way, the Borel function :

$$\tilde{A}_\varepsilon(x) = \sup \{ \frac{\|T_x f^{k+n}(v)\|}{\|T_x f^k(v)\|} \exp(-\mu n - \varepsilon |k|) \mid n \le 0, \, k \in \mathbb{Z}, \, v \in E^u_x \}$$

is also finite.

The function $A_\varepsilon(x)$ is defined by : $A_\varepsilon(x) = \max(\bar{A}_\varepsilon(x), \tilde{A}_\varepsilon(x))$. □

The following lemma is almost contained in the preceding proof.

LEMMA 3. Given $x \in B_{\lambda,\mu}$ and $\varepsilon > 0$, there exists a constant E such that, if $v = v_s + v_u$, $v_s \in E^s_{f^k(x)}$, $v_u \in E^u_{f^k(x)}$, we have :

$$\|v_s\|^2 + \|v_u\|^2 \le E \, \|v\|^2 \exp(\varepsilon |k|) \quad .$$

Proof. Applying inequality (1), we obtain :

$$\begin{aligned}\|v_s\|^2 + \|v_u\|^2 &\le m(\|v_s\|_k^2 + \|v_u\|_k^2) \\ &= m(\|v_s + v_u\|_k^2) \\ &\le m C^2 \|v\|^2 \exp(\varepsilon |k|) \quad .\end{aligned}$$

The constant E is defined by the well-known formula $E = mC^2$. □

We now define on $TM|B_{\lambda,\mu}$ a new norm which depends measurably on the point x and on which we can read directly the "hyperbolicity" of f.

If $v \in E_x^s$: $|||v||| = \sum_{n=0}^{+\infty} \exp(-\lambda n) \, ||T_x f^n(v)||$.

If $v \in E_x^u$: $|||v||| = \sum_{n=0}^{+\infty} \exp(\mu n) \, ||T_x f^{-n}(v)||$.

If $v = v_s + v_u$, $v_s \in E_x^s$, $v_u \in E_x^u$: $|||v||| = \max(|||v_s|||, |||v_u|||)$.

The properties of this norm are summarized in the following proposition.

PROPOSITION 4. (a) There exists a constant K, such that, for any point x in $B_{\lambda,\mu}$: $|||T_x f||| \le K$, $|||T_x f^{-1}||| \le K$,

$$|||T_x f | E_x^s||| \le \exp\lambda \ , \quad |||T_x f^{-1} | E_x^u||| \le \exp(-\mu) \ .$$

(b) Given $\varepsilon > 0$, there exists a Borel function B_ε, defined on $B_{\lambda,\mu}$ such that, for any $x \in B_{\lambda,\mu}$:

$$\forall v \in T_x M \ , \quad \tfrac{1}{2} ||v|| \le |||v||| \le B_\varepsilon(x) \, ||v|| \ ,$$

$$\forall n \in \mathbb{Z} \ , \quad B_\varepsilon(f^n(x)) \le B_\varepsilon(x) \exp(\varepsilon |n|) \ .$$

Proof. (a) The inequalities $|||T_x f | E_x^s||| \le \exp\lambda$ and $|||T_x f^{-1} | E_x^u||| \le \exp(-\mu)$ follow immediately from the definition of the norm $||| \ |||$.

The compactness of M implies that, for some constant K_o and any point x in M, we have : $||T_x f|| \le K_o$, $||T_x f^{-1}|| \le K_o$. It follows that, for $x \in B_{\lambda,\mu}$ and $v \in E_x^u$:

$$\begin{aligned} |||T_x f(v)||| &= \sum_{n=0}^{+\infty} ||T_x f^{-n+1}(v)|| \exp(n\mu) \\ &= ||T_x f(v)|| + |||v||| \exp\mu \\ &\le K_o ||v|| + |||v||| \exp\mu \\ &\le (K_o + \exp\mu) \, |||v||| \ . \end{aligned}$$

Putting $|||T_x f | E_x^s||| \le \exp\lambda$ and $|||T_x f | E_x^u||| \le K_o + \exp\mu$ together, we obtain that $|||T_x f|||$ is bounded uniformly in x. A similar proof gives the same statement for $|||T_x f^{-1}|||$.

(b) The inequality $\|v\| \le 2\,|||v|||$ is trivial. For $x \in B_{\lambda,\mu}$, we define $n_0(x)$ as the smallest integer n such that x belongs to $B_{\lambda-\frac{1}{n},\mu+\frac{1}{n}}$. Proposition 1, applied to $B_{\lambda-1/n_0(x),\mu+1/n_0(x)}$, gives a Borel function A_ϵ such that :

$$\forall p \ge 0,\ \forall v \in E^s_x,\quad \|T_x f^p(v)\| \le A_\epsilon(x)\,\|v\|\,\exp\left(p\left(\lambda - \frac{1}{n_0(x)}\right)\right).$$

$$\forall p \ge 0,\ \forall v \in E^u_x,\quad \|T_x f^{-p}(v)\| \le A_\epsilon(x)\,\|v\|\,\exp\left(-p\left(\mu + \frac{1}{n_0(x)}\right)\right)$$

$$\forall p \in \mathbf{Z},\qquad A_\epsilon(f^p(x)) \le A_\epsilon(x)\exp\left(\frac{\epsilon|p|}{2}\right)$$

From this, it follows that, if v belongs to E^s_x or E^u_x, we have :

$$|||v||| \le A_\epsilon(x)\left[1-\exp\left(-\frac{1}{n_0(x)}\right)\right]^{-1}\|v\|.$$

For any $k \in \mathbf{Z}$ and any vector $v \in T_{f^k(x)}M$, with $v = v_s + v_u$, $v_s \in E^s_{f^k(x)}$, $v_u \in E^u_{f^k(x)}$, we obtain from Lemma 3 and the above inequality :

$$|||v|||^2 \le |||v_s|||^2 + |||v_u|||^2$$

$$\le A_\epsilon(f^k(x))^2\left[1-\exp\left(-\frac{1}{n_0(x)}\right)\right]^{-2}\left(\|v_s\|^2 + \|v_u\|^2\right)$$

$$\le EA_\epsilon(x)^2\left[1-\exp\left(-\frac{1}{n_0(x)}\right)\right]^{-2}\|v\|^2\exp(2\epsilon|k|).$$

The measurable function :

$$B_\epsilon(x) = \sup\left\{\frac{|||v|||}{\|v\|}\exp(-|n|\epsilon)\ \middle|\ n \in \mathbf{Z},\ v \in T_{f^n(x)}M\right\}$$

is therefore finite on $B_{\lambda,\mu}$ and satisfies (b). □

§ 2. - MORE ESTIMATES

Let Exp denote the C^∞ exponential map associated to the C^∞ Riemannian metric $\| \ \|$ on M. By the compactness of M there exists $r > 0$ such that, for every $x \in M$, the map Exp_x is a C^∞ diffeomorphism from $B(0_x, r)$ the r-ball around 0_x in T_xM onto $B(x,r)$, the r-ball around x in M. Moreover, we have $d(x, \mathrm{Exp}_x(v)) = \|v\|$, for $v \in B(0_x, r)$.

From the uniform continuity of f, there exists a constant δ $(0 < \delta < r)$, such that :
$$d(x,y) < \delta \Rightarrow d(f(x), f(y)) < r \ .$$
One can then lift f to a map F on $\{v \mid v \in TM, \|v\| \le \delta\}$ defined by :
$$\forall v \in B(0_x, \delta), \quad F(v) = \mathrm{Exp}^{-1}_{f(x)} [f(\mathrm{Exp}_x(v))] \ .$$

The map F is as smooth as f. By definition, the following diagram is commutative :
$$\begin{array}{ccc} B(0,\delta) & \xrightarrow{\ F\ } & B(0,r) \\ \Big\downarrow \mathrm{Exp} & & \Big\downarrow \mathrm{Exp} \\ M & \xrightarrow{\ f\ } & M \end{array} \ .$$

Let F_x be the restriction of F to $B(0_x, \delta)$. One has : $T_{0_x} F_x = T_x f$.

We will extend F to a fiber map $\overline{F}$ defined on the whole of TM. Let $\theta : [0, +\infty[\ \to \mathbb{R}$ be a C^∞ function such that :
$$\theta(t) = 1 \quad \text{if } t \le \frac{\delta}{2} \ ;$$
$$\theta(t) = 0 \quad \text{if } t \ge \delta \ ;$$
$$0 \le \theta(t) \le 1 \quad \text{if } \frac{\delta}{2} \le t \le \delta \ .$$

Define $\overline{F}_x(v)$ for $v \in T_xM$, by :
$$\overline{F}_x(v) = T_x f(v) + (F_x - T_x f)(v)\, \theta\!\left(\frac{\|v\|}{\beta(x)}\right)$$
where $\beta(x) \in \ ]0,1[$ will be defined later.

Denote $(F_x - T_x f)$ by φ_x, and $(\overline{F}_x - T_x f)$ by $\overline{\varphi}_x$. One has $\varphi_x = \overline{\varphi}_x$ on $B(0_x, \delta \frac{\beta(x)}{2})$, $\varphi_x(0) = 0$ and $T_0 \varphi_x = 0$.

LEMMA 5. Suppose Tf satisfies a Hölder condition of order α $(0<\alpha<1)$. Then, there exists a constant C such that, for each x in M:

$$\mathrm{Lip}^1_{\|\ \|}(\bar{\varphi}_x) \leq C(\beta(x))^\alpha .$$

If $0<\alpha'<\alpha$, $\mathrm{Lip}^{\alpha'}_{\|\ \|}(T\bar{\varphi}_x) \leq C\beta(x)^{\alpha-\alpha'}$.

Proof. We are only going to give the estimates for the first inequality. The estimates for the second are similar.

From the Hölder condition on Tf, we find a constant C_0 such that $\|\varphi_x(v)\| \leq C_0\beta(x)^{1+\alpha}$, $\|T_v\varphi_x\| \leq C_0\beta(x)^\alpha$ for $\|v\| \leq \delta\beta(x)$ and $\mathrm{Lip}^1(\varphi_x|B(0,\delta\beta(x)) \leq C_0\beta(x)^\alpha$; moreover, if $0<\alpha'<\alpha$, $\mathrm{Lip}^{\alpha'}(T\varphi_x|B(0,\delta\beta(x)) \leq C_0\beta(x)^{\alpha-\alpha'}$.

Let Θ be defined for $v \in TM$ by : $\Theta(v) = \theta(\|v\|)$.

We estimate $\|\bar{\varphi}_x(v) - \bar{\varphi}_x(v')\|$. The following cases can occur :

- $v, v' \notin B(0,\delta\beta(x))$: $\bar{\varphi}_x(v') = \bar{\varphi}_x(v) = 0$.
- $v \in B(0,\delta.\beta(x))$, $v' \notin B(0,\delta.\beta(x))$:

$$\|\bar{\varphi}_x(v) - \bar{\varphi}_x(v')\| = \|\varphi_x(v)\|\left(\Theta\left(\frac{v}{\beta(x)}\right) - \Theta\left(\frac{v'}{\beta(x)}\right)\right)$$

$$\leq C_0\beta(x)^{1+\alpha}(\mathrm{Lip}^1_{\|\ \|}\Theta)\,\beta(x)^{-1}\|v-v'\| .$$

- $v, v' \in B(0,\delta.\beta(x))$:

$$\|\bar{\varphi}_x(v) - \bar{\varphi}_x(v')\| \leq \|\varphi_x(v)\|\left|\Theta\left(\frac{v}{\beta(x)}\right) - \Theta\left(\frac{v'}{\beta(x)}\right)\right| + \Theta\left(\frac{v'}{\beta(x)}\right)\|\varphi_x(v')-\varphi_x(v)\|$$

$$\leq C_0\beta(x)^\alpha\|v-v'\|\,\mathrm{Lip}^1(\Theta) + C_0\beta(x)^\alpha\|v-v'\| . \qquad \square$$

The preceding estimates can be translated in the norm $|||\ |||$, défined in § 1 on $TM|B_{\lambda,\mu}$, as follows.

LEMMA 6. For $\epsilon>0$ and $x \in B_{\lambda,\mu}$, we have :

1) $\mathrm{Lip}^1_{|||\ |||}(\bar{\varphi}_x) \leq 2C B_\epsilon(x)\,\beta(x)^\alpha \exp\epsilon$;

2) $\mathrm{Lip}^{\alpha'}_{|||\ |||}(T\bar{\varphi}_x) \leq 2^{1+\alpha'} C B_\epsilon(x)\beta(x)^{\alpha-\alpha'}\exp\epsilon$,

where $B_\epsilon(x)$ is the function introduced in proposition 4 .

Proof.

1) $$|||\bar\varphi_x(v) - \bar\varphi_x(v')||| \le B_\epsilon(f(x))\, ||\bar\varphi_x(v) - \bar\varphi_x(v')||$$
$$\le B_\epsilon(f(x))\, \mathrm{Lip}^1_{||\ ||}(\bar\varphi_x)\, ||v - v'||$$
$$\le 2B_\epsilon(f(x))\, \mathrm{Lip}^1_{||\ ||}(\bar\varphi_x)\, |||v - v'||| \quad .$$

But : $2B_\epsilon(f(x))\, \mathrm{Lip}^1_{||\ ||}(\bar\varphi_x) \le 2B_\epsilon(x)\, C\beta(x)^\alpha \exp\epsilon$.

2) $$|||T_u\bar\varphi_x(h) - T_v\bar\varphi_x(h)||| \le B_\epsilon(f(x))\, ||T_u\bar\varphi_x(h) - T_v\bar\varphi_x(h)||$$
$$\le B_\epsilon(f(x))\, ||T_u\bar\varphi_x - T_v\bar\varphi_x||\, ||h||$$
$$\le 2B_\epsilon(f(x))\, ||T_u\bar\varphi_x - T_v\bar\varphi_x||\, |||h||| \quad .$$

It follows :

$$|||T_u\bar\varphi_x - T_v\bar\varphi_x||| \le 2B_\epsilon(f(x))\, \mathrm{Lip}^{\alpha'}_{||\ ||}(T\bar\varphi_x)\, ||u - v||^{\alpha'}$$
$$\le 2^{1+\alpha'}\, \mathrm{Lip}^{\alpha'}_{||\ ||}(T\bar\varphi_x)\, |||u - v|||^{\alpha'}\, B_\epsilon(x)\exp\epsilon \quad .$$

Therefore $\quad \mathrm{Lip}^{\alpha'}_{|||\ |||}(T\bar\varphi_x) \le 2^{1+\alpha'}\, CB_\epsilon(x)\,\beta(x)^{\alpha-\alpha'}\exp\epsilon$. □

Fix $\alpha' \in\ ]0,\alpha[$. Let η be a given number > 0 . Define a strictly positive measurable function on $B_{\lambda,\mu}$ by :

$$\beta(x) = \min\left[\left(\frac{\eta}{2C\, B_\epsilon(x)\exp\epsilon}\right)^{1/\alpha}, \left(\frac{1}{2^{1+\alpha'}CB_\epsilon(x)\exp\epsilon}\right)^{\frac{1}{\alpha-\alpha'}} \right] ,$$

we obtain, for $x \in B_{\lambda,\mu}$: $\mathrm{Lip}^1_{|||\ |||}(\bar\varphi_x) \le \eta$, $\mathrm{Lip}^{\alpha'}_{|||\ |||}(T\bar\varphi_x) \le 1$.

Moreover, for $x \in B_{\lambda,\mu}$ and $n \in \mathbb{Z}$, $\beta(f^n(x)) \le \beta(x)\exp\left(\frac{\epsilon}{\alpha-\alpha'}|n|\right)$.

Since $\theta : \mathbb{R}^+ \to \mathbb{R}^+$ verifies $\theta(t) = 1$ for $|t| \le \frac{\delta}{2}$, we have $\bar\varphi_x(v) = \varphi_x(v)$ for $\frac{||v||}{\beta(x)} \le \frac{\delta}{2}$, which implies $\bar\varphi_x(v) = \varphi_x(v)$ for $|||v||| \le \frac{\delta\,\beta(x)}{4}$.

After slightly changing the notations, we summarize what we have obtained in the following proposition :

PROPOSITION 7. Suppose Tf satisfies a Hölder condition of order α and fix α' $(0 < \alpha' < \alpha)$. Given $\epsilon > 0$ and $\eta > 0$, there exists a measurable map

$\overline{F} : TM|B_{\lambda,\mu} \longrightarrow TM|B_{\lambda,\mu}$, fibered over f and a measurable function C_ϵ, defined on $B_{\lambda,\mu}$, such that, for $x \in B_{\lambda,\mu}$:

1) $\mathrm{Lip}^1_{|||\ |||}(\overline{F}_x - T_x f) \leq \eta$;

2) $\mathrm{Lip}^{\alpha'}_{|||\ |||}(T\overline{F}_x) \leq 1$;

3) $\overline{F}_x(v) = F_x(v)$ if $|||v||| \leq C_\epsilon(x)$;

4) $\forall n \in \mathbb{Z}$, $C_\epsilon(f^n(x)) \leq C_\epsilon(x) \exp(\epsilon|n|)$.

§ 3. - LOCAL STABLE MANIFOLDS

We keep the notations introduced above and suppose that Tf satisfies a Hölder condition of order α. We fix α', $0 < \alpha' < \alpha$ and λ, μ, with $\lambda < \mu < 0$.

Since we want to apply the stable manifold theorem, we begin by introducing a convenient Banach space.

Let $\Gamma_b = \Gamma_b(B_{\lambda,\mu})$ be the Banach space of sections σ of $TM|B_{\lambda,\mu}$, which are bounded for $|||\ |||$, i.e. $|||\sigma||| = \sup\{|||\sigma(x)|||\ |\ x \in B_{\lambda,\mu}\} < +\infty\}$. Consider Γ_{bb} the subspace of Γ_b consisting of Borel measurable sections. The next lemma shows that Γ_{bb} is a Banach subspace of Γ_b.

LEMMA 8. Let X and Y be two metric spaces and $(f_n)_{n\in\mathbb{N}}$ be a sequence of Borel maps from X to Y, converging at every point to a map f. Then, f is Borel.

Proof. It suffices to show that if A is a closed subset of Y, then $f^{-1}(A)$ is Borel. This follows from :

$$f^{-1}(A) = \bigcap_{n\geq 1} \bigcup_{N\geq 1} \bigcap_{k\geq N} f_k^{-1}[V_{\frac{1}{n}}(A)] ,$$

where $V_{\frac{1}{n}}(A) = \{y \in Y \mid d(y,A) \leq \frac{1}{n}\}$. □

The space Γ_b can be splitted as $\Gamma_b = \Gamma_b^s \oplus \Gamma_b^u$, where $\Gamma_b^s = \{\sigma \mid \forall s \in B_{\lambda,\mu}, \sigma(x) \in E_x^s\}$ and $\Gamma_b^u = \{\sigma \mid \forall x \in B_{\lambda,\mu}, \sigma(x) \in E_x^u\}$.

If $\sigma = \sigma_s + \sigma_u$, where $\sigma_s \in \Gamma_b^s$, $\sigma_u \in \Gamma_b^u$, then $|||\sigma||| = \max(|||\sigma_s|||, |||\sigma_u|||)$, because $|||\ |||$ verifies the same equality on each fiber T_xM . Since E_x^s and E_x^u depend measurably on x , one has : $\Gamma_{bb} = \Gamma_{bb}^s \oplus \Gamma_{bb}^u$ with $\Gamma_{bb}^s = \Gamma_{bb} \cap \Gamma_b^s$, and $\Gamma_{bb}^u = \Gamma_{bb} \cap \Gamma_b^u$.

We define maps :

$$\mathfrak{T} : \Gamma_b \to \Gamma_b , \ \sigma \to \mathfrak{T}(\sigma) , \text{ where } \mathfrak{T}(\sigma)(x) = T_{f^{-1}(x)} f[\sigma(f^{-1}(x))] ;$$

$$\mathfrak{F} : \Gamma_b \to \Gamma_b , \ \sigma \to \mathfrak{F}(\sigma) , \text{ where } \mathfrak{F}(\sigma)(x) = \bar{F}_{f^{-1}(x)} [\sigma(f^{-1}(x))] .$$

Here $\bar{F}$ is the map defined by proposition 7. From this same proposition, we obtain :

<u>LEMMA 9</u>. 1) <u>$\mathfrak{T}$ is a continuous linear map</u> ;

2) <u>Γ_b^s and Γ_b^u are invariant under $\mathfrak{T}$, moreover</u>

$$|||\mathfrak{T} \mid \Gamma_b^s||| \leq \exp\lambda , \ |||\mathfrak{T}^{-1} \mid \Gamma_b^u||| \leq \exp(-\mu) ;$$

3) <u>$\mathfrak{F} - \mathfrak{T}$ is Lipschitz, with</u> $\mathrm{Lip}^1_{|||\ |||}(\mathfrak{F} - \mathfrak{T}) \leq \eta$ <u>and</u> $\mathfrak{F}(0) = 0$;

4) <u>Γ_{bb} is invariant under $\mathfrak{F}$ and $\mathfrak{T}$, moreover Γ_{bb}^s and Γ_{bb}^u are invariant under $\mathfrak{T}$</u> .

Choose now ρ , such that $\lambda < \log\rho < \mu < 0$, and then η with $0 < \eta < \min(\exp(\mu) - \rho , \rho - \exp\lambda))$; then, we can apply the stable manifold theorem in a Banach space (see Appendix A, theorem A5). Hence we obtain :

PROPOSITION 10. There exists a Lipschitz map $\Psi : \Gamma_b^s \to \Gamma_b^u$, with $\Psi(0) = 0$, and, $\mathrm{Lip}^1_{\|\|\|\ \|\|\|}(\Psi) \leq 1$, such that the graph of Ψ is precisely the set $W^{s,\rho} = \{\sigma \in \Gamma_b \mid \sup_{n\geq 1} |||\rho^{-n}\mathfrak{F}^n(\sigma)||| < +\infty\}$. Moreover $\mathrm{Lip}^1_{\|\|\|\ \|\|\|}(\mathfrak{F} \mid W^{s,\rho}) \leq$ $\leq \exp(\lambda)+\eta < \rho$, in particular, $\rho^{-n}\mathfrak{F}^n(\sigma) \to 0$, if $\sigma \in W^{s,\rho}$.

One also has :

COROLLARY 11. $\Psi(\Gamma_{bb}^s) \subset \Gamma_{bb}^u$.

Proof. One can also apply the stable manifold theorem to $\mathfrak{F} \mid \Gamma_{bb}$ and $\mathfrak{F} \mid \Gamma_{bb}$. This gives a map $\tilde{\Psi} : \Gamma_{bb}^s \to \Gamma_{bb}^u$, such that each $\sigma \in \Gamma_{bb}^s$ satisfies $\rho^{-n}\mathfrak{F}^n(\sigma, \tilde{\Psi}(\sigma)) \to 0$, $n \to +\infty$, in particular $(\sigma, \tilde{\Psi}(\sigma)) \in W^{s,\rho}$, which yields $(\sigma, \Psi(\sigma)) = (\sigma, \tilde{\Psi}(\sigma))$, i.e. $\Psi \mid \Gamma_{bb}^s = \tilde{\Psi}$. □

PROPOSITION 12. There exists a fibered map $\psi : E^s|B_{\lambda,\mu} \to E^u|B_{\lambda,\mu}$, such that for every $\sigma \in \Gamma_b^s$, $\Psi(\sigma)(x) = \psi[\sigma(x)]$.

Proof. We will show that if $\sigma, \sigma' \in \Gamma_b^s$ are such that $\sigma(x_o) = \sigma'(x_o)$, then $\Psi(\sigma)(x_o) = \Psi(\sigma')(x_o)$.

Define $\tau \in \Gamma_b^u$ by $\tau(x) = \Psi(\sigma)(x)$ if $x \neq x_o$, and $\tau(x_o) = \Psi(\sigma')(x_o)$. One has : $|||\mathfrak{F}^n(\sigma,\tau)||| \leq \max(|||\mathfrak{F}^n(\sigma,\Psi(\sigma))|||, |||\mathfrak{F}^n(\sigma',\Psi(\sigma'))|||)$, because :

$$\mathfrak{F}^n(\sigma,\tau)(y) = \bar{F}_{f^{-1}(y)} \cdots \bar{F}_{f^{-n}(y)} [\sigma(f^{-n}(y)), \tau(f^{-n}(y))],$$

which implies :

$$\mathfrak{F}^n(\sigma,\tau)(y) = \mathfrak{F}^n(\sigma,\Psi(\sigma))(y), \text{ if } y \neq f^n(x_o)$$

and $\mathfrak{F}^n(\sigma,\tau)(f^n(x_o)) = \mathfrak{F}^n(\sigma',\Psi(\sigma'))(f^n(x_o))$.

The inequality above gives : $\rho^{-n}|||\mathfrak{F}^n(\sigma,\tau)||| \to 0$, $n \to \infty$, hence $\tau = \Psi(\sigma)$, in particular $\Psi(\sigma')(x_o) = \tau(x_o) = \Psi(\sigma)(x_o)$. □

Let $\psi_x : E_x^s \to E_x^u$ be the restriction of ψ to E_x^s (for $x \in B_{\lambda,\mu}$).

COROLLARY 13. For any $x \in B_{\lambda,\mu}$, ψ_x is Lipschitz with $\mathrm{Lip}^1_{\|\|\|\ \|\|\|}(\psi_x) < 1$.

Proof. The map ψ_x is the composition :

$$E_x^s \xrightarrow{i_x} \Gamma_b^s \xrightarrow{\Psi} \Gamma_b^u \xrightarrow{ev_x} E_x^u$$

$$v \longrightarrow \delta_x^v$$

$$\sigma \longrightarrow \Psi(\sigma)$$

$$\tau \longrightarrow \tau(x) \quad ,$$

where $\delta_x^v(y) = 0$ for $y \neq x$, $\delta_x^v(x) = v$. The maps i_x and ev_x are linear of norm 1, and Ψ is Lipschitz with $\mathrm{Lip}^1_{|||\,|||}\Psi < 1$. □

PROPOSITION 14. Denote by $C^o(E^s, E^u)$ the Borel measurable bundle with base B , whose fiber above x consists of the continuous maps from E_x^s to E_x^u . Then, the map $x \to \psi_x$ is a measurable section of $C^o(E^s, E^u)\,|B_{\lambda,\mu}$.

The proof of this proposition is given in appendix C, theorem C.6.

Let $x \in B_{\lambda,\mu}$, $v \in \mathrm{graph}\,(\psi_x)$, $y = \mathrm{Exp}_x(v)$. The section $\delta_x^v \in \Gamma_{bb}$ belongs to the graph of ψ . Moreover : $\mathfrak{F}^n(\delta_x^v)(f^n(x)) = \bar{F}_{f^{n-1}(x)}\,\bar{F}_{f^{n-2}(x)}\cdots\bar{F}_x(v)$.

By proposition 10, we obtain :

$$|||\bar{F}_{f^{n-1}(x)}\cdots\bar{F}_x(v)||| \leq \eta + \exp\lambda)^n\,|||v||| \quad .$$

Therefore, if for $n \geq 0$, $(\eta + \exp\lambda)^n\,|||v||| \leq C_\varepsilon(f^n(x))$, we have by induction from proposition 7, for $n \geq 0$:

$$\bar{F}_{f^{n-1}(x)}\,\bar{F}_{f^{n-2}(x)}\cdots\bar{F}_x(v) = F_{f^{n-1}(x)}\cdots F_x(v) = \mathrm{Exp}^{-1}_{f^n(x)}\,f^n(y) \; .$$

From $C_\varepsilon(f^{n'}(y)) \leq C_\varepsilon(y)\exp(\varepsilon\,|n'|)$, we obtain, with $x = f^{n'}(y)$, $n = -n'$:

$$\forall n \in \mathbb{Z}\,,\; C_\varepsilon(f^n(x)) \geq C_\varepsilon(x)\exp(-\varepsilon\,|n|) \; .$$

Now, the above inequality $(\eta + \exp\lambda)^n\,|||v||| \leq C_\varepsilon(f^n(x))$ is a consequence of :

$$\forall n \geq 0\,,\; (\eta + \exp\lambda)^n\,|||v||| \leq C_\varepsilon(x)\exp(-\varepsilon n) \; .$$

In the following, we suppose that $\varepsilon \in \,]0,-\mu[$, so we have $\rho < \exp(-\varepsilon)$.

Define now : $\widetilde{W}^{s,\rho}_{x,loc} = \mathrm{Exp}_x\,[\mathrm{graph}\,(\psi_x\,|\,\{v\,;\,|||v||| < C_\varepsilon(x)\})\,]$.

PROPOSITION 15.

1) $\widetilde{W}^{s,\rho}_{x,loc}$ is a measurable function of $x \in B_{\lambda,\mu}$;

2) For any y, z in $\widetilde{W}^{s,\rho}_{x,loc}$, we have :

$$|||\mathrm{Exp}^{-1}_{f^n(x)}(f^n(y)) - \mathrm{Exp}^{-1}_{f^n(x)}(f^n(z))||| \leq (\eta + \exp\lambda)^n \, |||\mathrm{Exp}^{-1}_x(y) - \mathrm{Exp}^{-1}_x(z)||| \;;$$

3) $f(\widetilde{W}^{s,\rho}_{x,loc}) \subset \widetilde{W}^{s,\rho}_{f(x),loc}$.

4) Suppose z satisfies $|||\mathrm{Exp}^{-1}_{f^n(x)}(f^n(z))||| \leq \rho^n C_\varepsilon(x)$, for $n \geq 0$, then : $z \in \widetilde{W}^{s,\rho}_{x,loc}$.

5) For any $v \in E^s_x$ with $|||v||| \leq C_\varepsilon(x)$, we have :

$$|||(v,\psi_x(v))||| = |||v||| \;.$$

Proof. 1) is a consequence of proposition 14 ;

2) follows from the considerations made above and from proposition 7.

3) It is sufficient to show that if $y \in \widetilde{W}^{s,\rho}_{x,loc}$, then :

$$|||\mathrm{Exp}^{-1}_{f(x)}(f(y))||| \leq C_\varepsilon(f(x)).$$

But this follows from $C_\varepsilon(f(x)) \geq C_\varepsilon(x)\exp(-\varepsilon)$, $\eta + \exp\lambda \leq \exp(-\varepsilon)$, and $|||\mathrm{Exp}^{-1}_{f(x)}(f(y))||| \leq (\eta + \exp\lambda)\, |||\mathrm{Exp}^{-1}_x(y)||| \leq (\eta+\exp\lambda)\, C_\varepsilon(x)$.

4) is proved in the same way, remarking that we have chosen $\rho < \exp(-\varepsilon)$.

Finally, 5) is an easy consequence of the inequality $\mathrm{Lip}\,\psi_x \leq 1$. □

We want to translate our results in the Riemannian metric $\| \ \|$. We recall the estimates :

a) $\forall v \in E^s_x$, $\|v\| \leq |||v|||$.

$\forall v \in E^u_x$, $\|v\| \leq |||v|||$.

b) $\forall v \in T_xM$, $\frac{1}{2}\|v\| \leq |||v||| \leq B_\varepsilon(x)\,\|v\|$.

c) $\forall n \in \mathbb{Z}$, $1 \leq B_\varepsilon(f^n(x)) \leq B_\varepsilon(x)\exp(\varepsilon|n|)$.

Define $\delta_\varepsilon(x) = \dfrac{C_\varepsilon(x)}{B_\varepsilon(x)}$, and

$$B^s(x,\delta_\varepsilon(x)) = \{v \in E^s_x \mid \|v\| \le \delta_\varepsilon(x)\} \subset \{v \mid |||v||| \le C_\varepsilon(x)\} .$$

$$W^{s,\rho}_{x,loc} = \mathrm{Exp}_x[\mathrm{graph}\,(\psi_x \mid B^s(x,\delta_\varepsilon(x)))]$$

$$= \mathrm{Exp}_x\,\{(v,\psi_x(v)) \mid v \in E^s_x,\ \|v\| \le \delta_\varepsilon(x)\} .$$

Here are some properties of $W^{s,\rho}_{x,loc}$.

1) $W^{s,\rho}_{x,loc} \subset \widetilde{W}^{s,\rho}_{x,loc}$.

Proof. This follows from b) and the definition of $\delta_\varepsilon(x)$. □

2) For $n \in \mathbb{Z}$, $\delta_\varepsilon(f^n(x)) \le \delta(x)\exp(2\varepsilon|n|)$.

Proof. This follows from analogous properties for B_ε and C_ε . □

3) $f(W^{s,\rho}_{x,loc}) \cap B(f(x), \dfrac{\delta_\varepsilon(f(x))}{B_\varepsilon(f(x))}) \subset W^{s,\rho}_{f(x),loc}$.

Proof. Let $z = \mathrm{Exp}_x(v,\psi_x(v))$, with $\|v\| \le \delta_\varepsilon(x)$ and $d(f(z),f(x)) \le \dfrac{\delta_\varepsilon(f(x))}{B_\varepsilon(f(x))}$.

Since $f(\widetilde{W}^{s,\rho}_{x,loc}) \subset \widetilde{W}^{s,\rho}_{f(x),loc}$, we can write $f(z)$ as some $\mathrm{Exp}_{f(x)}(v',\psi_{f(x)}(v'))$.

But : $\|v'\| \le |||v'||| = |||(v',\psi_{f(x)}(v'))||| \le B_\varepsilon(f(x))\,\|(v',\psi_{f(x)}(v')\|$

$$\le B_\varepsilon(f(x))\,d(f(z),f(x)) \le \delta_\varepsilon(f(x)) . \quad \square$$

Let $r > 0$ be the number introduced in § 2 such that Exp_x is defined on $B(x,r)$. The compactness of M implies that, for some constant C and any y , z such that $d(y,x) \le r$, $d(z,x) \le r$, we have :

$$\frac{1}{C}\,\|\mathrm{Exp}_x^{-1}(y) - \mathrm{Exp}_x^{-1}(z)\| \le d(y,z) \le C\,\|\mathrm{Exp}_x^{-1}(y) - \mathrm{Exp}_x^{-1}(z)\| .$$

Therefore we obtain :

$$\frac{1}{CB_\varepsilon(x)}\,|||\mathrm{Exp}_x^{-1}(y) - \mathrm{Exp}_x^{-1}(y)||| \le d(y,z) \le 2C\,|||\mathrm{Exp}_x^{-1}(y) - \mathrm{Exp}_x^{-1}(z)||| .$$

4) For $n \ge 0$ and $y,z \in W^{s,\rho}_{x,loc}$:

$$d(f^n(y),f^n(z)) \le 2C^2 B_\varepsilon(x)\,d(y,z)\,(\eta + \exp\lambda)^n .$$

Proof. From the above inequality and proposition 15, we have :

$$d(f^n(y),f^n(z)) \le 2C\,|||\mathrm{Exp}^{-1}_{f^n(x)}(f^n(y)) - \mathrm{Exp}^{-1}_{f^n(x)}(f^n(z))|||$$

$$\leq 2C(\eta + \exp\lambda)^n \, |||\mathrm{Exp}_x^{-1}(y) - \mathrm{Exp}_x^{-1}(z)|||$$

$$\leq 2C^2(\eta + \exp\lambda)^n B_\epsilon(x)\, d(x,y) \quad . \qquad \square$$

5) For $y \in W^{s,\rho}_{x,loc}$ and $n \geq 0$:

$$d(f^n(y), f^n(x)) \leq 2B_\epsilon(x)\, d(x,y)(\eta + \exp\lambda)^n \quad .$$

Proof. $d(f^n(y), f^n(x)) = \|\mathrm{Exp}^{-1}_{f^n(x)}(f^n(y))\|$

$$\leq 2\, |||\mathrm{Exp}^{-1}_{f^n(x)}(f^n(y))||| \leq 2(\eta + \exp\lambda)^n \, |||\mathrm{Exp}_x^{-1}(y)|||$$

$$\leq 2B_\epsilon(x)\, (\eta + \exp\lambda)^n\, d(x,y) \quad . \qquad \square$$

6) If z satisfies $d(f^n(x), f^n(z)) \leq \delta_\epsilon(x)\, \rho^n \exp(-n\epsilon)$ for every $n \geq 0$ and $d(x,z) < \dfrac{\delta_\epsilon(x)}{B_\epsilon(x)}$, then $z \in W^{s,\rho}_{x,loc}$.

Proof. Recall that $\delta_\epsilon(x) = \dfrac{C_\epsilon(x)}{B_\epsilon(x)}$. We have :

$$|||\mathrm{Exp}^{-1}_{f^n(x)}[f^n(z)]||| \leq B_\epsilon(f^n(x)) \frac{C_\epsilon(x)}{B_\epsilon(x)} \rho^n \exp(-\epsilon n) \leq \rho^n C_\epsilon(x) \quad ,$$

hence $z \in \widetilde{W}^{s,\rho}_{x,loc}$ by proposition 15. We can then write $z = \exp_x(v, \psi_x(v))$. We still have to show that $\|v\| \leq \delta_\epsilon(x)$; this follows from the inequalities :

$$\|v\| \leq |||v||| = |||(v, \psi_x(v))||| \leq B_\epsilon(x)\, \|(v, \psi_x(v))\| \leq B_\epsilon(x)\, d(x,z)$$

$$\leq B_\epsilon(x) \frac{\delta_\epsilon(x)}{B_\epsilon(x)} = \delta_\epsilon(x) \quad . \qquad \square$$

7) For every v in E_x^s , we have :

$$\frac{1}{2B_\epsilon(x)}\, d(x, \mathrm{Exp}_x(v, \psi_x(v))) \leq \|v\| \leq B_\epsilon(x)\, d(x, \mathrm{Exp}_x(v, \psi_x(v))) \quad .$$

Proof. We have : $d(x, \mathrm{Exp}_x(v, \psi_x(v))) = \|(v, \psi_x(v))\| \leq 2\, |||(v, \psi_x(v))|||$

$$\leq 2\, |||v||| \leq 2B_\epsilon(x)\, \|v\| \quad ,$$

and also :

$$\|v\| \leq |||v||| = |||(v, \psi_x(v))||| \leq B_\epsilon(x) \|(v, \psi_x(v))\| \leq B_\epsilon(x)\, d(x, \mathrm{Exp}_x(v, \psi_x(v)))$$

Let us define a Borel map γ_ϵ by $\gamma_\epsilon(x) = 2C^2 B_\epsilon(x) [\geq 2B_\epsilon(x) \geq B_\epsilon(x) \geq 1]$. After changing slightly the notations, we can restate what we obtained as follows.

THEOREM 16. Given $\epsilon > 0$ and $\rho \in]\exp\lambda, \exp\mu[$, there exists :

i) Two Borel functions $\delta_\epsilon : B_{\lambda,\mu} \to]0,\infty[$ and $\gamma_\epsilon : B_{\lambda,\mu} \to [1,\infty[$;

ii) For every x in $B_{\lambda,\mu}$, a Lipschitz map :

$$\psi_x : B^s(x,\delta_\epsilon(x)) = \{v \in E_x^s \mid \|v\| \leq \delta_\epsilon(x)\} \to E_x^u \quad ;$$

such that :

1) ψ_x depends measurably on x ;

1') $W^{s,\rho}_{x,loc} = \mathrm{Exp}_x[\mathrm{graph}\,\psi_x]$ is a small Lipschitz disc which depends measurably on x ;

2) $\forall y, z \in W^{s,\rho}_{x\,loc}$, $\forall n \geq 0$, $d(f^n(y), f^n(z)) \leq \gamma_\epsilon(x)\rho^n d(y,z)$;

3) If z satisfies $d(f^n(x), f^n(z)) \leq \rho^n \delta_\epsilon(x)$, $\forall n \geq 0$, and $d(z,x) \leq \dfrac{\delta_\epsilon(x)}{\gamma_\epsilon(f(x))}$, then $z \in W^{s,\rho}_{x,loc}$;

4) $\forall x \in B_{\lambda,\mu}$, $f(W^{s,\rho}_{x,loc}) \cap B\left(f(x), \dfrac{\delta_\epsilon(f(x))}{\gamma_\epsilon(f(x))}\right) \subset W^{s,\rho}_{x,loc}$;

5) $\forall x \in B_{\lambda,\mu}$, $\forall n \in \mathbb{Z}$, $\gamma_\epsilon(f^n(x)) \leq \gamma_\epsilon(x)\exp(\epsilon|n|)$ and
$\delta_\epsilon(f^n(x)) \leq e^{\epsilon|n|}\delta_\epsilon(x)$;

6) $\forall v \in E_x^s$, with $\|v\| \leq \delta_\epsilon(x)$, we have :

$$\frac{1}{\gamma_\epsilon(x)} d(x, \mathrm{Exp}_x(v, \psi_x(v))) \leq \|v\| \leq \gamma_\epsilon(x)\, d(x, \mathrm{Exp}_x(v, \psi_x(v))) .$$

COMPLEMENT TO THEOREM 16. Moreover $W^{s,\rho}_{x,loc}$ is $C^{1,\alpha'}$, for each $x \in B_{\lambda,\mu}$.

Proof. By the $C^{1,\alpha'}$ version of the stable manifold theorem in a Banach space, all we have to see is that the map $\mathfrak{F}$ defined in § 3 is $C^{1,\alpha'}$. Since the map $\mathfrak{T}$ defined in § 3 is linear, we show that $\Phi = \mathfrak{F} - \mathfrak{T}$ is $C^{1,\alpha'}$. Let us first compute the derivative of Φ . By definition :

$$\Phi(\sigma + \Delta\sigma)(x) = \bar{\varphi}_{f^{-1}(x)} [\sigma(f^{-1}(x)) + \Delta\sigma(f^{-1}(x))] ,$$

which gives :

$$|||\Phi(\sigma + \Delta\sigma)(x) - \Phi(\sigma)(x) - T_{\sigma(f^{-1}(x))} \bar{\varphi}_{f^{-1}(x)} (\Delta\sigma(f^{-1}(x)))|||$$

$$\leq \left[\int_0^1 |||T_{\sigma(f^{-1}(x)) + t\Delta\sigma(f^{-1}(x))} \bar{\varphi}_{f^{-1}(x)} - T_{\sigma(f^{-1}(x))} \bar{\varphi}_{f^{-1}(x)}||| \, dt\right] |||\Delta\sigma|||$$

$$\leq |||\Delta\sigma|||^{\alpha'} \, |||\Delta\sigma||| \quad .$$

We used part 2) of proposition 7 for the last estimate. This shows clearly that Φ is C^1 and identifies its derivative.

We check now that $T\Phi$ satisfies a Hölder condition of order α' :

$$|||T_\sigma \Phi - T_{\sigma'} \Phi||| \leq \sup_{x \in B_{\lambda,\mu}} |||T_{\sigma(f^{-1}(x))} \bar{\varphi}_{f^{-1}(x)} - T_{\sigma'(f^{-1}(x))} \bar{\varphi}_{f^{-1}(x)}|||$$

$$\leq |||\sigma - \sigma'|||^{\alpha'} \quad .$$

The last estimate is also from part 2) of proposition 7. □

§ 4. - THE GLOBAL STABLE MANIFOLD OF A POINT

In the following, we suppose that ρ and ε have been chosen so that $\log \rho + 2\varepsilon < 0$.

As the local stable manifold constructed above depends on the choice of $\varepsilon > 0$, we will sometimes denote it by $W^{s,\rho,\varepsilon}_{x,loc}$.

THEOREM 17. Let $x \in B$, the Borel set on which the conclusion of Osedelets' theorem holds ; then, the set $W^s_x = \{y, \limsup_{n \to +\infty} \frac{1}{n} \log d(f^n(x), f^n(y)) < 0\}$ is the image of a C^1 injective immersion of an euclidean space, such that :

1) $\dim W^s_x = \dim E^s_x$, where $E^s_x = \bigoplus_{\lambda_i(x)<0} W_i(x)$;

2) $T_x W^s_x = E^s_x$;

3) $W^s_x = \bigcup_{n\geq 0} f^{-n}(W^{s,\rho}_{f^n(x),loc})$, <u>where</u> $\log \rho \in]\lambda_{i_o}(x), 0[$ <u>and</u>

$\lambda_{i_o}(x) = \max\{\lambda_i(x) < 0\}$.

4) $W^s_x = \{y \mid \limsup \frac{1}{n} \log d(f^n(x), f^n(y)) \leq \lambda_{i_o}(x)\}$.

<u>Proof</u>. Let x be a fixed point in B . Take two choices (ρ, ε), (ρ', ε') , with $\lambda_{i_o}(x) < \log \rho \leq \log \rho'$, denote by W_ℓ and W'_ℓ the manifolds $W^{s,\rho,\varepsilon}_{f^\ell(x),loc}$ and $W^{s,\rho',\varepsilon'}_{f^\ell(x),loc}$ obtained by applying theorem 16.

<u>CLAIM 1</u>. For ℓ large enough, $W_o \subset f^{-\ell}(W'_\ell)$. In the same way, given $m \in \mathbb{N}$, $W_m \subset f^{-k}(W'_{m+k})$ for k large enough.

<u>Proof</u>. Let $d = \sup\{d(x,y) \mid y \in W_o\} < +\infty$, we have, for any $y \in W$ and $\ell \geq 0$:
$$d(f^\ell(x), f^\ell(y)) \leq \gamma_\varepsilon(x)\, \rho^\ell d \ .$$

The inequality $\dfrac{\delta_{\varepsilon'}(f^\ell(x))}{\gamma'_{\varepsilon'}(f^\ell(x))} \geq \gamma_\varepsilon(x) . \rho^\ell d$, which proves the claim by part 3) of theorem 16, now follows from $\dfrac{\delta_{\varepsilon'}(x)}{\gamma_{\varepsilon'}(x)} \exp(-2\ell\varepsilon') \leq \dfrac{\delta_{\varepsilon'}(f^\ell(x))}{\gamma_{\varepsilon'}(f^\ell(x))}$ and $d\gamma_\varepsilon(x)\, [\rho \exp(2\varepsilon')]^{\ell} \leq \dfrac{\delta_{\varepsilon'}(x)}{\gamma_{\varepsilon'}(x)}$, for ℓ large enough , since $2\varepsilon' + \log\rho \leq 2\varepsilon' + \log\rho' < 0$.

The second part of the claim is the first one applied to $f^m(x)$. □

Taking $\rho = \rho'$, $\varepsilon = \varepsilon'$, we obtain from the claim that $\bigcup_{\ell \geq 0} f^{-\ell}(W_\ell)$ is the increasing union of a subfamily ; but every $f^{-n}(W_n)$ is the image of a C^1 injective immersion of an open disk of dimension $\dim E^s_x$. It follows that the same is true for $\bigcup_{n\geq 0} f^{-n}(W_n)$ (see [Hi], chap. 8) .

It follows also from claim 1, applied to $\rho = \rho'$, ε, ε', that $\bigcup_{n\geq 0} f^{-n}(W_n)$ is independent of ε.

We now show that $\bigcup_{n\geq 0} f^{-n}(W_n) = \bigcup_{n\geq 0} f^{-n}(W'_n)$. By the previous remark, we can assume that $\varepsilon = \varepsilon'$. The inclusion $\bigcup_{n\geq 0} f^{-n}(W_n) \subset \bigcup_{n\geq 0} f^{-n}(W'_n)$ is deduced from the claim. To prove the other inclusion, it is sufficient to show that the inclusion $W'_o \subset f^{-\ell}(W_\ell)$ holds for some $\ell \geq 0$.

CLAIM 2. For $\|v\| \leq \delta(x) = \inf(\delta_\varepsilon(x), \delta'_\varepsilon(x))$, $\psi_x(v) = \psi'_x(v)$.

Proof. We know that $\overset{o}{W}_o$ and $\overset{o}{W}{}'_o$ are open submanifolds of $\bigcup_{n\geq 0} f^{-n}(W'_n)$; their intersection $\overset{o}{W}_o \cap \overset{o}{W}{}'_o$ is open in both $\overset{o}{W}_o$ and $\overset{o}{W}{}'_o$. Now $\mathrm{Exp}_x^{-1}(\overset{o}{W}_o \cap \overset{o}{W}{}'_o)$ is closed in $\overset{o}{B}{}^s(0,\delta(x)) \times E^u_x$, since $\mathrm{Exp}_x^{-1}(W_o)$ and $\mathrm{Exp}_x^{-1}(W'_o)$ are graphs. It follows easily that $\psi_x = \psi'_x$ on the connected set $\overset{o}{B}{}^s(0,\delta(x))$. □

Now let $y \in W'_o$; for $\ell \geq 0$, $f^\ell(y) = \mathrm{Exp}_{f^\ell(x)}(v_\ell, \psi'_{f^\ell(x)}(v_\ell))$, for some $v_\ell \in E^s_{f^\ell(x)}$. From claim 1, for ℓ large enough, $f^\ell(y) \in W'_\ell$, thus $\|v_\ell\| < \delta'_\varepsilon(f^\ell(x))$. By theorem 16, parts 3) and 6), we obtain, for such ℓ:

$$\|v_\ell\| \leq \gamma'_\varepsilon(f^\ell(x))\, d(f^\ell(x), f^\ell(y)) \leq \gamma'_\varepsilon(x)\, d(x,y)\, \rho'^{\ell} \exp(\ell\varepsilon)$$

$$\leq C(x)\, \rho'^{\ell} \exp(2\ell\varepsilon)\, \delta_\varepsilon(f^\ell(x))$$

with $C(x) = \gamma'_\varepsilon(x)\, d(x,y)\, (\delta_\varepsilon(x))^{-1}$.

Therefore, for ℓ large enough $\|v_\ell\| \leq \inf[\delta_\varepsilon(f^\ell(x)), \delta'_\varepsilon(f^\ell(x))]$. By the claim 2, we have $f^\ell(y) \in W_\ell$ for such ℓ.

We have proved $\bigcup_{n\geq 0} f^{-n}(W_n)$ does not depend on ρ and ε.

From theorem 16, it is easy to prove :

i) $y \in \bigcup_{n\geq 0} f^{-n}(W_n) \Rightarrow \limsup_{n\to+\infty} \{\frac{1}{n} \mathrm{Log}\, d(f^n(x), f^n(y))\} \leq \mathrm{Log}\, \rho$.

ii) $\limsup_{n\to+\infty} \{\frac{1}{n} \operatorname{Log} d(f^n(x), f^n(y))\} < \operatorname{Log} \rho \Rightarrow$ for n large enough, $f^n(y) \in W_n$.

This finishes the proof of theorem 17. □

§ 5. - HIGHER DIFFENTIABILITY OF STABLE MANIFOLDS

To prove the higher differentiability of the stable manifolds, it would have been convenient to know that the map $\mathfrak{F}$ of § 3 is as differentiable as f is. Unfortunately, this is not the case as can be seen from Lemma 19 below. We get rid of this minor point in the following way ; first we restrict to the orbit $\mathcal{O}(x) = \{f^n(x) \mid n \in \mathbb{Z}\}$; the estimates we get in lemma 19 suggest to look at the space of sections of $TM \mid \mathcal{O}(x)$ which decrease exponentially along the orbit. It is a matter of routine to show that the map induced by f on this space of sections is enough differentiable and to relate its stable manifold to the local stable manifolds of f along the orbit of x .

Keeping the notations introduced in § 1, § 2, we fix $\lambda < \mu < 0$. We suppose that f is a C^r diffeomorphism and that its C^r derivative satisfies a Hölder condition of order α $(0 < \alpha < 1)$. We will show that a local stable manifold at x , $W^{s,\rho,\epsilon}_{x,loc}$ is actually a $C^{r,\alpha}$ manifold. Without loss of generality, we can assume r finite.

We begin by recalling Lemma 5 and adding estimates for higher derivatives of $\bar{\varphi}_x$ which are proved in similar ways.

<u>LEMMA 18. Assuming f is $C^{r,\alpha}$, there exists a constant K such that the following estimates hold for any</u> $j = 1,\dots,r$, $x \in B_{\lambda,\mu}$, $v \in T_xM$:

i) $\|T^j_v \bar{\varphi}_x\| \leq K\beta(x)^{1+\alpha-j}$;

ii) $\mathrm{Lip}^{\alpha}_{\| \ \|}(T^r \bar{\varphi}_x) \leq K\,\beta(x)^{1-r}$.

We can translate these estimates in the norm $|\!|\!| \ |\!|\!|$, using Proposition 4, (b). An easy calculation gives :

LEMMA 19. Under the same assumptions as in Lemma 18, there exists a constant K' such that, for $x \in B_{\lambda,\mu}$ and $v \in T_x M$, we have :

i) $|\!|\!|T^j \bar{\varphi}_x|\!|\!| \leq K' B_\epsilon(x)\,\beta(x)^{1+\alpha-j}$, for $j = 1,\dots,r$.

ii) $\mathrm{Lip}^{\alpha}_{|\!|\!| \ |\!|\!|}(T^r \bar{\varphi}_x) \leq K' B_\epsilon(x)\,[\beta(x)]^{1-r}$.

We now fix $x \in B_{\lambda,\mu}$. From lemmas 6, 19 and proposition 4 , there exists a constant K'' (depending on x) such that, for $k \in \mathbf{Z}$ and $v \in T_{f^k(x)}M$, the following estimates hold : For $j = 1,2,\dots,r$

i) $|\!|\!|T^j_v \bar{\varphi}_{f^k(x)}|\!|\!| \leq K'' \exp((j-\alpha)|k|\epsilon)$;

ii) $\mathrm{Lip}^{\alpha}_{|\!|\!| \ |\!|\!|}(T^r \bar{\varphi}_{f^k(x)}) \leq K'' \exp(r|k|\epsilon)$;

iii) $|\!|\!|T_v \bar{\varphi}_{f^k(x)}|\!|\!| \leq \eta$;

iv) $|\!|\!|T_{f^k(x)} f|\!|\!| \leq K''$, $|\!|\!|T_{f^k(x)} f^{-1}|\!|\!| \leq K''$;

v) $|\!|\!|T_{f^k(x)} f \,|\, E^s_x|\!|\!| \leq \exp\lambda$, $|\!|\!|T_{f^k(x)} f^{-1} \,|\, E^u_x|\!|\!| \leq \exp(-\mu)$.

Now choose $\theta < 1$, such that $\theta^{-1}.\exp\lambda < \theta.\exp\mu$ (this is possible if θ is sufficiently close to 1). We define a space of sequences $\mathcal{S}$ as follows :

$$\mathcal{S} = \{(v_k)_{k\in\mathbf{Z}} \mid v_k \in T_{f^k(x)}M,\ \sup_{k\in\mathbf{Z}} (\theta^{-|k|}\,|\!|\!|v_k|\!|\!|) < +\infty\} .$$

This is a Banach space with the norm : $|\!|\!|(v_k)_{k\in\mathbf{Z}}|\!|\!| = \sup_{k\in\mathbf{Z}} (\theta^{-|k|}\,|\!|\!|v_k|\!|\!|)$.

We define, as in § 3, two maps $\mathcal{T}$ and Φ, from $\mathcal{S}$ to $\mathcal{S}$, by :

$$\mathcal{T}((v_n)_{n\in\mathbf{Z}}) = (v'_n)_{n\in\mathbf{Z}} , \quad \Phi((v_n)_{n\in\mathbf{Z}}) = (v''_n)_{n\in\mathbf{Z}} ,$$

where, for $n \in \mathbb{Z}$: $v'_n = T_{f^{n-1}(x)} f(v_{n-1})$, $v''_n = \bar{\varphi}_{f^{n-1}(x)} f(v_{n-1})$.

It is a consequence of relations (iii), (iv) that $\mathfrak{T}$ and Φ take their values in $\mathcal{S}$. Moreover, $\mathfrak{T}$ is linear, continuous, invertible and there exists an invariant splitting $\mathcal{S} = \mathcal{S}^s \oplus \mathcal{S}^u$, such that :

$$|||\mathfrak{T} \mid \mathcal{S}^s||| \leq \Theta^{-1} \exp\lambda \ , \qquad |||\mathfrak{T}^{-1} \mid \mathcal{S}^u||| < \Theta^{-1} \exp(-\mu) \ .$$

From Appendix B, lemma B.9 and relations (i), (ii) and (iii), we see that Φ is Lipschitz, with Lipschitz constant η/Θ , and that Φ is $C^{r,\alpha}$, as soon as we have, for $j = 2,3,\ldots,r,r+\alpha$: $\sup_{n\in\mathbb{Z}} (\Theta^{|n|(j-1)} \exp[(j-\alpha)|n|\varepsilon]) < +\infty$.

We therefore choose ε small enough to have : $(2-\alpha)\varepsilon < |\mathrm{Log}\,\Theta|$ which implies the above estimate .

The norm on $\mathcal{S}$ is the supremum of the norms on $\mathcal{S}^s$ and $\mathcal{S}^u$ (because it is so for $|||\ |||$ on T_xM , E^s_x , E^u_x). Let $\rho \in]\Theta^{-1}\exp\lambda , \Theta\exp\mu[$ and choose η (and then $\beta(x)$) so that :

$$\eta < \Theta.\inf[(\rho - \Theta^{-1}\exp\lambda), \Theta\exp(\mu) - \rho] \ .$$

We then have, from the choice of Θ :

$$|||\mathfrak{T} \mid \mathcal{S}^s||| \leq \Theta^{-1}\exp\lambda < \rho < \Theta\exp\mu \leq |||\mathfrak{T}^{-1} \mid \mathcal{S}^u|||^{-1}$$

$$\mathrm{Lip}(\Phi) < \inf(\rho - \Theta^{-1}\exp\lambda , \Theta\exp(\mu) - \rho) \ .$$

The stable manifold theorem holds under these assumptions and give a map $\Psi : \mathcal{S}^s \to \mathcal{S}^u$ such that :

$$x \in \text{graph}\,\Psi \iff \lim_{n\to+\infty} (\rho^{-n} |||(\mathfrak{T}+\Phi)^n(x)|||) = 0$$

As in § 3, it is easily proven that Ψ factorizes to maps

$\psi_{f^k(x)} : E^s_{f^k(x)} \longrightarrow E^u_{f^k(x)}$, for any $k \in \mathbb{Z}$.

It follows from the stable manifold theorem that Ψ , and therefore $\psi_{f^k(x)}$, are $C^{r,\alpha}$.

Choose now $\rho' \in]\exp\lambda , \Theta\rho[$. Let y be a point of the local stable manifold $W^{s,\rho',\varepsilon}_{x,loc}$. We have, for some $v \in T_xM$: $y = \mathrm{Exp}_x(v)$.

Let δ_v be the section of $\mathcal{S}$ defined by :

$$(\delta_v)_o = v \ , \quad (\delta_v)_k = 0 \ , \quad \text{if } k \neq 0 \ .$$

We now have, for $n \geq 0$:

$$|||\bar{F}_{f^{n-1}(x)} \cdots \bar{F}_x(v)||| \leq \rho'^n \, |||v|||$$

from which we derive :

$$|||(\mathcal{J} + \Phi)^n (\delta_v)||| \leq \left(\frac{\rho'}{\mathcal{G}}\right)^n |||v||| \quad .$$

This last estimate implies, as $\frac{\rho'}{\mathcal{G}} < \rho$, that δ_v belongs to the graph of Ψ and therefore that v belongs to the graph of ψ_x . Therefore $W^{s,\rho',\varepsilon}_{x,loc}$ is actually the image, under the exponential map, of the graph of a $C^{r,\alpha}$ map in the tangent bundle.

<u>Remark</u>. If f is C^r $(r \geq 2)$ but $T^r f$ does not satisfy an Hölder condition, the preceding results show that $W^{s,\rho,\varepsilon}_{x,loc}$ is a $C^{r-1,1-\varepsilon}$ manifold, for any $\varepsilon > 0$. (Take $r' = r-1$, $\alpha = 1-\varepsilon$ in the preceding discussion.) We lack, in the above discussion, some uniform estimate to show that the map $\mathcal{F} = \Phi + \mathcal{J}$ is actually C^r (what we obtain is C^{r-1} and Lipschitz).

APPENDIX A.

STABLE MANIFOLD OF A POINT IN A BANACH SPACE

We prove here the stable manifold theorem using the method of Irwin, see [Ir] . We give the version which we need and we will not consider the greatest generality.

§ 1. Some technical facts about contracting maps

LEMMA A.1. Let $\theta : X \times Y \to Y$ be a continuous map. Suppose that Y is a complete metric space and that θ is a uniform k-contraction $(k < 1)$ on the factor Y, i.e. :

$$\exists\, k < 1 , \quad \forall x \in X, \quad \forall y, y' \in Y, \quad d[\theta(x,y),\theta(x,y')] \leq k\, d(y,y').$$

Denote by θ_x the map : $Y \to Y$, $y \longmapsto \theta(x,y)$, and let $\varphi(x)$ be the fixed point of θ_x . The map φ is continuous. Moreover, if X is metric and θ is Lipschitz, φ is Lipschitz.

Proof. Let x, x' be in X ; we have :

$$\begin{aligned} d[\varphi(x),\varphi(x')] &= d[\theta(x,\varphi(x)),\ \theta(x',\varphi(x'))] \\ &\leq d[\theta(x,\varphi(x)),\ \theta(x',\varphi(x))] + d[\theta(x',\varphi(x)),\ \theta(x',\varphi(x'))] \\ &\leq d[\theta(x,\varphi(x)),\ \theta(x',\varphi(x))] + k\, d[\varphi(x),\varphi(x')] . \end{aligned}$$

Hence we obtain :

$$d[\varphi(x),\varphi(x')] \leq \frac{1}{1-k}\, d[\theta(x,\varphi(x)),\ \theta(x',\varphi(x))] .$$

The lemma follows easily from this inequality. □

LEMMA A.2. Suppose that in the situation above, X is metric and that on $X \times Y$ we choose the max metric, i.e. :

$$d[(x,y),(x',y')] = \max\,[d(x,x'),d(y,y')] .$$

If θ is Lipschitz with Lipschitz constant $\leq k$ (< 1), then φ is Lipschitz with Lipschitz constant $\leq k$.

Proof. We have : $d[\varphi(x),\varphi(x')] = d[\theta(x,\varphi(x)),\ \theta(x',\varphi(x'))]$

$$\leq k \max\,[d(x,x'),d(\varphi(x),\varphi(x'))] .$$

Since $k < 1$, we obtain easily : $d(\varphi(x),\varphi(x')) \leq k\, d(x,x')$. □

LEMMA A.3. With the hypothesis of A.1, suppose moreover that X and Y are Banach spaces. If θ is of class C^k $(k \in \mathbb{N})$, then φ is of class C^k with :

$$T\varphi(x) = [Id - T_2\theta(x,\varphi(x))]^{-1} T_1\theta(x,\varphi(x)) .$$

[Here T_1 (resp. T_2) is the partial derivative with respect to $x \in X$ (resp. $y \in Y$).]

Proof. We remark first that $\|T_2\,\theta(x,y)\| \leq k < 1$, since $\|\theta(x,y) - \theta(x,y')\| \leq k\|y - y'\|$. Consequently $Id - T_2\theta(x,y)$ is invertible and its inverse is given by :

$$[Id - T_2\,\theta(x,y)] = \sum_{i=0}^{\infty} [T_2\,\theta(x,y)]^i .$$

To show that φ is C^1 with derivative at x : $[Id - T_2\theta(x,\varphi(x))]^{-1} T_1\theta(x,\varphi(x))$, we have to show that $\|\varphi(x+v) - \varphi(x) - [Id - T_2\theta(x,\varphi(x))]^{-1} T_1\theta(x,\varphi(x))(v)\|$ is $o(\|v\|)$:

$$\|\varphi(x+v) - \varphi(x) - [Id - T_2\theta(x,\varphi(x))]^{-1}\, T_1\theta(x,\varphi(x))(v)\| \leq$$

$$\leq \|[Id-T_2\theta(x,\varphi(x))]^{-1}\|\,\|\varphi(x+v)-\varphi(x) - T_2\theta(x,\varphi(x))\,[\varphi(x+v)-\varphi(x)] - T_1\theta(x,\varphi(x))(v)\|$$

$$\leq \|[Id-T_2\theta(x,\varphi(x))]^{-1}\|\,\|\theta(x+v,\varphi(x+v))-\theta(x,\varphi(x))-T\theta(x,\varphi(x))\,[(v,\varphi(x+v)-\varphi(x))]\| .$$

This last expression is $o(\|(v,\varphi(x+v) - \varphi(x))\|)$ and is also $o(\|v\|)$ since φ is Lipschitz. Up to now, we have shown that φ is derivable if θ is. We remark that $T\varphi$ is equal to the following composition :

$$X \xrightarrow{(id,\varphi)} X \times Y \xrightarrow{(T_1\theta,T_2\theta)} L(X,Y) \times U \xrightarrow{(Id,\rho)} L(X,Y) \times L(Y,Y) \xrightarrow{comp} L(X,Y) ;$$

where $U = \{T \in L(Y,Y) \mid \|T\| < 1\}$, $\rho(T) = (Id - T)^{-1}$, $comp(Q,P) = P \circ Q$.

An easy induction argument shows that φ is C^k, if θ is. □

If $r \in \mathbb{N}$ and $\alpha \in]0,1]$, we will say that the map $f : X \to Y$ (X, Y Banach spaces) is $C^{r,\alpha}$, if $T^r f : X \to L^r(X,Y)$ satisfies locally a Hölder condition of order α, i.e. for every $x \in X$, there exists a neighborhood U of x such that $T^r f | U$ satisfies a Hölder condition of order α. It is a well-known fact that, with this definition, if we have a composition of maps such that one of them is $C^{r,\alpha}$ and the others are C^{r+1}, then the composition is $C^{r,\alpha}$.

LEMMA A.4. With the hypothesis of A.3, if θ is $C^{r,\alpha}$, then φ is $C^{r,\alpha}$.

Proof. We know that φ is C^r. Also $T\varphi$ is the composition of four maps, one of them (Id,φ) is C^r, the second one $(T_1\theta, T_2\theta)$ is $C^{r-1,\alpha}$ and the last two (Id,ρ) and comp are C^∞. Hence T_φ is $C^{r-1,\alpha}$, which means that φ is $C^{r,\alpha}$. □

§2. Stable manifold of a hyperbolic fixed point.

Let E_1 and E_2 be two Banach spaces. We define $E = E_1 \oplus E_2$ and take as a norm on E the max norm i.e. : $\|(x_1,x_2)\| = \max(\|x_1\|,\|x_2\|)$. We denote by p_i the projection : $E = E_1 \times E_2 \to E_i$, it is a linear map of norm 1.

Let $T : E \to E$ be a continuous linear map with $T(E_1) \subset E_1$ and $T(E_2) = E_2$, we define $T_i = T \mid E_i$. We suppose that T_2 is invertible and $\|T_1\| < \|T_2^{-1}\|^{-1}$. Let $\varkappa$ be such that $\|T_1\| < \varkappa < \|T_2^{-1}\|^{-1}$.

THEOREM A.5. Let $f : E \to E$ be a (global) Lipschitz map such that $f(0) = 0$ and $\mathrm{Lip}(f-T) = \ell < \min(\varkappa - \|T_1\|, \|T_2^{-1}\|^{-1} - \varkappa$. The set $W^{s,\varkappa} = \{x \in E \mid \sup_{n\geq 0} \|\varkappa^{-n} f^n(x)\| < +\infty\}$ is the graph of a Lipschitz map $g : E_1 \to E_2$, with $\mathrm{Lip}(g) < 1$. Moreover, $\mathrm{Lip}(f \mid W^{s,\varkappa} \leq \|T_1\| + \ell$, which implies that $\varkappa^{-n} f^n(x) \to 0$ for $n \to \infty$ if $x \in W^{s,\varkappa}$.

Proof. We define :

$$\mathcal{B}^\varkappa(E_1) = \{(\gamma_1(n))_{n\geq 1} \mid \gamma_1(n) \in E_1 \ \& \ \|\gamma_1\| = \sup_{n\geq 1} \|\varkappa^{-n}\gamma_1(n)\| < +\infty\}$$

and $\mathcal{B}^\varkappa(E_2) = \{(\gamma_2(n))_{n\geq 1} \mid \gamma_2(n) \in E_2 \ \& \ \|\gamma_2\| = \sup_{n\geq 1} \|\varkappa^{n+1}\gamma_2(n)\| < +\infty\}$.

These two spaces are Banach spaces. The norm on $\mathcal{B}^\varkappa(E_2)$ is $\sup_{n\geq 1} \|\varkappa^{-n+1}\gamma_2(n)\|$ and not $\sup_{n\geq 1} \|\varkappa^{-n}\gamma_2(n)\|$ for practical reasons.

On $E_1 \times \mathcal{B}^\varkappa(E_1) \times \mathcal{B}^\varkappa(E_2)$, we put the norm max of the norms on E_1, $\mathcal{B}^\varkappa(E_1)$, $\mathcal{B}^\varkappa(E_2)$.

We define $\mathcal{G} : E_1 \times \mathcal{B}^\varkappa(E_1) \times \mathcal{B}^\varkappa(E_2) \to \mathcal{B}^\varkappa(E_1) \times \mathcal{B}^\varkappa(E_2)$, $(x_1,\gamma_1,\gamma_2) \to (\nu_1,\nu_2)$ by :

$$\nu_1(1) = f_1(x_1,\gamma_2(1)) \ ;$$

$$\nu_1(n) = f_1(\gamma_1(n-1),\gamma_2(n)), \quad n \geq 2 \ ;$$

$$\nu_2(1) = T_2^{-1}[\gamma_2(2) + T_2\gamma_2(1) - f_2(x_1,\gamma_2(1))] \ ;$$

$$\nu_1(n) = T_2^{-1}[\gamma_2(n+1) + T_2\gamma_2(n) - f_2(\gamma_1(n-1),\gamma_2(n))], \quad n \geq 2 \ ;$$

where $f_1 = p_1 f$ and $f_2 = p_2 f$.

Suppose that $\mathcal{G}(x_1,\gamma_1,\gamma_2) = (\gamma_1,\gamma_2)$, it follows that :

$$\gamma_1(1) = f_1(x_1,\gamma_2(1)), \quad \gamma_1(n) = f_1(\gamma_1(n-1),\gamma_2(n))$$

and $\quad \gamma_2(1) = T_2^{-1}[\gamma_2(2) + T_2\gamma_2(1) - f_2(x_1,\gamma_2(1))]$, or $\gamma_2(2) = f_2(x_1,\gamma_2(1))$,

and in the same way $\gamma_2(n+1) = f_2(\gamma_1(n-1),\gamma_2(n))$. This means that $(\gamma_1(n),\gamma_2(n+1)) = f^n(x_1,\gamma_2(1))$ which implies that $(x_1,\gamma_2(1)) \in W^{s,\varkappa}$. Conversely, if $(x_1,x_2) \in W^{s,\varkappa}$ the point $(\gamma_1,\gamma_2) \in \mathcal{B}^{\varkappa}(E_1) \times \mathcal{B}^{\varkappa}(E_2)$, defined by $\gamma_1(n) = p_1f^n(x_1,x_2)$, $\gamma_2(n) = p_2f^{n-1}(x_1,x_2)$, is such that $\varsigma(x_1,\gamma_1,\gamma_2) = (\gamma_1,\gamma_2)$.

We will show that ς is a uniform contraction on the factor $\mathcal{B}^{\varkappa}(E_1)\times\mathcal{B}^{\varkappa}(E_2)$. This implies that $W^{s,\varkappa}$ is the graph of the map g which is the composition of φ with $\mathcal{B}^{\varkappa}(E_1) \times \mathcal{B}^{\varkappa}(E_2) \to E_2$, $E_2(\gamma_1,\gamma_2) \mapsto \gamma_2(1)$.

We check now that ς is well defined and compute its Lipschitz constant.

Remark first that $\varsigma(0,0,0) = (0,0)$ since $f(0) = 0$. Define then $\varsigma(x_1,\gamma_1,\gamma_2) = (\nu_1,\nu_2)$ and $\varsigma(x'_1,\gamma'_1,\gamma'_2) = (\nu'_1,\nu'_2)$; we have :

$$\nu'_1(1) - \nu_1(1) = f_1(x'_1,\gamma'_2(1)) - f_1(x_1,\gamma_2(1)) =$$

$$= T_1(x'_1-x_1) + [p_1(f-T)(x'_1,\gamma'_2(1)) - p_1(f-T)(x_1,\gamma_2(1))] ,$$

which gives :

$$\|\nu'_1(1)-\nu_1(1)\| \le \|T_1\|\|x'_1-x_1\| + \varepsilon \max(\|x_1-x'_1\|,\|\gamma'_2(1)-\gamma_2(1)\|) .$$

In the same way :

$$\|\nu_1(n)-\nu_1(n)\| \le \|T_1\|\|\gamma'_1(n-1)-\gamma_1(n-1)\| + \varepsilon \max(\|\gamma'_1(n-1)-\gamma_1(n-1)\|,\|\gamma'_2(n)-\gamma_2(n)\|) .$$

Since

$$\|\nu'_1 - \nu_1\| = \sup_{n\ge 1} \varkappa^{-n} \|\nu'_1(n) - \nu_1(n)\|$$

$$\|\gamma'_1 - \gamma_1\| = \sup_{n\ge 1} \varkappa^{-n} \|\gamma'_1(n) - \gamma_1(n)\|$$

$$\|\gamma'_2 - \gamma_2\| = \sup_{n\le 1} \varkappa^{-n+1} \|\gamma'_2(n) - \gamma_2(n)\| ,$$

we obtain :

$$\|\nu'_1 - \nu_1\| \le \varkappa^{-1}(\|T_1\| + \varepsilon) \max(\|x_1-x'_1\|,\|\gamma_1-\gamma'_1\|,\|\gamma_2-\gamma'_2\|) .$$

A similar computation shows that :

$$\|\nu'_2-\nu_2\| \le \|T_2^{-1}\|(\varkappa + \varepsilon) \max(\|x'_1-x_1\|,\|\gamma'_1-\gamma_1\|,\|\gamma'_2-\gamma_2\|) .$$

This shows that ς is well defined and that it is a Lipschitz map with $\mathrm{Lip}(\varsigma) \le \max[\varkappa^{-1}(\|T_1\| + \varepsilon), \|T_2^{-1}\|(\varkappa + \varepsilon)] < 1$.

An application of lemma A.1 of this appendix shows that the map $\varphi : E_1 \to \mathcal{B}^{\varkappa}(E_1) \times \mathcal{B}^{\varkappa}(E_2)$, $x_1 \mapsto \mathrm{fix}(\varsigma_{x_1})$ is a Lipschitz map. By lemma A.2,

$\mathrm{Lip}(\varphi) \leq \mathrm{lip}(\zeta) < 1$. If we define $g: E_1 \to E_2$ as the composition of φ with $\mathcal{B}^{\varkappa}(E_1) \times \mathcal{B}^{\varkappa}(E_2) \to E_2$, $(\gamma_1,\gamma_2) \mapsto \gamma_2(1)$ is linear of norm 1 .

As we have shown above, $W^{s,\varkappa}$ is the graph of g .

Remark that by definition of $W^{s,\varkappa}$, we have $f(W^{s,\varkappa}) = W^{s,\varkappa}$.

We compute now $\mathrm{Lip}(f \mid W^{s,\varkappa})$. First, since $\mathrm{Lip}(g) \leq 1$, we have : $\|(x_1,g(x_1)) - (x_1',g(x_1'))\| = \|x_1 - x_1'\|$. Since $f(x_1,g(x_1)) = [f_1(x_1,g(x_1)), g(f_1(x_1,g(x_1)))]$ and the same for x_1' , we obtain :

$$\|f(x_1,g(x_1)) - f(x_1',g(x_1')) \leq \mathrm{Lip}(f_1)\, \|x_1 - x_1'\| .$$

But $\mathrm{Lip}(f_1) \leq \|T_1\| + \mathrm{Lip}(f_1 - T_1) \leq \|T_1\| + \varepsilon$, hence we have :

$$\mathrm{Lip}(f \mid W^{s,\varkappa}) \leq \|T_1\| + \varepsilon .$$

Since $f(0) = 0$, if $x \in W^{s,\varkappa}$ we get : $\|f^n(x)\| \leq (\|T_1\| + \varepsilon)^n \|x\|$, which implies $\varkappa^{-n} f^n(x) \to 0$, since $\varkappa^{-1}(\|T_1\| + \varepsilon) < 1$. □

THEOREM A.6. *With the hypothesis of* A.5, *if* $\varkappa \leq 1$ *and* f *is* $C^{r,\alpha}$, $r \in \mathbb{N}$, $r \geq 1$, $\alpha \in [0,1]$, *then* g *is* $C^{r,\alpha}$. *If moreover* $Tf(0) = T$, *then* $Tg(0) = 0$, *hence* $W^{s,\varkappa}$ *is tangent to* E_1 .

Proof. Consider first the case $\varkappa < 1$. By Appendix B, lemma B.5, if f is $C^{r,\alpha}$ then ζ is $C^{r,\alpha}$. By lemma A.4 of this appendix, φ is also $C^{r,\alpha}$. Moreover if $Tf(0) = T$, a computation of $T\zeta$ shows that $T\zeta(0,0,0)(h_1,\alpha_1,\alpha_2) = (\beta_1,\beta_2)$, with $\beta_1(1) = T_1(h_1)$; $\beta_1(n) = T_1[\alpha_1(n-1)]$, $n \geq 2$; $\beta_2(1) = T_2^{-1}[\alpha_2(2)]$; $\beta_2(n) = T_2^{-1}[\alpha_2(n+1)]$, $n \geq 2$.

If we call $T_1\zeta$ the derivative of ζ with respect to $x_1 \in E_1$, we have :

$$T_1\zeta(0,0,0)(h_1) = [(T_1 h_1, 0, \ldots, 0, \ldots), 0] .$$

In the same way if $T_2\zeta$ is the derivative of ζ with respect to (γ_1,γ_2) in $\mathcal{B}^{\varkappa}(E_1) \times \mathcal{B}^{\varkappa}(E_2)$, we have :

$$T_2\zeta(0,0,0)(\alpha_1,\alpha_2) = [(0, T_1(\alpha_1(1)), T_1(\alpha_1(2)), \ldots), (T_2^{-1}(\alpha_2(2)), T_2^{-1}(\alpha_2(3)), \ldots)] .$$

A computation shows that $[\mathrm{Id} - T_2\zeta(0,0,0)]^{-1}(\alpha_1,\alpha_2) = (\gamma_1,\gamma_2)$ with $\gamma_1(n) = \sum_{i=1}^{n} T_1^{n-i}(\alpha_1(i))$, $\gamma_2(n) = \sum_{i=n}^{\infty} T_2^{-i+n}(\alpha_2(i))$. By lemma I.3, we have :

$$T\varphi(0)(h_1) = [(T_1 h_1, T_1^2 h_1, \ldots, T_1^n h_1, \ldots), 0] .$$

Since $g: E_1 \to E_2$ is the composite of φ with the linear map $\mathcal{B}^{\varkappa}(E_1) \times \mathcal{B}^{\varkappa}(E_2) \to E_2, (\gamma_1,\gamma_2) \to \gamma_2(1)$, we obtain that g is $C^{r,\alpha}$ if f is $C^{r,\alpha}$ and also $Tg(0) = 0$ if $Tf(0) = T$.

The case $\varkappa = 1$ can be done similarly once we replace $\mathcal{B}^1(E_i)$ by the space $C(E_i)$ of convergent sequences in order to apply Appendix B lemma B.3. □

APPENDIX B.

ON DIFFERENTIABILITY OF MAPS ON SPACES OF SEQUENCES

§ 1. One sided sequences

Let E be a Banach space and $\mu > 0$, we define $\mathcal{B}^{\mu}(E)$ as the Banach space :

$$\mathcal{B}^{\mu}(E) = \{\gamma = (\gamma(n))_{n\in\mathbb{N}} \mid \sup_{n\in\mathbb{N}} \|\mu^{-n}\gamma(n)\| < +\infty\} .$$

Remark that for $\mu < \mu'$, the inclusion $\mathcal{B}^{\mu}(E) \hookrightarrow \mathcal{B}^{\mu'}(E)$ is linear continuous of norm 1. Let us also define $\mathcal{C}(E) \subset \mathcal{B}^{1}(E)$ as the subspace consisting of convergent sequences. It is easy to see that $\mathcal{C}(E)$ is a closed linear subspace of $\mathcal{B}^{1}(E)$, hence a Banach space. If E and F are two Banach spaces and $f : E \to F$ is a continuous map, we define a map $\mathcal{C}(f) : \mathcal{C}(E) \to \mathcal{C}(F)$ by $\mathcal{C}(f)[(\gamma(n))_{n\in\mathbb{N}}] = (f[\gamma(n)])_{n\in\mathbb{N}}$.

LEMMA B.1. The map $\mathcal{C}(f)$ is well defined and continuous.

Proof. Since f is continuous, if $(\gamma(n))_{n\in\mathbb{N}}$ is convergent, $(f[\gamma(n)])_{n\in\mathbb{N}}$ is convergent ; this shows that $\mathcal{C}(f)$ is well defined. For continuity, remark that $\overline{\{\gamma(n) \mid n\in\mathbb{N}\}} = K$ is a compact subset of E. Hence by the uniform continuity of f at K, given $\varepsilon > 0$, there exists $\delta > 0$ such that if $x \in K$ and $v \in E$, with $\|v\| < \delta$, then $\|f(x+v) - f(x)\| < \varepsilon$. It follows that $\alpha \in \mathcal{C}(E)$ with $\|\alpha\| < \delta$, we have $\|\mathcal{C}(f)(\gamma+\alpha) - \mathcal{C}(f)(\gamma)\| < \varepsilon$. □

Suppose now that f is $C^{0,\alpha}$, this means that f satisfies locally a Hölder condition of order α, we have :

LEMMA B.2. If f is $C^{0,\alpha}$, then $\mathcal{C}(f)$ is $C^{0,\alpha}$.

Proof. Let $\gamma = (\gamma(n))_{n\in\mathbb{N}} \in \mathcal{C}(E)$, since the sequence is convergent and f is $C^{0,\alpha}$, it is easy to find $\varepsilon > 0$ and $K > 0$, such that if $\|x - \gamma(n)\| < \varepsilon$ and $\|y - \gamma(n)\| < \varepsilon$ for some n, then $\|f(x) - f(y)\| \le K\|x-y\|^{\alpha}$. It follows from this that $\mathcal{C}(f)$ satisfies a Hölder condition of order α on the neighborhood of γ. □

We have a continuous linear map $\mathcal{J}$ of norm one, $\mathcal{J} : \mathcal{C}(L(E,F)) \to L(\mathcal{C}(E),\mathcal{C}(F))$, $T = (T_n)_{n\in\mathbb{N}} \mapsto \mathcal{J}(T)$, with $[\mathcal{J}(T)(\gamma)](n) = T_n(\gamma(n))$.

We suppose that f is $C^{r,\alpha}$, which means that f is C^r and that the rth derivative of f satisfies a Hölder condition of order α (for $\alpha = 0$, $C^{r,\alpha}$ means C^r).

LEMMA B.3. Let $f: E \to F$ be $C^{r,\alpha}$ with $r \geq 1$, then $\mathcal{C}(f)$ is $C^{r,\alpha}$ and : $T\mathcal{C}(f) = \mathcal{J} \circ \mathcal{C}(Tf)$, which means $T_\gamma \mathcal{C}(f)(\nu) = (T_{\gamma(n)} f(\nu(n)))_{n \in \mathbb{N}}$.

Proof. Remark that $\mathcal{C}(Tf)$ is continuous, since $Tf: E \to L(E,F)$ is continuous. Hence $\mathcal{J} \circ \mathcal{C}(Tf)$ is continuous.

We show now the derivability of $\mathcal{C}(f)$. We have :

$$[\mathcal{C}(f)(\gamma+\nu) - \mathcal{C}(f)(\gamma) - \mathcal{J}\circ\mathcal{C}(Tf)(\gamma)(\nu)](n) = \int_0^1 [T_{\gamma(n)+t\nu(n)} f - T_{\gamma(n)} f](\nu(n))dt .$$

From which it follows that :

$$\|\mathcal{C}(f)(\gamma+\nu) - \mathcal{C}(f)(\gamma) - \mathcal{J}\circ\mathcal{C}(Tf)(\gamma)(\nu)\| \leq \|\nu\| \max_{\|\beta\| \leq \|\nu\|} \|\mathcal{C}(Tf)(\gamma+\beta) - \mathcal{C}(Tf)(\gamma)\| .$$

By lemma 1, $\max_{\|\beta\| \leq \|\nu\|} \|\mathcal{C}(Tf)(\gamma+\beta) - \mathcal{C}(Tf)(\gamma)\|$ goes to zero as $\|\nu\|$ goes to zero.

Up to now we have shown that $\mathcal{C}(f)$ is C^1 if f is C^1. An easy induction a argument, joined to lemma 2, shows that $\mathcal{C}(f)$ is $C^{r,\alpha}$, if f is. □

Suppose now that $f: E \to F$ is (uniformly) Lipschitz and that $f(0) = 0$. We define $\mathcal{B}^\mu(f) : \mathcal{B}^\mu(E) \to \mathcal{B}^\mu(F)$ by $(\gamma(n))_{n\in\mathbb{N}} \mapsto (f(\gamma(n)))_{n\in\mathbb{N}}$.

LEMMA B.4. The map $\mathcal{B}^\mu(f)$ is well defined and Lipschitz with $\mathrm{Lip}^1(\mathcal{B}^\mu(f)) \leq \mathrm{Lip}^1(f)$.

Proof. We have $\mathcal{B}^\mu(f)(0) \in \mathcal{B}^\mu(F)$ since $f(0) = 0$. If $\gamma, \gamma' \in \mathcal{B}^\mu(E)$, we have :

$$\|f(\gamma(n)) - f(\gamma'(n))\| \leq \mathrm{Lip}_1(f)\|\gamma(n) - \gamma'(n)\| \leq \mathrm{Lip}_1(f)\mu^n \|\gamma - \gamma'\| .$$

The rest of the proof is easy. □

We have a continuous linear map of norm one :

$$\mathcal{J}_\mu : \mathcal{B}^1(L(E,F)) \longrightarrow L(\mathcal{B}^\mu(E), \mathcal{B}^\mu(F)) , \; T = (T_n)_{n\in\mathbb{N}} \longmapsto \mathcal{J}_\mu(T) ,$$

with $[\mathcal{J}_\mu(T)(\gamma)](n) = T_n(\gamma(n))$.

We will denote by $\overline{\mathcal{J}_\mu}$ the restriction of $\mathcal{J}_\mu$ to $\mathcal{C}(L(E,F))$.

If μ is < 1, we have the inclusion $\mathcal{B}^\mu(E) \overset{k_\mu}{\hookrightarrow} \mathcal{C}(E)$. This inclusion is linear continuous of norm one.

LEMMA B.5. Suppose that f is lipschitzian and $C^{r,\alpha}$, with $r \geq 1$. If $\mu < 1$, the map $\mathcal{B}^{\mu}(f)$ is $C^{r,\alpha}$ and $T\mathcal{B}^{\mu}f = \overline{\mathcal{J}}_{\mu} \circ C(Tf) \circ k_{\mu}$, which means $[T_{\gamma}\mathcal{B}^{\mu}(f)(\nu)](n) = T_{\gamma(n)} f(\nu(n))$, $n \in \mathbb{N}$.

Proof. By lemma 3, the map $C(Tf)$ is $C^{r-1,\alpha}$. Since $\mathcal{J}_{\mu}$ and k_{μ} are linear and continuous, the map $\overline{\mathcal{J}}_{\mu} \circ C(Tf) \circ k_{\mu}$ is $C^{r-1,\alpha}$.

We have now to show that $\overline{\mathcal{J}}_{\mu} \circ C(Tf) \circ k_{\mu}$ is the derivative of $\mathcal{B}^{\mu}(f)$. We have :

$$[\mathcal{B}^{\mu}(f)(\gamma+\nu) - \mathcal{B}^{\mu}(f)(\gamma) - [\overline{\mathcal{J}}_{\mu} \circ C(Tf) \circ k_{\mu}(\gamma)](\nu)](n) =$$

$$= f(\gamma(n)+\nu(n)) - f(\gamma(n)) - T_{\gamma(n)} f(\nu(n)) = \int_0^1 [T_{\gamma(n)+t\nu(n)} f - T_{\gamma(n)} f](\nu(n))\, dt .$$

But :

$$\sup_{n \in \mathbb{N}} \left\| \mu^{-n} \int_0^1 [T_{\gamma(n)+t\nu(n)} f - T_{\gamma(n)} f](\nu(n))\, dt \right\| \leq$$

$$\leq \|\nu\| \max_{\|\beta\| \leq \|\nu\|} \|C(Tf) \circ k_{\mu}(\gamma+\beta) - C(Tf) \circ k_{\mu}(\gamma)\| .$$

Since $C(Tf)$ and k_{μ} are continuous, the quantity $\max_{\|\beta\| \leq \|\nu\|} \|C(Tf) \circ k_{\mu}(\gamma+\beta) - C(Tf) \circ k_{\mu}(\gamma)\|$ goes to zero as $\|\nu\|$ goes to zero. □

§ 2. Two sided sequences

Let E be a Banach space and $\lambda > 0$, we denote by $\mathcal{S}^{\lambda}(E)$ the Banach space :

$$\mathcal{S}^{\lambda}(E) = \{(\gamma(n))_{n \in \mathbb{Z}} \mid \|\gamma\| = \sup \|\lambda^{-|n|} \gamma(n)\| < +\infty\} .$$

Suppose F is another Banach space, $\mu > 0$, and $(T_n)_{n \in \mathbb{Z}}$ is a sequence of continuous linear maps $E \to F$; define $\mathcal{J} : \mathcal{S}^{\lambda}(E) \to \mathcal{S}^{\mu}(F)$ by $\mathcal{J}[(\gamma(n))_{n \in \mathbb{Z}}] = (T_n(\gamma(n)))_{n \in \mathbb{Z}}$. We have the following lemma :

LEMMA B.6. If $(T_n)_{n \in \mathbb{Z}} \in \mathcal{S}^{\mu/\lambda}(L(E,F))$, then $\mathcal{J}$ is well defined and continuous. Moreover the map $\mathcal{S}^{\mu/\lambda}(L(E,F)) \to L(\mathcal{S}^{\lambda}(E), \mathcal{S}^{\mu}(F)), (T_n)_{n \in \mathbb{Z}} \to \mathcal{J}$ is a continuous linear map of norm ≤ 1.

Proof. We have :

$$\|\mathcal{J}(\gamma)\| = \sup_{n \in \mathbb{Z}} \mu^{-|n|} \|T_n(\gamma(n))\| \leq \sup_{n \in \mathbb{Z}} \mu^{-|n|} \|T_n\| \|\gamma(n)\|$$

$$\leq \left(\sup_{n \in \mathbb{Z}} \left(\frac{\mu}{\lambda}\right)^{-|n|} \|T_n\|\right) \|\gamma\| \qquad □$$

Suppose now that we have a sequence of maps $f_n : E \to F$, $n \in \mathbb{Z}$, each satisfying a global Hölder condition of order $\alpha \in]0,1]$, with Hölder constant $\mathrm{Lip}^{\alpha}(f_n)$ i.e. $\forall x,y \in E$, $\|f_n(x) - f_n(y)\| \leq \mathrm{Lip}^{\alpha}(f_n)\|x-y\|^{\alpha}$.

We try to define $\mathcal{F} : \mathcal{S}^{\lambda}(E) \to \mathcal{S}^{\mu}(F)$ by $\mathcal{F}[(\gamma(n))_{n\in\mathbb{Z}}] = (f_n)(\gamma(n)))_{n\in\mathbb{Z}}$. We have the following lemma.

LEMMA B.7. The map $\mathcal{F}$ is well defined and continuous provided : $\sup_{n\in\mathbb{Z}} \mu^{-|n|}\|f_n(0)\| < +\infty$ and $\sup_{n\in\mathbb{Z}} [(\frac{\mu}{\lambda^{\alpha}})^{-|n|}\mathrm{Lip}^{\alpha}(f_n)] < +\infty$. Moreover the map $\mathcal{F}$ satisfies a global Hölder condition of order α and $\mathrm{Lip}^{\alpha}(\mathcal{F}) \leq \sup_{n\in\mathbb{Z}} [(\frac{\mu}{\lambda^{\alpha}})^{-|n|}\mathrm{Lip}^{\alpha}(f_n)]$.

Proof. We have :

$$\mu^{-|n|}\|f_n(\gamma(n)) - f_n(\gamma'(n))\| \leq \mu^{-|n|}\mathrm{Lip}^{\alpha}(f_n)\|\gamma(n)-\gamma'(n)\|^{\alpha}$$
$$\leq \mu^{-|n|}\lambda^{\alpha|n|}\mathrm{Lip}^{\alpha}(f_n)\|\gamma-\gamma'\|^{\alpha} .$$

The lemma is an easy consequence of this inequality. □

LEMMA B.8. Suppose that each f_n is $C^{1,\alpha}$ and that we have :

$$\sup_{n\in\mathbb{Z}}[\mu^{-|n|}\|f_n(0))\|] < +\infty, \quad \sup_{n\in\mathbb{Z}}[(\tfrac{\mu}{\lambda})^{-|n|}\|Tf_n\|_{C^0}] < +\infty \quad \text{and}$$

$\sup_{n\in\mathbb{Z}}[(\frac{\mu}{\lambda^{1+\alpha}})^{-|n|}\mathrm{Lip}^{\alpha}(Tf_n)] < +\infty$. Then the map $\mathcal{F}$ is $C^{1,\alpha}$. Moreover $T\mathcal{F}$ is equal to the composition :

$$T\mathcal{F} : \mathcal{S}^{\lambda}(E) \xrightarrow{(Tf_n)} \mathcal{S}^{\mu/\lambda}(L(E,F)) \longrightarrow L(\mathcal{S}^{\lambda}(E),\mathcal{S}^{\mu}(E)) ,$$

and

$$\|T\mathcal{F}\|_{C^0} \leq \sup_{n\in\mathbb{Z}}[(\tfrac{\mu}{\lambda})^{-|n|}\|Tf_n\|_{C^0}],\ \mathrm{Lip}^{\alpha}(T\mathcal{F}) \leq \sup_{n\in\mathbb{Z}}[(\tfrac{\mu}{\lambda^{1+\alpha}})^{-|n|}\mathrm{Lip}^{\alpha}(Tf_n)]$$

Proof. Since $\mathrm{Lip}^1(f_n) \leq \|Tf_n\|_{C^0}$, we already know by lemma B.7 that $\mathcal{F}$ is lipschitzian. We first check that the formula we have given for $T\mathcal{F}$ defines a map which is continuous, satisfies a Hölder condition of order α and that :

$$\|T\mathcal{F}\|_{C^0} \leq \sup_{n\in\mathbb{Z}}[(\tfrac{\mu}{\lambda})^{-|n|}\|Tf_n\|_{C^0}] ;$$

$$\mathrm{Lip}^{\alpha}(T\mathcal{F}) \leq \sup_{n\in\mathbb{Z}}[(\tfrac{\mu}{\lambda^{1+\alpha}})^{-|n|}\mathrm{Lip}^{\alpha}(Tf_n)] .$$

Since $T\mathcal{F}$ is the composite of $(Tf_n)_{n\in\mathbb{Z}} : \mathcal{S}^{\lambda}(E) \to \mathcal{S}^{\mu/\lambda}(L(E,F))$ and the linear map of norm ≤ 1, $\mathcal{S}^{\mu/\lambda}(L(E,F)) \to L(\mathcal{S}^{\lambda}(E),\mathcal{S}^{\mu}(F))$, it suffices to check the analogous statements for $(Tf_n)_{n\in\mathbb{Z}}$. But the statements on $(Tf_n)_{n\in\mathbb{Z}}$ are easy consequences of the last lemma and the inequalities assumed in the hypothesis. The fact that $T\mathcal{F}$ is indeed the derivative of $\mathcal{F}$ follows from the following computation :

$$\|f_n(\gamma(n)+h(n)) - f_n(\gamma(n)) - Tf_n(\gamma n)(h(n))\|$$
$$\leq \left[\int_0^1 \|Tf_n(\gamma(n)+th(n)) - Tf_n(\gamma(n))\| \, dt\right] \|h(n)\|$$
$$\leq \operatorname{Lip}^{\alpha}(Tf_n) \, \|h(n)\|^{\alpha} \, \|h(n)\| \quad .$$

Hence :

$$\|\mathcal{F}(\gamma+h) - \mathcal{F}(\gamma) - T\mathcal{F}(\gamma)(h)\| \leq \|h\| \left[\sup \left(\frac{\mu}{\lambda^{1+\alpha}}\right)^{-|n|} \operatorname{Lip}^{\alpha}(Tf_n)\right] \|h\|^{\alpha} \quad . \qquad \square$$

<u>LEMMA B.9</u>. <u>Suppose that each</u> f_n <u>is</u> $C^{r,\alpha}$, <u>with</u> :

$$\sup_{n\in\mathbb{Z}} [\mu^{-|n|} \|f_n(0)\|] < +\infty , \quad \sup_{n\in\mathbb{Z}} \left[\left(\frac{\mu}{\lambda^j}\right)^{-|n|} \|T^j f_n\|_{C^0}\right] < +\infty \quad , \; j = 1,\dots,r ,$$

$$\sup_{n\in\mathbb{Z}} \left\{\left(\frac{\mu}{\lambda^{r+\alpha}}\right)^{-|n|} \operatorname{Lip}_{\alpha}(T^r f_n)\right\} < +\infty .$$ Then $\mathcal{F}$ is $C^{r,\alpha}$ and

$$\|T^j\mathcal{F}\|_{C^0} \leq \sup \left[\left(\frac{\mu}{\lambda^j}\right)^{-|n|} \|T^j f_n\|_{C^0}\right] , \; j = 1,\dots,r ,$$

$$\operatorname{Lip}^{\alpha}(T^r\mathcal{F}) \leq \sup_{n\in\mathbb{Z}} \left(\frac{\mu}{\lambda^{r+\alpha}}\right)^{-|n|} \operatorname{Lip}^{\alpha}(T^r f_n) \quad .$$

<u>Proof</u>. The proof is by induction on r , using last lemma. $\square$

APPENDIX C.

ON MEASURABILITY

The goal of this appendix is to give a proof of proposition 14. We will in fact give a proof of a more general result. First, we have to recall some facts.

THEOREM C.1. Let X be a Polish space, i.e. separable metric complete and Y be a separable metric space. Let $A \subset X$ be a Borel subset and $f : A \hookrightarrow Y$ be a Borel injective map. Then $f(A)$ is a Borel subset of Y, and f is an isomorphism of the Borel structure of A on that of $f(A)$.

For a proof, see [Bo], [Co] or [Pa].

Let M and N be two C^∞-manifolds (separable, metrizable). On $C^r(M,N)$, $r = 0,1,\ldots,+\infty$, we put the topology of uniform convergence on compact sets of derivatives up to order r. The topology on $C^{r,\alpha}(M,N)$ is given by the C^r-topology plus the α-Hölder topology on the rth derivative on each compact subset. It is well known that this topology is metric complete, but it is not separable unless $\alpha = 0$, i.e. $C^r(M,N)$ is separable. For $\alpha' < \alpha$, $i : C^{r,\alpha}(M,N) \hookrightarrow C^{r,\alpha'}(M,N)$ and the image is a σ-compact subset of $C^{r,\alpha'}(M,N)$, in particular $i(C^{r,\alpha}(M,N))$ is a Borel subset of the complete separable metric space $i(C^{r,\alpha}(M,N))$.

THEOREM C.2.. A map $A \to C^r(M,N)$ is Borel if and only if $A \to C^r(M,N) \hookrightarrow C^0(M,N)$ is Borel.

Proof. This is a consequence of theorem 1 since $C^r(M,N) \hookrightarrow C^0(M,N)$ is continuous and both spaces are complete metric and separable. □

THEOREM C.3. Let $\alpha' < \alpha$ and $r \geq 0$, a map $A \to C^{r,\alpha}(M,N)$ is Borel for the $C^{r,\alpha'}$ topology on $C^{r,\alpha}(M,N)$, if and only if $A \to C^{r,\alpha}(M,N) \hookrightarrow C^0(M,N)$ is Borel.

Proof. We know that for the $C^{r,\alpha'}$ topology, $C^{r,\alpha}(M,N)$ is a Borel subset of a complete metrizable separable space, so we can apply theorem C.1. □

THEOREM C.4. Let A be a Borel set in a Polish space, and $f : A \to C^0(M,N)$, $x \mapsto f_x$ be a map. The following conditions are equivalent :

1) The map f is Borel ;

2) For each m in M , the map $A \to N$, $x \to f_x(m)$ is Borel ;

3) There exists a dense sequence $(m_i)_{i \in \mathbb{N}}$ of points in M , such that, for each $i \in \mathbb{N}$, the map $A \to N$, $x \to f_x(m_i)$ is Borel.

Proof. 1) $\Rightarrow$ 2) follows from the continuity of the map $C^0(M,N) \to N$, $f \mapsto f(m)$, for each $m \in M$.

2) $\Rightarrow$ 3) is clear.

3) $\Rightarrow$ 1) Consider the map $\theta : C^0(M,N) \to N^{\mathbb{N}}$, $f \mapsto (f(m_i))_{i \in \mathbb{N}}$. This map is continuous, it is also injective since $(m_i)_{i \in \mathbb{N}}$ is dense in M . Moreover, $N^{\mathbb{N}}$ is Polish. Now the implication 3) $\Rightarrow$ 1) is an easy consequence of theorem 1 . □

THEOREM C.5. Let Map (A,M) (resp. Map (A,N)) be the set of all maps from X to M (resp. N). Let $f : A \to C^0(M,N)$, $x \mapsto f_x$, be a map. We define $\mathfrak{F}$: Map (A,M) $\to$ Map (A,N) by $\mathfrak{F}(\theta)(x) = f_x(\theta(x))$, $\theta \in$ Map (A,M) , $x \in A$. If $\mathfrak{F}$ sends Borel maps to Borel maps, then f is Borel.

Proof. If $m \in M$, let θ_m be the map $A \to M$, $x \to m$. By hypothesis $\mathfrak{F}(\theta_m) : X \to N$, $x \mapsto f_x(m)$, is Borel. By theorem 4, the map f is Borel. □

Using the notations of proposition 14 and the fact that $E^s \to B$ and $E^u \to B$ are trivial as Borel bundles, we obtain the following

THEOREM C.6. For each $x \in B$, let ψ_x be a map $E^s_x \to E^u_x$, which is $C^{r,\alpha}$. Suppose that the induced map from sections of $E^s \to B$ to sections of $E^u \to B$ transforms Borel sections to Borel sections. Then ψ_x depends measurably of x for the $C^{r,\alpha'}$-topology for each $\alpha' < \alpha$. Moreover if $\alpha = 0$, the map $x \mapsto \psi_x$ depends measurably on x for the C^r-topology.

REFERENCES

[Bo] N. BOURBAKI, Topologie générale, chap. IX, Actualités scientifiques et industrielles 1045, Hermann (1958), Paris.

[Co] D.L. COHN, Measure theory, Birkhäuser (1980), Boston, Basel & Stuttgart.

[Hi] M.W. HIRSCH, Differential Topology, Graduate texts in Mathematics 33, Springer Verlag (1976), Berlin, Heideberg & New-York.

[Ir] M.C. IRWIN, On the smoothness of the composition map, Quart. J. of Math., 23 (1972), p. 113-133.

[Os] V.I. OSELEDEC, Multiplicative ergodic theorem, Ljapunov characteristic numbers for dynamical systems, Trudy Moskov. Mat. Obsc 19 (1968), p. 179-210, english translation, Trans. Mòscow Math. Soc. 19 (1968), p. 197-231.

[Pa] K.R. PARTHASARATHY, Probability measures on metric spaces, Academic Press (1967), New-York & London.

[Pe_1] Y.B. PESIN, Families of invariant manifolds corresponding to non zero characteristic exponents, Izv. Akad. Nauk SSSR, ser. mat. 40 (1976), p. 1332-1379, English translation Math. USSR Izvestija 10 (1976), p. 1261-1305.

[Pe_2] Y.B. PESIN, Characteristic Lyapunov exponents and smooth ergodic theory, Uspekhi Mat. Nauk. 32:4 (1977), p. 55-114, english translation Russian Math., Surveys 32:4 (1977), p. 55-112.

[PS] C. PUGH & M. SHUB,

[Ra] M.S. RAGHUNATHAN, Proof of Oseledec's multiplicative ergodic theorem, Israël J. of Math. 32 (1979), p. 356-362.

[Ru] D. RUELLE, Ergodic theory of differentiable dynamical systems, Publ. Math. IHES 50 (1979), p. 27-58.

M.R. HERMAN
Centre de Mathématique
de l'Ecole Polytechnique
91128 PALAISEAU - France

A. FATHI and J.C. YOCCOZ
Université Paris-Sud,
Mathématique, bâtiment 425
91405 ORSAY - France

PROPRIETÉS DE GÉNÉRICITÉ DES TRANSFORMATIONS CANONIQUES

Jean-Pierre FRANÇOISE

I. INTRODUCTION

Considérons $\mathbb{R}^{2m}$ sur lequel on définit les coordonnées linéaires (x_i,y_i), $i = 1,\ldots,m$, notées (x,y) en abrégé et la forme symplectique:

$$\omega = \sum_{i=1}^{m} dx_i \wedge dy_i = dx \wedge dy .$$

Introduisons les sous-espaces de codimension 2: $A_i = \{(x,y) \in \mathbb{R}^{2m} / x_i = y_i = 0\}$, que nous appellerons par abus de langage, les axes du système. Et soit $\Omega = \mathbb{R}^{2m} \setminus \bigcup_{i=1}^{m} A_i$.

Sur l'ouvert Ω on utilise le système de coordonnées "action angle" (p_i,q_i), noté (p,q) en abrégé, relié au précédent par les relations:

$$\text{(I.1)} \qquad p_i = (x_i^2+y_i^2)/2, \qquad x_i = (2p_i)^{\frac{1}{2}} \cos q_i, \quad y_i = (2p_i)^{\frac{1}{2}} \sin q_i .$$

Les coordonnées (p_i,q_i) définissent un difféomorphisme de Ω sur $\mathbb{R}_+^m \times \mathbb{T}^m$. Le type d'homotopie de Ω est donc celui de $\mathbb{T}^m$; $\pi_1(\Omega) \simeq \mathbb{Z}^m$.

En restriction à Ω, $\omega = \sum_{i=1}^{m} dp_i \wedge dq_i = dp \wedge dq$.

Dans $\mathbb{C}^{2m}$ rapporté à un système de coordonnées (x_i,y_i), on construit aussi les coordonnnées "action angle" sur l'ouvert constitué des points vérifiant, pour tout i, $\mathrm{Re}(x_i^2+y_i^2) > 0$;

sur cet ouvert le choix de la détermination de la racine carrée des p_i, obtenue après une coupure suivant l'axe réel négatif, donne, en effet, sens aux relations I.1. On définit la transformation $F_o|_\Omega : \Omega \to \mathbb{R}^{2m}$; en dehors des axes par les relations:

$$F_o|_\Omega : \begin{matrix} p_i \\ q_i \end{matrix} \longmapsto \begin{matrix} p_i^1 = p_i \\ q_i^1 = q_i + c_i + \sum_{j=1}^{m} Q_{ij}p_j \end{matrix} .$$

Avec $c = (c_i) \in \mathbb{R}^m$, $Q = (Q_{ij})$ matrice symétrique supposée inversible.

(Il convient de remarquer que la deuxième série de relations qui définit $F_o|_\Omega$ est à lire dans le revêtement universel du tore $\mathbb{T}^m : \mathbb{R}^m$.)

Ecrivons l'expression de $F_o|_\Omega$ dans les coordonnées linéaires (x_i, y_i)

$$(\text{I.2}) \qquad F_o|_\Omega \; \Big| \; \begin{matrix} x_i \\ \\ y_i \end{matrix} \longmapsto \begin{matrix} x_i^1 = (x_i \cos c_i - y_i \sin c_i)\cos(Q_i^j p_j) \\ \qquad - (y_i \cos c_i + x_i \sin c_i)\sin(Q_i^j p_j) \\ y_i^1 = (y_i \cos c_i + x_i \sin c_i)\cos(Q_i^j p_j) \\ \qquad + (x_i \cos c_i - y_i \sin c_i)\sin(Q_i^j p_j) \end{matrix} .$$

Cette seconde écriture de F_o est valable dans $\mathbb{R}^{2m}$ tout entier et (I.2) sera prise comme définition de F_o dans $\mathbb{R}^{2m}$.

Dans la suite, nous utiliserons une notation condensée dont voici un exemple:

$$(\text{I.3}) \qquad F_o|_\Omega : \begin{matrix} p \\ q \end{matrix} \longmapsto \begin{matrix} p^1 = p \\ q^1 = q + c + Qp \end{matrix} .$$

De la définition de F_o résulte immédiatement deux conséquences:

(1) F_o est symplectique: $F_o^*\omega = \omega$ (du fait que Q est symétrique)

(2) F_o s'étend à $\mathbb{C}^{2m}$ en une transformation analytique.

Remarquons que (I.2) a pour conséquence que la partie linéaire de F_o (que je désignerai par F_1 dans la suite) est la multi-rotation d'angles c_i.

F_1 se diagonalise donc dans les coordonnées isotropes et admet pour valeurs propres:

$$\lambda_i = e^{\sqrt{-1}\, c_i} \ (i=1,\dots,m), \qquad \lambda_{i+m} = e^{-\sqrt{-1}\, c_i} \ (i=1,\dots,m).$$

Je noterai D_1 le polydisque unité ouvert de $\mathbb{C}^{2m}$:

$$D_1 = \{(x,y) \in \mathbb{C}^{2m} / |x_i| < 1 \ \text{ et } \ |y_i| < 1\} \ ;$$

$\bar{D}_1$ le polydisque fermé.

DEFINITION - $\mathcal{F}_{c,Q,R}$ <u>sera l'ensemble des transformations symplectiques réelles tangentes à</u> F_o <u>modulo l'ordre</u> 4 <u>(c'est-à-dire que</u> F <u>étant un élément de</u> $\mathcal{F}_{c,Q,R}$, <u>la différence</u> $F-F_o$ <u>a pour fonctions coordonnées des éléments de la 4ième puissance de l'idéal maximal des fonctions analytiques nulles à l'origine) qui se prolongent à</u> $\mathbb{C}^{2m}$ <u>en des transformations analytiques dans</u> D_1 <u>et vérifiant</u>

$$F \in \mathcal{F}_{c,Q,R} \Rightarrow \|F-F_o\|_{\bar{D}_1} = \sup_{z\in \bar{D}_1} |F(z)-F_o(z)| < R .$$

Etant donné un élément F de $\mathcal{F}_{c,Q,R}$ on s'intéresse à ses itérés successifs $F, F^2, F^3, \dots, F^k, \dots$

$$F^k : \begin{matrix} x_i \\ y_i \end{matrix} \longmapsto \begin{matrix} x_i^k \\ y_i^k \end{matrix} .$$

Pour les transformations symplectiques de classe C^4, avec une condition arithmétique bénigne et la même condition d'inversibilité de la métrique Q, J. Moser démontre dans [4] l'existence de suites de points périodiques au voisinage du point fixe $\underline{o}$.

Nous allons donner une autre preuve de le résultat qui ne vaut que dans le cadre analytique et qui fait apparaître des points périodiques "munis d'un indice ν" suivant un schéma de démonstration élaboré par J. Vey dans [12] pour les champs de vecteurs qui utilise aussi de façon essentielle la méthode de la fonction génératrice prônée par Poincaré (cf [4]). Le rôle de cet indice ν se révele déterminant dans la preuve des propriétés de génericité qui viennent dans la suite.

Le premier résultat que l'on donc a en vue (théorème III.1) est de prouver que lorsqu'un élément ν de $2\pi\mathbb{Z}^m$ et un entier N satisfont certaines conditions d'association, un élément F de $\mathcal{F}_{c,Q,R}$ possède des points périodiques de période N et d'indice ν. De façon plus précise:

DEFINITION I.1. - Un élément ν de $2\pi\mathbb{Z}^m$ et un entier positif N sont dits associés par β et γ s'ils satisfont:

(i) $\nu - Nc \in QP'_o$

(ii) $|\nu-Nc| < \beta N^{1/5}$

(iii) $N > \gamma$

où P'_o désigne un cône convexe strictement contenu dans $\mathbb{R}^m_+$, β et γ sont deux constantes réelles.

DEFINITION I.2. - Un point M $(M=(p,q))$ de Ω est dit point périodique d'un élément F de $\mathcal{F}_{c,Q,R}$, de période N et d'indice ν si:

1º) <u>les itérés successifs de</u> M <u>par</u> F <u>sont tous dans</u> Ω <u>et, si</u> (p^i, q^i) <u>désignent les coordonnées action-angle de</u> $F^i(M)$;

2º) $p^N = p, \quad q^N = q+\nu$.

THEOREME III.4. - <u>On peut déterminer</u> β <u>et</u> γ <u>uniformes pour tous les éléments</u> F <u>de</u> $\mathfrak{F}_{c,Q,R}$ <u>telles que si</u> ν <u>et</u> N <u>sont associés par</u> β <u>et</u> γ, <u>tout élément</u> F <u>de</u> $\mathfrak{F}_{c,Q,R}$ <u>possède au moins un point périodique de période</u> N <u>et d'indice</u> ν.

Pour démontrer le théorème III.4, il est nécessaire d'obtenir un certain nombre de majorations qui sont fournies par l'extension complexe et l'utilisation de la formule de Cauchy; c'est l'objet du paragraphe II.

Les estimations du paragraphe II permettent la construction de la fonction génératrice et de la fonction séculaire au paragraphe III.

Le théorème III.4 résulte alors de l'étude de l'ensemble critique de la fonction séculaire.

Dans le IVème paragraphe, on utilise les points périodiques pour démontrer les trois propositions suivantes:

PROPOSITION IV.5. - <u>Si les valeurs propres</u> $\lambda_1, \lambda_2, \dots, \lambda_{m+1}, \dots, \lambda_{2m}$ <u>de</u> F_1 <u>satisfont</u>

$$\lambda_k \cdot \lambda_\ell = 1 \Leftrightarrow |k-\ell| = m \quad \text{(condition } A_1\text{)},$$

<u>un élément de</u> $\mathfrak{F}_{c,Q,R}$ <u>n'est pas génériquement l'exponentielle d'un champ hamiltonien</u>.

PROPOSITION IV.6. - Si les valeurs propres de F_1 ne vérifient aucune relation multiplicative sur $\mathbb{Z}$ autre que celles déduites des relations $\lambda_i \cdot \lambda_{i+m} = 1$, $i=1,\dots,m$ (ce que je noterai condition A_2 dans la suite), un élément F de $\mathfrak{F}_{c,Q,R}$ n'admet pas génériquement de fonction invariante non constante.

IV.6 implique la divergence générique de la forme normale des transformations de $\mathfrak{F}_{c,Q,R}$ définie par G.D. Birkhoff puis S. Sternberg dans [11] (voir également [2] à ce sujet). Dans le cas de la dimension deux, H. Rüssmann avait déjà obtenu ce résultat dans [7]. Pour la question des intégrales premières d'un champ hamiltonien on consultera les articles de J. Möser[5], J. Möser et W. Kyner [6].

Remarquons que IV.6, avec une hypothèse plus forte généralise le résultat de IV.5; puisque si une transformation est l'exponentielle d'un champ hamiltonien, elle admet l'hamiltonien pour fonction invariante.

PROPOSITION IV.7. - Si F_1 est l'identité, un élément de $\mathfrak{F}_{c,Q,R}$ n'est pas génériquement l'exponentielle d'un champ hamiltonien.

Le paragraphe V s'articule ainsi: à un élément F de $\mathfrak{F}_{c,Q,R}$ on associe la transformation formelle symplectique $\hat{F}$ de C^{2m} obtenue en écrivant les développements de Taylor des fonctions composantes de F au voisinage de l'origine.

En s'appuyant sur un théorème de Lewis démontré par Sternberg dans [10] :

THEOREME V.1. - Soit $\hat{F}$ une transformation formelle de $\mathbb{C}^{2m}$; $\lambda_1,\dots,\lambda_{2m}$ les valeurs propres de la partie linéaire F_1. S'il existe un logarithme des λ_i : $k_i = \log \lambda_i$ $(i=1,\dots,2m)$ satisfaisant la condition (S):

(S) "si $\sum_i n_i k_i = 2\pi\sqrt{-1}n$, $n_i \in \mathbb{Z}$, $n \in \mathbb{Z}$. Alors $n = 0$ ".

$\hat{F}$ est l'exponentielle d'un champ forme $\hat{X}$ (uniquement déterminé par le choix du logarithme des λ_i).

De plus, si $\hat{F}$ est symplectique, $\hat{X}$ est hamiltonien.

On peut alors écrire:

PROPOSITION V.2. - Soit F un élément de $\mathcal{F}_{c,Q,R}$: si les valeurs propres de F_1 satisfont la condition A_2 ou sont toutes égales à 1, $\hat{F}_1$ est l'exponentielle d'un champ formel hamiltonien (on prend pour logarithme des λ_i celui obtenu après coupure suivant l'axe réel négatif).

Le théorème V.1 permet de préciser l'énoncé des propositions IV.5 et IV.7 de la façon suivante:

PROPOSITION V.3. - Si les valeurs propres de F_1 satisfont A_2 ou sont toutes égales à 1; un élément F de $\mathcal{F}_{c,Q,R}$ est "formellement" l'exponentielle d'un champ hamiltonien $\hat{X}$. Et génériquement sur les termes de F d'ordres supérieurs ou égaux à 4, les séries entières associées à $\hat{X}$ ont des domaines de convergence d'intersection réduit à l'origine.

Pour ce qui est de la bibliographie, outre celle déjà indiquée on peut se reporter à l'article d'H. Rüssmann [8] pour une étude générale des formes normales de difféomorphismes en analogie avec la théorie de Brjuno [1] pour les champs de vecteurs.

Le résultat analogue à la proposition IV.6 pour les champs de vecteurs est un célèbre théorème de C.L. Siegel [9]. En relation directe avec ces sujets et pour une étude plus récente et systématique on consultera les articles de E. Zehnder [13].

Les résultats présentés ici furent annoncés par une note à l'Académie des Sciences [2] et ont fait partie d'une thèse à l'Université de Grenoble sous la direction de J. Vey.

II. MAJORATIONS

Dans ce paragraphe, on met à profit l'extension complexe d'un élément F de $\mathfrak{F}_{c,Q,R}$ pour effectuer diverses majorations.

§1. Précisons d'abord quelques notations:

Étant donné z un élément de $\mathbb{C}^{2m}$; $z = (z_1,\dots,z_{2m}) = (z_i)_{i=1,\dots,2m}$. On pose $|z| = \sup_{i=1,\dots,2m} |z_i|$, $\operatorname{Re} z = (\operatorname{Re} z_i)_{i=1,\dots,2m}$, $\operatorname{Im} z = (\operatorname{Im} z_i)_{i=1,\dots,2m}$.

Soit P_1 un cône convexe strictement contenu dans $\mathbb{R}_+^m$, a, ρ et δ trois nombres strictement positifs. Avec:

$$a < 1 \quad \text{et} \quad \rho+1 < \left(\frac{3}{2}\right)^{4/5}.$$

On désigne par $D(P_1,a,\rho,\delta)$ l'ouvert de $\mathbb{C}^{2m}$ ainsi constitué

$$D(P_1,a,\rho,\delta) = \{(x,y)\in\mathbb{C}^{2m};\ \operatorname{Re} p\in P_1, |\operatorname{Im} p| < \rho|\operatorname{Re} p|^{5/4}, |\operatorname{Im} q| < \delta, |p| < a\}.$$

Remarquons que P_1 étant inclus dans $\mathbb{R}_+^m$, um point (x,y) de $\tilde{D}(P_1,a,\rho,\delta)$ peut être repéré par ses coordonnées "action-angle" (p,q).

Je désignerai par p^k la coordonnée action (en fait condensé pour p_i^k, $i = 1,\dots,m$) du point $F^k(p,q)$; k-ième itéré par F du point (p,q) et lorsqu'elle existe par q^k la coordonnée angulaire de $F^k(p,q)$.

LEMME 1. - Soit P_2 un autre cône convexe contenu strictement dans $\mathbb{R}_+^m$ et contenant P_1. On se convainc facilement de l'existence d'une constante $k > 0$ ne dépendant que des deux cônes P_1 et P_2 telle que

$$\text{"}\forall\ (u,v) \in \mathbb{R}_+^m \times \mathbb{R}_+^m,\quad u \in P_1 \quad \text{et} \quad |u-v| < k|u|\text{"}$$

implique que v appartient à P_2.

LEMME 2. - Il existe une constante Γ_1 (resp. Γ_2) uniforme pour tous les éléments F de $\mathcal{F}_{c,Q,R}$ telle que pour tout point (p,q) de $D(P_1,a,\rho,\delta)$ (resp. $D(P_2,2a,2\rho,2\delta)$): $|p^1(p,q)-p| < \Gamma_1|p|^{5/2}$ (resp. $|p^1(p,q)-p| < \Gamma_2|p|^{5/2}$). Γ_1 est une fonction croissante de a, dans la suite, on suppose a suffisamment petit pour que $\Gamma_1|a|^{5/4} < k$.

LEMME 3. - Si $\operatorname{Re} p^1$ appartient à P_2, q^1 a un sens, $q^1-q-c-Qp$ est une fonction analytique sur $D(P_1,a,\rho,\delta)$ en les coordonnées linéaires et il existe une constante Γ' indépendante de F, élément de $\mathcal{F}_{c,Q,R}$ considéré, telle que:

$$|q^1-q-c-Qp| < \Gamma'|p|^{3/2}.$$

Les lemmes 2 et 3 sont impliqués par le fait que nous avons supposé F tangente à F_o modulo l'ordre 4, un calcul simple montre que p^1 est tangente à p à l'ordre 5 tandis que q^1 est tangente à $q+c+Qp$ à l'ordre 3. Γ_1 (resp. Γ_2) et Γ' sont uniformes pour tout les F de $\mathcal{F}_{c,Q,R}$ parce que $\|F-F_o\|_{\bar{D}_1} < R$. Il faut ajouter que dans les domaines $D(P_1,a,\rho,\delta)$ les quantités $\frac{|x_i|}{p}$ et $\frac{|y_i|}{p}$ sont bornées.

LEMME 4. - Si (p,q) est un point de $D(P_1,a,\rho,\delta)$, alors $|p| < (\frac{3}{2})^{4/5}|\operatorname{Re} p|$. En effet: $|p| \le |\operatorname{Im} p| + |\operatorname{Re} p|$, $|p| \le \rho|\operatorname{Re} p|^{5/4} +$

$+ |\mathrm{Re}\ p|$. <u>Mais</u> $|\mathrm{Re}\ p| \le |p| < a < 1$. <u>Donc</u>

$$|p| \le (\rho+1)|\mathrm{Re}\ p| < (\tfrac{3}{2})^{4/5}\ |\mathrm{Re}\ p|.$$

LEMME 5. - <u>On considère à nouveau un cône convexe</u> P_2 <u>strictement contenu dans</u> $\mathbb{R}_+^m$ <u>et contenant</u> P_1; <u>il existe une constante</u> ε <u>qui dépend de</u> a, ρ, P_1 <u>et</u> P_2 <u>telle que: pour tout point</u> (p,q) <u>de</u> $D(P_1, a, \rho, \delta)$ <u>si</u>

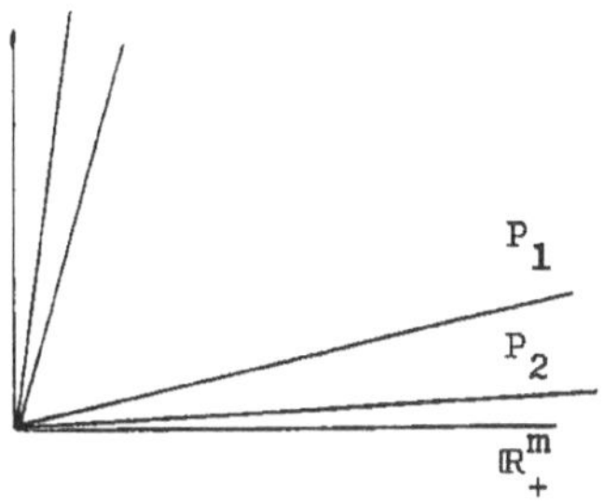

(figure 1)

i) $|\tilde{p}-p| < \varepsilon |p|^{5/4}$

ii) $|\mathrm{Im}\ \tilde{q}| < 2\delta$.

<u>Alors</u> $(\tilde{p},\tilde{q})$ <u>est un point de</u> $D(P_2, 2a, 2\rho, 2\delta)$.

Démonstration. - Introduisons la constante $k = k(P_1, P_2)$ du lemme 1 et remarquons que:

$$|\mathrm{Re}\ \tilde{p} - \mathrm{Re}\ p| \le |\tilde{p}-p| < \varepsilon |p|^{5/4} < \tfrac{3}{2}\ \varepsilon |\mathrm{Re}\ p|^{5/4}$$

d'après le lemme précédent, de sorte qu'en exigeant $\frac{3}{2}\varepsilon < k$ on est assuré que $\mathrm{Re}\ \tilde{p}$ appartient à P_2. $|\mathrm{Im}\ \tilde{p} - \mathrm{Im}\ p| \le |\tilde{p}-p| < \varepsilon |p|^{5/4}$ implique $|\mathrm{Im}\ \tilde{p}| < |\mathrm{Im}\ p| + \varepsilon |p|^{5/4}$ soit $|\mathrm{Im}\ \tilde{p}| < \rho |\mathrm{Re}\ p|^{5/4} + \varepsilon |p|^{5/4}$. Le lemme 4 entraîne $|\mathrm{Im}\ \tilde{p}| < (\rho + \frac{3}{2}\varepsilon)|\mathrm{Re}\ p|^{5/4}$, or

$$|\mathrm{Re}\ \tilde{p} - \mathrm{Re}\ p| < \tfrac{3}{2}\ \varepsilon |\mathrm{Re}\ p|^{5/4} \Rightarrow |\mathrm{Re}\ p| - \tfrac{3}{2}\ \varepsilon |\mathrm{Re}\ p|^{5/4} < |\mathrm{Re}\ \tilde{p}|$$

$|\mathrm{Re}\ p| - \frac{3}{2}\ \varepsilon |\mathrm{Re}\ p|^{5/4} > |\mathrm{Re}\ p|\{1 - \frac{3}{2}\ \varepsilon\ a^{1/4}\}$; imposons

$$\boxed{1 - \tfrac{3}{2}\ \varepsilon\ a^{1/4} > (\tfrac{2}{3})^{4/5}}\ .$$

Les deux inégalités ci-dessus donnent:

$$|\mathrm{Re}\ p| < (\tfrac{3}{2})^{4/5}\ |\mathrm{Re}\ \tilde{p}|,$$

et reportant dans une inégalité obtenue précédemment, il vient

$$|\mathrm{Im}\ \tilde{p}| < \tfrac{3}{2}\ (\rho + \tfrac{3}{2}\varepsilon)|\mathrm{Re}\ \tilde{p}|^{5/4}.$$

Une dernière contrainte sur ε sera que

$$\boxed{\frac{3}{2}\,(\rho+\frac{3}{2}\varepsilon) < 2\rho} \;.$$

Elle donne $|\mathrm{Im}\,\tilde{p}| < 2\rho|\mathrm{Re}\,\tilde{p}|^{5/4}$. Enfin, $|\tilde{p}| < |p| + \varepsilon|p|a^{1/4}$.

Imposons

$$\boxed{1 + \varepsilon a^{1/4} < 2}$$

$|\tilde{p}| < 2|p| < 2a$, le lemme en résulte ∎

Dans la suite, par la notation $O(|p|^\alpha)$, je désigne une quantité telle que $\dfrac{O(|p|^\alpha)}{|p|^\alpha}$ soit bornée uniformément pour tous les F de $\mathfrak{F}_{c,Q,R}$ sur les domaines dignes d'intérêt.

On est en mesure d'énoncer la proposition suivante:

PROPOSITION II.1 - <u>Il existe une constante</u> A_1 <u>uniforme pour tous les éléments de</u> $\mathfrak{F}_{c,Q,R}$ <u>telle que si</u> $(p,q) \in D(P_1,a,\rho,\delta)$ <u>et</u> $N|p|^{5/4} < A_1$.

1) <u>La coordonnée angulaire</u> q^k <u>du</u> k-<u>ième itéré</u> $F^k(p,q)$ <u>existe pour</u> $k = 0,\ldots,N$. <u>En fait:</u>

2) $(p^k,q^k) \in D(P_2,2a,2\rho,2\delta)$, $k = 0,\ldots,N$; <u>de plus</u>

3) $p^N = p + O(N|p|^{5/2})$.

<u>Démonstration</u>. - Les quatre contraintes imposées à A_1 sont les suivantes:

(i) $(1 + \frac{3}{2}\Gamma_2 A_1 a^{1/4})^{5/2} < \frac{3}{2}$;

(ii) $(\frac{3}{2})^{9/5}\, A_1\Gamma_2 a^{1/4} < k$;

(iii) $\{2\rho\|Q\|(\frac{3}{2})^{1/2} + \Gamma^1(\frac{3}{2})^{3/5}(2a)^{1/4}\}A_1 < \delta$;

(iv) $\frac{3}{2}\, A_1\Gamma_2 < \varepsilon$.

Nous allons supposer que pour $k = 0,\ldots,n-1$, (p^k,q^k) appartient à $D(P_2,2a,2\rho,2\delta)$ et montrer que si $n \le N$, (p^n,q^n) est encore dans ce domaine, la majoration arrive en prime.

a . - L'hypothèse de récurrence implique que pour $k = 0,\dots,n-1$, $|p^k|^{5/2} < \frac{3}{2}|p|^{5/2}$, en effet, d'après le lemme (2) $|p^k - p^{k-1}| < \Gamma_2 |p^{k-1}|^{5/2}$. Je vais supposer démontré que pour $\ell = 0,\dots,k-1$, $|p^\ell|^{5/2} < \frac{3}{2}|p|^{5/2}$. Pour $\ell = 0$ cette relation étant acquise, il vient:

$$|p^k| < |p^{k-1}| + \tfrac{3}{2}\Gamma_2|p|^{5/2}$$

$$|p^{k-1}| < |p^{k-2}| + \tfrac{3}{2}\Gamma_2|p|^{5/2}$$

$$|p^k| < |p^{k-2}| + 2\,\tfrac{3}{2}\Gamma_2|p|^{5/2}$$

et ainsi de suite:

$$|p^k| < |p| + \tfrac{3}{2}k\Gamma_2|p|^{5/2}$$

$$|p^k|^{5/2} < \{1 + \tfrac{3}{2}N\Gamma_2|p|^{5/4}a^{1/4}\}^{5/2}\,|p|^{5/2}$$

et le résultat $|p^k|^{5/2} < \frac{3}{2}|p|^{5/2}$ grâce à (i) pour $k = 0,\dots,n-1$.

b . - Examinons p^n:

$$|p^n - p| \le |p^n - p^{n-1}| + |p^{n-1} - p^{n-2}| + \dots + |p^1 - p|$$

$$|p^n - p| \le \Gamma_2|p^{n-1}|^{5/2} + \Gamma_2|p^{n-2}|^{5/2} + \dots + \Gamma_2|p|^{5/2}$$

$$|p^n - p| \le \tfrac{3}{2}n\Gamma_2|p|^{5/2}, \quad \text{donc} \quad |\mathrm{Re}\, p^n - \mathrm{Re}\, p| \le \tfrac{3}{2}A_1\Gamma_2 a^{1/4}|p|.$$

D'après le lemme 4

$$|\mathrm{Re}\, p^n - \mathrm{Re}\, p| \le (\tfrac{3}{2})^{9/5} A_1\Gamma_2 a^{1/4}|\mathrm{Re}\, p|$$

et (ii) implique, grâce au lemme 1, que $\mathrm{Re}\, p^n$ appartient à P_2.

c . - Je puis donc introduire la coordonnée q^n du n-ième de F:

$$|\mathrm{Im}\, q^n - \mathrm{Im}\, q| \le \sum_{k=1}^{n} |\mathrm{Im}\, q^k - \mathrm{Im}\, q^{k-1}|$$

pour $k = 0,\dots,n-1$, (p^k,q^k) est dans $D(P_2, 2a, 2\rho, 2\delta)$, $\mathrm{Re}\, p^n$ est dans P_2 donc le lemme 3 implique:

$$|\mathrm{Im}\, q^k - \mathrm{Im}\, q^{k-1}| < \|Q\|\,|\mathrm{Im}\, p^{k-1}| + \Gamma^1|p^{k-1}|.$$

Avec $\|Q\| = \underset{i,j}{\text{Max}} |Q_{i,j}|$ (c est réel).

$$|\text{Im } p^{k-1}| < 2\rho |\text{Re } p^{k-1}|^{5/4} < 2\rho |p^{k-1}|^{5/4} < 2\rho (\tfrac{3}{2})^{1/2} |p|^{5/4}.$$

Donc

$$\sum_{k=1}^{n} |\text{Im } q^k - \text{Im } q^{k-1}| < \{2\rho \|Q\| (\tfrac{3}{2})^{1/2} + \Gamma^1 (\tfrac{3}{2})^{3/5} (2a)^{1/4}\} N |p|^{5/4}$$

et (iii) implique $|\text{Im } q^n| < 2\delta$.

La dernière contrainte (iv) assure les deux autres conditions requises pour avoir que (p^n, q^n) appartient à $D(P_2, 2a, 2\rho, 2\delta)$ comme il est démontré dans le lemme 5. Puisqu'on peut écrire

$$|p^n - p| < \tfrac{3}{2} N |p|^{5/4} \Gamma_2 |p|^{5/4};$$

les majorations entreprises jusqu'à n sont valables pour tout $n \leq N$, de sorte que l'on obtient 1) et 2).

D'autre part, appliquons II.1 pour $n = N$; on obtient l'estimation:

$$|p^N - p| < \tfrac{3}{2} N\Gamma_2 |p|^{5/4} |p|^{5/4}.$$

Soit $|p^N - p| < \frac{3}{2} \Gamma_2 A_1 |p|^{5/4}$ et 3), $p^N = p + O(N|p|^{5/2}) =$
$= p + O(|p|^{5/4})$. ■

PROPOSITION II.2 - Avec les notations précédentes,

$$q^N - q - Nc - NQp = O(|p|^{1/4}) + O(N^2 |p|^{5/2}).$$

Preuve:

$$|q^N - q - Nc - NQp| \leq \sum_{k=1}^{N} |q^k - q^{k-1} - c - Qp|$$

$$|q^N - q - Nc - NQp| \leq \sum_{k=1}^{N} |q^k - q^{k-1} - c - Qp^{k-1}| + \|Q\| |p^{k-1} - p|.$$

Donc, d'après les majorations de la proposition II.2

$$|q^N - q - Nc - NQp| \leq \sum_{k=1}^{N} (\tfrac{3}{2})^{3/5} \Gamma_2 |p|^{3/2} + \sum_{k=1}^{N} \tfrac{3}{2} \|Q\| \Gamma_2 k |p|^{5/2} \quad ■$$

Dans toute la suite, je supposerai A_1 et a suffisamment petits pour pouvoir écrire: $|q^N-q-Nc-NQp| < \pi$ dès que $N|p|^{5/4} < A_1$ et $|p| < a$.

Cette condition m'assure que q^N-q est périodique en la coordonnée q lorsque l'on considère q^N comme une fonction de l'action p et de l'angle q du point.

§2. Dans ce paragraphe, on utilise la formule de Cauchy afin d'obtenir des estimations de dérivées.

PROPOSITION II.3 - On introduit un cône P_o convexe strictement contenu dans le cône P_1 et on utilise $D(P_o,\frac{a}{2},\frac{\rho}{2},\frac{\delta}{2})$ qui est strictement contenu dans $D(P_1,a,\rho,\delta)$.

Soit (p,q) un point variable de $D(P_o,\frac{a}{2},\frac{\rho}{2},\frac{\delta}{2})$ vérifiant $N|p|^{5/4} < A$ $(A = A_1/2)$; $p^N = p^N(p,q)$ et $q^N(p,q) = q^N$ les coordonnées du N-ième itéré par F de (p,q) considérées comme fonctions des coordonnées indépendantes (p,q):

$$\frac{\partial p^N}{\partial q} = O(N|p|^{5/2})$$

$$\frac{\partial p^N}{\partial p} = 1 + O(N|p|^{5/4})$$

$$\frac{\partial q^N}{\partial p} = NQ + O(N|p|^{1/4}) + O(N^2|p|^{5/4})$$

$$\frac{\partial q^N}{\partial q} = 1 + O(|p|^{1/4}) + O(N^2(|p|^{5/2})).$$

Pr Preuve. Posant $i = \sqrt{-1}$ on introduit les nombres complexes $z_j = e^{iq_j}$. Dans $D(P_o,\frac{a}{2},\frac{\rho}{2},\frac{\delta}{2})$, $e^{-\delta/2} < |z_j| < e^{\delta/2}$, $(j=1,\ldots,m)$. Les fonctions p^N et q^N-q se développent en série de Laurent dans la couronne

$$\Gamma(\delta) = (z \in \mathbb{C}^m, \quad z = (z_i)_{i=1,\ldots,m}, \quad e^{-\delta} < |z_i| < e^{\delta}),$$

les coefficients étant holomorphes pour

$$|p| < a, \quad (\mathrm{Re}(p_i))_{i=1,\dots,m} = \mathrm{Re}(p) \in P_1, \ |\mathrm{Im}\ p| < \rho|\mathrm{Re}\ p|^{5/4}.$$

En utilisant une représentation intégrale de la dérivée, on obtient pour les dérivées $\frac{\partial p^N}{\partial z_j}$, $\frac{\partial q^N}{\partial z_j}$ des estimations identiques à celles des fonctions p^N et q^N; de là, on passe aux dérivées par rapport à q en utilisant

$$\frac{\partial}{\partial q_j} = iz_j \frac{\partial}{\partial z_j}, \qquad e^{\delta/2} < |z_j| < e^{\delta/2}.$$

Pour les dérivées par rapport aux p_i; le lemme 3 assure l'existence d'une constante ε' telle que si $(p,q) \in D(P_o,\frac{a}{2},\frac{\rho}{2},\frac{\delta}{2})$, $N|p|^{5/4} < A$ et $|\tilde{p}-p| < \varepsilon'|p|^{5/4}$, $|\mathrm{Im}\ \tilde{q}| < \delta$, alors:

$$(\tilde{p},\tilde{q}) \in D(P_1,\frac{a}{2},\frac{\rho}{2},\frac{\delta}{2}).$$

On peut bien sûr exiger de ε' que

$$N|\tilde{p}|^{5/4} < A_1.$$

On introduit alors le cercle γ centré en (p,q) de rayon $\varepsilon'|p|^{5/4}$ tracé à $p_2,\dots,p_m$; $q_1,q_2,\dots,q_m$ constants et on écrit:

$$\frac{\partial(p^N-p)}{\partial p_1} = \frac{1}{2\pi j}\int_\gamma \frac{p^N(\zeta,p_2,\dots,p_m,q)d\zeta}{(\zeta-p_1)^2}$$

qui implique:

$$\left|\frac{\partial(p^N-p)}{\partial p_1}\right| \le \varepsilon'|p|^{5/4}\, \varepsilon'^{-2}|p|^{-5/2}\, O(N|p|^{5/2}),$$

soit

$$\frac{\partial p^N}{\partial p_1} = 1 + O(N|p|^{5/4});$$

il va de soi que l'on peut procéder de même pour n'importe quelle coordonnée action p_i, et on est conduit à remarquer que l'estimation de la dérivée par rapport à p_i d'une certaine fonction s'obtient en divisant par $|p|^{5/4}$ l'estimation de ladite fonction.

Par exemple:

$$\frac{\partial q^N}{\partial p} = NQ + O(N|p|^{1/4}) + O(N^2|p|^{5/4})$$

s'obtient ainsi à partir de

$$q^N = q + Nc + NQp + O(N|p|^{3/2}) + O(N^2|p|^{5/2}) \quad \blacksquare$$

§3. Dans ce paragraphe, je vais expliciter ce que les majorations obtenues dans le domaine complexe donnent en restriction au domaine réel et préciser les contraintes supplémentaires que j'exigerai dans la suite sur a et A.

Soit (p,q) un point réel tel que $p \in P_1$, $N|p|^{5/4} < A_1$, $|p| < a$, (p,q) est un point réel de $D(P_1,a,\rho,\delta)$. L'extension complexe (proposition II.1) nous précise que son N-ième itéré par un élément F de $\mathfrak{F}_{c,Q,R}$ est repéré par ses coordonnées action-angle p^N et q^N. De plus (p^N,q^N) est un élément de $D(P_2,2a,2\rho,2\delta)$.

Mais F est une transformation réelle, de sorte que (p^N,q^N) est un point réel de $D(P_2,2a,2\rho,2\delta)$ et les résultats 1 et 2 de II.1 se lisent:

$$p^N \in P_2, \qquad |p^N| < 2a.$$

On peut de plus profiter dans le domaine réel des estimations

$$p^N = p + O(N(|p|^{5/2}))$$

$$q^N - q - Nc - NQp = O(N|p|^{5/4}) + O(N^2|p|^{5/2})$$

$$\frac{\partial p^N}{\partial p} = 1 + O(N|p|^{5/4})$$

$$\frac{\partial p^N}{\partial q} = O(N|p|^{5/2})$$

$$\frac{\partial q^N}{\partial p} = NQ + O(N|p|^{1/4}) + O(N^2|p|^{5/4})$$

$$\frac{\partial q^N}{\partial q} = 1 + O(N|p|^{5/4}) + O(N^2|p|^{5/2}).$$

J'insiste une nouvelle fois sur le fait très important que ces estimations sont uniformes pour tous les éléments F de $\mathfrak{F}_{c,Q,R}$.

Quitte à diminuer a et A je puis affirmer:

PROPOSITION II.4 - <u>Il existe deux constantes</u> a <u>et</u> A <u>telles que si</u> $p \in P_0$, $|p| < a$, $N|p|^{5/4} < A$:

① <u>les itérés successifs</u> (p^k,q^k) <u>satisfont</u> $p^k \in P_2$ <u>pour</u> $k = 1,\dots,N$ <u>avec</u> $p^N = p + O(N|p|^{5/2})$.

② $|q^N - q - Nc - NQp| < \pi$

③ $|\frac{\partial p^N}{\partial p} - 1| < 1$, $|N^{-1}Q^{-1}\frac{\partial p^N}{\partial p} - 1| < 1$ et $|\frac{\partial q^N}{\partial q} - 1| < 1$.

Il importe de remarquer que 3 implique l'inversibilité de $\frac{\partial p^N}{\partial p}$, $\frac{\partial q^N}{\partial p}$ et $\frac{\partial q^N}{\partial q}$.

Dans la suite, j'abandonne l'extension complexe pour ne considérer que des points réels.

III. POINTS PERIODIQUES

Dans ce chapitre, P_o' désigne un cône convexe strictement contenu dans P_o. Je rappelle qu'un entier N et un élément ν de $2\pi Z^m$ sont dits associés (définition I.1) si:

i) $\nu - Nc \in QP_o'$

ii) $|\nu - Nc| < \beta N^{1/5}$

iii) $N > \gamma$

Je vais montrer que l'on peut choisir β et γ de façon que si ν et N sont associés, tout élément F de $\mathcal{F}_{c,Q,R}$ possède au moins deux points périodiques de période N et d'indice ν. C'est-à-dire deux solutions au moins aux équations en p et q:

$$\begin{cases} q^N(p,q) = q + \nu \\ p^N(p,q) = p. \end{cases}$$

Le problème de la recherche des points périodiques gagne à être présenté de la façon suivante: on introduit U_N l'ouvert de $\mathbb{R}^m$ ainsi défini:

$$U_N = \{p \in \mathbb{R}_+^m / p \in P_o,\ |p| < a,\ N|p|^{5/4} < A\}$$

où a et A sont les deux constantes de la proposition II.4.

Soit F un élément de $\mathcal{F}_{c,Q,R}$. D'après ce qui a été fait précédemment, on peut interpréter F^N comme une transformation de $U \times \mathbb{T}^m$ dans $P_2 \times \mathbb{T}^m$. Rapportons la variété $U_N \times \mathbb{T}^m$ à un système de coordonnées locales (p,q) et la variété $P_2 \times \mathbb{T}^m$ au système (p',q'). Dans la variété produit $(U_N \times \mathbb{T}^m) \times (P_2 \times \mathbb{T}^m)$ rapportée aux coordonnées (p,q,p',q'), on introduit les quatre sous-variétés suivantes:

Δ_N : le graphe de F^N paramétré par les deux séries de relations $p' = p^N(p,q)$, $q' = q^N(p,q)$.

$X_\nu^{(1)}$: la sous-variété paramétrée par $q' = q + \nu$.

$X_\nu^{(2)}$: la sous-variété constituée des points tels que $p' = p$;

$X_\nu = X_\nu^{(1)} \cap X_\nu^{(2)}$.

Il est évident que prouver l'existence de points périodiques de période N et d'indice ν revient à s'assurer que $X_\nu \cap \Delta_N$ est non vide.

Dans la suite, on suppose ν et N associés; β et γ seront précisées en cours de preuve.

On commence par étudier $X_\nu^{(1)} \cap \Delta_N$.

§1. LE TORE DE RETOUR DES ANGLES $\mathbb{T}^m_{\nu,N} = X_\nu^{(1)} \cap \Delta_N$.

PROPOSITION III.1 - <u>Il existe un choix des constantes</u> β <u>et</u> γ <u>qui permet la construction d'une fonction analytique</u> ψ_N^ν <u>définie sur le tore</u> $\mathbb{T}^m$ <u>telle que</u>

$$p = \psi_N^\nu(q) \iff \begin{cases} q^N(p,q) = q + \nu \\ p \in U_N \end{cases}$$

Posons $q^N(p,q) = q + Nc + NQp - G_N(p,q)$ et $\bar{\omega} = N^{-1}Q^{-1}(\nu - Nc)$.

Introduisons por les besoins de la preuve, l'application f définie sur U_N à valeurs dans $\mathbb{R}^m$ par:

$$f: p \mapsto f(p) = \bar{\omega} + N^{-1}Q^{-1}G_N(p,q).$$

LEMME - f <u>est strictement contractante sur un compact de</u> U_N.

Soi L un compact de U_N, $p, \tilde{p} \in L$;

$$|f(p)-f(\tilde{p})| \leq |p-\tilde{p}| \sup_L |N^{-1}Q^{-1} \frac{\partial G_N}{\partial p}|.$$

L est inclus dans U_N et la deuxième inégalité de II.4 (3) implique

$$\sup_L |N^{-1}Q^{-1} \frac{\partial G_N}{\partial p}| < 1 \quad ■$$

Démontrons maintenant III.1: on commence par examiner $\bar{\omega}$. Du fait des conditions d'association de ν et N: (i), (ii) et (iii) résulte:

$\bar{\omega} \in P'_o$ (à cause de (i))

$|\bar{\omega}| \leq N^{-1}\|Q^{-1}\| \, |\nu - Nc| < \beta N^{-4/5} \|Q^{-1}\|$

$|\bar{\omega}|^{5/4} < \beta^{5/4} N^{-1}\|Q^{-1}\|^{5/4}$

$N|\bar{\omega}|^{5/4} < \beta^{5/4} \|Q^{-1}\|^{5/4}.$

Une première contrainte sur β sera que $N|\bar{\omega}|^{5/4} < A/2$, d'autre part,

$|\bar{\omega}| < \beta\gamma^{-4/5} \|Q^{-1}\|.$

J'exige de γ que $\beta\gamma^{-4/5} \|Q^{-1}\| < a/2$, il vient

$|\bar{\omega}| < a/2.$

Afin d'appliquer le théorème du point fixe, je vais montrer que f laisse invariante un certain compact de U_N.

Soit k une constante telle que si $u \in P'_o$ et $|u-v| < k|u|$, on peut affirmer que $v \in P_o$.

Exigeons de plus $(1+k)^{5/4} < 2$.

Je noterai B la boule fermée centrée en $\bar{\omega}$ est dans P'_o et de la définition de k résulte immédiatement que si p appartient à B, p appartient à P_o; de plus, si p appartient à B:

$$|p| < (1+k)|\bar{\omega}| < (1+k)\frac{a}{2} < (1+k)^{5/4}\,\frac{a}{2} < a$$

$$N|p|^{5/4} < N(1+k)^{5/4}|\bar{\omega}|^{5/4} < \frac{A}{2}\,(1+k)^{5/4} < A.$$

De sorte que B est incluse dans U_N.

Montrons que $f(B) \subset B$ moyennant une dernière contrainte sur β.

$$f(p) - \bar{\omega} = N^{-1}Q^{-1}G_N(p,q)$$

pour p appartenant à U_N, une des estimations du §3 de II conduit à écrire:

$$|f(p)-\bar{\omega}| = O(|p|^{3/2}) + O(N|p|^{5/2}).$$

Lorsque p appartient à B on peut donc écrire

$$|f(p)-\bar{\omega}| = |\bar{\omega}|[O(|\bar{\omega}|^{1/2}) + O(N|\bar{\omega}|^{3/2}].$$

Dés lors, remarquons que

$$N|\bar{\omega}|^{3/2} = N|\bar{\omega}|^{5/4}\,|\bar{\omega}|^{1/4} < \frac{A}{2}\,|\bar{\omega}|^{1/4}$$

et rappelons que

$$|\bar{\omega}| < \beta\gamma^{-4/5}\,\|Q^{-1}\|.$$

De sorte qu'en choisissant β suffisamment petit, on obtient

$$|f(p)-\bar{\omega}| < k|\bar{\omega}|;$$

ainsi $f(B) \subset B$ ■

Il existe donc pour tout q un point fixe $p = \psi_N^\nu(q)$ et on construit de la sorte une fonction ψ_N^ν dont on peut conclure par transversalité qu'elle est analytique. De plus:

$$p = \psi_N^\nu(q) \quad \Leftrightarrow \quad \begin{cases} q^N(p,q) = q + \nu \\ p \in U_N \end{cases}.$$

Observons que ψ_N^ν est périodique en q et peut ainsi d'interpréter comme définie sur T^m. En effet, soit η un élément de $2\pi\mathbb{Z}^m$; si

$p \in U_N$, $q^N(p,q+\eta) = q^N(p,q)+\eta$ car (cf. prop. II.4) $|q^N-q-NQp-Nc| < \pi$. Donc

$$q^N(p,q) = q+\nu \Rightarrow q^N(p,q+\eta) = q+\eta+\nu$$

et

$$p = \psi_N^\nu(q) = \psi_N^\nu(q+\eta).$$

Dans la variété produit $(U_N \times \mathbf{T}^m) \times (P_2 \times \mathbf{T}^m)$, la sous-variété $X_\nu^{(1)} \cap \Delta_N$ est paramétrée par

$$X_\nu^{(1)} \cap \Delta_N : \begin{cases} q' = q+\nu \\ q' = q^N(p,q) \\ p' = p^N(p,q) \end{cases} .$$

La proposition III.1 suggère de la décrire ainsi:

$$X_\nu^{(1)} \cap \Delta_N : \begin{cases} q' = q+\nu \Leftrightarrow q = q'-\nu \\ p' = p^N(p,q) \Leftrightarrow p' = p^N(p,q') \\ p = \psi_N^\nu(q) \Leftrightarrow p = \psi_N^\nu(q') \end{cases} .$$

La seconde relation provenant du fait évident que p^N est périodique en q à p fixé. La coordonnée q' détermine un difféomorphisme de $X_\nu^{(1)} \cap \Delta_N$ sur $\mathbf{T}^m$, ce qui justifie d'appeler tore de retour des angles $X_\nu^{(1)} \cap \Delta_N$ et la notation

$$X_\nu^{(1)} \cap \Delta_N = \mathbf{T}^m_{\nu,N} .$$

§2. LA FONCTION GENERATRICE DE F^N.

Dans la variété produit $(U_N \times \mathbf{T}^m) \times (P_2 \times \mathbf{T}^m)$ introduisons la forme symplectique

$$\sigma = dp \wedge dq - dp' \wedge dq'.$$

Et remarquons que du fait que F est symplectique:

$$\sigma|_{\Delta_N} = 0.$$

Afim de paramétrer le graphe Δ_N j'ai utilisé comme coordonnées indépendantes p et q. Je prétends que l'on peut tout aussi bien utiliser p et q' comme coordonnées indépendantes pour repérer un point de Δ_N , ceci est justifié par le lemme qui suit:

PROPOSITION III.2 - _A p fixé dans U_N l'application de T^m dans T^m: $q \mapsto q^N(p,q)$ est un difféomorphisme._

p appartenant à U_N, la proposition II.4 donne que $\frac{\partial q^N}{\partial q}$ est inversible, donce $q \mapsto q^N$ est un étalement mais $\mathbb{T}^m$ est compact; donc c'est un revêtement.

Ecrivons $F = F_o + \tilde{F}$ (F_o définie en I.3); l'arc de transformations $F_\mu = F_o + \mu\tilde{F}$, $\mu \in [0,1]$ permet de s'assurer que $q \to q^N$ est homotope à l'application correspondante pour F_o. Mais pour F_o (cf. I.3)

$$q \mapsto q^N = q + Nc + NQp$$

et clairement à p fixé $q \mapsto q^N$ (pour F_o) est un difféomorphisme; $q \mapsto q^N$ est homotope à cette application donc est un difféomorphisme ■

On choisit p et q' comme coordonnées indépendantes sur le graphe Δ_N. Il nous faut maintenant préciser comment on paramétrise Δ_N au moyen de p et de q'. On se place dans le revêtement universel de Δ_N: $\tilde{\Delta}_N$. La 1-forme $qdp + p'dq'$ (où p et q' sont dans $\mathbb{R}^m$) est fermée en restriction à $\tilde{\Delta}_N$; il apparaît une fonction $S_N(p,q')$ définie sur $\tilde{\Delta}_N$ telle que

$$qdp + p'dq' = dS_N$$

S_N est appelée fonction génératrice de F^N et elle vérifie

$$q = \frac{\partial S_N}{\partial p}(p,q'), \qquad p^N = \frac{\partial S_N}{\partial q'}(p,q') = p'.$$

Considérons la fonction génératrice modifiée

$$S'_N(p,q') = S_N(p,q') - pq'.$$

PROPOSITION III.3 - <u>A</u> p <u>fixé</u>, $S'_N(p,q')$ <u>est périodique en</u> q'.

Dans Δ_N on trace, à p fixé, un lacet γ d'indice d'homotopie η: $\int_\gamma dq' = \eta$. Calculons $\int_\gamma dS'_N$.

$$\int_\gamma dS'_N = \int_\gamma \frac{\partial S'_N}{\partial q'} dq' = \int_\gamma (p^N - p)dq'$$

$$\int_\gamma dS'_N = \int_\gamma p^N dq' - p\eta .$$

Examinons $\int_\gamma (p^N dq' - pdq)$. La 1-forme $p^N dq' - pdq$ est fermée puisque F^N est symplectique. On peut facilement l'exprimer à l'aide des coordonnées linéaires et on constate qu'elle est analytique à l'origine. γ peut donc être contracté sur un point dans le domaine d'analycité. Il s'ensuit $\int_\gamma (p^N dq' - pdq) = 0$.

D'autre part, lorsque q' "tourne de η" q "tourne" aussi du même indice η (prop. II.4, $|q^N - q - NQp - Nc| < \pi$). Donc

$$\int_\gamma p^N dq' = \int_\gamma pdq = p\eta$$

et

$$\int_\gamma dS'_N = 0.$$

S'_N est de la sorte définie sur Δ_N et elle en fournit une paramétrisation:

$$q - p' = \frac{\partial S'_N}{\partial p}(p,q')$$

$$p' - p = \frac{\partial S'_N}{\partial q'}(p,q')$$

■

§3. LA FONCTION (ν-N) SECULAIRE.

On définit la fonction (ν-N) séculaire $R_N^\nu : \mathbb{T}^m_{\nu,N} \to \mathbb{R}$ par:

$$R_N^\nu(q') = S'_N(\psi_N^\nu(q'),q') + \nu\psi_N^\nu(q').$$

Remarquons que R_N^ν est périodique en q', puisque ψ_N^ν et S'_N le sont. Calculons $\dfrac{dR_N^\nu(q')}{dq'}$:

$$\frac{dR_N^\nu}{dq'}(q') = \frac{\partial S'_N}{\partial q'} + \frac{\partial \psi_N^\nu}{\partial q'}\frac{\partial S'_N}{\partial p}(\psi_N^\nu(q'),q') + \nu\frac{\partial \psi_N^\nu}{\partial q'}$$

$$\frac{\partial S'_N}{\partial p}(\psi_N^\nu(q'),q')\Big|_{\mathbb{T}^m_{\nu,N}} = q - q'\Big|_{\mathbb{T}^m_{\nu,N}} = -\nu.$$

Donc

$$\frac{dR_N^\nu}{dq'}(q') = p' - p\Big|_{\mathbb{T}^m_{\nu,N}} .$$

Et on obtient que $X_\nu^{(2)} \cap \mathbb{T}^m_{\nu,N}$ coincide avec l'ensemble critique de la fonction R_N^ν. Or

$$X_\nu \cap \Delta_N = X_\nu^{(2)} \cap \mathbb{T}^m_{\nu,N} .$$

D'autre part, la fonction R_N^ν est analytique sur $\mathbb{T}^m_{\nu,N}$ puisque ψ_N^ν- et S'_N le sont. Elle présente au moins deux points critiques sur le tore $\mathbb{T}^m_{\nu,N}$ correspondant à son maximum et à son minimum. Et on peut énoncer en définitive:

THÉORÈME III.4 - <u>Il existe deux constantes</u> β <u>et</u> γ <u>tels que si</u> ν <u>et</u> N <u>sont associés par</u> β <u>et</u> γ, <u>tout élément</u> F <u>de</u> $\mathcal{F}_{c,Q,R}$ <u>possède au moins un point périodique de période</u> N <u>et d'indice</u> ν.

§4. LA CONSTRUCTION DE SUITES DE POINTS PERIODIQUES.

Je vais préciser, dans ce paragraphe, quelques propriétés afférentes aux points périodiques du théorème III.4.

Examinons tout d'abord comment se présentent les points périodiques de F_o. Ils sont donnés par les deux stocks d'équations

$$\begin{cases} p^N = p \\ q^N = q+\nu \end{cases}$$

qui pour F_o se réduisent à

$$q + Nc + NQp = q + \nu,$$

soit $p = N^{-1}Q^{-1}(\nu-Nc)$. Les points périodiques de F_o apparaissent donc en tores centrés autour de l'origine. Dans ce cas, la fonction R_N^ν est constante sur tout le tore de retour des angles $\mathcal{T}_{\nu,N}^m$; comme le montre un calcul facile.

Il faut voir un élément F de $\mathcal{F}_{c,Q,R}$ comme une perturbation de F_o puisqu'il lui est tangent à l'ordre 4 (cf. définition de l'introduction).

Le théorème III.4 peut alors se comprendre par le fait que dans la perturbation les points périodiques de F_o ne disparaissent pas complètement si tant est que ν et N soient associées.

J'avais noté $\bar{\omega} = N^{-1}Q^{-1}(\nu-Nc)$ au paragraphe 1. Ici pour faire clairement apparaître la dépendance en ν et N, je poserai:

$$\bar{\omega}_N^\nu = N^{-1}Q^{-1}(\nu-Nc).$$

Cette façon de voir les points périodiques d'un élément de $\mathcal{F}_{c,Q,R}$, est renforcée par l'estimation rencontrée au paragraphe 1 de la coordonnée action de ces points périodiques, que je rappelle:

$$p = \omega_N^\nu + O(|\bar{\omega}_N^\nu|^{5/4}).$$

Cette estimation joint aux deux lemmes qui suivent a pour conséquence que l'on peut trouver des points périodiques aussi proches de l'origine qu'on le désire.

PROPOSITION III.5 - Il existe une suite $\bar{\omega}_N^\nu$ qui tend vers zéro (avec ν et N associés par β et γ du théorème III.4).

Il suffit d'examiner de près les conditions d'associations:

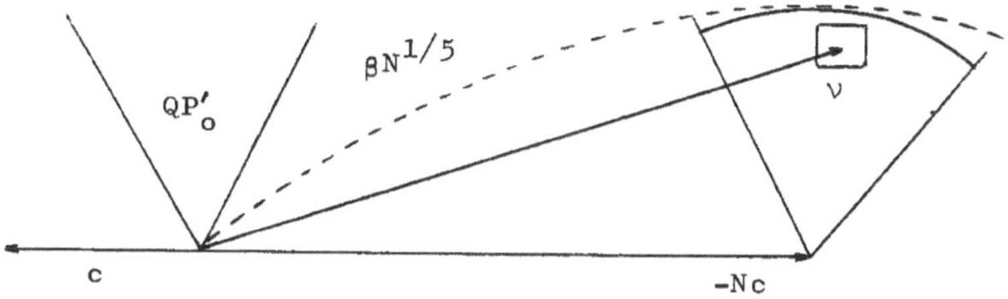

On introduit QP'_o et le translaté de QP'_o dans $\mathbb{R}^m$: $QP'_o + Nc$. Pour N suffisamment grand, on se convainc que l'on peut inscrire un hypercube de longueur 2π dans l'intersection de la boule centrée en $-Nc$, de rayon $\beta N^{1/5}$, et du translaté et donc on s'assure d'un ν (élément de $2\pi\mathbb{Z}^m$) tel que:

(i) $\nu - Nc \in QP'_o$

(ii) $|\nu - Nc| < \beta N^{1/5}$

il n'y a aucun obstacle à faire augmenter N indéfiniment

(iii) finit par être satisfaite et $|\omega_N^\nu| < \beta \| Q^{-1} \| N^{-4/5}$ tend vers zéro.

On peut améliorer le résultat précédent:

PROPOSITION III.6 - <u>Etant donné un cône convexe</u> χ <u>contenu</u> dans P'_o il existe une suite $\bar{\omega}_N^\nu$ qui tend vers zéro dans le cône χ.

La preuve est identique à celle de la proposition III.5.

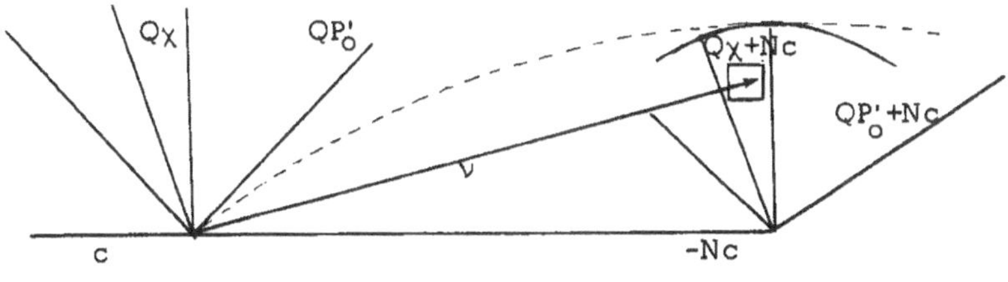

Il faut cette fois-ci placer l'hypercube dans $Q\chi + Nc$.

L'estimation de la coordonnée action des points périodiques que j'ai rappelée fournit immédiatement la

PROPOSITION III.7 - Etant donné un cône convexe χ contenu dans P'_o et F un élément de $\mathfrak{F}_{c,Q,R}$; il existe une suite de points périodiques, de période N et d'indice ν, de F; qui tend vers zéro dans le cône χ.

Remarque - Pour chacun des points de cette suite, je ne précise pas encore la coordonnée angulaire qu'on a la liberté de choisir parmi les points critiques de la fonction séculaire.

IV. PROPRIETES DE GENERICITE DE $\mathfrak{F}_{c,Q,R}$

La norme $\| \|_{\bar{D}_1}$ munit $\mathfrak{F}_{c,Q,R}$ d'une structure d'espace topologique complet. Nous désignerons par propriété de généricité une propriété vérifiée par tous les éléments d'une intersection dénombrable d'ouverts partout denses.

Avant d'établir les propriétés de généricité en vue; il est utile d'obtenir plus d'information sur la fonction séculaire et sur la fonction génératrice d'un élément de $\mathfrak{F}_{c,Q,R}$.

Je désigne toujours par F un élément de $\mathfrak{F}_{c,Q,R}$.

§1. LA TRANSFORMATION DE POINCARÉ-FLOQUET.

Comme il a été précisé dans le chapitre précédent F^N fait correspondre à un point (p^o,q^o) de $U_N \times \mathbb{T}^m$ un point (p^N,q^N) de $P_2 \times \mathbb{T}^m$. Je désignerai à nouveau par (p,q) les coordonnées locales de $U_N \times \mathbb{T}^m$ et (p',q') celles de $P_2 \times \mathbb{T}^m$.

L'application tangente TF^N en restriction à l'espace tangent $T_{(p^o,q^o)}(U_N \times \mathbb{T}^m)$ est une application linéaire définie sur $T_{(p^o,q^o)}(U_N \times \mathbb{T}^m)$ à valeurs dans $T_{(p^N,q^N)}(P_2 \times \mathbb{T}^m)$.

Supposons que (p^o,q^o) soit un point périodique pour F de période N, d'indice ν,

$$T_{(p^o,q^o)}(U_N \times \mathbb{T}^m) \text{ s'identifie à } T_{(p^N,q^N)}(P_2 \times \mathbb{T}^m).$$

Dans ce cas, l'application linéaire $TF^N|_{T_{(p^o,q^o)}}(U_N \times \mathbb{T}^m)$ est appelée transformation de Poincaré-Floquet associée au point périodique (p^o,q^o) et à F.

PROPOSITION IV.1 (cf. [12]) - <u>Soit</u> (p^o,q^o) <u>un point périodique de période</u> N <u>et d'indice</u> ν <u>de</u> F:

$$p^N = p^o, \qquad q^N = q^o + \nu.$$

<u>La multiplicité de la valeur propre</u> 1 <u>de sa transformation de Poincaré-Floquet est égale au co-rang de la Hessienne de la fonction</u> $(\nu\text{-}N)$ <u>séculaire calculée au point</u> q^N <u>de</u> $\mathbb{T}^m_{\nu,N}$ <u>correspondant</u>.

Précisons les notations: comme coordonnées locales de $T(U_N \times \mathbb{T}^m)$ on utilise $(p,q,\Delta p,\Delta q)$; sur $T(P_2 \times \mathbb{T}^m)$, $(p',q',\Delta p',\Delta q')$.

Un vecteur propre $(\Delta p,\Delta q')$ de $TF^N|_{T_{(p^o,q^o)}}(U_N \times \mathbb{T}^m)$ de valeur propre 1 vérifie

$$\begin{cases} \Delta p = \dfrac{\partial p^N}{\partial p} \Delta p + \dfrac{\partial p^N}{\partial q'} \Delta q' \\[2ex] \Delta q' = \dfrac{\partial q}{\partial p} \Delta p + \dfrac{\partial q}{\partial q'} \Delta q' \end{cases}$$

Il faut bien remarquer que $T_{(p^o,q^o)}(U_N \times \mathbb{T}^m)$ s'identifiant à $T_{(p^N,q^N)}(P_2 \times \mathbb{T}^m)$, il est loisible de mélanger les coordonnées et de choisir p et q' comme variables indépendantes. Et je rappelle à cette occasion que c'est le fait que l'application $q \mapsto q^N$ est un difféomorphisme (proposition III.2) qui permet de le faire.

Mais ces deux systèmes s'écrivent à l'aide de S'_N (cf. ch.III, §2) ainsi:

(IV.1) (IV.1)

$$\begin{cases} 0 = \dfrac{\partial^2 S'_N}{\partial p \partial q'} \Delta p + \dfrac{\partial^2 S'_N}{\partial q'^2} \Delta q' \\[2ex] 0 = \dfrac{\partial^2 S'_N}{\partial p^2} \Delta p + \dfrac{\partial^2 S'_N}{\partial p \partial q'} \Delta q' \end{cases}$$

et il est donc équivalent d'affirmer que $(\Delta p, \Delta q')$ est dans le noyau de la Hessienne de S'_N .

Supposons qu'un vecteur $(\Delta p, \Delta q')$ satisfasse (IV.1) et remarquons que pour (p,q) appartenant à $U_N \times \mathbf{T}^m$, $\dfrac{\partial^2 S'_N}{\partial p^2}$ est inversible. En effet: $\dfrac{\partial^2 S'_N}{\partial p^2} = \left(\dfrac{\partial q^N}{\partial q}\right)^{-1} \dfrac{\partial q^N}{\partial p}$ et $\dfrac{\partial q^N}{\partial p}$ est inversible (cf. proposition II.4 (3)). Donc (IV.1) implique

$$\Delta p = -\left(\frac{\partial^2 S'_N}{\partial p^2}\right)^{-1} \frac{\partial^2 S'_N}{\partial p \partial q'} \Delta q'$$

et

$$0 = -\frac{\partial^2 S'_N}{\partial p \partial q'} \left(\frac{\partial^2 S'_N}{\partial p^2}\right)^{-1} \frac{\partial^2 S'_N}{\partial p \partial q'} + \frac{\partial^2 S'_N}{\partial q'^2} \Delta q' . \tag{IV.2}$$

Je rappelle l'expression de la fonction $(\nu\text{-}N)$ séculaire R^{ν}_N

$$R^{\nu}_N(q') = S'_N(\psi^{\nu}_N(q'), q') + \nu \psi^{\nu}_N(q').$$

Un calcul rapide montre que (IV.2) écrite en un point qui vérifie: $p = \psi^{\nu}_N(q')$ est exactement l'équation (IV.3)

$$\frac{\partial^2 R^{\nu}_N(q')}{\partial q'^2} \Delta q' = 0. \tag{IV.3}$$

Inversement, étant donné un point q' de $\mathbf{T}^m_{\nu,N}$ et un vecteur $\Delta q'$ qui vérifie (IV.3), alors avec:

$$\Delta p = -\left(\frac{\partial^2 S'_N}{\partial p^2}\right)^{-1} \frac{\partial^2 S'_N}{d p \partial q'} \Delta q' ,$$

le vecteur $(\Delta p, \Delta q')$ vérifie (IV.1).

On détermine ainsi une bijection entre le noyau de la hessienne de S'_N, calculée en un point (p,q') avec $p = \psi_N^\nu(q')$, et le noyau de la Hessienne de R_N^ν. Le résultat annoncé en découle □

§2. PERTURBATION D'UN ELEMENT F DE $\mathcal{F}_{c,Q,R}$.

Rappelons qu'un champ hamiltonien X est un champ qui satisfait $L_X\omega = 0$ où L_X désigne la dérivée de Lie de X. La formule de Cartan $L_X = di_X + i_Xd$ donne $di_X\omega = 0$ et le lemme de Poincaré permet de conclure à l'existence locale d'une fonction H associée à X par $i_X\omega = -dH$; H est appelé l'hamiltonien de X (il est défini à une constante près). Le flot de X: $\exp \varepsilon X$ est symplectique puisque:

$$\frac{\partial}{\partial\varepsilon}(\exp \varepsilon X)^*\omega = (\exp \varepsilon X)^*L_X\omega = 0,$$

$(\exp \varepsilon X)^*\omega$ est indépendante de ε or pour $\varepsilon = 0$, elle est égale à ω, donc:

$$(\exp \varepsilon X)^*\omega = \omega.$$

Soit F un élement de $\mathcal{F}_{c,Q,R}$; la transformation $F_\varepsilon = \exp \varepsilon X \circ F$ est symplectique car la composée de deux transformations symplectiques est symplectique.

LEMME - <u>On peut choisir</u> H <u>de façon que</u> F_ε <u>soit encore un élément de</u> $\mathcal{F}_{c,Q,R}$ <u>pour</u> ε <u>réel suffisamment petit</u>.

Il suffit d'exiger de H d'être une fonction polynomiale qui appartient à la cinquième puissance de l'idéal maximal des fonctions analytiques nulles à l'origine.

En effet, F_ε s'étend à $\mathbb{C}^{2m}$ en une transformation analytique sur D_1 puisque H est polynomial; F_ε est encore tangente à l'ordre 4 à F_o et en prenant ε suffisamment petit, on s'assure $\|F_\varepsilon - F_o\|_{\bar{D}_1} < R$.

Dans toute la suite, on n'utilisera que des perturbations F_ε de F du type précédemment décrit, avec un hamiltonien H convenable.

Chaque transformation F_ε, étant un élément de $\mathcal{F}_{c,Q,R}$, possède une fonction génératice $S_N(\varepsilon,p,p')$ et une fonction séculaire $R_N^\nu(\varepsilon,q')$: il importe pour la suite de calculer la variation au premier ordre en ε de ces deux fonctions dans la perturbation $F \to F_\varepsilon$.

PROPOSITION IV.2 -

$$S_N(\varepsilon,p,q') = S_N(p,q') + \varepsilon \left.\frac{\partial S_N(\varepsilon,p,q')}{\partial\varepsilon}\right|_{\varepsilon=0} + O(\varepsilon^2).$$

Avec

$$\left.\frac{\partial S_N(\varepsilon,p,q')}{\partial\varepsilon}\right|_{\varepsilon=0} = -[H(p^1,q^1) + H(p^2,q^2) + \ldots + H(p^N,q^N)].$$

Dans le second membre de l'égalité, il faut lire les (p^k,q^k) comme des fonctions de p et de q'. En les interprétant d'abord, par exemple comme des fonctions de p et de q puis en inversant la relation $q' = q^N(p,q)$.

Démonstration - Elle repose sur la formule classique suivante: étant donné un paramètre réel ε et un champ X, une forme ω_ε dépendant de ε vérifie:

$$\frac{\partial(\exp \varepsilon X)^*\omega_\varepsilon}{\partial\varepsilon} = (\exp \varepsilon X)^* L_X\omega_\varepsilon + (\exp \varepsilon X)^* \frac{\partial\omega_\varepsilon}{\partial\varepsilon}.$$

Je noterai ainsi les itérés de F_ε:

$$F_\varepsilon^k : \begin{vmatrix} p \\ q \end{vmatrix} \longmapsto \begin{matrix} \tilde{p}^k \\ \tilde{q}^k \end{matrix}.$$

On commence par écrire $\frac{\partial}{\partial\varepsilon}\tilde{p}^N d\tilde{q}^N$

$$\frac{\partial}{\partial\varepsilon}(\tilde{p}^N d\tilde{q}^N) = \frac{\partial}{\partial\varepsilon}(F_\varepsilon^{N*}(pdq)) = \frac{\partial}{\partial\varepsilon}(F^*(\exp \varepsilon X)^*(\tilde{p}^{N-1}d\tilde{q}^{N-1}))$$

$$= F^* \frac{\partial}{\partial\varepsilon}(\exp \varepsilon X)^* \tilde{p}^{N-1}d\tilde{q}^{N-1}.$$

D'après la formule rappelée, le second membre de l'égalité vaut:

$$F_\varepsilon^* L_X \tilde{p}^{N-1}d\tilde{q}^{N-1} + F_\varepsilon^* \frac{\partial}{\partial\varepsilon}\tilde{p}^{N-1}d\tilde{q}^{N-1}.$$

Examinons:

$$F_\varepsilon^* L_X \tilde{p}^{N-1} d\tilde{q}^{N-1} = F_\varepsilon^* d i_X \tilde{p}^{N-1} d\tilde{q}^{N-1} + F_\varepsilon^* i_X (d\tilde{p}^{N-1} \wedge d\tilde{q}^{N-1}).$$

F_ε^{N-1} est symplectique donc $d\tilde{p}^{N-1} \wedge d\tilde{q}^{N-1} = dp \wedge dq$ et je rappelle que $i_X dp \wedge dq = -dH$ il vient

$$F_\varepsilon^* L_X \tilde{p}^{N-1} d\tilde{q}^{N-1} = d(\tilde{p}^N F_\varepsilon^* i_X d\tilde{q}^{N-1}) - F_\varepsilon^* dH.$$

On recommence la même opération sur $\frac{\partial}{\partial\varepsilon} \tilde{p}^{N-1} d\tilde{q}^{N-1}$ et etc..., il vient:

$$\begin{aligned}\frac{\partial}{\partial\varepsilon}(\tilde{p}^N d\tilde{q}^N) &= -F_\varepsilon^* dH - F_\varepsilon^{*2} dH - \dots - F_\varepsilon^{*N} dH \\ &\quad + d(\tilde{p}^N (F_\varepsilon^* i_X d\tilde{q}^{N-1} + F_\varepsilon^{*2} i_X d\tilde{q}^{N-2} + \dots + F_\varepsilon^{*N} i_X dq)).\end{aligned}$$

Je pose

$$A = d(\tilde{p}^N (F_\varepsilon^* i_X d\tilde{q}^{N-1} + \dots + F_\varepsilon^{*N} i_X dq)).$$

Examinons

$$\frac{\partial \tilde{q}^N}{\partial\varepsilon} = \frac{\partial F_\varepsilon^{*N} q}{\partial\varepsilon} = F^* \frac{\partial}{\partial\varepsilon} (\exp \varepsilon X)^* \tilde{q}^{N-1}$$

$$\frac{\partial \tilde{q}^N}{\partial\varepsilon} = F_\varepsilon^* L_X \tilde{q}^{N-1} + F_\varepsilon^* \frac{\partial \tilde{q}^{N-1}}{\partial\varepsilon}$$

toujours par le même procédé, il vient

$$\frac{\partial \tilde{q}^N}{\partial\varepsilon} = F_\varepsilon^* i_X d\tilde{q}^{N-1} + \dots + F_\varepsilon^{*N} i_X dq$$

et donc

$$A = d\left(\tilde{p}^N \frac{\partial \tilde{q}^N}{\partial\varepsilon}\right) = d\tilde{q}^N \frac{\partial \tilde{q}^N}{\partial\varepsilon} + \tilde{p}^N \frac{\partial}{\partial\varepsilon} d\tilde{q}^N$$

et on a obtenu

$$\frac{\partial \tilde{p}^N}{\partial\varepsilon} d\tilde{q}^N - \frac{\partial \tilde{q}^N}{\partial\varepsilon} d\tilde{q}^N = -F_\varepsilon^* dH - F_\varepsilon^{*2} dh - \dots - F_\varepsilon^{*N} dH.$$

On fait le changement de coordonnées locales:

$$p \mapsto p$$

$$p \mapsto q' = \tilde{q}^N(\varepsilon, p, q).$$

Un calcul bref montre que le premier membre de la dernière relation est:

$$\frac{\partial}{\partial\varepsilon}(\tilde{p}^N dq' + q dp) = \frac{\partial}{\partial\varepsilon} dS_N(\varepsilon,p,q') = d\frac{\partial}{\partial\varepsilon} S_N(\varepsilon,p,q')$$

et donc cette relation s'interprète par:

$$d\frac{\partial}{\partial\varepsilon} S_N(\varepsilon,p,q') = d[-F_\varepsilon^* H - F_\varepsilon^{*2} H - \dots - F_\varepsilon^{*N} H].$$

Profitant des constantes arbitraires qui figurent dans les définitions de S_N et de H et posant $\varepsilon = 0$, on obtient:

$$\frac{\partial}{\partial\varepsilon} S_N(\varepsilon,p,q')\Big|_{\varepsilon=0} = -F^* H - F^{*2} H - \dots - F^{*N} H$$

qui est le résultat annoncé à un changement de notation près ■

Je vais maintenant préciser la perturbation de la fonction séculaire $R_N^\nu(\varepsilon,p,q')$. L'exposé gagne en clarté par le changement de notation suivant: je pose

$$S_N(\varepsilon,p,q') = S_N(p,q') + \Delta S_N(p,q') + O(\varepsilon^2)$$

où $\Delta S_N(p,q')$ est linéaire en ε. Calculons $\Delta R_N^\nu(q')$ la partie linéaire en ε de $R_N^\nu(\varepsilon,q')$.

PROPOSITION IV.3 - $\Delta R_N^\nu(q') = \Delta S_N(p,q')\Big|_{p=\psi_N^\nu(q')}$.

<u>Preuve</u> - Introduisons $\Delta\psi_N^\nu(q')$ la partie linéaire en ε de $\psi_N^\nu(\varepsilon,q')$, la paramétrisation du tore de retour des angles de F_ε^*; et remarquons que $S_N'(\varepsilon,p,q') = S_N(\varepsilon,p,q') - pq'$ donc $\Delta S_N' = \Delta S_N$.

Par définition

$$R_N^\nu(\varepsilon,p,q') = S_N'(\varepsilon,\psi_N^\nu(\varepsilon,q'),q') + \nu\psi_N^\nu(\varepsilon,q').$$

Donc

$$\Delta R_N^\nu = \Delta S_N'(p,q')\Big|_{p=\psi_N^\nu(q')} + \frac{\partial S_N'(\varepsilon,p,q')}{\partial p}\Big|_{p=\psi_N^\nu(q')} \Delta\psi_N^\nu + \nu\Delta\psi_N^\nu .$$

Mais les deux derniers termes disparaissent par définition de S_N' et de ψ_N^ν ■

On a donc obtenu

$$\Delta R_N^{\nu}(q') = -\varepsilon \sum_{k=1}^{N} (F^k)^{*N}\Big|_{p=\psi_N^{\nu}(q')}, \qquad \text{IV.3.}$$

§3. PERTURBATION D'UN ELEMENT F DE $\mathcal{F}_{c,Q,R}$ ASSOCIEE A L'UN DE SES POINTS PERIODIQUES

Soit F un élément de $\mathcal{F}_{c,Q,R}$ et (p^o,q^o) un point périodique de F de période N et d'indice ν tel que $q'^o = q^o + \nu$ soit un minimum de R_N^{ν}.

PROPOSITION IV.4 - On peut construire $H \in \mathfrak{m}^5$ de telle façon qu'au moins pour ε assez petit (p^o,q^o) soit encore un point périodique de $F_{\varepsilon} = \exp \varepsilon X \circ F$ de période N et d'indice ν et que la fonction $R_N^{\nu}(\varepsilon, q')$ présente un point de Morse en q'^o.

D'après IV.3,

$$\text{Hess } R_N^{\nu}(\varepsilon,q') = \text{Hess } R_N^{\nu}(q') + \varepsilon \text{ Hess}\Big(- \sum_{k=1}^{N} F^{*k}H\Big|_{p=\psi_N^{\nu}(q')}\Big) + O(\varepsilon^2)$$

(le tout calculé au point q^N). Il reste à choisir H de manière adéquate:

dans $\mathbb{R}^{2m}$, les N points itérés successifs par F de (p^o,q^o) forment une variété algébrique de codimension $2m$; considérons $2m$ équations algébriques f_{ℓ} qui définissent cette variété et sont indépendantes dessus:

$$\begin{cases} f_1(p^k,q^k) = 0 \\ \vdots \\ f_{2m}(p^k,q^k) = 0, \quad k = 0,\dots,N-1 \end{cases}$$

On commence par les multiplier toutes par la première coordonnée action (par exemple) p_1 à la puissance trois: posons $h_{\ell} = p_1^3 f_{\ell}$ on a encore

$$\begin{cases} h_1(p^k,q^k) = 0 \\ \vdots \\ h_{2m}(p^k,q^k) = 0, \quad k = 0,\dots,N-1. \end{cases}$$

Par ailleurs, en un point (p^k,q^k):

$$dh_1 \wedge \ldots \wedge dh_{2m} = p_1^{6m} \, df_1 \wedge \ldots \wedge df_{2m} .$$

Donc au voisinage des points (p^k,q^k), les h_ℓ sont indépendantes (je rappelle que les points (p^k,q^k) sont dans $P_2 \times T^m$ donc p_1 ne peut s'annuler sur l'un des itérés). Je choisis

$$H = - \sum_{\ell=1}^{2m} h_\ell^2 ,$$

cet H convient, en effet:

- H est polynomial et appartient à la cinquième puissance de l'idéal maximal des fonctions analytiques nulles à l'origine (lemme IV.2).

- H est critique en les points (p^k,q^k), donc ces points sont laissés fixes par $\exp \varepsilon X$ et F_ε les admet encore pour points périodiques.

Comme H est critique sur les points itérés, la signature de la Hessienne de H est indépendante dy système de coordonnées choisi pour l'exprimer. En prenant les h_ℓ pour coordonnées locales, on se persuade que Hess H est définie négative. Il en est de même de

$$\mathrm{Hess}(F^*H + \ldots + F^{*N}H)\Big|_{p=\psi_N^\nu(q')} .$$

Il résulte de IV.3 que si on choisit ε positif suffisamment petit

$$\mathrm{Hess}\ R_N^\nu(\varepsilon, q') = \mathrm{Hess}\ R_N^\nu(q') - \varepsilon\ \mathrm{Hess}(F^*H + \ldots + F^{*N}H)\Big|_{p=\psi_N^\nu(q')} + O(\varepsilon^2)$$

sera définie positive en $q' = q^N$, ce qui achève la démonstration ■

Introduisons $Z_{\nu,N}$ l'ensemble des transformations F de $\mathcal{F}_{c,Q,R}$ telles que la fonction séculaire associée R_N^ν présente un point de Morse. $Z_{\nu,N}$ est un ouvert de $\mathcal{F}_{c,Q,R}$ à cause de la stabilité des points de Morse et de la continuité du passage $F \to S_N' \to R_N^\nu$ qui se fait par transversalité.

La proposition IV.4 implique que $Z_{\nu,N}$ est partout dense dans $\mathcal{F}_{c,Q,R}$.

Dans la suite, nous aurons à utiliser le fait que

$Z = \bigcap_{\nu,N \text{ associés}} Z_{\nu,N}$	est une intersection dénombrable d'ouverts partout denses.

§4. PROPRIETES DE GENERICITE

On suppose que F_1 (partie linéaire de F_o) a des valeurs propres λ_k, $k = 1,\dots,2m$ qui satisfont la condition

$$A_1: \lambda_k \cdot \lambda_\ell = 1 \Leftrightarrow |k-\ell| = m.$$

Soit F un élément de $\mathcal{F}_{c,Q,R}$ qui est l'exponentielle d'un champ analytique hamiltonien X. Je noterai H l'hamiltonien de X

$$F = \exp X \quad \text{avec} \quad i_X\omega = -dH.$$

LEMME - Dans le compact constitué des points (p,q): $|p| \leq r$ et $p \in P_o$. Avec r: rayon de convergence de H en tant que fonction analytique.

$$\frac{\partial H}{\partial p} = \alpha + O(|p|^{1/2})$$

où α est un vecteur constant non nul.

Preuve - H est défini à une constante arbitraire près de sorte qu'on peut imposer $H(0) = 0$; d'autre part, F laisse l'origine fixe et donc H est critique à l'origine. Le développement de Taylor de H commence par une partie quadratique H_2:

$$H(x,y) = H_2(x,y) + H_3(x,y), \quad \text{avec} \quad H_3 \in \mathfrak{m}^3 ;$$

$\mathfrak{m}$ désignant l'idéal maximal des fonctions analytiques nulles à l'origine. Remarquons que sur le domaine en question

$$H_3 = O(|p|^{3/2}).$$

Ecrivons que F admet H pour fonction invariante: $F^*H = H$. Le développement suivant les composantes homogènes donne au premier ordre:

$$F_1^* H_2 = H_2 .$$

Je rappelle que F_1 se diagonalise dans le complesifié C^{2m} suivant les axes isotropes.

Notons $X_i = x_i + \sqrt{-1} y_i$, $X_{i+m} = x_i - \sqrt{-1} y_i$ pour $i=1,\ldots,m$. H_2 est une certaine forme quadratique en les X_k $(k = 1,\ldots,2m)$,

$$H_2(x,y) = \sum_{k,\ell} \alpha_{k\ell} X_k X_\ell$$

mais $F_1^* H_2 = H_2$ implique

$$\sum_{k,\ell=1}^{2m} \alpha_{k\ell} \lambda_k \lambda_\ell X_k X_\ell = \sum_{k,\ell=1}^{2m} \alpha_{k\ell} X_k X_\ell$$

par conséquent, A_1 donne

$$H_2(x,y) = \sum_{i=1}^{m} (\alpha_{i,i+m} + \alpha_{i+m,i}) X_i X_{i+m}$$

$$H_2(x,y) = 2 \sum_{i=1}^{m} (\alpha_{i,i+m} + \alpha_{i+m,i}) p_i^2 .$$

Ce que j'écris en notation condensée $\frac{\partial H_2}{\partial p} = \alpha$ où α est un certain vecteur constant.

Remarquons que α ne peut être nul, s'il l'était, en effet, H_2 serait nul et F serait tangente à l'identité à l'ordre 2 ce qui contredirait l'hypothèse A_1.

L'utilisation de la formule de Cauchy donne

$$\frac{\partial H_3}{\partial p} = O(|p|^{1/2})$$

et le résultat du lemme $\frac{\partial H}{\partial p} = \alpha + O(|p|^{1/2})$ se trouve précisé.

Cela étant fait, on peut énoncer la première propriété de généricité.

PROPOSITION IV.5 - Si la partie linéaire de F_o satisfait A_1 un élément F de $\mathcal{F}_{c,Q,R}$ n'est pas génériquement l'exponentielle d'un champ hamiltonien analytique.

Preuve - Soit F un élément de $\mathcal{F}_{c,Q,R}$ qui est l'exponentielle d'un champ analytique X avec:

$$F = \exp X, \qquad i_X\omega = -dH.$$

Je vais montrer que F ne peut appartenir au résiduel $Z = \bigcap_{\nu,N} Z_N^\nu$. Supposons par l'absurde que F appartient à Z. Pour tout indice ν et période N associés on peut donc trouver un point périodique de F dont la coordonnée angulaire est un point de Morse de R_N^ν .

En fait, il existe une suite de tels points périodiques tendant vers l'origine; ainsi que le précise la proposition III.7 et la remarque qui la suit.

Insistons maintenant sur le fait que H est invariante par F. Il implique:

$$dH(TF^N - Id) = 0.$$

Puisque F appartient à Z, TF^N ne peut avoir 1 pour valeur propre (proposition IV.1) en conséquence H est critique au point périodique considéré.

Il est en fait nécessaire que H soit critique sur tout les points de la suite qui tend vers l'origine (proposition III.7). Or le lemme qui donne $\frac{\partial H}{\partial p} = \alpha + O(|p|^{1/2})$ avec α différent de zéro fournit une estimation qui finit par entrer en contradiction avec cette exigence.

La contradiction obtenue exprime que F ne peut appartenir à Z et prouve en cela la propriété de généricité ■

On suppose maintenant

A_2 : les valeurs propres λ_k de F_1 ne satisfont aucune relation multiplicative sur $\mathbb{Z}$ hormis celles déduites de $\lambda_i\lambda_{i+m} = 1, \quad i = 1,\dots,m.$

La même méthode prouve:

PROPOSITION IV.6 - <u>Si</u> F_1 <u>satisfait</u> A_2 <u>un élement</u> F <u>de</u> $\mathcal{F}_{c,Q,R}$ <u>ne peut admetre génériquement de fonctions invariantes analytiques non constantes</u>.

C'est-à-dire que, génériquement, il ne peut exister de fonction ψ analytique non constante avec $F^*\psi = \psi$. On peut toujours supposer $\psi(0) = 0$.

<u>Démonstration</u> - On développe ψ en série de Taylor au voisinage de l'origine. Soit ψ_ℓ sa première composante homogène non nulle

$$\psi = \psi_\ell + \psi_{\ell+1} \quad \text{où} \quad \psi_{\ell+1} \text{ appartient à } \mathfrak{m}^{\ell+1}.$$

Le développment suivant les composantes homogènes de $F^*\psi = \psi$ conduit à

$$F_1^*\psi_\ell = \psi_\ell \ .$$

On se convainc alors que A_2 implique que ℓ est forcément pair et que ψ_ℓ n'est fonction de x et y que par l'intermédiaire de p. Posons $\ell = 2h$, donc:

$$\psi = \psi_{2h}(p) + O(|p|^{h+1/2})$$

sur le compact $|p| < r'$, $p \in P_o$ (avec r': rayon de convergence de ψ en tant que fonction analytique).

$\dfrac{\partial\psi_{2h}(p)}{\partial p}$ est une certaine fonction homogène de degré $h-1$ en p. La sous-variété de $\mathbb{R}_+^{2m}$ constituée par les p tels que $\dfrac{\partial\psi_{2h}(p)}{\partial p} = 0$ ne peut contenir tout un cône P'_o. Prenons p_o dans P'_o tel que

$$\left|\frac{\partial\psi_{2h}(p)}{\partial p}\right|_{p=p_o} > \varepsilon > 0$$

par continuité sur tout un petit arc γ tracé à $|p|$ constant autour de p_o, on aura encore $\left|\dfrac{\partial\psi_{2h}(p)}{\partial p}\right| > \varepsilon > 0$ et dans le voisinage

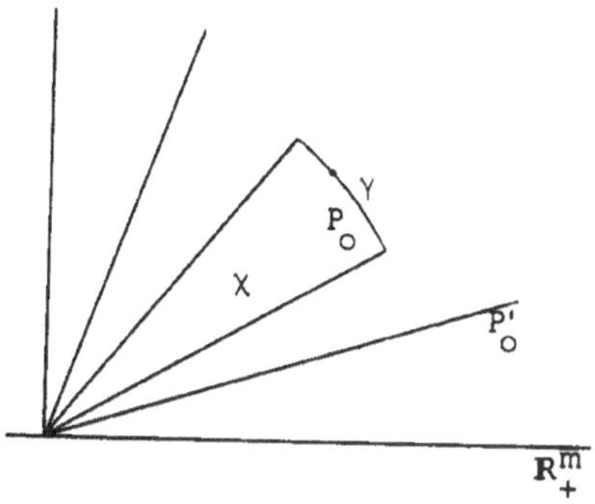

conique χ engendré par γ; du fait que ψ_{2h} est homogène de degré 2h

$$\left|\frac{\partial\psi_{2h}(p)}{\partial p}\right| > \varepsilon \frac{|p|^{h-1}}{|p_o|^{h-1}} .$$

De sorte que dans χ :

$$\frac{\partial\psi}{\partial p} = \frac{\partial\psi_{2h}}{\partial p}(p) + O(|p|^{h-3/4})$$

ne peut s'annuler dans un voisinage pointé de l'origine.

Un élément F qui appartient à Z ne peut avoir ψ pour fonction invariante.

Raisonnons à nouveau par l'absurde: on suppose qu'un élément F de Z admet ψ pour fonction invariante. ψ devrait être critique sur tous les points périodiques; il suffit de considérer une suite de points périodiques qui ont pour coordonnées angulaires des points de Morse de R_N^ν et qui tend vers l'origine dans χ (cf. prop. III.7) pour aboutir à une contradiction □

Cette proposition est à comparer avec le théorème d'absence générique d'intégrale première indépendante de l'hamiltonien démontré pour les champs hamiltoniens par J. Vey dans [12].

Elle a pour conséquence la divergence générique de la forme normale de Birkhoff des transformations analytiques.

Un élément de $\mathcal{F}_{c,Q,R}$ dont la forme normale converge possède, en effet, m fonctions invariantes non constantes.

PROPOSITION IV.7 - Si $F_1 = \mathrm{Id}$. Un élément F de $\mathfrak{F}_{c,Q,R}$ n'est pas génériquement l'exponentielle d'un champ analytique.

Démonstration - Ecrivons $F = \exp X$, $i_X\omega = -dH$ dans les coordonnées linéaires (x_i,y_i), $i = 1,\dots,m$. (I.2) devient (je rappelle que F est tangente à F_o à l'ordre 4)

$$F: \begin{cases} x_i^1 = x_i - y_i Q_i^j p_j + \eta_i \\ y_i^1 = y_i + x_i Q_i^j p_j + \zeta_i . \end{cases}$$

$\eta_i \in \mathfrak{m}^4$ et $\zeta_i \in \mathfrak{m}^4$; en développant $x_i^1 = x_i + X\cdot x_i + \frac{X^2}{2!}\cdot x_i + \dots$ et $y_i^1 = y_i + X\cdot y_i + \frac{X^2}{2!}\cdot y_i + \dots$ il vient

$$X_i = -\frac{\partial H}{\partial y_i} = -y_i Q_i^j p_j + h_i$$

$$X_{i+m} = \frac{\partial H}{\partial x_i} = x_i Q_i^j p_j + h_i'$$

avec h_i' et h_i dans $\mathfrak{m}^4$. Donc

$$\frac{\partial H}{\partial p_i} = [x_i \frac{\partial H}{\partial x_i} + y_i \frac{\partial H}{\partial y_i}]\frac{1}{p_i} = Q_i^j p_j + \frac{x_i h_i' - y_i h_i}{p_i} .$$

Et dans le compact $|p| < r''$, $p \in P_o$ (r'' désignant le rayon de convergence de H en tant que fonction analytique). En notation condensée:

$$\frac{\partial H}{\partial p} = Qp + o(|p|^{3/2}). \tag{IV.7}$$

On conclut tout comme pour (IV.6) en choisissant un p_o tel que $|Qp_o| > \varepsilon > 0$ puis, ayant tracé un arc γ à $|p|$ constant sur lequel on s'assure de $|Qp| > \varepsilon > 0$, on considère une suite de points périodiques qui tend vers l'origine dans le cône χ engendré par γ.

Là encore, un élément F de $\mathfrak{F}_{c,Q,R}$ qui est l'exponentielle d'un champ hamiltonien analytique ne peut appartenir à Z de par l'estimation (IV.7) ∎

V. COMPARAISON DU CAS FORMEL ET DU CAS ANALYTIQUE

A un élément F de $\mathcal{F}_{c,Q,R}$ considéré comme une transformation de $\mathbb{C}^{2m}$ on associe la transformation formelle de $\mathbb{C}^{2m}$: $\hat{F}$ obtenue en écrivant les développements de Taylor des fonctions composantes de F au voisinage de l'origine.

Dans [10], S. Sternberg démontre le résultat suivant qu'il attribue à Lewis:

THÉORÈME V.1 - Soit $\hat{F}$ une transformation formelle de $\mathbb{C}^{2m}$; $\lambda_1,\dots,\lambda_{2m}$ les valeurs propres de la partie linéaire F_1. S'il existe un logarithme des λ_i: $k_i = \log \lambda_i$ $(i = 1,\dots,2m)$ satisfaisant la condition (S):

(S) "si $\sum_i n_i k_i = 2\pi\sqrt{-1}n$, $n_i \in \mathbb{Z}$, $n \in \mathbb{Z}$, alors $n = 0$ ".

$\hat{F}$ est l'exponentielle d'un champ formel $\hat{X}$ (uniquement déterminé par le choix du logarithme des λ_i) de plus si $\hat{F}$ est symplectique $\hat{X}$ est hamiltonien.

Pour la démonstration, je renvoie à la référence [10].

Il est classique qu'une transformation formelle telle que les valeurs propres de la partie linéaire sont multiplicativement indépendantes sur Z peut être interpolée par un groupe à un paramètre de transformations formelles (donc est l'exponentielle d'un champ formel).

Les transformations symplectiques, par leur nature même, ont une partie linéaire qui ne satisfait pas cette condition d'indépendance.

L'intérêt du théorème précédent est qu'il peut s'appliquer à certaines transformations symplectiques:

PROPOSITION V.2 - Soit F un élément de $\mathcal{F}_{c,Q,R}$. Si les valeurs de F_1 satisfont la condition A_2 on sont toutes égales à 1; $\hat{F}$ est l'exponentielle d'un champ formel hamiltonien.

Démonstration - Si toutes les valeurs propres de F_1 sont égales à 1, on prend $k_i = 0$, $i = 1,\dots,2m$. Ce logarithme satisfait évidemment (S). Réécrivons la condition A_2 sous la forme

$$" \prod_{i=1}^{2m} \lambda_i^{n_i} = 1, \quad n_i \in \mathbb{Z} \Rightarrow n_j = n_{j+m}, \quad j = 1,\dots,m \; ".$$

F_1 a pour valeurs propres $\lambda_1,\dots,\lambda_{2m}$ avec $\lambda_j = \bar{\lambda}_{j+m}$, $|\lambda_i| = 1$. Aucune de ces valeurs propres ne peut être réelle négative car elle serait égale à -1 ce qui contredirait A_2.

Je prends pour logarithme des λ_i celeui obtenu après coupure suivant la demi-droite réelle négative. Posons $k_i = \log \lambda_i$, on vérifie $k_{j+m} = -k_j$ pour $j = 1,\dots,m$ car $\lambda_j \cdot \lambda_{j+m} = 1$.

Montrons que si F_1 satisfait A_2 les k_i satisfont (S). Supposons que

$$\sum_i n_i k_i = 2\pi \sqrt{-1} n \quad \text{avec} \quad n_i \in \mathbb{Z}, \quad n \in \mathbb{Z}.$$

Alors en prenant les exponentielles, on obtient

$$\prod_i \lambda_i^{n_i} = 1, \quad \text{donc} \quad n_j = n_{j+m} \quad (j=1,\dots,m)$$

$$\sum n_i k_i = \sum_j n_j (k_j + k_{j+m}) = 0$$

et par consequent

$$n = 0 \quad \blacksquare$$

Je dirai à la suite du théorème V.1 que les éléments de $\mathcal{F}_{c,Q,R}$, lorsque F_1 satisfait les conditions requises, sont "formellement" des exponentielles de champ hamiltonien.

On peut en conclusion préciser l'énoncé des propositions IV.5 et IV.7 de la façon suivante:

PROPOSITION V.3 - Si les valeurs propres de F_1 satisfont A_2 ou sont toutes égales à 1; un élément F de $\mathcal{F}_{c,Q,R}$ est "formellement" l'exponentielle d'un champ hamiltonien $\hat{X}$ et génériquement sur les termes de F d'ordres supérieurs ou égaux à quatre; les séries entières associées à $\hat{X}$ ont des domaines de convergence d'intersection réduite à l'origine.

REFERENCES

[1] A.D. BRJUNO

"Analytical form of differential equations" Transactions of the Moscow mathematical society t 25 p 131-288 (1971).

[2] F. BRUHAT

Séminaire Bourbaki, $13^{ém}$ année nº 217 p 1-18 (1960-61).

[3] J.P. FRANÇOISE

"Points périodiques des transformations canoniques" Comptes rendus à l'académie des Sciences de Paris t 286 p 1215-1217 (1978).

[4] J. MÖSER

"Proof of a generalized form of a fixed points theorem" p 464-495 Geometry and Topology - Rio de Janeiro July 1976 Lect. Notes in maths. nº 597.

[5] J. MOSER

"Non existence of Integrals" Comm. Pure and Apply. Math. Vol. 8 (1955) p 409-436.

[6] J. MÖSER and W. KYNER

Mem. of Ann. Math. Soc. nr 81 (1968).

[7] H. RÜSSMANN

"Über die Existenz einer Normal form inhaltstreuer elliptisher Transformationen" Math. Annalen Bd 137 p 64-77 (1969).

[8] H. RÜSSMANN

"On the convergence of power series transformations of analytic mappings near a fixed point into a normal form" Preprint of IHES, June 1977.

[9] C.L. SIEGEL

"Über die Existenz einer Normal form analytischer Hamiltonscher Differentialgleichungen in der Nähe einer Gleichzewischtslösung" Math. Ann. 128 p 144-170 (1954).

[10] S. STERNBERG

"Infinite Lie groups and the formal aspects of Dynamics" Journal of Math. and Mechanics vol 10 nº 3 p 451-476 (1961).

[11] S. STERNBERG

"The structure of local homeomorphisms III" Amer. Journal of Mathematics t 81 p 578-604 (1959).

[12] J. VEY

"Orbites périodiques d'un système hamiltonien au voisinage d'un point d'équilibre" Annali della Scuola Normale Superiore di Pisa, Serie IV vol V nº 4 p 757-787 (1978).

[13] E. ZEHNDER

"Generalized implicit function theorem with applications to some small divisor problems I" Comm. Pure Appl. Math. vol 28 p 91-140 (1975).

Institut Fourier
B.P. 116
Saint Martin d'Héres
38402 FRANCE

et

Instituto de Matemática Pura e Aplicada (IMPA)
Rua Luiz de Camões, 68
20.060 - Rio de Janeiro
BRASIL

PERIODIC POINTS AND TWISTED COEFFICIENTS

David FRIED*)
Mathematics Department
University of California
Santa Cruz, Ca. 95064 U.S.A.

We will develop in this paper certain topological methods for proving that a map has infinitely many periodic points. As our techniques are homological they are computable and they apply equally to all maps in an appropriate homotopy class.

Under various special hypotheses, there are a number of theorems that prove a map has infinitely many periodic points. The classical theorem is Poincaré's Last Theorem which applies to an area preserving twist map of the annulus [B] . A more general and topological theorem is that of Shub and Sullivan for a C^1 map whose iterates have arbitrarily large Lefschetz numbers [SS] . Halpern has determined certain homotopy classes on the torus in which all continuous maps have infinitely many periodic points by showing that the Nielsen numbers of the iterates are unbounded [H] . The investigations of this paper were motivated by these last two papers.

To prove topologically that a given continuous map has infinitely many periodic points it is not enough to show that the Lefschetz numbers are unbounded (see Shub's example in section 4 below). This suggests that one should compute the Nielsen numbers of the iterated map. But typically this will be impossible since one must solve an infinity of unrelated problems, each of which involves awkward computations in a non-abelian group. What is done in this paper is to present an abelian theory that carries partial information on the Nielsen classes of the iterates of a map and their respective fixed point indices. It is both computable and applicable to maps that aren't necessarily C^1 .

The main tool we use is an equivariant version of the Lefschetz fixed point theorem proven in section 1. Essentially we use the action of our map on the chains of an appropriate abelian cover (that is chains in the base with twisted coefficients) to compute the indices of certain collections of fixed points. One may also describe this as measuring how the lifts of fixed points move in this cover. This overlaps in part with a fixed point theorem of Fadell and Husseini [FH] but is better suited to dynamical applications. We apply our theorem to the iterates of a map and use a suitable zeta function $\tilde{\zeta}_H(t)$ to express the results efficiently. This "twisted Lefschetz zeta function" is developed in section 2 and compared there to other zeta

*) Partially supported by the National Foundation Grant N° MCS-8003622.

functions.

Corresponding to the two sides of the Lefschetz formula, the twisted Lefschetz zeta function has two forms. One form is a rational function found by viewing the Lefschetz numbers as homological invariants (Theorem 2). The other form is a product of power series, one for each periodic orbit. These power series are defined and discussed in section 3. The two forms of $\tilde{\zeta}_H(t)$ are equal by the twisted product formula of Theorem 5 below.

In section 4, we apply this product formula to a map with finitely many periodic points. For a C^1 map, we obtain rather sharp results (Theorem 6) that generalize the Shub-Sullivan theorem cited above. For a continuous map, we obtain nontrivial computable constraints whenever the induced map on first integral homology isn't nilpotent (Lemma 6). These constraints are on the irreducible factors of the rational function $\tilde{\zeta}_H$ (Theorem 10) or the highest coefficients in the numerator and denominator of $\tilde{\zeta}_H$ (Theorems 11 and 12). Our constraints on nondifferentiable maps depend on a technical algebraic theorem that is proved in the Appendix (section 6).

In section 5, we apply our theory to some maps of 2 dimensional complexes. For some of these maps, all their iterates have Lefschetz number zero and Nielsen number two yet any map of that homotopy type must have infinitely many periodic points.

We are preparing some related papers that build on this one. One paper will apply twisted cohomology to estimate the entropy of certain maps [Fr3] , generalizing Shub's entropy conjecture. The examples in that paper also illustrate our theorems on periodic points. Another paper will prove identities for the periodic orbits of certain flows by applying the equivariant Lefschetz formula to certain Poincaré maps [Fr2]. These results generalize Franks' work on Smale flows on S^3 [F 1]. We will apply this periodic point theory to compute some asymptotic growth rates of surface homeomorphisms [Fr4, cf. T].

We will conclude with an open question. Let X be a compact simply connected manifold and $f : X \to X$ a continuous map. The converse to the Lefschetz fixed point theorem implies that any iterate of f is homotopic to a map with at most one fixed point. It is conceivable (i.e., consistent with the theory of fixed point classes) that f is homotopic to a map with no more than one periodic point. If X is a sphere, one may indeed produce such a map in each homotopy class by collapsing an invariant interval joining the two periodic points in Shub's example (see section 4). These examples suggest

Question. Does every homotopy class of maps on a compact simply connected manifold contain a representative with a finite number of periodic points ?

An affirmative answer would show in a strong way that the use of fixed point classes and covering spaces in this paper is essential for the problem we study.

1. The Equivariant Lefschetz Fixed Point Formula.

We consider the following simple geometric situation. X is a finite complex and $\tilde{X}$ is a regular covering space of X with deck transformation group H, so that $\tilde{X}/H = X$. We take $f : X \to X$ a continuous self-map of X that lifts to $\tilde{X}$ and we assume that every lift $\tilde{f}$ commutes with every deck transformation $h \in H$.

Already, we may deduce something about H that will be useful in later sections

Lemma 1. H is abelian.

Proof. For any $h \in H$, $h\tilde{f}$ is a lift of f. By assumption, $h\tilde{f}$ commutes with g for any $g \in H$. Thus $g(h\tilde{f}) = (h\tilde{f})g = (\tilde{f}h)g = \tilde{f}(hg) = (hg)\tilde{f}$. Taking $y \in \tilde{X}$, we have $gh(\tilde{f}y) = hg(\tilde{f}y)$ so $gh = hg$. Q.E.D.

Our hypothesis assures that the fixed point set $\mathrm{Fix}(\tilde{f})$ is H-invariant. We associate an algebraic weight $i(\tilde{f}) \in \mathbb{Z}$ to this fixed point set by (essentially) summing the fixed point indices of all fixed points in a fundamental domain in $\tilde{X}$. As there may be an infinity of such points, however, we must proceed less hastily. First observe that to each $x \in \mathrm{Fix}(f)$ there is a unique $h \in H$ such that $\tilde{f}y = hy$ for all $y \in \tilde{X}$ over x. Thus $\mathrm{Fix}(f)$ is the disjoint union of the sets $S_h = \mathrm{Fix}(h^{-1}\tilde{f})/H$. Clearly S_h is closed for all h. As $\tilde{f}$ is H-equivariant and X is compact, $\tilde{f}$ is a bounded distance away form the identity and the sets $\mathrm{Fix}(h^{-1}\tilde{f})$ are empty for all but finitely many h. Thus the S_h form a finite partition of $\mathrm{Fix}(f)$ into open-closed sets. By standard index theory, there is thus a fixed point index $\mathrm{ind}(f,S_e) \in \mathbb{Z}$ attached to the closed-open subset S_e and we let $i(\tilde{f}) = \mathrm{ind}(f,S_e)$.

Now, regard H as a multiplicative group and form the integral group ring $\mathbb{Z}H$ of all finite formal sums of elements of H. We have that $i(h^{-1}\tilde{f}) = 0$ whenever S_h is empty, so we may define the generalized Lefschetz number $L_H(\tilde{f}) = \Sigma\, i(h^{-1}\tilde{f})\cdot h \in \mathbb{Z}H$. This agrees with a more general definition due to Fadell and Husseini; we will compare their results with ours after Theorem 1. Note that the coefficient of h in $L_H(\tilde{f})$ counts the algebraic number of solutions of $\tilde{f} = h$ per fundamental domain.

In terms of index theory, our sets S_h are unions of fixed point classes (this also shows that they are open in Fix f). It follows easily that $L_H(\tilde{f})$ is a homotopy invariant. More precisely if f_t is a homotopy and $\tilde{f}_t$ a lift, then $L_H(\tilde{f}_t)$ is independent of t. This suggests that one seeks a formula for $L_H(\tilde{f})$ in terms of the topology of $\tilde{X}$ and the action of $\tilde{f}$, generalizing the Lefschetz Fixed Point Formula for the usual Lefschetz number $L(f) = L_1(f)$ (i.e., the case $H = 1$).

Accordingly, we lift the triangulation of X to an H-equivariant triangulation of $\tilde{X}$. H permutes the i-simplices of X and so makes the i-chains $C_i(\tilde{X};\mathbb{Z})$ into a $\mathbb{Z}H$ module. Picking one simplex in $\tilde{X}$ over each simplex in X gives a free basis for $C_i(\tilde{X};\mathbb{Z})$ over $\mathbb{Z}H$. To discretize the action of $\tilde{f}$, we homotop f to a cellular map F and lift this homotopy to obtain a cellular map $\tilde{F}$ homotopic to $\tilde{f}$. As usual one associates chain maps $F_i : C_i(X,\mathbb{Z})\circlearrowleft$ and $\tilde{F}_i : C_i(\tilde{X},\mathbb{Z})\circlearrowleft$ to these cellular maps for $i = 0,1,2,\ldots$. Using the natural integral basis of $C_i(\tilde{X},\mathbb{Z})$ and the $\mathbb{Z}H$-basis chosen above for $C_i(\tilde{X},\mathbb{Z})$, we may regard F_i and $\tilde{F}_i$ as square matrices, with entries in $\mathbb{Z}$ and $\mathbb{Z}H$ respectively. Clearly $\tilde{F}_i$ is obtained from F_i by substituting 1 for h throughout, that is by applying the augmentation homomorphism $\varepsilon : \mathbb{Z}H \to \mathbb{Z}$. We obtain

<u>Theorem 1</u> (Equivariant Lefschetz Formula). Let X, $\tilde{X}$, H, f, and $\tilde{f}$ be as in the first paragraph of this section and let F_i , $\tilde{F}_i$ and ε be as in the preceding paragraph. Then

$$\text{(E.L.F.)} \qquad L_H(\tilde{f}) = \sum_{i\geq 0} (-1)^i \operatorname{Trace}(\tilde{F}_i)$$

and $\varepsilon(L_H(\tilde{f})) = L(f)$.

<u>Proof</u>. We note that the last formula follows immediately from the additivity of the fixed point index [Br] and the finite partition of $\mathrm{Fix}(f)$ into sets S_h , since it says precisely that $\sum_h \operatorname{ind}(f,S_h) = \operatorname{ind}(f,X) = L(f)$.

We now gradually reduce the first formula to the usual Lefschetz formula. For the first reduction, we observe that a component of X that isn't mapped to itself contributes to neither side of the formula and may be ignored. As the formula is additive in the invariant components, we may suppose X is connected (delete from $\tilde{X}$ all points not over this component).

Second, we will show that we may suppose $\tilde{f}$ preserves the components of $\tilde{X}$. For some h , $h\tilde{f}$ preserves one component of $\tilde{X}$ and hence all of them. But passing from $\tilde{f}$ to $h\tilde{f}$ affects both sides of our formula the same way, namely each side is multiplied by h . So if the formula holds for one lift, it holds for all.

Next, we reduce to the case when $\tilde{X}$ is also connected. If a deck transformation $h \in H$ leaves one component of $\tilde{X}$ invariant then it leaves them all invariant. Such h's form a subgroup $H' \subset H$. Clearly if S_h is nonempty then $h \in H'$, so $L_H(\tilde{f}) \in \mathbb{Z}H'$. But we have arranged that $\tilde{f}$ preserve the components of X , so that $\tilde{F}$ also preserves components and $\tilde{F}_i$ is defined over $\mathbb{Z}H'$. We pass from $\tilde{X}$ to some component, restrict $\tilde{f}$, and use H' instead of H .

We now have that X , $\tilde{X}$ are connected. Connected regular covers correspond to quotient groups of $\pi_1(X)$, so H is finitely generated. Suppose the formula fails and let $\Delta \in \mathbb{Z}H$ be the nonzero difference between the two sides, $\Delta = \Sigma\Delta_h h$. Let $S = \{h \in H \mid \Delta_h \neq 0\}$. Since H is finitely generated and abelian, it is easy to find a finite quotient group H_0 of H such that S maps 1-1 into H_0 . Using the obvious naturality of both terms in the formula with respect to intermediate covers, it it clear that the formula fails for the cover corresponding to H_0 as well.

Thus, we reduce to the case of H finite, say $|H| = n < \infty$. Since $\tilde{X}$ is an n-sheeted cover of X , it is a finite complex. So the usual Lefschetz formula applies to self-maps of $\tilde{X}$. Pass from our $\mathbb{Z}H$-basis B_i for $C_i(\tilde{X};\mathbb{Z})$ to the usual integral basis $\{g\beta \mid g \in H , \beta \in B_i\}$. Any $\sigma \in B_i$ such that $F\sigma$ wraps around $h\sigma$ with degree k contributes $k\cdot h \in \mathbb{Z}H$ to $\mathrm{Trace}_{\mathbb{Z}H}(\tilde{F}_i)$ but the translates $g\sigma$, $g \in H$, contribute a total of $nk \in \mathbb{Z}$ to $\mathrm{Trace}_{\mathbb{Z}}(h^{-1}\tilde{F}_i)$. Thus

$$\sum_i (-1)^i \mathrm{Trace}_{\mathbb{Z}H}(\tilde{F}_i) = \frac{1}{n}\sum_{h,i} (-1)^i \mathrm{Trace}_{\mathbb{Z}}(h^{-1}\tilde{F}_i)\cdot h \quad .$$

Since each fixed point $x \in S_h$ of f corresponds to n fixed points of the same index for the lift $h^{-1}\tilde{f}$, we have

$$L_H(f) = \sum_h \mathrm{ind}(f,S_h)\cdot h = \sum_h \frac{1}{n}\,\mathrm{ind}(h^{-1}\tilde{f},\tilde{X})\cdot h = \frac{1}{n}\sum_h L(h^{-1}\tilde{f})\cdot h \; .$$

So when H is finite, we see that equality holds by applying the usual Lefschetz formula to each lift $h^{-1}\tilde{f}$. Q.E.D.

Remarks. 1) In case H is finite this formula is a special case of results of Fadell and Husseini [FH] who did not make the assumption of equivariance. Their paper also considers infinite covers but they made assumptions on the lift $\tilde{f}$ that cannot hold when $|H| = \infty$ and $\tilde{f}$ is equivariant. Perhaps there is a common generalization of these theorems. Theorem 1 is ideally suited for the study of iterates of f , i.e., for the dynamical applications of this paper.

2) Given a finite complex X and a continuous self-map $f : X \circlearrowleft$, we may produce the covers $\tilde{X}$ and groups H in a purely algebraic manner. One readily reduces to the case when X is connected and then the following construction gives all the connected covers $\tilde{X}$ satisfying the hypothesis of this section. Consider the action of f on first integral homology $f_* : H_1(X;\mathbb{Z}) \circlearrowleft$. Let H be a quotient group of $H_1(X;\mathbb{Z})$ on which f acts and acts trivially. Then there is a connected cover $\tilde{X}$ with deck transformation group H defined by the usual construction, so that a loop γ lifts to $\tilde{X} \Leftrightarrow \gamma$ maps to $0 \in H$. As $f_* : H \circlearrowleft$ is the identity, all lifts of f commute with H .

We see that there is a natural choice for H , namely the cokernel H_{max} of $(f_*-\mathrm{id}) : H_1(X;\mathbb{Z}) \circlearrowleft$. Any other H satisfying the conditions of this section and associated to a connected cover is a quotient of H_{max} . The covering spaces obtained in this way correspond to all the intermediate covers $\tilde{X} = \tilde{X}_{max}/G$ with G a subgroup of H_{max} and $H = H_{max}/G$ .So the most information is obtained from applying the theorem to H_{max} , although it is often convenient to use it for other H . For instance, if $f^* : H^1(X;\mathbb{Z})\circlearrowleft$ fixes a cohomology class u then the infinite cyclic cover $\tilde{X}$ corresponding to $u : H_1(X,\mathbb{Z}) \to \mathbb{Z}$ satisfies our hypothesis. The cover $\tilde{X}$ is connected iff X is connected and u is indivisible (i.e., maps onto $\mathbb{Z}$) and then u specifies an infinite cyclic quotient H of H_{max} .

3) By standard approximation techniques [Br] one may extend the equivariant Lefschetz formula to the case when X is a compact ANR .

4) By using the generalized notion of trace due to Stallings [St] , one may replace $\tilde{F}_i$ in Theorem 1 by the $\mathbb{Z}H$-module map on twisted homology $\tilde{f}_* : H_i(\tilde{X},\mathbb{Z})\circlearrowleft$. One needs this extended definition because $H_i(\tilde{X};\mathbb{Z})$ is usually not free over $\mathbb{Z}H$. It is a well-known algebraic result that when generalized traces are used, the generalized Lefschetz number is the same on the chain level as on the homology level [St].

5) Suppose H is infinite cyclic and that the rational homology of $\tilde{X}$ is finite dimensional. Then the homology of $\tilde{X}$ over the quotient field of $\mathbb{Z}H$ vanishes It follows that $L_H(\tilde{f}) = 0$. By Theorem 1, $L(f) = 0$. This result was proven and used in Hirsch's study of the homology of Anosov diffeomorphisms [Hi].

6) While the Lefschetz number cannot be computed from the definition without detailed knowledge of the dynamics of f , one needs only a crude understanding of f to produce a cellular approximation F . One can then compute the action of f on $H_1(X;\mathbb{Z})$ and find an appropriate H (as in remark 4). Most importantly, one can proceed to compute the chain maps $\tilde{F}_i$ (and hence $L_H(f)$) without constructing $\tilde{X}$.

To simplify the computation, we assume X has a cell complex structure with only one vertex rather than a triangulation. Then one chooses F a cellular approximation to f . For each cell σ choose a basepath γ_σ from the vertex v to the center p_σ of σ . We may suppose $F^{-1}(p_\sigma) \cap \tau$ is finite whenever $\dim \tau = \dim \sigma = i$ by homotoping F further if need be. To each $x \in F^{-1}(p_\sigma) \cap \tau$, one may associate an integral weight and an element of H . The integer is the degree $d = d(x)$ of the map $F : S_\epsilon^{i-1}(x) \to (\sigma - p)$ where $S_\epsilon^{i-1}(x)$ is an ϵ sphere around x in τ . The element $h = h(x) \in H$ is the class of the loop $f(\gamma_\tau)\cdot f(\sigma_x)\cdot\gamma_\tau^{-1}$, where σ_x is a path in $\mathrm{int}(\tau)$ from p_τ to x . Intuitively, d is the weight of x and h is its position (see Figure 1). Since H is generated by the 1 cells of X , this class h is not hard to compute. Finally, the chain map $\tilde{F}_i$ is defined by $\tilde{F}_i(\tau) = \sum_\sigma(\Sigma d(x)\cdot h(x))\sigma$ where x varies over $F^{-1}(p_\sigma) \cap \tau$.

It is well known that these $\tilde{F}_i$ serve to compute the action of $\tilde{f}$ on twisted cohomology of X, and hence may be used in Theorem 1. The choices of basepath are tantamount to choosing a $\mathbb{Z}H$-basis for chains in $\tilde{X}$ and $\tilde{F}$ is implicitly lift of F that fixes the 0-cells (so $\tilde{F}_0$ is the identity map of $\mathbb{Z}H$).

There is an evident analog of this computation for a fitted diffeomorphism with one 0-handle, in which i-cells are replaced by i-handles [cf. F2].

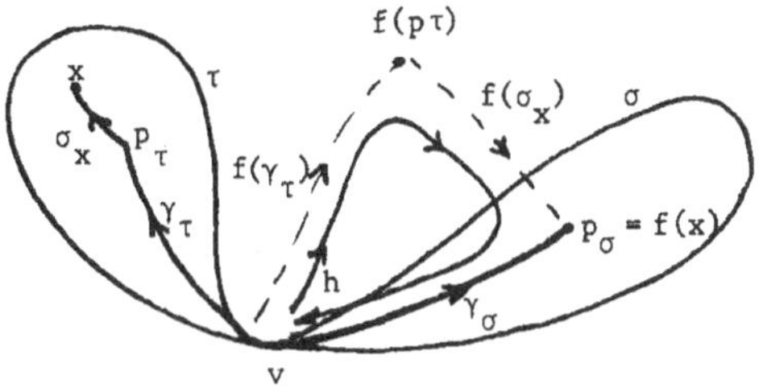

Figure 1. Constructing $h = h(x)$.

2. Twisted Zeta Functions for Dynamical Systems.

In this section, we will develop a zeta function $\tilde{\zeta}_H(t)$ that will later be used to interpret what the sequence $L_H(f^n)$ of generalized Lefschetz numbers tells us about the periodic point structure of f .

We will begin by summarizing the prior uses of zeta functions in dynamics, both to motivate our definitions and to differentiate the zeta function used in this paper from other related zeta functions. The following questions help to distinguish one zeta function from another. Is the dynamical system a flow or a map ? Is a periodic orbit counted geometrically (with weight one), algebraically (with weight given by index theory), or analytically (with weight given by summing a continuous function over an orbit) ? Does one distinguish orbits by their homology, homotopy or neither ? In cases when the zeta function is rational, what ring do its coefficients belong to ?

The original zeta function in dynamics was that of Artin and Mazur [AM]. One counts fixed points of a map geometrically, no topological invariants are mentioned and one obtains in certain favorable instances a rational function over $\mathbb{Q}$. Their definition holds for a map f with $N_n < \infty$ points of period n , $n = 1,2,\ldots$, namely

$$\zeta(t,f) = \zeta(t) = \exp\Big(\sum_{n>0} \frac{N_n}{n} t^n\Big) .$$

Smale and Ruelle replaced N_n by algebraic or analytic weights and extended the definitions to flows [R,Sm]. The use of $\mathbb{Z}_2$ coefficients for discrete dynamical systems was begun by Franks [F4]. Williams replaced t^n by a monomial in noncommuting variables to describe the homotopy classes of orbits in the Lorenz attractor [W] and this suggested the study of homology classes of closed orbits in [Fr1]. Both algebraic weights and homology classes are used in [F1,Fr2] for flows.

When counting with analytic weighting (geometric weighting being a special case) one makes strong hypotheses on the dynamics and seeks strong regularity assumptions on the zeta function (rationality, meromorphicity, etc.). When algebraic weights are used (such zeta functions are often denoted $\tilde{\zeta}$ and named after Lefschetz) one can apply an appropriate Lefschetz formula to see that the zeta function is rational and computable for very general dynamical systems, and one focuses instead on drawing dynamical conclusions from the zeta function.

Into this rough outline, the zeta function $\tilde{\zeta}_H(t)$ of this paper fits as follows It is defined for maps and uses algebraic weights. It is rational and has coefficients in the integral group ring $\mathbb{Z}H$ of an abelian group H (one may just as easily use FH for $F = \mathbb{Z}_2$, etc.). Also $\tilde{\zeta}_H(t)$ does take the homology classes of orbits into account and one may interpret this in two ways. One is to distinguish periodic points of f according to the homology classes of the corresponding orbits in the suspension flow. This viewpoint will be developed and applied in another paper [Fr2].

The other interpretation is more related to the definition of $L_H(\tilde{f})$ in the previous section. Consider the case when $\tilde{X}$ is connected so H is a quotient of $H_1(X;\mathbb{Z})$. We say $x,y \in \mathrm{Fix}(f^n)$ are in the same <u>H-fixed point class</u> for f^n if there is a path γ from x to y such that the closed loop $f^n\gamma\cdot\gamma^{-1}$ maps to zero in H. When $n = 1$, this gives the partition of $\mathrm{Fix}(f)$ into sets S_h as in section 1. When $n = 1$, $\pi_1(x)$ is abelian, and $H = H_{max}$ then H-fixed point class is the same as fixed point class but usually it is a weaker equivalence relation. Our second description of how homology is accounted for in $\tilde{\zeta}_H(t)$ is that is distinguishes H-fixed point classes of the iterates of f. We have now set the stage for the following.

<u>Definition</u>. Let X, $\tilde{X}$, H, $\tilde{f}$ and $\tilde{f}$ be as in the first paragraph of section 1. Then $\tilde{\zeta}_H(t) = \exp(\Sigma \frac{L_H(\tilde{f}^n)}{n} t^n)$ is the Lefschetz zeta function of $\tilde{f}$ with coefficients in $\mathbb{Z}H$ or (less precisely) the <u>twisted Lefschetz zeta function</u> of f.

Note that $\tilde{\zeta}_H(t)$ is a formal power series over $\mathbb{Q}H$ with constant term 1. We now present the standard proof that $\tilde{\zeta}_H(t)$ is rational.

<u>Theorem 2</u>. $\tilde{\zeta}_H(t)$ is the ratio of two polynomials over H with constant term 1. In the notation of Theorem 1, we have

$$\zeta_H(t) = \prod_{i\geq 0} \det(I-t\tilde{F}_i)^{(-1)^{i+1}}$$

<u>Proof</u>. Clearly it suffices to prove the formula. Using the formal identity $\det(I-tA)^{-1} = \exp(\sum_{n>0} \frac{\mathrm{Trace}\, A^n}{n} t^n)$ for $A = \tilde{F}_i$, $i = 0,1,2,\ldots,$ and taking the alternating product over all i, the right side becomes

$$\exp(\sum_{\substack{i\geq 0\\ n>0}} \frac{(-1)^i}{n} \mathrm{Trace}(\tilde{F}_i)^n t^n) .$$

Now apply Theorem 1 to $\tilde{f}^n$ to obtain

$$\exp(\sum_{n>0} \frac{L_H(f^n)}{n} t^n) = \tilde{\zeta}_H(t) . \quad \text{Q.E.D.}$$

<u>Remarks</u>. 1) It is crucial in the preceding formal computation that one work over a commutative ring, so Lemma 1 is needed.

2) Clearly when $H = 1$, $\tilde{\zeta}_H(t)$ is the Lefschetz-Smale zeta function $\tilde{\zeta}(t)$ and this theorem is then well-known [F3].

3) Remark 6 of section 1 indicates how $\tilde{\zeta}_H(t)$ may be computed from a nice cellular approximation to f, without explicit knowledge of $\tilde{X}$.

4) $\tilde{\zeta}_H(t)$ depends on the choice of lift $\tilde{f}$ in a simple manner. As mentioned previously, $L_H(h\tilde{f}) = hL_H(\tilde{f})$. Applying this to $\tilde{f}^n$, $n > 0$, one finds that (in an evident notation)

$$\tilde{\zeta}_H(t,h\tilde{f}) = \exp\left(\frac{L_H(h\tilde{f})^n}{n} t^n\right)$$

$$= \exp\left(\Sigma\frac{L_H(\tilde{f}^n)}{n} (ht)^n\right) = \tilde{\zeta}_H(ht,\tilde{f}) .$$

So f and H determine the zeta function $\tilde{\zeta}_H(t)$ up to a substitution of ht for t, $h \in H$.

5) One can modify the definition of $\tilde{\zeta}_H(t)$ by replacing $\mathrm{ind}(f,S_h)$ by $\#(S_h)$ throughout, provided that $\mathrm{Fix}(f^n)$ is finite for all $n > 0$. The resulting zeta function (of a map, using geometric weights, and keeping track of the H-fixed point class) will be denoted $\zeta_H(t)$ and called the twisted Artin-Mazur zeta function of f

Consider the case when X is a smooth manifold and f is an Axiom A diffeomorphism. In this case the rationality of the (untwisted) Artin-Mazur zeta function $\zeta(t)$ was proved by Manning, after partial results were obtained by Williams and Guckenheimer. It is clear that $\varepsilon(\zeta_H(t)) = \zeta(t)$, in the notation of Theorem 1. In fact, the more refined zeta function $\zeta_H(t)$ is also rational, as was essentially shown in [Fr 1]. There a rational zeta function denoted $\zeta_H(\varphi)$ was associated to the suspension flow φ of f (although "H" had no special significance in that paper, referring only to "Homology"). One can easily show that $\zeta_H(t) = \zeta_H(\varphi)$, so $\zeta_H(t)$ is rational.

The relationship between the twisted zeta functions $\zeta_H(t)$ and $\tilde{\zeta}_H(t)$ for Axiom A diffeomorphisms is like that found by Franks in the untwisted case $H = 1$. Then $\zeta(t)$ and $\tilde{\zeta}(t)$ are rational over $\mathbb{Z}$ and Franks showed that they agree after the coefficients are reduced mod 2 in standard form (i.e., when the numerator and denominator have constant term 1). Likewise, if one reduces the coefficients of $\tilde{\zeta}_H(t)$ mod 2, then one obtains a rational function over $\mathbb{Z}_2H$ that agrees with the reduction of $\zeta_H(t)$. This follows easily from the proof of rationality in [Fr 1] by passing from unsigned transition matrices to signed transition matrices. The former can be used to compute $\zeta_H(t)$, the latter to compute $\tilde{\zeta}_H(t)$ and inserting ± 1's has no effect mod 2.

3. Local Zeta Functions and the Product Formula.

One of the properties common to all the zeta functions that arise in dynamical systems is that they admit a product formula whenever the periodic orbits of the system are isolated. The terms in this possibly infinite product are some appropriate sort of "local zeta function" that measures the contribution of a single periodic orbit. We will develop the product formula for twisted Lefschetz zeta functions in this section.

Throuthout this section, we will consider the following local Lefschetz zeta function (cf.[F4]). Let x be a fixed point of a continuous map $f : X \circlearrowleft$ of a finite complex X such that x is isolated in $\mathrm{Fix}(f^n)$ for all $n > 0$. Then $\mathrm{ind}(f^n,x)$ is defined and we let $\tilde{\zeta}_x(t,f) = \tilde{\zeta}_x(t) = \exp(\sum_{n>0} \frac{\mathrm{ind}(f^n,x)}{n} t^n)$. It is a formal power series over $\mathbb{Q}$ with constant term 1. For f a C^1 diffeomorphism, $\tilde{\zeta}_x(t)$ was studied previously by Franks. When x is hyperbolic or when x is an isolated point of the chain recurrent set, Franks proved that $\tilde{\zeta}_x(t)$ is rational [F1].

We will generalize these results and show that $\tilde{\zeta}_x(t)$ is rational assuming only that f is a C^1 map. Clearly, one must analyze the sequence of indices $\mathrm{ind}(f^n,x)$, $n = 1,2,\dots$. This was done by Shub and Sullivan who proved that this is a bounded sequence [SS]. A close reading of their proof shows that there is a modulus M determined by the eigenvalues of Tf at x such that $\mathrm{ind}(f^n,x)$ depends only on the greatest common divisor of n and M. This was observed by Chow, Mallet-Paret, and Yorke [CMY] who moreover established that the sequence $\mathrm{ind}(f^n,x)$ is a finite integral linear combination of certain special sequences that we will denote $\alpha_i(n)$. For a fixed $i > 0$, the sequence $\alpha_i(n)$, $n = 1,2,3,\dots,$ takes the value 0 if $i \nmid n$ and i if $i|n$. We now show

Theorem 3. Let f be a C^1 self-map of a manifold X. Let $x \in \mathrm{Int}(X)$ be isolated in $\mathrm{Fix}(f^n)$ for all $n > 0$. Then there is a finite factorization $\tilde{\zeta}_x(t) = \prod_{i\geq 1} (1-t^i)^{a_i}$, $a_i \in \mathbb{Z}$, $a_i = 0$ for i large.

Proof. By the results of [CMY] cited above, $\log \tilde{\zeta}_x(t)$ is finite integral combination of series $\sum \frac{\alpha_i(n)}{n} t^n$. But $\sum_{n>0} \frac{\alpha_i(n)}{n} t^n = \sum_{m>0} \frac{it^{mi}}{mi} = \log(1-t^i)^{-1}$. Exponentiating gives the desired finite factorization of $\tilde{\zeta}_x(t)$. Q.E.D.

One can easily check that the bifurcation invariant in [CMY] is the negative of the sum of the exponents a_i in the preceding theorem. Actually, each a_i is itself a bifurcation invariant.

Next we show that $\tilde{\zeta}_x(t)$ may be irrational when f is merely continuous. To begin let $g : S^{n-1} \to S^{n-1}$ be a continuous map. We regard S^n as the union of two cones $S(S^{n-1}) = C_1S^{n-1} \cup_{S^{n-1}} C_2S^{n-1}$ and denote the poles (cone points) by P_1, P_2. Let $h : S^n \longrightarrow S^n$ be a diffeomorphism that preserves the system of cone lines $C_i(p) \subsetneq C_iS^{n-1}$, $p \in S^{n-1}$, that satisfies $\mathrm{Fix}(h) = S^{n-1} \cup \{P_1,P_2\}$ and that pushes points $z \notin \mathrm{Fix}(h)$ away from S^{n-1} toward the poles. Then $h \circ Sg : S^n \to S^n$ has the p_i as sinks and acts by g on the equator $S^{n-1} \subset S^n$. Let $I \subset S^{n-1}$ be a closed g-invariant set and C_1I the cone on I with vertex P_1.

<u>Lemma 2</u>. The quotient space $X = S^n/C_1I$ is a topological n-sphere, with a basepoint x corresponding to C_1I. There is a map $f : X \circlearrowleft$ induced by $h \circ Sg : S^n \circlearrowleft$ with $fx = x$. If $\mathrm{Fix}(g^n)$ is finite then $\mathrm{Fix}(f^n)$ is finite and $\mathrm{ind}(f^n,x) = 1 - \sum_y \mathrm{ind}(g^n,y)$, where $y \in I \cap \mathrm{Fix}(g^n)$.

<u>Proof</u>. One may isotop $S^n - C_1I$ to an open n-disc. Hence X is topologically the one point compactification of $\mathbb{R}^n$, i.e., a topological n-sphere [see Figure 2].

Clearly C_1I is invariant by h and Sg. Thus f is defined and $fx = x$.

$\mathrm{Fix}\,(f^n) = \{x,P_2\} \cup \{z \mid z \in \mathrm{Fix}\, g^n,\ z \notin I\}$. Thus $L(f^n) = \sum \mathrm{ind}((h\circ Sg)^n,z) + \mathrm{ind}(f^n,x)+1$. Since the equator is a repellor for $h\circ Sg$, we have $\mathrm{ind}((h\circ Sg)^n,z) = -\mathrm{ind}(g^n,z)$.

As f is homotopic to Sg, we have

$$\begin{aligned} L(f^n) = L(Sg^n) &= 1+(-1)^n\deg(Sg^n) \\ &= 1+(-1)^n\deg(g^n) \\ &= 2-L(g^n) \\ &= 2-\sum_z \mathrm{ind}(g^n,z) - \sum_y \mathrm{ind}(g^n,y) \end{aligned}$$

where $y \in I \cap \mathrm{Fix}\, g^n$.

Equating the two expressions for $L(f^n)$ gives the formula desired. Q.E.D.

We now show

<u>Theorem 4</u>. On any compact manifold $X \neq S^1$, there is a continuous self map f with finitely many points of each period and a point $x \in \mathrm{Fix}(f)$ with $\tilde{\zeta}_x(t)$ irrational.

<u>Proof</u>. It is not hard to see that it suffices to construct a 2 dimensional example. For this we take, in the setting of Lemma 2, $n = 2$ and $g : S^1 \circlearrowleft$ the map $g(z) = z^2$. The invariant set I will be chosen as follows. Let $\sigma : 2^{\mathbb{Z}^+} \circlearrowleft$ be the full one-sided 2-shift and $\pi : 2^{\mathbb{Z}^+} \to S^1$ the usual "binary expansion"

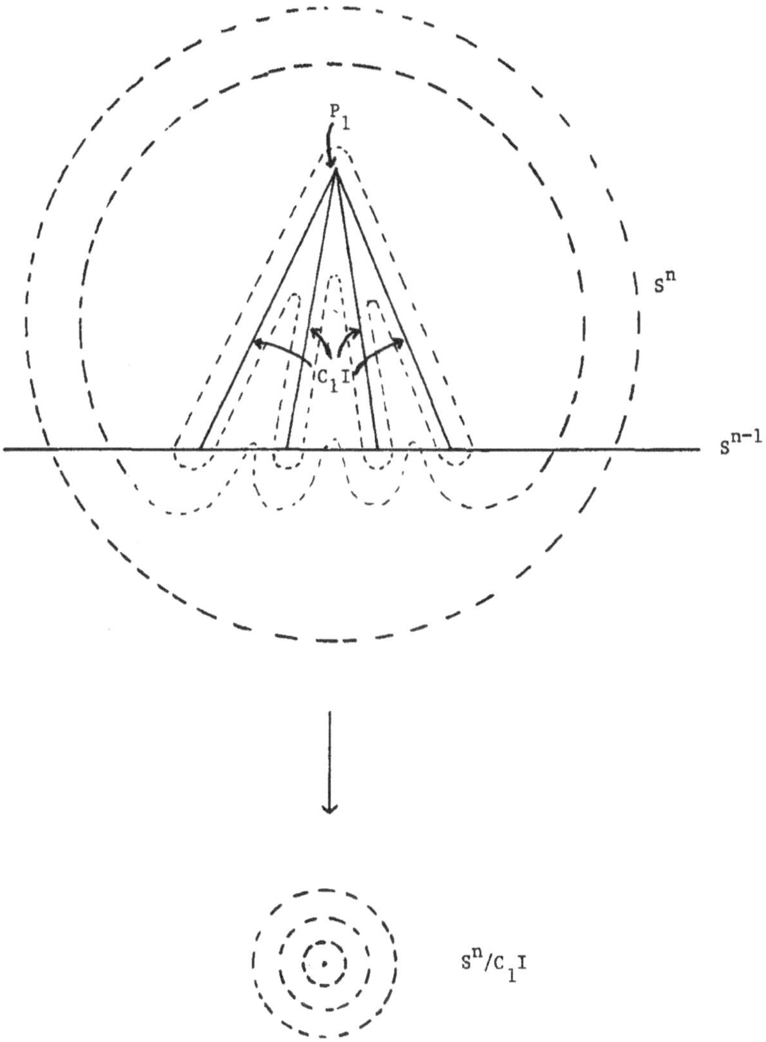

Figure 2. $S^n - C_1 I \simeq \mathbb{R}^n$ $\quad S^n/C_1 I \simeq S^n$.

semiconjugacy from σ to g . Let $J \subset 2^{\mathbb{Z}^+}$ be a subshift with $\zeta(t,\sigma|J)$ irrational [BL] and let $I = \pi J$.

Then lemma 2 gives, as all periodic points have index -1 ,

$$\tilde{\zeta}_x(t) = \exp\{\sum_{n>0} \frac{t^n}{n} [1- \sum_Y \mathrm{ind}(g^n,Y)]\}$$

$$= (1-t)^{-1}\zeta(t,g|I) \ .$$

Since π induces a 1-1 correspondence between $\mathrm{Fix}(g^n) - \mathrm{Fix}(g)$ and $\mathrm{Fix}(\sigma^n) - \mathrm{Fix}(\sigma)$, $\zeta(t,g|I)$ can only differ from $\zeta(t,\sigma|J)$ by one or two factors of $(1-t)$. Thus $\zeta(t,g|I)$ is irrational and so is $\tilde{\zeta}_x(t)$. Q.E.D.

Remarks. 1) If $X = S^1$ then $\tilde{\zeta}_x(t)$ is always rational.

2) Question (Gilbert Levitt): If X is a surface, $f : X \to X$ a homeomorphism and $x \in X$ an isolated point of $\mathrm{Fix}(f^n)$ for all $n > 0$ is $\tilde{\zeta}_x(t)$ rational ?

3) The f in the Theorem 4 may be taken a homeomorphism when $\dim X \geq 3$. For this, one starts with g the Plykin diffeomorphism of S^2 [P] and one proceeds, as before, to construct an example on S^3 .

4) For any continuous f , $\tilde{\zeta}_x(t)$ has integral coefficients. For by imbedding X in Euclidean space as a neighborhood retract, we may assume X is a manifold without changing the sequence of indices. Then the approximation argument in [CMY] shows that the sequence $\mathrm{ind}(f^n,x)$ is a (possibly infinite) integral combination of the sequences $\alpha_i(n)$ used above. This gives $\tilde{\zeta}_x(t) = \prod_{i>0} (1-t^i)^{a_i}$, $a_i \in \mathbb{Z}$. But the latter has integer coefficients.

We now turn to the product formula. We shall use the notation of the first paragraph of section 1 and add the hypothesis that $\mathrm{Fix}(f^n)$ is finite for all $n > 0$. For convenience, we assume that X and $\tilde{X}$ are connected so H is a quotient group of $H_1(X;\mathbb{Z})$. Let b be a basepoint for X and σ a basepath from b to fb.

Let x be any periodic point for f . We denote by $p = p(x)$ the least period of x . We define a class $h = h(x) \in H$ as follows. Let γ_x be a path from b to x. Then $f^p(\gamma_x)$ joins $f^p b$ to $f^p x = x$, so $\gamma_x^{-1} \cdot (\sigma \cdot f\sigma \cdot \ldots \cdot f^{p-1}\sigma) \cdot f^p(\gamma_x)$ is a loop based at x and determines a class $h(x) \in H$. If one changes from γ_x to another path γ'_x one sees that $h(x)$ changes by $f^p(g) \cdot g^{-1}$, where g is the class of $\gamma'_x \gamma_x^{-1}$. As f acts trivially on H , $h(x)$ is well defined. We can now state

Theorem 5. (The Twisted Product Formula). For f as above

$$\tilde{\zeta}_H(t) = \prod_x \tilde{\zeta}_x(t^p h, f^p)$$

where $p = p(x)$, $h = h(x)$ and x varies over one point from each periodic orbit of f . The lift $\tilde{f}$ used to compute $\tilde{\zeta}_H(t)$ is such that the basepath σ lifts to a path $\tilde{\sigma}$ from $\tilde{b}$ to $\tilde{f}\tilde{b}$ in $\tilde{X}$.

Proof. For fixed $n > 0$, consider the term $L_H(\tilde{f}^n)$ that arises in the definition of $\tilde{\zeta}_H(t)$. We have $L_H(\tilde{f}^n) = \sum_h i(\tilde{f}^n h^{-1})\cdot h$ and $i(\tilde{f}^n h^{-1}) = \mathrm{ind}(\tilde{f}^n, \mathrm{Fix}(f^n h^{-1})/H)$ $= \Sigma\, \mathrm{ind}(f^n,y)$ where $y \in \mathrm{Fix}(\tilde{f}^n h^{-1})/H$. Now y arises for some $h \overset{y}{\Longleftrightarrow} y \in \mathrm{Fix}(f^n) \Longleftrightarrow p(y)\,|\,n$. The relation between this h and $h(y)$ is found as follows. Let $\tilde{y} \in \tilde{X}$ be above y and let γ_y be a path from b to y . Then

Lemma 3. $h(y)^q = h$, where $q = n/p(y)$.

Proof of Lemma. One may find h by lifting the loop $\gamma_y^{-1}\cdot\sigma\cdot\ldots\cdot f^{n-1}\sigma\cdot f^n\gamma_y$ to $\tilde{X}$ to obtain a path from $\tilde{y}$ to $h\tilde{y} = \tilde{f}^n\tilde{y}$. But this loop may be factored into loops as follows (see Figure 3). $[\gamma_y^{-1}\cdot(\sigma\cdot\ldots\cdot f^{p-1}\sigma)f^p\gamma_y]\cdot[f^p\gamma_y^{-1}\cdot(f^p\sigma\cdot\ldots\cdot f^{2p-1}\sigma)\cdot f^{2p}\gamma_y]\cdot\ldots\cdot$ $[\ldots f^n\gamma_y]$ and so represents $h(y)\cdot f_*^p h(y)\cdot\ldots\cdot f_*^{p(q-1)}h(y)$. As f acts trivially on H, we have $h = h(y)^q$, proving Lemma 3 .

Thus $L_H(\tilde{f}^n) = \sum_y \mathrm{ind}(f^n,y)\cdot h(y)^q$ where y runs over $\mathrm{Fix}(f^n)$ and $q = n/p(y)$. Now choose one representative x from each periodic orbit and note that $\mathrm{ind}(f^n,x) = \mathrm{ind}(f^n, f^{ip}x)$ for all i . We obtain

$$L_H(\tilde{f}^n) = \sum_x \mathrm{ind}(f^n,x)\cdot p\cdot h^q$$

where $p = p(x)\,|\,n$, $h = h(x)$, and $q = n/p$. Then

$$\begin{aligned}\tilde{\zeta}_H(t) &= \exp\Big(\sum_{n>0} \frac{L_H(\tilde{f}^n)}{n}\, t^n\Big)\\ &= \exp\Big(\sum_{\substack{n>0\\ x}} \frac{1}{n}\, \mathrm{ind}(f^n,x)\cdot p\cdot h^q\cdot t^n\Big)\\ &= \prod_x \exp\Big(\sum_{\substack{n>0\\ p(x)|n}} \frac{1}{n}\, \mathrm{ind}(f^n,x)\cdot p\cdot h^q\cdot t^n\Big)\ .\\ &= \prod_x \exp\Big(\sum_{q>0} \frac{1}{q}\, \mathrm{ind}((f^p)^q,x)h^q\cdot(t^p)^q\Big)\\ &= \prod_x \tilde{\zeta}_x(ht^p, f^p)\end{aligned}$$

as desired. Q.E.D.

There are two main ideas in the preceding proof. The first is that $\tilde{\zeta}_H(t)$ is a product of factors arising from periodic orbits. This is obvious because the Lefschetz formula was additive over the periodic points and $\tilde{\zeta}_H(t)$ was defined using the exponential function. The other idea is that a given orbit of period p only contributes along a ray through $t^p h$ for some h. This holds because $\tilde{f}^p y = hy \Rightarrow \tilde{f}^{pi} y = h^i y$ for all $i > 0$.

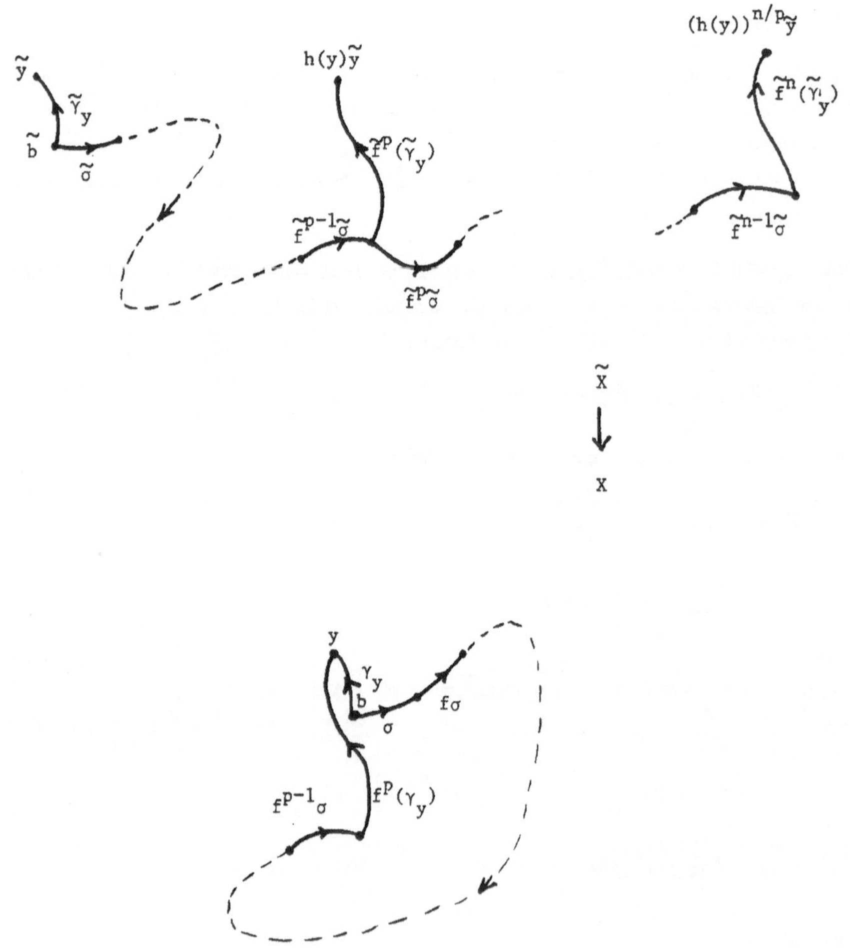

Figure 3.

4. Maps with Finitely Many Periodic Points.

Throughout this section, we will use the notation of the first paragraph of section 1. We will make the additional assumption that f has only finitely many periodic points and we will show that this imposes constraints on the twisted zeta function $\tilde{\zeta}_H(t)$. As $\tilde{\zeta}_H(t)$ is computable, these theorems furnish practical methods for proving that certain maps have infinitely many periodic points.

For the first theorem of this section, we assume extra smoothness properties for f .

Theorem 6. Let X be a compact manifold and $f : X \to \mathrm{Int}(X)$ a C^1 map with finitely many periodic points. Let $\tilde{X}$, H and $\tilde{f}$ be as in section 1 . Then the twisted zeta function $\tilde{\zeta}_H(t)$ has a finite factorization into terms of the form $(1-t^n h)^{\pm 1}$, $n > 0$, $h \in H$.

Proof. One easily reduces to the case when X and $\tilde{X}$ are connected. The twisted product formula of Theorem 5 gives $\tilde{\zeta}_H(t) = \prod_x \tilde{\zeta}_x(t^p h, f^p)$, a finite factorization over the periodic orbits of f . By Theorem 3, applied to $x \in \mathrm{Fix}(f^p)$, $\tilde{\zeta}_x(t^p h, f^p) = \prod_{i=1}^{n} (1-(t^p h)^i)^{a_i}$ for some $n = n(x)$, $a_i \in \mathbb{Z}$. The terms in this factorization are of the desired form. Q.E.D.

Note that letting $H = 1$ in Theorem 6 gives nontrivial constraints on $\tilde{\zeta}(t)$, i.e., the untwisted Lefschetz zeta function. This is just a reworking of the main theorem of [SS] using the sharper local results of [CMY] and the language of zeta functions.

We remark that for continuous maps there is no such constraint on $\tilde{\zeta}(t)$. Consider, for instance, the following counterexample of M. Shub. Let $Y = S^1$ and let $g : Y \to Y$ be the map $g(z) = z^2$. Let $X = SY = S^2$ and let $h : X \to X$ be a Morse-Smale diffeomorphism with north pole P as source, south pole P' as sink and such that all other points are mapped "south" from P to P' . Then, we let $f = h \circ Sg$. Clearly f has only 2 periodic points, P and P'. One easily sees that $L(f^n) = 2^n + 1$ so that $\tilde{\zeta}(t) = (1-t)^{-1}(1-2t)^{-1}$. This rational function does not have a finite factorization into terms $(1-t^n)^{\pm 1}$. As $\tilde{\zeta}_P(t) = (1-2t)^{-1}$, we see also that Theorem 3 fails for continuous maps.

We will, however, obtain certain nontrivial constraints on $\tilde{\zeta}_H(t)$ for a continuous f with finitely many periodic points when $H \neq 1$. We cannot proceed as in Theorem 6, since the local Lefschetz zeta functions may be irrational (Theorem 4), so we must analyze the algebra more deeply.

Let R be a commutative ring with unity, e.g., $R = \mathbb{Z}H$, H an abelian group •

We denote the multiplicative group of all formal power series in the variable t over R with constant term 1 by $R_1[[t]]$. Thus $\tilde{\zeta}_H(t) \in R_1[[t]]$ for $R = \mathbb{Z}H$. When $R = \mathbb{Z}H$, we will call $\xi(t) \in (\mathbb{Z}H)_1[[t]]$ a <u>ray series</u> if it is expressible as $\eta(t^n h)$, for some $\eta \in \mathbb{Z}[[t]]$, $n > 0$, $h \in H$. The name is motivated by the case when H is free abelian with generators $x_1,\dots,x_r$. One calls the terms $t^a x_1^{a_1} \dots x_r^{a_r} \in \mathbb{Z}^+ \times H$ whose coefficient in ξ is nonzero the <u>support</u> of $\xi(t)$. Then $\xi(t)$ is a ray series $\Longleftrightarrow \{(a,a_1,\dots,a_r) \in \mathbb{Z}^{r+1} \mid t^a x_1^{a_1} \dots x_r^{a_r} \in \operatorname{Supp} \xi(t)\}$ lies in a geometric ray through the origin in $\mathbb{R}^{r+1}$. In the general case, we call $\{(t^n h)^i \mid i > 0\} \subset \mathbb{Z}^+ \times H$ the <u>ray</u> through $t^n h$. We see that $\xi(t)$ is a ray series $\Longleftrightarrow$ its support lies on a ray.

The following is an immediate consequence of the Twisted Product Formula and remark 4 after Theorem 4.

<u>Theorem 7</u>. In the notation of section 1, suppose f has m periodic points, $m < \infty$. Then $\tilde{\zeta}_H(t)$ is a product of m ray series.

It is perhaps not obvious that this is a nontrivial constraint on $\tilde{\zeta}_H(t)$ for $H \neq 1$ (it is, in fact, trivial when $H = 1$). To see what the conclusion of Theorem 7 means, consider $\xi(t) \in (\mathbb{Z}H)_1[[t]]$. We may factor $\xi(t) = \prod_{n>0} \prod_{h \in H} (1-t^n h)^{a_{nh}}$ as follows. Assume a_{nh} are known for $n < N$ so that

$$\xi_{N-1}(t) = \prod_{n=1}^{N-1} \prod_{h \in H} (1-t^n h)^{a_{nh}}$$

agrees with $\xi(t)$ up to terms of order $< N$. Then $\xi(t)/\xi_{N-1}(t) = 1 - B_N t^N + \dots$ where $B_N = \Sigma b_n h \in \mathbb{Z}H$. We let $a_{Nh} = b_h$ and see that $\xi_N(t)$ agrees with $\xi(t)$ in terms of order N as well. It is clear from this construction that the exponents a_{Nh} are uniquely determined, so the above factorization is unique.

We will call the above infinite product the <u>canonical factorization</u> of $\xi(t)$ and the a_{nh} the <u>canonical exponents</u> of $\xi(h)$. Note that for fixed n only finitely many a_{nh} are nonzero. Conversely, a collection of integers a_{nh} with this finiteness property determines an element $\xi(t) \in (\mathbb{Z}H)_1[[t]]$ with canonical exponents a_{nh}. The points $(n,h) \in \mathbb{Z}^+ \times H$ for which the canonical exponent a_{nh} is nonzero form the <u>logarithmic support</u> of ξ. We have the following more intuitive reformulation of Theorems 6 and 7.

<u>Theorem 8</u>. If f is as in Theorem 7, then the logarithmic support of $\tilde{\zeta}_H(t)$ lies in a finite union of rays. If f is C^1 then this logarithmic support is finite.

<u>Proof</u>. The second statement is immediate from Theorem 6. To show the first, we need only show the following algebraic result.

Lemma 4. $\xi(t) \in \mathbb{Z}H_1[[t]]$ is a finite product of ray series $\Longleftrightarrow$ the logarithmic support of $\xi(t)$ lies in a finite union of rays.

Proof of Lemma 4. Suppose $\eta(t) \in \mathbb{Z}_1[[t]]$. Then we may factor $\eta(t) = \Pi(1-t^i)^{a_i}$ and see that the ray series $\eta(t^n h) = \Pi(1-t^{ni}h^i)^{a_i}$ has logarithmic support on the ray through $t^n h$. This proves the forward implication and the converse follows similarly. Q.E.D.

It is now clear that Theorem 7 had nontrivial content for $H \neq 1$, since then $\mathbb{Z}^+ \times H$ is not the union of finitely many rays. But since canonical exponents are difficult to compute it is not clear how to tell in a finite number of steps whether a given $\xi(t)$ satisfies the conclusion of Theorem 7. This must be done to obtain a practical method for showing maps have infinitely many periodic orbits, such as the method given by Theorem 6 for C^1 maps. Essentially, we must find a criterion that depends on the rational function aspect of $\tilde{\zeta}_H(t)$ (which is how it is computed as in Theorem 2) and not on its power series aspect (which is how it was defined). We begin by introducing terminology for rational functions over $\mathbb{Z}H$ analogous to that we have for power series.

Let R be a commutative ring with unit. We let $R_1[t]$ be the cancellative semigroup of polynomials over R with constant term 1. The group generated by $R_1[t]$ consists of rational functions over R with numerator and denominator having constant term 1 and is denoted $R_1(t)$. There is a natural inclusion $R_1(t) \subset R_1[[t]]$ given by expansion into formal power series.

When $R = \mathbb{Z}H$, we say $\xi(t) \in R_1(t)$ is a ray function if $\xi(t) = \eta(t^n h)$ for some $\eta(t) \in \mathbb{Z}_1(t)$, $n > 0$, $h \in H$. Clearly the formal power series of a ray function is a ray series. The converse does not always hold, however, that is for some H (with nontrivial torsion; see Theorem 9) and $\xi(t) \in (\mathbb{Z}H)_1(t)$, the power series expansion is a ray series and $\xi(t)$ is not a ray function. Such an example is given in the appendix.

A modified converse does hold, however.

Theorem 9. Let H be an abelian group and $\xi(t) \in (\mathbb{Z}H)_1(t)$. Assume that the power series expansion of $\xi(t)$ factors into m ray series, $m < \infty$ (e.g., $\xi = \tilde{\zeta}_H(t)$ in Theorem 7). Then some power $\xi^N(t)$, $N \geq 1$, factors into m ray functions. One may take $N = 1$ if H is torsionfree.

We present the algebraic proof of Theorem 9 in section 6.

We consider further the case when H is torsionfree and finitely generated. Then H is free abelian on generators $x_1,\ldots,x_n$. We see that

$(\mathbb{Z}H)_1[t] \subset \mathbb{Z}[x_1,x_1^{-1},\ldots,x_n,x_n^{-1},t] = R$. R has unique factorization and if $p(t) \in (\mathbb{Z}H)_1[t]$ factors in R then (for appropriate choice of signs) the factors also belong to $(\mathbb{Z}H)_1[t]$. Thus a function $\xi(t) \in (\mathbb{Z}H)_1(t)$ has a unique factorization into powers of irreducible polynomials in $(\mathbb{Z}H)_1[t]$ that we call the irreducible factors of $\xi(t)$. As these factors don't change when one passes from H to a larger abelian group, we don't need the assumption that H is finitely generated to define these irreducible factors. We have then

Theorem 10. With the conventions of section 1 , assume that f has only finitely many periodic points and that H is torsionfree. Then the irreducible factors of $\tilde{\zeta}_H(t)$ are ray functions.

Proof. By Theorems 7 and 9, $\tilde{\zeta}_H(t)$ is a product of ray functions $\xi_1(t),\ldots,\xi_m(t) \in (\mathbb{Z}H)_1(t)$. It suffices to show that each irreducible factor of each $\xi_i(t)$ is also a ray function, as these include the irreducible factors of $\tilde{\zeta}_H(t)$. So we only need

Lemma 5. The irreducible factors of a ray function $\xi(t) \in (\mathbb{Z}H)_1[t]$ are also ray functions

Proof of Lemma 5. We may suppose H is finitely generated, since only finitely many elements of H are involved. We know that $\xi(t) = \eta(t^n h)$, $n > 0$, $h \in H$, $\eta \in \mathbb{Z}_1(t)$. Considering the numerator and denominator of η separately, we may suppose $\eta \in \mathbb{Z}_1[t]$. We will then show that any factor in $(\mathbb{Z}H)_1[t]$ of $\xi(t)$ is a ray function.

Suppose $\xi(t) = p(t)q(t)$, $p(t), q(t) \in (\mathbb{Z}H)_1[t]$. We have $\xi(t) = \eta(t^n h)$. Choosing an integral basis $x_1,\ldots,x_r$ for H and writing $h = x_1^{a_1}\ldots x_r^{a_r}$, we have that $\xi(t)$ is a polynomial in t , $x_1,x_1^{-1},\ldots,x_r,x_r^{-1}$ all of whose terms lie on the geometric ray through $(n,a_1,\ldots,a_r) \in \mathbb{Z}^+ \times \mathbb{Z}^r \subset \mathbb{R}^{r+1}$. We suppose for some term in $p(t)$, $t^b x_1^{b_1},\ldots,x_r^{b_r}$ say, $(b,b_1,\ldots,b_r)$ lies off this ray and obtain a contradiction. There is an integral functional $\ell: \mathbb{Z}^{r+1} \to \mathbb{Z}$ that annihilates the ray but $\ell(b,b_1,\ldots,b_r) > 0$. Let $p_1(t)$ be the sum of those terms in $p(t)$ for which ℓ is largest and likewise define $q_1(t)$. Then $p_1(t)q_1(t) \neq 0$ and all its terms occur (without cancellation) in $p(t)q(t)$. Thus $p(t)q(t)$ has terms with $\ell > 0$ whereas $\xi(t)$ does not. Q.E.D.

Since the irreducible factors of elements of $\mathbb{Z}[x_1,x_1^{-1},\ldots,x_r,x_r^{-1},t]$ can be computed (e.g., there is an algorithm in the computer language MACSYMA for factoring integral polynomials), Theorem 10 gives a computable algebraic criterion that may be used to show a map has infinitely many periodic points.

We now seek such a criterion when H is not assumed to be torsionfree. There is the additional difficulty that then $(\mathbb{Z}H)_1[t]$ may not have unique factorization : for instance when $H = \{1,x\}$, $x^2 = 1$, we have $(1-xt)(1+xt) = (1-t)(1+t)$. We also must contend with the unknown exponent N arising in Theorem 9. Nevertheless, we obtain in Theorems 11 and 12 useful criteria, that (when H is torsionfree) are weaker than the preceding ones but easier to apply.

Theorem 11. In the setting of section 1, let $\tilde{\zeta}_H(t) = \frac{p(t)}{q(t)}$, $p,q \in (\mathbb{Z}H)_1[t]$. Let the highest nonzero coefficients of p and q be α and β, respectively.

If f has only finitely many periodic points, then there is $N \geq 1$, $q \in \mathbb{Q}(q \neq 0)$, and $h \in H$ such that

$$\alpha^N = qh\beta^N \quad .$$

If H is torsionfree, this holds for $N = 1$.

Proof. By theorems 7 and 9, we have an $N \geq 1$ and ray functions $\xi_1(t),\ldots,\xi_m(t) \in (\mathbb{Z}H)_1(t)$ for which $\xi_1(t) \ldots \xi_m(t) = [\tilde{\zeta}_H(t)]^N$.

We write $\xi_i(t) = p_i(t^{n_i} h_i)/q_i(t^{n_i}h_i)$ for $p_i,q_i \in \mathbb{Z}_1[t]$, $n_i > 0$, $h_i \in H$, $i = 1,\ldots,m$. We obtain

$$\prod_{i=1}^{m} p_i(t^{n_i}h_i)\cdot q(t)^N = \prod_{i=1}^{m} q_i(t^{n_i}h_i)\cdot p(t)^N \quad .$$

We now analyze the coefficients of the highest nontrivial terms in this equation in $(\mathbb{Z}H)[t]$.

It is known that $\mathbb{Z}H$ has no nilpotent elements. To see this it is enough to consider the case when H is finitely generated : but then for any nonzero $x \in \mathbb{Z}H$ there is a representation $\rho : \mathbb{Z}H \to \mathbb{C}$ with $\rho(x) \neq 0$. Hence the highest term in $q(t)^N$ has coefficient β^N.

For any i, the highest term in $p_i(t^{n_i}h)$ has the form $a\cdot t^n h$, $a \in \mathbb{Z}$, $a \neq 0$, $n>0$, $h \in H$. So we see that $\prod_i p_i(t^{n_i}h)\cdot q(t)^N$ has highest coefficient of the form $m\cdot g\beta^N$, $m \in \mathbb{Z}$, $m \neq 0$, $g \in H$. Reasoning similarly for the other side of our equation gives the desired identity. Q.E.D.

We found computable choices for p,q in Theorem 2, so the above criterion is computable, at least if N is fixed. To eliminate this uncertainty regarding N, we further weaken the criterion of Theorem 11 as follows.

Let $\rho : H \to S^1$, $i = 1,\ldots,n$, be a character of the abelian group H. Let the associated homomorphism $\mathbb{Z}H \to \mathbb{C}$ also be denoted ρ. For $\alpha,\beta \in \mathbb{Z}H$ as in Theorem 11, we let $v_\rho = (|\rho(\alpha)|, |\rho(\beta)|) \in \mathbb{R}^2$.

We have

Theorem 12. If f has only finitely many periodic points then the vectors

$v_\rho \in \mathbb{R}^2$ are proportional for all characters $\rho \in \hat{H}$.

Proof. By Theorem 11, there is an equation $m\cdot\alpha^N = n\cdot h\cdot\beta^N$, $m , n \in \mathbb{Z} - \{0\}$, $h \in H$, $n \geq 1$ where $q = \frac{n}{m}$.

Consequently, $\rho(m\alpha^N) = \rho(n\cdot h\cdot\beta^N)$. Thus $m\rho(\alpha)^N = n\rho(h)\ \rho(\beta)^N$. Taking absolute values gives $|m|\ |\ \rho(\alpha)|^N = |n|\ |\ \rho(\beta)|^N$, since $\rho(h) \in S^1$. Taking Nth roots gives $|\ \rho(\alpha)| = |q|^{1/N}|\ \rho(\beta)$. As the factor $|q|^{1/N}$ is nonzero and independent of ρ , we see that the vectors v_ρ have the same slope, as desired. Q.E.D.

The criterion of Theorem 12 may be easily applied from the computation of $p(t)$, $q(t)$ in Theorem 2, for any choice of ρ 's .

5 . Examples.

We begin with the complex $X = S^1 \vee S^2$, the space obtained by joining a circle and a 2-sphere at one point. All we say will apply to other finite complexes of the homotopy type of X, for instance the compact 3-manifold X' obtained by removing an open ball from the interior of a solid torus. In the usual embedding of X in $\mathbb{R}^3$, X' is its regular neighborhood. When a theorem is stated for C^1 maps, one should regard them as defined on X' instead of X.

We shall assume that $f_* : H_1(X;\mathbb{Z}) \circlearrowleft$ is the identity. Then f is homotopic to a map F that fixes S^1. This F is a cellular map for the usual cell complex structure on X with one vertex ρ, one one-cell τ and one two-cell σ.

Now we choose $H = H_1(X;\mathbb{Z})$ and construct the cover $\tilde{X}$ associated to H. $\tilde{X}$ is obtained by gluing a copy of the 2-sphere S^2 to each integer point in $\mathbb{R}$. The deck transformation group $\{g^i : i \in \mathbb{Z}\}$ acts on $\mathbb{R}$ by integer translations, so $g(y) = y+1$ for $y \in \mathbb{R} \subset X$. We see that $\tilde{X}$ is simply connected, hence the universal cover of X. The path $\gamma(t) = t$, $0 \leq t \leq 1$, in X determines a loop in X that generates $H_1(X;\mathbb{Z})$ and gives the isomorphism $H \simeq \{g^i : i \in \mathbb{Z}\}$.

We lift the cells ρ, τ, σ to $\tilde{\rho}, \tilde{\tau}, \tilde{\sigma}$ in $\tilde{X}$. Then the translates $g^i\tilde{\rho}$, $g^i\tilde{\tau}$, $g^i\tilde{\sigma}$ give a cell structure on $\tilde{X}$ (see Figure 4). We have $C_i(\tilde{X};\mathbb{Z})$ a free $\mathbb{Z}H$ module for $i = 0,1,2$ with generators $\tilde{\rho}, \tilde{\tau}, \tilde{\sigma}$ respectively. The boundary operators are $\partial\tilde{\sigma} = 0$, $\partial\tilde{\rho} = 0$ and $\partial\tilde{\tau} = (g-1)\tilde{\rho}$, where $\tilde{\tau}$ is oriented as a subset of $\mathbb{R}$. Thus $H_0(\tilde{X};\mathbb{Z}) \simeq \mathbb{Z}$; $H_1(\tilde{X};\mathbb{Z}) = 0$, and $H_2(X;\mathbb{Z})$ is the free $\mathbb{Z}H$ module generated by $\tilde{\sigma}$.

We now take $\tilde{F}$ to be the lift of F that fixes $\mathbb{R}$ and $\tilde{f}$ the lift of f found by lifting the homotopy from F to f. We find that the chain matrices are $\tilde{F}_0 = (1)$ and $\tilde{F}_1 = (1)$, but we haven't specified f enough to know $\tilde{F}_2$. In fact, $\tilde{F}_2$ is arbitrary. That is, we may choose an element $A \in \mathbb{Z}H$, say $A = \Sigma a_i g^i$, and construct a map F with $\tilde{F}_2 = (A)$ as follows. For each i for which $a_i \neq 0$, choose a disc $D_i \subset \sigma$ with the D_i closed and disjoint. F should map $\sigma-(\cup D_i)$ to ρ. F should wrap D_i i times around τ and then around σ with degree a_i. Moreover, <u>any</u> map f satisfying our asssumption is homotopic to such an F by elementary homotopy theory.

Thus, we have the 1 by 1 chain matrices (1), (1) and (A) in dimension 0, 1, and 2 respectively. The generalized Lefschetz number of section 1 is $L_H(\tilde{f}) = 1-1+A = A$ using Theorem 1. If we pass to powers of $\tilde{f}$, we obtain $L_H(\tilde{f}^n) = 1^n - 1^n + A^n = A^n$. Thus by definition $\tilde{\zeta}_H(t) = \exp(\Sigma \frac{A^n}{n} t^n) = (1-At)^{-1}$. This agrees with Theorem 4 : $\tilde{\zeta}_H(t) = \dfrac{\det(I-(t))}{\det(I-(t))\det(I-(At))} = (1-At)^{-1}$.

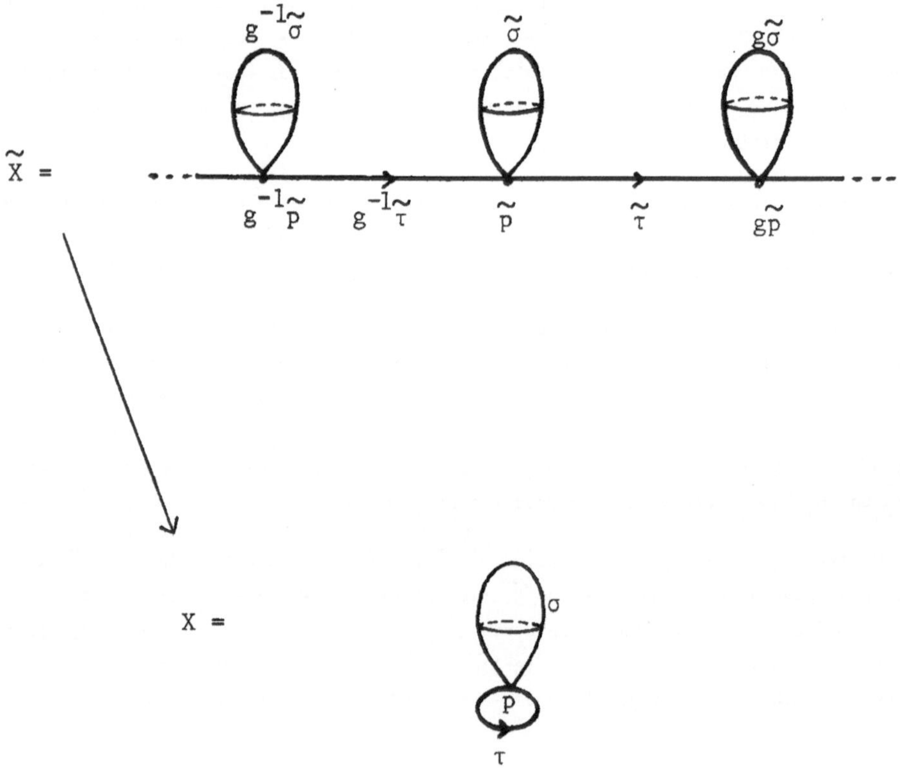
$\tilde{X}$ =
$g^{-1}\tilde{\sigma}$
$\tilde{\sigma}$
$g\tilde{\sigma}$
$g^{-1}\tilde{p}$
$g^{-1}\tilde{\tau}$
$\tilde{p}$
$\tilde{\tau}$
$g\tilde{p}$
X =
σ
p
τ

Figure 4.

We now determine those values of A for which our theorems show that f has infinitely many periodic points. Theorem 10 applies since H is torsionfree and it implies that the irreducible factors of $1-At$ must be ray functions whenever f has only finitely many periodic points. As $1-At$ is clearly irreducible, this means A must be a monomial, i.e. a_i is nonzero for at most one i. So if $a_i \neq 0$, $a_j \neq 0$ nd $i \neq j$ then f has infinitely many periodic points.

If f is a C^1 map of the above homotopy type with finitely many fixed points, Theorem 6 gives more information. We see that not only is $\tilde{\zeta}_H(t) = (1-a_i g^i t)^{-1}$ but $1-a_i t$ must be a product of terms $(1-t^j)^{\pm 1}$. Thus roots of $1-a_i t$ are roots of unity, so $a_i = -1$, 0 or 1.

When $A = a_i g^i$, it is not hard to construct an f, based on the Shub example of section 4, that has finitely many periodic points. When $A = 0$ or $A = \pm g^i$, one can construct Morse-Smale f's of this homotopy type. Thus, due to the simple nature of the example, our results are sharp in this case.

By contrast, consider what information the ordinary Lefschetz formula and the untwisted Lefschetz zeta function gives. The sequence of Lefschetz numbers is B^n, $B = \Sigma a_i$. If $|B| \leq 1$ then this sequence is bounded and the Shub-Sullivan criterion gives no information. Indeed if $A = 1-g$ then $B = 0$, so all the iterates of f have zero Lefschetz number: but actually f must have infinitely many periodic points.

To illustrate the theory when H has nontrivial torsion, we modify the preceding example as follows. Replace S^1 by $\mathbb{RP}^2$, i.e. let $X = \mathbb{RP}^2 \vee S^2$, let $F = X \circlearrowleft$ fix $\mathbb{RP}^2$ and let f be homotopic to F. Then the usual cell decompositions of $\mathbb{RP}^2$ and S^2 give a cell decomposition of X with one 0-cell, one 1-cell and two 2-cells. As before, one shows that the chain matrices of F are (1), (1), $\begin{pmatrix} 1 & * \\ 0 & A \end{pmatrix}$ over H, $H = \{1,g\}$, $g^2 = 1$, where g is a generator of $H_1(X;\mathbb{Z}) \simeq \mathbb{Z}_2$. We find $\zeta_H(t) = (1-t)^{-1}(1-At)^{-1}$, $A = a_o + a_1 g \in \mathbb{Z}H$.

We now assume f has finitely many periodic points and we apply our criteria from section 4 to find constraints on A. We begin with Theorem 12. Then we have $\alpha = 1$ and $\beta = A$ as the highest coefficients of the numerator and denominator of $\tilde{\zeta}_H(t)$. We take the characters of H with $\rho_1(g) = 1$, $\rho_2(g) = -1$, respectively. Then $v_1(\alpha) = (|\rho_1(\alpha)|, |\rho_1(\beta)|) = (1, |a_0+a_1|)$ and $v = (|\rho_2(\alpha)|, |\rho_2(\beta)|) = (1, |a_o - a_1|)$. So v_1 and v_2 are proportional $\Longleftrightarrow |a_0+a_1| = |a_0-a_1|$, i.e., one of a_o and a_1 is zero. For such $A = a_i g^i$, we have in fact that $\tilde{\zeta}_H(t)$ satisfies the conclusion of Theorem 7 and no futher information will be gained by applying our more stringent criteria of Theorems 7 or 11.

It is rather surprising that so much more information is obtained in this example by just passing to a double cover. When $A = a-ag$, $a \neq 0$, the Nielsen numbers of every iterate of f are two and the Lefschetz numbers of every iterate are zero, yet f has infinitely many periodic points.

If f is a C^1 map of the preceding homotopy type, then Theorem 6 says more : $1 - At$ must be a product of terms $(1-ht^n)^{\pm 1}$. Thus $1-At = \prod_{n>0} (1-t^n)^{b_n}(1-t^n g)^{c_n}$. We solve for $A = a_i g^i$ as follows. Substituting 1 for g (i.e., applying the homomorphism $\rho_1 : \mathbb{Z}H \to \mathbb{Z}$) gives $(1-a_i t) = \prod(1-t^n)^{b_n+c_n}$. Thus, as in the original example, $a_i = 1$, 0 , or -1 . So A is $\pm g$, ± 1 or 0 .

In a less artificial example, one may be given an $f : X \circlearrowleft$ with $f_* : H_1(X;\mathbb{Z}) \circlearrowleft$ not the identity. Then one should take H to be a quotient group of $H_1(X;\mathbb{Z})$ on which f_* induces the identity, i.e., a quotient of the group $H_{max} = H_{max}(f_*)$ considered in section 1 . If H_{max} is trivial (or if the criteria of Section 4 don't apply) one may try to pass to a power f^M of f before computing H and applying section 4. As f has finitely many periodic points $\Leftrightarrow f^M$ does, this may work. To see the extent to which this may help, let $G = H_1(X;\mathbb{Z})$ and $\phi = f_*$ in the following.

<u>Lemma 6.</u> Let G be a finitely generated abelian group and ϕ an endomorphism of G . If ϕ is not nilpotent then there is an $M \geq 1$ and a nontrivial quotient group $H \neq 0$ of G in which ϕ induces the identity.

<u>Proof.</u> By passing to $G/(\cup \ker \phi^i)$, we may assume that ϕ is 1-1 . If G has torsion, then $H = \text{Torsion}(G)$, $M = |H|$ suffice. Otherwise, we may take $G = \mathbb{Z}^n$, $n \geq 1$. Then let p be a prime that doesn't divide $\det(\phi)$. Let $H = G/pG$. We see that ϕ induces an automorphism of H . Again we let $M = |H|$. Q.E.D.

Thus, section 4 gives nontrivial conditions for a continuous map f to have finitely many periodic points unless the map induced by f on first homology is nilpotent.

6. Algebraic Appendix.

We present here the proof of Theorem 9.

Proof of Theorem 9. We are given $p(t)$, $q(t) \in (\mathbb{Z}H)_1[t]$ and $\eta_i(t) \in \mathbb{Z}_1[[t]]$, $i \in I$, I finite, $n_i > 0$, $h_i \in H$ for which $p(t)/q(t) = \prod_{i \in I} \eta_i(t^{n_i} h_i)$. By passing to the subgroup generated by the terms of p and q and the h_i, we may suppose H is finitely generated.

We need two lemmas. The first is a form of Gauss' Lemma for integral power series series. It follows from results of Fatou [Fa] exposed in [BL, Prop. 1]

Lemma 7. $\mathbb{Q}(t) \cap \mathbb{Z}_1[[t]] = \mathbb{Z}_1(t)$.

The next lemma gives a factorization of certain rational functions into simpler rational functions.

Lemma 8. Let F be a field and $R = F(x_1,\ldots,x_n)$ the rational functions over F in indeterminates x_i. Let $G \subset R^*$ be the multiplicative group generated by the x_i. Suppose there are $\sigma_i(t) \in F_1[[t]]$, $m_i > 0$ and $g_i \in G$, $i = 1,\ldots,k$, and $\xi(t) \in R(t)$ such that

$$\xi(t) = \prod_{j=1}^{k} \sigma_j(t^{m_j} g_j)$$

If $g_i^{m_j} \neq g_j^{m_i}$ for $i \neq j$ then $\sigma_i(t) \in F(t)$, $i = 1,\ldots,k$.

Proof. We let $V = G \otimes \mathbb{R} = \mathbb{R}^n$ and identify G with $\mathbb{Z}^n$ in the obvious way. The points $v_i = m_i^{-1} g_i \in \mathbb{R}^n$ span a convex polyhedron $C \subset \mathbb{R}^n$. Suppose, reordering the indices if necessary, that v_k is a vertex of C. Then there is a linear functional $\alpha : V \to \mathbb{R}$ for which $\alpha(v_i) > \alpha(v_k)$ for all $i < k$.

Now, write $\xi(t) = p(t)/q(t)$, $p(t)$, $q(t) \in R[t]$, $p(0) = q(0) \neq 0$. Then $q(t) \prod_{i=1}^{n} \sigma_i(t^{m_i} g_i) = p(t)$. Let $\beta(t^n h) = n\alpha(v_k) - \alpha(h)$ for all $n > 0$, $h \in G$, so that $\beta(t^{m_i} g_i) = m_i \alpha(v_k) - \alpha(g_i) \leq 0$ with equality only for $i = k$. Let $p_*(t)$ (respectively $q_*(t)$) be the terms in $p(t)$ (respectively $q(t)$) on which β is largest. Then $q_*(t)\sigma_k(t^{m_k} h_k) = p_*(t)$ so $\sigma_k(t^{m_k} h_k) \in R(t)$ and $\sigma_k(t) \in F(t)$.

Now $\xi(t)/\sigma_k(t^{m_k} h_k)$ satisfies the hypothesis of the theorem for a smaller value of k. An obvious induction on k concludes the proof of Lemma 8.

We now prove Theorem 9 when H is torsionfree. We take $x_1,\ldots,x_n$ an integral basis for H. We group all the factors $\sigma_i(t^{n_i} h_i)$ lying on one ray to obtain factors $\sigma_j(t^{m_j} h_j)$ with distinct values of $h_j \otimes \frac{1}{m_j} \in H \otimes \mathbb{R}$. Then Lemma 8 with $F = \mathbb{Q}, G = H$

gives a factorization of $\xi(t)$ into rational functions $\sigma_j(t^{m_j}h_j)$. Lemma 7 gives $\sigma_j(t) \in \mathbb{Z}_1(t)$ for all j so that the factors $\sigma_j(t^{m_j}h_j)$ are indeed ray functions.

We now handle the case when H has a nontrivial torsion subgroup T. We may write $H = G \times T$ where G is free abelian on $n \geq 0$ generators $x_1,\ldots,x_n$ and T is finite. Given $h \in H$, we have $h = h_G \cdot h_T$, where $h_G \in G$, $h_T \in T$. As above, we partition the factors η_i according to the points v_i in $H \otimes \mathbb{R} = G \otimes \mathbb{R} = \mathbb{R}^n$ they determine by the formula $h_i \otimes \frac{1}{n_i} = v_i$. We denote by $\hat{T}$ the character group $\mathrm{Hom}(T,S^1)$ of T. Each $\rho \in \hat{T}$ determines a power series $\rho_*\xi(t)$ over $\mathbb{C}[x_1,x_1^{-1},\ldots,x_n,x_n^{-1}]$ that one can prove is a finite product of ray functions with complex coefficients, with one factor for each value of v_i. The problem is that the various ray functions thus arising depend on ρ in a complicated way and need not come from ray functions over H under ρ_*. To circumvent this pathology, one must construct the ray functions in a more controlled way.

So let M be the product of the least common multiple of all the n_i's with the least common multiple of all the orders of elements of T. Let λ be a primitive Mth root of unity and $F = \mathbb{Q}(\lambda)$. We let C be the power series group $R_1[[t]]$ where $R = F(x_1,\ldots,x_n)$. There are two natural group actions on C. The Galois group $\Gamma = \mathrm{Gal}(F/\mathbb{Q})$ has a natural action and the cyclic group $\mathbb{Z}_M$ acts by substituting $t\lambda^j$ for t, $j \in \mathbb{Z}_M$. Together these generate a group Φ of automorphisms of C. Φ is a semi-direct product of Γ and $\mathbb{Z}_M$, hence finite.

Now fix a $v \in H \otimes \mathbb{R}$ and let $I_v \subset I$ consist of all indices i with $v_i = v$. Let $\xi_v(t) = \prod_{i\in I_v} \eta_i(t^{n_i}h_i)$. We distinguish two Φ-invariant subgroups $A_v \subset B_v \subset C$. B_v is the subgroup generated by the series $\eta_i(t^{n_i}\lambda^j h_{iG})$, $j \in \mathbb{Z}_M$, $i \in I_v$. A_v is the smallest Φ-invariant subgroup of B_v containing the series $\rho_*\xi_v(t)$, $\rho \in \hat{T}$. Since logarithmic supports of elements of B_v lie on $\{(n,h) \mid h \otimes \frac{1}{n} = v\}$, one sees that the subgroups B_v are independent in C. So we have $B = \oplus B_v \subset C$ and $A = \oplus A_v \subset B$. As there are only finitely many v_i, B is a finitely generated group.

Now we pass to additive notation. B is a torsionfree finitely generated abelian group with a linear action of Φ and A is an invariant subgroup. Tensoring with $\mathbb{Q}$ and applying Maschke's Theorem, we obtain a Φ-equivariant projection $\pi : B \otimes \mathbb{Q} \to A \otimes \mathbb{Q}$ that we may choose so that $\pi(B_v \otimes \mathbb{Q}) \subset A_v \otimes \mathbb{Q}$ for all v. Choose $r > 0$ so that $r\cdot\pi(B \otimes 1) \subset A \otimes 1$, $r \in \mathbb{Z}$.

We return to multiplicative notation and let $P : B \to A$ be the map corresponding to $r\pi$. Then P is Φ-equivariant, $P(x) = x^r$ for $x \in A$, and $P(B_v) \subset A_v$ for all v.

Now fix $i \in I$. Let $\emptyset_j(t) = P(\eta_i(t^{n_i}\lambda^j h_{iG})) \in A_v$, $j \in \mathbb{Z}_M$. As P is $\mathbb{Z}_M$ equivariant, we have $\emptyset_j(\lambda t) = \emptyset_{j+n_i}(t)$, for all $j \in \mathbb{Z}_M$. It follows that for $d = M/n_i$, $\emptyset_o(\lambda^d t) = \emptyset_o(t)$. As λ^{d^i} is a primitive n_ith root of unit, the coefficient of t^p in $\emptyset_o(t)$ must be zero when p is not divisible by n_i . As P is Γ-equivariant and $\eta_i(t^{n_i}h_{iG})$ has integer coefficients, $\emptyset_0(t)$ has integer coefficients as well. Thus $\emptyset_0(t) = \tau_i(t^{n_i}h_{iG})$, for some uniquely determined series $\tau_i(t) \in \mathbb{Z}_1[[t]]$.

We now show that a power of $\xi(t)$ factors into ray series belonging to A .

<u>Lemma 9.</u>

$$(\xi(t))^r = \prod_{i \in I} \tau_i(t^{n_i}h_i) \quad .$$

<u>Proof.</u> It suffices to show that this equation over $(RT)_1[[t]]$ gives a correct formula after applying ρ_* to both sides for all $\rho \in \hat{T}$. So we fix $\rho \in \hat{T}$ and show $(\rho_*\xi(t))^r = \prod_{i \in I} \tau_i(t^{n_i}\rho_*(h_{iT})h_{iG})$.

Now fix $i \in I$. Then $P\rho_*(\eta_i(t^{n_i}h_i)) = P\eta_i(t^{n_i}\rho_*(h_{iT})h_{iG})$ and by the choice of λ , $\rho_*(h_{iT}) = \lambda^j$ where $n_i|j$, say $j = n_i\sigma$. Thus

$$\begin{aligned} P(\eta_i(t^{n_i}\lambda^j h_{iG})) &= \emptyset_j(t) \\ &= \emptyset_0(\lambda^\sigma t) \\ &= \tau_i((\lambda^\sigma t)^{n_i}h_{iG}) \\ &= \tau_i(t^{n_i}\lambda^j h_{iG}) \quad . \end{aligned}$$

Combining these formulas gives

$$P\rho_*(\eta_i(t^{n_i}h_i)) = \tau_i(t^{n_i}\rho_*(h_{iT})h_{iG}) \quad .$$

Now fix v . Multiplying over $i \in I_v$ and using $\xi_v(t) \in A$ gives

$$\begin{aligned} (\rho_*\xi_v(t))^r &= P\rho_*\xi_v(t) \\ &= \prod_{i \in I_v} P\rho_*(\eta_i(t^{n_i}h_i)) = \prod_{i \in I_v} \tau_i(t^{n_i}\rho_*(h_{iT})h_{IG}) \ . \end{aligned}$$

Finally, we multiply over all v's to get the desired formula for $(\rho_*\xi(t))^r$. This proves Lemma 9.

By applying Lemma 8 to $\rho_*\xi(t)$ with $\sigma_j(t^{m_j}g_j) = \rho_*\xi_v(t)$, we see that for any $\rho \in \hat{T}$, and any v, $\rho_*\xi_v(t) \in R(t)$. Thus $A \subset R(t)$ and in particular $\tau_i(t) \in F(t)$.

We write $\tau_i(t) = \frac{p_i(t)}{q_i(t)}$, with $p_i(t)$, $q_i(t) \in F[t]$. As $\tau_i(t) \in \mathbb{Z}_1[[t]]$, we have $\gamma\tau_i(t) = \tau_i(t)$ for all $\gamma \in \Gamma$. Thus for $\ell = |\Gamma|$, as γ runs over Γ we have

$$\tau_i(t)^\ell = \Pi\ \gamma\tau_i(t) = \frac{\Pi\ \gamma p_i(t)}{\Pi\ \gamma q_i(t)} = \frac{P_i(t)}{Q_i(t)},$$

where P_i, $Q_i \in \mathbb{Q}[t]$. If $\psi_i(t) = \tau_i(t)^\ell$, we have $\psi_i(t) \in \mathbb{Q}(t) \cap \mathbb{Z}_1[[t]]$. Now Lemma 7 gives $\psi_i(t) \in \mathbb{Z}_1(t)$ for all i.

Setting $N = \ell\cdot r$ and applying Lemma 9 gives $\xi(t)^N = \prod_i \psi_i(t^{n_i}h_i)$, the desired factorization of a power of $\xi(t)$ into ray functions. Q.E.D.

We now present our example of a $\xi(t)$ satisfying the hypothesis of Theorem 9 for which one must take $N > 1$ in the conclusion.

Let $H = \{1,x\}$, $x^2 = 1$. Let $\xi(t) = 1-(3+x)t+5t^2 \in (\mathbb{Z}H)_1[t]$. Let $\eta_0(t) = (1+2t+5t^2)^{-1}(1+4t+5t^2)^{-1}$, $\eta_1(t) = (1+2t+5t^2)(1-4t+5t^2)$ and $\eta_2(t) = 1-6t+25t^2$, so $\eta_i(t) \in \mathbb{Z}_1(t)$, $i = 0,1,2$. Then one can easily check that $\xi(t)^2 = \eta_0(t)\eta_1(xt)\eta_2(xt^2)$ (as below it suffices to check this for $x = \pm 1$). It follows that the logarithmic support of $\xi(t)$, which is clearly the same as the logarithmic support of $\xi(t)^2$, lies on the rays through t, xt and xt^2. Thus $\xi(t)$ satisfies the hypothesis of Theorem 9 (and the conclusion with $N = 2$).

We now suppose $\xi(t)$ is itself a product of finitely many ray functions and we will find a contradiction. Simplifying this product, we may write $\xi(t) = \alpha_0(t)\alpha_1(xt)\alpha_2(xt^2)\ldots\alpha_{n+1}(xt^{2^n})$ for some n, with $\alpha_i(t) \in \mathbb{Z}_1(t)$.

Given any $\eta(t) \in (\mathbb{Z}H)_1(t)$, we define $\eta_+(t)$ (respectively $\eta_-(t)$) to be the image of $\eta(t)$ under the homomorphism associated to the trivial representation $\rho_+ : H \to \{1\}$ (respectively the nontrivial representation $\rho_- : H \to \{1,-1\}$)

$$(*)\qquad \begin{aligned} \xi_+(t) &= \alpha_0(t)\alpha_1(t)\alpha_2(t^2)\ldots\alpha_{n+1}(t^{2^n}) \\ \xi_-(t) &= \alpha_0(t)\alpha_1(-t)\alpha_2(-t^2)\ldots\alpha_{n+1}(-t^{2^n}) \end{aligned}$$

over $\mathbb{Z}_1(t)$.

We define a homomorphism $F : \mathbb{Z}_1(t) \circlearrowleft$ by the rule $F(\eta(t))\circ t^2 = \eta(t)\eta(-t)$. Since $\eta(t)\eta(-t)$ is invariant under $t \to -t$ it is a function of t^2, so F is well defined. If we take $\eta(t) = \xi_+(t)/\xi_-(t)$, we have

$$\eta(t) = (1-4t+5t^2)(1-2t+5t^2)^{-1}$$

$$F\eta(t) = (1-6t+25t^2)(1+6t+25t^2)^{-1}$$

$$F^i\eta(t) = 1, \quad \text{for all} \quad i \geq 2 \ .$$

Now, we rewrite $\eta(t)$ using (*) and apply F n times. We may use the rules $F(\beta(t)\beta(-t)^{-1}) = 1$ and $F(\beta(t^2)) = \beta(t)^2$ to find that $F^n\eta(t) = \alpha_{n+1}(t)^m \cdot \alpha_{n+1}(-t)^{-m}$, $m = 2^n$.

We compare our 2 expressions for $F^n\eta(t)$. If $n \geq 2$ then we see $\alpha_{n+1}(t)^m = \alpha_{n+1}(-t)^m$ so that $\alpha_{n+1}(t) = \alpha_{n+1}(-t)$. If follows that $\alpha_{n+1}(xt^{2n})$ has no terms containing x and may be absorbed into $\alpha_0(t)$. Continuing in this way, we may suppose $n = 1$ in (*).

For $n = 1$, we find $F\eta(t) = \dfrac{\alpha_2(t)^2}{\alpha_2(-t)^2} = \beta(t)^2$, for some $\beta(t) \in \mathbb{Z}_1(t)$. Thus the zeroes and poles of $F\eta(t)$ must have even order. By inspection, however, all the zeroes and poles of $F\eta(t)$ are of order 1. Thus, no factorization of $\zeta(t)$ of the above type can exist.

We remark that the factorization of $\xi(t)^2$ given above can be obtained by studying the action of F on $\xi_+(t)$ and $\xi_-(t)$. In fact when $H = \mathbb{Z}_2$ there is a constructive proof of Theorem 9 based on this operator F. This other proof shows that one may take the exponent N in Theorem 9 to be a power of 2 when $\# H = 2$.

REFERENCES

[AM] Artin, M and B. Mazur, On periodic points, Annals of Math. (2) 81 (1965), 82-99.

[B] Birkhoff G. D., Dynamical Systems, AMS Colloq. Publ. IX, Providence, 1966.

[BL] Bowen, R. and O.E. Lanford III., Zeta functions of the restrictions of the shift transfromation, Proc. Symp. Pure Math. XIV, 43-49.

[Br] Brown, Robert F.. The Lefschetz Fixed Point Theorem, Scott Foresman and Co., 1971

[CMY] Chow, S. N., J. Mallet-Paret and J. A. Yorke, A periodic orbit index which is a bifurcation invariant, preprint.

[FH] Fadell, E. and S. Husseini, Fixed point theory for nonsimply connected manifolds, Topology 20 (1981), 53-92.

[Fa] Fatou, P., Series trigonometrique et series de Taylor, Acta Math. 30 (1906), 335-400.

[F1] Franks, J.,Knots, links and symbolic dynamics, Annals of Math. 113 (1981), 529-552.

[F2] Franks, J., Homology and Dynamical Systems. Notes on NSF-CBMS Reg. Conf. at Emory University, preprint.

[F3] Franks, J., Morse inequalities for zeta functions, Annals of Math. 102 (1975), 143-157.

[F4] Franks, J., A reduced zeta function for diffeomorphisms, Amer. Journal of Math. 100 (1978), 217-243.

[Fr1] Fried, D., Flow equivalence, hyperbolic systems and a new zeta function for flows, Commentarii Math. Helv., to appear.

[Fr2] Fried, D., Homological identities for closed orbits, Inv. Math., to appear.

[Fr3] Fried, D., Entropy and twisted cohomology, to appear.

[Fr4] Fried, D., Growth rates of surface homomorphisms, to appear.

[H] Halpern, B., Periodic points on tori, Pac. J. Math. 83 (1979), 117-133.

[Hi] Hirsch.,M. Anosov maps, polycyclic groups and homology, Topology 10 (1971), 177-183.

[P] Plykin, R. V., Sources and sinks of Axiom A diffeomorphisms of surfaces, Mat. Sb. (N.S.) 94 (136) (1974), 243-264.

[R] Ruelle, D., Thermodynamic Formalism. Encyc. of Math. and its Appl., Vol. 5, Addison-Wesley, 1978.

[S] Shub, M., Dynamical systems, filtrations and entropy. Bull. Amer. Math. Soc. 80 (1974), 27-41.

[SS] Shub, M. and Sullivan., A remark on the Lefschetz fixed point formula for differentiable maps, Topology 13 (1974) 189-191.

[SM] Smale, S., Differentiable dynamical systems, Bull. Amer. Math. Soc. 73 (1967) 747-817.

[St] Stallings, J., Centerless groups - an algebraic formulation of Gottlieb's theorem, Topology 4 (1965), 129-134.

[T] Thurston, W.P., On the geometry and dynamics of diffeomorphisms of surfaces, preprint.

[W] Williams, R.F., The structure of Lorenz attractors, Publ. Math. IHES 50 (1980), 73-79.

STATISTICAL PROPERTIES OF FOLIATIONS

by Lucy Garnett

A nonsingular flow on a manifold corresponds to a foliation of a manifold by one dimensional leaves where the leaves are provided with Riemannian metrics and directed. The Birkhoff Ergodic Theorem is a tool for studying the recurrence patterns of the orbits of the flow as they wind around in the manifold We want to find a similar result for foliations with higher dimensional leaves in order to study such questions as how often does a leaf intersect a given transverse section? To apply the ideas behind the Birkhoff Ergodic Theorem a group action is needed to move averages around. The naive idea of starting at a point on a leaf, counting the number of intersection points with the transverse section within a distance R of the starting point, dividing by the area within that distance and then taking the limit as R goes to infinity doesn't provide the needed semi-group action. Moreover to apply the Birkhoff theorem an invariant measure is required. Although there is a concept of a (invariant) transverse measure, that is a measure on each transverse section which is invariant under motion along paths in leaves, these measures don't always exist. The two difficulties; the lack of a semi-group and measure, are solved with one stroke by replacing the naive average as described above by a diffused average over the whole leaf.

Let $(M, \mathcal{F})$ be as smoothly foliated manifold of finite dimension. Provide the tangent bundle to with a C^3 Riemannian structure. We require the induced geometry on the leaves to be uniformly bounded. In particular this insures that on each leaf:

(1) the sectional curvature is bounded from above and below

(2) the injectivity radius is bounded from below

(3) the volume of a leaf grows at most exponentially.

Whenever M is compact these conditions are automatically insured. Under these conditions each leaf inherits a Riemannian structure from the global structure making the leaf into a connected Riemannian manifold complete for diffusion. Solving the heat equation on each leaf gives a heat kernel $p_t(x,y)$ which represents the amount of heat that flows from point x to another point y on the same leaf in time t. A global function $f{:}M \longrightarrow R$ diffuses with time by the operator

$$D_t f(x) = \int_L f(y) p_t(x,y)\, dm\,(y)$$

where m is the Riemannian measure on the leaf L containing x. We can show that if f is a continuous function on M then so is $D_t f$. This diffusion operator provides the needed one parameter semi-group of operators.

A measure on M is said to be invariant if the integral of f with respect to that measure equals the integral of Df with respect to the measure for any continuous function f. It is easy to show, either as a fixed point theorem or by direct construction , that every compact M has at least one invariant measure. A measure (class) is ergodic if the manifold cannot be split into two disjoint measurable leaf saturated sets with intermediate measure. The stage is now set for the :

FOLIATION ERGODIC THEOREM

Let μ be a finite invariant measure on M. For all $f \in L^1(\mu)$ there exists a unique $\tilde{f} \in L^1_\mu$ such that

① $\lim\limits_{N\to\infty} \frac{1}{N} \sum\limits_{t=0}^{N-1} D_t f(x) = \tilde{f}(x)$

② $\tilde{f}$ is constant on leaves

③ $\langle \tilde{f}, \mu \rangle = \langle f, \mu \rangle$ $\qquad$ ($\langle\ ,\ \rangle$ represents the integral)

④ If μ is ergodic then $\tilde{f} = \langle f, \mu \rangle$

The proof of ①, ②, and ④ follows directly from a standard ergodic theorem as found in Yosida's book (5). The proof of 2 which is easy for one dimensional orbits requires more work in the higher dimensional case.

An unexpected bonus is a nice local characterization of the invariant measures. This generalizes the notion of a transverse measure. A measure is invariant if and only if it disintegrates locally into positive harmonic functions on the plaques of a flowbox times the Riemannian measures on these plaques. Given any transverse invariant measure, a global measure may be formed by locally integrating the Riemannian leaf measures with respect to the transverse invariant measure. In this case we say that the (global) measure on M is obtained from a transverse measure. Any such (global) measure distingrates locally to a constant function times the Riemannian leaf measures and thus, by the previous characterization, is an invariant measure for the foliation. However, the concept of an invariant measure is more general than transverse measure as is seen in the example of the geodesic flow on the surface of genus two with constant negative curvature. Grouping together the geodesic flow lines which are asymptotic at infinity produces a codimension one foliation of this three dimensional manifold. The leaves have exponential growth and work by Plante shows that such foliations have no transverse invariant measures. However, the measure obtained by integrating the visual (Poisson) measure over the Poincare metric on the leaf is invariant. Moreover, this measure is the unique invariant measure on this foliation.

There are several consequences of this local characterization on an invariant measure, If the leaves are simply connected and have no positive nonconstant harmonic functions then any invariant measure is obtained fron a transverse measure. (Moser has shown that any quasi-Euclidean space has no positive nonconstant harmonic functions.) In the case of a two dimensional foliation of te solid three torus, Rosenberg has shown that the leaves are conformal to R^2 if there are no Reeb components. This would necessitate that any invariant measure must come from a transverse measure.

This ergodic theorem for foliations has an interesting interpretation in terms of paths in the leaves. With respect to any invariant measure on a foliaiton the time limit of any L function along almost (in the sense of Wiener) path starting at any point in a leaf exists and equals the leaf diffused average given in the Foliaiton Ergodic Theorem.

Let us now restrict attention to the case of a compact foliated manifold. Extending the standard ergodic theory arguments to leaf saturated sets we can construct a measurable leaf saturated set R of regular points having the following properties:

1 For any $x \in R$ the diffused dirac measure $\tilde{\delta}_x$ exists, is ergodic and contains x in its support. The diffused dirac measure is defined by $\langle f, \tilde{\delta}_x \rangle = \langle \tilde{f}, \delta_x \rangle$

for any continuous $f: M \longrightarrow R$.

2 Any two points on the same leaf in R have equivalent diffused dirac measures.

3 R is given measure one by any invariant probability measue.

From the discussion of regular points it follows that any invariant measure is a convex combination of ergodic ones coming from diffused dirac masses.

ERGODIC DECOMPOSTION THEOREM

If μ is an invariant probalitiy measure then

① $\mu = \int_{x \in R} \tilde{\delta}_x \, d\mu(x)$

and if μ is obtained from a transverse measure then

② $\mu = \int_{x \in R_0} \tilde{\delta}_x \, d\mu(x)$

where R_0 is the subset of R whose diffused dirac masses come from transverse measures.

A pure foliation fact follows easily from this analysis. Namely the set of leaves in a compact manifold which are non-recurrent(the leaf interesects an open transversal in one point) is given measure zero by any finite transverse measure.

We end with a couple of remarks. Although this theory is stated for foliations there is no obstruction to applying it to laminations. The foliation ergodic theorem only states that the limit of the Cesaro sums of the diffused function exists. It seems likely that the limit of the diffused function should exist directly without taking Cesaro sums, but I've been unable to prove that. The proof and references for these statements can be found in Garnett(1).

BIBLIOGRAPHY

[1] Garnett,L. ""Functions and Measures Harmonic along the Leaves of a Foliation and the Ergodic Theorem" Journal of Functional Analysis (to appear)

[2] Moser,J. "Harnach Inequalities" Comm of Pure and Applied Math 1973

[3] Plante,J. "Foliations with Measure Preserving Holonomy" Annals Math 102 1975

[4] Ruelle and Sullivan,"Currents,Flows and Diffeomorphisms" Topology Vol.14

[5] Yosida,K. "Functional Analysis" Springer-Verlag, Berlin, 1965

Dehn surgery on Anosov flows

Sue Goodman

Several new and interesting examples of Anosov flows on 3-manifolds have been found recently. In [2], Handel and Thurston give the first examples of non-algebraic Anosov flows. In their construction a geodesic flow is cut apart along an incompressible torus and reglued with a Dehn twist, preserving an Anosov structure, but the resulting manifold is neither an S^1-bundle or a torus-bundle over S^1. It is a graph manifold, and the flow is volume-preserving.

Franks and Williams [1] have recently given an example of an Anosov flow which is not volume-preserving; hence, also not algebraic. The manifold is two copies of the complement of the figure eight knot, identified along the boundary torus. Each piece has a hyperbolic structure, but not the entire manifold.

In this paper, we produce more new examples of closed 3-manifolds which support Anosov flows. This is accomplished by doing a Dehn surgery, preserving longitudes, on a neighborhood of a periodic orbit of any Anosov flow. Many of these examples are also non-algebraic. Further, some of the manifolds produced are hyperbolic, and hence atoroidal, answering a question posed in [2].

Although the construction is somewhat different from Handel-Thurston's, the proof that the resulting flow is again Anosov is verysimilar to theirs. The author is grateful to them for several suggestions concerning the proof.

Let ψ_t be an Anosov flow on a compact 3-manifold M and let $g(t)$, $0 \leq t \leq 2\pi$, be a periodic orbit of ψ_t. Since $g(t)$ is hyperbolic, the induced diffeomorphism on a small cross-sectional disk is hyperbolic. Enclose $g(t)$ with a torus $T^2 \cong S^1 \times S^1$, parametrized by (θ_1, θ_2), $0 \leq \theta_i \leq 2\pi$, bounding $S^1 \times D^2 = M_1$ so that each $\{\theta_1\} \times D^2$ is as shown:

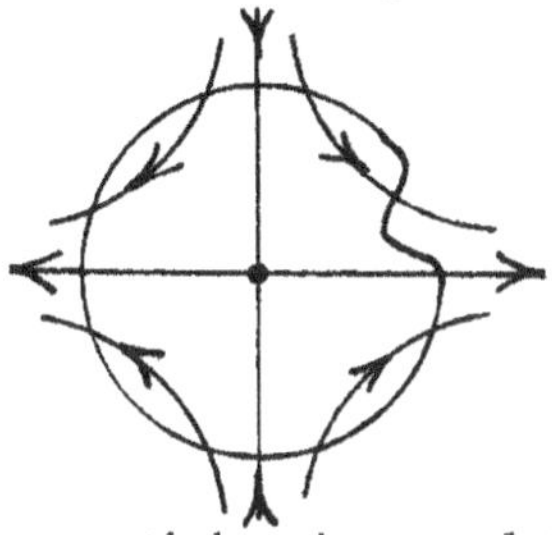

i.e., $\partial D^2 = S^1$ is smooth but is normal to the flow on a narrow strip, say $\{\theta_1\} \times (\varepsilon, 2\varepsilon)$. Let M_2 be the remaining component of M.

There are natural coordinates on this torus. Let the longitude be determined by weak unstable manifold of $g(t)$ intersected with T^2, and the meridian a curve bounding a disk $\{\theta_1\} \times D^2$ which meets $g(t)$ once. We will perform a Dehn surgery on this torus, doing all the twisting in the normal strip, and we will show that the resulting flow, on a new manifold M^1, is again Anosov.

To define the surgery gluing map $F : \partial M_1 \to \partial M_2$, let $h(\theta)$ be a smooth function on $[0, 2\pi]$ so that

$$h(\theta) = \begin{cases} 0 & \text{for } \theta \in [0, \varepsilon] \\ 2\pi & \text{for } \theta \in [2\varepsilon, 2\pi] \\ \text{is strictly increasing on} & (\varepsilon, 2\varepsilon). \end{cases}$$

Define F by $F(\theta_1, \theta_2) = (\theta_1 + h(\theta_2)k, \theta_2)$ where k is a positive integer. Notice that longitudes are taken to longitudes and that F is the identity outside the strip.

To prove that the flow Φ_t is Anosov we want to find the stable and unstable directions. The idea is, as in Handel-Thurston, that a half-open pair of scissors in the vertical plane for ψ_t (i.e., the plane determined by the strong stable and unstable directions) will in forward time approach the slope of the unstable foliation of Φ_t and in backward time, the stable. The scissors close at an exponential rate under the action of $D\psi_t$, since it is Anosov. Under the action of DF, DF^{-1}, we will see that they do not open (Proposition 1). Since the flow ϕ_t is a composition of these actions, with $D\psi_t$ predominating, this gives a well-defined continuous slope function for the intersection of the weak stable (and unstable) foliations with the vertical plane. Hence this slope together with the flow direction determines the 2-dimensional Anosov foliation associated with Φ_t. One then does a similar argument in each of these 2-dimensional plane fields to determine strong stable and unstable line fields (Proposition 2). The proof follows.

The flow ψ_t being Anosov means that there is a $D\psi_t$ - invariant splitting of the tangent bundle to M, TM, into line bundles $E^t \oplus E^{uu} \oplus E^{ss}$ where E^t is tangent to the flow and E^{uu}, E^{ss} have the following property. For any $p \in M$, there are constants $A_u, A_s > 0$, $\lambda_u, \lambda_s > 1$ so that if $v_p \in E^{uu}$ and $t \geq 0$ then $\| D\psi_t(v_p) \| \geq A_u \lambda_u^t \| v_p \|$ and if $v_p \in E^{ss}$ and $t \leq 0$, then $\| D\psi_t (v_p) \| \geq A_s \lambda_s^{-t} \| v_p \|$. Let t_p, u_p, s_p be a basis for these line bundles at p.

For $p \in T^2$, there is also a natural orthonormal basis L_p, M_p, N_p where L_p is a unit vector parallel to the longitude of T^2, M_p is a unit vector parallel to the meridian, and N_p is a unit vector normal to T^2 and pointing out from M_1. We also choose M_p and L_p so that the u-coordinates of both (denoted m_u and ℓ_u) are positive.

It will be useful to write the map $DF : TM|_{\partial M_1} \to TM|_{\partial M_2}$ in terms of both types of coordinates. If v_p is an arbitrary tangent vector to M at $p = (\theta_1, \theta_2) \in T^2$, $v_p = at_p + bu_p + cs_p = xL_p + yM_p + zN_p$. The map DF is given by $DF(v_p) = [x + h'(\theta_2) k y]L_{F(p)} + yM_{F(p)} + zN_{F(p)}$, or in flow coordinates, $at_{F(p)} + bu_{F(p)} + cs_{F(p)} + h'(\theta_2) k v_p \cdot M_p L_{F(p)}$. (Note: $v_p \cdot M_p = y$.)

Also notice that $DF(t_p) = t_{F(p)}$, since if p is in the strip, $t_p \cdot M_p = 0$, and otherwise $h'(\theta_2) = 0$. Hence, there is a smooth flow Φ_t induced on M'. We wish to show that this flow is Anosov. To do this, we first find F^s and F^u, the weak stable and unstable plane fields for Φ_t.

<u>Proposition 1</u>: There exist $D\Phi_t$-invariant plane fields F^u, F^s on M', intersecting along the line field tangent to Φ_t.

Proof for F^u (F^s case is similar):

We will produce a plane field on each component M_i .

For each p in the interior of M_1 , $v_p \in T_pM_1$ and $\Phi_t(v_p) \in \operatorname{int} M_1$, we can write $D\Phi_t(v_p) = D\psi_{t_n} \circ DF^{-1} \circ D\psi_{t_{n-1}} \circ DF \circ \ldots \circ DF \circ D\psi_{t_1}(v_p)$ where $t = \Sigma t_i$ and the t_{2k} are uniformly bounded above 0 . Similarly for $p \in M_2$.

The actions of $D\psi_t$ and DF on the u,s-plane at a point p are shown below.

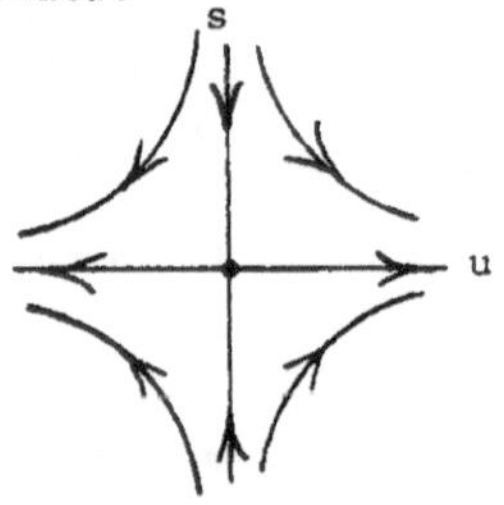

$D\psi_t$, $t \geq 0$

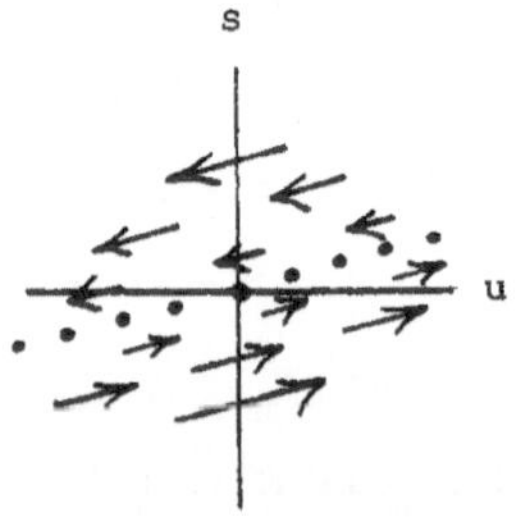

DF for $p \in$ strip

The slope of the line of fixed points for DF is $-\dfrac{m_u}{m_s}$ (since it is perpendicular to M). Also notice that DF is the identity for p outside the strip and that DF^{-1} is always the identity, since the flow is directed out of M_1 on the strip.

For each p , there is a $D\Phi_t$-invariant ($t \geq 0$) neighborhood K_p of the u-axis, as shown below.

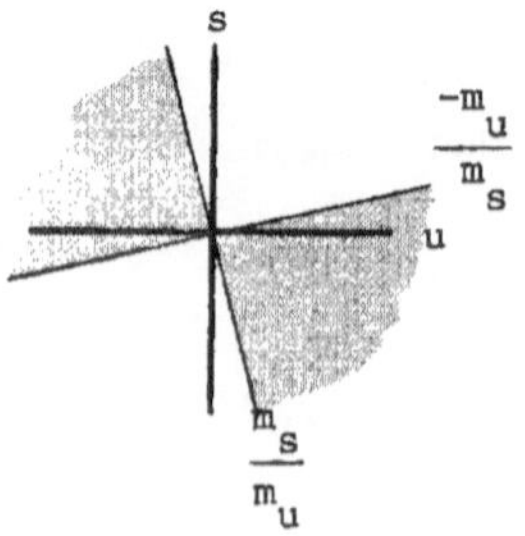

For any two vectors v_1, v_2 in this neighborhood, we will see that the difference in slopes of these vectors in the u,s-plane approaches zero exponentially in t under the action of $D\Phi_t$. This is true for $D\psi_t$, since it is Anosov. For DF, at p in the strip, if $v_\ell = a_\ell t + b_\ell u + c_\ell s$ is in K, a straightforward calculation shows that the difference in slopes of $DF(v_2)$ and $DF(v_1)$ is $\left| \frac{c_2 + h'(\theta_2)ky_2\ell_s}{b_2 + h'(\theta_2)ky_2\ell_u} - \frac{c_1 + h'(\theta_2)ky_1\ell_s}{b_1 + h'(\theta_2)ky_1\ell_u} \right| =$

$$\left|\frac{c_2}{b_2} - \frac{c_1}{b_1}\right| \cdot \frac{1}{1 + h'(\theta_2)k\left(\frac{y_1}{b_1} + \frac{y_2}{b_2} + h'(\theta_2)\frac{y_1y_2}{b_1b_2}\ell_u\right)} \le \left|\frac{c_2}{b_2} - \frac{c_1}{b_1}\right| .$$

The difference is left unchanged for DF^{-1}. Since $D\Phi_t$ is a composition of these maps, the result follows. So $\alpha(p) = \lim_{t\to\infty} (\text{slope } D\psi_t(v))$ for $v \in K_p$ is a well-defined continuous function with values between $\frac{m_s}{m_u}$ and $-\frac{m_u}{m_s}$. This function determines the slope of the line of intersection of F_p^u with the u,s-plane, i.e., $F_p^u = \{v_p \mid v_p = at_p + bu_p + \alpha(p)\, bs_p\}$. These match up on the boundary and are $D\Phi$-invariant.

F_p^s is defined similarly from a slope function $\beta(p)$ with values greater than $-\frac{m_u}{m_s}$ or less than $\frac{m_s}{m_u}$ (or vice versa if $\frac{m_s}{m_u}$ is positive). F_p^s and F_p^u are therefore transverse. □

We now find F^{uu} and F^{ss}, the strong unstable and stable line bundles for Φ_t by finding the slope of the line F_p^{uu} in F_p^u and F_p^{ss} in F_p^s.

Proposition 2: There are continuous $D\Phi$-invariant line bundles F^{uu}, F^{ss} on M' which, along with the flow, provide an Anosov structure for Φ_t.

Proof for F^{uu}:

A vector v_p in F_p^u has form $at_p + bu_p + \alpha(p)bs_p$. As in the proof of Proposition 1, it will be shown that there is a continuous slope function $\delta(p)$ in the u,t-plane defined by $\lim_{t\to\infty}$ (slope $D\Phi_t(v)$) for v in F^u with $b > 0$. For two vectors $(v_\ell)_p \in F_p^u$, $\ell = 1,2$, the difference in slopes of $D\psi_t(v_\ell)$ decreases exponentially in t by the Anosov structure of ψ_t.

For DF in the strip, this difference is given by

$$\left|\frac{a_2 + h'(\theta_2)\,k\,(b_2m_u + \alpha b_2m_s)\ell_t}{b_2 + h'(\theta_2)\,k(b_2m_u + \alpha b_2m_s)\ell_u} - \frac{a_1 + h'(\theta_2)\,k\,(b_1m_u + \alpha b_1m_s)\ell_t}{b_1 + h'(\theta_2)\,k\,(b_1m_u + \alpha b_1m_s)\ell_u}\right| =$$

$$\left|\frac{a_2}{b_2} - \frac{a_1}{b_1}\right| \cdot \frac{1}{(1 + h'(\theta_2)\,k\,(m_u + \alpha m_s)\ell_u}\,.$$

It is easy to show that if m_u, ℓ_u have the same sign, $1 + h'(\theta_2)\,k\,(m_u + \alpha m_s)\,\ell_u \geq 1$, hence DF does not increase the difference in slopes. Nor does DF^{-1}; hence, $\delta(p)$ exists.

Finally, we must show that $D\Phi_t$ expands vectors in F^{uu} exponentially. A vector in F_p^{uu} has form $\delta(p)b\,t_p + b\,u_p + \alpha\,(p)b\,s_p$, so it is enough to show that the u-coordinate b is expanded exponentially. For $D\psi_t$, this is clear. For DF, the u-coordinate of $DF(v_p)$ is $b[1 + h'(\theta_2)\,k\,(m_u + \alpha m_s)\,\ell_u]$ which is no less than b in absolute value. Hence $D\Phi_t$ expands b exponentially in t. □

<u>Example</u>. Let ψ_t be the suspension of the Anosov diffeomorphism of T^2 induced by the map $\begin{pmatrix} 2 & 1 \\ 1 & 1 \end{pmatrix}$. Let the periodic orbit $g(t)$ be the suspension of the fixed point $(0,0)$. The complement of this orbit is a fiber bundle over S^1 with fiber T^2-point, and it is homeomorphic to the complement of the figure eight knot in S^3 . [See Rolfsen [3], p. 337, for the fibering of this knot complement.]

Thurston [4] has shown that this manifold has a hyperbolic structure (i.e., a Riemannian metric with constant curvature -1) and further that all but a finite number of Dehn surgeries on the figure eight knot yield a closed 3-manifold with a hyperbolic structure. The construction given in this paper produces Anosov flows on many of these manifolds.

Further, Thurston points out that all but two of these surgeries produce manifolds which are not sufficiently large. Thus, we also produce the first examples of Anosov flows on non-sufficiently large manifolds which are not Seifert manifolds (since they are hyperbolic).

List of References

1. J. Franks and R. Williams, "Anomolous Anosov Flows," Proceeding of Northwestern Conference, Lecture Notes in Math 819, pp. 158-174.

2. M. Handel and W. Thurston, "Anosov Flows on new 3-manifolds," Inv. Math. 59 (1980), 95-103.

3. D. Rolfsen, <u>Knots and Links</u>, Math. Lecture Series 7, Publish or Perish (1976).

4. W. Thurston, "The Geometry and Topology of 3-Manifolds," Lecture Notes, Princeton University.

Smoothability of Cherry Flows on Two-Manifolds

Carlos Gutierrez*

§1 Introduction

Let M and $\tilde{M}$ denote compact, manifolds. We say that two continuous flows (M,ϕ) and $(\tilde{M},\tilde{\phi})$ are (topologically) equivalent if there is a homeomorphism of M onto $\tilde{M}$ which takes orbits of ϕ onto orbits of $\tilde{\phi}$ preserving sense (but not necessarily parametrization). A continuous flow is smoothable if it is equivalent to a smooth one. A particular case of our theorem is:

Proposition - Any continuous flow ϕ on a compact two-manifold M, with finitely many fixed points and a dense positive semi-trajectory is smoothable.

In 1938 T.M. Cherry [1] introduced and studied an interesting class of flows on the torus. We continue his work.

Definition - A Cherry flow is a continuous flow ϕ on a compact, two-manifold M provided:

(a) ϕ has only finitely many fixed points

(b) Let $p_1, p_2, \ldots, p_m$ be the source-fixed-points of ϕ and let $\lambda_1, \lambda_2, \ldots, \lambda_m$ be their basins of repulsion (= their unstable manifolds). Then, each λ_j contains a unique trajectory θ_j con-

* Partially supported by CNPq Brazil.

necting p_j to another (unique) fixed point $q_j \in \partial\lambda_j$ (See Fig.1).

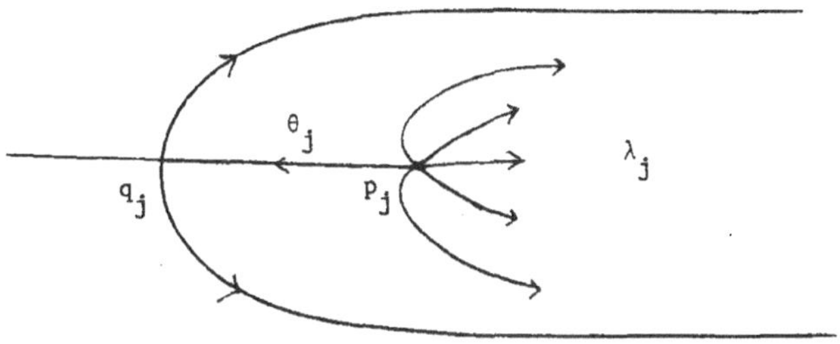

Fig. 1

(c) There are finitely many positive semi-trajectories $\gamma_1^+, \gamma_2^+, \ldots, \gamma_n^+$ such that $(\bigcup_{i=1}^{m} \lambda_i) \cup (\bigcup_{j=1}^{n} \gamma_j^+)$ is dense in M.

We prove the following:

Theorem 1 - Any continuous Cherry flow is smoothable.

It will be proved that, in general, Cherry flows have non-trivial ω-recurrent trajectories. Flows with this sort of trajectory are not always smoothable; for example, Denjoy [2] constructed on the torus a C^1 flow which is not topologically equivalent to any C^2-flow; his example has no fixed points, however, it does not satisfy the condition (c) of the definition of Cherry flows. Recent examples of unsmoothable flows in higher dimensions have been given by J. Harrison [5].

Cherry [1] proved theorem 1 on the torus for a flow with two fixed points: one source and one saddle point. We remark that his example is analytical and we conjecture that all Cherry flows are equivalent to analytic ones.

To explain briefly what we do in this paper we need the following definition. An injective differentiable map, $T: \mathbb{R}/\mathbb{Z} \to \mathbb{R}/\mathbb{Z}$, defined everywhere except possibly at finitely many points is said to be a (generalized) interval exchange transformation if there exists a constant $\beta \in (0,1]$ such that for every x in the domain of definition of T, $|T'(x)| = \beta$. If $\beta = 1$ (resp. $\beta < 1$) T is called <u>standard</u> (resp. <u>contractive</u>).

In section 3 we introduce the concept of basic Cherry flow and prove, in sections 3 and 4, that they are (modulo topological equivalence) smooth suspensions of interval exchange transformations. In section 5 we see that the smoothability of Cherry flows can be reduced to that of basic Cherry flows. The proof of the proposition mentioned above is in section 4.

Before continuing I wish to thank C. Pugh for his helpful and stimulating conversations. I also wish to thank Gilbert Levitt for his useful comments.

§2 Local Study of the Fixed Points

Let ϕ be a continuous flow on a two-manifold. An isolated fixed point p of ϕ is said to be a multisaddle (Fig. 2) if there are only finitely many trajectories such that their α- or ω-limit set is precisely $\{p\}$ (see [9, pp.24] for the definition of α- and ω-limit set).

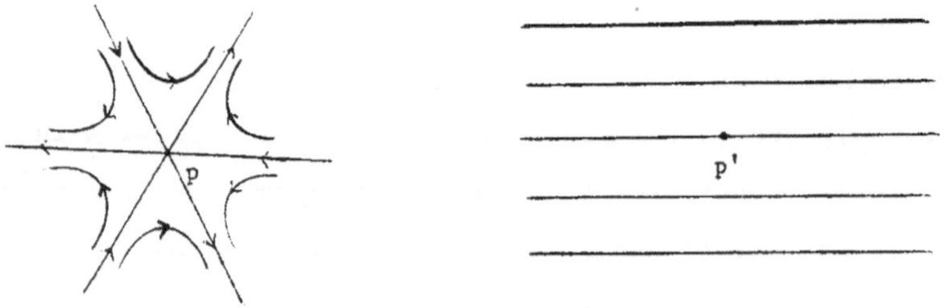

Multisaddle points p and p'

Fig. 2

Lemma 1 - Let ϕ be a Cherry flow. Any fixed point p of ϕ is either a source or a multisaddle.

Proof: Let p be a fixed point of ϕ which is not a source. By the local sector-classification of isolated fixed points [6, pp.161], because finitely many semi-trajectories together with finitely many unstable manifolds of sources are dense, and the fact that there is only one trajectory connecting a source-fixed point to other fixed point, we see that at p there are no elliptic sectors, nor parabolic sectors, nor else elliptic parts of hyperbolic sectors. Since any isolated fixed point has finite index we must merely consider each of the finitely many hyperbolic sectors. Each hyperbolic sector S has its two boundary curves as separatries, each separatrix in S comes from or goes to the fixed point p. □

§3 Basic Cherry Flows with Sources

In this section we define basic Cherry flows and prove that the ones having sources are, modulo topological equivalence, smooth suspensions of contractive interval exchange transformations.

Let (N,ψ) be a continuous flow (on a two-manifold N) and γ be a trajectory of ψ which is neither a fixed point nor a closed orbit. Let $(\mathbb{R}^2,\varphi)$ be a continuous flow with precisely two fixed points, a source p and a saddle point q, connected by a separatrix λ_4 of q. Denote by V the subset of $\mathbb{R}^2$ formed by the union of q, its separatrices λ_1, λ_2, λ_3 and λ_4 and $W^u(p)$. Notice that $W^u(p) \cap \{\lambda_1 \cup \lambda_2 \cup \lambda_3\} = \phi$.

Certainly $\tilde{N} = (N-\gamma) \cup V$ admits a structure of manifold and there exists a continuous flow $(\tilde{N},\tilde{\psi})$ such that

(A) $\tilde{N}$ is homeomorphic to N,

(B) The trajectories of $\tilde{\psi}|_{(N-\gamma)}$ (i.e. $\tilde{\psi}$ restricted to $N-\gamma$) and $\tilde{\psi}|_V$ are the same as those of $\psi|_{(N-\gamma)}$ and $\varphi|_V$ respectively.

(C) Let $\widetilde{N-\gamma}$ (resp. $N-\gamma$) denote the subspace $N-\gamma$ of $\tilde{N}$ (resp. N). The identity map $h\colon \widetilde{N-\gamma} \mapsto N-\gamma$ is a homeomorphism and extends to a continuous map $h\colon \tilde{N} \to N$ such that $h(V) = \gamma$.

Under these conditions we will say that either $(\tilde{N},\tilde{\psi})$ has been obtained by <u>blowing up</u> γ, or γ has been <u>blown up</u>. Roughly speaking what we do is to cut N along γ and insert V; the way is indicated in figure 3a, before blowing up γ, and in figure 3b, after blowing up γ.

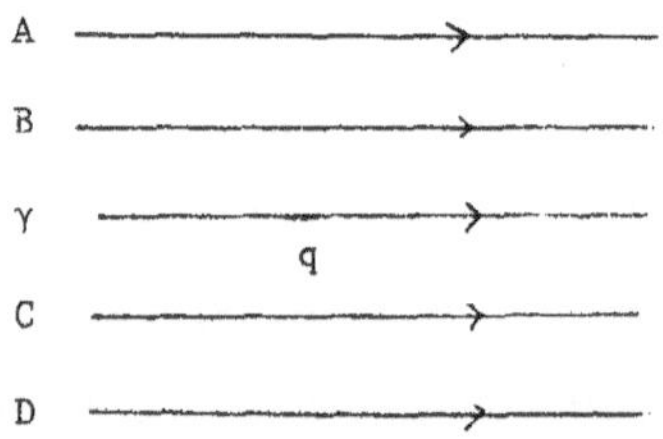

Before blowing up γ

Fig. 3a

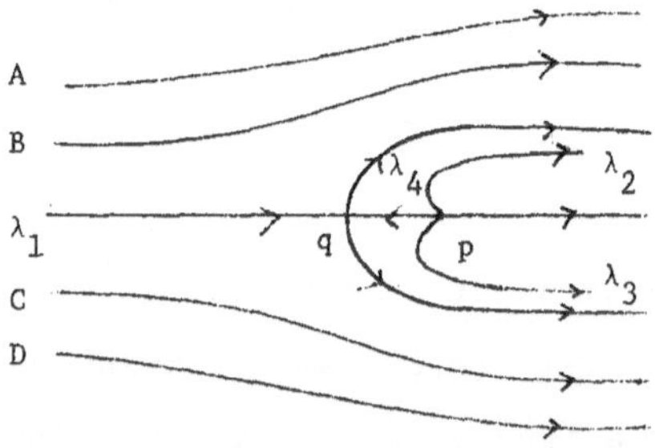

After blowing up γ

Fig. 3b

If we have a multisaddle fixed point of $2n$ separatrices $A_1, A_2, \ldots, A_{2n-1}, \gamma$ $(n \geq 1)$, we can also blow up γ as indicated (for $n=3$) in Figs. 4a and 4b.

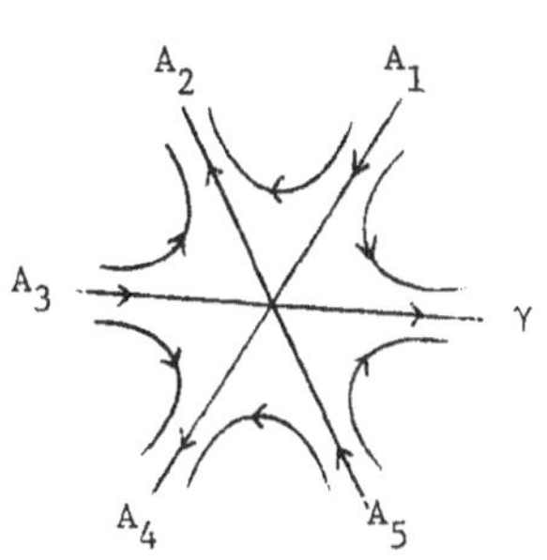

Before blowing up γ

Fig. 4a

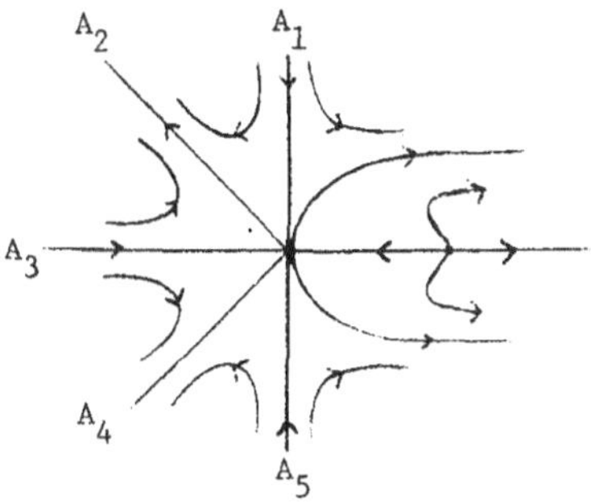

After blowing up γ

Fig. 4b

The reverse process to that of blowing up a trajectory is called the <u>blowing down</u> of the unstable manifold of a source.

A <u>basic Cherry flow</u> is a continuous flow defined on a compact connected two-manifold and such that either:

(a) it has finitely many fixed points and a dense positive semi-trajectory, or

(b) it has been obtained by blowing up finitely many dense positive semi-trajectories of a flow which satisfies (a) above.

Let γ_p (resp. $\gamma_p^{\pm}$) denote a trajectory (resp. positive/negative semi-trajectory) of ϕ passing through the point p. γ_p^+ and γ_p^- are taken to contain p. If p is a source-fixed-point of ϕ, let us denote by $W^u(p)$ the unstable manifold of p.

<u>Lemma 2</u> - (Peixoto). If ψ is a flow on a compact two-manifold M, having only finitely many fixed points and C is a circle transverse to ψ, then the domain of definition of the Poincaré map $T: C \to C$ induced by ψ is the finite union of intervals (a,b). The endpoints a, b lie on trajectories tending to fixed points as $t \to \infty$.

<u>Proof</u>: (See [10, Lemma 3, pp. 106] or [9, pp. 255]). □

We notice that if ϕ is a basic Cherry-flow, then we can construct a circle C transverse to ϕ (See [4, Lemma 2]).

<u>Lemma 3</u> - Let ϕ be a basic Cherry flow with sources $q_1, q_2, \dots, q_n$, $n \geq 1$. Let C be a circle transverse to ϕ. Then, there exists a homeomorphism $h: C \to \mathbb{R}/\mathbb{Z}$ such that $\tilde{T} = h \circ T \circ h^{-1}$ is a contractive interval exchange transformation.

<u>Proof</u>: Recall that $\alpha(x)$ denotes the α-limit set of the point x.

We define $\forall\, i \in \mathbb{N}$, $\forall\, j \in \{1,2,\dots,n\}$, $A_{ij} = \{x \in C / \alpha(x) = q_j$ and γ_x^- crosses C exactly i times$\}$ (See Fig. 5). Notice that $x \in \gamma_x^-$.

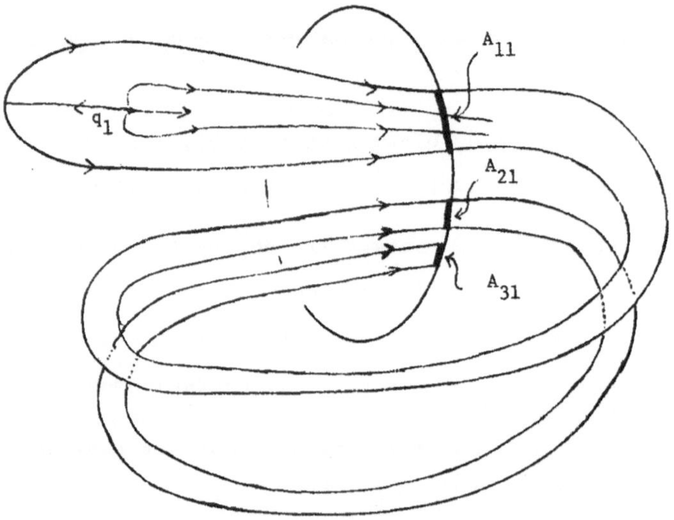

Fig. 5

Since the unstable manifolds of the sources of ϕ intersect C infinitely many times, $\forall\, i \in \mathbb{N}$, $\forall\, j \in \{1,2,\dots,n\}$, A_{ij} is an open non-empty interval.

Let θ denote either a fixed orientation in the circle C or the usual positive orientation of $\mathbb{R}/\mathbb{Z}$. Let $\Gamma \in \{C,\mathbb{R}/\mathbb{Z}\}$, $a,b \in \Gamma$, $z \neq b$. We define the interval $(a,b) = \{z \in \Gamma - \{a\}$ / if "$<$" denotes the linear order induced by the orientation θ in $\Gamma - \{a\}$, then $z < b\}$. Observe that $(a,b) \cap (b,a) = \emptyset$ and $(a,b) \cup (b,a) = \Gamma - \{a,b\}$.

Fix $\beta \in (0,1/n)$. Given $j \in \{1,2,\dots,n\}$, we define a measure μ in the Borel subsets of $A_{1j} = (a_j,b_j)$ such that

(3.1) the map $x \to \mu((a_j,x))$ is a homeomorphism from A_{1j} onto $(0,\beta)$.

Notice that $\{A_{ij}\}$, $i \in \mathbb{N}$, $j \in \{1,2,\dots,n\}$ are pairwise disjoint. Let us assume that we have defined μ in $\bigcup_{i=1}^{k-1}\bigcup_{j=1}^{n} A_{ij}$ and proceed inductively to define μ in $\bigcup_{j=1}^{n} A_{kj}$. If (x,y) is a sub-interval of A_{kj}, for some $j \in \{1,2,\dots,n\}$, we define

$$\mu((x,y)) = (1-n\beta)\,\mu(T^{-1}(x,y)). \tag{3.2}$$

Finally we extend μ to the Borel subsets of C in such a way that.

$$\mu(C - \bigcup_{i,j} A_{ij}) = 0. \tag{3.3}$$

Now we define $h\colon C \to \mathbb{R}/\mathbb{Z}$. Fix $\lambda_0 \in C$. Given $p \in C$ we let

$$h(p) = \begin{cases} 0, & \text{if } p = \lambda_0 \\ \mu((\lambda_0,p)), & \text{if } p \neq \lambda_0 \end{cases}$$

It follows from (3.1), (3.2) and (3.3) that h is continuous and strictly monotonic and also that $\mu(A_{ij}) = \beta(1-n\beta)^{i-1}$, $\forall\, i \in \mathbb{N}$, $\forall\, j \in \{1,2,\dots,n\}$. This implies that $\mu(C) = \sum_{i,j} \mu(A_{ij}) = 1$ and therefore that h is onto and a homeomorphism.

By Lemma 2, T is defined everywhere except possibly at finitely many points. Therefore $\tilde{T} = h\circ T\circ h^{-1}$ has the same property.

We claim that, for every x in the domain of definition of $\tilde{T}$, $|\tilde{T}'(x)| = 1-n\beta$. In fact, if $\tilde{T}$ is defined in (x,y), $x \neq y$, and J denotes the subinterval of C such that $h(J) = (x,y)$, we have that:

$$|\tilde{T}(x)-\tilde{T}(y)| = |h\circ T(h^{-1}(x)) - h\circ T(h^{-1}(y))| = \mu(T(J)) = \text{(Def. of } h)$$
$$= (1-n\beta)\mu(J) \qquad ((3.1),(3.2) \text{ and } (3.3))$$
$$= (1-n\beta)\mu(h(J)) = (1-n\beta)|y-x|. \qquad \text{(Def. of } h).$$

This proves the lemma. □

Let C_1 be a circle and $S: C_1 \to C_1$ be a continuous injective map defined everwhere except possibly at finitely many points. We say that a continuous flow (N,φ) is a <u>suspension</u> of S if N is a two-manifold containing C_1, φ is transversal to C_1 and the Poincaré map from C_1 into itself, induced by φ, is precisely S. For example, if (M,ϕ), C and $T: C\mapsto C$ are as in Lemma 3, then (M,ϕ) is a suspension of T. Notice that Image(T) is not dense and that two suspensions of T may not be equivalents.

<u>Proposition 1</u> - If ϕ is a basic Cherry flow with sources $q_1,q_2,\ldots,q_n$, $n \geq 1$, then ϕ is smoothable.

<u>Proof</u>: We need some notation. If ψ is a continuous flow on a manifold N, we denote by Σ_ψ (resp. Λ_ψ) the set of fixed points (resp. separatrices) of ψ. If A is a subset of N, $\psi|_A$ denotes the restriction of ψ to A. Let C, h, T and $\tilde{T}$ be as in Lemma 3. We consider a smooth circle $\tilde{C} \subset M$ close to C and disjoint from the fixed points of ϕ. Certainly, via a coordinate system, we can assume that $\tilde{C} = \mathbb{R}/\mathbb{Z} \times \{0\}$ and that $\mathbb{R}/\mathbb{Z} \times [-2,2] \subset M - \Sigma_\phi$.

First, we will prove that there exists a smooth flow $\tilde{\phi}$ defined on M such that

(1) $\tilde{\phi}$ is a suspension of $\tilde{T}: \mathbb{R}/\mathbb{Z} \times \{0\} \to \mathbb{R}/\mathbb{Z} \times \{0\}$. Moreover, there exists an auxiliary homeomorphism $\tilde{h}: M \circlearrowleft$ satisfying $\tilde{h}|_C = h: C \to \mathbb{R}/\mathbb{Z} \times \{0\}$, $\tilde{h}(\Sigma_\phi) = \Sigma_{\tilde{\phi}}$ and $\tilde{h}(\Lambda(\phi|_{M-C})) = \Lambda(\tilde{\phi}|_{M-(\mathbb{R}/\mathbb{Z}\times\{0\})})$.

The proof of (1) will be obtained by using (2) and (3) below. We assert:

(2) Let $\Gamma \in \{\mathbb{R}/\mathbb{Z}, (0,1)\}$ and $f\colon \Gamma\times\{-1\} \to \Gamma\times\{1\}$ be a smooth orientation preserving diffeomorphism. Then there exists a smooth vector field X defined in $\Gamma \times [-1,1]$ such that

(2.1) X has no singularities,

(2.2) X restricted to a neighborhood of $\Gamma \times \{-1,1\}$ is equal to $(0,1)$ and

(2.3) The Poincaré map $\Gamma\times\{-1\} \mapsto \Gamma\times\{1\}$ induced by X is precisely f.

The assertion (2) is a particular case of either [9, Prop. 3.7] or [11, II.1].

Recall that the fixed points of ϕ are either sources or multisaddles. Now we will prove that

(3) There exists a smooth vector field Y defined in $M - (\mathbb{R}/\mathbb{Z} \times (-1,1))$ such that

(3.1) $\Sigma_{\phi} = \{\text{singularities of } Y\} = \Sigma_Y$

(3.2) There exists an (auxiliary) homeomorphism $g\colon M-C \to M-(\mathbb{R}/\mathbb{Z} \times [-1,1])$ satisfying $g(\Sigma_{\phi}) = \Sigma_Y$ and $g(\Lambda(\phi|_{M-C})) = \Lambda(Y|_{M-\mathbb{R}/\mathbb{Z}\times[-1,1]})$.

(3.3) Y restricted to $\mathbb{R}/\mathbb{Z} \times ([-2,-1] \cup [1,2])$ is equal to $(0,1)$.

(3.4) Given $p \in \Sigma_Y$ there exists a compact neighborhood B_p of p such that if p is a multisaddle of $2n$ separatrices (resp. a source) then, via a coordinate system: $p = (0,0)$, $B_p = \{(x,y) \in \mathbb{R}^2 \mid x^2+y^2 \leq 1\}$ and the vector field Y is defined in B_p by $Y(x+iy) = (x-iy)^n$ (resp. $Y(x+iy) = x+iy$), where $\mathbb{R}^2$ is identified with $\mathbb{C}$ in the canonical way. Moreover the sets $\{B_p\}_{p\in\Sigma_Y}$ are pairwise disjoint.

(3.5) The Poincaré map $P\colon \mathbb{R}/\mathbb{Z}\times\{1\} \to \mathbb{R}/\mathbb{Z}\times\{-1\}$ induced by Y satisfies $\text{Domain}(P) = \text{Domain}(\tilde{T})$ and $\text{Image}(P) = \text{Image}(\tilde{T})$.

(3.6) If $x \in \text{Domain}(P)$ is sufficiently close to a point of $\mathbb{R}/\mathbb{Z}\times\{1\} - \text{Domain}(P)$, then $|P'(x)| = 1$.

Certainly, it is obvious that there exists such vector field Y satisfying (3.1)-(3.5).

We will prove that Y can be taken to satisfy also (3.6). Let p be a multisaddle point of Y. If we use (3.4) to identify B_p with $\{(x,y) \mid x^2+y^2 \le 1\}$ and to have a expression of $Y|_{B_p}$, we have that the phase portrait of $Y|_{B_p}$ has several symmetries. As a consequence of this, if $p_1, p_2 \in B_p \cap \{(x,y) \in \mathbb{R}^2 \mid x^2+y^2 = 1/2\}$ are in different separatrices of p and L_{p_1} (resp. L_{p_2}) is a small line segment orthogonal to Y at p_1 (resp. at p_2) then the Poincaré map $\tilde{P}: L_{p_1} \mapsto L_{p_2}$ is an isometry in any interval where it is defined.

Therefore, as a consequence of (2), the construction of Y in a neighborhood of $\Lambda(Y)$ can be made to satisfy (3.6). That is, we can construct Y satisfying not only (3.1)-(3.5) but also (3.6).

Let (a,b) be a connected component of Domain(P). We define f in $P(a,b)$ by $f = \tilde{T} \circ P^{-1}$. In this way we define f in Image(P). Because of (3.5) and (3.6), f can be extended to a smooth orientation preserving diffeomorphism $f: \mathbb{R}/\mathbb{Z} \times \{-1\} \mapsto \mathbb{R}/\mathbb{Z} \times \{1\}$. By using (2) we get a smooth vector field X defined in $\mathbb{R}/\mathbb{Z} \times [-1,1]$ which suspends f and can be glued together with Y (see (3.3)) to give a smooth vector field Z defined on the whole manifold M. Notice that the Poincaré map $\mathbb{R}/\mathbb{Z} \times \{1\} \to \mathbb{R}/\mathbb{Z} \times \{1\}$ induced by Z is precisely $\tilde{T}$. The flow $\tilde{\phi}$ induced by Z satisfies (1).

Next, we prove that ϕ and $\tilde{\phi}$ are topologically equivalent. A canonical region of M corresponding to the flow ϕ and the circle C (which will be called simply a canonical region of the pair (ϕ, C)) is the closure in M of a connected component of $M-(C \cup \Sigma_\phi \cup \Lambda(\phi|_{(M-C)}))$. By a result of D. Neumann and T. O'Brien [8, Theorem 1 and its proof, p. 97], given a canonical region R of ϕ, there exists a homeomorphism $h_R: R \to \tilde{h}(R)$ which is a topological equivalence between $\phi|_R$ and $\tilde{\phi}|_{\tilde{h}(R)}$ and such that h_R restricted to

$R \cap (\Sigma_\phi \cup \Lambda(\phi|_{(M-C)}) \cup C)$ is equal to $\tilde{h}$. Therefore, we can patch together equivalences on the (finitely many) canonical regions of (ϕ, C) and get a topological equivalence between ϕ and $\tilde{\phi}$. □

§4 Basic Cherry Flow with a Dense Semitrajectory

Throughout this section, we are going to assume that ϕ is a basic Cherry flow which has a dense positive semitrajectory $\gamma^+_{p_1}$. (In consequence, ϕ is continuous and has finitely many fixed points). In this section we prove that ϕ is, modulo topological equivalence, a smooth suspension of a standard interval exchange transformation.

Let C be a transversal circle to ϕ and $T: C \to C$ be the Poincaré map induce by ϕ. We denote by p_i the i-th intersection of $\gamma^+_{p_1}$ with C. For $0 < \beta < 1$ we define an atomic measure $\mu_\beta: C \to \mathbb{R}$ in the following way: $\mu_\beta(p_i) = \beta(1-\beta)^{i-1}$, and if $A \subset C$ and $P = A \cap \{p_1, \ldots, p_n, \ldots\}$, we write $\mu_\beta(A) = \sum_{p \in P} \mu_\beta(p)$. Certainly $\mu_\beta(C) = 1$. Let θ denote both a fixed orientation in C as well as the positive orientation of $\mathbb{R}/\mathbb{Z}$. Let $\Gamma \in \{C, R/Z\}$, $a, b \in \Gamma$, $a \neq b$. <u>We define the interval</u> $(a,b) = \{z \in \Gamma - \{a\} \mid$ if "<" denotes the linear order induced by the orientation θ in $\Gamma - \{a\}$, then $z < b\}$. The notation $a < c < b$ will mean that $c \in (a,b)$.

<u>Lemma 4</u> - Let $a, b \in C$, $a \neq b$. Then $\inf\{\mu_\beta((a,b)) \,/\, 0 \le \beta \le \frac{1}{2}\} > 0$.

<u>Proof</u>: Let $x, y \in (a,b)$ such that $a < x < y < b$ and

(4.1) $\{x,y\} \cap \{p_1, p_2, \ldots, p_n, \ldots\} = \phi$.

Certainly, $\forall \beta \in (0,1/2], \mu_\beta((x,y)) < \mu_\beta((a,b))$. In consequence, we only have to prove that $\inf\{\mu_\beta((a,b))/0 \le \beta \le \frac{1}{2}\} > 0$.

Let $\sigma_1, \sigma_2, \ldots, \sigma_{n-1} \in (x,y)$, $x = \sigma_0 < \sigma_1 < \ldots < \sigma_n = y$, such that

(4.2) Given $p \in (x,y)$, $p \in \{\sigma_1, \sigma_2, \dots, \sigma_{n-1}\}$ if and only if γ_p^+ goes to the set {fixed points of ϕ} $\cup \{x\} \cup \{y\} \cup \{p_1\}$ before reintersecting $(x,y) - \{p_1\}$ (recall that $p \in \gamma_p^+$).

By Lemma 2, the fact that $\gamma_{p_1}^+$ is dense and (4.2) the Poincaré map $S\colon (x,y) \longmapsto (x,y)$ induced by ϕ is defined and is continuous at each (σ_i, σ_{i+1}), $\forall\, i \in \{0,1,\dots,n-1\}$. Therefore:

(4.3) $\forall\, i \in \{0,1,\dots,n-1\}$, there exists $\ell_i \in \mathbb{N}$ such that, $\forall\, s \in \{1,2,\dots,\ell_i\}$, $T^s((\sigma_i,\sigma_{i+1})) \cap (x,y) = \emptyset$, but $T^{\ell_i+1}((\sigma_i,\sigma_{i+1})) \subset (x,y)$ (this implies that $S|_{(\sigma_i,\sigma_{i+1})} = T^{\ell_i+1}|_{(\sigma_i,\sigma_{i+1})}$). And

(4.4) There exists $\ell_n \in \mathbb{N}$ such that $\{p_1, p_2, \dots, p_{\ell_n}\} \cap (x,y) = \emptyset$, but $p_{\ell_n+1} \in (x,y)$.

Observe that if $p_k \in (\sigma_i, \sigma_{i+1})$, for some $k \in \mathbb{N}$ and $i \in \{0,1,\dots,n-1\}$, then, by (4.3), $p_{k+\ell_i+1} \in (x,y)$ which implies (by (4.1) and (4.3)) that $p_{k+\ell_i+1} \in (\sigma_j, \sigma_{j+1})$ for some $j \in \{0,1,\dots,n-1\}$. Thus, using (4.3) and (4.4) we obtain that, $\forall\, \tilde{s} \in \mathbb{N}$,

$$(4.5) \qquad p_{\ell_n+1+\tilde{s}} \in \bigcup_{i=0}^{n-1} \bigcup_{s=0}^{\ell_i} T^s((\sigma_i,\sigma_{i+1})).$$

In consequence, since $\{p_{\ell_n+1+\tilde{s}}\}_{\tilde{s}\in\mathbb{N}}$ is dense in C, we conclude that

$$(4.6) \qquad \text{closure}\,\Big(\bigcup_{i=0}^{n-1} \bigcup_{s=0}^{\ell_i} T^s((\sigma_i,\sigma_{i+1}))\Big) = C.$$

Now, we claim that $\forall\, i \in \{0,1,\dots,n-1\}$ and $\forall\, s \in \{0,1,\dots,\ell_i\}$,

$$(4.7) \qquad \mu_\beta(T^s(\sigma_i,\sigma_{i+1})) = (1-\beta)^s\, \mu_\beta((\sigma_i,\sigma_{i+1})).$$

In fact, $\mu_\beta(T((\sigma_i,\sigma_{i+1}))) = \sum_{p_j \in (\sigma_i,\sigma_{i+1})} \mu_\beta(p_{j+1}) =$

$$= (1-\beta) \sum_{p_j \in (\sigma_i,\sigma_{i+1})} \mu_\beta(p_j) = (1-\beta)\, \mu_\beta(\sigma_i,\sigma_{i+1}).$$

It follows from (4.4), (4.5), (4.6) and (4.7) that:

$$\mu_\beta(C) = \sum_{i=0}^{n-1} \mu_\beta((\sigma_i,\sigma_{i+1}))(1+(1-\beta)+\dots+(1-\beta)^{\ell_i}) + \sum_{i=1}^{\ell_n} \mu_\beta(p_i). \tag{4.8}$$

If we assume that there is a sequence $\beta_1,\beta_2,\dots,\beta_j,\dots$ such that $\lim_j \mu_{\beta_j}((x,y)) = 0$, we have that $\lim_j \mu_{\beta_j}((\sigma_i,\sigma_{i+1})) = 0$, $\forall\, i \in \{0,1,\dots,n+1\}$. Therefore by (4.8) $\lim_j \mu_{\beta_j}(C) = 0$. This is a contradiction because $\mu_\beta(C) = 1$, $\forall\, \beta \in (0,1/2]$.

<u>Remark 1</u>: Using the ideas contained in Lemma 4, in particular (4.6), we can prove that given any trajectory γ of ϕ, we have that $\omega(\gamma)$ (Resp. $\alpha(\gamma)$) is either a fixed point or the whole manifold.

Fix $\lambda_0 \in C-\gamma_{p_1}$. Since the set $\{\mu_\beta((\lambda_0,p_i))/\beta \in (0,1/2],\ i\in\mathbb{N}\}$ is bounded we can find a sequence $\{\beta_j\}$, $\beta_j \in (0,1/2]$, such that $\lim_j \beta_j = 0$ and, $\forall\, i \in \mathbb{N}$, $\lim_j \mu_{\beta_j}((\lambda_0,p_i)) = h(p_i)$ exists. Given $x \in C-\lambda_0$ we define $h(x) = \sup_{\lambda_0<p_i<x} h(p_i)$. We also define $h(\lambda_0) = 0$.

<u>Lemma 5</u> - The map $h: C \to \mathbb{R}/\mathbb{Z}$ is a homeomorphism and $\tilde{T} = h\circ T\circ h^{-1}$ is a standard interval exchange transformation.

<u>Proof</u>: By Lemma 4, $\lambda_0 < p_n < p_m$, implies that $0 < h(p_n) < h(p_m)$. So, h is monotonic and continuous from below. Suppose h is not continuous. Thus, there exist $x \in C$ and sequences $\{p_{n_j}\}$, $\{p_{m_j}\}$, $j \in \mathbb{N}$, such that

$$p_{n_1} < p_{n_2} <\dots< p_{n_j} <\dots< p_{m_j} <\dots< p_{m_2} < p_{m_1}, \tag{5.1}$$

$$\lim_j p_{n_j} = x = \lim_j p_{m_j}, \quad \text{and} \tag{5.2}$$

$$\lim_j (h(p_{m_j}) - h(p_{n_j})) > \delta. \tag{5.3}$$

We will only consider the case where x does not belong to a separatrix of ϕ. Let N be a positive integer satisfying $N \geq \frac{2}{\delta}$. It follows from (5.1) and (5.2) that there exist $n \in \{n_1, n_2, \ldots\}$ and $m \in \{m_1, m_2, \ldots,\}$ such that $\{(p_n, p_m), T((p_n, p_m)), \ldots, T^N((p_n, p_m))\}$ are pairwise disjoint intervals contained in the domain of T. We observe that, $\forall\, s \in \{0, 1, \ldots, N\}$, $\lim_i \mu_{\beta_i}(T^s((p_n, p_m))) =$ $= \lim_i (1-\beta_i)^s \mu_{\beta_i}((p_n, p_m)) > \delta$. (See (4.7) in Lemma 4). Therefore, $\lim_i \mu_{\beta_i}(\bigcup_{s=0}^{N} T^s((p_n, p_m))) \geq N\delta > 2$. This is a contradiction, because $\lim_i \mu_{\beta_i}(C) = 1$. Consequently h is a homeomorphism.

Let (a,c) be an interval contained in the domain of definition of T. We claim that

$$(5.4) \qquad |h(a) - h(c)| = |hT(a) - hT(c)|.$$

Since $\{p_k\}_{k \in \mathbb{N}}$ is dense in C and h and T are continuous, we only have to prove (5.4) when $a = p_i$ and $c = p_j$, for some $i, j \in \mathbb{N}$, $i \neq j$. Notice that $\lambda_o < p_i < p_j$. We only consider the case $T(p_j) < T(p_i) < T(\lambda_o)$. Hence:

$$\begin{aligned}
|h(p_i) - h(p_j)| &= \\
&= |\lim_k \mu_{\beta_k}((\lambda_o, p_i)) - \lim_k \mu_{\beta_k}((\lambda_o, p_j))| = \\
&= \lim_k \mu_{\beta_k}((p_i, p_j)) = \\
&= \lim_k \frac{1}{(1-\beta_k)} \cdot \lim_k \mu_{\beta_k}((T(p_j), T(p_i))) = \text{(See (4.7) in proof of Lemma 4)} \\
&= |hT(p_i) - hT(p_j)|.
\end{aligned}$$

If $(\tilde{a}, \tilde{c})$ is an interval of the domain of definition of $\tilde{T} = h \circ T \circ h^{-1}$, then, by (5.4), $|\tilde{T}(\tilde{a}) - \tilde{T}(\tilde{c})| = |\tilde{a} - \tilde{c}|$. This proves the lemma.

<u>Proposition 2</u> - Any basic Cherry flow with a dense positive trajectory is smoothable.

Proof: As in Proposition 1 of §3, we can suspend the map $\tilde{T}$, of Lemma 5, to a smooth flow $\tilde{\phi}$ topologically equivalent to ϕ. □

§5 The Main Result

In this section we prove that the problem of smoothability of Cherry flows can be reduced to that of smoothability of basic Cherry flows.

A Cherry flow ϕ is said to be reduced if

(a) Any trajectory γ of ϕ which satisfy $\alpha(\gamma) \cup \omega(\gamma) \subset$ $\subset$ {multisaddle points of ϕ} is necessarily a multisaddle point, and

(b) for any two source-fixed-points of ϕ, say p and p', which are connected to the same multisaddle point q by means of trajectories θ and θ' respectively, there is no separatrix of q which is adjacent to both θ and θ' (see in Fig. 6 the situation which does not occur).

Lemma 6 (Reduction lemma) - Let ϕ a Cherry flow on a compact two-manifold M. Then, there exist a reduced Cherry flow $\tilde{\phi}$ such that ϕ is smoothable if and only if $\tilde{\phi}$ is.

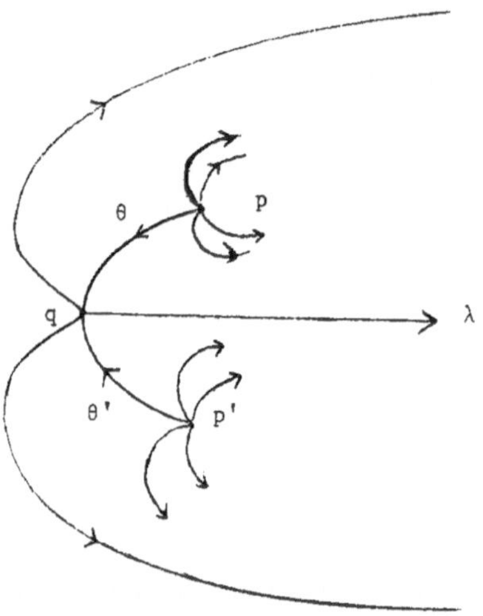

The separatrix λ of q is adjacent to both θ and θ'

Fig. 6

To prove this lemma, we will construct a finite sequence of Cherry flows $(M_o,\phi_o),(M_1,\phi_1),\ldots,(M_k,\phi_k)$ such that:

(a) $(M,\phi) = (M_o,\phi_o)$ and, $\forall\, i \in \{1,2,\ldots,k\}$, ϕ_i is smoothable if and only if ϕ is, and

(b) (M_k,ϕ_k) is a reduced Cherry flow.

Let us suppose that (M_i,ϕ_i) has been defined and proceed inductively to define (M_{i+1},ϕ_{i+1}) in one of the three following ways:

(1) If ϕ_i has a trajectory, say γ, connecting two different multisaddles points p and q, we obtain (M_{i+1},ϕ_{i+1}) by blowing down γ to a point; that is, we define $M_{i+1} = M_i/\{p\} \cup \{q\} \cup \gamma$ and a flow ϕ_{i+1} on M_{i+1} in such a way that the trajectories of ϕ_{i+1} coincide with those of ϕ_i in the set $M_i - \{\{p\} \cup \{q\} \cup \gamma\}$, and moreover $\{p\} \cup \{q\} \cup \gamma$ is a fixed point of ϕ_{i+1}. See Fig. 7a before blowing down γ and Fig. 7b after blowing down γ.

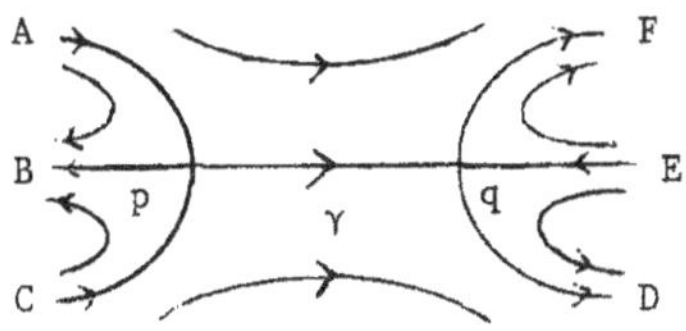

the flow ϕ_i

Fig. 7a

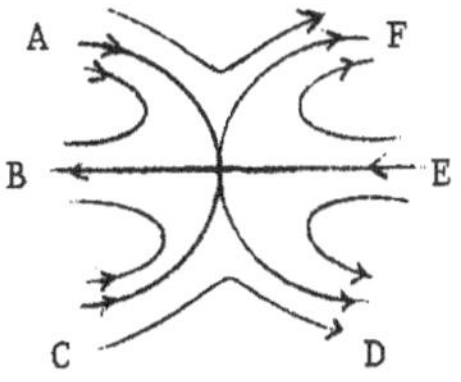

the flow ϕ_{i+1}

Fig. 7b

(2) If ϕ_i has no trajectories connecting two distinct multisaddle points but it has two source-fixed-points, say p and p', which are connected to the same multisaddle point q by means of trajectories θ and θ' respectively, and moreover q has a separatrix which is adjacent to both θ and θ' (see Fig. 8a). Then, we make $M_i = M_{i+1}$ and define ϕ_{i+1} on M_{i+1} in such a way that: (A) The trajectories of ϕ_{i+1} coincide with those of ϕ_i in $M_i - \{\lambda \cup W^u_{\phi_i}(p) \cup W^u_{\phi_i}(q)\}$ ($W^u_{\phi_i}(p)$ denote the unstable manifold of p for the flow ϕ_i) and (B) ϕ_{i+1} has a source-fixed-point r connected to q be a unique trajectory and $W^u_{\phi_{i+1}}(r) = \lambda \cup \cup W^u_{\phi_i}(p) \cup W^u_{\phi_i}(q)$ (See Figs. 8a and 8b).

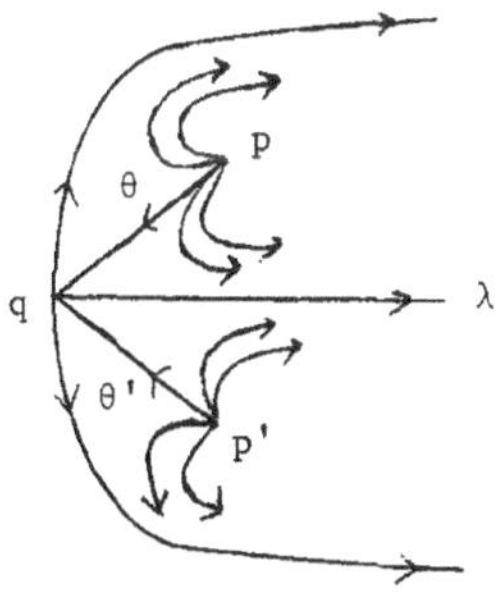

the flow ϕ_i

Fig. 8a

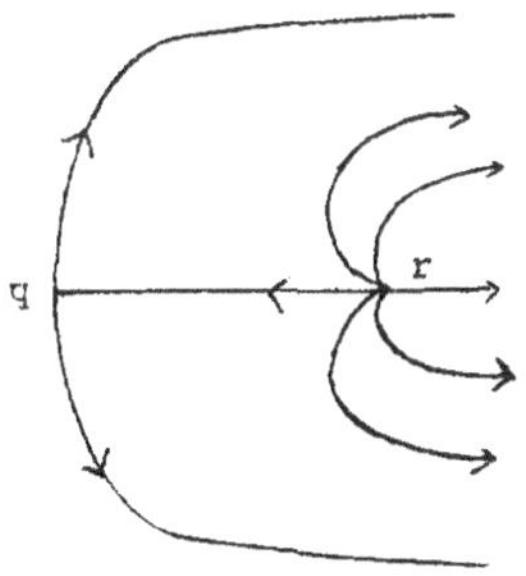

the flow ϕ_{i+1}

Fig. 8b

(3) If ϕ_i does not satisfy the conditions required in (1) or (2) above to define (M_{i+1}, ϕ_{i+1}), but it has a (non-fixed point) trajectory, say γ, such that both $\alpha(\gamma)$ and $\omega(\gamma)$ are the same multisaddle point q. Then, in case that $C = \gamma \cup \{\alpha(\gamma)\}$ is a one-sided (resp. two-sided) closed curve, we define the manifold M_{i+1} as the compactification of $M_i - C$ by means of a point x (resp. two points y and z) at infinity. When C is two-sided y comcatifies one side of C and z the other. Let (M_{i+1}, ϕ_{i+1}) be any of the continuous flows induced by (M_i, ϕ_i) in such a way that x (resp. y and z) is a fixed point of ϕ_{i+1}. Se Figs. 9a and 9b.

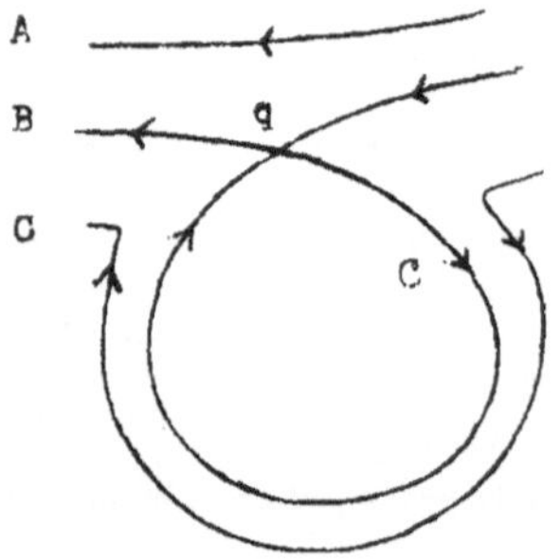

the flow ϕ_i

Fig. 9a

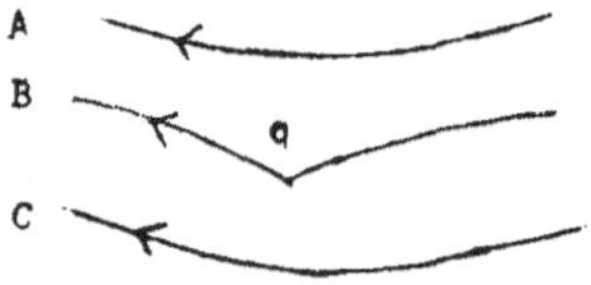

the flow ϕ_{i+1}

Fig. 9b

If ϕ_i does not satisfy any one of the conditions required in (1), (2) or (3) above to define (M_{i+1}, ϕ_{i+1}), then ϕ_i is already a reduced Cherry flow. Certainly, given a Cherry flow (M,ϕ), there exists $k \in \mathbb{N}$ such that (M_k, ϕ_k) is a reduced Cherry flow.

<u>Proof of Theorem 1</u>: Because of Lemma 1, we only need to consider reduced Cherry flows. Let ϕ be a reduced Cherry flow on a connected manifold M. As in its definition, let $\lambda_1, \lambda_2, \ldots, \lambda_n$ be the unstable manifolds of the source-fixed-points of ϕ and $\lambda_{n+1}, \lambda_{n+2}, \ldots, \lambda_{n+s}$ be the positive semi-trajectories of ϕ such that

(1.1) $$\bigcup_{i=1}^{n+s} \lambda_i \text{ is dense in } M.$$

Let γ be a non-trivial ω-recurrent trajectory of ϕ (See [4, §1]) and $p \in \gamma$. We consider a small open segment S transverse to ϕ, passing through p and such that;

(1.2) $\forall\, i \in \{1,2,\dots,n+s\}$, either λ_i intersects S infinitely many times or $\lambda_i \cap S = \phi$.

We want S to have the following property:

(1.3) Let $i_1, i_2, \dots, i_k \in \{1,2,\dots,n+s\}$ such that λ_i intersects S infinitely many times if and only if $i \in \{i_1, i_2, \dots, i_k\}$. If $r \in S - \bigcup_{j=1}^{k} \lambda_{i_j}$, then $r \in \overline{\lambda_{i_j}}$, $\forall\, j \in \{1,2,\dots,k\}$.

We assume that S has property (1.3) because otherwise by recurrence, using the fact that ϕ is a reduced Cherry flow, we can find an open subinterval of S which has property (1.3). Now we construct a circle C_1 transverse to ϕ and such that (See [4, Lemma 4 and its proof]):

(1.4) A trajectory of ϕ intersects C_1 only if it intersects S.

Let $T_1 \colon C_1 \to C_1$ be the Poincaré map induced by ϕ. It follows from Lemma 2, (1.1), (1.3) and (1.4) that:

(1.5) The domain of definition $\mathcal{D}_1$ of T_1 is everywhere except possibly finitely many points. Moreover, if $q \in C_1 - \mathcal{D}_1$, then the possitive semi-trajectory starting at q goes directly to a fixed point without intersecting C again.

Now, we claim that

(1.6) If (x,y) is a connected component of $\overline{C_1\text{-image }(T_1)}$, then (x,y) is contained in the unstable manifold of exactly one source.

In fact, as a consequence of (1.5), if a positive semi-trajectory crosses (x,y), then it never comes back to (x,y).

Therefore, since ϕ is a Cherry flow, the unstable manifolds of the source-fixed-points of ϕ must be dense in (x,y). Finally, because ϕ is a reduced Cherry flow, (x,y) have to be contained in the unstable manifold of exactly one source.

As a consequence of (1.3), (1.5) and (1.6), we have that $F_1 = \overline{\bigcup_{t\in\mathbb{R}} \phi_t(C_1)}$ is a submanifold of M and that ϕ restricted to F_1 is a basic Cherry flow. Since M is a two-dimensional compact manifold, there exist finitely many submanifolds $F_1, F_2, \ldots, F_m$ of M such that

(1.7) $\forall\, i \in \{1,2,\ldots,m\}$, ϕ resctricted to F_i is a basic Cherry flow and if $i,j \in \{1,2,\ldots,m\}$, $i \neq j$, then $\underline{\text{interior}}\ (F_i) \cap \cap\ \text{interior}\ (F_j) = \emptyset$. Moreover, ϕ restricted to $M - \bigcup_{i=1}^{m} F_i$ has no non-trivial ω-recurrent trajectories.

Certainly, by [9, Prop. 2.3, pp. 244] or [10, Proof of Lemma 3, pp. 107] and the fact that ϕ has no non-trival ω-recurrent trajectories in $\overline{M - \bigcup_{i=1}^{m} F_i}$, we know that the ω-limit set, say L, of any trajectory in $\overline{M - \bigcup_{i=1}^{m} F_i}$ must be either a fixed point, or a closed orbit, or else a graph made up of fixed points and trajectories connecting them. By using the fact that ϕ is a reduced Cherry flow, we can see that if L is not a fixed point, then L bounds a disc D which contains a source-fixed-point and a multi-saddle point with exactly two separatrices. See Fig. 10.

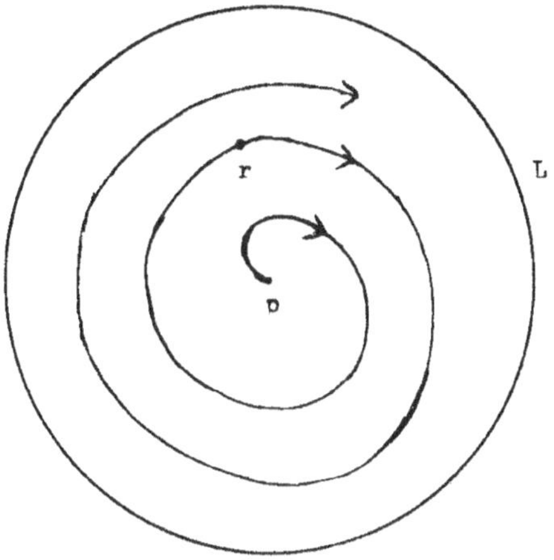

the disc D bounded by L

Fig. 10

Obviously, ϕ is smoothable in $M - \overline{\bigcup_{i=1}^{m} F_i}$. By Propositions 1 and 2, ϕ is smoothable in $\bigcup_{i=1}^{m} F_i$. Therefore, ϕ is smoothable everywhere. □

Remark 2: Let $T: \mathbb{R}/\mathbb{Z} \to \mathbb{R}/\mathbb{Z}$ be a standard interval exchange transformation, then there exists a continuous flow ϕ (called a suspension of T) defined on a two manifold M which contains $\mathbb{R}/\mathbb{Z}$ and such that

(a) $\bigcup_{t\in\mathbb{R}} \phi_t(\mathbb{R}/\mathbb{Z}) = M$ and the Poincaré map $\mathbb{R}/\mathbb{Z} \to \mathbb{R}/\mathbb{Z}$ induced by ϕ is precisely T.

(b) ϕ has finitely many fixed points and all of them are multisaddles (See Lemma 1).

(c) If T has a positive dense orbit, then the ω-limit set of any trajectory of ϕ is either a multisaddle point or the whole manifold M (See remark 1).

Therefore, the problem of the existence of basic Cherry flows (with or without sources)) can be transfered to that of the existence of standard interval exchange transformations with a dense semi-orbit. Keany [7] studied this last problem when the standard interval exchange transformation is orientation preserving on every interval of its domain.

Remark 3: There is no Cherry flow with a dense positive semi-trajectory on the Torus with a cross-cap: The proof of this is essentially contained in [4, Proposition 2].

Remark 4: On any non-orientable two-manifold of genus ≥ 4, there are Cherry flows with a dense non-orientable trajectory [3]. A dense trajectory, say γ, is non-orientable if given $p \in \gamma$ and a segment S which contains p and is transverse to the flow, there exist connected components $\overline{ab} \subset \gamma_p^+ - S$ and $\overline{cd} \subset \gamma_p^- - S$ such that both $\overline{ab} \cup S$ and $\overline{cd} \cup S$ contain a one-sided simple closed curve.

References

[1] T.M. Cherry - Analytic quasi-periodic curves of discontinuous type on a torus. Proc. Lond. Math. Soc. v. 44 (1938), 175-215.

[2] A. Denjoy - Sur les courbes définies par les équations différentielles à la surface du tore. J. de Math. Pures et Appliquées (9) 11 (1932), pp. 333-375.

[3] C. Gutierrez - Smooth non-orientable non-trivial recurrence on two-manifolds. Jour. Diff. Eq. (v. 29) 3 (1978) pp. 388-395.

[4] C. Gutierrez - Structural stability for flows on the torus with a cross-cap. Trans. AMS 241 (1978), 311-320.

[5] J. Harrison - Unsmoothable diffeomorphisms on higher dimensional manifolds. To be published in Proc. Amer. Math. Soc.

[6] P. Hartman - Ordinary Differential Equations, John Wiley & Sons, Inc. (1973).

[7] M. Keane - Interval Exchange Transformation, Math. Zeits., 141, (1975), pp. 25-31.

[8] D. Neumann and T. O'Brien - Global Structure of Continuous Flows on Two-manifolds. Journal of Diff. Equations 22, (1976) pp. 89-110.

[9] J. Palis and W. de Melo - Introdução aos Sistemas Dinâmicos, Instituto de Matemática Pura e Aplicada do CNPq, Rio de Janeiro, Brazil.

[10] M. Peixoto - Structural Stability on two-dimensional manifolds. Topology, vol. 1 (1962) pp. 101-120.

[11] S. Smale - Differentiable dynamical systems, Bull. AMS 73 (1967), 747-817.

Instituto de Matemática Pura e Aplicada, Brazil
University of California at Berkeley.

Carlos Gutierrez
I.M.P.A.
Rua Luiz de Camões, 68
20.060 - Rio de Janeiro, RJ
BRAZIL

AN APPROXIMATION THEOREM FOR IMMERSIONS WITH STABLE CONFIGURATIONS OF LINES OF PRINCIPAL CURVATURE

by

C. Gutiérrez and J. Sotomayor

Abstract

It is proved that every immersion of a compact oriented two-dimensional smooth manifold into R^3 can be arbitrarily C^2-approximated by smooth immersions β whose principal configurations $\mathcal{P}_\beta = (\mathcal{U}_\beta, \mathcal{F}_\beta, f_\beta)$, defined by umbilical points and families of lines of principal curvature, are stable under C^3-sufficiently small perturbations of β. Actually, the elements β are found in the class $\mathcal{S}^r$, $r \geq 4$, of C^3-principally structurally stable immersions, introduced in [3].

Examples of immersions with recurrent lines of principal curvature are also given.

1. Preliminaries

Let M be a compact connected, oriented, two dimensional smooth (i.e. C^∞) manifold. An immersion α of M into R^3 is a map such that $D\alpha_p: TM_p \to R^3$ is one-to-one for every $p \in M$. Denote by $\mathcal{J}^r = \mathcal{J}^r(M,R^3)$ the set of immersions of class C^r, $\infty \geq r \geq 4$, of M into R^3. When endowed with the C^s-topology, $s \leq r$, this set is denoted by $\mathcal{J}^{r,s} = \mathcal{J}^{r,s}(M,R^3)$.

Associated to every $\alpha \in \mathcal{J}^r$ is defined the Gaussian normal map $N_\alpha: M \to S^2$:

$$N_\alpha(p) = \alpha_u(p) \wedge \alpha_v(p) \ / \ |\alpha_u(p) \wedge \alpha_v(p)|,$$

where $(u,v): (M,p) \to (R^2,0)$ is a positive chart of M around p, $\alpha_u(p) = \frac{\partial\alpha}{\partial u}(p) = D\alpha_p(\frac{\partial}{\partial u}(p))$, $\wedge$ denotes the exterior product of vectors in R^3 determined by a once for all fixed orientation of R^3 and $| \; | = \langle \; , \; \rangle^{1/2}$ is the euclidean norm in R^3.

Since $DN_\alpha(p)$ has its image contained in the image of $D\alpha(p)$ the endomorphism $\mathfrak{w}_\alpha : TM \to TM$ is well defined by

$$D\alpha \cdot \mathfrak{w}_\alpha = DN_\alpha \; .$$

It is well known that $\mathfrak{w}_\alpha$ is a self adjoint endomorphism, when TM is endowed with the metric $\langle \; , \; \rangle_\alpha = \alpha^*\langle \; , \; \rangle$, induced by α from the euclidean metric $\langle \; , \; \rangle$ in R^3. Clearly N_α is well defined and of class C^{r-1} in M.

Let $\mathcal{K}_\alpha = \det \mathfrak{w}_\alpha$ and $\mathcal{H}_\alpha = -\frac{1}{2}$ trace $\mathfrak{w}_\alpha$ be the <u>Gaussian</u> and <u>mean curvatures</u> of the immersion α.

A point $p \in M$ is called an <u>umbilical point of</u> α if $\mathcal{H}_\alpha^2(p) - \mathcal{K}_\alpha(p) = 0$. This means that the eigenvalues of $\mathfrak{w}_\alpha$ are equal at p. The set of umbilical points of α will be denoted by $\mathcal{U}_\alpha$.

Outside $\mathcal{U}_\alpha$ the eigenvalues of $\mathfrak{w}_\alpha$ are distinct and given by $-\mathcal{H}_\alpha \pm \sqrt{\mathcal{H}_\alpha^2 - \mathcal{K}_\alpha}$. Their oposite values $K_\alpha = \mathcal{H}_\alpha + \sqrt{\mathcal{H}_\alpha^2 - \mathcal{K}_\alpha})$ and $k_\alpha = \mathcal{H}_\alpha - \sqrt{\mathcal{H}^2 - \mathcal{K}_\alpha})$ are called respectively <u>maximal</u> and <u>minimal principal curvatures of</u> α. The eigenspaces associated to the principal curvatures define two C^{r-2} line fields $\mathcal{L}_\alpha$ and ℓ_α, mutually orthogonal in TM (with the metric $\langle \; , \; \rangle_\alpha$), called <u>the principal line fields of</u> α. They are characterized by Rodrigues' equations [10]

$$\mathfrak{w}_\alpha \mathcal{L}_\alpha + K_\alpha \mathcal{L}_\alpha = 0$$

$$\mathfrak{w}_\alpha \ell_\alpha + k_\alpha \ell_\alpha = 0.$$

The integral curves of $\mathcal{L}_\alpha$ (resp. ℓ_α) are called <u>lines of maximal</u> (resp. <u>minimal</u>) <u>principal curvature</u>. The family of such curves i.e. the integral foliation of $\mathcal{L}_\alpha$ (resp. ℓ_α) in $M - \mathcal{U}_\alpha$ will be denoted by $\mathcal{F}_\alpha$ (resp. f_α) and called the <u>maximal</u> (resp.

minimal) principal foliation of α.

The triple $\rho_\alpha = (\mathcal{U}_\alpha, \mathcal{F}_\alpha, f_\alpha)$ will be called the principal configuration of α.

An immersion $\alpha \in \mathcal{J}^r$ is said to be C^s-principally structurally stable, $s \leq r$, if there is a neighborhood $\mathcal{V}_\alpha$ of α in $\mathcal{J}^{r,s}$ such that for every $\beta \in \mathcal{V}(\alpha)$ it is possible to find a homeomorphism $h = h_\beta : M \to M$ such that $h(\mathcal{U}_\alpha) = \mathcal{U}_\beta$ and $h|M - \mathcal{U}_\alpha$ is simultaneously a topological equivalence between $\mathcal{F}_\alpha$ and $\mathcal{F}_\beta$ and between f_α and f_β. Shortly it is said that h maps ρ_α to ρ_β or that h is a topological equivalence between ρ_α and ρ_β.

In [3] were provided sufficient conditions for an immersion $\alpha \in \mathcal{J}^r$, $r \geq 4$, to be C^s-principally structurally stable, $s \geq 3$. The reader is also refered to this paper for a discussion on the general background for the study of principal configurations.

2. Sufficient Conditions for Structural Stability [3].

These conditions are expressed in terms of the umbilical points $\mathcal{U}_\alpha$, the principal cycles of α i.e. the compact principal lines of either $\mathcal{F}_\alpha$ or f_α and the assymptotic behaviour of non compact principal lines, specially of umbilical separatrices.

2.1 Definition. An immersion $\alpha \in \mathcal{J}^r$, $r \geq 3$ is said to satisfy Condition D (for Darboux [2]) at a point $p \in \mathcal{U}_\alpha$ provided there is a chart $(u,v): (M,p) \to (R^2,0)$ and an isometry Γ of R^3 with $\Gamma(\alpha(p)) = 0$ such that

$$(\Gamma \circ \alpha)(u,v) = \left(u, v, \frac{k}{2}(u^2+v^2) + \frac{a}{6}u^3 + \frac{b}{2}uv^2 + \frac{c}{6}v^3 + O((u^2+v^2)^2)\right),$$

where

γ) $b(b-a) \neq 0$, and

δ) either one of the following inequalities hold

D_1: $a/b > (c/2b)^2 + 2$;

D_2: $\quad (c/2b)^2 + 2 > a/b > 1, \quad a \neq 2b;$

D_3: $\quad 1 > a/b.$

The local principal configurations under Condition D are illustrated in Fig. 2.1, [2 , 3].

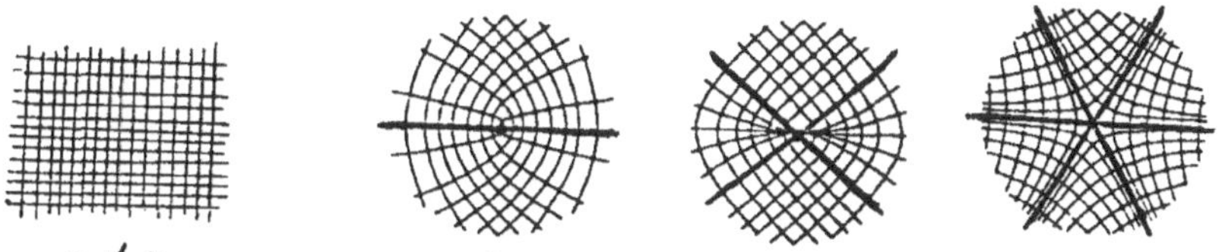

$p \notin \mathcal{U}_\alpha$ $\qquad D_1 \qquad D_2 \qquad D_3$

Fig. 2.1

The index i refers to the number of <u>umbilical separatrices</u> of D_i, $i = 1,2,3$ in each of the foliations $\mathcal{F}_\alpha$, f_α. Separatrices are drawn in heavy lines in the picture.

2.2 <u>Definition</u>. It is said that $\alpha \in \mathcal{J}^r$, $r \geq 3$, satisfies <u>Condition H</u> at a principal cycle c provided either one of the following conditions, (which are equivalent [3]), is satisfied

H_1: $$\int_c \frac{d\,\mathcal{H}_\alpha}{(\mathcal{H}_\alpha^2 - K_\alpha)^{1/2}} \neq 0.$$

H_2: The cycle c is a <u>hyperbolic cycle</u> of the foliation to which it belongs. That is, the Poincaré first return map h associated to a transversal to c at a point q is such that $h'(q) \neq 1$.

H_3: $$\int_c \frac{dk_\alpha}{K_\alpha - k_\alpha} \neq 0.$$

H_4: $$\int_c \frac{d\,K_\alpha}{K_\alpha - k_\alpha} \neq 0.$$

2.3 <u>Definition</u>. Let $\mathcal{S}^r = \mathcal{S}^r(M)$ denote the set of $\alpha \in \mathcal{S}^r$, $r \geq 4$, such that

a) α satisfies condition D of 2.1 at each of its umbilical points.

b) α satisfies condition H of 2.2 at each of its principal cycles.

c) The limit set of every principal line of α is the union of umbilical points and principal cycles.

d) There is no umbilical separatrix of α which is a separatrix of two different umbilical points or twice a separatrix of the same umbilical point.

Umbilical separatrices which violate condition d) of 2.3 are called umbilical connections.

In Section 6 have been given examples of immersions with principal lines which are contained in their limit sets and which are not principal cycles. They are called non-trivial recurrent principal lines. Such principal lines certainly violate condition c) of 2.3.

2.4 Theorem [3]. Let $r \geq 4$. The set $\mathcal{S}^r$ defined in 2.3 is open in $\mathcal{J}^{r,3}$ and every $\alpha \in \mathcal{S}^r$ is C^3-principally structurally stable.

3. The Approximation Theorem.

3.1 Main Theorem. Let $r \geq 4$. The set $\mathcal{S}^r$ defined in 2.3 is dense in $\mathcal{J}^{r,2}$, the space of C^r-immersions endowed with the C^2-topology.

The discovery of a proof of the density of $\mathcal{S}^r$ in $\mathcal{J}^{r,s}$ for $s \geq 3$ would easily lead to the characterization of the set of C^s-principally structurally stable immersions. Such proof however involves serious difficulties for the elimination of non-trivial recurrent principal lines by means of C^s-small perturbations of immersions. Difficulties of this kind were found and solved by the first time by M. Peixoto [7], for orbits of smooth vector fields on compact two-dimensional manifolds. See also the work of C. Pugh [9]. The methods of Peixoto and Pugh cannot be directly applied to the present case due to the integrability conditions of surface theory that must be respected while the perturbations of the principal line fields are made and due to the presence of oscillatory recurrences which are not possible for vector fields.

3.2 Proof of 3.1. The proof of Theorem 3.1 is given below. It is based on several preliminary propositions. With the exception of the more lenghty, Proposition 3.2.6, which deal with the elimination of non-trivial recurrences and umbilical connections, the other propositions are proved here. These propositions are used also in Sections 4 and 5, where an adjustment of the main steps of the methods developed by Peixoto and Pugh [7 , 9] for orbits of vector fields is made in order to be applicable to principal lines.

Since $\mathcal{J}^{\infty}$ is dense in $\mathcal{J}^{r,s}$, for any r and s, it sufficient to prove 3.1 for $r = \infty$. For this reason all the propositions below are formulated for C^{∞} immersions.

3.2.1 Proposition

a) Let $\mathcal{S}_1^{\infty}$ be the set of $\alpha \in \mathcal{J}^{\infty}$ all of whose umbilical points satisfy condition D (2.1).

b) Let $\mathcal{S}_2^{\infty}$ the set of $\alpha \in \mathcal{J}^{\infty}$ such that for every $\lambda \in \mathbb{R}^3$, the function $g_{\lambda,\alpha}: M \to R$, defined by $g_{\lambda,\alpha}(x) = |\alpha(x) - \lambda|^2$, has only isolated critical points.

c) Let $\mathcal{S}_3^2$ the set of $\alpha \in \mathcal{J}^{\infty}$ such that for every unitary vector $v \in S^2$, the function $p_{v,\alpha}: M \to R$, defined by

$$p_{v,\alpha}(x) = \langle \alpha(x), v \rangle$$

has only isolated critical points.

Then $\mathcal{S}_i^{\infty}$, $i = 1,2,3$ are dense in $\mathcal{J}^{\infty}$. More precisely, $\mathcal{S}_1^{\infty}$ also open and $\mathcal{S}_2^{\infty}$ and $\mathcal{S}_3^{\infty}$ contain open sets which are dense in $\mathcal{J}^{\infty}$.

Proof. The proof follows from Thom's Transversality Theorem for jet spaces and is essentially known. For instance condition $\mathcal{T}$) of 2.1 amounts to the transversality of $j^2\alpha$ to the submanifold (of codimension 2) of umbilical 2-jets. From this follows the openeness and density of the immersions which satisfy $\mathcal{T}$) of 2.2. The openess and density of those which also satisfy δ) of 2.1 is obvious.

Case b) follows from the openess and density of the set of α's for which $g_{\lambda,\alpha}$ present as critical points only the elementary catastrophes of codimension ≤ 3 (folds (A_2), cusps (A_3), swallowtails (A_4) and hyperbolic and elliptic umbilics (D_4) [5 , 8 ,11].

Case c) follows from the openess and density of the set of α's for which $N_\alpha : M \to S^2$ is a Whitney map i.e. has only folds and cusps as singularities, [5 ,12].

It must also be mentioned that cases b) and c) appear in a unified context in [5], which applies also to higher dimensional manifolds. See also [1] for a proof in case b), based in the interesting fact that for any $\alpha_o \in \mathcal{J}^\infty$, the set of affine motions a of R^3 for which $g_{\lambda, a\circ\alpha_o}$ has a critical point of codimension ≥ 4, has null Lebesgue measure.

3.2.2 Remark

Let $\alpha \in \mathcal{S}_2^\infty \cap \mathcal{S}_3^\infty$ then along any minimal (resp. maximal) principal line c, $dk_\alpha|\ell_\alpha$ (resp. $dK_\alpha|\mathcal{L}_\alpha$) does not vanish on any open arc of c.

Proof. If $k_\alpha|c$ were a constant k_o on an arc $c_o \subset c$, the map $\frac{1}{k_o} N_\alpha + \alpha$, when $k_o \neq 0$, would be a constant λ_o on c_o, by Rodrigues' formula. Therefore, the function $g_{\lambda_o,\alpha}$ of 3.2.1,b would have all the points of the arc c_o as critical points. When $k_o \equiv 0$; N_α would be a constant v_o on c_o. Therefore, the function $p_{v_o,\alpha}$ of 3.2.1.c would have all the points of the arc c_o as critical points.

The case of K_α is similar.

3.2.3 Proposition. Let $\alpha \in \mathcal{J}^\infty$ have a non-hyperbolic minimal principal cycle c, i.e. $\int_c \frac{dk_\alpha}{K_\alpha - k_\alpha} = 0$, by 2.2. Let $v: M \to R$ be a C^∞ function such that $v/c = 0$ and $|dv| = 1$ at all points of c.

a) For any C^∞ function $w: M \to R$

$$\alpha_\varepsilon = \alpha + \frac{\varepsilon}{2} v^2 w N_\alpha$$

belongs to $\mathcal{J}^\infty$, if ε is small, and has c as a minimal principal cycle.

Furthermore, on c, $K_{\alpha_\varepsilon} = \varepsilon w + K_\alpha$ and $k_{\alpha_\varepsilon} = k_\alpha$.

b) If $k_\alpha | c$ is not constant, which happens for instance if $\alpha \in \mathcal{S}_2^\infty \cap \mathcal{S}_3^\infty$ (3.2.1), and if the support of $w|c$ is an arc c_o where $dk_\alpha | \ell_\alpha \neq 0$, then c is a hyperbolic minimal principal cycle of α_ε, when ε is small.

Analogous result holds for maximal principal cycles.

Proof. Part a) follows from a direct calculation of K_{α_ε} and k_{α_ε}. Part b) follows from 2.2 and from the fact that the integral

$$I(\varepsilon) = \int_c \frac{dk_{\alpha_\varepsilon}}{K_{\alpha_\varepsilon} - k_{\alpha_\varepsilon}} = \int_c \frac{dk_\alpha}{K_\alpha + \varepsilon w - k_\alpha}$$

does not vanish for $|\varepsilon|$ small, since

$$\frac{dI}{d\varepsilon}(0) = \int_{c_o} \frac{-w dk_\alpha}{(K_\alpha - k_\alpha)^2} \neq 0.$$

3.2.4 Corollary. Suppose that the minimal cycle c of 3.2.3 has multiplicity $n < \infty$; i.e. $h^{(n)}(0) \neq 0$ and $h^{(i)}(0) = 0$, $i = 2, \ldots, n-1$. Here, $h: L_o \to L$ is the Poincaré first return map of c, on a transversal arc $L_o \subset L$ which meets L_o at 0, defined by the minimal foliation of α.

Then, if w is like in b) of 3.2.3, there is a neighborhood U_o of $0 \in R$ and V_c of c in M such that for any $\varepsilon \in U_o - \{0\}$ all the principal minimal cycles of α_ε contained in V_c are hyperbolic.

Proof. Denote by h_ε the Poincaré map of α_ε. For ε small it can be also defined from L_o to L. By the Malgrange Preparation Theorem [6] there are C^∞ functions $g, a_1, \ldots, a_n$ such that

$$h_\varepsilon(x) - x = g(x,\varepsilon)(x^n + a_1(\varepsilon)x^{n-1} + \ldots + a_{n-1}(\varepsilon)x + a_n(\varepsilon)),$$

with $g(0,0) \neq 0$. The fact that $h_\varepsilon(0) = 0$ implies that $a_n \equiv 0$ and the multiplicity n of c implies that $a_i(0) = 0$ $i=1,\ldots,n-1$.

Also $h'_\varepsilon(0) = 1+g(0,\varepsilon)\cdot a_{n-1}(\varepsilon)$ which is equalt to $\exp[-I(\varepsilon)]$, by the formula derived in [3] for the derivative of the Poincaré map. Therefore by the proof of b), 2.3.3 $a'_{n-1}(0) = \frac{1}{g(0,0)} \int_{c_0} \frac{w dk_\alpha}{(K_\alpha - k_\alpha)^2} \neq 0$.

Let $\bar{\varepsilon} = \bar{\varepsilon}(x)$ be the implicit function defined by

$$(*) \qquad x^{n-1} + a_1(\varepsilon)x^{n-2} + \ldots + a_{n-1}(\varepsilon) = 0.$$

It follows that all the fixed points of h_ε, for ε and x small are given by $\{x=0\}$, and the graph of $\varepsilon = \bar{\varepsilon}(x)$. The fact that $\bar{\varepsilon}$ has the form $\bar{\varepsilon}(x) = x^{n-1}(1 + xa(x))$ with a of class C^∞, which follows from (*), implies that $\bar{\varepsilon}'(x)$ is not zero for small $x \neq 0$. This means that at the points $x \neq 0$ for which $\bar{\varepsilon}(x) = \varepsilon$, $h'_\varepsilon(x) \neq 1$. This finishes the proof of the Corollary.

3.2.5 <u>Definition</u>. Let $\alpha \in \mathcal{S}_1^\infty$, p be an umbilical point of α and γ be a minimal separatrix of p. Orient γ in such a way that the α-limit set of γ is p. The separtrix γ is said to be <u>stabilized</u> if:

1) it is not an umbilical connection
2) the ω-limit set of γ is either an umbilical point (of type D_2) or a minimal principal cycle
3) the properties (1) and (2) above of γ presist for small C^3-perturbations of α.

3.2.6 <u>Proposition</u>. Any immersion $\alpha \in \mathcal{S}_1^\infty$ can be arbitrarily C^2-approximated by a smooth immersion $\beta \in$ interior $(\mathcal{S}_1^\infty \cap \mathcal{S}_2^\infty \cap \mathcal{S}_3^\infty)$ such that:

a) their (maximal and minimal) separatrices are stabilized
b) the ω-limit set of any oriented minimal (resp. maximal) principal line of β is either an umbilical point or a minimal (resp. maximal) principal cycle
c) any $\tilde{\beta} \in \mathcal{J}^\infty$ close enough to β in the C^∞-topology, satisfies a) and b) above.

The proof of this proposition is given in Section 5.

3.2.7 Proof of Theorem 3.1

By Proposition 3.2.1, α can be assumed to be actually in $\mathcal{S}_1^\infty$. By Proposition 3.2.6, α can be arbitrarily C^2-approximated by $\alpha_1 \in$ interior $(\mathcal{S}_1^\infty \cap \mathcal{S}_2^\infty \cap \mathcal{S}_3^\infty)$ which verifies a), b) and c) of that proposition. Approximate α_1, in the C^∞-topology, by $\alpha_2 \in \mathcal{J}^\omega$ [13]. The immersion $\alpha_2 \in$ interior $(\mathcal{S}_1^\infty \cap \mathcal{S}_2^\infty \cap \mathcal{S}_3^\infty)$, it satisfies a), b) and c) of Proposition 3.2.6, and has finitely many principal cycles all of which have finite multiplicity. Using 3.2.4, α_2 can be C^∞ approximated by β whose principal cycles are all hyperbolic. Because of 3.2.6 (c) β can be taken in $\mathcal{S}^\infty$.

4. Elimination of nontrivial recurrent principal lines.

This section is devoted to the proof of the following proposition.

4.1 Proposition. Let $\alpha \in \mathcal{S}_1^\infty \cap \mathcal{S}_2^\infty \cap \mathcal{S}_3^\infty$ and $\mathcal{G}$ be a closed nowhere dense subset of M formed by principal lines of α none of which is non-trivial recurrent. Then α can be arbitrarily C^2-approximated by a $\beta \in \mathcal{S}_1^\infty$, without non-trivial recurrent minimal principal lines and such that the support of $\beta-\alpha$ is arbitrarily small and disjoint of $\mathcal{G} \cup \mathcal{U}_\alpha$.

The proof depends on Proposition 4.9 proved below after some preliminary lemmas.

A chart (u,v) on whose domain of definition holds that $\ell_\alpha = R\frac{\partial}{\partial u}$ and $\mathcal{L}_\alpha = R\frac{\partial}{\partial v}$ will be called a principal chart of α.

4.2 Lemma. Let $\alpha \in \mathcal{J}^\infty$ and $(u,v)\colon M \to 2I\times 2I$ be a principal chart of α, where $I = [-1,1]$. Let $A = \alpha + \varepsilon\varphi N_\alpha$ where N_α is the normal to α and φ is a real valued function on M, with support contained in $(u,v)^{-1}(I\times I) = D$. Then for ε small, the minimal foliation $f_{A(\cdot,\varepsilon)}|D$ defines a map $T(\cdot,\varepsilon)$ from $u^{-1}(-1)$ to $u^{-1}(1)$ such that $v(T(\cdot,0)) = v(\cdot)$ and

$$D(v\circ T)\,\frac{\partial}{\partial\varepsilon}(p,0) = -\int_{-1}^{1} \varphi\,\frac{\partial}{\partial u}\Big(\frac{\frac{\partial}{\partial v}(k_\alpha)}{(K_\alpha-k_\alpha)^2\langle\frac{\partial\alpha}{\partial v},\frac{\partial\alpha}{\partial v}\rangle}\Big)du\,+$$

$$+\int_{-1}^{1}\frac{\frac{\partial\varphi}{\partial v}\cdot\frac{\partial}{\partial u}(k_\alpha)du}{(K_\alpha-k_\alpha)^2\langle\frac{\partial\alpha}{\partial v},\frac{\partial\alpha}{\partial v}\rangle}\,.$$

Where the integrands are evaluated on the segment $I\times\{v(p)\}$.

Proof. The following notation will be used: $\theta = \theta_\alpha\circ(u,v)^{-1}$, where $\theta_\alpha = K_\alpha, k_\alpha, N_\alpha$. Moreover $A_u = DA\cdot(\frac{\partial}{\partial u})$, $A_{u\varepsilon} = D(A_u)\cdot\frac{\partial}{\partial\varepsilon},\ldots,$ etc. Let $E = \langle A_u,A_u\rangle$, $F = \langle A_u,A_v\rangle$, $G = \langle A_v,A_v\rangle$, $e = \langle A_{uu}, \frac{A_u\wedge A_v}{|A_u\wedge A_v|}\rangle$, $f = \langle A_{uv}, \frac{A_u\wedge A_v}{|A_u\wedge A_v|}\rangle$ and $g = \langle A_{vv}, \frac{A_u\wedge A_v}{|A_u\wedge A_v|}\rangle$. If ε small is fixed, then $E(\cdot,\varepsilon)$, $F(\cdot,\varepsilon)$, $G(\cdot,\varepsilon)$ and $e(\cdot,\varepsilon)$, $f(\cdot,\varepsilon)$, $g(\cdot,\varepsilon)$ are respectively, the coefficients of the first and second fundamental forms of the immersion $A(\cdot,\varepsilon)$.

Direct calculation shows that

(1)
$$A_u = (1-\varepsilon\ k\varphi)\alpha_u + \varepsilon\ \varphi_u\ N$$
$$A_v = (1-\varepsilon\ K\varphi)\alpha_v + \varepsilon\ \varphi_v\ N$$
$$\alpha_u\wedge\alpha_v = |\alpha_u|\,|\alpha_v|N$$
$$\begin{aligned}|\alpha_u|\,|\alpha_v|\ A_u\wedge A_v = {}&(1-\varepsilon\ k\varphi)(1-\varepsilon\ K\varphi)E(\cdot,0)G(\cdot,0)N\\ &-\varepsilon\ \varphi_u(1-\varepsilon\ K\varphi)G(\cdot,0)\alpha_u\\ &-\varepsilon\ \varphi_v(1-\varepsilon\ k\varphi)E(\cdot,0)\alpha_v\end{aligned}$$
$$e(\cdot,0) = \langle N,\alpha_{uu}\rangle = kE(\cdot,0)$$
$$g(\cdot,0) = \langle N,\alpha_{vv}\rangle = KG(\cdot,0)$$
$$E = E(\cdot,0)(1-\varepsilon\ k\varphi)^2 + \varepsilon^2\varphi_u^2$$
$$F = \varepsilon^2\varphi_u\varphi_v$$
$$G = G(\cdot,0)(1-\varepsilon\ k\varphi)^2 + \varepsilon^2\varphi_v^2$$

The following relation holds:

(2)
$$\alpha_{uv} = \frac{k_v}{K-k}\,\alpha_u - \frac{K_u}{K-k}\,\alpha_v\,.$$

In fact, the equality $((\alpha+N)_u)_v = ((\alpha+N)_v)_u$ means that $\alpha_{uv} - k_v\alpha_u - k\alpha_{uv} = \alpha_{uv} - K_u\alpha_v - K\alpha_{uv}$. This implies (2).

It follows from (2) that

(3) $$A_{uv} = \frac{(k_v - \varepsilon K k_v \varphi - \varepsilon k K \varphi_v + \varepsilon k^2 \varphi_v)}{K - k} \alpha_u + $$
$$+ \frac{(-K_u + \varepsilon k K_u \varphi + \varepsilon k K \varphi_u - \varepsilon K^2 \varphi_u)}{K - k} \alpha_v + $$
$$+ \varepsilon \varphi_{uv} N.$$

From (1), (2) and (3) follows that

$$|\alpha_u||\alpha_v||A_u \wedge A_v| \, \mathbf{f} = |\alpha_u||\alpha_v| \langle A_u \wedge A_v, A_{uv} \rangle =$$
$$= -\varepsilon \varphi_u \frac{1-\varepsilon K\varphi}{K-k} (k_v - \varepsilon K k_v \varphi - \varepsilon k K \varphi_v + \varepsilon k^2 \varphi_v) E(\cdot,0) G(\cdot,0)$$
$$-\varepsilon \varphi_v \frac{1-\varepsilon k\varphi}{K-k} (-K_u + \varepsilon k K_u \varphi + \varepsilon k K \varphi_u - \varepsilon K^2 \varphi_u) E(\cdot,0) G(\cdot,0)$$
$$+\varepsilon \varphi_{uv} (1-\varepsilon k\varphi)(1-\varepsilon K\varphi) E(\cdot,0) G(\cdot,0).$$

The equation of the lines of principal curvature of $A(\cdot,\varepsilon)$ is given by:

(5) $$\begin{vmatrix} (dv)^2 & -dudv & (du)^2 \\ E(\cdot,\varepsilon) & F(\cdot,\varepsilon) & G(\cdot,\varepsilon) \\ \tilde{e}(\cdot,\varepsilon) & \tilde{f}(\cdot,\varepsilon) & \tilde{g}(\cdot,\varepsilon) \end{vmatrix} \equiv 0$$

where $(\tilde{e}(\cdot,\varepsilon), \tilde{f}(\cdot,\varepsilon), \tilde{g}(\cdot,\varepsilon)) = |\alpha_u||\alpha_v||A_u \wedge A_v|(e(\cdot,\varepsilon), f(\cdot,\varepsilon), g(\cdot,\varepsilon))$, [10].

Let $p \in u^{-1}(-1)$. Denote by $\tilde{v} = \tilde{v}(u,v(p),\varepsilon)$ the principal minimal arc of $A(\cdot,\varepsilon)$ which is a solution of (5), with initial condition $\tilde{v}(-1,v(p),\varepsilon) = v(p)$. Since φ is bounded, assuming $\varepsilon > 0$ small, $\tilde{v}(u,v(p),\varepsilon)$ will be defined for all $u \in I$, moreover

(6) $$v(T(p,\varepsilon)) = \tilde{v}(1,v(p),\varepsilon) = v(p) + \int_{-1}^{1} \frac{\partial \tilde{v}}{\partial u} (u,v(p),\varepsilon) du.$$

This implies that

(7) $$D(v \circ T) \frac{\partial}{\partial \varepsilon}(p,0) = \int_{-1}^{1} \frac{\partial^2 \tilde{v}}{\partial u \partial \varepsilon} (u,v(p),0) du.$$

Using (5), it follows that $\frac{\partial^2 \tilde{v}}{\partial u \partial \varepsilon} (u,v(p),0) =$

$$= - \frac{\frac{\partial}{\partial\varepsilon}[(E\tilde{f}-F\tilde{e})](u,v(p),0)}{[(Eg-Ge)(EG)](u,v(p),0)} .$$ Notice that, by (1) and (4),

$\frac{\partial}{\partial\varepsilon}(E\tilde{f}-F\tilde{e})(u,v(p),0) = (E\frac{\partial\tilde{f}}{\partial\varepsilon})(u,v(p),0)$ and $(Eg-Ge)(u,v(p),0) = (K-k)(u,v(p))\cdot(EG)(u,v(p),0)$. Therefore, using (4) again, it follows that

$$(8) \qquad \frac{\partial^2\tilde{v}}{\partial u\partial\varepsilon}(u,v(p),0) = - \frac{(-\varphi_u k_v + \varphi_v K_u + \varphi_{uv}(K-k))(u,v(p))}{(K-k)^2(u,v(p))\cdot G(u,v(p),0)}$$

Using (8) and the fact that φ vanishes outside $I\times I$, the following expression is obtained integrating (7) by parts.

$$(9) \qquad D(v\circ T)\frac{\partial}{\partial\varepsilon}(p,0) =$$
$$= - \int_{-1}^{1} \varphi\left(\frac{k_v}{(K-k)^2G(\cdot,0)}\right)_u du + \int_{-1}^{1} \varphi_v\left[\frac{\partial}{\partial u}\left(\frac{1}{(K-k)G(\cdot,0)}\right) - \frac{K_u}{(K-k)^2G(\cdot,0)}\right]du,$$

where the integrands are evaluated on the segment $I \times \{v(p)\}$.

From (2) and the relation $\langle\alpha_{uv},\alpha_v\rangle = \frac{1}{2}(G(\cdot,0))_u$, follows that

$$(10) \qquad \frac{(G(\cdot,0))_u}{G(\cdot,0)} = - \frac{2K_u}{K-k} .$$

Substituting (10) in the expression that results calculating $\frac{\partial}{\partial u}\left[\frac{1}{(K-k)G(\cdot,0)}\right]$, the second integral of (9) reduces to $\int_{-1}^{1}\frac{\varphi_v\, k_u\, du}{(K-k)^2G(\cdot,0)}$. This gives the expression for $D(v\circ T)\frac{\partial}{\partial\varepsilon}(p,0)$ required in the conclusion of the lemma.

4.3 <u>Lemma</u>. Let $\alpha \in \mathcal{I}^\infty$ and $p \in M$ be such that $dk_\alpha/\ell_\alpha(p) \neq 0$. Let $(u,v): M \to 2I\times 2I$ be a principal chart of α around p, where $I = [-1,1]$. Then given any $\varepsilon > 0$ and any sequence of C^r norms $\|\ \|_r$, $r = 2,3,\ldots,$ on $\mathcal{I}^\infty$, there are numbers $\delta = \delta(\varepsilon,\|\ \|_2) > 0$ and $c = c(\varepsilon,\|\ \|_2) > 0$ such that for any $\rho \in (0,\delta]$ and any $p_0 \in u^{-1}(-1) \cap v^{-1}((1-\delta)I)$ it is possible to construct a continuous family $\{\alpha_\mu\}$, $\mu \in [0,1]$, of C^∞ immersions which satisfy the following conditions

i) The support of $\alpha_\mu - \alpha$ is contained in $D = v^{-1}(v(p_0)+2\rho I)$

and $\alpha_o = \alpha$.

ii) For all $\mu \in [0,1]$, $\|\alpha_u - \alpha\|_2 < \varepsilon$

iii) The minimal principal arc of $\alpha_\mu|D$ which passes through p_o meets the segment $u^{-1}(1)$ in a point denoted by $r_\mu(p_o)$. The range of the map $\mu \to v(r_\mu(p_o))$, $\mu \in [0,1]$, contains the interval $[v(p_o), v(p_o) + \rho c]$. See Fig. 4.1.

iv) There exists $\mu_o = \mu_o(r) > 0$ such that $\|\alpha_{\mu_o} - \alpha\|_r < \varepsilon$ and $v(r_{\mu_o}(p_o)) > v(p_o)$.

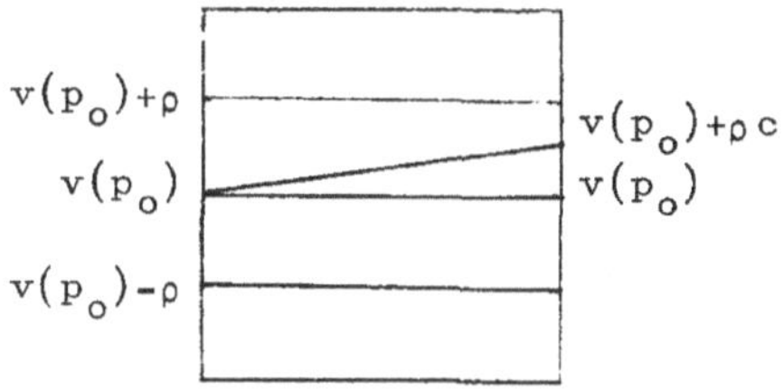

Fig. 4.1

Proof. For a C^s function $\varphi: R^2 \to R$ with compact support $|\varphi|_s$ will denote the C^s-norm of φ, defined by

$$|\varphi|_s = \sup_{\substack{0 \le k+\ell \le s \\ (x,y) \in R^2}} \left| \frac{\partial^{k+\ell}}{\partial x^k \partial y^\ell} (x,y) \right| .$$

Let $(u,v): M \to 2I \times 2I$ be a principal chart around p, $k = k_\alpha \circ (u,v)^{-1}$ and $K = K_\alpha \circ (u,v)^{-1}$. Suppose that the domain of definition of (u,v) is so small that for some constant $\tilde{c} > 0$ and for all $(u,v) \in 2I \times 2I$:

$$3\tilde{c} \ge \left| \left(\frac{\frac{\partial k}{\partial u}}{G(K-k)^2} \right)(u,v) \right| \ge 2\tilde{c} \tag{1}$$

where $G = \langle D\alpha \frac{\partial}{\partial v}, D\alpha \frac{\partial}{\partial v} \rangle$.

Let $\varepsilon > 0$ be given. Certainly there exists a $\lambda > 0$ such that

(2) If $\varphi: M \to R$ is a smooth function whose support is contained in $(u,v)^{-1}(I\times I)$ and satisfies $|\varphi\circ(u,v)^{-1}|_2 \le \lambda$; then $\|\alpha - (\alpha+\varphi N_\alpha)\|_2 = \|\varphi N_\alpha\|_2 \le \varepsilon$.

Let $\theta: R \to [0,1]$ be a smooth function whose support is contained in $2I$ and such that $\theta^{-1}(1) \supset I$. Using (1) it is easy to see that there is a smooth function $\tau: 2I \to [0,1]$ with support contained in I and satisfying:

(3.1) $$|\tau|_2 \le \frac{\lambda}{16|\theta|_2}, \quad \text{with } \lambda \text{ as in (1)}$$

and

(3.2) There is some constant $c \in (0,1/8)$ such that for any $(u_o,v_0) \in I\times I$

$$\int_{-1}^{1} \frac{\tau(u)\,\frac{\partial k}{\partial u}(u,v_0)}{[G(K-k)^2](u,v_0)}\,du \ge 2c$$

and

$$\int_{-1}^{u_o} \frac{\tau(u)\,\frac{\partial k}{\partial u}(u,v_0)}{[G(K-k)^2](u,v_0)}\,du \le 3c.$$

Let $\delta \in (0,\frac{1}{2})$, $N = N_\alpha\circ(u,v)^{-1}$ and $A: 2I \times I \times I \times \delta I \to R^3$ be defined by $A(u,v,v_0,\sigma) = \alpha(u,v) + \sigma(v-v_0)\ (u)\ N(u,v)$. Suppose $\delta > 0$ is so small that for any given $(v_0,\sigma) \in I\times\delta I$, the map $A(\cdot,\cdot,v_0,\sigma)$ which takes $(u,v) \in 2I\times I$ to $A(u,v,v_0,\sigma)$ is a smooth embedding. Let $\tilde{v}(u,v_0,\sigma)$, $u \in 2I$, be the principal minimal arc of $A(\cdot,\cdot,v_0,\sigma)$ satisfying the inicial condition $\tilde{v}(-1,v_0,\sigma) = v_0$. It follows from Lemma (4.2) and inequalities (3.2) that for all $(u,v_0) \in I\times I$

(4) $$\frac{\partial\tilde{v}}{\partial\sigma}(1,v_0,0) \ge 2c \quad \text{and} \quad \frac{\partial\tilde{v}}{\partial\sigma}(u,v_0,0) \le 3c.$$

Since I is compact and $\frac{\partial\tilde{v}}{\partial\sigma}(u,v_0,\sigma)$ depends continuously of $(u,v_0,\sigma) \in I \times I \times \delta I$, taking $\delta > 0$ small enough and using (4), it holds that, for all $(u,v_0,\sigma) \in I \times I \times \delta I$,

(5) $$\frac{\partial\tilde{v}}{\partial\sigma}(1,v_0,\sigma) \ge c \quad \text{and} \quad \frac{\partial\tilde{v}}{\partial\sigma}(u,v_0,\sigma) \le 4c.$$

Notice that $8c < 1$ and that for all $(u,v_o) \in I \times I$, $\tilde{v}(u,v_o,0) = v_o$. Using this, (5) and the Mean Value Theorem it is seen that, for all $(u,v_o,\sigma) \in I \times I \times \delta I$,

(6a) $$\tilde{v}(1,v_o,\sigma) \geq v_o + c\sigma$$

(6b) $$|\tilde{v}(u,v_o,\sigma) - v_o| \leq \sigma/2$$

Given $\rho \in (0,\delta)$ and $p_o \in M$ with $u(p_o) = -1$ and $v(p_o) \in (1-\delta)I$, let $\varphi: M \to \mathbb{R}$ be the smooth function, with support contained in $(u,v)^{-1}(2I \times 2I)$, such that $\hat{\varphi}(u,v) = \rho\theta(\frac{v-v(p_o)}{\rho})(v-v(p_o))\tau(u)$, where $\hat{\varphi} = \varphi \circ (u,v)^{-1}$. Consider the continuous family $\alpha_\mu = \alpha + \mu\varphi N_\alpha$, $\mu \in [0,1]$, of smooth mappings on M. Using (3.1) in can be verified that $|\mu\hat{\varphi}|_2 \leq \lambda$. This and (2) imply that each α_μ is an immersion which is ϵ-C^2-close to α. Certainly each α_μ has support contained in $v^{-1}(v(p_o) + 2\rho I)$. It remains to show that $\{\alpha_\mu\}$ satisfies (iii) of this lemma. Notice that for all $v \in (v(p_o) + \rho I)$, $\theta(\frac{v-v(p_o)}{\rho}) = 1$. Therefore for all $\mu \in I$ $\alpha_\mu \circ (u,v)^{-1}$ and $A(\cdot,\cdot,v(p_o),\mu\rho)$ coincide when restricted to $2I \times (v(p_o) + \rho I)$. Relation (6b) implies that the minimal principal arc $\tilde{v}(u,v(p_o),\mu\rho)$ of $A(\cdot,\cdot,v(p_o),\mu\rho)$ is also a minimal principal arc of $\alpha_\mu \circ (u,v)^{-1}$. Assertion (iii) of the lemma follows from this and from (6a). Assertion (iv) is obvious from the above discussion. The proof of the lemma is in this way finished.

For future reference, the following lemma due to Pugh [9] is stated.

4.4 <u>Lemma</u>. Let $\{p_n\}$ be a sequence in $I = [-1,1]$ of which $0 \in I$ is a cluster point. Given $\epsilon > 0$, small, there are points p_i and p_{i+n} such that

1) $\theta = |p_i - p_{i+n}| < \epsilon$

2) If $\sigma_1 = \min\{p_i,p_{i+n}\} - \theta/2$ and $\sigma_2 = \max\{p_i,p_{i+n}\} + \theta/2$

then

$$[\sigma_1,\sigma_2] \cap \{p_i,p_{i+1},\ldots,p_{i+n}\} = \{p_i,p_{i+n}\}.$$

4.5 <u>Lemma</u>. Let $\alpha \in \mathcal{J}^\infty$, γ be a minimal non-trivial recurrent principal line of α and $p \in \gamma$. Then there exists a C^1 circle C which contains p and is transverse to the minimal foliation f_α of α.

<u>Proof</u>. Similar to that for smooth vector fields on two-dimensional manifolds [4, Lemma 2].

4.6 <u>Lemma</u>. Let $\alpha \in \mathcal{S}_2^\infty \cap \mathcal{S}_3^\infty$ of 3.2.1, $\| \ \|_2$ be a C^2 norm on $\mathcal{J}^\infty$ and $\tilde{\gamma}$ be an oriented minimal principal line of α. Assume that $\omega(\tilde{\gamma})$ (the ω-limit set of $\tilde{\gamma}$) contains a minimal non-trivial recurrent principal line γ. Let $\mathcal{G}$ be a closed nowhere dense subset of M formed by principal lines of α none of which is non-trivial recurrent. Then there exists a point $p \in \gamma - \mathcal{G}$ and a C^1-circle C, transverse to the minimal foliation f_α, passing through p, such that for any $\varepsilon > 0$ and for any given neighborhood V of p, there is a continuous family $\{\alpha_\mu\}$, $\mu \in [0,1]$, in $\mathcal{J}^\infty$, with $\alpha_o = \alpha$ and $\|\alpha_\mu - \alpha\|_2 < \varepsilon$, $\mu \in [0,1]$, and such that for some principal minimal chart of α: $(s,t)\colon M \to I \times I$, where $I = [-1,1]$ (i.e. $\frac{\partial}{\partial s}$ is tangent to the minimal foliation), it holds that

i) The principal line $\tilde{\gamma}$ contains an arc $\overrightarrow{AB}$, with endpoints A and B in $s^{-1}(-1)$ such that if $\mathcal{D}$ denotes the domain of definition of the chart (s,t), then either.

i.1) $\overrightarrow{AB} \cap \mathcal{D} = \{A\} \cup t^{-1}(t(B))$ or

i.2) $\overrightarrow{AB} \cap \mathcal{D} = t^{-1}(t(A)) \cup t^{-1}(t(B))$ and $C \cap (\overrightarrow{AB} - \mathcal{D})$ contains at least two points (See Fig. 4.2).

ii) The circle C is disjoint of $s^{-1}(-1) \cup s^{-1}(1)$.

iii) The support of each $\alpha_\mu - \alpha$, $\mu \in [0,1]$, is contained in $V \cap \mathcal{D}$.

iv) The minimal principal arc of $\alpha_\mu|\mathcal{D}$ which passes through A meets the segment $s^{-1}(1)$ at a point r_μ. The range of the map $\mu \to r_\mu$, $\mu \in [0,1]$, contains the points $\tilde{A} = s^{-1}(1) \cap t^{-1}(t(A))$ and $\tilde{B} = s^{-1}(1) \cap t^{-1}(t(B))$.

v) If the α-limit set of $\tilde{\gamma}$ is an umbilical point, then it can be assumed that $\tilde{\gamma}$ intersects $\mathcal{D}$ for the first time at B.

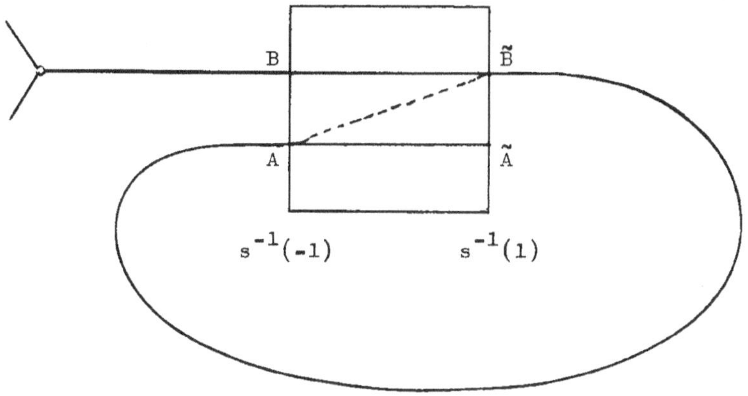

Case i.1

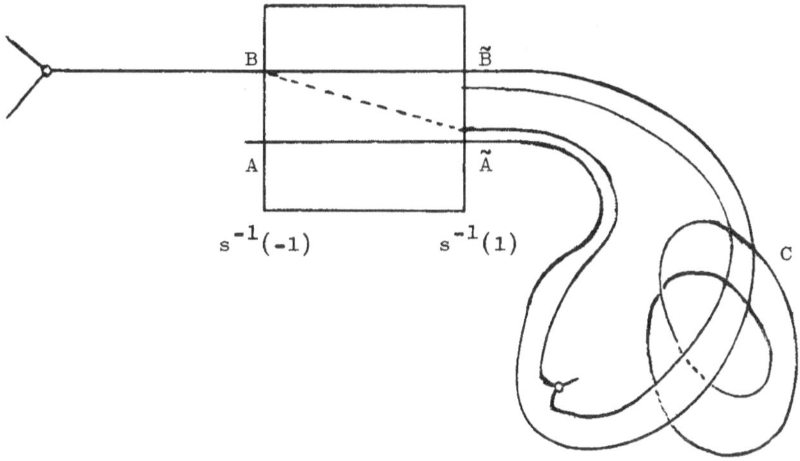

Case i.2

Fig. 4.2

<u>Proof</u>. Except for the statements about the circle C, whose existence is stated in Lemma 4.5, this lemma can be proved following the methods of [9]. For this reason, only an outline of the proof is given below.

By 3.2.3 a point $p \in \gamma - \mathcal{G}$ such that $dk_p|\ell_\alpha \neq 0$ can be chosen. Given a neighborhood V of p, there is a principal chart $(u,v)\colon M \to 2I\times 2I$ such that

(1) The domain of definition $\mathcal{D}(u,v) = (u,v)^{-1}(2I\times 2I)$ of (u,v) is a neighborhood of p contained in V, the hypothesis of Lemma 4.3 are satisfied and C intersects $\mathcal{D}(u,v)$ in a segment which is disjoint of $u^{-1}(-1) \cup u^{-1}(1)$.

Given $\epsilon > 0$, let $\delta = \delta(\epsilon,\| \|_2) > 0$ and $c = c(\epsilon,\| \|_2) > 0$ be as in Lemma 4.3. Let $n \in N$ be such that $(n-1)c \geq 4$. Choose n consecutive points $p_1,p_2,\ldots,p_n$ through which γ crosses $u^{-1}(-1) \cap$ $\cap\, v^{-1}((1-\delta)I)$. The simplifying assumption below does not change the essence of the argument.

(2) Given $i \in \{1,2,\ldots,n-1\}$, the open sub-arc $\overset{\frown}{p_i p_{i+1}}$ of γ with endpoints p_i and p_{i+1} intersects $\mathcal{D}(u,v)$ in an interval having p_i as an endpoint. See Fig. 4.3.

Consider a minimal chart $(\tilde{s},\tilde{t})\colon M \to I\times I$ for which $\tilde{t}^{-1}(0)$ is the principal minimal arc joining p_1 with p_n, $\tilde{s}(p_1) = -1$ and $\tilde{s}^{-1}(1) \cup \tilde{s}^{-1}(-1) \subset u^{-1}(-1)$.

Let $D(\tilde{s},\tilde{t}) = (\tilde{s},\tilde{t})^{-1}(I\times I)$. Given $i \in \{1,2,\ldots,n\}$. Let $\tilde{\Sigma}_i$ be the connected component of $u^{-1}(-1) \cap \mathcal{D}(\tilde{s},\tilde{t})$ which contains p_i. Suppose $(\tilde{s},\tilde{t})$ thin enough in the $\tilde{t}$ direction so that $v(\tilde{\Sigma}_i) \cap$ $\cap \{-1,1\} = \phi$. Certainly it holds that

(3) For $i = 1,2,\ldots,n-1$, the Poincaré map $T_i\colon v(\tilde{\Sigma}_i) \mapsto v(\tilde{\Sigma}_{i+1})$ induced by the minimal principal arcs of $\alpha|\mathcal{D}(\tilde{s},\tilde{t})$ has the form $T_i(v(p_i)+x) = v(p_{i+1}) + \lambda_i x + r_i(x)$, where $\lambda_i \in \mathbb{R}-\{0\}$ and $r_i(x)$ is the Taylor rest. It satisfies: $r_i(0) = r_i'(0) = 0$.

To simplify the exposition suppose at this point that

(4) For $i = 1,2,\ldots,n-1$, $r_i(x) \equiv 0$ and $\lambda_i > 0$.

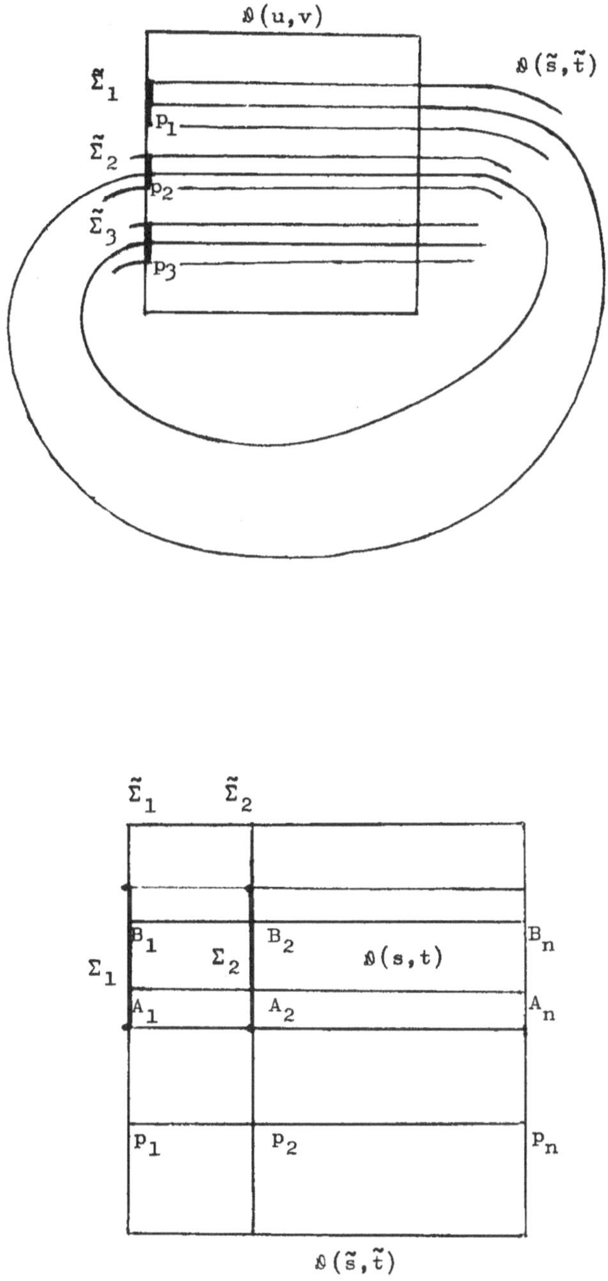

$\mathfrak{D}(u,v)$
$\mathfrak{D}(\tilde{s},\tilde{t})$
$\tilde{\Sigma}_1$
p_1
$\tilde{\Sigma}_2$
p_2
$\tilde{\Sigma}_3$
p_3
$\tilde{\Sigma}_1$
$\tilde{\Sigma}_2$
B_1
B_2
B_n
Σ_1
Σ_2
$\mathfrak{D}(s,t)$
A_1
A_2
A_n
p_1
p_2
p_n
$\mathfrak{D}(\tilde{s},\tilde{t})$

Fig. 4.3

Since $\omega(\tilde{\gamma}) \supset \gamma$, p_1 is an accumulation point of $\tilde{\gamma} \cap u^{-1}(-1)$. Thus, by Lemma 4.4, there exists an arc of trajectory $\overset{\frown}{AB}$ of $\tilde{\gamma}$ with endpoints in $u^{-1}(-1)$, such that $A,B \in \tilde{\Sigma}_1$ and if $v(A) < v(B)$ and $2\theta = v(B) - v(A)$, then

(5) The subinterval $[v(A)-\theta, v(B)+\theta]$ of I is contained in $v(\tilde{\Sigma}_1)$ and intersects $v(\overset{\frown}{AB} \cap u^{-1}(-1))$ only at $v(A)$ and $v(B)$.

Let $(s,t): M \mapsto \mathbb{R}^2$ be the restriction of $(\tilde{s},\tilde{t})$ to $\mathcal{D} = \tilde{t}^{-1}(\tilde{t}(v^{-1}([v(A)-\theta, v(B)+\theta])))$. By (5), either

(6.1) $\overset{\frown}{AB} \cap \mathcal{D} = \{A\} \cup t^{-1}(t(B))$ or

(6.2) $\overset{\frown}{AB} \cap \mathcal{D} = t^{-1}(t(A)) \cup t^{-1}(t(B))$

Let Σ_i be the subinterval $\tilde{\Sigma}_i \cap \mathcal{D}$ of $u^{-1}(-1)$ and let A_i (resp. B_i) be the point of $\Sigma_i \cap t^{-1}(t(A))$ (resp. $\Sigma_i \cap t^{-1}(t(B))$. When (6.2) happens, the subarc $\overset{\frown}{A_nB_n}$ of $\overset{\frown}{A_1B_1}$ meets $u^{-1}(-1)$ from the same side at A_n and B_n. Thus $\overset{\frown}{A_nB_n}$ contains the arcs $v^{-1}(v(A_n))$ and $v^{-1}(v(B_n))$ which, by (1), meet C. Moreover, in this case, $\overset{\frown}{A_1B_1} - \mathcal{D}(s,t) = \overset{\frown}{A_nB_n}$. Therefore:

(6.3) When (6.2) happens, $C \cap \{\overset{\frown}{A_nB_n} - \mathcal{D}(s,t)\}$ contains at least two points.

Because $v(\Sigma_1) = [v(y)-\theta, v(z)+\theta]$ and $T_i: v(\Sigma_i) \to v(\Sigma_{i+1})$ satisfies $T_i(x) - T_i(y) = \lambda_i(x-y)$, it follows that

(7) (a) $v(\Sigma_i)$ is equal to $[v(A_i) - \lambda_1 \cdot \lambda_2 \cdot \ldots \cdot \lambda_{i-1}\theta$, $v(B_i) + \lambda_1 \cdot \lambda_2 \cdot \ldots \cdot \lambda_{i-1}\theta]$ and $\Sigma_i \cap t^{-1}(t(B)) = \{B_i\}$, $\Sigma_i \cap t^{-1}(t(A)) = A_i$.

(b) $v(B_i) = v(A_i) + 2\lambda_1 \cdot \lambda_2 \cdot \ldots \cdot \lambda_{i-1}\theta$.

Let $x_i = v(A_i) + \frac{2(i-1)}{n-1} \lambda_1 \cdot \ldots \cdot \lambda_{i-1}\theta$, $i = 1,2,\ldots,n$. Obviously $v(A_i) \leq x_i \leq v(B_i)$ and therefore.

(8) $[x_i - \lambda_1 \cdot \lambda_2 \cdot \ldots \cdot \lambda_{i-1}\theta,\ x_i + \lambda_1 \cdot \lambda_2 \cdot \ldots \cdot \lambda_{i-1}\theta] \subset v(\Sigma_i)$.

Using (4) and the fact that $\frac{c}{2} \geq \frac{2}{n-1}$ it follows that

(9) $T_i(x_i) = v(A_{i+1}) + \frac{2(i-1)}{n-1}\lambda_1\cdot\lambda_2\cdot\ldots\cdot\lambda_i\theta$ and

$x_{i+1} - T_i(x_i) \leq \lambda_1\cdot\lambda_2\cdot\ldots\cdot\lambda_i\theta c/2.$

Therefore, by (4), (7) and (9), Lemma 4.3 can be applied to construct a continuous family α_μ, $\mu \in [0,1]$, of smooth immersions such that.

(10.i) $\alpha_o = \alpha$ and the support of α_μ is contained in $\bigcup_{i=1}^{n} \Omega_i$, where $\Omega_i = v^{-1}\{x_i + \lambda_1\cdot\lambda_2\cdot\ldots\cdot\lambda_{i-1}\theta I\}$.

(10.ii) For all $\mu \in [0,1]$, $\|\alpha_\mu - \alpha\|_2 \leq \epsilon$.

(10.iii) Given $i \in \{1,2,\ldots,n\}$, the minimal principal arc of $\alpha_\mu|\Omega_i$ which passes through $u^{-1}(-1) \cap v^{-1}(x_i)$ meets the segment $u^{-1}(1)$ at the point $\tilde{r}_\mu(x_i)$. The following inclusion holds $\{v(\tilde{r}_\mu(x_i)),\ \mu \in [0,1]\} \supset [T_i(x_i), x_{i+1}]$.

Let $\tau: R^2 \to R^2$ be the affine map which takes the image of (s,t) onto $I\times I$ and makes $\tau\circ(s,t)$ a minimal chart. If the chart $\tau\circ(s,t)$ is still denoted by (s,t) it will follow, by (1), (5), (6) and (10.i)-(10.iii), that all the conditions of this lemma are verified.

Notice that the whole argument does not change in the absence of assumption (2). To extend this proof to the general case (3), i.e., omitting the restriction (4), it must be observed that in this case the fact that λ_i be either positive or negative does not change the essence of the argument. On the other hand, observe that taking the minimal chart $(\tilde{s},\tilde{t})$ (introduced in the proof above) thin enough in the $\tilde{t}$ direction, it will follow that each function $r_i(x)$ has a Lipschitz constant which is very small in relation to the positive numbers $\frac{|\lambda_1|}{2n}, \frac{|\lambda_2|}{2n}, \ldots, \frac{|\lambda_{n-1}|}{2n}$. Besides, taking $n \in N$ satisfying $(n-1)c \geq 8$ instead of $(n-1)c \geq 4$, it will result that the alterations introduced in (7), (8) and (9), by the consideration of the rest $r_i(x)$, are negligible and the proof of (10) and of this lemma will follow.

4.7 Lemma. Let $\alpha \in \mathcal{S}_1^\infty$ and γ be a minimal non-trivial recurrent principal line. Either α has no umbilical points or there exists an umbilical separatrix of α whose limit set contains γ.

Proof. Similar to that for the case of saddle separatrices of smooth vector fields on two-manifolds [7, Lemma 5].

4.8 Lemma. Let $\alpha \in \mathcal{S}_1^\infty$. Let $T: [a,b] \to [c,d]$ be a homeomorphism between C^1 closed segments which are transverse to the minimal foliation of α. Suppose that when $y \in [a,b)$, y and $T(y)$ determine a continuous family of principal minimal arcs $\overset{\frown}{yT(y)}$ which intersect $[a,b] \cup [c,d]$ onty at y and $T(y)$. Then the map $y \to \overset{\frown}{yT(y)}$ extends continuously to b and $\overset{\frown}{bT(b)}$ is either a principal minimal arc or an arc made up of finitely many umbilical points and principal arcs of minimal umbilical separatrices. See Fig. 4.4.

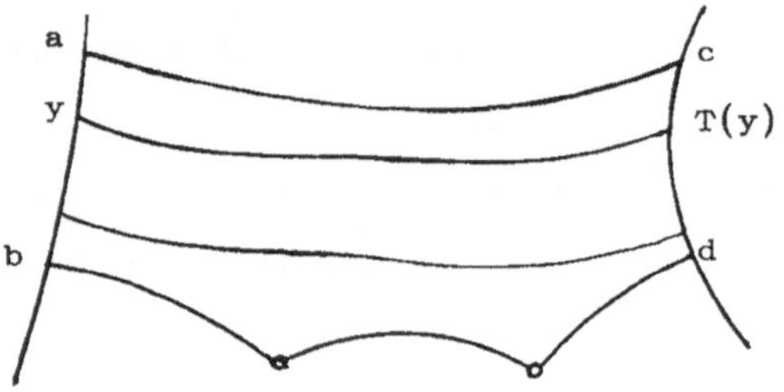

Fig. 4.4

Proof. Similar to that for the case of smooth vector fields on two-manifolds [7 , Lemma 3].

4.9 Proposition. Let $\alpha \in \mathcal{S}_1^\infty \cap \mathcal{S}_2^\infty \cap \mathcal{S}_3^\infty$ of 3.2.1 have a non-trivial recurrent minimal separatrix γ. Let $\mathcal{C}$ be a closed nowhere dense subset of M formed by principal lines of α none of which is non-trivial recurrent. Then there is a point $p \in \gamma - \mathcal{C}$ such that given $\epsilon > 0$ and a neighborhood V of p, arbitrarily small, there is a

$\beta \in \mathcal{S}_1^\infty$ ε-C^2-close to α, with support contained in V and which satisfies one of the following alternatives.

a) β has a minimal principal cycle φ through V. Moreover if some minimal cycle of β disjoint of V, say $\tilde{\varphi}$, which together with φ bound a cylinder, then this cylinder contains an umbilical point of β.

b) β has at least one connection of minimal separatrices more than α.

Proof. Assume that α has umbilical points. Call $\tilde{\gamma}$ an oriented minimal separtrix of an umbilical point $\tilde{\mu}$ such that $\omega(\tilde{\gamma}) \supset \gamma$. The existence of such a $\tilde{\gamma}$ follows from Lemma 4.7. Let $p \in \gamma - \mathcal{G}$ and C be (a C^1 circle, transverse to f_α) as in Lemma 4.6. Given $\varepsilon > 0$ and V, neighborhood of p, consider the minimal chart $(s,t): M \to I \times I$ with domain of definition $\mathcal{D}$, the family $\{\alpha_\mu,\ 0 \le \mu \le 1\}$, the arc $\overrightarrow{AB}$ and all the notations and conclusions stated in Lemma 4.6. There are two cases to study.

Case i.1: $\overrightarrow{AB} \cap \mathcal{D} = \{A\} \cup t^{-1}(t(B))$.

In this case an immersion β satisfying a) of this proposition can be found. In fact, for some $\mu_0 \in [0,1]$, the immersion $\alpha_{\mu_0} = \beta$ has a principal minimal arc, contained in $\mathcal{D}$ connecting A with the point $\tilde{B} = s^{-1}(1) \cap t^{-1}(t(B))$. Since the subarc $\overrightarrow{A\tilde{B}}$ of $\overrightarrow{AB}$ is a principal minimal arc of both α and β, the immersion β has a principal minimal cycle φ passing through $\tilde{B}$. Now, let $\tilde{\varphi}$ be a minimal cycle of β disjoint of V which together with φ bound a cylinder. By the proof of Lemma 4.6, it can be seen that if a principal minimal line of α meets $\mathcal{D}$ it also meets V. Therefore, $\tilde{\varphi}$ does not meet $\mathcal{D}$ and is a minimal cycle of both α and β. Under these circumstances if it is assumed that $\tilde{\varphi}$ and φ bound a cylinder without umbilical points of β, then, considering the sub-segment AB of $s^{-1}(-1)$, it will be obtained that both $\tilde{\varphi}$ and $AB \cup \overrightarrow{AB}$ will bound a cylinder free of umbilical points of α. In this cylinder the minimal foliation f_α is orientable. Therefore, by the Poincaré-

Bendixson theorem, $\overset{\frown}{AB}$ cannot be contained in a non-trivial recurrent minimal principal line. This contradiction proves this lemma in case i.1.

Case i.2: $\overset{\frown}{AB} \cap \mathcal{S} = t^{-1}(t(A)) \cap t^{-1}(t(B))$.

In this case an immersion satisfying the alternative b) of this proposition can be found. In fact, denote by D the set of $x \in s^{-1}(1)$ such that each y belonging to the closed subinterval $[p,x]$ of $s^{-1}(1)$ determines a point $T(y) \in s^{-1}(1)$ by the following condition

(1) y and $T(y)$ are the endpoints of a principal minimal arc $\overset{\frown}{yT(y)}$ which depends continuously on y, coincides with $\overset{\frown}{AB}$ when $y = A$ and whose intersection with $s^{-1}(1)$ is $\{y,T(y)\}$.

Certainly, the assignment $x \to T(x)$ defines a continuous map T on D. Since M is orientable and $\overset{\frown}{AT(A)}$ meets $\{1\}\times I$, at A and $B = T(A)$, for the same side:

(2) T has no fixed point and D and $T(D)$ are disjoint half-open intervals of the form $[A,\tau)$ and $(\sigma,B]$, respectively.

Extend T to the closed interval $[A,\tau]$ by defining $T(\tau) = \sigma$. It follows from Lemma 4.8 that

(3) There is a continuous map on $[A,\tau]$ taking y in $\overset{\frown}{yT(y)}$, where $\overset{\frown}{\tau T(\tau)}$ is either a principal minimal arc on an arc made up of umbilical points and arcs of umbilical minimal separatrices.

Suppose that

(4) Each α_μ, $\mu \in [0,1]$, is so C^2-close to α that, for all $\mu \in [0,1]$, C is transverse to the minimal foliation f_{α_μ} of α_μ.

It will be proved that

(5) $\overset{\frown}{AT(A)}$ and $\overset{\frown}{\tau T(\tau)}$ are disjoint.

In fact, given $y \in [A,\tau]$, let $N_1(y)$ be the cardinal number of the set $\overset{\frown}{yT(y)} \cap C$. Certainly $N_1(y)$ depends continuously on y

and therefore

(6) $N_1(y) \equiv N_1$ is constant for all $y \in [A,\tau]$.

Given $y \in [A,\tau]$ let $\overset{\frown}{yp(y)}$ and $\overset{\frown}{q(y)T(y)}$ be the subarcs of $\overset{\frown}{yT(y)}$ which meet C only at $p(q)$ and $q(y)$, respectively. The existence of $p(y)$ and $q(y)$ and the fact that $p(y) \neq q(y)$ is guaranteed by (6) and Lemma 4.6, i.2). Certainly, (1) implies that the maps $y \to q(y)$ and $y \to p(y)$, defined in $[A,\tau]$ are homeomorphism when restricted to $[A,\tau)$ with disjoint images. Therefore $p(\tau) \neq p(A)$. This implies (5) because otherwise $\overset{\frown}{\tau T(\tau)}$ would necessarily contain $\overset{\frown}{AT(A)}$ and since $p(\tau) \neq p(A)$ it would have that $N_1(\tau) > N_1(A)$ which would contradict (6).

Using (3), (5) and (6) it can be concluded that

(7) There are an umbilical point $U(0) \in \overset{\frown}{\tau T(\tau)}$ and an umbilical minimal separatrix $\lambda(0)$ of $U(0)$ which contains τ and

such that, denoting by $\overset{\frown}{U(0)\tau}$ the subarc of $\lambda(0) \cap \overset{\frown}{\tau T(\tau)}$, the cardinal number N_2 of the set $\overset{\frown}{U(0)\tau} \cap C$ is less that N_1.

When μ varies along $[0,1]$, there are natural continuations $U(\mu)$ and $\lambda(\mu)$ of $U(0)$ and $\lambda(0)$, respectively. $U(\mu)$ and $\lambda(\mu)$ are an umbilical point and an umbilical minimal separatrix of α_μ, respectively. Suppose V small enough so that

(8) There is no minimal umbilical connection of α intersecting V.

It will be proved that

(9) For some $\mu_o \in [0,1]$, $\lambda(\mu_o)$ is a minimal umbilical connection.

Suppose that (9) is not true. Let G be the connected set formed by $\mu \in [0,1]$ such that there is a natural continuation $\tau(\mu)$ of $\tau(0) = \tau$ which belong to $s^{-1}(-1) - (t^{-1}(-1) \cup t^{-1}(1))$. Given $\mu \in G$, denote by $\overrightarrow{U(\mu)\tau(\mu)}$ the subarc of $\lambda(\mu)$ with endopoints $U(\mu)$ and $\tau(\mu)$. Using Lemma 4.8, the assumption that (9) is false and the fact that, when $\mu \in G$, $\tau(\mu)$ never reaches $\{s^{-1}(1) \cap t^{-1}(1)$,

$s^{-1}(1) \cap t^{-1}(-1)\}$ it can be proved that the map $\mu \to \overrightarrow{U(\mu)\tau(\mu)}$ extends continuously to the non-zero endpoint $\tilde{\mu}$ of $\mathcal{G}$ and that

(10) $\overrightarrow{U(\tilde{\mu})\tau(\tilde{\mu})} \subset \lambda(\tilde{\mu})$.

Now, for $\mu \in \bar{\mathcal{G}} = \mathcal{G} \cup \{\tilde{\mu}\}$, denote by $N_3(\mu)$ the cardinal number of the set $\overrightarrow{U(\mu)\tau(\mu)} \cap C$. Since $N_3(\mu)$ depends continuously on μ, $N_3(\mu) \equiv N_3$ is constant. It follows from (7) that $N_3 < N_1 < N_2$ and therefore that $\overrightarrow{U(\tilde{\mu})\tau(\tilde{\mu})}$ does not contain $\overset{\frown}{AT(A)}$. This implies that

(11) $\tau(\tilde{\mu}) \notin \{s^{-1}(1) \cap t^{-1}(1), s^{-1}(1) \cap t^{-1}(-1)\}$.

By (10) and (11) it is concluded that $\tilde{\mu} \in \mathcal{G}$ which implies that $\mathcal{G} = [0,1]$. Under these conditions, using (iv) of Lemma 4.6, it is seen that there exists $\mu_o \in [0,1]$ such that the minimal umbilical separatrix of α_{μ_o}, passing through B, also passes through $\tau(\mu_o)$, i.e. $\lambda(\mu_o)$ is the minimal umbilical connection required to prove (b) of this lemma.

Finally, assume that α has no umbilical point. In this situation take $\gamma = \tilde{\gamma}$. The same argument above shows that case i.2) cannot occur. Thus, it can be proved that the alternative a) of this lemma holds. This ends the proof of this lemma.

<u>Proof of 4.1</u>. Apply 4.9 to $(\alpha,\mathcal{G})$ and obtain an immersion α_1 with a minimal cycle $\tilde{\varphi}$ or a minimal umbilical connection $\tilde{\gamma}$, for which a set $\mathcal{G}_1$ can be defined as follows

a) In case $\tilde{\varphi}$ bounds a disk in M, $\mathcal{G}_1$ is the union of $\mathcal{G}$ and $\tilde{\varphi}$.

b) In case $\tilde{\varphi}$ together with another minimal cycle φ bound a cylinder, $\mathcal{G}_1$ is the union of $\mathcal{G}$, φ and $\tilde{\varphi}$.

c) In case $\tilde{\varphi}$ together with no minimal cycle of $\mathcal{G}$ bound a cylinder, $\mathcal{G}_1$ is the union of $\mathcal{G}$ and $\tilde{\varphi}$.

d) In case α_1 contains no cycle such as $\tilde{\varphi}$, $\mathcal{G}_1$ is the union of $\mathcal{G}$ and $\tilde{\gamma}$.

In case α_1 has nontrivial recurrent minimal lines, apply 4.9 to $(\alpha_1,\mathcal{G}_1)$, and obtain $(\alpha_2,\mathcal{G}_2)$ as above. This process would produce a sequence $(\alpha_n,\mathcal{G}_n)$, which must be finite, because in each step it diminishes de number of available umbilical points or increases the number of independent cycles in M.

This means that after a finite number of applications of 4.9 one obtains an immersion β as required in 4.1.

5. Proof of Proposition 3.2.6

5.1 Lemma. Let $\alpha \in S_1^\infty \cap S_2^\infty \cap S_3^\infty$. Suppose that M contains a cylinder D, free of umbilical points of α, one of whose boundary components is a graph C which consists of umbilical points and minimal umbilical connections and such that either C is accumulated by minimal principal cycles contained in D or the minimal foliation $f_\alpha | D$ can be oriented so that the ω-limit set of any minimal principal line of $f_\alpha | D$ is C. (See Fig. 5.1). Then α can be arbitrarily C^∞-approximated by an immersion β which has one more stabilized separatrix than α.

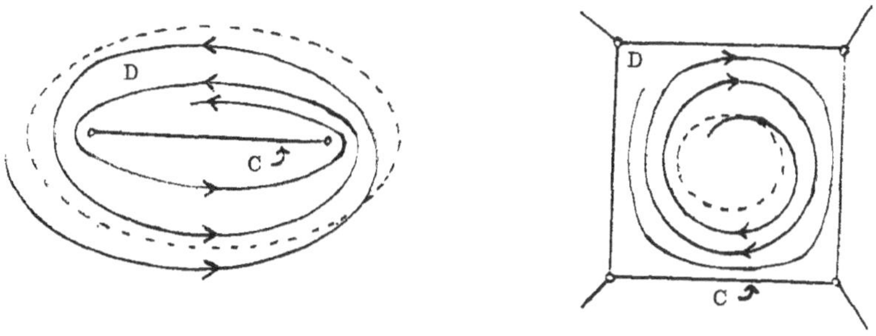

Fig. 5.1

Proof. The same argument of [7 , Lemma 8] works in this case. Let $p \in (M - \mathcal{U}_\alpha) \cap C$ and θ be the minimal umbilical connection passing through p. Because of iv) of Lemma 4.6, there is a perturbation β of α, localized in a small neighborhood of p, which breaks up θ into two minimal umbilical separatrices, say λ and γ.

Notice that the orientation of $f_\alpha | D$ induces an orientation of $f_\beta | D$ and also on $\lambda \cap D$. Choose β so that the ω-limit set of λ is contained in D. By the Poincaré-Bendixson Theorem λ will be stabilized. This finishes the proof of this lemma because β can be taken equal to α in the closure of the stabilized minimal separatrices of α.

5.2 Lemma. Let $\alpha \in \mathcal{S}_1^\infty$ without non-trivial recurrent minimal principal lines. Then the ω-limit set of any oriented minimal principal line is either an umbilical point or a minimal principal cycle or else a graph which consists of umbilical points and minimal umbilical connections.

Proof. Similar to that of the case of vector fields [7 , Lemma 7].

5.3 Proof of Proposition 3.2.6. The immersion α can be arbitrarily C^2-approximated by an immersion $\alpha_1 \in$ interior $(\mathcal{S}_1^\infty \cap \mathcal{S}_2^\infty \cap \mathcal{S}_3^\infty)$ such that:

(1) No small C^2-perturbation of α_1, far away from umbilical points and stabilized separatrices can increase the number of its stabilized separatrices.

In fact, approximate α in $\mathcal{I}^\infty$ by $\beta_1 \in$ interior$(\mathcal{S}_1^\infty \cap \mathcal{S}_2^\infty \cap \mathcal{S}_3^\infty)$. Call $\nu(\beta_1)$ the number of stabilized umbilical separatrices of β_1. Notice that $\nu(\beta_1) \geq \nu(\alpha)$. Either it is possible to arbitrarily approximate β_1 in $\mathcal{I}^{\infty,2}$ by $\beta_2 \in \mathcal{S}_1^\infty$ such that support $(\beta_2 - \beta_1)$ is disjoint from $\mathcal{U}_{\beta_1}$ and stabilized separatrices of β_1 and $\nu(\beta_2) > \nu(\beta_1)$ or $\alpha_1 = \beta_1$ satisfies (1).

If the first possibility holds, approximate β_2 in $\mathcal{I}^\infty$ by $\beta_3 \in$ interior $(\mathcal{S}_1^\infty \cap \mathcal{S}_2^\infty \cap \mathcal{S}_3^\infty)$. Certainly β_3 has the same number of umbilical separatrices as β_1 and $\nu(\beta_3) > \nu(\beta_2)$.

If $\alpha_1 = \beta_3$ does not satisfy (1), one can continue to produce $\beta_4 \in$ interior $(\mathcal{S}_1^\infty \cap \mathcal{S}_2^\infty \cap \mathcal{S}_3^\infty)$ with $\nu(\beta_4) > \nu(\beta_3)$, and so on. This process however must be finite since the number of umbilical separatrices remains constant in each step.

It follows from Proposition 4.1 and Lemma 5.2 that α_1 can be C^2-aproximated by $\alpha_2 \in \mathcal{S}_1^\infty$ without non-trivial recurrent principal lines, sharing property (1) and such that

(2) The support of $\alpha_2 - \alpha_1$ is far away from umbilical connections and from the closure of the separatrices of α_1 whose limit set is made up of umbilical points or principal cycles.

It will be proved that

(3) The immersion α_2 has no umbilical connnection.

Suppose by contradiction that α_2 has a minimal (or maximal) umbilical connection γ. Then, by Lemmas 5.1 and 5.2, γ is neither accumulated by minimal principal cycles nor it is contained in the limit set of any minimal principal line. Therefore a principal chart $(u,v)\colon M \to I \times I$ of α_2, with small domain of definition, centered in a point of γ can be found so that if Ω is the set formed by the principal minimal lines of α_2 crossing the domain of definition $\mathcal{D}$ of the chart (u,v), it holds that

(4.1) Any $\lambda \in \Omega - \gamma$ in not a minimal umbilical connection and it intersects $\mathcal{D}$ in a segment; and

(4.2) If the minimal foliation $f_{\alpha_2}|\Omega$ is oriented; then for all λ in a connected component of $\Omega - \gamma$, $\theta = \omega(\lambda)$ does not depend on the particular λ taken.

Moreover $\omega(\lambda)$ is either a principal minimal cycle or an umbilical point.

Taking (2) into account and using iv) of Lemma 4.6, given any $r \in N$, by means of a small C^r-perturbation of α_2 localized in $\mathcal{D}$, the umbilical minimal connection γ can be broken up and a new stabilized separatrix having θ as limit set can be produced. This contradiction with (1) proves (3).

By Lemma 5.2, in order to prove that $\alpha_2 \in \mathcal{S}_1^\infty$ satisfies a) and b), it remains to show that if γ is an oriented minimal separatrix such that its ω-limit set $\omega(\gamma)$ is a principal cycle c, then γ is a stabilized separatrix. Otherwise, because of (2), iv) of Lemma 4.6 could be used to make a C^∞-small perturbation around a point of c so as to transform γ into a new stabilized separatrix for the perturbed immersion. This would be a contradiction with (1).

Now by Proposition 3.2.1, α_2 can be arbitrarily C^∞-approximated by a $\beta \in$ interior $(\mathcal{S}_1^\infty \cap \mathcal{S}_2^\infty \cap \mathcal{S}_3^\infty)$. It follows from Lemma 4.7, Lemma 5.2 and from the fact that α_2 satisfies a) and b) that β satisfies not only a), but also b) and c).

6. Examples of non-trivial recurrent principal lines.

In this section are given examples of immersions of T^2 and S^2 into R^3, which are actually embeddings, and which have non-trivial recurrent principal lines.

In 6.1 is constructed an embedding of T^2 one of whose principal foliations is an irrational foliation of T^2. This embedding is obtained as a small deformation of an embedding of T^2 different from the usual one, but whose principal configuration is also given by the parallels and meridians of T^2.

In 6.2 is constructed an embedding of S^2 which is a deformation of the ellipsoidal embedding of S^2 and which has non-trivial recurrent principal lines of oscillatory type.

6.1 An embedding of T^2 with non-trivial recurrent principal line.

(1) Let $\gamma = (\gamma_1, \gamma_2): R \to R^2 - \{0\}$ be C^∞ curve of period 2π, enclosing the origin $(0,0)$ and such that

(1.1) $|\gamma'(u)| = (\gamma_1'(u)^2 + \gamma_2'(u)^2)^{1/2} = 1$, that is, u is the arc lenght of γ.

(1.2) For each $u \in (\frac{3\pi}{2} - \frac{\pi}{10}, \frac{3\pi}{2} + \frac{\pi}{10})$, $\gamma''(u) = 0$.

(1.3) For each $u \in [0,\pi]$, $\gamma(u) = (\cos u, \text{sen } u)$.

(2) Let θ be a C^∞ function of period 2π such that

$$(2.1) \qquad \theta(u) = \begin{cases} 0, & \text{if } u \in [\frac{3\pi}{2} + \frac{\pi}{20}, 2\pi] \cup [0, \frac{\pi}{2}]; \\ 1, & \text{if } u \in [\frac{3\pi}{4}, \frac{3\pi}{2} - \frac{\pi}{20}]. \end{cases}$$

(2.2) For each $u \in [0,\pi]$, $\theta'(u) \geq 0$ and for each $u \in [\pi,2\pi]$, $\theta'(u) \leq 0$.

Let $\tilde{N}(u) = (\gamma_2'(u), -\gamma_1'(u))$. Clearly

(3) $\gamma''(u) = k(u)\tilde{N}(u)$ and $\tilde{N}'(u) = -k(u)\gamma'(u)$, where $|k|$ is the curvature of γ.

Define $\alpha: R \times R \times [-1,1] \to R^3$ by $\alpha(u,v,s) = (\gamma_1(u) + \delta\gamma_2'(u) \cos v, \gamma_2(u) - \delta\gamma_1'(u) \cos v, s\varepsilon\theta(u) + \delta \sin v)$, where $\varepsilon,\delta \in (0,1/4]$ are constants. Certainly ε and δ can be chosen so that

(4) The map $s \to \alpha(\cdot,\cdot,s)$, with $s \in [-1,1]$, is a continuous curve of smooth embeddings of the torus $T^2 = R^2/(2\pi\mathbb{Z})^2$ into R^3.

It will be used the following notation:
$E_s = \langle\alpha_u,\alpha_u\rangle$, $F_s = \langle\alpha_u,\alpha_v\rangle$, $G_s = \langle\alpha_v,\alpha_v\rangle$, $e_s = \langle\alpha_u\wedge\alpha_v, \alpha_{uu}\rangle$, $f_s = \langle\alpha_u\wedge\alpha_v, \alpha_{uv}\rangle$ and $g_s = \langle\alpha_u\wedge\alpha_{vv}\rangle$, where $\alpha_u,\alpha_{uv},\ldots,$ denote respectively $\frac{\partial\alpha}{\partial u}(u,v,s)$, $\frac{\partial^2\alpha}{\partial u\partial v}(u,v,s),\ldots$. Consider the equation of the principal minimal lines of $\alpha(\cdot,\cdot,s)$ [10]

$$(5) \qquad \begin{vmatrix} (\frac{dv}{du})^2 & -\frac{dv}{du} & 1 \\ E_s & F_s & G_s \\ e_s & f_s & g_s \end{vmatrix} = 0$$

The calculation performed in (9) will show that $F_o \equiv 0$, $f_o \equiv 0$, which implies that the parallels $u = u_o$, and the meridians $v = v_0$ are principal lines of $\alpha(\cdot,\cdot,0)$. Therefore, $\varepsilon > 0$ can be taken so small that for any given $v_0 \in R/2\pi\mathbb{Z}$, $s \in [-1,1]$, the solution $v(u,v_0,s)$ of (5) satisfying the initial condition $v(0,v_0,s) = v_0$ is defined for all $u \in \mathbb{R}$. Given $s \in [-1,1]$, consider the diffeomorphism T_s defined by the principal lines of $\alpha(\cdot,\cdot,s)$, from the circle $C = \{(u,v);\ u = 0\}$ onto itself by $T_s(v_0) = v(2\pi,v_0,s)$, with $T_o = Id_C$. It will be proved that for all $v_0 \in C$

(6) $$\frac{dT_s}{ds}(v_0) = \frac{\partial v}{\partial s}(2\pi,v_0,0) > 0.$$

This will imply that the rotation number of T_s changes with s around $s = 0$. Therefore, for some $s_o \in [-1,1]$, the rotation number of T_{s_o} will be irrational.

(7) The principal foliation of $\alpha(\cdot,\cdot,s_o)$, which induces T_{s_o}, is an irrational foliation on the torus T^2 and therefore all its lines are dense in T^2.

For all $(v_0,s) \in C \times [-1,1]$, $v(2\pi,v_0,s) = v_0 + \int_0^{2\pi} \frac{\partial v}{\partial u}(u,v_0,s)du$. Hence for all $v_0 \in C$,

(8) $$\frac{\partial v}{\partial s}(2\pi,v_o,0) = \int_0^{2\pi} \frac{\partial^2 v}{\partial u \partial s}(u,v_0,0)du.$$

Using (3), it follows that:

(9) $$\alpha_u(u,v,s) = (\gamma_1'(u)-\delta k(u)\cos v\ \gamma_1'(u),\ \gamma_2'(u)-\delta k(u)\cos v\ \gamma_2'(u), s\varepsilon\theta'(u))$$

$$\alpha_v(u,v,s) = (-\delta\gamma_2'(u)\sin v,\ \delta\gamma_1'(u)\sin v,\ \delta\cos v)$$

$$\alpha_{uu}(u,v,s) = (k(u)\gamma_2'(u)-\delta(k(u)\gamma_1'(u))'\cos v,\ -k(u)\gamma_1'(u) - \delta(k(u)\gamma_2'(u))'\cos v,\ s\varepsilon\theta''(v))$$

$$\alpha_{uv}(u,v,s) = (\delta k(u)\gamma_1'(u)\sin v,\ \delta k(u)\gamma_2'(u)\sin v, 0)$$

$$\alpha_{vv}(u,v,s) = (-\delta\gamma_2'(u)\cos v,\ \delta\gamma_1'(u)\cos v,\ -\delta\sin v)$$

$$\alpha_u(u,v,s) \wedge \alpha_v(u,v,s) = (\delta(\cos v)\gamma_2'(u)-\delta^2 k(u)\cos^2 v\ \gamma_2'(u) - s\varepsilon\gamma_1'(u)\theta'(u)\sin v,$$
$$-\delta(\cos v)\gamma_1'(u) + \delta^2 k(u)(\cos^2 v)\gamma_1' +$$

$$+ \; s\varepsilon\delta\gamma_2'(u)\;\theta'(u)\;\sin v,$$
$$,\; \delta \sin v - \delta^2 k(u)\sin v \cos v).$$

Derivating with respect to s equation (5) evaluated on $v = \tilde{v}(u,v_o,s)$, one obtains that

$$(9.1)\quad 2\left(\frac{\partial v}{\partial u}(u,v_o,0)\right)\cdot\frac{d}{ds}\left(\frac{\partial v}{\partial u}(u,v_o,s)\right)\Big|_{s=0}\cdot[F_o g_o - G_o f_o] +$$
$$+ \left(\frac{\partial u}{\partial u}(u,v_o,0)\right)^2 \frac{d}{ds}[F_s g_s - G_s f_s]\Big|_{s=0} +$$
$$+ \frac{\partial^2 v}{\partial u \partial s}(u,v_o,0)\cdot(E_o g_o - e_o G_o) + \frac{\partial v}{\partial u}(u,v_o,0)\cdot\frac{d}{ds}[E_s g_s - e_s G_s]\Big|_{s=0} +$$
$$+ \frac{\partial}{\partial v}[E_o f_o - F_o e_o]\cdot\frac{\partial v}{\partial s}(u,v_o,s)\Big|_{s=0} + \frac{\partial}{\partial s}[E_s f_s - e_s F_s]\Big|_{s=0} = 0.$$

From the expressions in (9), follows that $f_o(u,v) \equiv 0$ and $F_o(u,v) \equiv 0$, which implies that $\frac{\partial v}{\partial u}(u,v_o,0) \equiv 0$. Therefore (9.1) reduces to

$$(10)\quad \frac{\partial^2 v}{\partial u \partial s}(u,v_o,0) = - \frac{\frac{\partial}{\partial s}[(E_s f_s - e_s F_s)(u,v_o)]\big|_{s=0}}{[E_o g_o - G_o e_o](u,v_o)}$$

Taking into account separately the particular form that (9) and (10) take on the intervals $[0,\pi]$, $[\pi, \frac{3\pi}{2} - \frac{\pi}{2}]$, $[\frac{3\pi}{2} - \frac{\pi}{20}, \frac{3\pi}{2} + \frac{\pi}{20}]$ and $[\frac{3\pi}{2} + \frac{\pi}{20}, 2\pi]$ (See (1) and (2)) it follows that:

(11.1) when $u \in [0,\pi]$, $\frac{\partial^2 \tilde{v}}{\partial u \partial s}(u,v_o,0) = \frac{\theta'(u)}{-1 + 2\delta \cos v_o}$, and

(11.2) when $u \in [\pi,2\pi]$, $\frac{\partial^2 \tilde{v}}{\partial u \partial s}(u,v_o,0) \equiv 0.$

Since $\delta \in (0,\frac{1}{4}]$, (8) and (11) imply that

$$\frac{\partial \tilde{v}}{\partial s}(2\pi,v_o,0) = - \int_0^{\pi} \frac{\theta'(u)du}{-1 + 2\delta \cos v_o} > 0.$$

This proves (6) and therefore also (7).

6.2 An embedding of S^2 with oscillatory recurrent principal lines.

Let φ and ψ be C^∞ real functions on R, defined as follows: φ is even, non-negative, $\varphi(z) = 0$ for $|z| \leq 1/4$ and $\varphi(z) = 1$ for $|z| \geq 1$.
ψ is non-negative, $\psi(z) = 0$ for $z \leq 1/4$ and $\psi(z) = 1$ for $z \geq 1/2$.

Let $S^2 = \{(x,y,z);\ x^2+y^2+z^2 = 1\}$. Denote by α the restriction to S^2 of the map

$$A_{b,c,\theta} : (x,y,z) \to (x_1,y_1,z_1),$$

where

$$x_1 = x\cos[\psi(z)\theta] - [1+b\varphi(z)]y\sin[\psi(z)\theta]$$

$$y_1 = x\sin[\psi(z)\theta] + [1+b\varphi(z)]y\cos[\psi(z)\theta]$$

$$z_1 = (1+c)z.$$

It will be verified that for $0 < b << c$, b small and θ/π irrational, the minimal foliation of α has all its lines dense in S^2.

In fact the minimal foliation of α can be explained in three steps, as follows:

For $b = 0$, $c > 0$, $\theta = 0$, this foliation is the radial foliation on each hemisphere; since the embedding $A_{(0,c,0)}|S^2$ has as image the ellipsoid of revolution

$$(1) \qquad x^2 + y^2 + \frac{z^2}{(1+c)^2} = 1.$$

For $0 < b << c$, the foliation is the same in the northern and southern hemispheres. The illustration below follows from the fact that $A_{(b,c,0)}|S^2$ has its image on the ellipsoid

$$(2) \qquad x^2 + \frac{y^2}{(1+b)^2} + \frac{z^2}{(1+c)^2} = 1$$

for $|z| \geq 1/2$, while for $|z| \leq 1/4$,

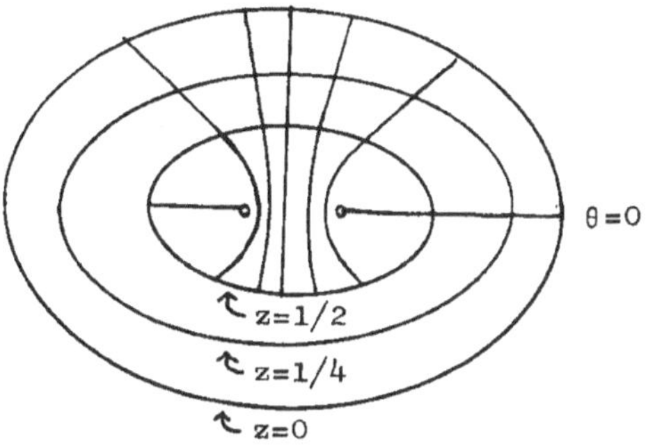

Fig. 6.1

coincides with the ellipsoid of revolution (1).

The foliation is symmetric with respect to reflections about the coordinate axis. The first return map is given by $P(e^{iu}) = e^{-iu}$ on $S^1 = \{z=0\}$. The principal configuration on $|z| \geq 1/2$ illustrated in Fig. 6.1 is well known [3 ,10].

For $0 < b << c$, $\theta > 0$. The foliation in the southern hemisphere coincides with that of Fig. 6.1. In the northern hemisphere, is given in Fig. 6.2.

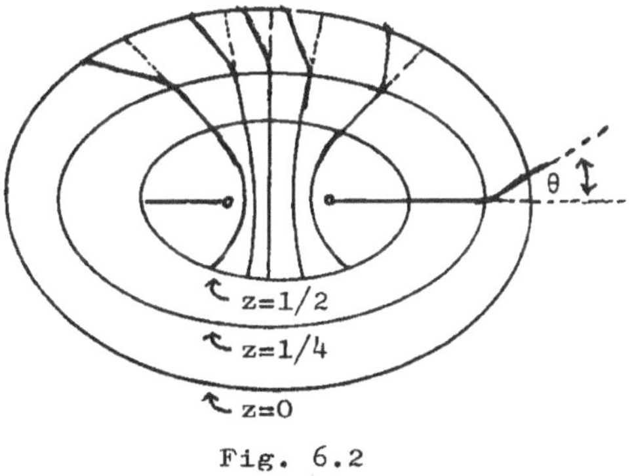

Fig. 6.2

Therefore, the first return map defined by the foliation in the northern hemisphere in this case is given by $Q(e^{iu}) = e^{i(-u+2\theta)}$.

Therefore the second return map defined by the foliation is given by $P\circ Q(e^{i(u)}) = e^{i(u-2\theta)}$, which is an irrational rotation when θ/π is irrational. In this case, the minimal foliation of α has all its lines dense in S^2.

References

[1] Bruce, J.W., Giblin, P.J. - Generic curves and surfaces, J. London Math. Soc. (2) Vol. 24, 1981.

[2] Darboux, G. - Sur la forme des lignes de courbure dans le voisinage d'un ombilic, Note VII; Leçons sur la théorie générale des surfaces, Vol. IV. Gauthier-Villars, 1896.

[3] Gutiérrez, C., Sotomayor, J. - Structurally stable configurations of lines of principal curvature. To appear in Astérisque.

[4] Gutiérrez, C. - Structural stability for flows on the torus with a cross - cap. Trans. AMS, Vol. 241, 1978.

[5] Looijenga, E. - Structural stability of smooth families of C^∞ functions, Thesis, Univ. of Amsterdam, 1974.

[6] Malgrange, B. - Ideals of Differentiable Functions, Oxford University Press, 1966.

[7] Peixoto, M. - Structurally stable vector fields on two-dimensional manifolds, Topology, Vol. 1, 1962.

[8] Porteous, I.R. - The normal singularities of a submanifold, J. Diff. Geom. Vol. 5, 1971.

[9] Pugh, C. - The closing lemma. Am. J. Math. Vol. 89, 1967.

[10] Struik, D. - Lectures on classical differential geometry, Addison Wesley, 1950.

[11] Thom, R. - Stabilité Structurelle et Morphogénèse, Benjamin, 1972.

[12] Wilson, L., Bleecker, D. - Stability of Gauss Maps, Ill. Journ. of Math. Vol. 22, 1978.

[13] Whitney, H. - Differentiable manifolds, Annals of Math., Vol. 37, 1936.

Currents on a circle invariant by a Fuchsian group

by A. Haefliger and Li Banghe

Introduction

Let Γ be a finitely generated group of diffeomorphisms of the circle S^1 with generators $\gamma_1, \gamma_2, \dots, \gamma_K$. We consider the following problem.

Given a C^∞ function f on S^1, when is it possible to find C^∞ 1-forms $\alpha_1, \dots, \alpha_K$ such that

$$*) \qquad df = \sum_{i=1}^{K} \alpha_i - \gamma_i^* \alpha_i$$

In other words, if we denote by $\Omega^d(S^1)$ the vector space of C^∞ forms of degree d on S^1, when is df homologous to zero in the group $H_0(\Gamma, \Omega^1(S^1))$, which is by definition the quotient of $\Omega^1(S^1)$ by the vector space generated by elements of the form $\alpha - \gamma^* \alpha$, where $\alpha \in \Omega^1(S^1)$ and $\gamma \in \Gamma$.

Our interest in this question comes from a problem connected with foliations with all leaves minimal (cf. [2]). Suppose Γ is finitely generated. Then Γ is isomorphic to the holonomy group of a foliation F transversal to the fibers of a circle bundle X over a compact surface S. Indeed choose a surface of high enough genus g such that there is a surjective homomorphism ϕ of $\pi_1(S)$ on Γ. If $\tilde{S}$ is the universal covering of S, let X be the quotient of $\tilde{S} \times S^1$ by $\pi_1(S)$ acting diagonally on the first factor S by covering translations and on the second one through the representation ϕ. It is a circle bundle over S. The horizontal foliation $\tilde{F}$ on $\tilde{S} \times S^1$ whose leaves are of the form $\tilde{S} \times \{y\}$ is invariant by the action of $\pi_1(S)$ and is transversal to the fibers $\{x\} \times S^1$. So on the quotient X, we get a foliation F tranversal to the fibers and with global holonomy group Γ.

Suppose g_0 is a Riemannian metric defined along the leaves of F with volume form ω_0. By a process of integration along the leaves of F described in [2], one gets a C^∞ function f on S^1 whose class in $H_0(\Gamma, \Omega^0(S^1))$ is well defined. The metric g_0 extends to a metric on X for which all the leaves are minimal iff $df \sim 0$ in $H_0(\Gamma,\Omega^1(S^1))$.

An obvious necessary condition is the following.

__For any__ Γ-__invariant__ 1-__current__ T (i.e., a continuous linear form $\omega \mapsto \langle T,\omega\rangle$ on the space of 1-forms such that $\langle T,\omega\rangle = \langle T,\gamma^*\alpha\rangle$ for any $\alpha \in \Omega^1(S^1)$ and $\gamma \in \Gamma$), __then__ $\langle T,df\rangle = 0$.

If all those conditions are satisfied, then F is the limit in the C^∞-topology of a sequence of functions f_n with $df_n \sim 0$ in $H_0^*(\Gamma,\Omega^1(S^1))$. This follows easily from the Hahn-Banach theorem.

So one might consider two steps. First determine all the Γ-invariant currents. Second find out when the vector subspace of $\Omega^1(S^1)$ generated by elements of the form $\alpha - \gamma^*\alpha$, $\alpha \in \Omega^1(S^1)$ and $\gamma \in \Gamma$, is a closed subspace of $\Omega^1(S^1)$.

This second step seems to be much more difficult and we know almost nothing about this question. Let us illustrate the situation by considering a simple example.

Suppose Γ is generated by a single rotation γ of angle $2\pi\rho$. When ρ is irrational, the space of Γ-invariant 1-currents is generated by the current $\omega \mapsto \int_{S^1} \omega$ so that any Γ-invariant 1-current vanishes on df for any f. But the subspace of $\Omega^1(S^1)$ generated by elements of the form $\alpha - \gamma^*\alpha$ is closed iff ρ satisfies a Diophantine condition, namely if there are constants $a > 0$ and k such that

$$|\rho - n/m| \geq \frac{a}{m^k}$$

for all integers $n, m \neq 0$ (cf. [2]).

In this talk we shall determine the space of Γ-invariant currents in the case Γ is the restriction to the boundary of the unit disk $D \subset \mathbb{C}$ of a Fuchsian group Γ such that $\Gamma\backslash D$ is a Riemann surface with a finite number of punctures.

The authors took benefit of a conversation with B. Malgrange and S. J. Patterson.

§1. Currents and harmonic functions

A 1-current T on the circle S^1 is a linear form on the space $\Omega^1(S^1)$ of complex valued 1-forms on S^1 continuous with respect to the C^∞-topology: T is continuous if there is an integer K and a constant $a > 0$ such that

$$|\langle T,\omega\rangle| \leq a\ \|\omega\|_K$$

where $\|\omega\|_K$ is the norm of ω in the C^K-topology.

The Fourier coefficients $c_n(T)$ of T are the numbers $c_n(T) = \langle T, e^{-in\theta}d\theta\rangle$. The sequence $c_n(T)$ has slow growth, namely there is an integer K and a constant $b > 0$ such that

$$|c_n(T)| < b|n|^K \qquad n \neq 0.$$

Conversely, if c_n is a sequence of complex numbers, $n \in \mathbb{Z}$, with slow growth, then the sequence $\sum_{-N}^{+N} c_n e^{in\theta}d\theta$ converges to a 1-current whose Fourier coefficients are the c_n.

The series

$$c_0 + \sum_1^\infty c_n(T)z^n + \sum_1^\infty c_{-n}(T)\bar{z}^n = \sum_{-\infty}^{+\infty} c_n(T)r^n e^{in\phi}$$

where $z = re^{i\phi}$, converges in the unit disk $D = \{z : |z| < 1\}$ to a harmonic function $\tilde{T}(z)$.

One can also define $\tilde{T}(z)$ using the Poisson Kernel by the formula:

$$\tilde{T}(z) = \langle T, \frac{1-|z|^2}{|e^{i\theta}-z|^2}\, d\theta\rangle .$$

<u>Definition</u>: A function f on D has <u>slow growth</u> if there is an integer K and a constant $a > 0$ such that

$$|f(z)| \leq \frac{a}{(1-|z|)^K} .$$

In terms of the hyperbolic distance d on D, this is equivalent to

$$|f(z)| \leq a' e^{Kd(z,z_0)}$$

where $d(z,z_0)$ is the distance of z to a fixed point $z_0 \in D$.

Theorem. a) The map $T \mapsto \tilde{T}$ associating to the 1-current T on the circle S^1 the harmonic function $\tilde{T}(z)$ in the unit disk D is an isomorphism of the vector space of 1-current on the circle on the vector space of harmonic functions on D with slow growth. This map commutes with the natural action on $S^1 = \partial D$ and D of the group G of isometries of D with the hyperbolic metric.

b) There is also a G-equivariant isomorphism between the space of 0-currents T on S^1 such that $\langle T,1\rangle = 0$ and the vector space of differential of harmonic functions on D will slow growth (equivalently such a 0-current T is represented by a gradient vector field ξ on the hyperbolic space D which is divergence free and such that $|\xi|$ has slow growth).

Proof. a) Assume $|c_n(T)| < a|n|^K$, $n \neq 0$. Let $z = re^{i\phi}$.

Then

$$|\tilde{T}(z)| \leq c_0 + 2a \sum_{n=1}^{\infty} n^k r^n \leq \frac{b}{(1-r)^{k+1}}$$

because $\sum_{n=1}^{\infty} n^k r^n = \frac{P(r)}{(1-r)^{k+1}}$ where $P(r)$ is a polynormal in r.

Conversely, let $f(z)$ be a harmonic function in D. Then

$f(re^{i\phi}) = \sum_{-\infty}^{+\infty} c_n r^{|n|} e^{in\phi}$, where the c_n are defined by

$$c_n = \frac{1}{2\pi} \int_0^{2\pi} \frac{f(re^{i\theta})e^{-in\theta}}{r^{|n|}} d\theta$$

for any $r \in]0,1[$.

Assume that $|f(re^{i\phi})| \leq \frac{a}{(1-r)^k}$, then $|c_n| \leq \frac{a}{(1-r)^k r^{|n|}}$. Choose

$r = \frac{|n|-1}{|n|}$, $|n| > 1$. Then $|c_n| \leq \frac{a|n|^k |n|^{|n|}}{(|n|-1)^{|n|}} \leq b|r|^k$.

Hence the c_n are the Fourier coefficients of a 1-current T on S^1 such that $\tilde{T}(z) = f(z)$.

b) A 0-current S such that $\langle S,1\rangle = 0$ is the boundary of a 1-current T, i.e., for any C^∞ function f on S^1, we have

$$\langle S,f\rangle = -\langle T,df\rangle .$$

To T correspond a harmonic function $\tilde{T}(z)$ with slow growth on D. The differential of $\tilde{T}(z)$ is independent of the choice of T and will be the harmonic 1-form associated to S.

Note that for a 1-form $\alpha = a(\theta)d\theta$ on S^1,

$$\langle T,\alpha\rangle = \lim_{r\to 1} \int_0^{2\pi} \tilde{T}(re^{i\theta})a(\theta)d\theta .$$

Examples: The Poisson Kernel is the harmonic function corresponding to the 1-current associating to the 1-form $a(\theta)d\theta$ on S^1 the number $a(0)$. This curren is represented by the tangent vector $\partial/\partial\theta$ at the point 1.

If S is a 0-current on S^1 such that $\langle S,1\rangle = 1$, then S can be considered as a distribution on S^1 with total charge 0 and the corresponding harmonic 1-form can be represented by a divergence free gradient vector field.

§2. Currents invariants under the action of a Fuchsian group with finite volume

Let Γ be a discrete subgroup of the group of conformal maps of the unit disk $D = \{z : |z| < 1\}$ such that $\Gamma\backslash D$ has finite volume. Then $\Gamma\backslash D$ is isomorphic to a compact Riemann surface S of genus g with k points removed.

Theorem: *Assume $\Gamma\backslash D$ is a Riemann surface S of genus g with k punctures at $x_1,x_2,\ldots,x_k$. Then*

(a) The space of Γ-invariant 0-currents on S^1 which vanish on constant functions is isomorphic to the space of harmonic 1-forms on S having at most poles of order 1 at the x_i. Its dimension is $\max(2g,2g+2k-2)$.

(b) The space of Γ-invariant 1-currents on S^1 is isomorphic to the space of harmonic functions on S whose differential has at most poles of order 1 at the x_i. Its dimension is $\max(1,k)$.

<u>Proof</u>. If Γ is a discrete subgroup of the group of conformal maps of D, then $\Gamma\backslash D$ is naturally a Riemann surface, the projection $p : D \to \Gamma\backslash D$ being holomorphic.

Considered as a quotient of the hyperbolic plane D by the subgroup Γ of the group of isometries of D, it is a hyperbolic surface with conical singularities at the image by p of points of D whose isotropy subgroups are not trivial. A metric on $\Gamma\backslash D$ is defined by

$$d(x,y) = \min_{\gamma\in\Gamma} d(\tilde{x},\gamma\tilde{y})$$

where $x = p(\tilde{x})$, $y = p(\tilde{y})$.

The map $f \mapsto p\circ f = \tilde{f}$ is a bijection of the space of harmonic functions on the Riemann surface $\Gamma\backslash D$ on the space of Γ-invariant harmonic functions on D.

$\tilde{f}$ is of slow growth iff there are positive constants a and k such that, for a fixed $x_0 \in \Gamma\backslash D$,

$$|f(x)| \leq ae^{kd(x,x_0)} .$$

Also the map $\alpha \mapsto p^*\alpha = \tilde{\alpha}$ is a bijection between the space of harmonic 1-forms on $\Gamma\backslash D$ and the space of Γ-invariant harmonic forms on D. Moreover, $\tilde{\alpha}$ is of slow growth iff there are constants a and $k > 0$ such that

$$\|\alpha(x)\| \leq ae^{kd(x,x_0)}$$

where the norm $\|\alpha(x)\|$ is defined as the supremum of $|\langle\alpha(x),\xi\rangle|$ for $\xi \in T_x(\Gamma\backslash D)$, $|\xi| = 1$.

If $\Gamma\backslash D$ has finite volume, then there is compact Riemann surface S such that $\Gamma\backslash D$ is isomorphic to the complement in S of k points, called the cusps of $\Gamma\backslash D$. If $x_i \in S$ is a cusp, then a disk neighborhood of x_i intersected with $\Gamma\backslash D$ (so a punctured disk centered in x_i) is isometric to the quotient of a half-plane $H = \{z : \operatorname{Im} z > c > 0\}$ by a translation $\gamma : z \mapsto z + 2\pi$, via the map $z \mapsto e^{iz} = w$, where w is a suitable local coordinate around x_i.

Suppose $\tilde{\alpha}(z)$ is a holomorphic 1-form on H invariant by γ. Then

$$\tilde{\alpha}(z) = \sum_{-\infty}^{+\infty} c_n e^{inz} dz .$$

As $\|dz\| = y$ at the point $z = x + iy$, $\tilde{\alpha}(z)$ has slow growth in H iff $|\Sigma c_n e^{inz}|y \leq ay^k$ for some a and $k > 0$.

This is possible iff $c_n = 0$ for $n < 0$, because

$$|c_n e^{-ny}| = \frac{1}{2\pi}\left|\int_0^{2\pi} (\sum_m c_m e^{imz}) e^{-inx} dx\right| \leq ay^{k-1} .$$

Any harmonic 1-form $\tilde{\alpha}$ invariant by γ is the sum of a holomorphic 1-form and the conjugate of a holomorphic 1-form, both invariant by γ, so

$$\tilde{\alpha} = \sum_0^\infty c_n e^{inz} dz + \sum_0^\infty c_{\bar{n}}\, e^{-in\bar{z}}\, d\bar{z} .$$

In the coordinate $w = e^{iz}$ around x_i, the 1-form α corresponding to $\tilde{\alpha}$ is of the form

$$\alpha = f(w)dw + \overline{g(w)}\, d\bar{w}$$

where f and g are holomorphic functions with a pole of order at most 1 at $w = 0$. (Note that $dz = -\frac{idw}{w}$.)

Now it follows from Riemann-Roch theorem, for instance, that the dimension of the space of holomorphic 1-forms on S having at most poles of order 1 at the punctures $x_1, x_2, \ldots, x_k$ is equal to $g + k - 1$. So the dimension of the space of harmonic 1-forms having poles of order at most 1 at the x_i is $2g + 2k - 2$ (such a real form α is characterized by its $2g$ periods on a basis of the homology of S and by the values of the residues of α and its conjugate α^*, remembering that the sum of the residues must vanish, cf. [1]).

The differential df of a harmonic function f on $\Gamma\backslash D$ with slow growth will correspond to a harmonic form on S having at most poles of order 1 at the x_i and whose periods and residues are zero. The dimension of the space of such 1-forms is $k - 1$ (near a puncture f will be of the form $c \log|w|$ + bounded function). Taking account of the constant functions, this proves b).

§3. Remarks and problems

1) In case $\Gamma\backslash D$ is a compact Riemann surface S (so $k = 0$), then there are no Γ-invariant 0-current (i.e., distribution) taking the value 1 on the

constant function 1.

Indeed, as pointed out to us by Morris Hirsch, this would imply that the Euler class of the circle bundle over S, quotient of $D \times S^1$ by the action $\gamma(z,x) = (\gamma z, \gamma x)$ of Γ, would be zero, as follows from Theorem 2.1, p.376 of Hirsch-Thurston (cf. [3], the proof works for any distribution non zero on the constant functions). But this bundle is isomorphic to the unit tangent circle bundle of S, so its Euler class is non zero.

For $k > 0$, we don't know if there are Γ-invariant 0-currents non zero on the constant functions.

2) It would be interesting to determine the space of Γ-invariant currents for cocompact discrete subgroups Γ of the group of isometries of the hyperbolic n-space acting on the sphere at infinity (our method works only for (n-1)-currents and their boundaries).

For a group Γ acting on a compact smooth n-manifold M, one can define the double complex

$$C^{p,q} = C^p(\Gamma, \mathcal{C}_{n-q}(M))$$

of p-cochains on Γ with values in the space $\mathcal{C}_{n-q}(M)$ of (n-q)-currents on M considered as a Γ-module. The differential $\delta : C^{p,q} \to C^{p+1,q}$ is the usual one and the differential $\partial : C^{p,q} \to C^{p,q+1}$ is induced by $\partial : \mathcal{C}_{n-q}(M) \to \mathcal{C}_{n-q-1}(M)$.

One can show that the cohomology of the total complex associated to this double complex is the cohomology of the bundle $E\Gamma \times_\Gamma M$ associated to the universal bundle $E\Gamma$ over $B\Gamma$, the classifying space for Γ. Using this remark, one could also get the conclusion of Remark 1.

One could also consider the double complex

$$C_{p,q} = C_p(\Gamma, \Omega^{n-q}(M))$$

of p-chains of Γ with values in the space $\Omega^{n-q}(M)$ of (n-q)-forms on M. The homology of the associated total complex should be the homology with real coefficients of $E\Gamma \times_\Gamma M$.

The group one would like to compute in relation with [2] is $H_0(\Gamma, \Omega^1(M))$.

It would also be useful to consider the more general case where Γ is replaced by a pseudogroup H acting on M (cf. [2]). For instance, what are the invariant currents for the pseudogroup generated by the map $z \mapsto z^2$ of S^1 onto itself which appears as the holonomy pseudogroup of a foliation of codimension 1 ?

3) As we have pointed out, it seems difficult to know when the subspace of $\Omega^1(S^1)$ (resp. $\Omega^0(S^1)$) generated by elements of the form $\alpha - \gamma^*\alpha$, $\alpha \in \Omega^1(S^1)$, $\gamma \in \Gamma$ (resp. $f - f\circ\gamma$, $f \in \Omega^0(S^1)$) is closed or not.

We have already mentioned that, when γ is generated by a rotation of $2\pi\rho$, then those spaces are closed iff ρ is not a Liouville number. In fact, they are also closed for a group Γ generated by one hyperbolic or parabolic element γ.

Proposition: Suppose Γ is generated by an element γ which has only two fixed points x_+ and x_- such that the derivative of γ at x_+ (resp. x_-) is a number $\lambda_+ > 1$ (resp. a number $0 < \lambda_- < 1$). Then a C^∞-function f on S^1 is of the form $f = g - g\circ\gamma$, where g is a C^∞-function, iff $f(x_+) = f(x_-) = 0$ and $\sum_{-\infty}^{+\infty} f(\gamma^n x) = 0$. One can then choose $g(x) = \sum_0^\infty f(\gamma^n x)$.

A 1-form ω can be written as $\alpha - \gamma^*\alpha$ iff $\sum_{-\infty}^{+\infty} (\gamma^n)^*\omega = 0$ on $S^1 - \{x_+, x_-\}$.

Corollary: The space generated by elements of the form $g - g\circ\gamma$ (resp. $\alpha - \gamma^*\alpha$) is closed in $\Omega^0(S^1)$ (resp. in $\Omega^1(S^1)$).

We first note the following lemma.

Lemma: Let f be a C^∞-function on an open interval $U \subset \mathbf{R}$ containing 0, such that $f(0) = 0$. Let λ be a number such that $0 < \lambda < 1$. Then the function $x \mapsto \sum_0^\infty f(\lambda^n x)$ is C^∞ in U.

Indeed let K be a compact interval in U containing 0. For $x \in K$, we have

$$|f(\lambda^n x)| \leq \lambda^n |x| \; \|f'\|_K$$

where $\|f'\|_K = \sup_{k \in K} |f'(x)|$.

So we have uniform convergence on K. The same is true for the derivatives, because if $f_n(x) = f(\lambda^n x)$, then $f_n'(x) = f'(\lambda^n x)\lambda^n$. ■

Proof of the proposition: For any f which vanishes at x_+ and x_-, the series $\sum_{-\infty}^{+\infty} f(\gamma^n x)$ converges to a C^∞-function on $S^1 - \{x_+, x_-\}$ invariant by γ (apply the lemma to $\sum_0^\infty f(\gamma^n x)$ near x_- and to $\sum_1^\infty f(\gamma^{-n} x)$ near x_+).

If $f(x_+) = f(x_-) = 0$, let $g(x) = \sum_0^\infty f(\gamma^n x)$. By the lemma g is C^∞ at x_- and also at x_+ because $\sum_0^\infty f(\gamma^n x) = -\sum_1^\infty f(\gamma^{-n} x)$.

The proof for 1-forms is similar. ■

Along the same lines, one can prove that if γ is a parabolic diffeomorphism of S^1 with fixed point x_0, then a C^∞-function f can be written as $f(x) = g(x) - g(\gamma x)$ iff f and df vanish at x_0 and $\sum_{-\infty}^{+\infty} f(\gamma^n x) = 0$.

References:

[1] Ahlfors and Sario: Riemann Surfaces, Princeton University Press, 1960.

[2] A. Haefliger: Some remarks on foliations with minimal leaves, J. Differential Geometry 15(1980), 269-284.

[3] M. Hirsch and W. Thurston: Foliated bundles, Inveriant measures and flat manifolds, Ann. of Math. 101(1975), 369-390.

Institute for Advanced Study
Princeton, New Jersey
and
Institut mathematique
Université de Genève

INFINITE DIMENSIONAL DYNAMICAL SYSTEMS

by

Jack K. Hale

The purpose of this paper is to outline an approach to the development of a theory of dynamic systems in infinite dimensions which is analogous to the theory for finite dimensions. The first problem is to find a class for which there is some hope of classification and yet general enough to include some interesting applications. My goal has been to discover something about a class which includes retarded functional differential equations (RFDE), certain types of neutral functional differential equations (NFDE), parabolic partial differential equations (parabolic PDE) and some other special PDE's. The underlying theory of RFDE's and NFDE's can be found in Hale [7] and parabolic PDE's in Henry [14]. For some details of how these equations fit into the abstract framework below, see Hale [8].

Let X, Y, Z be Banach spaces and let $\mathscr{X}^r = C^r(Y,Z)$, $r \geq 1$, be the set of functions from Y to Z which are bounded and uniformly continuous together with their derivatives up through order r. We impose the usual topology on $\mathscr{X}^r$. (In applications, other topologies may be needed; for example, the Whitney topology.) For each $f \in \mathscr{X}^r$, let $T_f(t) : X \to X$, $t \geq 0$, be a strongly continuous semigroup of transformations on X. For each $x \in X$, we suppose $T_f(t)x$ is defined for $t \geq 0$ and is C^r in x.

We say a point $x_0 \in X$ has a <u>backward extension</u> if there is a $\varphi : (-\infty,0] \to X$ such that $\varphi(0) = x_0$ and $T_f(t)\varphi(\tau) = \varphi(t+\tau)$ for $0 \leq t \leq -\tau$, $\tau \leq 0$. If there is a backward extension φ through x_0, we define $T_f(t)x_0 = \varphi(t)$, $t \leq 0$. A set $M \subset X$ is <u>invariant</u> if, for each $x \in M$, $T_f(t)x$ is defined and belongs to M for $t \in (-\infty,\infty)$. The <u>orbit</u> $\gamma^+(x)$ through x is defined as $\gamma^+(x) = \bigcup_{t\geq 0} T_f(t)x$.

Let

$$A_f = \{x \in X : T_f(t)x \text{ is defined and bounded} \\ \text{for } t \in (-\infty,\infty)\}.$$

The set A_f contains much of the interesting information about the semigroup $T_f(t)$. In fact, it is very easy to verify the following result.

Proposition 1. If A_f is compact, then A_f is maximal, compact, invariant. If, in addition, all orbits have compact closure, then A_f is a global attractor. Finally, if $T_f(t)$ is one-to-one on A_f, then $T_f(t)$ is a continuous group on A_f.

The first difficulty in infinite dimensional systems is to decide how to compare two semigroups $T_f(t)$, $T_g(t)$. It seems to be almost impossible to make a comparison of any systems on all or even an arbitrary bounded set of X. If A_f is compact, Proposition 1 indicates that all essential information is contained in A_f. Thus, we define equivalence relative to A_f.

Definition 2. We say f is equivalent to g, $f \sim g$, if there is a homeomorphism $h: A_f \to A_g$ which preserves orbits and the sense of direction in time We say f is structurally stable if there is a neighborhood V of f in $\mathcal{X}^r$ such that $g \in V$ implies $g \sim f$. We say f is a bifurcation point if f is not structurally stable.

The basic problem is to discuss detailed properties of the set A_f and to determine how A_f and the structure of the flow on A_f change with f.

If A_f is not compact, very little is known at this time. It becomes important therefore to isolate a class of semigroups for which A_f is compact.

If $T_f(t)$ is an α-contraction for $t > 0$ and $T_f(t)$ is compact dissipative, then it was proved by Hale and Lopes [11] (see, also, Hale [9], Massatt [27,28]) that A_f is compact, uniformly asymptotically stable and attracts bounded sets if orbits of bounded sets are bounded. We do not define an α-contraction, but a special case which is very important in the applications is

$$T_f(t) = S_f(t) + U_f(t), \ t \geq 0,$$

where $S_f(t)$ is a strict contraction for $t > 0$ and $U_f(t)$ is completely continuous for $t \geq 0$. Compact dissipative means there is a bounded set B in X such that for any compact set K in X, there is a $t_0 = t_0(K,B)$ such that $T_f(t)\, K \subset B$, $t \geq t_0$.

If $T_f(t)$ is completely continuous for $t \geq r$ for some $r > 0$, then it was shown by Billotti and LaSalle [1] that A_f is compact if $T_f(t)$ is point dissipative. Other conditions for A_f to be compact which are very usefu in the applications have been given recently by Massatt [27,28].

Before proceeding further, we give two examples of semigroups which can be used as models to illustrate several of the ideas.

Suppose $u \in \mathbb{R}^k$, $x \in \mathbb{R}^n$, Ω is a bounded, open set in $\mathbb{R}^n$ with smooth boundary, D is a $k \times k$ constant diagonal, positive matrix, Δ is the Laplacian operator, and consider the equation

$$u_t - D\Delta u = f(x,u,\text{grad } u) \quad \text{in } \Omega$$
$$u = 0 \quad \text{on } \partial\Omega .$$

Other boundary conditions could also be used. Let $W = W_0^{1,2}(\Omega) \cap W^{2,2}(\Omega)$ be the domain of $-\Delta$ and let $X = W^\alpha$, $0 \leq \alpha \leq 1$, be the domain of the fractional power $(-\Delta)^\alpha$ of $-\Delta$ with the graph norm. Under appropriate conditions on f, α, this equation generates a strongly continuous semigroup $T_f(t)$ on X which is compact for $t > 0$. In this case $\mathscr{X}^r = C^r(\Omega \times \mathbb{R}^k \times \mathbb{R}^{kn}, \mathbb{R}^k)$. If f is independent of x, then $\mathscr{X}^r = C^r(\mathbb{R}^k \times \mathbb{R}^{kn}, \mathbb{R}^k)$. If f depends only on u, then $\mathscr{X}^r = C^r(\mathbb{R}^k, \mathbb{R}^k)$. In each of these cases, the theory will be different.

As another example, suppose $r > 0$, $C = C([-r,0], \mathbb{R}^n)$, $\mathscr{X}^r = C^r(C, \mathbb{R}^n)$, $r \geq 1$, and consider the RFDE,

$$\dot{x}(t) = f(x_t)$$

where, for each fixed t, x_t designates the restriction of a function x as $x_t(\theta) = x(t+\theta)$, $-r \leq \theta \leq 0$. For any $\varphi \in C$, let $x(\varphi)(t)$, $t \geq 0$, designate the solution with $x_0(\varphi) = \varphi$ and define $T_f(t)\varphi = x_t(\varphi)$. If this function is defined for $t \geq 0$, then $T_f(t) : C \to C$ is a strongly continuous semigroup and $T_f(t)$ is completely continuous for $t \geq r$ if it takes bounded sets to bounded sets.

For differential difference equations

$$\dot{x}(t) = f(x(t), x(t-r))$$
$$\dot{x}(t) = f(x(t-r))$$

the space $\mathscr{X}^r$ is respectively, $C^r(\mathbb{R}^n \times \mathbb{R}^n, \mathbb{R}^n)$, $C^r(\mathbb{R}^n, \mathbb{R}^n)$.

We now begin our discussion of the set A_f. There are few general results which are independent of f. However, there is an important one concerning the dimension.

Theorem 3. (Mallet-Paret, Mañé) _If_ A_f _is compact and_ $D_x T_f(t)x$ _is the sum of a contraction and a completely continuous operator, then_ A_f _has finite limit capacity. In particular,_ A_f _has finite Hausdorff dimension._

Mallet-Paret [21] proved the part on finite Hausdorf dimension for the case in which X is a Hilbert space. Mañé [24] proved the general case, but a different type of analysis was required.

This result has some important implications if one uses the results of Cartwright [3,4] on almost periodic functions. In particular, one can prove the following result.

Corollary 4. If A_f _is compact and the hypothesis of Theorem_ 3 _is satisfied, then there is an integer_ $N = N_f$ _such that, if_ $T_f(t)x$ _is almost periodic in_ t, _then_ $T_f(t)x$ _is quasiperiodic with frequency base of dimension_ $\leq N$.

Landau and Lifschitz have proposed a principle for the onset of turbulence which consists in the successive introduction through bifurcation of independent frequencies in the oscillatory motion. If the motion is known to be described by the Navier-Stokes equations, then the results of Ladyzenskaya [17] show that A_f is compact and the hypothesis of Theorem 3 is satisfied. Thus, the Landau-Lifschitz principle cannot be valid as a consequence of Corollary 4.

Other than Theorem 3, there are no general results on A_f. The research has followed along the lines of considering special types of equations which lead to the semigroup $T_f(t)$; in particular, functional differential equations and parabolic partial differential equations. On the other hand, to explain some of the results that are known, it is convenient to pose general questions.

Q.1. _Is_ $T_f(t)$ _one-to-one on_ A_f _generically in_ f?

Q.2. If f is structurally stable, is $T_f(t)$ one-to-one on A_f?

Q.3. When is A_f a manifold or a finite union of manifolds?

Q.4. Can A_f be embedded in a finite dimensional manifold generically in f?

Q.5. For each $x \in A_f$, is $T_f(t)x$ continuously differentiable in $t \in \mathbb{R}$?

Q.6. Are Kupka-Smale semigroups generic?

Q.7. Are Morse-Smale systems open and structurally stable?

Before discussing specific results, one important observation must be made. All of the above questions are posed for A_f. This set is much smaller than X and, thus, the questions have a better chance of being answered. Also, Q.5 is not even meaningful on the whole space for several important applications.

For Q.1, one-to-oneness of $T_f(t)$ on A_f, there is no general result known. However, for retarded functional differential equations and special types of neutral functional differential equations, it follows from Nussbaum [33], Hale [7] that $T_f(t)$ is one-to-one on A_f if f is analytic. This is proved by showing that $T_f(t)x$ is an analytic function of t for each $x \in A_f$. Since $(T_f(t)x)(\theta) = (T_f(t+\theta)x)(0)$ for all θ, this implies $x \in C([-r,0],\mathbb{R}^n)$ is also analytic. It is easy to construct examples where f is analytic in a retarded functional differential equation and $T_f(t)$ is not one-to-one on C. A trivial example is $\dot{x}(t) = 0$, $t \geq 0$, $x(t) = \varphi(t)$, $t \in [-1,0]$, $\varphi \in C$. Non-trivial examples may be found in Hale [7].

For some types of parabolic equations, the results of Henry [14], Miller [32] (see Manselli and Miller [25] for further references) imply that $T_f(t)$ is one-to-one on all of X. For these equations, it would be interesting to study more detailed properties of the solutions on A_f. For example, if f is analytic, when are the solutions of $u_t = \Delta u + f(u)$ in a bounded domain, with some boundary conditions, analytic in t and the space variable? A personal communication from D. Henry for one space variable shows that, in one space dimension, this analyticity holds for all solutions on the unstable manifold of a hyperbolic equilibrium point. The same conclusion is probably true for A_f.

In a personal communication, J. Mallet-Paret has given an example where $T_f(t)$ is not one-to-one on A_f. However, it is not structurally stable and, thus, the question Q.2 is posed.

For retarded functional differential equations defined on a compact manifold M without boundary, which are close in some sense to an ordinary differential equation (for example, a differential difference equation with one delay which is small), Kurzweil [16] has shown that A_f is diffeomorphic to M. Oliva [35] has generalized these results giving other conditions which imply A_f is diffeomorphic to M. The corresponding problems for parabolic equations have not been discussed. However, there should be some analogue.

For certain gradient systems of parabolic equations, Henry [14,15] has shown that A_f is the union of a finite number of manifolds.

If it is known that the number of critical points is finite and the ω-limit set of every orbit is a hyperbolic equilibrium point, then A_f will be the union of a finite number of manifolds; namely, $A_f = \bigcup_i W^u_f(\alpha_i)$ where $W^u_f(\alpha_i)$ is the unstable manifold (necessarily finite dimensional) of the equilibrium point α_i.

For Q.5, the differentiability of $T_f(t)x$ in t for $x \in A_f$ is known for some special cases. For RFDE's, this is obviously true since $T_f(t)x$ is defined for $t \leq 0$. We remark that this is true for $x \in A_f$ and not for every $x \in C$. For certain NFDE's, it is also known to be true (see Hale [7]).

The results in PDE's generally relate to the differentiability of $T_f(t)x$ for x in a very large class (see Marsden and McCracken [26]). It should be possible to obtain better results if one restricts x to be in A_f.

In a personal communication, O. Lopes has shown that periodic orbits are always continuously differentiable for the abstract semigroups $T_f(t) = S_f(t) + U_f(t)$ above.

In studying the properties of semigroups $T_f(t)$ which are generic in f, the "size" of the space of functions f plays an important role. For example, if one is attempting to prove that a periodic orbit may be assumed to be hyperbolic generically in f, then the space of functions must be large enough to move the characteristic multipliers that are on the unit circle in any direction whatsoever by an appropriate variation of f. The same difficulty arises with any other property being discussed. When there are more restrictions on the vector field, the characterization of properties which are generic becomes more difficult. We are certainly familiar with similar difficulties in finite dimensional problems; for example, restrictions to vector fields corresponding to Hamiltonian systems, electric circuits, learning models, population models, etc. In infinite dimensional systems, there is even more flexibility in the choice of the vector field. These restrictions sometimes may be natural or may be imposed to make the problem easier to discuss. Also, each system has a specific role to play in applications.

For retarded functional differential equations, for example, one could be considering either of the following equations

$$\dot{x}(t) = f(x_t)$$
$$\dot{x}(t) = f(x(t), x(t-1))$$
$$\dot{x}(t) = f(x(t-1))$$

with, respectively, $f \in C^r([-1,0], \mathbb{R}^n)$, $f \in C^r(\mathbb{R}^n \times \mathbb{R}^n, \mathbb{R}^n)$, $f \in C^r(\mathbb{R}^n, \mathbb{R}^n)$. All of these equations are retarded functional differential equations $\dot{x}(t) = F(x_t)$ with F, respectively, being $F(\varphi) = f(\varphi)$, $F(\varphi) = f(\varphi(0), \varphi(-1))$, $F(\varphi) = f(\varphi(-1))$ For the first two cases, Mallet-Paret [21,22] has shown that the Kupka-Smale systems are generic.

For the third case, $\dot{x}(t) = f(x(t-1))$, nothing is known except that one may assume the equilibrium points are hyperbolic generically. This is especially interesting since this equation is certainly the one that is most often discussed in the literature as far as the existence of periodic orbits is concerned (see Nussbaum [33] for references). On the other hand, in many of the applications, this equation arose through a transformation of variables of an equation of the form

$$\dot{y}(t) = ay(t) + g(y(t-1)).$$

Perhaps one can prove that the Kupka-Smale systems are generic in this class.

For parabolic equations, very little is known about Kupka-Smale systems. There are even several technical difficulties that arise in the discussion of hyperbolicity of the equilibrium solutions. To be more specific, consider the scalar, one-dimensional parabolic equation

$$u_t = u_{xx} + g(x,u), \qquad 0 < x < 1$$
$$u = 0 \quad \text{at} \quad x = 0,\ x = 1,$$

where $g \in C^r(\mathbb{R} \times \mathbb{R}, \mathbb{R})$. In this case, it is not difficult to show that the equilibrium points are hyperbolic generically in g. The reason for this is the fact that the function g is allowed to depend upon x.

On the other hand, if $g(x,u) = f(u)$ where $f \in C^r(\mathbb{R},\mathbb{R})$ is independent of x and $u^0(x)$ is an equilibrium point for f^0, then the linear variational equation about u^0 depends on x. One must now prove that it is possible to move the spectrum of this linear operator by choosing $f(u)$ in a neighborhood of $f^0(u)$ independent of x. This is a nontrivial problem. Smoller and Wasserman [37] have given an example where hyperbolicity can be determined in a class of $f(u)$. Brunovsky and Chow [2] have shown that it always occurs generically in f. More precisely, suppose $u(x,\eta,f)$ is the solution of $u_{xx} + f(u) = 0$ with $u(0) = 0$, $u_x(0) = \eta$ and let $D_f = \{\eta : u(x,\eta,f)$ is zero for some $x > 0\}$. For any $\eta \in D_f$, let $T(\eta,f) > 0$ be the first positive zero of $u(x,\eta,f)$. Brunovsky and Chow [2] prove the following result.

Theorem 4. *There is a residual set $\mathcal{G} \subset$ {space of C^2 functions f endowed with the Whitney topology} such that, for any $f \in \mathcal{G}$, the function $T(\eta,f)$ is a Morse function. Furthermore, there is a residual set $\mathcal{G}' \subset \mathcal{G}$ such that, for each $f \in \mathcal{G}'$, if $T(\eta,f) = 1/n$ for some integer $n = 1,2,\ldots,$ then*

$\partial T(\eta,f)/\partial\eta \neq 0$. <u>In particular, the equilibrium points of</u>

$$u_t = u_{xx} + f(u) \qquad 0 < x < 1$$
$$u = 0 \quad \text{at} \quad x = 0,1 \tag{1}$$

<u>are hyperbolic generically in</u> f.

This theorem also implies that, generically in f, the bifurcation of equilibrium points occur as saddle-node bifurcations for Eq. (1)

It does not seem to be possible to extend the proof of Brunovsky and Chow [2] to several space dimensions. Also, in several space dimensions, one probably should include the domain Ω with the vector field f in the discussion of generic properties since the shape of Ω sometimes determines the multiplicity of eigenvalues.

The discussion of the genericity of the property of the transversal intersection of stable and unstable manifolds for general parabolic equations has not been considered. Henry [14] has given specific examples where he has shown that this property does hold. These examples are also structurally stable

For the above scalar one-dimensional parabolic equations on a bounded interval, no one has given an example where the equilibrium points are hyperbolic and the stable and unstable manifolds intersect nontransversally. After spending some time trying, unsuccessfully, to construct such an example, I conjecture that no such example exists. If this is the case, then these equations are Kupka-Smale if and only if the equilibrium points are hyperbolic. Furthermore, the only way the topological structure of the flow can change on a generic set of f is through saddle-node bifurcations from the result of Brunovsky and

Chow. Even if the above conjecture is generally false, it would be interesting to characterize those f for which it is true.

A vector field f is Morse-Smale if it is Kupka-Smale with a finite number of critical points and periodic orbits with the set of nonwandering points $\Omega(f)$ equal to the set of critical points and periodic orbits. In this case, A_f should be the union of the unstable manifolds for the critical points and periodic orbits. It is not known if such a system is structurally stable.

One can define Axiom A as in finite dimensions and it would be very interesting to prove that the analogue of the Smale decomposition theorem holds.

Let us now give some simple examples to illustrate several of the remarks above. Consider the scalar one-dimensional parabolic equation

$$\begin{aligned} &u_t = u_{xx} + \lambda(u-u^3), \qquad 0 < x < \pi, \quad t \geq 0 \\ &u = 0 \quad \text{at} \quad x = 0,\pi \end{aligned} \tag{2}$$

where λ is a real parameter. This equation defines a strongly continuous semigroup $T_\lambda(t)$, $t \geq 0$, on $H_0^1(0,\pi)$ (see, for example, Henry [14]).

If

$$V(\varphi) = \int_0^\pi \left[\frac{1}{2}\varphi_x^2 - \lambda\left(\frac{\varphi^2}{2} - \frac{\varphi^4}{4}\right)\right] dx$$

for $\varphi \in H_0^1(0,\pi)$, then $\dot{V}(u(t,\cdot))$, the derivative of V along the solutions of (2), satisfies

$$\dot{V}(u(t,\cdot)) = -\int_0^\pi u_t^2 dx \leq 0.$$

This implies every solution of (2) is bounded for $t \geq 0$. Also, every bounded orbit has compact closure which implies the ω-limit set exists. By the invariance principle, the ω-limit set of each orbit belongs to the set of equilibrium points; that is, a solution in $H_0^1(0,\pi)$ of the equation

$$0 = u_{xx} + \lambda(u-u^3), \qquad 0 < x < \pi, \qquad\qquad (3)$$
$$u = 0 \quad \text{at} \quad x = 0,\pi.$$

This result was essentially proved by Chafee and Infante [5]. It is also known that the ω-limit set of each bounded solution (even for general vector field $f(u)$) consists of only one equilibrium point (see Matano [30], Hale and Massatt [12]).

The next step is to analyze how A_λ, the maximal compact invariant set, depends upon λ. The eigenvalues of the linear variational equation for the zero solution are $\lambda_n = n^2$, $n = 1,2,\ldots$. At each λ_n, two hyperbolic solutions φ_n^+, φ_n^- bifurcate from zero. Henry [14] proves that these are the only points of bifurcation and, for $\lambda_n < \lambda < \lambda_{n+1}$, there are exactly $2n+1$ equilibrium solutions with the unstable manifolds $W^u(\varphi_j^+)$, $W^u(\varphi_j^-)$ of φ_j^+, φ_j^- having dimension $j - 1$. In this interval of λ, the unstable manifold $W^u(0)$ of 0 has dimension n.

Let E_λ be the set of equilibrium points for a given λ. For $\lambda_n < \lambda < \lambda_{n+1}$, the set A_λ is given by $A_\lambda = \bigcup_{\varphi\in E_\lambda} W^u(\varphi)$ = closure of $W^u(0)$ and has dimension n. If the stable and unstable manifolds of the equilibrium points always intersect transversally, then Eq. (2) is structurally stable. Henry proves this is the case for $0 < \lambda < 1$, $1 < \lambda < 4$, $4 < \lambda < 9$, $9 < \lambda < 16$. For the latter interval, the oddness of $f(u) = u - u^3$ was exploited. The other cases hold for general $f(u)$.

It is the belief of the author that the transversal intersection property always holds and the proof of this fact must exploit more detailed properties of the solutions of Eq. (2). The recent results of Matano [31] on the change of complexity of solutions with increasing time perhaps could play a role and do, in fact, make it easier to obtain the results of Henry [14]. It is reasonable to refer to Eq. (2) as a gradient flow since formally $\dot{u}_t = -\text{grad } V(u)$ in $H_0^1(0,\pi)$. One can define gradient flows in several space dimensions and vector functions u (see Henry [14]). It would be very interesting to know if gradient Morse-Smale systems are open and dense. No results in this direction seem to be available.

The scalar one-dimensional equation (2) behaves qualitatively as a scalar ordinary differential equation with the analogy being complete if we knew that stable and unstable manifolds for hyperbolic equilibria always intersect transversally. In several respects, retarded functional differential equations generate semigroups which have several properties in common with parabolic equations. However, in detail, these equations behave quite differently with the retarded equations having a more complicated orbit structure. We give an example to illustrate these remarks.

Consider the scalar equation

$$\dot{x}(t) = -\int_{-1}^{0} a(-\theta)g(x(t+\theta))d\theta \tag{4}$$

where $g \in C^2(\mathbb{R},\mathbb{R})$, $a \in C^2([0,1],\mathbb{R})$, $G(x) = \int_0^x g \to \infty$ as $|x| \to \infty$, $a(1) = 0$, $a(s) \geq 0$, $\dot{a}(s) \leq 0$, $\ddot{a}(s) \geq 0$. We consider Eq. (3) for initial data in $C([-1,0],\mathbb{R})$

For Eq. (4), Levin and Nohel [20] exhibited a Liapunov function

$$V(\varphi) = G(\varphi(0)) - \frac{1}{2}\int_{-r}^{0} \dot{a}(-\theta)\left[\int_{\theta}^{0} g(\varphi(s))ds\right]^2 d\theta$$

whose derivative $\dot{V}$ along the solutions of (4) is given by

$$\dot{V}(\varphi) = \frac{1}{2}\dot{a}(r)\left[\int_{-r}^{0} g(\varphi(\theta))\,d\theta\right]^2 - \frac{1}{2}\int_{-r}^{0} \ddot{a}(-\theta)\left[\int_{\theta}^{0} g(\varphi(s))\,ds\right]^2 d\theta \le 0.$$

Using the invariance principle, one can then prove the following result (see Hale [7]).

Theorem 5. Every orbit of Eq. (4) is bounded and has an ω-limit set. If g has isolated zeros, then

(i) If there is an s such that $\ddot{a}(s) > 0$, then the ω-limit set of any orbit is a constant function corresponding to a zero of g;

(ii) If $\ddot{a}(s) = 0$ for all s and $a \not\equiv 0$, then the ω-limit set of any orbit is a periodic orbit of period 1 generated by a 1-periodic solution of the ordinary equation

$$\ddot{x} + a(0)g(x) = 0.$$

Let us consider first a special case of (i); namely, $\ddot{a}(s) > 0$ for $s \in (0,1)$, and the zeros $\alpha_1 < \alpha_2 < \cdots < \alpha_{2k+1}$ of g are simple. Let $A_{a,g}$ be the attractor for Eq. (4). We want to study the dependence of $A_{a,g}$ on g, keeping a fixed. One can show that each equilibrium point is hyperbolic, α_{2j+1} are stable, $j = 0,1,\ldots,k$, α_{2j} are saddle points with unstable manifold $W^u(\alpha_{2j})$ having dimension one, $j = 1,2,\ldots,k$. Thus, the attractor $A_{a,g} = \bigcup_{s=1}^{2k+1} W^u(\alpha_s)$ is one dimensional.

The basic question is the following: to which equilibrium points do the orbits on an unstable manifold $W^u(\alpha_{2j})$ tend? Is it always true that these orbits tend to α_{2j-1} and α_{2j+1}? Surprisingly, the answer to the latter question is negative and there can exist saddle-connections. This means the topological structure of the flow can change without having a bifurcation of equilibrium in contrast to what we believe is happening in the scalar one-dimensional parabolic Eq. (1). To state a precise result, suppose $k = 2$, that is, there are five simple zeros of g, $\alpha_1 < \alpha_2 < \alpha_3 < \alpha_4 < \alpha_5$. If the symbol $\alpha_{2j}(\alpha_k,\alpha_\ell)$ designates that the saddle point α_{2j} is connected by its unstable manifold to α_k,α_ℓ, then the orbit structure on $A_{a,g}$ is determined by a pair $[\alpha_2(\alpha_k,\alpha_\ell), \alpha_4(\alpha_m,\alpha_n)]$. Hale and Rybakowski [13] have shown that each of the following orbit structures can be attained by choosing g appropriately in the above class:

$$[\alpha_2(\alpha_1,\alpha_3), \alpha_4(\alpha_3,\alpha_5)]$$

$$[\alpha_2(\alpha_1,\alpha_4), \alpha_4(\alpha_3,\alpha_5)]$$

$$[\alpha_2(\alpha_1,\alpha_5), \alpha_4(\alpha_3,\alpha_5)]$$

$$[\alpha_2(\alpha_1,\alpha_3), \alpha_4(\alpha_2,\alpha_5)]$$

$$[\alpha_2(\alpha_1,\alpha_3), \alpha_4(\alpha_1,\alpha_5)].$$

The first case corresponds to the natural order of the reals on $A_{a,g}$; the second and fourth cases correspond to saddle connections; the third and fifth cases reverse the natural order of the reals. The first, third and fifth cases remain for small perturbations of g. Although the second and fourth cases seem as if they should not remain after appropriately small perturbations of g, but the authors have been unable to prove this fact and thus, the question is open: <u>Is the set of</u> g <u>for which saddle-connections exist nongeneric?</u>

Let us now suppose that $a(s) = a_0(s)$ is linear and, in particular, that $a_0(s) = 4\pi^2(1-s)$. Also, suppose g is restricted to the class of functions such that $xg(x) > 0$ for $x \neq 0$, $g'(0) = 1$. Then the linear variational equation about $x = 0$ has two eigenvalues on the imaginary axis with the remaining ones having negative real parts. In this case, it is natural to discuss the bifurcation of periodic orbits from zero which arise by small variations in a_0 or g. To do this, one must compute the bifurcation function at (a_0, g) and determine the first nonvanishing coefficients in the Taylor expansion. The generic Hopf bifurcation corresponds to the coefficient $\alpha_{a_0,g}$ of the cubic terms being $\neq 0$. It has been shown by Hale [10] that $\alpha_{a_0,g} = 0$ for all g in the above class and, therefore, a <u>generic Hopf bifurcation can never occur</u>. It is not known if there ever exist any coefficient in the Taylor series of the bifurcation function which is not zero.

The two preceding examples illustrate clearly that important questions in the qualitative theory of infinite dimensional systems arise in the simplest of systems. It is not easy to outline a general direction of research that should be pursued. On the other hand, it is clear that one should first determine the extent to which infinite dimensional systems share the general qualitative properties that are known for ordinary differential equations; for example, the genericity of Kupka-Smale systems, the structural stability properties of Morse-Smale systems, the decomposition theorem of Smale for systems that satisfy Axiom A, etc.

The preceding discussion shows that these questions are difficult in the most elementary examples. Examples of this type need to be discussed in great detail in order to develop the methods and intuition to proceed to more general cases. In scalar one-dimensional parabolic equations above, for instance, important properties of the system were not being used extensively - in particular, the maximum principle. What role, if any, does it play in these qualitative investigations?

The example of the retarded equation exhibited flows with a more complicated qualitative behavior by the introduction of saddle connections. Perhaps this is to be expected since the equation must define a flow in infinite dimensions. On the other hand, so did the parabolic equation. What makes the retarded equation more complicated? What types of flows can be generated by a scalar retarded equation?

Very complicated chaotic type of motions have been observed in applications of scalar retarded equations (see, for example, Lasota and Wazewska-Czyzewska [19], Lasota [18], Glass and Mackey [6]). Peters [36], Walther [38] have actually proved that chaos occurs in equations $\dot{x}(t) = f(x(t-1))$ for some nonlinear functions f. It should not be surprising that complicated dynamics might occur since mappings on the interval exhibit such behavior. However, this is pure speculation and the process involved is not understood. For example, consider the equation

$$\dot{x}(t) = x(t) - f_{\lambda}(x(t-1)) \tag{5}$$

where $f_{\lambda}(x)$ depends on a real parameter λ. Suppose that the interval map $x(t) = f_{\lambda}(x(t-1))$ has successive bifurcations of periodic points through period doubling as λ increases and, for some value of λ, "chaotic" motion occurs. Under what conditions do these periodic points correspond to periodic orbits of (5) and when does a chaotic motion occur in (5)? A version of this problem presently is being studied by Chow and Mallet-Paret for the equation

$$\mu\dot{x}(t) = x(t) - f_\lambda(x(t-1)) \tag{6}$$

where μ is a small parameter. Preliminary investigations indicate that the bifurcation phenomena in (6) is much more complicated than the corresponding one for the interval map.

REFERENCES

[1] Billotti, J. and J.P. LaSalle, Periodic dissipative processes. Bull. Am. Math. Soc. 6(1971), 1082-1089.

[2] Brunovsky, P. and S.N. Chow, Generic properties of stationary states of reaction diffusion equations. J. Differential Equations. To appear.

[3] Cartwright, M.L., Almost periodic flows and solutions of differential equations. Proc. London Math. Soc. (3) 17(1967), 355-380, Corrigenda (3) 17(1967), 769.

[4] __________, Almost periodic differential equations and almost periodic flows. J. Differential Eqns. 5(1969), 167-181.

[5] Chafee, N. and E.F. Infante, A bifurcation problem for a nonlinear partial differential equation of parabolic type. J. Math. Anal. Appl. 4(1974), 17-37.

[6] Glass, L. and M.C. Mackey, Pathological conditions arising from instabilities in physiological control systems. Annals N.Y. Acad. Sci. 316(1979), 214-235.

[7] Hale, J.K., Theory of Functional Differential Equations, Applied Math. Sci., Vol. 3, 2nd Edition, Springer-Verlag, 1977.

[8] __________, Topics in Dynamic Bifurcation Theory, CBMS Regional Conference Series in Math., No. 47(1981), Am. Math. Soc., Providence, R. I.

[9] __________, Some recent results on dissipative systems. Functional Differential Equations and Bifurcation (Ed. A.F. Ize), Lecture Notes in Math. Vol 799(1980), Springer-Verlag, pp. 152-172.

[10] __________, Generic properties of an integro-differential equation. Am. J. Math. To appear.

[11] Hale, J.K. and O. Lopes, Fixed point theorems and dissipative processes. J. Differential Eqns. 13(1973), 391-402.

[12] Hale, J.K, and P. Massatt, Asymptotic behavior of gradient-like systems. Proc. 2nd Fla. Conf. on Dyn. Sys. Feb. 1981. To appear.

[13] Hale, J.K. and K. Rybakowski, On a gradient-like integro-differential equation. Proc. Royal Soc. Edinburgh, Ser. A. To appear.

[14] Henry, D., Geometric Theory of Semilinear Parabolic Equations. Lecture Notes in Math., Vol.840(1981), Springer-Verlag.

[15] _______, Gradient flows defined by parabolic equations, pp. 122-128 in Nonlinear Diffusion (Eds. Fitzgibbon and Walker), Pitman, 1977.

[16] Kurzweil, J., Global solutions of functional differential equations, in Lecture Notes in Math., Vol. 144(1970), Springer-Verlag.

[17] Ladyzenskaya, O.A., A dynamical system generated by the Navier-Stokes equation, J. Sov. Math., 3(1975), 458-479.

[18] Lasota, A., Ergodic theorems in biology, Asterique 50(1977), 239-250.

[19] Lasota, A. and M. Wazewska-Czyzewska, Matematyczne problemy dynamiki ukladu krwinek czerwonych, Mat. Stosowana 6(1976), 23-40.

[20] Levin, J.J. and J. Nohel, On a nonlinear delay equation. J. Math. Ana. Appl. 8(1964), 31-44.

[21] Mallet-Paret, J., Negatively invariant sets of compact maps and an extension of a theorem of Cartwright. J. Differential Equations 22(1976), 331-348.

[22] _______________, Generic properties of retarded functional differential equations. Bull. Am. Math. Soc. 81(1975), 750-752.

[23] _______________, Generic periodic solutions of functional differential equations, J. Differential Equations 25(1977), 163-183.

[24] Mañé, R., On the dimension of the compact invariant set of certain nonlinear maps. Preprint.

[25] Manselli, P. and K. Miller, Dimensionality reduction methods for efficient numerical solution, backward in time, of parabolic equations with variable coefficients. SIAM J. Math. Anal. 11(1980), 147-159.

[26] Marsden, J.E. and M. McCracken, The Hopf Bifurcation and Its Applications, Appl. Math. Sci. Vol.19, Springer-Verlag, 1976.

[27] Massatt, P., Stability and fixed points of dissipative systems. J. Differential Equations 40(1981), 217-231.

[28] __________, Attractivity properties of α-contractions. J. Differential Equations. To appear.

[29] __________, Asymptotic behavior of a strongly damped nonlinear wave equation. J. Differential Equations. To appear.

[30] Matano, H., Convergence of solutions of one-dimensional semi-linear parabolic equations, J. Math. Kyoto Univ. 18(1978), 221-227.

[31] ________, Nonincrease of the lap number of a solution for a one-dimensional semilinear parabolic equation. Pub. Fac. Sci. Univ. Tokyo. To appear.

[32] Miller, K., Nonunique continuation for certain ODE's in Hilbert space and for uniformly parabolic and elliptic equations in self-adjoint divergence form, p.85-101, Lecture Notes in Math., Vol.316(1973), Springer-Verlag.

[33] Nussbaum, R., Periodic solutions of analytic functional differential equations are analytic. Mich. Math. J. 20(1973), 249-255.

[34] __________, Periodic solutions of nonlinear autonomous functional differential equations, p.283-325 in Functional Differential Equations and Approximation of Fixed Points, (Eds. Peitgen and Walther), Lecture Notes in Math., Vol.730, 1979.

[35] Oliva, W.M., The behavior at infinity and the set of global solutions of retarded functional differential equations. Symposium on Functional Differential Equations, São Carlos, Brisil (1975), Coleção ATAS, Vol.8 Soc. Brasileira de Mat.

[36] Peters, H., Comportement chaotique d'une équation différéntielle retardée. C.R. Acad. Sci. Paris 290(1980), Ser. A., 1119-1122.

[37] Smoller, J. and A. Wasserman, Global bifurcation of steady state solutions. J. Differential Equations.

[38] Walther, H.O., Homoclinic solution and chaos in $\dot{x}(t) = f(x(t-1))$. Nonlin. Anal. 5(1981), 775-788.

Lefschetz Center for Dynamical Systems
Division of Applied Mathematics
Brown University
Providence, Rhode Island 02912

FLOWS ON S^3 AND $\mathbb{R}^3$ WITHOUT PERIODIC ORBITS

J. Harrison

James A. Yorke

Dedicated to José Massera

§1. Introduction

A number of authors have studied the problem of destroying periodic orbits. We restrict ourselves to flows on a three dimensional manifold M with metric d.

A C^k flow plug is a C^k vector field X defined on a neighborhood of the unit cube I of $\mathbb{R}^3$ satisfying

(i) $X(p) = -dz$ (a constant vector field) for p in a neighborhood of I, $p \notin I$.

(ii) X has no zeroes.

(iii) X has no periodic orbits.

(iv) If an orbit of X enters at $(x,y,1)$ in the top face of I and exits I at $(x',y',-1)$ in the bottom face of I, then $x = x'$ and $y = y'$.

(v) At least one orbit enters I, at $(0,0,1)$, say, and never exits.

C^k flow plugs can be used to replace a C^k flow f_t having only finitely many periodic orbits and finitely many fixed points by a C^k flow g_t having no periodic orbits and the same set of fixed points. Furthermore, $f_t = g_t$ outside a finite set of flow

This research was partially supported by National Science Foundation grant MCS 7818221A02.

boxes U_i, and there exists $V_i \subset U_i$ and a diffeomorphism $Q_i : I \xrightarrow{\text{onto}} V_i$ such that if Y_i is the restriction of the vector field of g_t to V_i, then the "pull-back" $Q_i^*(Y_i) = X$.

Wilson [7] showed how to construct a C^∞ flow plug in dimension > 3. Schweitzer [6] produced the first C^1 flow plug in dimension three. In [2,3] C^2 flow plugs are produced. It is not known if a C^3 flow plug exists.

The main application in [2,3,6] is the presentation of Seifert counterexamples; that is, C^k flows on S^3, $k = 1,2$, that have neither fixed points nor periodic orbits. Since it is easy to display flows on S^3 with precisely two periodic orbits and no fixed points, the existence of a C^k flow plug implies the existence of a C^k Seifert counterexample.

The purpose of this paper is to describe two other applications.

In our Example 1, we use the existence of a C^k flow plug, $k = 1,2$, to prove the existence of a C^k flow f_t on $\mathbb{R}^3$ having neither periodic orbits nor fixed points, and yet for each trajectory $f_t(x)$

$$\sup_{t \in \mathbb{R}} \|f_t(x)\| < \infty$$

that is, each trajectory is positively and negatively bounded.

In our Example 2 we describe a related construction. Suppose f_t is a flow on $\mathbb{R}^3$ for which 0 is an isolated fixed point. We say 0 is <u>totally stable</u> if it is positively and negatively Liapunov stable, that is, if for any $\theta_1 > 0$ there is a $\theta_2 \varepsilon (0,\theta_1)$ such that if $|q| \le \theta_2$ then the trajectory $f_t(q)$ satisfies $\sup_{t \in \mathbb{R}} \|f_t(q)\| \le \theta_1$. It then follows that if $|q| \ge \theta_1$, then $\inf_{t \in \mathbb{R}} \|f_t(q)\| \ge \theta_2$. This is equivalent to showing 0 is the intersection of nested invariant neighborhoods.

Using our Example 1 we construct a C^2 example of a flow on $\mathbb{R}^3$ having an isolated zero that is totally stable.

§2. Examples

Example 1

The construction of a flow with no fixed points and all orbits bounded positively and negatively is based on [4], though in that example there is an uncountable infinity of periodic orbits. The construction calls for a solid torus S_1 in $\mathbb{R}^3$ embedded in a larger solid torus S_2, and the sequence continues with infinitely many ever larger tori. See Figure 2.1. Figure 2.2 shows how the flow behaves in the solid torus S_{n+1}. The trajectories rotate about the center circle in S_n only in the vicinity of S_n, while elsewhere they dive away from the orbit A on the surface of S_n toward a corresponding orbit B on the opposite side of the surface of S_n. Clearly all orbits are bounded. It is not difficult to arrange the flow so that the only periodic orbits in each $S_{n+1}-S_n$ are the two on the surface of S_{n+1}, and S_1 has no periodic orbits in its interior. Since the periodic orbits are isolated, applying the C^k flow plug $(k = 1,2)$ to each is possible and the result is a C^k flow g_t with no fixed points and no periodic orbits. The flow plug is applied in a small neighborhood U(p) of

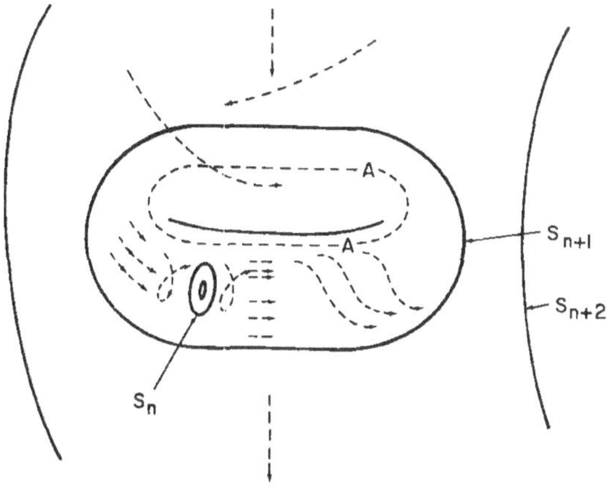

Figure 2.1

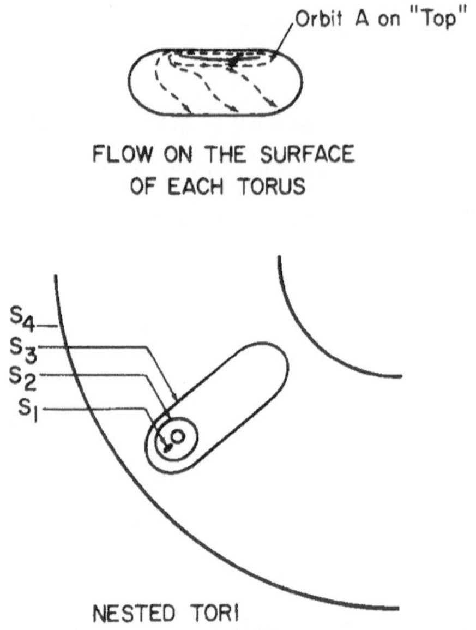

Figure 2.2

one point p on each orbit of f_t so if an orbit of the altered flow g_t is inside a torus S_n when it enters the flow plug neighborhood $U(p)$, it will again be inside if it leaves $U(p)$; so each orbit must still be positively and negatively bounded.

The embedding of a torus in a smaller torus is reminiscent of Fuller's embedding an orbit in a torus in [1].

Example 2

Given the flow on $S_2 - S_1$, there is a smooth map ψ (linear for example) of S_2 onto S_1 which carries the vector field on ∂S_2 onto the vector field ∂S_1. The image of S_1 is a torus we may denote S_0. Iterating the process we can produce a sequence of ever decreasing invariant tori S_{-n}, the intersection of which is a point, which we will call 0. The resulting vector field is C^∞ smooth except possibly at 0. Now each periodic orbit has a neighborhood which contains no other orbits, and a C^2 flow plug can be

construct examples in which there are no repellers, for example by having two attracting orbits separated by a repelling torus. (In our count, we ignore tori). The interchangeability of periodic orbits and zeroes is elementary and can be seen from Figure 3.1 top.

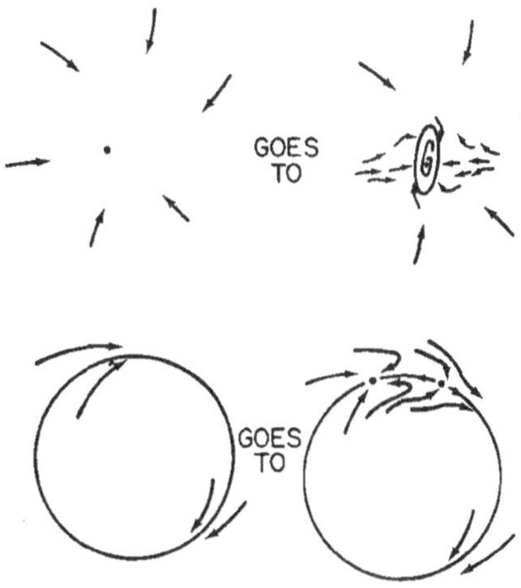

Figure 3.1

In the top an attractor point is changed into an attractor orbit, while in the bottom the reverse occurs. Both changes require changes only in a small neighborhood of the attracting point (top) or periodic orbit (bottom). Other examples can be given which show how to convert a sink into a source and vice versa using changes in the vector field in a neighborhood of the sink or source.

A sink stationary point can be changed into an attracting periodic orbit with a stationary solution that has a two dimensional unstable manifold. Conversely an attracting orbit can be destroyed by the insertion of a point attractor with a saddle point. Hence a sink (or source) stationary point can be replaced by a sink (or source) periodic orbit and vice versa.

Essentially the case of vector fields X on the disk D^3 with applied to each. We choose the U_i neighborhoods to be disjoint, none of them intersecting the boundary of more than one S_n. Our new flow h_t has no fixed points except possibly 0, which is also possibly a point of discontinuity of the flow. Note 0 is the intersection of invariant neighborhoods C_n, namely, the closure of the orbits that intersect S_n minus the flow plug balls U_i which intersect the boundary of S_n. Choosing a function $\rho : \mathbb{R}^3 \to [0,1]$ such that $\rho(q) = 0$ if and only if $q = 0$, we examine the vector field $X_2(q) = \rho(q)X_1(q)$ (defining $X_2(0) = 0$). Choosing ρ so that its derivatives go to zero sufficiently fast, the vector field X_2 is C^2 in $\mathbb{R}^3$. This completes the construction of Example 2.

§3. Speculation

In this section we summarize the current state of the question of what kinds of vector fields exist on S^3 by stating a strengthened form of the Seifert conjecture that permits the existence of finitely many hyperbolic fixed points and periodic orbits.

By a <u>sink</u> of a vector field X we will mean either a zero of the vector field X that is an attractor or a periodic orbit that is an attractor. By a <u>source</u> of X we mean a sink for the vector field $-X$.

<u>Conjecture</u>: If X is a C^3 vector field on S^3 having only finitely many periodic orbits and having only hyperbolic periodic orbits and hyperbolic zeroes, then the number of sources plus the number of sinks is at least 2.

<u>Remarks</u>: The restriction to finitely many orbits is necessary because the Guckenheimer-Lorenz flow has a ball on whose boundary the vector field is pointing in and yet there are no sources or sinks. The strange attractor (with an infinity of periodic orbits) takes over this role. Such a ball could be carefully glued to another having the reverse flow with a strange repeller, and thus there would be a counterexample to the conjecture. It is easy to

X transverse to the boundary is a special case, but we state this case explicitly.

D^3 Conjecture: If X is a C^3 vector field on D^3 and is transverse to the boundary of D^3 and has only finitely many periodic orbits, and if periodic orbits and stationary points are hyper bolic, then there is either a source or sink.

Plykin [5] has proved a related result for Axion A diffeomorphisms on S^2 or T^2, namely that if there exists a one dimensional basic set, there exists a periodic sink or source.

References

1. F. B. Fuller, "Notes on trajectories on a solid torus," Annals of Math (2) 59 (1952), 438-439.

2. J. Harrison, "Denjoy Fractals in the plane," preprint.

3. J. Harrison, "A C^2 Counterexample to the Seifert conjecture," preprint.

4. G. S. Jones and J. A. Yorke, "The existence and nonexistence of critical points in bounded flows," J. Differential Equations, 6 (1969), 238-246.

5. R. V. Plykin, "Sinks and sources for A-diffeomorphisms on surfaces," (Russian), Math. of U.S.S.R. Sbornik 94 (1974), 243-264.

6. P. A. Schweitzer, "Counterexamples to the Seifert conjecture and opening closed leaves of foliations," Annals of Math. (1971), 386-400.

7. F. W. Wilson, Jr., "On the minimal sets of non-singular vector fields," Annals of Math., 84 (1966), 529-536.

J.HARRISON
Somerville College, Oxford
England
and
University of California
Berkeley - California
USA

JAMES A.YORKE
Univ. of Maryland
College Park
Maryland USA

Generalizations of some theorems of small divisors

to non archimedean fields

M. Herman and J.C. Yoccoz*

0. Introduction

We propose to generalize theorems of C.L. Siegel [12], [13], and V.I. Arnold [2] to non archimedean fields, the main difficulty being the presence of small divisors. More precisely, the content of this paper is the following:

In part I.1, we show that a formal diffeomorphism $f(x) = Ax + \ldots \in (k[[x_1 \ldots x_n]])^n$, with k a commutative field and $A \in GL(n,k)$, is formally conjugate to A as soon as A satisfies the non-resonance condition $*$. In I.2, we introduce a diophantine condition (C) that will enable us to prove the convergence of the conjugacy (i.e. Siegel's theorem), assuming that k is a complete non-trivial valued field. We give examples of non archimedean fields k for which condition (C) can be satisfied in the Siegel domain. Unfortunately, this usuallv implies that the characteristic of $k = 0$. We also give examples of matrices A on $\mathbb{Q}_p^2$ not satisfying condition (C), and associated $\mathbb{Q}_p$-analytic diffeomorphisms of $\mathbb{Q}_p^2$, $f(x) = Ax + \ldots$, such that the formal conjugacy diverges. Part I.3 is devoted to various lemmas on composition.

* The authors wish to thank the Instituto de Matemática Pura e Aplicada for their hospitality.

In part I.4, we prove Siegel's theorem. The proof is almost the same as Rüssmann's [8], as generalized by Zehnder [15], in the case of $\mathbb{C}$. We don't suppose in this proof that char $k = 0$ or that k is locally compact. These results generalize results of Sibuya and Sperber [11] on vector fields (cf. I.5). In I. Appendix, we prove the stable manifold theorem (with almost the same proof as given by Meyer [6]).

In part II, we generalize a theorem of Arnold on diffeomorphisms of the unit circle of $\mathbb{C}$. For k a commutative field with non trivial non archimedean complete norm, we define diffeomorphisms of $S_k = \{x \in k, |x|_k = 1\}$ by $f: z \to \alpha z + \varphi(z)$, with $\alpha \in k$, $|\alpha|_k = 1$, and φ is a "small" Laurent series. (We propose Laurent series as a generalization of $\mathbb{R}$-analytic Fourier series). We show that, if α is not a root of unity and φ is small enough, f is conjugate to a rotation $z \to \alpha_1 z$, $|\alpha_1|_k = 1$, by a diffeomorphism $z \to z + \psi(z)$, with ψ a small Laurent series. The proof is almost the same as the one in [3, Appendice].

In part III, we study homeomorphisms of the ring of p-adic integers $\mathbb{Z}_p$ (p prime, $p \geq 2$) of the form $f(x) = x + u + \varphi(x)$, with $|u|_p = 1$, φ a Lipschitz map from $\mathbb{Z}_p$ to $\mathbb{Z}_p$ with $\mathrm{Lip}(\varphi)$ small. We prove, using interpolation series, that f is Lipschitz conjugate to $R_u: x \to x+u$, provided $\mathrm{Lip}(\varphi)$ is small enough.

The parts I and II are joint work, part III is due to the first author.

I. Generalization of a theorem of Siegel

1. The group of formal diffeomorphisms of k^n.

1.1 Let k in the following be a commutative field (of arbitrary characteristic).

We denote by $G^F(k^n)$ the group of formal diffeomorphisms of k^n leaving fixed 0:

$$G^F(k^n) = \{f = (f_1,\dots,f_n) \in k[[x_1,\dots,x_n]]^n,\ f(x) = Ax + \dots\ A \in GL(n,k)\}$$

$G^F(k^n)$ is a group under the composition of formal power series "leaving 0 fixed".

There is a canonical surjective group homomorphism, which splits:

$$f \in G^F(k^n) \to Df(0) \in GL(n,k).$$

We call $Df(0)$ the linear part of f.

1.2 The question we want to study now is to conjugate $f \in G^F(k^n)$ in the group $G^F(k^n)$ to its linear part A.

We say that a matrix $A \in GL(n,k)$ satisfies the condition $*$, if the eigenvalues $\lambda_1,\ldots,\lambda_n$ of A, in an algebraic closure of k, satisfy no relation:

$$\lambda_1^{i_1} \ldots \lambda_n^{i_n} = \lambda_j$$

with $i_k \in \mathbb{N}$, $\Sigma\, i_k \geq 2$, $1 \leq j \leq n$.

The condition does not depend on the algebraic closure of k.

<u>Proposition 1</u>. <u>If the linear part A of</u> f <u>satisfies condition</u> $*$, <u>then there exists a unique</u> $h \in G^F(k^n)$, $h(x) = x + \ldots$, <u>such that</u>:

$$f \circ h = h \circ A.$$

<u>Proof</u>: Let $A \in GL(n,k)$ be a matrix satisfying condition $*$, we define for $p \geq 2$ a linear operator L_A^p from the space Q^p of homogeneous polynomials of degree p, in n-variables and with values in k^n, into itself:

$$L_A^p(\varphi) = \varphi \circ A - A\varphi.$$

<u>Lemma 1</u>: L_A^p <u>is invertible</u>.

<u>Proof</u>: By replacing k, if necessary, by a finite field extension, we can suppose that A is upper triangular. We define an ordered basis of Q^p by:

for $1 \le j \le n$, $i = (i_1,\dots,i_n) \in \mathbb{N}^r$, $|i| = \Sigma i_k = p$

$$e_{i,j} = \begin{pmatrix} 0 \\ \vdots \\ x^i \\ \vdots \\ 0 \end{pmatrix} \leftarrow j^{th} \text{ row}, \quad x^i = x_1^{i_1} \dots x_n^{i_n},$$

the order being defined by $(i,j) > (i',j')$ if $j > j'$ or $j = j'$, $i > i'$ in lexicographic order. In this basis, the operator L_A^p has upper triangular matrix, with (i,j) eigenvalue $\lambda^i - \lambda_j$. ■

<u>Remark 1</u>: If A is a diagonal matrix, so is L_A^p (in the chosen basis).

<u>Proof of proposition</u>: We write $h(x) = x + \sum_{p\ge2} h_p(x)$, $h_p \in Q^p$.

We solve $f\circ h = h\circ A$ by induction on p: for $p \ge 2$, $L_A^p(h_p)$ = terms depending on f, $h_1,\dots,h_{p-1}$.

Uniqueness follows from the proof of existence. ■

<u>Remark 2</u>: If the linear part A of $f \in G^F(k^n)$ is diagonal, but does not satisfy condition $*$, the same method allows to obtain for f a (non unique) normal form in $G^F(k^n)$:

$$N(x) = Ax + \sum_{\substack{i\in\mathbb{N}^r \\ |i|\ge2}} b_i x^i ,$$

$$b_i = \begin{pmatrix} b_i^1 \\ \vdots \\ b_i^n \end{pmatrix}, \qquad b_i^j \neq 0 \Rightarrow \lambda_j = \lambda^i.$$

<u>Example</u>: For $n = 1$, and $\lambda \in k$, $\lambda^q = 1$, $q \in \mathbb{N}^*$, let $f(x) = \lambda x + x^2 \in k[[x]]$. Then f is not formally conjugate to its linear part. For otherwise, f^q would be the identity in $G^F(k)$; but it is a polynomial of degree 2^q. (It has a normal form $\lambda x + x \sum_{k\ge1} b_k x^{kq}$.)

1.3 Conjugation of formal vector fields.

We consider $X^F(k^n) = (k[[X_1,\dots,X_n]])^n$, the space of formal vector fields at 0. The group $G^F(k^n)$ acts on $X^F(k^n)$:

$$(h,X) \to Dh\circ h^{-1} \cdot X\circ h^{-1} = h_*(X).$$

We suppose for the following lemma that k has 0 characteristic.

Lemma 2: Let $X \in X^F(k^n)$ be a vector field with no constant term: $X = Ax + \dots$. Then there exists a unique $f_t(x)$ such that $(f_t(x),t) \in G^F(k^{n+1})$ with $(x,t) = (x_1,\dots,x_n,t) \in k^{n+1}$ and satisfying formally:

$$\frac{\partial f_t}{\partial t}(x) = X\circ f_t(x), \quad f_o(x) = x.$$

Moreover, $f_t(x)$ has the form:

$$f_t(x) = A_t x + B_t x^2 + \dots , \quad A_t = \exp(tA) = \sum_{n\geq 0} \frac{t^n}{n!} A^n$$

is a formal power series in t, as well as B_t ..., and $B_t|_{t=0} = 0$.

For a proof, see [10].

Under the conditions of the lemma, if $h \in G^F(k^n)$, consider $g_t(x) = h\circ f_t\circ h^{-1}(x)$; we have $\frac{\partial g_t}{\partial t}(x) = (h_*X)(g_t(x))$. It follows that, in a certain sense, the action of $G^F(k^n)$ on $X^F(k^n)$ is the adjoint action of $G^F(k^n)$.

Normal forms of vector fields.

Let k have 0 characteristic, $X \in X^F(k^n)$ satisfying $X(0) = 0$, $X(x) = Ax + \dots$. We say that a matrix $A \in GL(n,k)$ satisfies the condition ** if the eigenvalues $\lambda_1,\dots,\lambda_n$ of A in an algebraic closure of k satisfy no relation:

$$\sum_{j=1}^{n} k_j\lambda_j = \lambda_i$$

with $k_j \in \mathbb{N}$, $\sum_{j=1}^{n} k_j \geq 2$, $1 \leq i \leq n$.

Under the preceding conditions, the following proposition holds:

Proposition 2: Given $X \in X^F(k^n)$ such that its linear part A satisfy condition **, there exists a unique $h \in G^F(k^n)$, $h(x) = x+\ldots$, such that $(h_*X)(x) = Ax$.

The proof is similar to the proof of Proposition 1.

Remark 3: If the condition ** is not fulfilled, there exist non unique normal forms analogous to remark 2.

This could be of some interest when k has non-zero characteristic.

2. Diophantine approximations in k^n.

2.1 We suppose that k is a commutative valued field, with absolute value $|\ |_k : k \to R_+$, $|xy|_k = |x|_k \cdot |y|_k$, $|x+y|_k \leq |x|_k + |y|_k$, $|0|_k = 0$, $|1|_k = 1$.
If $|x+y|_k \leq \mathrm{Max}(|x|_k, |y|_k)$, we say that $(k, |\ |_k)$ is non archimedean. In the following, we suppose that the norm is non trivial: there exists $x \in k$ with $|x|_k \neq 0,1$.

2.2 Let $\lambda_1,\ldots,\lambda_n \in k^*$. We say that $(\lambda_1,\ldots,\lambda_n)$ satisfies a diophantine condition (condition (C)) if there exists a strictly positive real number C and a real number β, such that

$$|\lambda_1^{i_1} \ldots \lambda_n^{i_n} - \lambda_j|_k \geq C|i|^{-\beta}, \tag{C}$$

for any $1 \leq j \leq n$, $(i_1,\ldots,i_n) \in \mathbb{N}^n$ such that $|i| = \Sigma\, i_k \geq 2$.

Completion does not affect condition (C), so we can assume that k is complete. By Gelfand-Mazur's theorem, if k is archimedean, $k = R$ or $k = C$; these cases being well-known, we suppose in the following that k is non-archimedean.

2.3 Case $n = 1$.

If $|\lambda|_k \neq 1$, λ satisfies condition (C) (with $\beta = 0$). Suppose now that $|\lambda|_k = 1$.

Lemma 3. If k is locally compact, there exists a strictly increasing sequence of integers such that $\lim_{i\to+\infty} \lambda^{n_i} = 1$.

Proof: We can suppose that λ is not a root of unity. The group $G = \{\lambda^n \mid n \in \mathbb{Z}\}$ is infinite and included in the unit circle which is compact. So 1 is a cluster point of G. ■

So condition (C), $|\lambda^n - 1|_k \geq C|n|^{\beta}$, $n \in \mathbb{N}^*$ cannot be satisfied with $\beta = 0$ when k is locally compact and $|\lambda|_k = 1$.

1) Car $k = 0$.

Lemma 4. If $\lambda - 1$ is small enough, $\lambda \neq 1$, λ satisfies condition (C).

Proof: For $|x|_k$ small enough, $\exp x$ and $\mathrm{Log}(1+x)$ are defined and satisfy:

$|\mathrm{Log}(1+x)|_k = |x|_k$, $|\exp(x)-1|_k = |x|_k$, $\exp(\mathrm{Log}(1+x))-1 = \mathrm{Log}(\exp(x)) = x$.

Write $\lambda = \exp\alpha$, $\alpha = \mathrm{Log}\,\lambda$; we have:

$$|\lambda^n - 1|_k = |\exp(n\alpha)-1|_k = |n\alpha|_k = |n|_k\,|\alpha|_k.$$

But, for any p-adic norm on $\mathbb{Q}$, $|n|_p \geq |n|^{-1}$, and we have from Ostrowski's theorem $|x|_k = |x|_p^{\gamma}$ for some prime p, $\gamma \in \mathbb{R}_+$, and any $x \in \mathbb{Q}$. □

Corollary. If $\lambda \in k^*$ is not a root of unity, λ satisfies condition (C).

Proof: If 1 is not a cluster point of $\{\lambda^n \mid n \geq 1\}$, λ satisfies condition (C) with $\beta = 0$. Otherwise, we find $r \in \mathbb{N}$, $r \geq 1$, such that lemma 2 holds for λ^r, and $|\lambda^s - 1|_k > |\lambda^r - 1|_k$ for $1 \leq s < r$. We can write $\lambda^r = \exp\alpha$, so that, for $i \in \mathbb{N}$, $i \geq 1$:

$$|\lambda^{ir} - 1|_k = |\exp(i\alpha)-1|_k = |i\alpha|_k = |i|_k\,|\alpha|_k \leq |\alpha|_k.$$

We write any integer $m \geq 1$, $m = ir + s$, $i \in N$, $0 \leq s < r$.

a) If $s \neq 0$, we have:

$$|\lambda^{ir}-1|_k \leq |\lambda^r-1|_k < |\lambda^s-1|_k .$$

This implies:

$$|\lambda^m-1|_k = |\lambda^s(\lambda^{ir}-1) + (\lambda^s-1)|_k = |\lambda^s-1|_k.$$

b) If $s = 0$, we have:

$$|\lambda^m-1|_k = |i|_k \; |\lambda^r-1|_k = \left|\frac{\lambda^r-1}{r}\right|_k \; |m|_k ,$$

and we conclude as in Lemma 2. ■

<u>Remark 4</u>: Lemma 4 shows that the roots of unity form a discrete group in k.

2) <u>char $\mathbf{k} = p \neq 0$</u>.

a) If $|\lambda-1|_k < 1$, we have, putting $x = \lambda-1$, for $n \geq 1$:

$$\lambda^{p^n} = (1+x)^{p^n} = 1 + x^{p^n}.$$

Therefore $|\lambda^{p^n}-1|_k = |x|_k^{p^n}$, and λ cannot satisfy condition (C).

b) If k is locally compact, and $|\lambda|_k = 1$, using Lemma 1 we can find $r \in N$, $r \geq 1$, such that $|\lambda^r-1|_k < 1$. We have then, for $r \geq 1$:

$$|\lambda^{r\cdot p^n} - 1|_k = |\lambda^r - 1|_k^{p^n},$$

thus λ cannot satisfy condition (C).

2.4 <u>Case $n \geq 2$</u>.

In this case we only give examples.

a) <u>Poincaré domain</u>

It is the set $\{(\lambda_1,\ldots,\lambda_n),\ |\lambda_i|_k < 1,\ 1 \leq i \leq n\}\ \cup$ $\cup\ \{(\lambda_1,\ldots,\lambda_n),\ |\lambda_i|_k > 1,\ 1 \leq i \leq n\}$.

In this domain, a n-tuple $(\lambda_1,\dots,\lambda_n)$ can only satisfy a finite number of relations, if not, it satisfies the condition (C), with $\beta = 0$.

b) Siegel domain.

It is the complementary in k^n of the Poincaré domain. We study the condition (C'): there exists $C > 0$, $\beta \geq 0$, such that $|\lambda_1^{i_1} \dots \lambda^{i_n} - 1|_k \geq \frac{C}{|i|^\beta}$ for $(i_1,\dots,i_n) \in \mathbb{Z}^n - \{0\}$ with $|i| = |i_1| + \dots + |i_n|$.

It is clear that the condition (C') implies the condition (C). We will give examples when $k = \mathbb{Q}_p$ is the field of p-adic numbers (with p a prime number) with the normalized p-adic absolute value ($|p|_p = \frac{1}{p}$). We denote by $\mathbb{Z}_p$ the ring of p-adic integers.

(1) We suppose that $\lambda_i \in \mathbb{Q}_p$, $|\lambda_i - 1|_p \leq \frac{1}{p}$.

Then $\lambda_i \in \mathbb{Z}_p$ and there exists $\alpha_i \in \mathbb{Z}_p$ such that $\exp \alpha_i = \lambda_i$, so condition (C') reduces to linear approximations: there exists $C > 0$, $\beta \geq 0$ such that

$$|k_1\alpha_1 + \dots + k_n\alpha_n|_p \geq C|k|^{-\beta}$$

for

$$k = (k_1,\dots,k_n) \in \mathbb{Z}^n - \{0\}.$$

We have the following:

Proposition 3: (E. Lutz [4]) *Let μ be the normalized Haar measure on $(\mathbb{Z}_p)^n$. Let $\beta > n$. Then for μ-almost $(\alpha_1,\dots,\alpha_n) \in (\mathbb{Z}_p)^n$:*

$$\sum_{j \in \mathbb{Z}^n - \{0\}} \frac{1}{|j|^\beta |\langle j,\alpha\rangle|_p} < +\infty ,$$

with $\langle j,\alpha\rangle = \sum j_i\alpha_i$.

To prove the proposition, we use the following lemmas.

Lemma 5. *Let $(\alpha_i)_{i\in I}$ be a countable family of real positive numbers*

<u>such that</u> $\sum_{i\in I} \alpha_i < +\infty$. <u>Then, for any</u> $\delta > 1$, $\sum_{i\in I} \alpha_i^{\delta} < +\infty$.

<u>Proof:</u> Obvious. ∎

<u>Lemma 6.</u> <u>Let</u> $\delta > 1$; <u>then</u>

$$\int_{\mathbb{Z}_p} |x|_p^{-1/\delta} \, d\mu(x) < +\infty .$$

<u>Proof:</u> $\mu(p^{\ell}Z_p - p^{\ell+1}Z_p) = \frac{1}{p^{\ell}} - \frac{1}{p^{\ell+1}} = \frac{p-1}{p^{\ell+1}}$. Therefore,

$$\int_{\mathbb{Z}_p} |x|_p^{-1/\delta} \, d\mu(x) \leq \sum_{\ell=0}^{\infty} p^{\ell(\frac{1}{\delta}-1)} \frac{p-1}{p} < +\infty . \quad ∎$$

<u>Lemma 7.</u> <u>Let</u> $\delta > 1$, $j = (j_1,\dots,j_n) \in \mathbb{Z}^n - \{0\}$, <u>then</u>

$$\int_{(\mathbb{Z}_p)^n} |j_1x_1+\dots+j_nx_n|_p^{-1/\delta} \, d\mu(x_1)\dots d\mu(x_n) \leqslant C(\delta) \, [\max_{1\leq i\leq n} |j_i|_p]^{-1},$$

<u>with</u> $C(\delta)$ <u>a constant independent of</u> j.

<u>Proof:</u> Suppose for example that $|j_1|_p \geq |j_i|_p$, $2 \leq i \leq n$. We do the following change of variables:

$$F(x_1,\dots,x_n) = (j_1x_1 +\dots+ j_nx_n, \; x_2,\dots,x_n),$$

then $F((\mathbb{Z}_p)^n) \subset (\mathbb{Z}_p)^n$, $|\det F|_p = |j_1|_p$, therefore:

$$\int_{(\mathbb{Z}_p)^n} |j_1x_1+\dots+j_nx_n|_p^{-1/\delta} \, d\mu(x_1)\dots d\mu(x_n) \leq$$

$$\int_{(\mathbb{Z}_p)^n} |x_1|_p^{-1/\delta} \, |j_1|_p^{-1} \, d\mu(x_1)\dots d\mu(x_n) = C(\delta) \, |j_1|_p^{-1}. \quad ∎$$

<u>Proof of Proposition 3:</u> Let $\beta > n$, choose δ such that $1 < \delta < \frac{\beta}{n}$. By Lemma 5 and Lebesgue dominated convergence theorem, it is enough to prove that

$$\sum_{\mathbb{Z}^n-\{0\}} \int_{(\mathbb{Z}_p)^n} \frac{d\mu(\alpha)}{|j|^{\beta/\delta} \, |\langle j,\alpha\rangle|_p^{1/\delta}} < +\infty .$$

We estimate the partial sum

$$S_r = \sum_{p^r\mathbb{Z}^n - p^{r+1}\mathbb{Z}^n} \int_{(\mathbb{Z}_p)^n} |j|^{-\beta/\delta} \, |\langle j,\alpha\rangle|_p^{-1/\delta} \, d\mu(\alpha),$$

with $r \in \mathbb{N}$. By Lemma 7 we have:

$$S_r \le \sum_{p^r\mathbb{Z}^n - p^{r+1}\mathbb{Z}^n} |j|^{-\beta/\delta}\, C(\delta) p^r$$

$$\le C(\delta) \sum_{j' \in \mathbb{Z}^n - \{0\}} |j'|^{-\beta/\delta}\, p^{r(1-\frac{\beta}{\delta})}$$

$$\le C'(\beta,\delta)\, p^{r(1-\frac{\beta}{\delta})}, \quad \text{since } \frac{\beta}{\delta} > n.$$

Therefore $\sum_{r=0}^{+\infty} S_r$ is finite and the proposition is proved. ■

Let us show, for $n \ge 2$, examples where (C) is not satisfied, but the condition $*$ of §1 is satisfied. First we prove the following proposition

Proposition 4. *Let* $\beta \in \mathbb{Z}_p$, $|\beta|_p = 1$. *Let* φ *be a strictly positive function:* $\mathbb{N} \to \mathbb{R}_+$ *with* $\varphi(n) \to +\infty$ *as* $n \to +\infty$. *Then for a dense* G_δ *of* α's *in* $\mathbb{Z}_p$ *we have:*

$$\psi(\alpha) = \operatorname*{Inf}_{n\in\mathbb{N}} |\alpha - n\beta|_p\, \varphi(n) = 0.$$

We leave to the reader the following lemma.

Lemma 8. *Let* $\beta \in \mathbb{Z}_p$, $|\beta|_p = 1$, *then* $\{n\beta,\ n \in \mathbb{N}\}$ *is dense in* $\mathbb{Z}_p$.

Proof of the proposition 4: ψ is upper semi-continuous on $\mathbb{Z}_p$, so $\psi^{-1}(0)$ is a G_δ; it contains $\{n\beta,\ n \in \mathbb{N}\}$, thus is dense. ■

We have also the following lemma:

Lemma 9. *Let* $\beta \in \mathbb{Z}_p$; *for a dense* G_δ *of* α's *in* $\mathbb{Z}_p$, *there is no relation* $n_1\alpha + n_2\beta = 0$ *with* $(n_1,n_2) \in \mathbb{Z}^2 - \{0\}$.

To construct $(\lambda_1,\lambda_2) \in (\mathbb{Z}_p)^2$, satisfying condition $*$ but not condition (C), we choose $\alpha,\beta \in \mathbb{Z}_p$ rationally independent such that $\inf_{n\in\mathbb{N}} |\alpha - n\beta|\varphi(n) = 0$ and $\varphi(n) \ge \exp(n)$. Then $\lambda_1 = \exp(p\alpha)$, $\lambda_2 = \exp(p\beta)$ is a convenient choice.

Moreover we have:

Proposition 5. *There exists* λ_1,λ_2 *in* $\mathbb{Z}_p$ *satisfying condition* $*$

<u>and</u> $\theta(z_2) = \sum_{k\geq 2} a_k z_2^k$ <u>an entire function such that the diffeomorphism of</u> $(\mathbb{Q}_p)^2$ <u>given by</u>:

$$(z_1,z_2) \to (\lambda_1 z_1 + \theta(z_2), \lambda_2 z_2)$$

<u>is not</u> $\mathbb{Q}_p$<u>-analytically conjugate to its linear part</u>.

<u>Proof</u>. By proposition 1, the formal conjugacy is unique (up to its linear part) and given by $H(z_1,z_2) = (z_1 + \eta(z_2), z_2)$, with $\eta(\lambda_2 z_2) - \lambda_1 \eta(z_2) = \theta(z_2)$, and therefore we have formally:

$$\eta(z_2) = \sum_{k\geq 2} \frac{a_k z_2^k}{\lambda_2^k - \lambda_1}.$$

We choose, $\theta(z_2) = \sum_{k\geq 2} a_k z_2^k$ any entire function with $a_k \neq 0$, for $k \geq 2$, $\mathrm{Log}\,\lambda_1 = \exp(p\alpha)$, $\lambda_2 = \exp(p\beta)$, $\alpha,\beta \in \mathbb{Z}_p$ rationally independent and satisfying proposition 4 with $\varphi(n) = |a_n|_p^{-2}$ for $n \geq 2$. ■

② Let us give hyperbolic examples in the Siegel domain for $k = \mathbb{Q}_p$.

(a) <u>n = 2</u>. Let $\lambda \in \mathbb{Q}_p$, $|\lambda|_p > 1$, and $\alpha \neq 0$, $|\alpha|_p \leq \frac{1}{p}$.

The matrix $A = \begin{pmatrix} \lambda & 0 \\ 0 & \lambda^{-1}\exp(\alpha) \end{pmatrix}$ satisfies condition (C) with $\beta = 1$.

(b) <u>n = 3</u>. Let $\varphi(n) = e^n$; from proposition 4 and Lemma 9, we can find α and β such that $|\alpha|_p \leq 1/p$, $|\beta|_p \leq 1/p$, α and β are rationally independent, but $\inf_{n\in\mathbb{N}} \varphi(n)|\alpha - n\beta|_p = 0$. Take $\lambda_1 = \lambda^{-1}\exp\alpha$, $\lambda_2 = \lambda\exp(-\beta)$, $\lambda_3 = \lambda^{-1}$, $A = \mathrm{diag}(\lambda_1,\lambda_2,\lambda_3)$. Then A satisfies condition $*$. However, for $n \in \mathbb{N}$:

$$|\lambda_1\lambda_2^n\lambda_3^n - \lambda_3|_p = |\lambda^{-1}|_p \, |\exp(\alpha - n\beta)|_p .$$

Thus, A cannot satisfy condition (C).

3. <u>Analytic functions</u>.

3.1 We suppose that k is a commutative complete valued field with non trivial absolute value $|\ \ |$. As the case of $\mathbb{R}$ or $\mathbb{C}$ is well known, we suppose that k is non archimedean and recall the main facts of the theory of analytic functions on such fields [1], [7].

For $n \geq 1$, we put on k^n the norm $|(x_1,\dots,x_n)| = \sup_{1\leq i\leq n} |x_i|$.

3.2 Definition. Let $r > 0$. We define:

$$A_r(k^n) = \{f \in \sum_{j\in\mathbb{N}^n} a_j x^j \in k[[x_1,\dots,x_n]],\ \sup|a_j|r^{|j|} = \|f\|_r < +\infty\}.$$

Proposition 6. $(A_r(k^n), \|\ \|_r)$ is a non-archimedean Banach k-algebra, i.e. $\|f+g\|_r \leq \mathrm{Max}(\|f\|_r, \|g\|_r)$.

Proof: See [1], [7].

Remark 5. For $f \in A_r(k^n)$, f defines a function $\{|x| < r\} \to k$, which satisfies $|f(x)| \leq \|f\|_r$.

Lemma 10 (Cauchy's formula). Let $f \in A_r(k^n)$. Then, for $1 \leq i \leq n$, the formal partial derivative $\frac{\partial f}{\partial x_i}$ belongs to $A_r(k^n)$, and satisfies:

$$\left\|\frac{\partial f}{\partial x_i}\right\|_r \leq \frac{\|f\|_r}{r}.$$

Proof: Trivial. ■

3.3 The space of n-tuples $f = (f_1,\dots,f_n) \in (A_r(k^n))^n$ is a non-archimedean k-Banach space with the norm $\|f\|_r = \sup_i \|f_i\|_r$. It behaves well with respect to composition:

Lemma 11. Let $r,s > 0$, $g \in (A_r(k^n))^n$, $f \in (A_s(k^n))^n$, and suppose that $\|f\|_s < r$. Then, the composite $g\circ f$ is defined, belongs to $(A_s(k^n))^n$, and satisfies $\|g\circ f\|_s \leq \|g\|_r$.

Proof: Let E a non archimedean Banach space, $(x_i)_{i\in I}$ a family of elements in E. Then, the family $(x_i)_{i\in I}$ is summable if and only if 0 is the limit point of the family for the filter of sets $J \subset I$ such that $I - J$ is finite.

Now it is sufficient, if $g = (g_1,\dots,g_n)$, to define $g_i\circ f$ for $1 \leq i \leq n$, and to prove it belongs to $A_s(k^n)$ and satisfies

$\|g_i\circ f\|_s \le \|g_i\|_r$. Let $1 \le i \le n$, $g_i = \sum_{\ell\in\mathbb{N}^n} a_\ell x^\ell$, and consider the family $(a_\ell f^\ell)_{\ell\in\mathbb{N}^n}$ in the Banach k-algebra $A_s(k^n)$. We have the following estimate:

$$\|a_\ell f^\ell\|_s \le \|f\|_s^{|\ell|}\, |a_\ell| \le (\|f\|_s\cdot r^{-1})^{|\ell|}\, \|g\|_r .$$

Therefore this family is summable in $A_s(k^n)$; its sum is defined to be $g_i\circ f$, and satisfies:

$$\|g_i\circ f\|_s \le \sup_{\ell\in\mathbb{N}^n} (\|a_\ell f^\ell\|_s) \le \|g\|_r . \qquad \blacksquare$$

Remark 6. It follows from remark 5 that the function $\{|x| < s\} \to k^n$ associated with the composite $g\circ f$ is the composite of the functions associated with g and f.

3.4 We put on $\mathrm{End}_k(k^n)$ the Banach algebra norm $\|(a_{ij})\| = \sup|a_{ij}|$. We define, for $r > 0$, $A_r(k^n, \mathrm{End}(k^n)) = \{f = \sum_{\ell\in\mathbb{N}^n} a_\ell x^\ell,\ a_\ell \in \mathrm{End}(k^n),\ \sup_{\ell\in\mathbb{N}^n} \|a_\ell\| r^{|\ell|} = \|f\|_r < +\infty\}$.

Lemma 12. With the matrix multiplication, $A_r(k^n, \mathrm{End}(k^n))$ is a k-Banach algebra. If $f \in (A_s(k^n))^n$ satisfies $\|f\|_s < r$, and $A \in A_r(k^n, \mathrm{End}(k^n))$, the composite $A\circ f$ can be defined as in 3.3, belongs to $A_s(k^n, \mathrm{End}(k^n))$ and satisfies $\|A\circ f\|_s \le \|A\|_r$.

Proof: Trivial. $\blacksquare$

Lemma 13. Let $\varphi \in A_r(k^n, \mathrm{End}(k^n))$, such that $\|\varphi\|_r < 1$. Then $A = \mathrm{Id} + \varphi$ is invertible in $A_r(k^n, \mathrm{End}(k^n))$, and its inverse satisfies $\|A^{-1}-\mathrm{Id}\|_r = \|\varphi\|_r$.

Proof: The result is true in any non archimedean Banach algebra, and follows from the identity $(1-x)^{-1} = \sum_{\ell=0}^{+\infty} x^\ell$ for $\|x\| < 1$. $\blacksquare$

3.5 Proposition 7 (Taylor's formula). Let $r > 0$, $s > 0$, $f \in (A_r(k^n))^n$, $h, \Delta h \in (A_s(k^n))^n$ with $\|h\|_s < r$, $\|\Delta h\|_s < r$. Then the following estimate holds:

$$\|f\circ(h+\Delta h) - f\circ h - Df\circ h\cdot\Delta h\|_s \le \frac{\|f\|_r}{r^2}\|\Delta h\|_s^2 .$$

Proof: Let $f = (f_1 \dots f_n)$, $f_i = \sum_{\ell\in\mathbb{N}^n} a_\ell x^\ell$ for some $1 \le i \le n$.
According to 3.3, $f_i\circ(h+\Delta h)$ is in $A_s(k^n)$ the sum of a family of terms of the form $a_{k+\ell}\, h^k \Delta h^\ell$ $(k,\ell \in \mathbb{N}^n)$.

It is easy to check that the sum of those terms, for which $|\ell| = 0$, is just $f_i\circ h$, and the sum of those for which $|\ell| = 1$, is just $(Df\circ h\cdot\Delta h)_i = Df_i\circ h\cdot\Delta h$. Therefore we must estimate the sum of those terms for which $|\ell| \ge 2$; for such an expression, we have:

$$\|a_{k+\ell} h^k \Delta h^\ell\|_s \le \|\Delta h\|_s^2 \frac{\|f\|_r}{r^{k+\ell}} \cdot \|h\|_s^k \|\Delta h\|_s^{\ell-2}$$

$$\le \|\Delta h\|_s^2 \frac{\|f\|_r}{r^2} \cdot \left(\frac{\mathrm{Max}(\|h\|_s, \|\Delta h\|_s)}{r}\right)^{k+\ell-2} .$$

This implies the estimation of the propostion. ■

With similar (but easier) estimates, we prove the following lemma:

Lemma 14. Let $r,s > 0$, f, h, Δh as above; the following estimate holds:

$$\|f\circ(h+\Delta h) - f\circ h\|_s \le \frac{\|f\|_r}{r}\|\Delta h\|_s .$$

4. Siegel's theorem.

4.1 We suppose that k is a complete non archimedean commutative valued field, with non trivial absolute value $|\ |_k$.

For $r > 0$, we define $G_r^\omega(k^n)$ as the intersection $G^F(k^n) \cap \cap (A_r(k^n))^n$.

Let $f \in G_r^\omega(k^n)$, $A = Df(0)$ its linear part. We suppose in the following that A satisfies the condition (C) defined in 2.2: it is diagonal, and its eigenvalues $\lambda_1 \dots \lambda_n$ are such that, for some $C > 0$ and $\beta \ge 0$:

$$|\lambda_1^{i_1} \dots \lambda_r^{i_r} - \lambda_j|_k \ge \frac{C}{|i|^\beta}$$

for any $1 \le j \le n$ and $i = (i_1 \dots i_n) \in \mathbb{N}^n$ such that $|i| \ge 2$.

Then, the weaker condition $*$ (cf 1.2) is also satisfied by A; therefore, by proposition 1, there exists a unique $h \in G^F(k^n)$ such that $Dh(0) = \mathrm{Id}$ and $f \circ h = h \circ A$.

Theorem 1. Under the above diophantine condition, there exists a positive real number s, depending on r, $\|f\|_r$, C, β, such that the formal series h, satisfying $Dh(0) = \mathrm{Id}$ and $f \circ h = h \circ A$, belongs to $G_s^{\omega}(k^n)$.

For the complex case, see Siegel [12] and Zehnder [15].

Remarks 7: a) From proposition 5 we see that some kind of diophantine condition must be necessary for the conclusion of the theorem to hold.

b) If A is not diagonalizable, the theorem is not true, even if the eigenvalues of A satisfy condition (C). In fact, for $\lambda \in \mathbb{Z}_p$, $|\lambda - 1|_p < 1$, one can construct examples with $A = \begin{pmatrix} \lambda & 1 & 0 \\ 0 & \lambda & 0 \\ 0 & 0 & \lambda \end{pmatrix}$ where the formal conjugacy has zero radius of convergence.

4.2 The linearized equation.

For $r > 0$, we define the spaces $A_r^2(k^n) = \{f \in A_r(k^n) \mid f(0) = 0,\ Df(0) = 0\}$ and $B_r^2(k^n) = \{f \in A_r^2(k^n) \mid f \circ A \in A_r^2(k^n)\}$.

We put on $(B_r^2(k^n))^n$ the max norm:

$$\|f\|_r = \mathrm{Max}(\|f\|_r, \|f \circ A\|_r).$$

A linear operator $L: (B_r^2(k^n))^n \to (A_r^2(k^n))^n$ is defined by:

$$Lw = w \circ A - A \circ w.$$

We are interested in inverting L.

Lemma 15. Let $r > 0$, $g \in (A_r^2(k^n))^n$, L defined as above. There exists a unique formal series w satisfying $w(0) = 0$, $Dw(0) = 0$, $Lw = g$. Moreover, for any $\delta > 0$, $w \in (B_{r-\delta}^2(k^n))^n$ and satisfies:

$$\|w\|_{r-\delta} \le C_1 \frac{\|g\|_r}{\delta^{\beta}} r^{\beta}$$

$$\|Dw\|_{r-\delta} \le C_1 \frac{\|g\|_r}{\delta^\beta} \frac{r^\beta}{r-\delta},$$

$$\|Dw\circ A\|_{r-\delta} \le C_1 \frac{\|g\|_r}{\delta^\beta} \frac{r^\beta}{r-\delta},$$

<u>where</u> C_1 <u>is a constant which depends only on</u> C, β, $\|A\|$.

<u>Proof</u>: Write $g = (g_1 \ldots g_n)$, $w = (w_1 \ldots w_n)$, $g_i = \sum_{|\ell|\ge 2} b^i_\ell x^\ell$, $w_i = \sum_{|\ell|\ge 2} a^i_\ell x^\ell$; direct identification gives the unique formal solution $a^i_\ell = \dfrac{b^i_\ell}{\lambda^\ell - \lambda_i}$.

For convergence, we must estimate $\sup_{i,\ell} (|a^i_\ell|_k |r-\delta|^{|\ell|})$; we have:

$$\sup_{\substack{1\le i\le n\\ |\ell|>2}} (|a^i_\ell|_k |r-\delta|^{|\ell|}) \le C^{-1}\|g\|_r \sup_{|\ell|\ge 2} (|\ell|^\beta (1-\tfrac{\delta}{r})^{|\ell|}).$$

For $0 < a < 1$, the positive real valued function $s(x) = x^\beta a^x$ attains on the positive real line its maximum value at $x_o = -\beta(\mathrm{Log}\, a)^{-1}$, for which:

$$s(x_o) = e^{\beta(\mathrm{Log}\,\beta - 1)}(-\mathrm{Log}\, a)^{-\beta} = K(\beta)(-\mathrm{Log}\, a)^{-\beta}.$$

Taking $a = 1 - \frac{\delta}{r}$, $x = |\ell|$, we obtain:

$$s(x) = |\ell|^\beta (1-\tfrac{\delta}{r})^{|\ell|} \le K(\beta)(-\mathrm{Log}\, a)^{-\beta} < \frac{C}{\delta^\beta} r^\beta,$$

since $-\mathrm{Log}(1-\frac{\delta}{r}) > \frac{\delta}{r}$.

Therefore we obtain that $w \in (A^2_{r-\delta}(k^n))^n$, and

$$\|w\|_{r-\delta} \le C_o \frac{\|g\|_r}{\delta^\beta} r^\beta.$$

The estimate for $w\circ A$ follows from $Lw = g$, and the two other estimates from Cauchy's formula. ■

4.3 <u>Newton's method</u>.

Let $s,\delta > 0$ such that $1/2 \le s-\delta < s \le 1$. We suppose that $f \in (A_1(k^n))^n$, and take $\hat{h} \in (B^2_s(k^n))^n$ such that $\|\hat{h}\|_s < 1/2$, $\|D\hat{h}\circ A\|_s < 1/2$.

Denote the expression $f\circ h - h\circ A$ by $F_f(h)$, where $h(z) = z + \hat{h}(z)$. It belongs to $(A_s(k^n))^n$.

For $\Delta h \in (A^2_{s-\delta}(k^n))^n$, $\|\Delta h\|_{s-\delta} < 1/2$, we estimate $\|F_f(h+\Delta h)\|_{s-\delta}$ in the following way:

$$F_f(h+\Delta h) = F_f(h) + Df\circ h\cdot\Delta h - \Delta h\circ A + [f\circ(h+\Delta h) - f\circ h - Df\circ h\cdot\Delta h].$$

Using $DF_f(h) = Df\circ h\cdot Dh - (Dh\circ A)\cdot A$ and the auxiliary function $E = (Dh)^{-1}\cdot\Delta h$, we obtain:

$$F_f(h+\Delta h) = F_f(h) + [DF_f(h)\cdot E + f\circ(h+\Delta h) - f\circ h - Df\circ h\cdot\Delta h] + (Dh\circ A)(A\cdot E - E\circ A).$$

We estimate the two terms in the bracket by Cauchy's formula and Taylor's formula:

$$\|DF_f(h)E\|_{s-\delta} \le 2\|F_f(h)\|_{s-\delta}\ \|E\|_{s-\delta}\ ;$$

$$\|f\circ(h+\Delta h) - f\circ h - Df\circ h\cdot\Delta h\|_{s-\delta} \le 4\|f\|_1\ \|\Delta h\|^2_{s-\delta}\ .$$

By Lemma 15, we can solve $F_f(h) + (Dh\circ A)(A\cdot E - E\circ A) = 0$ for $E \in (B^2_{s-\delta}(k^n))^n$, with the estimates:

$$\|E\|_{s-\delta} \le C_2\frac{\|F_f(h)\|_s}{\delta^\beta}\|(Dh\circ A)^{-1}\|_s \le C_2\frac{\|F_f(h)\|_s}{\delta^\beta},$$

$$\|DE\|_{s-\delta} \le C_2\frac{\|F_f(h)\|_s}{\delta^\beta},$$

$$\|DE\circ A\|_{s-\delta} \le C_2\frac{\|F_f(h)\|_s}{\delta^\beta}.$$

We have used $\|Dh\circ A-\mathrm{id}\| < 1/2$ and Lemma 13. It follows from $\|Dh\|_s \le 1$, $\|Dh^{-1}\|_s \le 1$, $\|Dh\circ A\|_s \le 1$, $\|(Dh\circ A)^{-1}\|_s \le 1$ that $\Delta h = Dh\cdot E$ belongs to $(B^2_{s-\delta}(k^n))^n$ and satisfy the same estimates as E.

To obtain the initial assumption $\|\Delta h\|_{s-\delta} < 1/2$, we assume that $C_2\frac{\|F_f(h)\|_s}{\delta^\beta} < 1/2$. We can then choose Δh so that:

$$\|F_f(h+\Delta h)\|_{s-\delta} \le \frac{K}{\delta^{2\beta}}\|F_f(h)\|^2_s\ ,$$

where $K > C_2^2$ is a constant which depends on C, β, $\|A\|$, $\|f\|_1$.

Moreover, $h + \Delta h$ satisfies, with respect to $s-\delta$, the same hypothesis that h with respect to s.

4.4 Iteration process.

Let $s_o = 1$, $s_n = \frac{1}{2} + \frac{1}{2^{n+1}}$, $s_\infty = \frac{1}{2}$, $\delta_n = s_n - s_{n+1} = \frac{1}{2^{n+2}}$. Replacing, if necessary, f by $tf(t^{-1}z)$ ($t \in k$, $|t|_k$ large), we can suppose that $f \in (A_1(k^n))^n$, and that $\|f-A\|_1 = \varepsilon$ is as small as we want.

Suppose that, for some $p \geq 0$, we can find h_p satisfying the hypothesis of 4.3 and the estimate:

$$\|F_f(h_p)\|_{s_p} \leq \varepsilon 2^{-2\beta p}$$

For $\varepsilon < (C_2\ 2^{2\beta+1})^{-1}$, this implies:

$$C_2 \frac{\|F_f(h_p)\|_{s_p}}{\delta_p^\beta} < 1/2.$$

Therefore we can carry out the construction of 4.3 to obtain $\Delta h_p \in (B^2_{s_{p+1}}(k^n))^n$, such that $h_{p+1} = h_p + \Delta h_p$ satisfy the hypothesis of 4.3, and the estimate:

$$\|F_f(h_{p+1})\|_{s_{p+1}} \leq K\ 2^{2(p+2)\beta}\ (\varepsilon 2^{-2\beta p})^2 \leq \varepsilon e^{-2\beta(p+1)}[\varepsilon K 2^{6\beta}].$$

If we assume that $\varepsilon K 2^{6\beta} < 1$, i.e. take ε small enough, the induction step has been carried out.

For $p = 0$, $h_o = id$, the induction hypothesis is just the definition of ε; therefore we can construct a sequence h_p satisfying the hypothesis of 4.3 and:

$$\|F_f(h_p)\|_{s_\infty} \leq \varepsilon 2^{-2\beta p}.$$

Moreover, from 4.3 we have the estimates:

$$\|\Delta h_p\|_{s_\infty} \leq C_2 \frac{\|F_f(h_p)\|_{s_p}}{\delta_p^\beta} \leq 2^{-p\beta}.$$

Therefore the sequence h_p converges in $(A_{s_\infty}(k^n))^n$ to a limit h

which satisfy $h(0) = 0$, $Dh(0) = \mathrm{id}$. The continuity of F_f follows from Lemma 14; we can therefore conclude that $F_f(h) = 0$; this ends the proof of the theorem. ■

5. Differential equations.

We suppose in the following that k is a commutative field of characteristic 0, complete for a non discrete non archimedean absolute value $|\ |_k$.

5.1 Existence of solutions of differential equations.

Proposition 8. *Let* $\delta > 0$, $X \in (A_\delta(k^n))^n$. *There exists* $\delta_1 > 0$ *and a unique* $f_t(x) \in (A_{\delta_1}(k \times k^n))^n$ *satisfying for* $|t|_k < \delta_1$, $x \in k^n$, $|x|_k < \delta_1$:

$$\frac{\partial f_t}{\partial t}(x) = X \circ f_t(x), \qquad f_t(f_{t'}(x)) = f_{t+t'}(x), \qquad f_o(x) = x.$$

We begin by the following lemma.

Lemma 16. *Let* A *be an algebra containing* $\mathbb{Q}$; *for* $\tilde{X} \in (A[[x_1,\dots,x_n]])^n$, $\tilde{X} = \sum_{r \in \mathbb{N}^n} a_r x^r$, *consider the equation in* $v \in (A[[t]])^n$, $v = \sum_{r \geq 1} \frac{c_r}{r!} t^r$:

$$\tilde{X}(v(t)) = \frac{\partial v}{\partial t}(t)$$

It has a unique solution v; *the coefficients* $c_r \in A^n$ *are given by induction formulas* $c_{r+1} = Q_{r+1}(c_o,\dots,c_r,a_i)$, $|i| \leq r$ Q_{r+1} *being a polynomial with integral coefficients.*

Proof: The equation is:

$$\sum_{r \geq 0} c_{r+1} \frac{t^r}{r!} = \sum_{\ell \in \mathbb{N}^n} a_\ell \Big(\sum_{r \geq 1} c_r \frac{t^r}{r!} \Big)^\ell .$$

It gives $c_1 = a_o$, and c_{r+1} as a polynomial in $c_1,\dots,c_r$ and those a_i for which $|i| \leq r$. The coefficients are sums of terms of the form $\frac{r!}{p_1!\ldots p_s!}$, with $p_1+\ldots+p_s = r$, and thus are integers. ∎

<u>Proof of proposition</u>: We first remark that if f_t is a solution for X, f_{rt} is a solution for rX and $\frac{1}{r} f_t(rx)$ is a solution for $X(rx)$ $(r \neq 0)$. Therefore we can assume that X can be written

$$X(x+v) = \sum_{\ell \in \mathbb{N}^r} a_\ell(x) v^\ell \quad \text{with} \quad v \in k^n, \quad |v|_k < \delta_2,$$

$a_\ell(x) \in (A_{\delta_2}(k^n))^n$, $\|a_\ell\|_{\delta_2} < 1$.

Looking for $v_t(x) = f_t(x)-x$, we obtain the differential equation $X(x+v_t(x)) = \frac{\partial v_t}{\partial t}(x)$. Let $v_t(x) = \sum_{r \geq 1} c_r(x) \frac{t^r}{r!}$. By the lemma (with $A = k[[x_1 \ldots x_n]]$), we can solve for $c_1(x),\ldots,c_r(x)$, in formal series of x. The coefficients of the polynomials Q_r being integers, we can even show by induction, beginning with $c_1 = a_o$, that $c_r \in (A_{\delta_2}(k^n))^n$ and $\|c_r\|_{\delta_2} < 1$ for any $r > 0$.

We may assume, by Ostrowski's theorem, that the absolute value $|\ |_k$ on $\mathbb{Q}$ is some p-adic norm. Then $|\frac{1}{n!}|_k \leq p^{\frac{n}{p-1}}$, therefore we conclude that $v_t(x)$ are convergent for $|t|_k < \delta_1$, $|x|_k < \delta_1$. By construction, $f_o(x) = x$, $\frac{\partial f_t}{\partial t}(x) = X(f_t(x))$. The lemma gives the unicity, which is already valid in formal series. To check that $f_t \circ f_{t'} = f_{t+t'}$, we fix $t = t_o$ and consider the equation in $v_t(x) = \sum_{\ell \geq 1} c_\ell(x) \frac{t^\ell}{\ell!}$:

$$X(f_{t_o}(x) + v_t(x)) = \frac{\partial v_t}{\partial t}(x), \qquad v_0(x) = 0.$$

From above, we know two solutions $v_t^1 = f_{t+t_o} - f_{t_o}$, $v_t^2 = f_t \circ f_{t_o} - f_{t_o}$. Taking into the lemma $\tilde{X} = X(f_{t_o}(x)+v) \in (A[[v_1 \ldots v_n]])^n$, $A = k[[x_1 \ldots x_n]]$, we see that v_t^1 and v_t^2 must be formally equal, and we conclude that $f_{t+t_o} = f_t \circ f_{t_o}$. ∎

5.2 <u>Remark 8</u>. Let $X \in (A_\delta(k^n))^n$ be a vector field, $X(x) = Ax + \ldots$, then $f_t(x) = \exp(tA)x + \ldots$ $(t,x) \in k \times k^n$, $|t|_k < \delta_1$, $|x|_k < \delta_1$. Then A satisfies condition ** iff $\exp(tA)$ satisfies condition * for $0 < |t| < \delta_o$; furthermore A is diagonal iff $\exp(tA)$ is diagonal.

5.3 For $\lambda_1,\ldots,\lambda_n \in k$, we say that $(\lambda_1\ldots\lambda_n)$ (or the matrix $\operatorname{diag}(\lambda_1,\ldots,\lambda_n)$) satisfies the condition (C'') if there exists $C > 0$, $\beta \geq 0$ such that:

$$|i_1\lambda_1 + \ldots + i_n\lambda_n - \lambda_j|_k \geq C|i|^{-\beta}$$

for any $1 \leq j \leq n$, $(i_1\ldots i_n) \in \mathbb{N}^n$ such that $|i| = \Sigma\, i_\ell \geq 2$.

Let A be a diagonal matrix; it satisfies condition (C'') iff there exists $\delta_o > 0$ such that $\exp(tA)$ satisfies condition (C) for $0 < |t| < \delta_o$.

5.4 <u>Theorem 2</u>. <u>Let</u> $X = Ax + \ldots$ <u>be a vector field in</u> $(A_\delta(k^n))^n$, $\delta > 0$, <u>with</u> A <u>a diagonal matrix satisfying condition</u> (C''). <u>There exists</u> $s > 0$ <u>and a unique</u> $h \in G_s^\omega(k^n)$ <u>with</u> $Dh(0) = id$ <u>such that</u>:

$$h_*X(x) = Dh\circ h^{-1}(x)\cdot X(h^{-1}(x)) = Ax.$$

For the complex case, see Siegel [13].

<u>Proof</u>: Let $f_t(x)$ be the solution of the differential equation $\frac{\partial f_t}{\partial t}(x) = X(f_t(x))$, $f_o(x) = x$. By 1.3, there exists a formal $h_1(x) = x + \ldots \in G^F(k^n)$ satisfying $h_1^*(X)(x) = Ax$, $h_1\circ f_t\circ h_1^{-1} = \exp(tA)\cdot x$, the last equality holding between formal series of (n+1) variables. From 5.1, we obtain the convergence in t, therefore for $|t_o| \neq 0$ small, we have the equality $h_1\circ f_{t_o}\circ h_1^{-1} = \exp(tA)\cdot x$ in $G^F(k^n)$. By Siegel's theorem, h_1 is in fact in some $G_s^\omega(k^n)$ and the convergence is proved. ∎

Appendix: The stable manifold theorem.

Let k be a commutative field, complete for a non archimedean non trivial absolute value $|\ |_k$. The notations being those of I.3, let $f \in G_r^{\omega}(k^p)$ with $p \in \mathbb{N}$, $p \geq 1$, and $r > 0$. Assume that $Df(0)$ is of the form $\begin{pmatrix} A & 0 \\ 0 & B \end{pmatrix}$, with $A \in GL(n,k)$, $B \in GL(m,k)$, $m+n = p$ and $\|A\| < 1$, $\|B^{-1}\| < 1$ for the max norm on matrices; thus we can write, for $(x,y) \in k^n \times k^m = k^p$:

$$f(x,y) = (f_1(x,y),\ f_2(x,y)) = (Ax+g_1(x,y),\ By+g_2(x,y))$$

with $g_1 \in (A_r^2(k^p))^n$, $g_2 \in (A_r^2(k^p))^m$.

Theorem A1. For some neighborhood N of the origin in k^p, the set $W^s = \{u \in N \mid f^q(u) \in N \text{ for } q > 0\}$ is the intersection with N of the graph of a k-analytic function $h \in (A^2(k^n))^m$ and thus a n-dimensional f-invariant analytic manifold. Moreover, for $u \in W^s$, $f^q(u) \to 0$ as $q \to +\infty$.

Proof: The proof is similar to [Me].

The equation $h(f_1(x,h(x))) = f_2(x,h(x))$ of the f-invariance of the graph of h is equivalent to:

$$Bh - h(Ax + g_1(x,h(x))) + g_2(x,h(x)) = 0$$

This means that h is a fixed point of the map Φ defined if possible by:

$$\Phi(h)(x) = B^{-1}(h(Ax + g_1(x,h(x))) - g_2(x,h(x))).$$

For $0 < \delta < r$, define $B_\delta = \{h \in (A_r^2(k^n))^m,\ \|h\|_\delta \leq \delta\}$.

Lemma A2. For δ sufficiently small, Φ is well-defined on B_δ, sends B_δ into B_δ and is a contraction on B_δ.

Proof: For $0 < \delta < r$, we have, as g_1 and g_2 begin by quadratic terms, $\|g_1\|_\delta < \frac{\delta^2}{r^2}\|g_1\|_r$, $\|g_2\|_\delta < \frac{\delta^2}{r^2}\|g_2\|_r$. For $h \in B_\delta$, let $H = (\mathrm{id},\ h) \in (A_\delta(k^n))^p$; then $\|H\|_\delta \leq \delta$, thus by lemma 11, $\|g_i \circ H\|_\delta \leq \frac{\delta^2}{r^2}\|g_i\|_r$, $i = 1,2$. This implies, for δ sufficiently

small, $\|A + g_1 \circ h\|_\delta = \delta \|A\|$, the norm in the right term being the matrix norm. As $\|A\| < 1$, $\Phi\colon B_\delta \to (A_\delta^2(k^n))^m$ is well defined.
We now show that Φ is Lipschitz and estimate its Lipschitz constant. For $h, h' \in B_\delta$, $i = 1,2$, we have by Lemma 14:

$$\|g_i \circ H - g_i \circ H'\|_\delta \le \frac{\|g_i\|_\delta}{\delta} \|H-H'\|_\delta \le \frac{\delta}{r^2} \|g_i\|_r \|h-h'\|_\delta \ ;$$

$$\|h \circ (A+g_1 \circ H) - h' \circ (A+g_1 \circ H')\|_\delta$$
$$\le \operatorname{Max}(\|(h-h') \circ (A+g_1 \circ H')\|_\delta, \|h \circ (A+g_1 \circ H) - h \circ (A+g_1 \circ H')\|_\delta);$$

By Lemma 11, $\|(h-h') \circ (A+g_1 \circ H)\|_\delta \le \|(h-h')\|_\delta$. On the other hand, by Lemma 14:

$$\|h \circ (A+g_1 \circ H) - h \circ (A+g_1 \circ H')\| \le \frac{\|h\|_\delta}{\delta} \|g_1 \circ H - g_1 \circ H'\|_\delta$$
$$\le \frac{\delta}{r^2} \|g_1\|_r \|h-h'\|_\delta \ .$$

Therefore, for δ sufficiently small, we obtain:

$$\|\Phi(h) - \Phi(h')\|_\delta \le \|B^{-1}\| \|h-h'\|_\delta \ .$$

Moreover, we have $\Phi(0) = -B^{-1}(g_2 \circ (\mathrm{id},0))$, therefore $\|\Phi(0)\|_\delta \le$ $\le \frac{\delta^2}{r^2} \|g_2\|_r$. Putting this together with the Lipschitz estimate, we obtain that Φ sends B_δ to B_δ and is a contraction on B_δ. ■

<u>Proof of the theorem</u>: Fix δ sufficiently small for the conclusion of lemma A2 to hold. Then Φ has a unique fixed point h in B_δ. Let $K \in G_\delta^\omega(k^p)$ be defined by $K(x,y) = (x, y - h(x))$; K takes the graph of h on $\{y=0\}$, therefore $K \circ f \circ K^{-1}$ leaves $\{y=0\}$ invariant; therefore, it is of the form $K \circ f \circ K^{-1}(x,y) = (Ax + \tilde{g}_1(x,y), By + \tilde{g}_2(x,y))$ with $\tilde{g}_2(x,0) = 0$. Making now the same construction for $K \circ f^{-1} \circ K^{-1}$, we can conjugate it by some $L \in G_{\delta'}^\omega(k^n)$, $0 < \delta' < \delta$, of the form $L(x,y) = (x - \tilde{h}(y), y)$ so that the resulting diffeomorphism $\tilde{f} = L \circ K \circ f \circ K^{-1} \circ L^{-1} \in G_{\tilde{\delta}}^\omega(k^p)$ for $\tilde{\delta}$ sufficiently small and is of the form $\tilde{f}(x,y) = (Ax + \tilde{\tilde{g}}_1(x,y), By + \tilde{\tilde{g}}_2(x,y))$ with $\tilde{\tilde{g}}_1(0,y) = 0$, $\tilde{\tilde{g}}_2(x,0) = 0$. Let $\theta < 1$ such that $\|A\| < \theta$, $\|B^{-1}\| < \theta$. As $\tilde{\tilde{g}}_1$, $\tilde{\tilde{g}}_2$ begin by second order terms, we deduce from above, for (x,y) in

some sufficiently small neighborhood $\tilde{N}$ of 0 in k^p:

$$|Ax + \tilde{\tilde{g}}_1(x,y)|_k \leq \theta |x|_k$$

$$|By + \tilde{\tilde{g}}_2(x,y)|_k \geq \theta^{-1} |y|_k$$

From this, we see that the following properties for $(x,y) \in \tilde{N}$ are equivalent:

i) $y = 0$.

ii) $\tilde{f}^q(x,y) \in \tilde{N}$ for all $q > 0$.

iii) $\tilde{f}^q(x,y) \to 0$ as $q \to +\infty$.

Terefore we obtain the theorem with $N = K^{-1} \circ L^{-1}(\tilde{N})$. ∎

II. Local conjugacy of diffeomorphisms of the circle.

1.1 In the following, k denotes a complete non archimedean commutative field with a non trivial absolute value $| \ |_k$. Let $S_k = \{x \in k, |x|_k = 1\}$ be the unit circle. We propose to study diffeomorphisms of S_k which are perturbations of rotations, $x \to \alpha x$, $|\alpha|_k = 1$, and of the form $f(x) = \alpha x + \varphi(x)$.

One possible generalization of analytic diffeomorphisms of $S_{\mathbb{C}}$ is that φ is a Laurent series (a generalization of real analytic Fourier series).

1.2 Laurent series.

Let $r > 1$; we define $L_r = \{f(x) = \sum_{n \in \mathbb{Z}} a_n x^n, \ \|f\|_r = \sup_{n \in \mathbb{Z}} |a_n|_k \, r^{|n|} < +\infty\}$.

Lemma 17. L_r is a non archimedean Banach k-algebra, the product of $f = \sum a_n x^n$ and $g = \sum b_n x^n$ being defined as $fg = \sum c_n x^n$ with $c_n = \sum_{\ell \in \mathbb{Z}} a_{n-\ell} \, b_\ell$.

Proof: For $p,q \in \mathbb{Z}$, we have:

$$|a_p b_q|_k \; r^{|p+q|} \le \|f\|_r \; \|g\|_r \; r^{|p+q|-|p|-|q|}.$$

Therefore the series defining c_ℓ converges, its sum satisfying $|c_\ell|_k \le \|f\|_r \; \|g\|_r \; r^{-|\ell|}$. ■

Remarks 9. a) Any $f \in L_r$ defines a function $\{r > |x|_k > 1/r\} \to k$, which satisfies $|f(x)|_k \le \|f\|_r$.

b) If $f \in L_r$, the formal derivative Df of f belongs to L_r, and Cauchy's formula reads $\|Df\|_r \le r\|f\|_r$.

c) Any $f = \Sigma\, a_n x^n \in L_r$ decomposes into $f = f_1 + f_2$ where $f_1 = \sum_{n\ge 0} a_n x^n \in A_r(k)$ and $f_2(\frac{1}{x}) = \sum_{n<0} a_n x^{|n|} \in A_r(k)$.

d) Let f be defined by $f(z) = \alpha z + \varphi(z)$, $|\alpha|_k = 1$, $\varphi \in L_r$, $\|\varphi\|_r < \frac{1}{r}$; then f defines a function from the annulus $\{r > |x|_k > 1/r\}$ into itself.

1.3 Proposition 9. Let $1 < \nu < r$, $\psi \in L_r$, $\varphi \in L_\nu$, with $\|\varphi\|_\nu < 1/\nu$. For $\alpha \in k$, $|\alpha|_k = 1$ define $f(z) = \alpha z + \varphi(z)$; then $\psi \circ f \in L_\nu$ and $\|\psi \circ f\|_\nu \le \|\psi\|_r$.

We first prove the following lemma.

Lemma 18. Under the preceding conditions, f is invertible and $\|\frac{1}{f}\|_\nu \le \nu$.

Proof: We have $\|\frac{\varphi}{\alpha z}\|_\nu < 1$, therefore we can write the following power series in the Banach algebra L_ν:

$$\frac{1}{f(z)} = \frac{1}{\alpha z} \, \frac{1}{1+\frac{\varphi}{\alpha z}} = \frac{1}{\alpha z}\, \Sigma(-1)^i \left(\frac{\varphi}{\alpha z}\right)^i,$$

and the estimate follows. ■

We now prove the proposition and for this we estimate $a_n f^n$ with $\psi = \sum_{n\in\mathbb{Z}} a_n x^n$.

$$\|a_n f^n\|_\nu < \frac{\|\psi\|_r}{r^{|n|}} \operatorname{Max}(\|f\|_\nu, \|\tfrac{1}{f}\|_\nu)^{|n|} \le \|\psi\|_r .$$

This implies the proposition. ■

1.4. Proposition 10. Let $1 < \nu < r^{1/3}$, $\psi \in L_r$, $\varphi, \Delta\varphi \in L_\nu$ with $\|\varphi\|_\nu < 1/\nu$, $\|\Delta\varphi\|_\nu < 1/\nu$. For $\alpha \in k$, $|\alpha|_k = 1$ define $f(z) = \alpha z + \varphi(z) \in L_\nu$. Then the following estimate holds in L_ν:

$$\|\psi\circ(f+\Delta\varphi) - \psi\circ f - D\psi\circ f\,\Delta\varphi\|_\nu \le \|\psi\|_r\,\|\Delta\varphi\|_\nu^2 .$$

Proof: It is sufficient to consider the case $\psi = a_n z^n$. The case $n = 0$ or 1 is trivial.

1) $n \ge 2$:

$$\|a_n(f+\Delta\varphi)^n - a_n f^n - n a_n f^{n-1}\,\Delta\varphi\|_\nu \le \|\Delta\varphi\|_\nu^2\,\frac{\|\psi\|_r}{r^n}\,\nu^{n-2}$$
$$\le \|\Delta\varphi\|_\nu^2\,\frac{\|\psi\|_r}{r^2} .$$

2) $n \le -1$:

$$\left\|\frac{a_n}{(f+\Delta\varphi)^{|n|}} - \frac{a_n}{f^{|n|}} + \frac{|n|a_n}{f^{|n|+1}}\Delta\varphi\right\|_\nu \le$$

$$\le \|f a_n[f^{|n|} - (f+\Delta\varphi)^{|n|}] + (f+\Delta\varphi)^{|n|}\,a_n|n|\Delta\varphi\|_\nu\,\left\|\frac{1}{f}\right\|_\nu^{|n|+1}\left\|\frac{1}{f+\Delta\varphi}\right\|_\nu^{|n|} .$$

As the monomial terms of degree 0 or 1 in $\Delta\varphi$ vanish, the estimate follows from $\|f\|_\nu < \nu$, $\|\frac{1}{f}\|_\nu < \nu$, $\|\frac{1}{f+\Delta\varphi}\|_\nu < \nu$ and $\|\Delta\varphi\|_\nu < \nu$. ■

1.5 Proposition 11. Let $r > 1$, $\psi \in L_r$, $\alpha, \Delta\alpha \in k$, $|\alpha|_k = 1$, $|\Delta\alpha|_k < 1$. The following estimate holds:

$$\|\psi((\alpha+\Delta\alpha)z) - \psi(\alpha z) - D\psi(\alpha z)\Delta\alpha z\|_r \le \|\psi\|_r\,|\Delta\alpha|_k^2 .$$

Proof: It is again sufficient to study only the cases $\psi = a_n x^n$, $n \ge 2$ or $n \le -1$.

Case $n \ge 2$. We have to estimate $(\alpha+\Delta\alpha)^n - \alpha^n - \alpha^{n-1} n\,\Delta\alpha$:

$$|(\alpha+\Delta\alpha)^n - \alpha^n - n\alpha^{n-1}\,\Delta\alpha|_k \le |\Delta\alpha|_k^2 .$$

Case $n < 0$. We have:

$$\left|\frac{1}{(\alpha+\Delta\alpha)^{|n|}} - \frac{1}{\alpha^{|n|}} + \frac{|n|\Delta\alpha}{\alpha^{|n|+1}}\right|_k = \left|\frac{\alpha^{|n|+1} - \alpha(\alpha+\Delta\alpha)^{|n|} + |n|\Delta\alpha(\alpha+\Delta\alpha)^{|n|}}{\alpha^{|n|+1}(\alpha+\Delta\alpha)^{|n|}}\right|_k$$

$$\leq |\Delta\alpha|_k^2 . \quad \blacksquare$$

1.6 Proposition 12. Let $r > 1$, $\varphi \in L_r$ with $\|\varphi\|_r < 1/r$. For any $\alpha \in k$, $|\alpha|_k = 1$, and $1 < \nu < r^{1/3}$, there exists $\psi \in L_\nu$, with $\|\psi\|_\nu < 1/\nu$ such that, if $f = \alpha z + \varphi$, $g = \alpha^{-1}z + \psi$ we have $f\circ g = g\circ f = \mathrm{Id}$.

Proof: It is sufficient to consider the case $\alpha = 1$. Then ψ satisfies the following functional equation:

$$\psi = -\varphi\circ(\mathrm{Id} + \psi).$$

We define $B_\nu = \{\psi \in L_\nu,\ \|\psi\|_\nu \leq \frac{1}{\nu}\}$ and $\Phi\colon B_\nu \to L_\nu$ by $\Phi(\psi) = -\varphi\circ(\mathrm{Id} + \psi)$. It follows from Proposition 8 that Φ is well defined and $\Phi(B_\nu) \subset B_\nu$. By Proposition 9, Φ satisfies the Lipschitz condition:

$$\|\Phi(\psi_1)-\Phi(\psi_2)\|_\nu \leq \mathrm{Max}(\|D\varphi\|_r\|\psi_1-\psi_2\|_\nu, \|\varphi\|_r\|\psi_1-\psi_2\|_\nu^2).$$

But $\|D\varphi\|_r \leq r\|\varphi\|_r < 1$ and $\|\varphi\|_r\|\psi_1-\psi_2\|_\nu < 1$, therefore Φ is a contraction and fixes some ψ in B_ν. $\blacksquare$

2.1 Let us remark that the rotations $R_\alpha\colon z \to \alpha z$, $|\alpha|_k = 1$ preserve S_k. The main theorem is the following that generalizes the theorem of Arnold and Moser [3, App.]

Theorem 3. We suppose that the characteristic of k is 0. Let $\alpha \in k$, $|\alpha|_k = 1$, and not a root of unity, and $r > 1$. There exists $\varepsilon > 0$, $c > 0$, $1 < \nu < r$ depending on α and r such that if $\psi \in L_r$ satisfies $\|\psi\|_r < \varepsilon$, then there exists $\lambda \in k$, $|\lambda|_k < c < 1$, and $\varphi \in L_\nu$, such that $g = \mathrm{id} + \varphi$ satisfies $\|\varphi\|_\nu < 1/\nu$, $\|D\varphi\|_\nu < 1$, and:

$$f\circ g - g\circ R_{\alpha+\lambda} = 0$$

We assume in the following that char $k = 0$, and that $|\ |_k$ is normalized on the prime ring of k.

2.2 Lemma 19. Let $\alpha \in k$ be not a root of unity. There exist constants $B, C > 0$, depending on α, such that for any $\Delta\alpha \in k$, $|\Delta\alpha|_k < C$, and $n \in \mathbb{Z}$, $n \neq 0$, we have:

$$|(\alpha+\Delta\alpha)^n - 1|_k \geq B/|n|.$$

Proof: If $|\alpha|_k \neq 1$, $C = |\alpha|_k$ is a convenient choice. Assume $|\alpha|_k = 1$; if 1 is not adherent to $\{\alpha^n, n \in \mathbb{Z}-\{0\}\}$, taking $C = \inf\{|\alpha^n-1|_k, n \in \mathbb{Z}-\{0\}\}$, we obtain for $n \neq 0$:

$$|(\alpha+\Delta\alpha)^n - \alpha^n|_k \leq |\Delta\alpha|_k < C,$$

$$|(\alpha+\Delta\alpha)^n - 1|_k = |\alpha^n - 1|_k \geq C.$$

Otherwise, we can find $\beta \in k$, $n_o \in \mathbb{N}$ such that $\alpha^{n_o} = \exp\beta$, $|\alpha^n-1|_k > |\alpha^{n_o}-1|_k$ for $1 \leq n < n_o$. Take $C = |\alpha^{n_o}-1|_k$; then for $1 \leq n \leq n_o$, we have:

$$|(\alpha+\Delta\alpha)^n - \alpha^n|_k \leq |\Delta\alpha|_k < C \leq |\alpha^n-1|_k$$

$$\Rightarrow |(\alpha+\Delta\alpha)^n - 1|_k = |\alpha^n-1|_k .$$

Therefore $(\alpha+\Delta\alpha)^{n_o} = \exp\gamma$, for some $\gamma \in k$ with $|\gamma|_k = |\exp\gamma - 1|_k = |\alpha^{n_o} - 1|_k$. For $m \in \mathbb{Z}$, $m = in_o + s$, $i \in \mathbb{Z}$, $0 \leq s < n_o$, we estimate $|(\alpha+\Delta\alpha)^m - 1|_k$:

a) $s = 0$.

$$|\gamma|_k \geq |i\gamma|_k = |(\alpha+\Delta\alpha)^{in_o} - 1|_k = |\gamma|_k \left|\frac{m}{n_o}\right|_k \geq \left|\frac{\gamma}{n_o}\right|_k \frac{1}{|m|}.$$

b) $s \neq 0$.

$$|(\alpha+\Delta\alpha)^{in_o+s} - 1|_k = |(\alpha+\Delta\alpha)^s((\alpha+\Delta\alpha)^{in_o}-1) + (\alpha+\Delta\alpha)^s - 1|_k ;$$

but

$$|(\alpha+\Delta\alpha)^s-1|_k = |\alpha^s-1|_k > |\alpha^{n_o}-1|_k = |\gamma|_k ,$$

$$|(\alpha+\Delta\alpha)^{in_o}-1|_k \leq |\gamma|_k \quad \text{by case a).}$$

Therefore $|(\alpha+\Delta\alpha)^{in_o+s} - 1|_k > |\gamma|_k$, ending the proof of the lemma in the case b). ■

2.3 The linearized equation.

Proposition 13. *Let* $\alpha \in k$, $|\alpha|_k = 1$, *satisfying, for some* $B > 0$ *and any* $n \in \mathbb{Z} - \{0\}$, $|\alpha^n - 1|_k > B/|n|$. *Let* $r > 1$; *then, for any* $\psi \in L_r$, *there exist unique* $\lambda \in k$, $\eta \in \bigcap_{\delta>0} L_{r-\delta}$, *such that:*

$$\eta(\alpha z) - \alpha\eta(z) - \lambda z = \psi(z), \qquad D\eta(0) = 0.$$

Moreover, the following estimates hold:

$$|\lambda|_k \le \frac{\|\psi\|_r}{r}, \quad \|\eta\|_{r-\delta} \le \frac{r}{B\delta}\|\psi\|_r .$$

Proof: Let $\psi = \sum_{\mathbb{Z}} a_n z^n$, then $\lambda = a_1$, $b_n = \dfrac{a_n}{\alpha^n - \alpha}$ for $n \neq 1$, $\eta = \sum_{n\neq 1} b_n z^n$; we have $|\lambda|_k \le \dfrac{\|\psi\|_r}{r}$ and, for $n \neq 1$:

$$|b_n|_k (r-\delta)^{|n|} \le \frac{\|\psi\|_r}{B}\left(1-\frac{\delta}{r}\right)^{|n|}|n-1| .$$

Recalling the proof of Lemma 15, we have $\left(1-\frac{\delta}{r}\right)^{|n|}|n| \le \frac{1}{e}\frac{r}{\delta}$, and this implies, for $n \neq 0,1$:

$$|b_n|_k\,(r-\delta)^{|n|} \le \|\psi\|_r \frac{r}{B\delta}\left|\frac{n-1}{n.e}\right| \le \|\psi\|_r \frac{r}{B\delta} .$$

For $n = 0$, $|b_0|_k = \left|\dfrac{a_0}{\alpha - 1}\right|_k \le \dfrac{\|\psi\|_r}{B} \le \|\psi\|_r \dfrac{rB}{\delta}$.

The estimates of the proposition follow. ■

2.4 For $\lambda \in k$, $|\lambda|_k < 1$ and $g \in L_\nu$, $1 < \nu < r$, $|g - id|_\nu < 1/\nu$, we write $F_f(g,\lambda) = f\circ g - g\circ R_{\alpha+\lambda}$.

For $\Delta g \in L_\nu$, $\Delta\lambda \in k$, $|\Delta g|_\nu < 1/\nu$, $|\Delta\lambda|_k < 1$, we want to estimate $F_f(g+\Delta g, \lambda+\Delta\lambda) - F_f(g,\lambda)$, and to do this we perform the same kind of estimates that in I.4.3; more precisely, we have $F_f(g+\Delta g, \lambda+\Delta\lambda) - F_f(g,\lambda) = L + R_1 - R_2 - R_3$ with:

$$L = Df\circ g\,\Delta g - \Delta g\circ R_{\alpha+\lambda} - (Dg\circ R_{\alpha+\lambda})\cdot\Delta\lambda z ;$$

$$R_1 = f\circ(g+\Delta g) - f\circ g - Df\circ g\cdot\Delta g ;$$

$$R_2 = g\circ R_{\alpha+\lambda+\Delta\lambda} - g\circ R_{\alpha+\lambda} - Dg\circ R_{\alpha+\lambda}\cdot\Delta\lambda z ;$$

$$R_3 = \Delta g\circ R_{\alpha+\lambda+\Delta\lambda} - \Delta g\circ R_{\alpha+\lambda} .$$

On the other hand:

$$DF_f(g,\lambda) = Df\circ g\cdot Dg - (\alpha+\lambda)Dg\circ R_{\alpha+\lambda} .$$

Thus, putting $\Delta g = Dg\cdot E$, we obtain $L = R_o + \Phi_f(g,\lambda)\cdot(E,\Delta\lambda)$, with:

$$R_o = DF_f(g,\lambda)\cdot E$$

$$\Phi_f(g,\lambda)(E,\Delta\lambda) = -Dg\circ R_{\alpha+\lambda}\,[E\circ R_{\alpha+\lambda} - (\alpha+\lambda)E - \Delta\lambda z].$$

2.5 We now suppose that $\alpha \in k$ has absolute value 1 and is not a root of the unity; Lemma 8 determines two constants $B, C > 0$ such that, for $\Delta\alpha \in k$, $|\Delta\alpha|_k < C$, and $n \in \mathbb{Z} - \{0\}$, we have $|(\alpha+\Delta\alpha)^n - 1|_k \geq B/|n|$.

We now assume that $|\lambda|_k < C$; then, using proposition 9, we can solve $\Phi_f(g,\lambda)(E,\Delta\lambda) = -F_f(g,\lambda) = -F$, for $\Delta\lambda \in k$ and E belonging to any $L_{\nu-\delta}$, $0 < \delta < \nu - 1$, with the following estimates:

$$|\Delta\lambda|_k \leq \frac{1}{\nu}\left\|\frac{F}{Dg\circ R_{\alpha+\lambda}}\right\|_\nu ;$$

$$\|E\|_{\nu-\delta} \leq \frac{\nu}{B\delta}\left\|\frac{F}{Dg\circ R_{\alpha+\lambda}}\right\|_\nu .$$

Recalling that $\|g-id\|_\nu < 1/\nu$ implies $\|Dg-1\|_\nu < 1$, $\|Dg\|_\nu = 1$, and $\left\|\frac{1}{Dg\circ R_{\alpha+\lambda}}\right\|_\nu = \left\|\frac{1}{Dg}\right\|_\nu = 1$, we obtain therefore for E, Δg, $\Delta\lambda$ the estimates:

$$|\Delta\lambda|_k \leq \frac{1}{\nu}\|F\|_\nu ,$$

$$\|E\|_{\nu-\delta} \leq \frac{\nu}{B\delta}\|F\|_\nu ,$$

$$\|\Delta g\|_{\nu-\delta} \leq \frac{\nu}{B\delta}\|F\|_\nu ,$$

$$\|D\Delta g\|_{\nu-\delta} \leq \frac{\nu^2}{B\delta}\|F\|_\nu .$$

2.6 We now proceed to estimate R_i, $i = 0\ldots3$, using for E, Δg, $\Delta\lambda$ the values determined above. We obtain:

$$\|R_o\|_{\nu-\delta} \leq \|DF\|_{\nu-\delta}\,\|E\|_{\nu-\delta} \leq \frac{\nu^2}{B\delta}\|F\|_\nu^2 .$$

Assume in the following that $1 < \nu < r^{1/3}$; then, using propositions 9 and 10, we have:

$$\|R_1\|_{\nu-\delta} \le \|f\|_r \|\Delta g\|^2_{\nu-\delta} \le \|f\|_r \frac{\nu^2}{B^2\delta^2} \|F\|^2_\nu ;$$

$$\|R_2\|_{\nu-\delta} \le \|g\|_\nu |\Delta\lambda|^2_k \le \|F\|^2_\nu ;$$

$$\|R_3\|_{\nu-\delta} \le \|\Delta g\|_{\nu-\delta} |\Delta\lambda|_k \le \frac{1}{B\delta} \|F\|^2_\nu ;$$

provided that $\|\Delta g\|_{\nu-\delta} < \frac{1}{\nu}$, $|\Delta\lambda|_k < 1$. Thus, with $F_f(g+\Delta g, \lambda+\Delta\lambda) = R_o + R_1 - R_2 - R_3$, we can summarize as follows:

Proposition 14. Let $1 < \nu < r^{1/3}$, $0 < \delta < \nu-1$. Assume that $g \in L_\nu$, $\lambda \in k$ verify $|\lambda|_k < C \le 1$, $\|g-id\|_\nu < 1/\nu$; then, if $\|F_f(g,\lambda)\|_\nu < \inf(C, \frac{B\delta}{\nu^2})$, we can find $\Delta\lambda \in k$, $\Delta g \in L_{\nu-\delta}$, such that:

$$|\Delta\lambda|_k \le \frac{\|F\|_\nu}{\nu} < C,$$

$$\|\Delta g\|_{\nu-\delta} \le \|F\|_\nu \frac{\nu}{B\delta} < \frac{1}{\nu} ,$$

$$\|F_f(g+\Delta g, \lambda+\Delta\lambda)\|_{\nu-\delta} \le \|f\|_r \frac{\nu^2}{B^2\delta^2} \|F_f(g,\lambda)\|^2_\nu .$$

Thus, $g+\Delta g$, $\lambda+\Delta\lambda$ satisfy the same hypothesis that g, λ.

2.7 The iteration process.

Choose $1 < \nu_o < r^{1/3}$ small enough to have $\frac{(\nu_o-1)B}{4} = K \le 1$. Define, for $n \ge 0$, $\nu_n - 1 = (\nu_o-1)(\frac{1}{2} + \frac{1}{2^{n+1}})$, $\delta_n = \nu_n - \nu_{n+1} = \frac{K}{B} 2^{-n}$. Let $\nu_\infty = \frac{\nu_o+1}{2}$ be the limit of ν_n. Assume that $\|f(z)-\alpha z\|_r < \epsilon = \inf(C, \frac{K^2}{4r^3})$. We claim that, using proposition 14, we can construct two sequences $\lambda_n \in k$, $g_n \in L_{\nu_n}$ such that $\lambda_{n+1} - \lambda_n = \Delta\lambda_n$, $g_{n+1} - g_n = \Delta g_n \in L_{\nu_{n+1}}$ and $F_f(g_n,\lambda_n)$ satisfy the following estimates:

(1_n) $\|F_f(g_n,\lambda_n)\|_{\nu_n} < \epsilon\, 2^{-2n}$, $\|g_n-id\|_{\nu_n} < \frac{1}{\nu_n}$, $|\lambda_n|_k < C$;

(2_n) $\|\Delta g_n\|_{\nu_{n+1}} \le \frac{1}{r} 2^{-n}$;

(3_n) $\quad |\Delta\lambda_n|_k < C\, 2^{-2n}$.

Indeed, we take $g_o = \mathrm{id}$, $\lambda_o = 0$, so that (1_o) holds; assuming (1_n), we have $\|F_f(g_n,\lambda_n)\|_{\nu_n} < \varepsilon 2^{-2n} \le \inf(C, \frac{B\delta_n}{\nu_n^2})$, thus we can apply proposition 14 and obtain Δg_n, $\Delta\lambda_n$ with:

$$\|\Delta g_n\|_{\nu_{n+1}} \le \|F_f(g_n,\lambda_n)\|_{\nu_n} \le \frac{\varepsilon r}{K} 2^{-n} \le \frac{1}{r} 2^{-n} ;$$

$$|\Delta\lambda_n|_k \le \frac{1}{\nu_n} \|F_f(g_n,\lambda_n)\|_{\nu_n} < C\, 2^{-n};$$

$$\|F_f(g_{n+1},\lambda_{n+1})\|_{\nu_{n+1}} < \|f\|_r \frac{\nu_n^2}{B^2\delta_n^2}(\varepsilon 2^{-2n})^2 \le \frac{r^3}{K^2}\varepsilon^2\, 2^{-2n}$$

$$\le (\frac{4r^3}{K^2}\varepsilon)\,\varepsilon\; 2^{-2(n+1)} \le \varepsilon\; 2^{-2(n+1)}.$$

Therefore (1_{n+1}), (2_n), (3_n) are satisfied. By (2_n) and (3_n), the sequences g_n and λ_n are converging, respectively in L_{ν_∞} and k, to some limits g and λ. The continuity of F_f follows from propositions 9 and 10. We conclude that, in L_{ν_∞}, $f\circ g - g\circ R_{\alpha+\lambda} = 0$, and theorem 2 is proved. ∎

III - Lipschitz perturbations of minimal translations of $\mathbb{Z}_p$.

In the following, p is a prime number ≥ 2.

1. We put on $\mathbb{Q}_p$ (the field of p-adic numbers) the standard p-adic absolute value: $|p|_p = \frac{1}{p}$. The ring of p-adic integers $\mathbb{Z}_p$ is the unit ball of $\mathbb{Q}_p$ and the norm $|\ |_p$ defines a complete metric on $\mathbb{Z}_p$.

If $u \in \mathbb{Z}_p$, the translation of $\mathbb{Z}_p$, $R_u(x) = x+u$, is an isometry of $\mathbb{Z}_p$. It is minimal (i.e. for every $x \in \mathbb{Z}_p$, $\{R_u^n(x) \mid n \in \mathbb{Z}\}$ is dense in $\mathbb{Z}_p$) if and only if $|u|_p = 1$. For $|u|_p = 1$, all translations R_u are conjugate by linear maps of the form $x \in \mathbb{Z}_p \to \alpha x \in \mathbb{Z}_p$, $|\alpha|_p = 1$. Furthermore, the translations R_u ($|u|_p = 1$) are strictly ergodic: the unique probability measure invariant by R_u is the Haar measure of $\mathbb{Z}_p$.

We put on the group of homeomorphisms of $\mathbb{Z}_p$ the compact-open topology. The following proposition is a special case of the easy C^o-closing lemma.

Proposition 14. Given $u \in \mathbb{Z}_p$, $|u|_p = 1$ and $\varepsilon > 0$, we can find a homeomorphism f of $\mathbb{Z}_p$ having a periodic point and such that:

$$\|f-R_u\| = \sup_{x\in\mathbb{Z}_p} |f(x)-R_u(x)|_p < \varepsilon .$$

Proof: Choose n such that $|p^n|_p < \varepsilon/2$. Let φ be a homeomorphism of $\mathbb{Z}_p$ with support in $\{x \in \mathbb{Z}_p \mid |x|_p \leq |p^n|_p\}$ such that $\varphi(p^n u) = 0$. The homeomorphism $f = \varphi\circ R_u$ is such that $f^{p^n}(0) = 0$ and satisfies $\|f-R_u\| < \varepsilon$. ∎

We propose to prove that the conclusion of the proposition is not possible with Lipschitz perturbations. We denote by $C^o(\mathbb{Z}_p,\mathbb{Q}_p)$ the Banach $\mathbb{Q}_p$-algebra of continuous mappings from $\mathbb{Z}_p$ to $\mathbb{Q}_p$ with the norm $\|\varphi\| = \sup_{x\in\mathbb{Z}_p} |\varphi(x)|_p$, $\|\varphi\| \in \mathbb{R}_+$.

2. Lipschitz and C^1-mappings of $\mathbb{Z}_p$.

2.1 Definition. We say that $\varphi: \mathbb{Z}_p \to \mathbb{Q}_p$ is Lipschitz if

$$\sup_{\substack{x\neq y\\ x,y\in\mathbb{Z}_p}} \frac{|\varphi(x)-\varphi(y)|_p}{|x-y|_p} = \mathrm{Lip}(\varphi) < +\infty .$$

One easily checks that $\mathrm{Lip}(\mathbb{Z}_p) = \{\varphi: \mathbb{Z}_p \to \mathbb{Q}_p, \mathrm{Lip}(\varphi) < +\infty\}$ is a non-archimedean Banach $\mathbb{Q}_p$-algebra with the norm $\|\varphi\|_{\mathrm{Lip}} = \mathrm{Max}(\|\varphi\|, \mathrm{Lip}(\varphi))$.

Proposition 15. Let $\varphi \in \mathrm{Lip}(\mathbb{Z}_p)$ with $\|\varphi\| \leq 1$ and $\mathrm{Lip}(\varphi) < 1$, then $f(x) = x + \varphi(x)$ is a homeomorphism and an isometry of $\mathbb{Z}_p$.

Proof: As $\|\varphi\| \leq 1$, f sends $\mathbb{Z}_p$ to $\mathbb{Z}_p$, and we have for $x \neq y$:

$$|\varphi(x)-\varphi(y)|_p < |x-y|_p \Rightarrow |f(x)-f(y)|_p = |x-y|_p .$$

Let us show that f is surjective: if $y \in \mathbb{Z}_p$, the mapping $x \xrightarrow{\Phi} y-\varphi(x)$ sends $\mathbb{Z}_p$ into itself, and $|\Phi(x)-\Phi(y)|_p \leq \mathrm{Lip}(\varphi)|x-y|_p$;

by the contraction mapping theorem, Φ has a fixed point x_o, thus $f(x_o) = y$ and f is surjective. ■

Remark 9. Actually, any isometry of $\mathbb{Z}_p$ is an homeomorphism.

Caution. The mapping $x \to px$ sends $\mathbb{Z}_p$ into itself, is injective but not surjective.

One checks that if $\varphi, f \in \mathrm{Lip}(\mathbb{Z}_p)$, $\|\varphi\|_{\mathrm{Lip}} \leq 1$, then $f\circ\varphi \in \mathrm{Lip}(\mathbb{Z}_p)$ and $\mathrm{Lip}(f\circ\varphi) \leq \mathrm{Lip}(f)\mathrm{Lip}(\varphi)$.

The space of continuous mappings from $\mathbb{Z}_p$ to $\mathbb{Z}_p$, $C^o(\mathbb{Z}_p,\mathbb{Z}_p)$, is a complete metric space with the norm $\|\ \|$.

Proposition 16. Let $h \in \mathrm{Lip}(\mathbb{Z}_p)$ with $\|h\|_{\mathrm{Lip}} = k < 1$. Then the mapping:

$$\varphi \in C^o(\mathbb{Z}_p,\mathbb{Z}_p) \xrightarrow{\Phi} h\circ\varphi \in C^o(\mathbb{Z}_p,\mathbb{Z}_p)$$

satisfies $\|\Phi(\varphi_1) - \Phi(\varphi_2)\| \leq k\,\|\varphi_1-\varphi_2\|$.

The proof is immediate. ■

2.2 Definition. We say that $\varphi \in C^o(\mathbb{Z}_p,\mathbb{Q}_p)$ is C^1 or continuously strictly differentiable if the mapping:

$$(u,v) \in \mathbb{Z}_p^2 - \Delta \to \frac{\varphi(u)-\varphi(v)}{u-v}$$

extends to a continuous function on $\mathbb{Z}_p^2$, where $\Delta = \{(v,v) \in \mathbb{Z}_p^2 \mid v \in \mathbb{Z}_p\}$ is the diagonal.

We define the $\mathbb{Q}_p$-vector space:

$$C^1(\mathbb{Z}_p,\mathbb{Q}_p) = \{\varphi \in C^o(\mathbb{Z}_p,\mathbb{Q}_p) \mid \varphi \text{ is } C^1\}$$

and put on it the norm $\|\ \|_{C^1}$ defined by:

$$\|\varphi\|_{C^1} = \mathrm{Max}(\|\varphi\|, |\varphi|_{C^1}),$$

$$|\varphi|_{C^1} = \sup_{u\neq v} \left|\frac{\varphi(u)-\varphi(v)}{u-v}\right|_p .$$

One checks (cf [9]) that $C^1(\mathbb{Z}_p,\mathbb{Q}_p)$ is a non-archimedean Banach $\mathbb{Q}_p$-algebra with this norm.

It is immediate that any C^1 function φ is Lipschitz and satisfies $\mathrm{Lip}(\varphi) = |\varphi|_{C^1}$.

<u>Caution</u>. A function $\varphi \in C^o(\mathbb{Z}_p,\mathbb{Q}_p)$ with a continuous derivative (in the usual sense) on all $\mathbb{Z}_p$ is not necessarily C^1, not even Lipschitz.

Let $\varphi \in C^1(\mathbb{Z}_p,\mathbb{Q}_p)$, $\|\varphi\| \le 1$, $|\varphi|_{C^1} < 1$; then, we have seen that $h(x) = x + \varphi(x)$ defines an isometric homemorphism of $\mathbb{Z}_p$. As φ is C^1, h is C^1; one easily checks that h^{-1} is C^1 and satisfies:

$$Dh^{-1} = (Dh)^{-1}\circ h^{-1}, \qquad |h^{-1}|_{C^1} = 1.$$

<u>Example</u>. By proposition 7, every locally $\mathbb{Q}_p$-analytic function on $\mathbb{Z}_p$ is C^1 and therefore lipschitzian. Examples of such functions are polynomials.

3. Linear difference equations.

3.1 Interpolation.

Let $\binom{x}{n} = x(x-1)\dots(x-n+1)/n!$ for $n \in \mathbb{N}^*$, $\binom{x}{0} = 1$. Thus, for $x \in \mathbb{N}$, $0 \le x < n$, $\binom{x}{n} = 0$. By [5], any $\varphi \in C^o(\mathbb{Z}_p,\mathbb{Q}_p)$ can be written uniquely as $\varphi(x) = \sum_{n\ge 0} a_n\binom{x}{n}$ with $a_n = \sum_{i=0}^{n} (-1)^{n-i}\binom{n}{i}\varphi(i)$, $\lim_{n\to+\infty} |a_n|_p = 0$, $\|\varphi\| = \sup_{n\ge 0} |a_n|_p$. Thus, $\lim_{n\to+\infty} \|a_n\binom{x}{n}\| = \lim_{n\to+\infty} |a_n|_p = 0$, the series $\sum_{n\ge 0} a_n\binom{x}{n}$ are normally convergent.

<u>Proposition 17</u>. <u>The function</u> $\varphi \in C^o(\mathbb{Z}_p,\mathbb{Q}_p)$ <u>belongs to</u> $\mathrm{Lip}(\mathbb{Z}_p)$ <u>if and only if</u> $\sup_{n\ge 0} n|a_n|_p < +\infty$, <u>and one has</u>:

$$\frac{1}{p} \sup_{n\ge 0} n|a_n|_p \le \mathrm{Lip}(\varphi) \le \sup_{n\ge 0} n|a_n|_p .$$

<u>Proof</u>: Suppose that φ is Lipschitz; define, for $x \ne y$, $g(x,y) = [\varphi(x)-\varphi(y)](x-y)^{-1}$, so that $\mathrm{Lip}(\varphi) = \sup_{x\ne y} |g(x,y)|_p$. For

$i,j \in \mathbb{N}$, let $\beta_{i,j} = \sum_{m=0}^{i} \sum_{n=0}^{j} (-1)^{i+j-m-n}\binom{i}{m}\binom{j}{n} g(m,m+n+1)$; it is well defined because $m \neq m+n+1$, and satisfies $|\beta_{i,j}|_p \leq \mathrm{Lip}(\varphi)$ because $|\binom{i}{m}|_p \leq 1$.

But Weisman ([14]) proves that $\beta_{i,j} = a_{i+j+1}(j+1)^{-1}$ (see also [5, p 208]). This implies that:

$$\sup_n \gamma_n^{-1} |a_n|_p = \sup_{i,j} |a_{i+j+1}(j+1)^{-1}|_p \leq \mathrm{Lip}(\varphi),$$

with $\gamma_n^{-1} = \max\{|j|_p^{-1},\ 1 \leq j \leq n\}$.

From $n/p \leq \gamma_n^{-1} \leq n$, we deduce $\sup_n n|a_n|_p < +\infty$, and $\frac{1}{p} \sup_n n|a_n|_p \leq$ $\leq \mathrm{Lip}_1 \varphi$.

Conversely, we have for $x \neq y$:

$$g(x,y) = \sum_{n=1}^{+\infty} \frac{a_n(y)}{n} \binom{x-y-1}{n-1},$$

with $a_n(y) = \sum_{k=0}^{+\infty} a_{k+n}\binom{y}{k}$ (cf [5, p 209]).

Therefore, as $\|\binom{x}{n}\| = 1$:

$$|g(x,y)|_p \leq \sup_{\substack{k\geq 0\\ n\geq 1}} \left|\frac{a_{k+n}}{n}\right|_p \leq \sup_{n\geq 1} n|a_n|_p ,$$

and this proves the other part of the proposition. ■

<u>Remark 10</u>. Weisman ([14]) shows that φ is C^1 if and only if $\lim_{n\to+\infty} n|a_n|_p = 0$.

3.2 For $u \in \mathbb{Z}_p$, $|u|_p = 1$, define:

$$L_u: \varphi \in C^0(\mathbb{Z}_p,\mathbb{Q}_p) \to \varphi - \varphi\circ R_u \in C^0(\mathbb{Z}_p,\mathbb{Q}_p).$$

As the translations R_u are conjugate by linear operators: $x \to \alpha x$, $|\alpha|_p = 1$, it is enough to prove the following proposition for $u = 1$.

<u>Proposition 18</u>. <u>There are continuous linear operators</u>

$$S_u: C^0(\mathbb{Z}_p,\mathbb{Q}_p) \to C^0(\mathbb{Z}_p,\mathbb{Q}_p)$$

such that $\|S_u\eta\| \le \|\eta\|$, $L_u S_u \eta = \eta$; *furthermore if* $\eta \in \mathrm{Lip}(\mathbb{Z}_p)$ (*resp.* η *is* C^1), *then* $S_u\eta \in \mathrm{Lip}(\mathbb{Z}_p)$ (*resp. is* C^1) *and* $\mathrm{Lip}(S_u\eta) \le p\,\mathrm{Lip}(\eta)$.

Proof: Write $\eta = \sum_{n\ge 0} a_n \binom{x}{n}$; we have $L_1(\binom{x}{n}) = -\binom{x}{n-1}$ for $n \ge 1$, $L_1(1) = 0$, so $S_1(\eta) = \sum_{n\ge 1} -a_{n-1}\binom{x}{n}$ is the desired operator. The inequalities follow from 3.1. ∎

4. We propose to prove the main theorem.

Theorem 3. *Let* $u \in \mathbb{Z}_p$, $|u|_p = 1$. *If* $\varphi \in \mathrm{Lip}(\mathbb{Z}_p)$ *satisfies* $\|\varphi\|_{\mathrm{Lip}} < p^{-1}$, *then* $f: \mathbb{Z}_p \to \mathbb{Z}_p$, *defined by* $f(x) = x+u+\varphi(x)$ *is conjugate by an isometry* h *of* $\mathbb{Z}_p$ *to* R_u:

$$f = h\circ R_u \circ h^{-1}$$

with h of the form $\mathrm{id} + \eta$, $\eta \in \mathrm{Lip}(\mathbb{Z}_p)$, $\|\eta\|_{\mathrm{Lip}} < 1$.

Remarks 11. a) Example: For $\varepsilon \in \mathbb{Z}_p$, let $f_\varepsilon: \mathbb{Z}_p \to \mathbb{Z}_p$ be defined by $f_\varepsilon(x) = x + 1 + \varepsilon x^2 = x + 1 + \varphi_\varepsilon(x)$. We have $\|\varphi_\varepsilon\|_{\mathrm{Lip}} \le \varepsilon$, so that for $|\varepsilon|_p < p^{-1}$, f_ε is minimal, and conjugate to R_1 by an isometry h_ε of the form $h_\varepsilon(x) = x + \eta_\varepsilon(x)$, $\|\eta_\varepsilon\|_{\mathrm{Lip}} < 1$.

b) It is easy to prove that any minimal isometry of $\mathbb{Z}_p$ is conjugate to R_1 by an isometry of $\mathbb{Z}_p$.

c) Define $h(x) = x + 1 + \varphi(x)$, with $\varphi(x) = -p$ if $|x|_p < 1$, $\varphi(x) = 0$ if $|x|_p = 1$. Then $\|\varphi\|_{\mathrm{Lip}} = p^{-1}$ and $h^p(0) = 0$, therefore h is not conjugate to R_1; this shows that the condition $\|\varphi\|_{\mathrm{Lip}} < p^{-1}$ in the theorem is sharp.

Proof of theorem 3: We look for $h = \mathrm{id} + \eta$ with $\|\eta\|_{\mathrm{Lip}} < 1$, such that:

$$f\circ h = h\circ R_u \Leftrightarrow \varphi\circ h = \eta\circ R_u - \eta \Leftrightarrow S_u(\varphi\circ h) = \eta.$$

We put on $B = \{\eta \in C^0(\mathbb{Z}_p,\mathbb{Q}_p) \mid \|\eta\| \le p^{-1}\}$ the uniform distance for which B is complete, and define $\Phi: B \to B$ by $\Phi(\eta) = S_u(\varphi\circ(\mathrm{id}+\eta))$. By proposition 16, Φ is a contraction, so it has a unique fixed point η_∞ in B, satisfying $\eta_\infty = \lim_{n\to+\infty} \Phi^n(0)$.

Assume that $\mathrm{Lip}(\eta) < 1$; then $\mathrm{Lip}(id+\eta) = 1$, and from proposition 18 we obtain:

$$\mathrm{Lip}(\Phi(\eta)) \leq p \,\mathrm{Lip}[\varphi \circ (id+\eta)] = p \,\mathrm{Lip}\, \varphi = C < 1.$$

Taking $\eta_o = 0$, $\eta_n = \Phi^n(0)$, we have for $x,y \in \mathbb{Z}_p$ and any $n \in \mathbb{N}$ $|\eta_n(x) - \eta_n(y)|_p \leq C|x-y|_p$, thus $|\eta_\infty(x) - \eta_\infty(y)|_p \leq C|x-y|_p$ with $C < 1$. Therefore $\|\eta_\infty\|_{Lip} < 1$, $\Phi(\eta_\infty) = \eta_\infty$ and $h = id + \eta_\infty$ is the conjugation we were looking for. ■

<u>Question</u>: If f is C^1, C^1-near R_u, the theorem gives an isometric conjugation h of f to R_u; does h have to be C^1? A similar question holds for smoother classes.

<u>References</u>

[1] Y. Amice, Les nombres p-adiques, Presses Universitaires de France, Paris, 1975.

[2] V.I. Arnold, Small denominators I, Mappings of circumference onto itself, Amer. Math. Soc. Transl. Ser. 2, <u>46</u> pp. 213-284.

[3] M.R. Herman, Sur la conjugaison différentiable des difféomorphismes du cercle à des rotations, Publ. I.H.E.S. <u>49</u> pp. 5-233.

[4] E. Lutz, Sur les approximations diophantiennes linéaires p-adiques, Hermann, Paris, 1955.

[5] K. Mahler, P-adic numbers and their functions, Camb. Univ. Press, Cambridge, second edition, 1981.

[6] K. Meyer, The implicit function theorem and analytic differential equations, Springer Lect. Notes in Math. 468, Proc. Symp. on Dynamical Systems at Warwick (1974).

[7] A.C.M. Van Rooij, Non archimedean function analysis, Marcel Dekker, New York, 1978.

[8] H. Rüssmann, Kleine Nenner II: Bemerkungen zur Newton'schen Methode. Nachr. Akad. Wiss. Göttingen Math. Phys. Kl. (1972) pp. 1-20.

[9] W.H. Schikhof, Non archimedean calculus, Lecture notes, Nijmegean, Katholieke Univ., 1978.

[10] J.P. Serre, Lie algebras and Lie groups, Benjamin, New York, 1965.

[11] Y. Sibuya & S. Sperber, Convergence of power series of p-adic non linear differential equations, preprint Univ. Minnesota (1978).

[12] C.L. Siegel, Iteration of analytic functions, Ann. Math. 43 (1942), pp. 607-612.

[13] C.L. Siegel, Über die Normalform analytischer Differentialgleichungen in der Nähe einer Gleichawichtslösung, Nachr. Akad. Wiss. Göttingen, Math. Phys. K. (1952), pp. 21-30.

[14] C.S. Weisman, On p-adic differentiability, J. Number th. 9 (1977), pp. 79-86.

[15] E. Zehnder, A simple proof of a generalization of a theorem by C.L. Siegel, Lec. Notes in Math. 597, Springer-Verlag, Berlin, (1977), pp. 855-866.

M. HERMAN

IMPA and
Centre de mathematiques de
l'Ecole Polytechnique

91128 Palaiseau CEDEX

FRANCE

Y.C. YOCCOZ

Instituto de Matemática Pura e Aplicada (IMPA)

Estrada Dona Castorina 110,

CEP 22460 - Rio de Janeiro, RJ

BRASIL

THE TOPOLOGY OF LINEAR $\mathbb{C}^m$-FLOWS ON $\mathbb{C}^n$

Nicolaas H. Kuiper

I. Introduction and theorem

An action

$$\varphi: \mathbb{C}^m \times \mathbb{C}^n \to \mathbb{C}^n, \qquad 1 \leq m \leq n,$$

of the additive group $\mathbb{C}^m$ on $\mathbb{C}^n$ is called a (linear) $\mathbb{C}^m$-flow also called $\mathbb{C}^m$-action in case it is generated over $\mathbb{C}$ by m commuting (linear) vector fields. In the linear case they can be given in terms of $n \times n$-matrices σ_i by vectors

$$\sigma_i z \in \mathbb{C}^n \quad \text{at} \quad z \in \mathbb{C}^n, \quad \sigma_i \sigma_j = \sigma_j \sigma_i \qquad i,j = 1,\ldots,m. \tag{1}$$

The set $\varphi(\mathbb{C}^m, z)$ is the orbit of z.

Assuming that at least one linear combination of the matrices σ_i has all eigenvalues different, then the commutativity permits the unique choice of linear coordinates for $\mathbb{C}^n$ with respect to which all matrices σ_i are diagonal. The action is then obtained by integration from differential equations of the kind

$$\begin{aligned} dz_i &= z_i \sum_{j=1}^{m} \alpha_{ij}\, ds_j \qquad \alpha_{ij} \in \mathbb{C}, \\ z_i, s_j &\in \mathbb{C}, \quad i = 1,\ldots,n;\ j=1,\ldots,m, \end{aligned} \tag{2}$$

with solutions

$$z_i = z_i^o \exp \sum_{j=1}^{m} \alpha_{ij} s_j \qquad i = 1,\dots,n. \tag{2}$$

Let $\alpha_{i_1,\dots,i_m}$ for $1 \leqq i_1 < i_2 \dots < i_m \leqq n$ be the determinant of the square matrix obtained from the matrix $\{\alpha_{ij}\}$ by deleting the rows $(\alpha_{i1},\dots,\alpha_{im})$ for all $i \notin \{i_1,\dots,i_m\}$.

In order to avoid certain degeneracies we assume not only

$$\alpha_{i_1,\dots,i_m} \neq 0 \quad \text{for all} \quad 1 \leqq i_1 < \dots < i_m \leqq n \tag{3a}$$

but even

$$\alpha_{i_1,\dots,i_m} \notin \mathbb{R}\, \alpha_{j_1,\dots,j_m} \tag{3b}$$

for $\{i_1,\dots,i_m;\ j_1,\dots,j_m\}$ a set of exactly $m+1$ different numbers, and $(i_1,\dots,i_m) \neq (j_1,\dots,j_m)$. With a suitable coordinate transformation on C^m we find the normal form

$$\left.\begin{aligned} dz_i &= z_i\, ds_i && i = 1,\dots,m \\ dz_{m+j} &= z_{m+j} \sum_{k=1}^{m} \lambda_{jk}\, ds_k && j = 1,\dots,p, \quad n = m+p \end{aligned}\right\} \tag{4}$$

with solutions

$$\begin{aligned} z_i &= z_i^o \exp s_i && i = 1,\dots,m \\ z_{m+j} &= z_{m+j}^o \exp \sum_{k=1}^{m} \lambda_{jk}\, s_k && i = 1,\dots,p. \end{aligned} \tag{4}$$

Note the matrix of coefficients:

$$\begin{pmatrix} 1 & & & \\ & 1 & & \\ & & \ddots & \\ & & & 1 \\ \lambda_{11} & \cdots & \cdots & \lambda_{1m} \\ \cdot & \cdot & \cdot & \cdot \\ \lambda_{p1} & \cdots & \cdots & \lambda_{pm} \end{pmatrix} \tag{5}$$

From (32) follows that all determinants of square submatrices of any size of the submatrix of elements λ_{ij} are non zero. In particular $\lambda_{ij} \neq 0$ for all (i,j).

Two linear $\mathbb{C}^m$-flows φ and φ' are called conjugate if there exist automorphisms

$$\tau \in GL(m,\mathbb{C}), \quad \eta \in GL(n,\mathbb{C}), \quad \iota: i \to \pm i \quad \text{of the field } \mathbb{C},$$

sending φ to φ' in a commutative diagram:

$$\begin{array}{ccc} \mathbb{C}^m \times \mathbb{C}^n & \xrightarrow{\varphi} & \mathbb{C}^n \\ \downarrow \iota\tau \quad \downarrow \iota\eta & & \downarrow \iota\eta \\ \mathbb{C}^m \times \mathbb{C}^n & \xrightarrow{\varphi'} & \mathbb{C}^n \end{array}$$

They are called orbit-equivalent if there exists a homeomorphism $h: \mathbb{C}^n \to \mathbb{C}^n$ sending the orbits of φ onto those of φ':

$$h\varphi(\mathbb{C}^m, z) = \varphi'(\mathbb{C}^m, hz) \quad \text{for} \quad z \in \mathbb{C}^n \tag{6}$$

They corresponding foliations with some singular leaves are then homeomorphic.

The are called topologically conjugate in case the homeomorphism h can moreover be chosen to preserve the parameterspace, that is the additive real structure $\mathbb{R}^{2m}$ of $\mathbb{C}^m$: $\exists\ \eta: \mathbb{C}^m \to \mathbb{C}^m$ $\mathbb{R}$-linear invertible such that $h\varphi(x,z) = \varphi'(\eta x, hz)$.

The linear $\mathbb{C}^m$-action φ is called stable in case any sufficiently near linear $\mathbb{C}^m$-action φ' is orbit equivalent to φ.

The linear $\mathbb{C}^m$-action φ is called orbit rigid in case any sufficiently near orbit-equivalent $\mathbb{C}^m$-action φ' is conjugate to φ. This means that homeomorphism of the foliations of φ and φ' implies conjugacy.

The case $m=1$: We can simply write, instead of (2),

$$dz_i = z_i \alpha_i \, ds \qquad i = 1,\dots,n. \tag{7}$$

We assume again

$$\alpha_i \notin \mathbb{R}\,\alpha_j \quad \text{for} \quad i \neq j,$$

and we use the notation

$$\mathcal{H}(\alpha_1,\ldots,\alpha_n)$$

for the convex hull in $\mathbb{C}$ of the points $\alpha_1,\ldots,\alpha_n$. Then we can recall and formulate two known theorems. The theorem of Guckenheimer: The linear $\mathbb{C}$-action φ that obeys (5) is stable if and only if

$$0 \in \mathcal{H}(\alpha_1,\ldots,\alpha_n)$$

and

The theorem of Camacho-Kuiper-Palis [1] and Ladis [2]. Two linear $\mathbb{C}$-actions φ and φ' for which (3) holds and for which

$$0 \in \mathcal{H}(\alpha_1,\ldots,\alpha_n),$$

are orbit equivalent if and only if, after reordering of coordinates, there exists $g \in GL(2,\mathbb{R})$ acting on $\mathbb{R}^2 = \mathbb{C}$ such that

$$g(\alpha_i^{-1}) = (\alpha_i')^{-1} \qquad i = 1,\ldots,n.$$

φ is neither stable nor orbit-rigid in this case. However, any φ' near to φ and orbit equivalent to φ is topologically conjugate to φ as well.

The case $m = n-1 \geqq 2$ was solved by B. Klares in his Strasbourg thesis 1980. See [3]. Also by Camacho-Lins Neto [5] but presented in terms of differential forms.

Let $\alpha_{[j]} = \alpha_{12\,\ldots\,\hat{j}\,\ldots\,n}$ be the determinant of the square matrix obtained by deleting the j-th row of the matrix of elements α_{ij}.

Theorem of Klares and Camacho-Neto: Two linear $\mathbb{C}^{n-1}$-actions on $\mathbb{C}^n$, φ and φ', with determinants $\alpha_{[j]}$ and $\alpha'_{[j]}$, $j=1,\ldots,n$, for which

$$\alpha_{[i]} \notin \mathbb{R}\,\alpha_{[j]} \quad \text{if} \quad i \neq j \tag{3}$$

are orbit equivalent, if and only if, after suitable reordering of coordinates, there exists $g \in GL(2,R)$ acting on $\mathbb{R}^2 = \mathbb{C}$, such that

$$g(\alpha_{[j]}) = \alpha'_{[j]} \qquad j = 1,\dots,n \tag{8}$$

φ is neither stable nor orbit-rigid. However, again, any φ' near to φ and orbit equivalent to φ is topologically conjugate to φ as well.

In this note we use this theorem to prove the intermediate remaining cases:

Main Theorem: The orbit-rigid linear $\mathbb{C}^m$-flows on $\mathbb{C}^n$ form for $2 \leqq m \leqq n-2$ an open dense set in the space of all linear $\mathbb{C}^m$-actions.

Remark: It seems likely that the nondegeneracy condition (3) is sufficient for orbit-rigidity.

Corollary: The same conclusion holds for $\mathbb{C}^m$-flows on $\mathbb{C}P^n$ for $2 \leqq m \leqq n-2$.

II. Proof of Klares' theorem

For our purpose a proof slightly different from Klares' solution suits us. It rests however, on the same crucial topological invariant $\eta(\varphi)$. We assume (3) and use the normal form (4) (5) of the linear $\mathbb{C}^m$-flow on $\mathbb{C}^n$ for $m = n-1$:

$$\begin{aligned} dz_j &= z_j \, ds_j \qquad j = 1,\dots,m = n-1 \\ dz_n &= z_n \left(\sum_{j=1}^{m} \lambda_j \, ds_j \right) \end{aligned} \tag{9}$$

with solutions

$$\begin{aligned} z_j &= z_j^o \exp s_j \qquad j = 1,\dots,m \; ; \\ z_n &= z_n^o \exp(\lambda_1 s_1 + \dots + \lambda_m s_m) \end{aligned} \tag{9}$$

and matrix

$$\begin{pmatrix} 1 & & & \\ & 1 & & \\ & & \ddots & \\ & & & 1 \\ \lambda_1 & \lambda_2 & & \lambda_m \end{pmatrix} \qquad (9)$$

<u>Proof of necessity</u>: We prove that the equivalence class

$$\eta(\varphi) = (\lambda_1,\ldots,\lambda_m,1) \quad \text{modulo} \quad GL(2,\mathbb{R})$$

where $GL(2,R)$ acts on $\mathbb{R}^2 = \mathbb{C}$, is a topological invariant of the orbit structure of φ.

Let us call the set of points

$$\{(z_1,\ldots,z_n) \in \mathbb{C}^n : z_{i_1} = \ldots = z_{i_{n-p}} = 0, \quad 1 \leq i_1 < i_2 < \ldots < i_{n-p} \leq n\}$$

a <u>coordinate</u> p-<u>plane</u>. The union of all coordinate p-planes is called the p-<u>skeleton</u> $Sk_p = Sk_p(\mathbb{C}^n)$ of $\mathbb{C}^n$. If the oribt of a $\mathbb{C}^m$-flow through a point $z^o \in \mathbb{C}^n$ is a m-manifold at z^o, then z^o is called <u>regular</u>, othwerwise <u>singular</u>.

It is easy to see in the equations (9) (and (4) in the general case) that $Sk_{m-1}(\mathbb{C}^n)$ is precisely the set of singular points of φ, and therefore can be defined in terms of the topology of the orbit-structure. Consider for that also, the analogous normal forms obtained with other orderings of the n coordinates.

It then follows that coordinate m-1-planes can be distinguished topologically by the orbitstructure.

Now if the numbers $w_1,\ldots,w_m$, $z_1^o,\ldots,z_n^o$ are nonzero, $m+1 = n$, then we easily read from (9) that the closure in $\mathbb{C}^n$ of the orbit of $z^o = (z_1^o,\ldots,z_n^o)$ contains $(w_1,\ldots,w_m,0)$. Indeed if we put

$$z_j^o = w_j \exp a_j \qquad j = 1,\ldots,m$$

then the orbit of z^o is

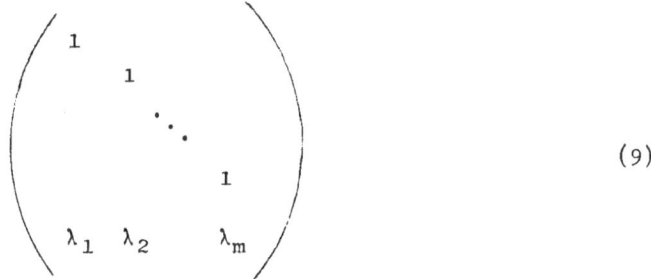

$$z_j = w_j \exp(a_j+s_j) \qquad j = 1,\dots,m$$

$$z_n = z_n^o \exp(\lambda_1 s_1 +\dots+ \lambda_m s_m)$$

and in the lattice

$$\{(s_1,\dots,s_m) : a_j+s_j \equiv 0 \ \text{mod.}\ 2\pi i, \quad j = 1,\dots,m\}$$

we easily find a sequence of points $(s_1,\dots,s_m) \in \mathbb{C}^m$ in this orbit for which z_n tends to zero. Hence the closure of the orbit of a point

$$z^o \in \mathbb{C}^{m+1}\setminus Sk_m , \quad m+1 = n,$$

contains all of $z_n = 0$ and (by reordering of coordinates) all of every coordinate hyperplane, hence their union Sk_m , and in particular $Sk_{m-1} \subset Sk_m$.

On the other hand if $z^o \in Sk_m\setminus Sk_{m-1}$ then this closure does not contain all of Sk_{m-1} (obvious by taking the point z^o in $z_n = 0$ for example).

Now Sk_{m-1} was defined in topological terms already. Any point z^o in $\mathbb{C}^{m+1}\setminus Sk_{m-1}$ lies in $Sk_m\setminus Sk_{m-1}$ if and only if the closure of its orbit does not contain all of Sk_{m-1}.

Therefore Sk_m is also defined topologically in terms of the orbit structure (we will give another topological definition of Sk_m in the next paragraph).

Let $V = \mathbb{C}^{m+1}\setminus Sk_m$, $z^o \in V$ and $U(z^o) \subset V$ a small open neighborhood in V homeomorphic to an open ball. In the class of all curves $z(t)$, $0 \leq t \leq 1$ in V, that begin at z^o and end inside $U(z^o)$, every curve is homotopic to a closed curve, beginning and ending at z^o, and gives rise to unique rotation numbers $N_j \in \mathbb{Z}$, concerning $z_j \neq 0$, for $j = 1,\dots,m+1$, around each of the (topologically defined) coordinate hyperplanes in Sk_m of real codimension 2. These integers are invariants of the homotopy classes of curves beginning at z^o and ending in $U(z^o)$. Next suppose the curve $z(t)$ lies in the leaf (9). For small $U(z^o)$ the end point $z(1)$ is near to the point

$$(z_1^o \exp 2\pi i\ N_1, \ldots, z_m^o \exp 2\pi i\ N_m,\ z_n^o \exp 2\pi i\ N_{m+1})$$

as well as near to the leaf point

$$(z_1^o \exp 2\pi i\ N_1, \ldots, z_m^o \exp 2\pi i\ N_m,\ z_{m+1}^o \exp 2\pi i\ (\lambda_1 N_1 + \ldots + \lambda_m N_m))$$

for $(s_1, \ldots, s_m) = (N_1, \ldots, N_m)$.

Therefore these last two points are near to each other, and so are the corresponding points in the universal covering space of V, for which the given arguments in the exponential functions are valid! Hence for $U(z^o)$ small, also

$$|\lambda_1 N_1 + \ldots + \lambda_m N_m - N_{m+1}| \quad \text{is small (bounded)} \tag{11}$$

Assuming for a moment the existence of such leaf curves for $U(z^o)$ decreasing and converging to z^o, we find a sequence of m+1-tuples of numbers:

$$\left(\frac{N_1}{N}, \ldots, \frac{N_{m+1}}{N}\right) : N = \sum_{j=1}^{m+1} |N_j|$$

with increasing integers N, going to ∞, any limit of which, $(c_1, \ldots, c_{m+1}) \in R^{m+1}$, obeys by substitution in (11)

$$\lambda_1 c_1 + \ldots + \lambda_m c_m - c_{m+1} = 0, \qquad \sum_{j=1}^{m+1} |c_j| = 1 \tag{12}$$

The set $\Delta(\varphi)$ of all such limits is by its definition a topological invariant of the orbit structure φ.

Next take any m+1-tuple of numbers obeying (12). By the first theorem on diophantine approximations (see the book of Minkowski or Cassels) there exist infinitely many natural numbers N and integers $N_1, \ldots, N_n$, $n = m+1$, such that

$$|c_j - N_j/N| < N^{-1-\frac{1}{n}}, \quad N = \sum_{j=1}^{n} |N_j|, \quad j=1,\ldots,n, \quad N \to \infty. \tag{13}$$

Look at the leaf curves

$$z(t) = (z_1^o \exp 2\pi i\ N_1 t, \ldots, z_m^o \exp 2\pi i\ N_m t, z_{m+1}^o \exp 2\pi i(\lambda_1 N_1 + \ldots + \lambda_m N_m)t) \tag{14}$$

For $t=1$ the last component equals

$$z^{o}_{m+1} \exp 2\pi i(\lambda_1 N_1 + \ldots + \lambda_m N_m - N_{m+1}).$$

But the <u>difference</u> between

$$\lambda_1 N_1 + \ldots + \lambda_m N_m - N_{m+1}$$

and

$$(\lambda_1 c_1 + \ldots + \lambda_m c_m - c_{m+1})N = 0 \qquad (!) ,$$

<u>is</u> in absolute value

$$\leqq |c_1 N - N_1| \cdot |\lambda_1| + \ldots + |c_m N - N_m| \cdot |\lambda_m| + |c_{m+1} N - N_{m+1}|$$

$$\leqq (|\lambda_1| + \ldots + |\lambda_m| + 1) \cdot N^{-\frac{1}{n}}, \qquad \text{by (13)}.$$

This tends to zero for $N \to \infty$, so that the endopoints of the curves indeed tend to z^{o}. Therefore the <u>topological invariant</u> $\Delta(\varphi)$ <u>consists precisely of all solutions of</u> (12). But this set of solutions $\Delta(\varphi)$ determines geometrically completely the set $(\lambda_1, \ldots, \lambda_m, 1)$ modulo $GL(2,\mathbb{R})$, that is the invariant $\eta(\varphi)$ in (10). The necessity is now established.

<u>The sufficiency of Klares' condition</u>: For φ and φ' with $\eta(\varphi) = \eta(\varphi')$ we will present an explicit homeomorphism h, which will even be a topological conjugacy. We start from φ as in (9) and try to get φ' by a proposed homeomorphism, which clearly will be a topological conjugacy if it is given as $h: \mathbb{C}^n \to \mathbb{C}^n$, $h(z_1, \ldots, z_n) = (z'_1, \ldots, z'_n)$, by formulas

$$z_j = \exp w_j, \quad z'_j = \exp w'_j, \quad w'_j = g_j(w_j), \quad j=1,\ldots,n. \tag{15}$$

Here $g_j \in GL(2,\mathbb{R})$ acts on $\mathbb{R}^2 = \mathbb{C}$, and we must assume

$$g_j(2\pi i) = 2\pi i, \quad g(i) = i. \tag{16}$$

Call

$$g_j(1) = \tau_j \notin i\mathbb{R} \qquad j = 1,\ldots,n \tag{17}$$

Substitution of (9) and the analogous equations in λ'_j, s'_j concerning

φ', yields equations

$$\left.\begin{aligned} s'_j &= g_j(s_j) \qquad\qquad j = 1,\dots,m \\ \sum_{j=1}^{m} \lambda'_j s'_j &= \sum_{j=1}^{m} \lambda'_j g_j(s_j) \equiv g_{m+1}\Big(\sum_{j=1}^{m+1} \lambda_j s_j\Big) \end{aligned}\right\} \tag{18}$$

For $s_j = i$ or 1 and $s_h = 0$ for $h \neq j$, we obtain in particular

$$\left.\begin{aligned} g_{m+1}(\lambda_j i) &= \lambda'_j\, i \\ g_{m+1}(\lambda_j) &= \lambda'_j \tau_j \end{aligned}\right\} \quad j = 1,\dots,m \tag{18}$$

Recall: $g_{m+1}(i) = i$, $g_{m+1}(1) = \tau_{m+1}$.

Once g_{m+1} (or τ_{m+1}) is given, as well as $\lambda_1,\dots,\lambda_m$, then $\lambda'_1,\dots,\lambda'_m$, $\tau_1,\dots,\tau_m$, $g_1,\dots,g_m$ are uniquely determined. If we put

$$g = -i \cdot g_{m+1} \cdot i \in GL(2,\mathbb{R}), \tag{19}$$

then

$$g(\lambda_j) = \lambda'_j, \quad j = 1,\dots,m, \quad \text{and} \quad g(1) = 1. \tag{20}$$

For given g in (20) we obtain g_{m+1} by (19) and the homeomorphism h, which is <u>unique</u> if we claim it to be a <u>topological conjugacy</u> between φ and φ' as well.

This ends the proof of Kalres' theorem.

III. The case $2 \leq m \leq n-2$

We assume the normal forms (3) (4) (5) of the equations and we recall that other such normal forms are obtained with other orderings of the n coordinates. Such considerations we use regularly to make conclusions about orbits in Sk_q, from facts about orbits in the coordinate q-plane

$$z_{q+1} = \dots = z_n = 0.$$

The singular set $Sk_{m-1} \subset \mathbb{C}^n$ is topologically defined, as we saw already, and so are the coordinate m-1-planes, as well as the coordinate q-planes for $1 \leq q \leq m-1$, by intersection. We now give topological definitions of Sk_m and Sk_{m+1}. Assuming this, the coordinate m+1-planes are topologically defined as well, and Klares' invariant is valid in each of them. Such Klares' invariants together lead to the orbit rigidity and our main theorem.

The orbits in $\mathbb{C}^n \setminus Sk_{m-1}$ form a foliation of dimension 2m without singularities. The set $Sk_q \setminus Sk_{q-1}$ for $q \geq m$ is a union of open sets in coordinate q-planes, with typical example the set

$$V^q = \{z_1, \ldots, z_q, 0, \ldots, 0) : z_1, z_2, \ldots, z_q \neq 0\} \tag{21}$$

It is an <u>immersion</u> of a fibre-bundle space with fibre C^m as we see from equations (4), $q = m+p$.

Any leaf may have periods forming a lattice

$$\Gamma \subset (2\pi i\, \mathbb{Z})^m \subset \mathbb{C}^m ,$$

consisting of the solutions (in (4)) of

$$\begin{cases} s_j \equiv 0 \mod 2\pi i, & j = 1, \ldots, m, \\ \lambda_{j1} s_1 + \ldots + \lambda_{jm} s_m \equiv 0 \mod 2\pi i & \text{for } j = 1, \ldots, p . \end{cases}$$

For $p = 0$, $\Gamma = (2\pi i\, \mathbb{Z})^m$, and each leaf is an immersed and even embedded manifold $\mathbb{C}^m/\Gamma$ homeomorphic to $T^m \times \mathbb{R}^m$, fibre of an embedded bundle. For $p > 0$, Γ is a, possibly empty, lattice, which has less than m generators, as otherwise some natural number M would exist for which

$$(M\, 2\pi i\, \mathbb{Z})^m \subset \Gamma \subset (2\pi i\, \mathbb{Z})^m$$

and then all λ_i ought to be real in contradiction to (3). So no leaf in $\mathbb{C}^n \setminus Sk_m$ is homeomorphic to $T^m \times \mathbb{R}^m$. This accomplishes the <u>topological distinction and definition</u> of the subspace $Sk_m \setminus Sk_{m-1}$ hence of Sk_m.

Next consider a point $z^o \in V^q$, $q = m+p \geq m+1$, typical point of $Sk_q \setminus Sk_{q-1}$. We look for endpoints of leaf curves in the orbit of

z^o, that are near to z^o. We let z^o be near to coordinate m-plane V^m of the first m coordinates:

$$\max_{1\leq j\leq p} |z^o_{m+j}| << \min_{1\leq i\leq m} |z^o_i| .$$

Any point near to z^o admits a small holonomic move in its leaf to a position where it has the same first m coordinates as z^o. So the leaf curve $z(t)$ can be assumed to have coordinates for $0 \leq t \leq 1$ in the form

$$\begin{cases} z^o_j \exp 2\pi i\, N_j t & j = 1,\dots,m \\ z^o_{m+\ell} \exp 2\pi i (\sum_{k=1}^{m} \lambda_{\ell k} N_k) t & \ell = 1,\dots,p \end{cases}$$

with end point coordinates

$$\begin{cases} z^o_j \exp 2\pi i\, N_j & j = 1,\dots,m \\ z^o_{m+\ell} \exp 2\pi i \sum_{k=1}^{m} \lambda_{\ell k} N_k & \ell = 1,\dots,p \end{cases}$$

which must be near (for some values $N_{m+\ell} \in \mathbb{Z}$) and even inside the exponential arguments to:

$$\begin{cases} z^o_j \exp 2\pi i\, N_j & j = 1,\dots,m \\ z^o_{m+\ell} \exp 2\pi i\, N_{m+\ell} & \ell = 1,\dots,p. \end{cases}$$

If we divide the exponential arguments by $N = \sum_{j=1}^{m} |N_j|$ (not the same as in §2!) then limit values

$$c_j \quad \text{of} \quad \frac{N_j}{N} \quad \text{and} \quad c_{m+\ell} \quad \text{of} \quad \frac{N_{m+\ell}}{N}$$

will obey relations of the kind

$$\lambda_{11}c_1 + \lambda_{12}c_2 +\dots+ \lambda_{1m}c_m = c_{m+1}$$

$$\dots\dots\dots\dots\dots\dots$$

$$\lambda_{p1}c_1 + \lambda_{p2}c_2 +\dots+ \lambda_{pm}c_m = c_{m+p}$$

with $\sum_{j=1}^{m} |c_j| = 1$.

For $p=1$, $m \geq 2$ the set of solutions $(c_1,\dots,c_m)$ has a

higher dimension then for $p > 1$, unless the complex numbers λ_{jk} obey special conditions. We express this by saying that <u>in general</u> this is the case. These solutions $(c_1,\dots,c_m)$ represent asymptotic directions to which sequences of points $(N_1,\dots,N_m) \in \mathbb{Z}^m \subset \mathbb{R}^m$ converge in $\mathbb{R}^m$.
Here $\mathbb{Z}^m$ is the fundamental group of the leaf in V^m, which guides the holonomy near $z^o \in V^m$. All this together implies that in general $p=1$ and $p > 1$ can be distinguished by topological means and so can therefore the sets

$$Sk_{m+1} \setminus Sk_m \quad \text{and} \quad \mathbb{C}^n \setminus Sk_{m+1}$$

As Sk_{m+1} is in this way topologically defined, so are all coordinate m+1-planes. Then Klares' invariant in each coordinate m+1-plane is an invariant of the topological orbitstructure of the $\mathbb{C}^m$-flow φ.

We still have to conclude to orbit rigidity. If there is a homeomorphism h giving rise to φ' not conjugate to φ, then we find that this homeomorphism inside the coordinate m+1-plane V^{m+1} determines (by composition with a φ-orbit preserving homeomorphism to) a unique topological conjugacy, such that the values λ'_{jk} and λ_{jk} are related by (see (18))

$$\left.\begin{aligned} g_{m+1}(\lambda_{1j}\, i) &= \lambda'_{1j}\, i \\ g_{m+1}(\lambda_{1j}) &= \lambda'_{1j}\, \tau_j \end{aligned}\right\} \quad j = 1,\dots,m$$

Therefore:

$$\tau_j = \frac{i\, g_{m+1}(\lambda_{1j})}{g_{m+1}(\lambda_{1j}\, i)} \qquad j = 1,\dots,m .$$

We can make the same calculation for the coordinate m+1-plane of the coordinates $z_1,\dots,z_m$ and $z_{m+\ell}$ and find (for the same τ_j):

$$\tau_j = \frac{i\, g_{m+\ell}(\lambda_{\ell j})}{g_{m+\ell}(\lambda_{\ell j}\, i)} \qquad j = 1,\dots,m \quad \ell = 1,\dots,p$$

Let us temporarily <u>omit</u> <u>various</u> <u>indices</u> and call

$$g_{m+\ell}(1) = 1+w, \quad \lambda_{\ell j} = \rho e^{i\varphi}, \quad \rho > 0,$$

and recall $g_{m+\ell}(i) = i$. We then obtain

$$\tau = \frac{ig(\cos\varphi + i\sin\varphi)}{g(i(\cos\varphi + i\sin\varphi)} = \frac{i(\cos\varphi\cdot(1+w)+\sin\varphi\cdot i)}{\cos\varphi\cdot i - \sin\varphi\cdot(1+w)} = \frac{\cos\varphi + i\sin\varphi + \cos\varphi\cdot w}{\cos\varphi + i\sin\varphi + i\sin\varphi\cdot w}$$

and

$$\tau - 1 = \frac{e^{-2i\varphi}w}{1+i\sin\varphi\cdot e^{-i\varphi}\cdot w}$$

Revival of indices gives

$$\tau_j - 1 = \frac{(\exp 2i\varphi_{\ell_j})w_\ell}{1+i\sin\varphi_{\ell_j}(\exp i\varphi_{\ell_j})w_\ell}$$

For $j = 1$ as well as for $j = 2$ we get a quadratic curve in the complex (w_1, w_2)-plane by putting the right hand sides equal for $\ell = 1$ and $\ell = 2$. If these curves have more than four points in common then they coincide completely and this is necessary for getting a nearby non conjugate φ'. Putting coefficients proportional in the resulting quadratic equations we find after some calculation:

$$\varphi_{22} - \varphi_{21} - \varphi_{12} + \varphi_{11} = 0 \mod \pi$$

or

$$0 \neq \frac{\lambda_{22}\lambda_{11}}{\lambda_{12}\lambda_{21}} \in \mathbb{R}.$$

This condition is <u>in general</u> not fulfilled and therefore orbit-rigidity follows in general.

<u>Remark</u>: With the powerful work on linearization near a singularity of a flow, of Mark Chaperon [6], it follows (modulo some small-divisor cases left open) that two near non linear $\mathbb{C}^m$-actions on $\mathbb{C}^n$, $2 \leq m \leq n-2$, φ and φ' with "hyperbolic" singularities at $0 \in \mathbb{C}^n$, have homeomorphic foliations near 0 if and only if the linear parts $L(\varphi)$ and $L(\varphi')$ are conjugate.

References

[1] C. Camacho, N.H. Kuiper, J. Palis: "The topology of holomorphic flows with singularity" Publ. Math. I.H.E.S., 48 (1979), p. 5-38.

[2] Ju. S. IL' iašenko: "Global and local aspects of the theory of complex differential equations." Proc. I.C. Helsinki, p. 821-826.

[3] B. Klares: "Classification topologique des n-uples de champs de vecteurs holomorphes commutatifs sur $P_{n+1}(\mathbb{C})$." Lecture Notes in Mathematics 1015, Springer-Verlag 1983.

[4] N.H. Kuiper: "La topologie des singularités hyperboliques des actions de $\mathbb{R}^2$." Asterisque 59-60 (1978) p. 131-150.

[5] C. Camacho, A. Lins Neto: "The topology of integrable differential forms near a singularity". To appear in Publ. Math. de l'IHES - 1982.

[6] M. Chaperon: "Propriétés génériques des germes d'actions différentiables de groupes de Lie commutatifs élémentaires", thèse, Université Paris 7, octobre 1980.

Institut des Hautes Etudes Scientifiques
91440 Bures-sur-Yvette (France)

LYAPUNOV FUNCTIONS AND STABILITY OF GEODESIC FLOWS

Jorge Lewowicz.

0. INTRODUCTION

Geodesic flows on manifolds of negative curvature have been studied extensively. They are known to be Anosov flows and hence structurally stable (see, for instance, [1]) and topologically stable [2]. Many of their ergodic properties are known, particulary for surfaces, also in the case when the curvature is only assumed to be non-positive [5].

In this article we consider, from the topological point of view, geodesic flows ϕ on two-dimensional compact reimannian manifolds M of non-positive curvature. When the subset of M where the curvature vanishes does not contain whole geodesics, the geodesic flow is again Anosov as may easily be shown (see, section 1). On the other hand, when that subset does contain an entire geodesic, it is never Anosov. Hence such a geodesic flow may present new topological aspects only in this case; nevertheless as we essentially show in this paper, unless the set of zeroes of the curvature contains an open invariant set of ϕ, the geodesic flow still has strong stability properties.

We show, among other things, that when the curvature K is such that the interior of $K^{-1}(\{0\})$ is void, the geodesic flow is topologically stable (Theorem 4.2.) and that it is C^1-structurally stable within the class of geodesic flows produced by riemannian metrics whose curvature (besides being everywhere non-positive) satisfies the above mentioned property (Corollary 4.3). We also show that if $K^{-1}(\{0\})$ is the union of a finite collection of pairwise disjoint closed geodesics such that the second derivative of the curvature in the direction normal to these geodesics is different from zero at each point of $K^{-1}(\{0\})$, then, any vector field Y on $S(M)$ (the bundle of unit vectors tangent to M) that is close enough in the C^4-topology to the vector field X of the geodesic flow, and that has the same 3 - jet as X at each point $(x,u)\in S(M)$, $u\in T_xM$, on those geodesics, produces a flow topologically conjugate to ϕ (Theorem 4.4). I am grateful to A. Katok for having mentioned this type of problem to me, during the symposium.

As the reader will realize the above mentioned results admit generalizations to both the case where the curvature may be positive somewhere, and the case of higher dimensional M. This last possibility of generalization rests cheefly upon the n-dimensional character of the topological methods we use. These methods are essentially those of [4]; however some new features appear due to the lack of an invariant natural family of local sections of the geodesic flow (see section 2).

1. PRELIMINARIES

M will be throughout a two dimensional smooth oriented compact connected riemannian manifold and $S(M)$ the circle bundle of M. As usual, we shall denote by $T(S(M))$ the tangent bundle of $S(M)$ and by $\pi: S(M) \to M$, the canonical projec-- tion. Elements of $S(M)$ will be denoted either by (x, u), (y, v), etc., $(x,y \in M$, $u \in T_xM$, $v \in T_yM)$ or by ξ, η, etc.

Let $\theta: T(S(M)) \to R$ be the reimannian connection 1-form (see, for instance, [3]), ω the 1-form defined by $\omega_{(x,u)}(\vec{t}) = (\pi'\vec{t}, u)$ and $\omega^{\perp}$ the 1-form defined by $\omega^{\perp}_{(x,u)}(\vec{t}) = (\pi'\vec{t}, iu)$. Here, $\vec{t} \in T(S(M))$, $(.,.)$ denotes the scalar product given by the riemannian metric, and $iu \in T_xM$ is the unit vector perpendicular to u and such that the basis $\{u, iu\}$ of T_xM agrees with the orientation of M. $K: M \to R$ will be the curvature function and X and V will stand, respectively, for the vector field defining the geodesic flow ϕ, and the vertical vector field on $S(M)$. Also iX will denote the horizontal vector field such that $\pi'_{(x,u)}(iX) = (x, iu)$; in general, if $\xi \in S(M)$, $\xi = (x, u)$ we shall write $i\xi$ for (x, iu) and $e^{i\varphi}\xi$ for $(x, \cos\varphi u + \sin\varphi\, iu)$, $\varphi \in R$.

Let us recall the structural equations relating the forms ω and $\omega^{\perp}$ to the riemannian connection form θ:

$$d\theta = -(K\circ\pi)\ \omega \wedge \omega^{\perp}$$

$$d\omega = \theta \wedge \omega^{\perp}$$

$$d\omega^{\perp} = -\theta \wedge \omega$$

From this formulae it is easy to calculate the Lie derivatives $\mathcal{L}_X$ of these forms along the geodesic flow:

$$\mathcal{L}_X\theta = -(K\circ\pi)\omega^{\perp},\ \mathcal{L}_X\omega = 0,\ \mathcal{L}_X\omega^{\perp} = \theta .$$

It follows, in particular, that the bundle $W = \{\vec{t} \in T(S(M)): \omega(\vec{t}) = 0\}$, transversal to X, is invariant under the geodesic flow.

This paragraph concerns Anosov vector fields ([1][2]). Let N be a compact smooth riemannian manifold and X a C^1-vector field on N such that there is a continuous vector sub-bundle $W \subset T(N)$, transversal to X, and invariant under the flow it generates. Then we may prove, with the same arguments as those used in the proof of Theorem 2.1 in [4], the following

Theorem 1.1. X *is an Anosov vector field if and only if there exists a non-degenerate continuous quadratic form* $B: W \to R$ *such that* $\mathcal{L}_X(B): W \to R$ *is positive definite.*

Let us return to our geodesic flow on M, and let $B(\vec{t}) = \theta(\vec{t}) \,.\, \omega^{\perp}(\vec{t})$, $\vec{t} \in T(S(M))$; we have that $\mathcal{L}_X(B) = \mathcal{L}_X(\theta) \,.\, \omega^{\perp} + \theta \,.\, \mathcal{L}_X(\omega^{\perp}) = -(K\circ\pi)(\omega^{\perp})^2 + \theta^2$ which is positive definite on W if $K(x) < 0$, $x \in M$. Obviously, B is non-degenerate, and therefore, by the previous theorem, the geodesic flow is Anosov if the curvature is negative.

Let $t_0 > 0$ and let us define $\hat{B}$ by $\hat{B} = \int_0^{t_0} \phi_s^{\#}(B)ds$, where, as usual, $(\phi_s^{\#}(B))_\xi(\vec{t}) = B_{\phi(\xi,s)}(\phi_s'(\xi)\vec{t})$, $\vec{t} \in T_\xi S(M)$. Then $\mathcal{L}_X(\hat{B}) = \int_0^{t_0} \phi_s^{\#}(\mathcal{L}_X(B))ds$ which is positive definite on W if $K \leq 0$ provided that for each $\xi \in S(M)$, there exists t, $0 \leq t \leq t_0$, such that $K\circ\pi\ (\phi(\xi, t)) < 0$. Since $\hat{B}(\vec{t}) > 0$ when $B(\vec{t}) > 0$ and $\hat{B}(\phi'_{-t_0}\ \vec{t}) < 0$ if $B(\vec{t}) < 0$, $\hat{B}$ is non-degenerate, and consequently, the geodesic flow is Anosov also in the case of non-positive curvature, unless $K^{-1}(\{0\})$ contains a whole geodesic. In that case it is never Anosov since if $\pi(\phi(\xi,t)) \in K^{-1}(\{0\})$, $t \in R$, we have that $\theta(\phi_t'(\xi)iX) = \theta_\xi(iX) = 0$ for every $t \in R$, and therefore $\phi_t'(\xi)(iX_\xi) = iX_{\phi(\xi,t)}$, $t \in R$, which would be impossible if ϕ were Anosov. Thus, for manifolds of non-positive curvature, new situations appear only in case $\pi^{-1} \circ K^{-1}(\{0\})$ contains a non void invariant set of the geodesic flow. To face these situations we shall have to consider, instead of Lyapunov forms such as B, Lyapunov functions defined on certain pairs of nearby points of $S(M)$.

2. LYAPUNOV FUNCTIONS

Let (m, n, φ) belong to a suitable neighbourhood of 0 in R^3; for each $\xi \in S(M)$ we define p_ξ on that neighbourhood by

$$p_\xi(m, n, \varphi) = e^{-(\frac{\pi}{2}-\varphi)i}\, \phi(i\phi(\xi, m), n).$$

Obviously p_ξ is smooth and $p_\xi(0, 0, 0) = \xi$. Also

$$p_\xi(m, n, \varphi) = p^0_{\phi(\xi,m)}(n, \varphi) \tag{2.1}$$

where $p^0_\xi(n, \varphi) = p_\xi(0, n, \varphi)$. It is easy to check that

$$p'_\xi(0)\, \frac{\partial}{\partial m} = X_\xi, \quad p'_\xi(0)\, \frac{\partial}{\partial n} = iX_\xi, \quad p'_\xi(0)\, \frac{\partial}{\partial \varphi} = V_\xi \tag{2.2}$$

which implies that p_ξ is a diffeomorphis on some neighbourhood of 0.

Let us calculate for further use $p'_\xi(0, n, \varphi)$, $\xi = (x, u)$, $x \in M$, $u \in T_xM$, $\|u\| = 1$. Call $\eta = (y, v) = p^0_\xi(n, \varphi)$; since $\pi o p^0_\xi = \pi o \hat{p}^0_\xi$, $\hat{p}^0_\xi(n, \varphi) = \phi(i\xi, n)$, we have that

$$\omega(p'_\xi(0, n, \varphi)\, \frac{\partial}{\partial m}) = (\pi'_\eta p'_\xi(0, n, \varphi)\, \frac{\partial}{\partial m}, v) = ((\pi o \hat{p}^0_\xi)'\, \frac{\partial}{\partial m}, v).$$

Let $\phi(i\xi, n) = (y, i\bar{u})$ and let $E \in T_{i\xi}S(M)$ be the horizontal vector such that $\pi'_{i\xi}(E) = (x, u)$. Then $\omega(p'_\xi(0, n, \varphi)\, \frac{\partial}{\partial m}) = (\pi'\phi'_n(i\xi), v)$, and since $v = \cos\varphi\bar{u} + \operatorname{sen}\varphi\, i\bar{u}$ we get that $\omega(p'_\xi(0, n, \varphi)\, \frac{\partial}{\partial m}) = -\cos\varphi\, \omega^\perp_{(y,i\bar{u})}(\phi'_n(i\xi)E)$ on account of the fact that $\omega_{(y,i\bar{u})}(\phi'_n(i\xi)E) = \omega_{i\xi}(E) = 0$. Similarly, $\omega^\perp(p'_\xi(0,n,\varphi)\frac{\partial}{\partial m}) = \sin\varphi\, \omega^\perp_{(y,i\bar{u})}(\phi'_n(i\xi)E)$ and $\theta(p'_\xi(0, n, \varphi)\frac{\partial}{\partial m}) = \theta_{(y,i\bar{u})}(\phi'_n(i\xi)E)$. On the other hand $p'_\xi(0, n, \varphi)\frac{\partial}{\partial n} \in T_\eta S(M)$ is a horizontal vector and $\pi' o\, p'_\xi(0, n, \varphi)\frac{\partial}{\partial n} = (y, i\bar{u})$; $\omega(p'_\xi(0, n, \varphi)\frac{\partial}{\partial n}) = (i\bar{u}, v) = \sin\varphi$; $\omega^\perp(p'_\xi(0,n,\varphi)\frac{\partial}{\partial n}) = (i\bar{u}, iv) = \cos\varphi$; $\theta(p'_\xi(0, n, \varphi)\frac{\partial}{\partial n}) = 0$. Since, clearly, $p'_\xi(0, n, \varphi)\frac{\partial}{\partial\varphi} = V_\eta$, we have proved

Lemma 2.1. (i) $p'_\xi(0, n, \varphi)\frac{\partial}{\partial m} = -\cos\varphi\, \omega^\perp_{(y,i\bar{u})}(\phi'_n(i\xi)E)X_\eta + \sin\varphi\, \omega^\perp_{(y,i\bar{u})}(\phi'_n(i\xi)E)iX_\eta$

$\underline{+\theta_{(y,i\bar{u})}(\phi_n'(i\xi)E)V_\eta}$

(ii) $\underline{p_\xi'(0,\ n,\ \varphi)\ \frac{\partial}{\partial n} = \sin\varphi\ X_\eta + \cos\varphi\ iX_\eta}$

(iii) $\underline{p_\xi'(0,\ n,\ \varphi)\ \frac{\partial}{\partial\varphi} = V_\eta}$

Now we are ready to go into Lyapunov functions. Let $N = \{n, \varphi\}$ be an open ball centered at $0 \in R^2$ so that p_ξ^0/N is an embedding and $p_\xi^0(N)$ is transversal to X for every $\xi \in S(M)$. Thus the $p_\xi^0(N)$, $\xi \in S(M)$ constitute a family of local sections of the geodesic flow which, in general, is not invariant under it.

Remark. Since, as it is easy to show,

$$-\frac{1}{2} = \theta \wedge \omega^{\perp}(iX,\ V) = d\omega(iX,\ V) = -\ \omega([\,iX,\ V]),$$

the distribution defined by $\omega = 0$ is never integrable. This justifies some of the constructions that will follow.

On the set

$$\{(\xi,\ \eta) \in S(M) \times S(M):\ p_\xi^0(n\ ,\ \varphi) = \eta,\ (n,\ \varphi) \in \frac{1}{2}\,N\}$$

we define $\mathcal{V}(\xi,\ \eta)$ by $\mathcal{V}(\xi,\ \eta) = \sin^2\varphi + n^2$, and $\dot{\mathcal{V}}(\xi,\ \eta)$ by

$$\dot{\mathcal{V}}(\xi,\ \eta) = \lim_{t\to 0} \frac{1}{t}\,(\mathcal{V}(\phi_\tau(\xi),\ \phi_t(\eta)) - \mathcal{V}(\xi,\ \eta))$$

where $\tau = \tau_\eta(t)$ is the unique (by (2, 2)) real number close to 0 such that $\phi_t(\eta) \in p^0_{\phi(\xi,\tau)}(N)$. Then

$$\dot{\mathcal{V}}(\xi,\ \eta) = \lim_{t\to 0} \frac{1}{t}(\sin^2\varphi(t) + n^2(t) - \sin^2\varphi - n^2)$$

where

$$\varphi(0) = \varphi,\ n(0) = n \quad \text{and} \quad p_\xi(\tau_\eta(t),\ n(t),\ \varphi(t)) = \phi(\eta,\ t)$$

for t close to 0. Consequently

$$\dot{\mathcal{V}}(\xi,\ \eta) = 2(\sin\varphi\cos\varphi\ \dot{\varphi}(0) + n\ \dot{n}(0))$$

where $p'_\xi(0, n, \varphi)\ (\dot{\tau}_\eta(0)\frac{\partial}{\partial m} + \dot{n}(0)\frac{\partial}{\partial n} + \dot{\varphi}(0)\frac{\partial}{\partial\varphi}) = X_\eta$

From Lemma 2.1 we get

$$\dot{n}(0) = \sin\varphi,\ \dot{\varphi}(0) = -\ \dot{\tau}_\eta(0)\ \theta_{(y.i\bar{u})}(\phi'_n(i\xi)E)$$

$$\dot{\tau}_\eta(0) = -\cos\varphi(\omega^{\perp}_{(y.i\bar{u})}(\phi'_n(i\xi)E))^{-1}$$

Let $\chi(n) = \theta_{(y,i\bar{u})}(\phi'_n(i\xi)E)\ (\omega^{\perp}_{(y.i\bar{u})}(\phi'_n(i\xi)E))^{-1}$

Lemma 2.2. $\frac{1}{2}\dot{V}(\xi, \eta) = \sin\varphi\cos^2\varphi\ \chi(n) + n\sin\varphi = (1-K\circ\pi(\xi))n\varphi + o(n^2+\varphi^2)$

Proof. The first formula follows from the previous computation. As for the second one, it is enough to show that $\chi(n) = -K\circ\pi(\xi)n + o(n)$ and this follows from the fact that $\omega^{\perp}_{i\xi}(E) = -1$ and

$$\theta_{(y,i\bar{u})}(\phi'_n(i\xi)E) = \int_0^n -K\circ\pi(\phi(i\xi, s))\omega^{\perp}(\phi'_s(i\xi)E)ds.$$

Ley us remark for further use, that since $K\circ\pi(\xi) \leq 0$, for every $\xi \in S(M)$, it is always possible to find $\rho > 0$ and subspaces S_ξ, $U_\xi \subset W_\xi$ such that $\dot{V}(\xi, \eta) < 0$ (> 0) if $\eta = p^0_\xi(n, \varphi)$, $n\ iX_\xi + \varphi V_\xi \in S_\xi$ (respectively, $n\ iX_\xi + \varphi V_\xi \in U_\xi$) and $0 < n^2 + \varphi^2 \leq \rho^2$. Moreover, the subspaces S_ξ, U_ξ can be chosen to vary continuously with ξ.

Now we define $\ddot{V}(\xi, \eta)$ with the same procedure used above, as $(\dot{V}(\xi, \eta))^{\cdot}$ and prove

Lemma 2.3. $\frac{1}{2}\ddot{V}(\xi, \eta) =$

$\cos^2\varphi\ n\ \chi(n) + \sin^2\varphi - K\circ\pi(\eta)\sin^2\varphi\cos^2\varphi + \chi^2(n)(\cos^4\varphi - 3\sin^2\varphi\cos^2\varphi)$.

Proof. The equality follows from the values for $\dot{n}(0)$, $\dot{\varphi}(0)$, $\dot{\tau}_\eta(0)$ obtained previously and the fact that $\frac{d\chi(n)}{dn} = K\circ\pi(\eta) - \chi^2(n)$.

Notice that if η lies in a suitably small neighbourhood of ξ, $\frac{1}{2}\ddot{V}(\xi, \eta) \geq$

$\sin^2\varphi + (\omega^{\perp}(\phi_n'(i\xi)E))^{-1}\cos^2\varphi\ n\int_0^n -K\circ\pi(\phi(i\xi, s)\omega^{\perp}(\phi_s'(i\xi)E)ds$ which is positive if $n^2+\varphi^2>0$ unless K vanishes on $\pi(\phi(i\xi, s))$ for every s on an open nieghbourhood of 0. Let T be a small positive number and define $\mathcal{W}(\xi, \eta)$ by

$$\mathcal{W}(\xi, \eta) = \int_0^T \mathcal{V}(\phi(\xi, \tau_\eta(s)), \phi(\eta, s))ds.$$

Then $\mathcal{W}(\phi(\xi,\tau_\eta(t)), \phi(\eta, t)) = \int_0^T \mathcal{V}(\phi(\phi(\xi,\tau_\eta(t)),\tau_{\phi(\eta,t)}(s)), \phi(\phi(\eta, t), s))ds$

and since $\tau_\eta(t)+\tau_{\phi(\eta,t)}(s) = \tau_\eta(t+s)$, if t, s are small, we have that

$$\mathcal{W}(\phi(\xi,\tau_\eta(t)), \phi(\eta, t)) = \int_0^T \mathcal{V}(\phi(\phi(\xi,\tau_\eta(s)), \tau_{\phi(\eta,s)}(t)), \phi(\phi(\eta, s),t))ds$$

Hence $\dot{\mathcal{W}}(\xi, \eta) = \int_0^T \dot{\mathcal{V}}(\phi(\xi, \tau_\eta(s)), \phi(\eta, s))ds$, and analogously, $\ddot{\mathcal{W}}(\xi, \eta) = \int_0^T \ddot{\mathcal{V}}(\phi(\xi, \tau_\eta(s)), \phi(\eta, s))ds$.

Now we choose any riemannian metric for the compact manifold $S(M)$ and state the following proposition that summarizes our previous discussion.

Proposition 2.4. Assume that M has non-positive curvature and that $K^{-1}(\{0\})$ has an empty interior. Then there exist a $\rho > 0$ and a smooth function $\mathcal{U}$ defined on pairs $\xi, \eta \in S(M)$ such that $p_\xi^0(n, \varphi) = \eta$, $\|n\ iX_\xi + \varphi\ V_\xi\| < \rho$ with the properties:

(i) $\mathcal{U}(\xi, \eta) > 0$ if $\xi \neq \eta$.

(ii) There exist subspaces S_ξ, U_ξ, $S_\xi \oplus U_\xi = W_\xi$, depending continuously on ξ and such that $\dot{\mathcal{U}}(\xi, \eta) < 0\ (> 0)$ if $0 < \|n\ iX_\xi + \varphi V_\xi\| < \rho$ and $n\ iX_\xi + \varphi V_\xi \in S_\xi$ (respectively, $n\ iX_\xi + \varphi V_\xi \in U_\xi$).

(iii) $\ddot{\mathcal{U}}(\xi, \eta) > 0$ if $\xi \neq \eta$.

Proof. Choose a small $T > 0$ and take $\mathcal{U}$ to be equal to our previous $\mathcal{W}$. From the remark after the proof of Lemma 2.3. it follows that if $\ddot{\mathcal{U}}(\xi, \eta) = 0$ and $\xi \neq \eta$, then $K\circ\pi(\phi(i\phi(\xi, t), s)) = 0$ for $0 \leq t \leq T$ and s in some neighbourhood of 0. K would then vanish on an open set, which is absurd.

As the reader will have noticed, the previous result may be generalized so as to include certain cases where $K \geq 0$ on some subset of M of non-void interior.

3. PERSISTENCE PROPERTIES

Let $\mathcal{U}$ and ρ be as in Proposition 2.4. and let ε be a positive number. Choose $h > 0$ so that $\mathcal{U}(\xi, \eta) > 0$ implies $\mathrm{dist}(\xi, \eta) < \varepsilon$ and $\| n\, iX_\xi + \varphi V_\xi \| \leq \rho$; here, as before, $p_\xi^0(n, \varphi) = \eta$. Let Y be a C^1-vector field on $S(M)$ producing the flow ψ. Assume that Y is so close to X in the C^1-topology, that $p_\xi^0(N)$ is transversal to Y for every $\xi \in S(M)$, and that the mixed derivatives

$$\dot{\mathcal{U}}^*(\xi, \eta) = \lim_{t \to 0} \frac{1}{t} (\mathcal{U}(\phi(\xi, \tau_\eta(t)), \psi(\eta, t)) - \mathcal{U}(\xi, \eta))$$

and

$$\ddot{\mathcal{U}}^*(\xi, \eta) = \lim_{t \to 0} \frac{1}{t} (\dot{\mathcal{U}}^*(\phi(\xi, \tau_\eta(t)), \psi(\eta, t)) - \dot{\mathcal{U}}^*(\xi, \eta))$$

where, as always, $\tau_\eta(t)$ is the only real number close to 0 such that $\psi(\eta, t) \in$ $\in p^0_{\phi(\xi, \tau_\eta(t))}(N)$, have the following properties when $\mathcal{U}(\xi, \eta) = h$.

$$\dot{\mathcal{U}}^*(\xi, \eta) < 0 \ (> 0) \quad \text{if} \quad n\, iX_\xi + \varphi V_\xi \in S_\xi \quad (\text{resp. } U_\xi) \tag{3.1}$$

and

$$\ddot{\mathcal{U}}^*(\xi, \eta) > 0 \tag{3.2}$$

Let $\xi \in S(M)$ and for $t \in R$ define

$$H_t(\xi) = \{\eta \in p^0_{\phi(\xi, t)}(N) \colon \mathcal{U}(\phi(\xi, t), \eta) \leq h\} .$$

Lemma 3.1. <u>For each</u> $\xi \in S(M)$ <u>there exists</u> $\eta \in H_0(\xi)$ <u>and a smooth increasing and surjective function</u> $\tau_\eta \colon R \to R$, $\tau_\eta(0) = 0$, <u>such that</u> $\psi(\eta, t) \in H_{\tau_\eta(t)}(\xi)$, $t \in R$.

Proof. We argue by contradiction as in [4]. If there is no such an η, then, for every $\eta \in H_0(\xi)$ there exists $t(\eta) \geq 0$ so that

i) There exists a smooth increasing function $\tau_\eta : [-t(\eta), t(\eta)] \to R$, $\tau_\eta(0) = 0$ such that $\psi(\eta, t) \in H_{\tau_\eta}(t)$, $t \in [-t(\eta), t(\eta)]$, and

ii) Either $\psi(\eta, -t(\eta)) \in \partial H_{\tau_\eta(-t(\eta))}(\xi)$ or

$$\psi(\eta, t(\eta)) \in \partial H_{\tau_\eta(t(\eta))}(\xi).$$

To prove this assertion take $t(\eta)$ to be the maximum of those $T \geq 0$ such that i) is satisfied on $[-T, T]$. It is clear that if $\tau : [-T, T] \to R$ and $\tau' : [-T', T']$ satisfy i), then $\tau' = \tau$ on $[-T, T] \cap [-T', T']$. Also, on account of the fact that both Y and X are transversal to $p_\xi^0(N)$ we may deduce that neither $t(\eta)$ nor $\tau(t(\eta))$ could be infinite. On the other hand, ii) is clear if we define $t(\eta)$ in this way.

Now, using (3.2) we prove as in [4] that $t(\eta)$ is a continuous function of η and that $t(\eta) = 0$ if $\mathfrak{U}(\xi, \eta) = h$. We need to show also that $\tau_\eta(t(\eta))$ is a continuous function of η, but this follows easily from the following remark: if $\eta_\nu \to \eta_\infty$ and $|t| \leq t(\eta_\nu)$, $t(\eta_\infty)$, then $\tau_{\eta_\nu}(t) \to \tau_{\eta_\infty}$, which may be proved on account of (2.2) by taking the supremum of those t_0 for which this is true for every t, $|t| \leq t_0$.

Let $G_\xi = \{n\, iX_\xi + \varphi V_\xi : \mathfrak{U}(\xi, p_\xi^0(n, \varphi)) \leq h\} \subset W_\xi$ and for each $\xi \in S(M)$ let $q_\xi : H_0(\xi) \to W_\xi$ be defined by $q_\xi(\eta) = n\, iX_\xi + \varphi V_\xi$, $\eta = p_\xi^0(n, \varphi)$. Consider the functions $\alpha, \beta : G_\xi \to W_\xi$ defined by

$$\alpha(\vec{t}) = (P_\eta \circ \phi'_{-\tau}(\phi_\tau(\xi)) \circ q_{\phi(\xi,\tau)} \circ \psi_{t(\eta)})(\eta)$$

$$\beta(\vec{t}) = (Q_\eta \circ \phi'_{-\bar{\tau}}(\phi_{\bar{\tau}}(\xi)) \circ q_{\phi(\xi,\bar{\tau})} \circ \psi_{-t(\eta)})(\eta)$$

where $q_\xi(\eta) = \vec{t}$, $\tau = \tau_\eta(t(\eta))$, $\bar{\tau} = \tau_\eta(-t(\eta))$ and $P_\eta : W_\xi \to W_\xi$ is the projection onto $\phi'_{-\bar{\tau}}(\phi_{\bar{\tau}}(\xi))\ U_{\phi(\xi,\bar{\tau})}$ along $\phi'_{-\tau}(\phi_\tau(\xi))\ S_{\phi(\varepsilon,\tau)}$, whereas Q_η is the complementary projection. This subspaces of W_ξ are transversal since $\theta\,\omega^\perp$ is positive in one of them and negative in the other one. Then, on account of (3.1) it follows as in [4], page 198, that $\gamma = \alpha + \beta : G_\xi \to W_\xi - \{0\}$ is a continuous function such that $\gamma(\vec{t}) = \vec{t}$ if $\vec{t} \in \partial G_\xi$, which is absurd. This completes the proof.

Lemma 3.2. For each $\eta \in S(M)$ there exists $\xi \in S(M)$ and a smooth increasing surjective function $\tau: R \to R$, $\tau(0) = 0$ such that $\psi(\eta, t) \in H_{\tau(t)}(\xi)$, $t \in R$.

Proof. Let $B = \{\xi \in S(M): \eta \in H_0(\xi)\}$. On account of the fact that the exponential mapping is a local diffeomorphism it is easy to show that the mapping $r: B \to \{(n, \varphi) \in R^2: \sin^2\varphi + n^2 \leq h\}$ defined by $r(\xi) = (n,\varphi)$ where $p^0_\xi(n, \varphi) = \eta$ is a surjective homeomorphism if h is small enough.

Let us argue again by contradiction. If the thesis is not true, for each $\xi \in B$ there exists $t(\xi)$, $0 \leq t(\xi) < \infty$ such that

i) There exists a smooth increasing function $\tau_\xi: [-t(\xi), t(\xi)] \to R$, $\tau(0) = 0$ with the property that $\psi(\eta, t) \in H_{\tau_\xi(t)}(\xi)$, $|t| \leq t(\xi)$,

ii) Either $\psi(\eta, t(\xi)) \in \partial H_{\tau_\xi(t(\xi))}(\xi)$ or

$$\psi(\eta, -t(\xi)) \in \partial H_{\tau_\xi(-t(\xi))}(\xi).$$

We show as above that $t(\xi)$, $\tau_\xi(t(\xi))$ are continuous and that $t(\xi) = 0$ if $\xi \in \partial B$. Take $q_{\phi(\xi,\tau(t(\xi)))}(\psi(\eta, t(\xi))) \in W_{\phi(\xi,\tau(t(\xi)))}$ and apply to this vector $\phi'_{-\tau(t(\xi))}$ obtaining a vector in W_ξ.

Proceed similarly with $q_{\phi(\xi,\bar{\tau}(\xi))}(\psi(\eta, -t(\xi))) \in W_{\phi(\xi,\bar{\tau}(\xi))}$ where $\bar{\tau}(\xi)$ stands for $\tau(-t(\xi))$, getting another vector in W_ξ. As in the proof of the previous lemma, we project the first one onto $\phi'_{-\bar{\tau}(\xi)} U_{\phi(\xi,\bar{\tau}(\xi))}$ along $\phi'_{-\tau(t(\xi))} S_{\phi(\xi,\tau(t(\xi)))}$ and apply the complementary projection to the second one. Finally we add up getting for each $\xi \in B$ a vector $\vec{t}(\xi) \in W_\xi$ that depends continuously on ξ and that never vanishes as it may be shown as in [4] on account of (3.1). Since $t(\xi) = 0$ when $\xi \in \partial B$, we have that for those ξ, $\vec{t}(\xi) = q_\xi(\eta)$. Now let $f: B \to R^2 - \{0\}$ be defined by $f(\xi) = (\omega^\perp(\vec{t}(\xi)), \theta(\vec{t}(\xi)))$. Then

$$f o r^{-1}: \{(n, \varphi) \in R^2: \sin^2\varphi + n^2 \leq h\} \to R^2 - \{0\}$$

is a continuous function such that, if $\sin^2\varphi + n^2 = h$, $f o r^{-1}(n, \varphi) = (n, \varphi)$ since $r^{-1}(n, \varphi) \in \partial B$; as this is absurd, the proof is complete.

If Y is only C^0-close to X, (3.2) is not necesarily true; nevertheless the same conclusions may be obtained on the basis of the remark at the end of the proof of Proposition 3.1. in [4], page 200. We summarize the previous conclusions in the following.

Proposition 3.3. If Y is a vector field on $S(M)$ producing a flow ψ and close enough to X in the C^0-topology, then for each $\xi(\eta) \in S(M)$ there exists $\eta(\xi) \in S(M)$ and a smooth increasing surjective function $\tau_\eta(\tau_\xi)\colon R \to R$, $\tau_\eta(0) = 0$ $(\tau_\xi(0) = 0)$ such that

$$\psi(\eta, t) \in H_{\tau_\eta(t)}(\xi)(H_{\tau_\xi(t)}(\xi)); \quad t \in R.$$

4. STABILITY

Lemma 4.1. Let $\eta \in H_0(\xi)$ be such that there exists a smooth increasing and surjective function $\tau\colon R \to R$, $\tau(0) = 0$ with the property that $\phi(\eta, t) \in H_{\tau(t)}(\xi)$, $t \in R$. Then $\eta = \xi$.

Proof. The arguments are analogous to those of Lyapunov's direct theorem. Consider $\mathcal{U}(\phi(\xi, \tau(t), \phi(\eta, t))$, and assume that $\dot{\mathcal{U}}(\xi, \eta) > 0$; since $\dot{\mathcal{U}}(\phi(\xi, \tau(t)), \phi(\eta, t))$ is increasing, we have that for $t \geq 0$ $\text{dist}(\phi(\xi, \tau(t)), \phi(\eta, t))$ is bounded away from zero, and on account of iii) in Proposition 2.4, this leads to a contradiction. In case $\dot{\mathcal{U}}(\xi, \eta) \leq 0$ we argue in a similar way for $t \leq 0$. This completes the proof.

Assume now that for some $\eta \in S(M)$ there exists $\xi . \xi' \in S(M)$ such that

$$\psi(\eta, t) \in \overline{H}_{\tau(t)}(\xi), \quad \psi(\eta, t) \in \overline{H}_{\tau'(t)}(\xi'), \ t \in R,$$

where τ and τ' have the properties mentioned in Lemma 4.1 and where $\overline{H}$ is defined as H but replacing h by $\overline{h}$, $0 < \overline{h} < h$. If $\overline{h}$ is small enough $\phi(\xi', t^*) \in H_0(\xi)$ for some t^* with $|t^*|$ small, and $\phi(\phi(\xi',t^*),t) \in H_{s(t)}(\xi)$ for some smooth increasing surjective function $s\colon R \to R$, $s(0) = 0$. By Lemma 4.1 this implies $\phi(\xi',t^*) = \xi$, and since $|t^*|$ is small we may conclude, recalling that p_ξ is a diffeomorphism that $\xi' = \xi$.

Theorem 4.2. If $K \leq 0$ and $K^{-1}(\{0\})$ has empty interior, the geodesic flow is topologically stable [2]

Proof. The previous results imply that if Y is close enough to X in the C^0-topology, to each $\eta \in S(M)$ we may associate a unique $\xi = \mu(\eta)$ and a function $\tau_\eta(t)$, with the known properties, such that $\psi(\eta, t) \in H_{\tau_\eta(t)}(\xi)$, $t \in R$. Moreover, μ is surjective and both μ and $\tau_\eta(t)$ are continuous as it is easy to check. Since $\mu(\psi_t(\eta)) = \phi(\mu(\eta), \tau_\eta(t))$ this completes the proof.

Let $\mathcal{H}$ be the class of vector fields Y on $S(M)$ producing geodesic flows on M and so that $K_Y(x) \leq 0$, $x \in M$, and that the interior of $K_Y^{-1}(\{0\})$ is void.

Corollary 4.3. If $X \in \mathcal{H}$, there is a C^1-neighbourhood $\mathcal{N}$ of X, such that if $Y \in \mathcal{N} \cap \mathcal{H}$, then the flows generated by Y and X are topologically conjugated.

Proof. Choose $\mathcal{N}'$ such that each $Y \in \mathcal{N}'$ is transversal to every $p_\xi^0(N)$, $\xi \in S(M)$ and also in order to have a fixed $h > 0$ such that for each $Y \in \mathcal{N}' \cap \mathcal{H}$, the Lyapunov function $\mathcal{U}_Y$ satisfies $\dot{\mathcal{U}}_Y(\xi, \eta) > 0$ if $0 < \sin^2\varphi + n^2 \leq h$, $p_\xi^0(Y, n, \varphi) = \eta$. Take $\mathcal{N} \subset \mathcal{N}'$ so small that for each $Y \in \mathcal{N} \cap \mathcal{H}$ and each $\eta \in S(M)$ there exist $\xi \in S(M)$ and $\tau(t)$ such that $\psi(\eta, t) \in \overline{H}_{\tau(t)}(\xi)$, $t \in R$, where

$$\overline{H}_t(\xi) = \{\eta \in p^0_{\phi(\xi,t)}(N) \colon \mathcal{U}_X(\phi(\xi, t), \eta) \leq \overline{h}\},$$

and where $\overline{h}$ is chosen so small that

$$\psi(\eta, t) \in \overline{H}_{\tau(t)}(\xi), \quad \psi(\eta', t) \in \overline{H}_{\tau'(t)}(\xi), \quad t \in R,$$

imply that

$$\psi(\eta', t) = p^0_{\psi(\eta,\tau''(t))}(Y, n, \varphi), \ \sin^2\varphi + n^2 \leq h;$$

here τ'' is a smooth increasing, surjective function, $\tau''\colon R \to R$, with $|\tau''(0)|$ small. Then Lemma 4.1 applies and we get, as before, $\eta = \psi(\eta', t_0)$ for some t_0 of small absolute value. Since Y is transversal to $p_\xi^0(N)$ this implies $t_0 = 0$, and consequently, the function μ defined in the proof of Theorem 4.2 is injective. This completes the proof.

We shall now assume that $K^{-1}(\{0\})$ is the union of a finite collection of pairwise disjoint closed geodesics and that

$$\left(\frac{d^2}{ds^2} K\circ\pi(\phi(i\,\xi_0, s))\right)_{s=0} < 0$$

for every ξ_0 that belongs to one of those geodesics, and, of course, we shall keep the assumption that K is non-positive everywhere. Let X be the vector field of the geodesic flow ϕ.

<u>Theorem 4.4. Assume that the riemannian manifold M satisfies the above conditions, and that Y is a vector field on S(M) that has the same 3-jet as X at each point ξ_0, such that</u> $K(\pi(\phi(\xi_0,t))) = 0$, <u>$t \in R$. If furthermore, Y is sufficiently close to X in the C^4-topology, then the flow produced by Y is topologically conjugated to</u> ϕ.

Proof. Let ξ,η be close to $\xi_0 \in (K\circ\pi)^{-1}(\{0\})$, $\xi = p^0_{\xi_0}(n_0, \varphi_0)$, $\eta = p^0_{\xi}(n, \varphi)$ and define the Lyapunov function $\mathcal{V}$ by

$$\mathcal{V}(\xi,\eta) = n\varphi + n^2(n\varphi + n_0\varphi_0);$$

clearly $\mathcal{V}(\xi,\eta) = 0$ if $\xi = \eta$. Then $\dot{\mathcal{V}}(\xi,\eta) = (\dot{n}\varphi + n\dot{\varphi})(1+n^2) + 2n^2\varphi\dot{n} + 2n\,\dot{n}\,n_0\varphi_0 + n^2(\dot{n}_0\varphi_0 + n_0\dot{\varphi}_0)$.

From section 2 we get that:

$$\dot{n} = \sin\varphi,\ \dot{\varphi} = \cos\varphi\ \chi(n),\ \dot{n}_0 = -\sin\varphi_0\ \cos\varphi\,(\omega^{\perp}(\phi'_n(i\xi)E))^{-1},$$

$$\dot{\varphi}_0 = -\cos\varphi\cos\varphi_0\chi(n_0)(\omega^{\perp}(\phi'_n(i\xi)E))^{-1}.$$

Since

$$\omega^{\perp}(\phi'_n(i\xi)E) = -1 + \int_0^n \theta(\phi'_s(i\xi)E)ds$$

and

$$\theta(\phi'_n(i\xi)E) = \int_0^n -(K\circ\pi)(\phi(i\xi,s))\omega^{\perp}(\phi'_s(i\xi)E)ds$$

we have that

$$\theta(\phi_n'(i\xi)E) = a_0 n + \frac{1}{2} b_0 n^2 + \frac{1}{6}(c_0 - a_0^2)n^3 + O(n^4),$$

$$\omega^{\perp}(\phi_n'(i\xi)E) = -1 + \frac{1}{2} a_0 n^2 + \frac{1}{6} b_0 n^3 + O(n^4),$$

$$\chi(n) = -a_0 n - \frac{1}{2} b_0 n^2 - \frac{1}{6}(c_0 - 2a_0^2)n^3 + O(n^4),$$

where

$$a_0 = K(\pi(\xi)), \quad b_0 = \left(\frac{\partial}{\partial n} K \circ \pi(\phi(i\xi, n))\right)_{n=0}, \quad c_0 = \left(\frac{\partial^2}{\partial n^2} K \circ \pi(\phi(i\xi, n))\right)_{n=0}$$

Let $k(m_0, n_0) = K \circ \pi(\phi(i\phi(\xi_0, m_0), n_0))$. Then

$$a_0 = k(0, n_0) = \frac{\partial^2 k}{\partial n_0^2}(0, 0)n_0^2 + o(n_0^2),$$

$$b_0 = -\frac{\partial k}{\partial m_0}(0, n_0)\sin\varphi_0 + \frac{\partial k}{\partial n_0}(0, n_0)\cos\varphi_0 = \frac{\partial^2 k}{\partial n_0^2}(0, 0)n_0^2 + o(n_0^2),$$

$$c_0 = \left(\frac{\partial^2}{\partial n_0^2} K \circ \pi(\phi(i\bar{\xi}_0, n))\right)_{n=0} + o(1); \quad \bar{\xi}_0 = e^{i\varphi_0}\xi_0$$

If $\pi(\phi(i\bar{\xi}_0, s)) = \pi(\phi(i\phi(\xi_0, m_0(s)), n_0(s)))$, we have that $\dot{m}_0(0) = -\sin\varphi_0$, $\dot{n}_0(0) = \cos\varphi_0$, and finally since

$$\left(\frac{\partial^2}{\partial n^2} K \circ \pi(\phi(i\bar{\xi}_0, n))\right)_{n=0} = \frac{\partial^2 k}{\partial n_0^2}(0, 0)(\dot{n}_0(0))^2,$$

we get that $c_0 = \frac{\partial^2 k}{\partial n_0^2}(0, 0)\cos^2\varphi_0 + o(1)$.

So, $\dot{\mathcal{V}}(\xi, \eta) = \varphi^2(1+o(1)) + n^2(2\varphi^2 + \varphi_0^2 - k(n_0^2 + \frac{1}{2} n_0 n + \frac{1}{6} n^2) + o(n^2 + \varphi^2 + n_0^2 + \varphi_0^2)) +$

$$+ \; O(n^2 + \varphi^2 + n_0^2 + \varphi_0^2)n\varphi,$$

where $k = \frac{\partial^2 k}{\partial n^2}(0, 0) < 0$ by assumption, and hence the coefficient of n^2 is greater than a positive definite quadratic form in the four variables, n, φ, n_0, φ_0. Consequently if we calculate the derivative of the <u>same</u> function $\mathcal{V}$ with respect to vector fields Y and if we had that the corresponding $\dot{n}_Y$, $\dot{\varphi}_Y$, $\dot{n}_{0Y}$, $\dot{\varphi}_{0Y}$ differ from the previous ones only by terms bounded by $C(n^4 + \varphi^4 + n_0^4 + \varphi_0^4)$, where C is a positive constant independent of n, φ, n_0, φ_0, ξ_0, we could find for those Y a fixed $h > 0$, and a fixed $h_0 > 0$, such that if $0 < n^2 + \varphi^2 \leq h$, $0 \leq n_0^2 + \varphi_0^2 \leq h_0$, then $\dot{\mathcal{V}}_Y(\xi, \eta)$, the derivative of the previous $\mathcal{V}$ with respect to the flow generated by Y, would be positive since the quadratic form in n, φ we would obtain in those cases would have a positive coefficient of φ^2

and a positive discriminant.

Now, let $\mathcal{U}$ be the previously defined Lyapunov function and let $\lambda: S(M) \to R$ be a C^∞ non-negative function which vanishes outside of a small neighbourhood of $(K\circ\pi)^{-1}(\{0\})$ and equals 1 on a smaller neighbourhood of the same subset of $S(M)$. Let c be a positive constant, and let $\mathcal{W} = \dot{\mathcal{U}} + c\lambda\mathcal{V}$. Then $\dot{\mathcal{W}} = \dot{\mathcal{U}} + c\dot{\lambda}\mathcal{V} + c\lambda\dot{\mathcal{V}}$ and since $\dot{\mathcal{U}}(\xi, \eta) \geq \alpha(n^2 + \varphi^2)$ for some $\alpha > 0$, provided ξ is outside some neighbourhood of $(K\circ\pi)^{-1}(\{0\})$ and $n^2 + \varphi^2$ is small, it is clear that we may choose c so small that $\dot{\mathcal{W}}(\xi, \eta) \geq \beta(n^2 + \rho^2)$, $\beta > 0$, if ξ lies outside a suitably small neighbourhood of $(K\circ\pi)^{-1}(\{0\})$ and that $\dot{\mathcal{W}}(\xi, \eta) \geq c\dot{\mathcal{V}}(\xi, \eta)$ on that neighbourhood.

Consider now a vector field Y producing a flow ψ on $S(M)$ and such that it coincides with X on $(K\circ\pi)^{-1}(\{0\})$. If $p^0_{\psi(\xi,t)}(n(t), \varphi(t)) = \psi(\eta, \tau(t))$, ($p^0_\xi(n, \varphi) = p^0_\xi(X, n, \varphi) = e^{-i(\pi/2-\varphi)}\phi(i\xi, n)$ as before) we have that

$$p'_{(n,\varphi)}(\xi)\, X_\xi\, \dot{m}_1(0) + p'_{(n,\varphi)}(\xi)\, E^\perp_\xi\, \dot{n}_1(0) + p'_{(n,\varphi)}(\xi)\, V_\xi\, \dot{\varphi}_1(0) =$$

$$= Y_\eta\, \dot{\tau}(0) - \dot{n}(0) p'_\xi(0, n, \varphi)\frac{\partial}{\partial n} - \dot{\varphi}(0)\, p'_\xi(0, n, \varphi)\frac{\partial}{\partial\varphi}\,.$$

Here $p_{(n,\varphi)}$ maps $\xi \in S(M)$ on $e^{-i(\pi/2-\varphi)}\phi(i\xi, n)$, $E^\perp_\xi \in T_\xi S(M)$ is the horizontal vector such that $\pi' E^\perp_\xi = i\xi$, and $p_\xi(m_1(t), n_1(t), \varphi_1(t)) = \psi(\xi, t)$, which means that

$$\dot{m}_1(0)\, X_\xi + \dot{n}_1(0)\, i\, X_\xi + \dot{\varphi}_1(0)\, V_\xi = Y_\xi.$$

Using the same procedure with ξ_0 instead of ξ and ξ instead of η, we get from $p^0_{\psi(\xi_0,t)}(n_0(t), \varphi_0(t)) = \psi(\xi, \tau_0(t))$ that

$$p'_{(n_0,\varphi_0)}(\xi_0) X_{\xi_0} = -\dot{n}_0(0) p'_{\xi_0}(0, n_0, \varphi_0)\frac{\partial}{\partial n} - \dot{\varphi}_0(0) p'_{\xi_0}(0, n_0, \varphi_0)\frac{\partial}{\partial\varphi} + \dot{\tau}_0(0)\, Y_\xi.$$

This formulae permit to calculate the $\dot{n}_Y$, $\dot{\varphi}_Y$, $\dot{n}_{0Y}$ and $\dot{\varphi}_{0Y}$ needed to obtain $\dot{\mathcal{W}}_Y(\xi, \eta)$ for ξ close to $(K_o\pi)^{-1}(\{0\})$, and to show that if the 3-jet of Y coincides with that of X on the set $\{\xi \in S(M): K(\pi(\phi(\xi, t))), t \in R\}$, we will have that, on a neighbourhood of this set, the differences $\dot{n}_Y - \dot{n}_X$, $\dot{\varphi}_Y - \dot{\varphi}_X$, $\dot{n}_{0_Y} - \dot{n}_{0_X}$, $\dot{\varphi}_{o_Y} - \dot{\varphi}_{o_X}$ are bounded by $C(n^4 + \varphi^4 + n_0^4 + \varphi_o^4)$, $C>0$. For Y close enough to X in the C^4-

topology, C can be chosen in such a way that the bound is uniform in Y. By previous remarks this implies the existence of an $h > 0$, also independent of Y, such that $\dot{W}_Y(\xi, \eta) > 0$ for any pair $\xi,\eta \in S(M)$, $p^0_\xi(n, \varphi) = \eta$, $0 < n^2 + \varphi^2 \leq h$. Then the arguments in the proof of Corollary 4.3 may be applied to get again that the function μ (defined in the proof of Theorem 4.2) is injective. This completes the proof.

REFERENCES

[1] V. Arnold et V. Avez. Problèmes Ergodiques de la Mécanique Classique, Gauthier Villars, Paris, 1967.

[2] K. Kato and A. Morimoto. Topological Stability of Anosov flows and their Centralizers. Topology 12 (1973), 255-273.

[3] S. Kobayashi and K. Nomizu. Foundations of Differential Geometry. Vol. I. Interscience, New York, 1969.

[4] J. Lewowicz. Lyapunov Functions and Topological Stability. J. Differential Equations. 38 (1980), 192-209.

[5] A. Manning. Curvature Bounds for the Entropy of the Geodesic Flow on a Surface. Mathematics Institute University of Warnick, (1980) 1-11.

UNIVERSIDAD SIMON BOLIVAR
DEPARTAMENTO DE MATEMATICAS
Y CIENCIA DE LA COMPUTACION
Apartado Postal 80.659
Caracas.

FINITE DETERMINACY OF GERMS OF INTEGRABLE 1-FORMS IN DIMENSION 3 (A SPECIAL CASE)

by

A. Lins Neto

§1. Introduction

In this paper we study a special class of singularities of germs of integrable 1-forms in $\mathbb{C}^3$ or $\mathbb{R}^3$. A 1-form ω, defined in an open set U ($U \subset \mathbb{R}^n$ or $\mathbb{C}^n$) is integrable if $\omega \wedge d\omega = 0$. A point $p \in U$ is a singularity of ω if $\omega_p = 0$. The forms considered here will be C^∞ if $U \subset \mathbb{R}^3$ or holomorphic if $U \subset \mathbb{C}^3$. For simplicity we use the notation K for $\mathbb{R}$ or $\mathbb{C}$.

Let $\omega = \sum_{i=1}^{3} \omega_i dx_i$ be an integrable 1-form, expressed in some coordinate system $x = (x_1, x_2, x_3)$ in a neighborhood of $0 \in K^3$. Let us suppose that $x = 0$ is a singularity of ω. In this coordinate system $d\omega = \sum_{1 \le i < j \le 3} \omega_{ij}\, dx_i \wedge dx_j$ has three components $\omega_{ij} = \frac{\partial \omega_j}{\partial x_i} - \frac{\partial \omega_i}{\partial x_j}$, $1 \le i < j \le 3$, therefore it makes sense to consider the case where 0 is an algebraically isolated singularity of $d\omega$, namely if

$$(1) \qquad \mu(d\omega) = \dim_{\mathbb{C}}(\hat{\mathcal{O}}(3)/[\tilde{\omega}_{ij}]) < \infty$$

where $\hat{\mathcal{O}}(3)$ is the ring of formal power series in three variables and $[\tilde{\omega}_{ij}]$ is the ideal generated by $\tilde{\omega}_{ij}$, $1 \le i < j \le 3$, where $\tilde{\omega}_{ij}$ is the complexified formal power series of the ∞-jet of ω_{ij}.

The following proposition is easy and shall not be proved here.

<u>Proposition</u>. Condition (1) is independent of the (formal) coordinate system.

In other words, the fact that 0 is an algebraically isolated singularity of $d\omega$, does not depend on the coordinate system, therefore we can give the following definition.

Definition. We say that $p \in U$ is a simple singularity of ω if $\omega_p = 0$ and either $d\omega_p \neq 0$ or there exists a coordinate system $x = (x_1,x_2,x_3)$ around p, such that $x(p) = 0$ and 0 is an algebraically isolated singularity of $d\omega$ (expressed in this coordinate system).

Another way to express this condition is the following. Let us fix some coordinate system $x = (x_1,x_2,x_3)$. Then, since $d\omega$ is a 2-form, there exists an unique vector field Y such that $d\omega = i_Y(\Omega)$, where $\Omega = dx_1 \wedge dx_2 \wedge dx_3$ and $i_Y(\Omega)$ is the interior product defined by

$$i_v(\mu)(v_1,\dots,v_p) = \mu(v,v_1,\dots,v_p)$$

where μ is $(p+1)$-form and $v,v_1,\dots,v_p$ are vectors. In fact, if $d\omega = \sum_{i<j} \omega_{ij}\, dx_i \wedge dx_j$ then $Y = \omega_{23}\, \partial/\partial x_1 + \omega_{31}\, \partial/\partial x_2 + \omega_{12}\, \partial/\partial x_3$, so that 0 is a simple singularity of ω if and only if either $Y(0) \neq 0$ or 0 is an algebraically isolated zero for Y. We use the notation $Y = \mathrm{rot}(\omega)$.

Before stating our results let us give some examples.

Example 1. Let us suppose $\omega_o = 0$ and $d\omega_o \neq 0$. This case was studied by Reeb [7], Kupka [4] and Medeiros [6]. It is not difficult to prove that, in this case, there exists a germ of diffeomorphism $f: (K^n,0) \to (K^n,0)$ such that $\omega^* = f^*(\omega) = \omega_1^*(x_1,x_2)dx_1 + \omega_2^*(x_1,x_2)dx_2$, that is $f^*(\omega)$ depends on 2 variables only and the foliation defined by ω, is locally the product of a singular foliation in K^2 by planes K^{n-2}.

Example 2. Let $Z = \sum_{i=1}^{3} Z_i dx_i$ be a homogeneous vector field of degree $k-1 \geq 1$, $R = \sum_{i=1}^{3} x_i\, \partial/\partial x_i$ be the radial vector field and $\Omega = dx_1 \wedge dx_2 \wedge dx_3$. Then $\omega = i_R i_Z(\Omega)$ is an integrable 1-form, homogeneous of degree k. (The integrability of ω follows from the relation $[R,Y] = (k-2)Y$. It is not difficult to verify that $\mathrm{rot}(\omega) = (k+1)Z - \mathrm{div}(Z)R$, where $\mathrm{div}(Z) = \sum_{i=1}^{3} \frac{\partial Z_i}{\partial x_i}$. If we take $Y = \mathrm{rot}(\omega)$ then $\omega = i_S i_Y(\Omega)$, where $S = \frac{1}{k+1} R$, Y is homogeneous of degree $k-1$ and $\mathrm{div}(Y) = 0$. Therefore 0 is a simple singularity of ω if and only if 0 is an algebraically isolated zero of Y. Let $\mathcal{H}_k$ be the set of forms ω obtainned as above and $\mathcal{S}_k$ be the set of forms $\omega \in \mathcal{H}_k$ for which 0 is a simple singularity. Then it is not difficult to prove that $\mathcal{S}_k$ is open and dense in $\mathcal{H}_k$. In [2] is proved the following result (theorem 2):

Let ω be a germ at $0 \in K^3$ of holomorphic integrable 1-form and let $\omega = \omega_k + \omega_{k+1} + \dots$ be its Taylor series at 0. If $\omega_k \in \mathcal{S}_k$, $k \geq 3$, then there exists a germ of diffeomorphism $h: (K^3,0) \to (K^3,0)$ (C^∞ if $K = \mathbb{R}$ or holomorphic if $K = \mathbb{C}$) such that $h^*(\omega) = \omega_k$.

Here we generalize this result in the following theorem:

Theorem 1. Let ω be a germ of integrable 1-form at $0 \in K^3$, such that 0 is a simple singularity of ω and the order of $d\omega$ at 0 is $k \geq 2$. Then there exists a germ of diffeomorphism $h: (K^3,0) \to (K^3,0)$ such that $h^*(\omega) = \omega^*$ is polynomial. Furthermore there exist vector fields S and $Y^* = \mathrm{rot}(\omega^*)$ such that $\omega^* = i_S i_{Y^*}(\Omega)$, where S is linear diagonal with rational positive eigenvalues, Y^* is polynomial and $[S,Y^*] = \alpha Y^*$, where $\alpha > 0$. The set of singularities of ω^* (outside of zero) consists of a finite number of orbits of S. This number $\ell \leq \mu(d\omega)-1$ where $\mu(d\omega)$ is as in (1). The order of $d\omega$ at 0 is the order of the first non zero jet of $d\omega$ at 0.

As an application of Theorem 1 we get the following

Corollary 1. Let ω be as in the hypothesis of Theorem 1. Then ω embeds in a local action of the group of affine transformations of K.

By putting together Theorem 4 of [2] and Theorem 1, it is not difficult to prove the following

Corollary 2. Let ω be a germ of holomorphic integrable 1-form at $0 \in \mathbb{C}^n$, $n \geq 4$. Suppose that there exists a 3-plane F such that $0 \in F$ is a simple singularity of the restriction ω/F. Then there exists a germ of holomorphic diffeomorphism $h: (\mathbb{C}^n,0) \to (\mathbb{C}^n,0)$ such that $h^*(\omega)$ is polynomial.

Another fact that we shall prove is the following.

Theorem 2. Let $S = \sum_{i=1}^{n} \lambda_i x_i \, \partial/\partial x_i$ and Y be vector fields in C^n such that $[S,Y] = \alpha Y$, where $\lambda_1,\ldots,\lambda_n \in \mathbb{N}$, $\alpha \geq 0$ and 0 is an isolated singularity for Y. Let $\mu(Y) = \dim_{\mathbb{C}} \mathcal{O}(n)/[Y^i]$ where $Y = \sum_{i=1}^{n} Y^i \, \partial/\partial x_i$ and $[Y^i]$ is the ideal of $\mathcal{O}(n)$ generated by $Y^1,\ldots,Y^n$. Then $\mu(Y) = f(\alpha,\lambda_1,\ldots,\lambda_n)$ where

$$(2) \qquad f(\alpha,\lambda_1,\ldots,\lambda_n) = \prod_{i=1}^{n} \left(\frac{\lambda_i+\alpha}{\lambda_i}\right).$$

In particular if this number is not integer there is no Y satisfying $[S,Y] = \alpha Y$ and such that 0 is an isolated singularity.

Observe that this result implies that if we have S and Y as in Theorem 1 then $\mu(d\omega) = \mu(Y)$ can be computed if we know $\lambda_1, \lambda_2, \lambda_3$ and α.

Remarks. 1) It is not difficult to construct examples of forms satisfying the hypothesis of Theorem 1 and which are not included in Example 2. For example if we take $S = \sum_{i=1}^{3} i x_i \, \partial/\partial x_i$ and $Y = \sum_{i=1}^{3} Y^i \, \partial/\partial x_i$ where

$$Y^1 = a x_2^2, \quad Y^2 = b x_1^5 - 2c x_2 x_3, \quad Y^3 = c x_3^2, \quad a,b,c \neq 0$$

then $\omega = i_S i_Y(\Omega)$ satisfies the hypothesis of the theorem.

2) It seems that the condition $\text{order}_0(d\omega) \geq 2$ is not the more general to prove Theorem 1 (see Lemma G). We think that a more general condition can be given in terms of $\mu(d\omega)$. On the other hand it is possible to construct examples of forms ω such that 0 is isolated for $d\omega$ but the eigenvalues of S are not positive (cf. [5]).

There are some natural problems which arise:

Problem 1. Let $\alpha, \lambda_1, \ldots, \lambda_n \in \mathbb{N}$ be such that $f(\alpha, \lambda_1, \ldots, \lambda_n)$ is integer (see (2)). Is there a vector field Y such that $[S,Y] = \alpha Y$ and 0 is an isolated singularity of Y ?

Problem 2. Are the singularities considered here stable? The word stable here means the following: Let ω be as in Theorem 1 and let $\eta = \omega + \Delta$ be a small perturbation of ω which is integrable. Is there a point p near the origin such that $d\eta$ has an isolated zero at p and $\mu(d\omega, 0) = \mu(d\eta, p)$? Notice that this fact is true in the case of Example 2, that is $S = \sum_{i=1}^{3} \lambda_i x_i \partial/\partial x_i$ with $\lambda_1 = \lambda_2 = \lambda_3$. (cf. [2]).

§2. Proofs

2.1 - Notations and preliminary results

Here we fix some notations and results which shall be used.

Notations

$\Lambda_A^p(n)$ - The set of germs at $0 \in K^n$ of differential p-forms of class C^∞ or holomorphic ($A = \infty$ if $K = \mathbb{R}$ and $A = H$ if $K = \mathbb{C}$).

$\mathfrak{X}^A(n)$ - The set of germs at $0 \in K^n$ of C^∞ or holomorphic vector fields.

$\mathrm{Diff}_0^A(n)$ - The set of germs at $0 \in K^n$ of C^∞ or holomorphic diffeomorphisms f, such that $f(0) = 0$.

If $\mathcal{N}^A$ or $\mathcal{N}_A$ is one of the above sets then $\hat{\mathcal{N}}$ is the set of corresponding formal power series (for example $\hat{\Lambda}^1(n) = \{ \sum_{i=1}^{n} \omega_i dx_i \mid \omega_i$ is a formal power series in n variables$\}$).
We use also the notation $J^k(\mathcal{N})$ for the set of k-jets of elements of $\mathcal{N}^A$ or $\mathcal{N}_A$. If $\alpha \in \mathcal{N}^A$ or $\mathcal{N}_A$ then $j^k(\alpha)$ is the k-jet section of α.

Definition. We say that $Y \in \mathfrak{X}^A(n)$ has the division property if for any $\alpha \in \Lambda_A^p(n)$, $1 \le p \le n-1$, such that $i_Y(\alpha) = 0$, then there exists $\beta \in \Lambda_A^{p+1}(n)$ such that $\alpha = i_Y(\beta)$.

Theorem A (De Rham). Let $Y \in \mathfrak{X}^A(n)$ be such that 0 is an algebraically isolated singularity of Y. Then Y has the division property.

The proof can be found in [2].

Theorem B (Brjuno [1]). Let $X \in \hat{\mathfrak{X}}(n)$. Then there exists $f \in \widehat{\mathrm{Diff}}_o(n)$ such that $f^*(X) = (df)^{-1}\cdot X \circ f = S+N$, where $S = \sum_{i=1}^{n} \lambda_i x_i\, \partial/\partial x_i$ and $N \in \hat{\mathfrak{X}}(n)$ are such that $[S,N] = 0$ and the linear part of N is nilpotent.

Remark. S and N are called the semi-simple and nilpotent parts of $f^*(X)$, respectively. The reason is the following: Let $Z \in \hat{\mathfrak{X}}(n)$ and let us consider the linear transformation in $\hat{\Lambda}^p(n)$ given by the Lie derivative

$$(3) \qquad \eta \in \hat{\Lambda}^p(n) \longmapsto L_z(\eta) = i_z(d\eta) + d(i_z y) \in \hat{\Lambda}^p(n).$$

For each $k \in \mathbb{N}$, (3) induces a corresponding linear transformation in $J^k(\Lambda^p(n)) = J^k$ by

$$(3)_k \qquad L_z^k(j^k(\eta)) = j^k(L_z(\eta)).$$

Lemma C. Let $Z = f^*(X) = S+N$. Then for any $k \ge 0$ and $p \in \{1,\dots,n\}$ $L_Z^k = L_S^k + L_N^k$ is the Jordan decomposition of L_Z^k, that is L_S^k is semi-simple, L_N^k is nilpotent and $L_S^k L_N^k = L_N^k L_S^k$. Furthermore the set $\mathcal{B} = \{x^\sigma\, dx_I \mid |\sigma| \le k,\ I = (i_1,\dots,i_p),\ i_1 < \dots < i_p\}$ is a basis of J^k formed by eigenvectors of L_S^k with $L_S^k(x^\sigma\, dx_I) = \lambda(\sigma,I)x^\sigma\, dx_I$, where $\sigma = (\sigma_1,\dots,\sigma_n)$, $x^\sigma = x_1^{\sigma_1}\dots x_n^{\sigma_n}$, $|\sigma| = \sigma_1 + \dots + \sigma_n$, $dx_I = dx_{i_1} \wedge \dots \wedge dx_{i_p}$ and $\lambda(\sigma,I) = \sum_{j=1}^{n} \sigma_j\lambda_j + \sum_{\ell=1}^{p} \lambda_{i_\ell}$.

Proof. The fact that L_S^k and L_N^k commute follows from $L_S L_N - L_N L_S = L_{[S,N]}$ and $[S,N] = 0$. The relation $L_S^k(x^\sigma dx_I) =$

$\lambda(\sigma,I)x^{\sigma}\, dx_I$ can be easily checked. Since $\mathcal{B}$ is a basis for J^k it follows that L_S^k is semi-simple. It remains to prove that L_N^k is nilpotent, namely that zero is the unique eigenvalue of L_N^k.

Let $N = N_1 + \tilde{N}$ where N_1 is the linear part of N at 0. Put $T = L_N^k$, $T_1 = L_{N_1}^k$ and $\tilde{T} = L_{\tilde{N}}^k$. If $\alpha \in J^k - \{0\}$ define

$$\text{order}(\alpha) = \min\{\ell \leq k \mid j^{\ell}(\alpha) \neq 0\}.$$

Using (3) and the fact that N_1 is nilpotent, it is not difficult to verify the following:

a) T_1 is nilpotent.

b) If $T_1(\alpha) \neq 0$ then $\text{order}(T_1(\alpha)) = \text{order}(\alpha)$

c) If $\tilde{T}(\alpha) \neq 0$ then $\text{order}(\tilde{T}(\alpha)) > \text{order}(\alpha)$.

Now suppose that α is an eigenvector of T with eigenvalue λ. We have $T_1(\alpha) + \tilde{T}(\alpha) = \lambda\alpha$ or $\tilde{T}(\alpha) = \lambda\alpha - T_1(\alpha)$. Using b) and c) we get $\tilde{T}(\alpha) = 0$ so that $T_1(\alpha) = \lambda\alpha$. Since T_1 is nilpotent we have $\lambda = 0$. Q.E.D.

<u>Theorem D</u>. Let $X \in \mathfrak{X}^A(n)$ and $X_1 = DX(0)$ be the linear part of X at 0. Suppose that all eigenvalues of X_1 are rational and positive. Then there exists $f \in \text{Diff}_0^A(n)$ such that $f^*(X) = S+N$, where S and N are as in Theorem B and N is polynomial.

In the holomorphic case the proof can be found in [1] and in the C^∞ case in [8].

2.2. Proof of Theorem 1

Let ω be as in the hypothesis of the theorem and $Y = \text{rot}(\omega)$.

<u>Lemma E</u>. There exists $X \in \mathfrak{X}^A(3)$ such that $\omega = i_X i_Y(\Omega)$ and $L_X\omega = \omega$.

<u>Proof</u>. Since $d\omega = i_Y(\Omega)$, from the integrability condition $\omega \wedge d\omega = 0$ we get $i_Y(\omega) = 0$ and so by Theorem A there exists $\alpha \in \Lambda_A^2(3)$ such that $\omega = i_Y(\alpha)$. Since α is a 2-form, there exists $X \in \mathfrak{X}^A(3)$ such that $\alpha = -i_X(\Omega)$. Hence $\omega = i_X i_Y(\Omega)$. On the other hand we have

(4) $L_X(\omega) = i_X(d\omega)+d(i_X\omega) = i_X(d\omega) = i_X i_Y(\Omega) = \omega$ Q.E.D.

Now let $\hat{X}$ and $\hat{\omega}$ be the ∞-jets of X and ω respectively. Equation (4) implies $L_{\hat{X}}(\hat{\omega}) = \hat{\omega}$. Let $\hat{f} \in \widehat{\mathrm{Diff}}_o(3)$ be as in Theorem B, so that $Z = \hat{f}^*(\hat{X}) = S+N$, where $S = \sum_{i=1}^{3} \lambda_i x_i\, \partial/\partial x_i$, $[S,N] = 0$ and N is nilpotent. Let $\omega^* = \hat{f}^*(\hat{\omega}) \in \hat{\Lambda}^1(3)$.

<u>Lemma F</u>. $L_S\omega^* = \omega^*$, $L_N\omega^* = 0$, $\lambda_1,\lambda_2,\lambda_3 \in \mathbb{Q}_+$ (rational positive numbers) and $\sum_{i=1}^{3} \lambda_i < 1$.

<u>Proof</u>. It is clear that

(5) $$\omega^* = L_Z(\omega^*) = L_S(\omega^*) + L_N(\omega^*).$$

Since S and N are the semi-simple and nilpotent parts of Z respectively, by Lemma C and (5) we have $L_S(\omega^*) = \omega^*$ and $L_N(\omega^*) = 0$. In the proof that $\lambda_1,\lambda_2,\lambda_3 \in \mathbb{Q}_+$ we use the hypothesis, namely 0 is an algebraically isolated zero for $d\omega^*$, order $(d\omega^*) \geq 2$ and the relation

(6) $$L_S(d\omega^*) = d\omega^* .$$

Let $d\omega^* = \sum_{1\leq i<j\leq 3} \sum_{|\sigma|\geq 2} a_\sigma^{ij} x^\sigma dx_i \wedge dx_j$. Then (6) is equivalent to

(6') $$a_\sigma^{ij} = \lambda(\sigma,i,j)a_\sigma^{ij}, \quad |\sigma| \geq 2, \quad 1 \leq i < j \leq 3.$$

<u>Assertion</u>. For any $k \in \{1,2,3\}$ there exists $1 \leq i,j \leq 3$ and $r \geq 2$ such that $a_\sigma^{ij} \neq 0$ where $\sigma_k = r$ and $\sigma_m = 0$ if $m \neq k$.

<u>Proof</u>. If the assertion is not true it is not difficult to see that there exists $m,n \in \{1,2,3\}$ such that

$$d\omega^* = x_m\alpha + x_n\beta$$

where $\alpha,\beta \in \hat{\Lambda}^2(3)$. This implies (see (1)) that 0 is not an algebraically isolated zero for $d\omega^*$, contradiction. Since order $(d\omega^*) \geq$ ≥ 2 it is clear that $r \geq 2$.

From the assertion and (6') it is not difficult to see that

$\lambda_1, \lambda_2, \lambda_3$ must satisfy a system of equations of the form

(7) $$r_k\lambda_k + \lambda_{i_k} + \lambda_{j_k} = 1, \quad k=1,2,3$$

where $r_k \geq 2$ and $1 \leq i_k < j_k \leq 3$. The following lemma implies that $\lambda_1, \lambda_2, \lambda_3 \in \mathbb{Q}_+$.

Lemma G - Let $n \geq 3$ and $\lambda_1 \geq \lambda_2 \geq \lambda_3 \geq \ldots \geq \lambda_n$ satisfy the system

(7') $$\sum_{j=1}^{n} a_{ij}\lambda_j = 1, \quad i=1,\ldots,n$$

where the matrix $A = (a_{ij})$ satisfy the following conditions:

a) $a_{ii} \geq n-1, \quad i=1,\ldots,n$

b) $0 \leq a_{ij} \leq 1, \quad$ if $i \neq j$.

Then A is non singular and $\lambda_i > 0$ for all $i = 1,\ldots,n,$ unless A is of the form

(8) $$A = \begin{pmatrix} n-1 & 0 & \ldots & 0 & a_{1n} \\ 0 & n-1 & \ldots & 0 & a_{2n} \\ \vdots & \vdots & & \vdots & \vdots \\ 0 & 0 & \ldots & n-1 & a_{n-1n} \\ 1 & 1 & \ldots & 1 & a_{nn} \end{pmatrix}$$

and in this case $\lambda_1 = \ldots = \lambda_{n-1} = 1/n-1$ and $\lambda_n = 0$. In particular if A is not as in (8) and the $a_{ij} \in \mathbb{Q}$ then all $\lambda_i \in \mathbb{Q}_+$ and $\sum_{i=1}^{n} \lambda_i < 1$ if at least one $a_{ij} = 1, \; i \neq j$.

Observe that in the case of the system (7), A is never like in (8) and, of course, we can suppose $\lambda_1 \geq \lambda_2 \geq \lambda_3$.

Proof. Let $D = \text{diag}(a_{11},\ldots,a_{nn})$ and write $A = D+E = D(I+D^{-1}E) = D(I+F)$, where $F = (f_{ij})$, $f_{ij} = a_{ii}^{-1}a_{ij}$ if $i \neq j$ and $f_{ii} = 0$.

In particular $0 \leq f_{ij} \leq 1/n-1$, if $i \neq j$.

Let us suppose by contradiction that -1 is an eigenvalue of F. In this case we shall prove that $f_{ij} = 1/n-1$ if $i \neq j$. In fact, let us suppose that there exist $i_o \neq j_o$ such that $f_{i_oj_o} < 1/n-1$ and $u = (u_1,\ldots,u_n)$ satisfying

$$-u_i = \sum_{j=1}^{n} f_{ij}u_j = \sum_{j\neq i} f_{ij}u_j, \quad i=1,\dots,n.$$

In this case we have

$$|u_i| \le (1/n-1) \sum_{j\neq i} |u_j|, \quad i=1,\dots,n$$

where the inequality is strict if $i = i_o$ or $u_{j_o} = 0$. If we sum these inequalities we get $\sum_{i=1}^{n} |u_i| < (1/n-1) \sum_i \sum_{j\neq i} |u_j| = \sum_{i=1}^{n} |u_i|$ or $\sum_{i\neq j_o} |u_i| \le \frac{n-2}{n-1} \sum_{i\neq j_o} |u_i|$ in the second case. In both cases we have a contradiction and so

$$(n-1)(I+F) = \begin{bmatrix} n-1 & 1 & \dots & 1 \\ 1 & n-1 & \dots & 1 \\ \vdots & \vdots & & \vdots \\ 1 & 1 & & n-1 \end{bmatrix} = G$$

Now, $\det G = 2(n-1)(n-2)^{n-1} > 0$, as can be verified directly. This proves the first assertion of the lemma.

Let $\lambda_1 \ge \dots \ge \lambda_n$ satisfy (7'). At least one of the λ's is positive. We can suppose $\lambda_1,\dots,\lambda_k > 0$ and $\lambda_{k+1},\dots,\lambda_n \le 0$. Let $\alpha = \sum_{j=1}^{k} \lambda_j > 0$ and $\beta = \sum_{j=k+1} |\lambda_j| \ge 0$. For $i \le k$ we have

(10) $$1 = \sum_{j=1}^{k} a_{ij}\lambda_j - \sum_{j=k+1}^{n} a_{ij}|\lambda_j| \ge (n-1)\lambda_i - \beta.$$

If we sum these inequalitties we obtain

(10') $$k \ge (n-1)\alpha - k\beta.$$

On the other hand, for $i \ge k+1$ we have

(11) $$1 = \sum_{j=1}^{k} a_{ij}\lambda_j - \sum_{j=k+1}^{n} a_{ij}|\lambda_j| \le \alpha - (n-1)|\lambda_i|$$

and so

(11') $$n-k \le (n-k)\alpha - (n-1)\beta.$$

If we multiply (11') by $n-1$, (10') by $n-k$ and subtract, we get

$$(n-k-1)(n-k) \le \beta(k(n-k)-(n-1)^2).$$

Now suppose $k < n$. In this case we have $(n-k-1)(n-k) \geq 0$ and so $k(n-k) - (n-1)^2 \geq 0$ or $\beta = 0$ in which case $k = n-1$. Let us consider the first possibility:

$$(n-1)^2 - k(n-k) = n^2 - (k+2)n + k^2 + 1 \leq 0. \tag{12}$$

If (12) is satisfied for some n we must have

$$\Delta = (k+2)^2 - 4(k^2+1) = 4k - 3k^2 \geq 0.$$

Since $k \geq 1$ we get $k \leq 4/3$ and so $k = 1$. If we substitute $k = 1$ in (12) we have $n^2 - 3n + 2 = (n-1)(n-2) \leq 0$ and so $n = 1$ or 2, contradicting the hypothesis.

Let us suppose $\beta = 0$, $k = n-1$. In this case we have $\lambda_1,\dots,\lambda_{n-1} > 0$, $\lambda_n = 0$ and from (10) and (11) we get $0 < \lambda_i \leq 1/n-1$ for $i = 1,\dots,n-1$ and $1 \leq \alpha = \sum_{0=1}^{n-1} \lambda_i$. This implies that $\lambda_1 = \dots = \lambda_{n-1} = 1/n-1$ and $\lambda_n = 0$. These values for the λ_i's are possible only if A is like in (8). This finishes the proof of Lemma G and hence of Lemma F either.

Now let ω, X, Y, S and N be as before. It follows from Lemma F that $\lambda_1,\lambda_2,\lambda_3 \in \mathbb{Q}_+$, and so from Theorem D there exists $f \in \mathrm{Diff}^A(3)$ such that $f^*(X) = S+N$. Let $\omega^* = f^*(\omega)$ and $\tilde{\omega} = \sum_{i=1}^{3} \sum_{|\sigma|\geq 1} \omega_\sigma^i x^\sigma dx_i$, $x^\sigma = x_1^{\sigma_1} x_2^{\sigma_2} x_3^{\sigma_3}$, be the ∞-jet of ω^* (in the holomorphic case $\tilde{\omega} = \omega^*$). Since $L_X\omega = \omega$, we get $L_{S+N}(\omega^*) = \omega^*$ and so by Lemma F we have

$$\begin{cases} L_S\tilde{\omega} = \tilde{\omega} \\ L_N\tilde{\omega} = 0 \end{cases} \tag{13}$$

From (13) we get

$$\omega_\sigma^i \left(\sum_{j=1}^{3} \lambda_j\sigma_j + \lambda_i\right) = \omega_\sigma^i, \quad i=1,2,3, \quad |\sigma| \geq 1. \tag{13'}$$

Hence $\omega_\sigma^i \neq 0$ implies

$$\sum_{j=1}^{3} \lambda_j\sigma_j + \lambda_i = 1.$$

Since $\lambda_1,\lambda_2,\lambda_3 \in \mathbb{Q}_+$ the set $\{(i,\sigma) \mid \sum_{j=1}^{3} \lambda_j\sigma_j + \lambda_i = 1\}$

is finite, therefore $\tilde{\omega}$ is polynomial. In the holomorphic case this implies already that ω^* is polynomial. In the C^∞ case this implies that $\omega^* = \tilde{\omega} + \eta$ where η is flat at $0 \in \mathbb{R}^3$ and $L_{S+N}\,\eta = \eta$. Let $Z = S+N$ and Z_t be its flow. Then $L_Z\eta = \eta$ is equivalent to

(14) $$\eta_x(v) = e^{-t}\,\eta_{Z_t(x)}\,(DZ_t(x)\cdot v)$$

for any $x,v \in \mathbb{R}^3$ (x sufficiently small), and $t \leq 0$. Since $\lim_{t\to-\infty} Z_t(x) = \lim_{t\to-\infty} DZ_t(x) = 0$ and η is flat at the origin, (14) implies that $\eta \equiv 0$ and so ω^* is polynomial in any case. It remains to prove that $N \equiv 0$. In fact if this is true, we have $f^*(X) = S$ and so

$$\omega^* = f^*(i_X i_Y \Omega) = i_S i_{f^*(Y)} f^*(\Omega) = i_S i_{Y^*}\Omega$$

where $Y^* = J(f)f^*(Y)$, $J(f)$ being the determinant of the Jacobian of f. We also have

$$d\omega^* = f^*(d\omega) = f^*(i_Y\Omega) = i_{Y^*}\Omega.$$

Hence $Y^* = \mathrm{rot}(\omega^*)$. Furthermore (13) implies that

$$i_{Y^*}\Omega = d\omega^* = L_S d\omega^* = L_S(i_{Y^*}\Omega) = i_{[S,Y^*]}\Omega + \mathrm{tr}(S)i_{Y^*}\Omega.$$

Therefore $[S,Y^*] = \alpha Y^*$ where $\alpha = 1-\mathrm{tr}(S) > 0$.

Let us prove that $N \equiv 0$. Since all objects we are considering are polynomials we can suppose that we are in the complex case. In order to simplify the notations, from now on we omit the stars (so instead of ω^* we use simply ω). Let us suppose $\lambda_1 \geq \lambda_2 \geq \lambda_3$.

<u>Some Remarks</u>:

1) Since N is nilpotent and $[S,N] = 0$, there exists a coordinate system (x_1,x_2,x_3) such that $S = \sum_{i=1}^{3} \lambda_i x_i\, \partial/\partial x_i$ and

(15) $$N = N^1(x_2,x_3)\partial/\partial x_1 + N^2(x_3)\partial/\partial x_2\ .$$

In fact in the case where $\lambda_1 > \lambda_2 > \lambda_3$ it is not necessary to change the coordinate system and we have

$$(15') \quad \begin{cases} N^2(x_3) = ax_3^r \\ N^1(x_2,x_3) = \sum_{i,j} b_{ij}x_2^i x_3^j \end{cases}$$

where $r = \lambda_2/\lambda_3 > 1$ and if $b_{ij} \neq 0$ then $\lambda_1 = i\lambda_2 + j\lambda_3$. The proof in the general case can be found in [1].

2) Since $L_N d\omega = 0$ we have

$$0 = L_N(i_Y\Omega) = i_{[N,Y]}\Omega + i_Y L_N\Omega .$$

It follows from (15) that $L_N\Omega = 0$ and so $[N,Y] = 0$.

3) Let $a,b,c \in \mathbb{C}$ and

$$P(a,b,c) = \{x \in C^3 \mid aN(x) + bS(x) + cY(x) = 0\}.$$

Then $P(a,b,c)$ is invariant by the flow of N, which we denote by N_t.

In fact, from $[S,N] = [N,Y] = 0$ we get

$$(16) \quad \begin{aligned} &aN(N_t(x)) + bS(N_t(x)) + cY(N_t(x)) = \\ &= DN_t(x)\cdot[aN(x)+bS(x)+cY(x)] = 0 \end{aligned}$$

<u>Lemma H</u>. If b or $c \neq 0$ then $P(a,b,c)$ is finite. Furthermore we have:

i) If $b \neq 0$ then $P(a,b,0) = \{0\}$

ii) If $c \neq 0$ then $\#P(a,b,c) \leq k$ where $k = \dim_{\mathbb{C}} \mathcal{O}(3)/[Y_i]$

iii) If b or $c \neq 0$ then $P(a,b,c) \subset \{x \mid N(x) = 0\}$.

<u>Proof</u>. Suppose first $c = 0$, $b \neq 0$. Then $aN(x) + bS(x) = 0$ is equivalent to

$$\begin{aligned} -b\lambda_1 x_1 &= aN^1(x_2,x_3) \\ -b\lambda_2 x_2 &= aN^2(x_3) \\ -b\lambda_3 x_3 &= 0 \end{aligned}$$

It is not difficult to see that $x = 0$ is the unique solution of this system.

Let us suppose $c \neq 0$. Since $\dim \mathfrak{G}(3)/[Y_i] = k$ there exist $r > 0$ and $\delta > 0$ such that if $|a_1|, |a_2| < \delta$ then the number of solutions of

$$Y(x) = a_1 N(x) + a_2 S(x)$$

with $\|x\| < r$ is at most k (cf. [3]) (In fact is k if we count the solutions with multiplicity). Suppose by contradiction that

$$Y(x) = -\frac{a}{c} N(x) - \frac{b}{c} S(x) = AN(x) + BS(x)$$

has more than k solutions, say $x = x_1, \ldots, x_n$, $n \geq k+1$. Since $[S,Y] = \alpha Y$, $\alpha > 0$, we have

$$(17) \qquad Y(e^{tS}x) = e^{t(\alpha I+S)}\, Y(x)$$

for any $t \in C$ and $x \in \mathbb{C}^3$. Hence

$$(17') \qquad \begin{aligned} Y(e^{tS}x_j) &= e^{\alpha t}e^{tS}\, Y(x_j) = e^{\alpha t}e^{tS}[AN(x_j)+BS(x_j)] = \\ &= e^{\alpha t}\, AN(e^{tS}x_j) + e^{\alpha t}\, BS(e^{tS}x_j) \end{aligned}$$

for $j = 1,\ldots,n$. Since $\alpha > 0$ and $\lim\limits_{t\to-\infty} e^{tS} = 0$, there exists t such that $|e^{\alpha t}A|, |e^{\alpha t}B| < \delta$ and $\|e^{tS}x_j\| < r$, $j = 1,\ldots,n$, which is a contradiction. This proves ii).

Now suppose b or $c \neq 0$. If there exists $x \in P(a,b,c)$ such that $N(x) \neq 0$ then $\{N_t(x) \mid t \in C\}$ is infinite. In this case, by (16), $P(a,b,c)$ is infinite which contradicts i) and ii).

Q.E.D.

<u>Corollary 1</u>. Let $D = \{x \in C^3 \mid N(x), S(x) \text{ and } Y(x) \text{ are linearly dependent}\}$. Then $D = \{x \mid N(x) = 0\}$

<u>Corollary 2</u>. Let $E = \{x \in C^3 \mid S(x) \text{ and } Y(x) \text{ are linearly - dependent}\}$. Then E consists of a finite number of orbits of S. If ℓ is this number then $\ell \leq k$ and one of the orbits is the origin.

<u>Proof</u>. From Lemma H it is sufficient to prove that E is invariant by e^{tS}. Let us suppose that $Y(x) = aS(x)$, $a \neq 0$. Then

$$Y(e^{tS}x) = e^{\alpha t}e^{tS}\, Y(x) = ae^{\alpha t}S(e^{tS}x).$$

Hence $e^{tS}x \in E$. Q.E.D.

Let us finish the proof that $N \equiv 0$. Define

$$(18) \qquad f = i_N i_S i_Y \Omega = \det \begin{vmatrix} N^1 & N^2 & 0 \\ \lambda_1 x_1 & \lambda_2 x_2 & \lambda_3 x_3 \\ Y^1 & Y^2 & Y^3 \end{vmatrix}$$

where $N^1 = N^1(x_2,x_3)$ and $N^2 = ax_3^r$, $r = \lambda_2/\lambda_3$. By Corollary 1 of Lemma H we have

$$f^{-1}(0) = \{x \mid N^1(x_2,x_3) = N^2(x_3) = 0\}$$

and so it is sufficient to prove that $f \equiv 0$. Suppose by contradiction that $f \not\equiv 0$. If $N^2(x_3) = ax_3^r \not\equiv 0$ then

$$f^{-1}(0) = \{x \mid N(x) = 0\} = \{x_3 = 0\}$$

because $f^{-1}(0)$ has complex codimension 1 and $f^{-1}(0) \subset \{x_3=0\}$. Hence $f(x) = bx_3^s \not\equiv 0$. On the other hand we have $Y(f) = 0$, because

$$Y(f) = L_Y(i_N i_S i_Y \Omega) = L_Y(i_N i_{S+N} i_Y \Omega) = L_Y(i_N \omega) =$$
$$= i_{[Y,N]}\omega + i_N L_Y \omega = i_N(L_Y \omega) = i_N(i_Y d\omega + d(i_Y(\omega))) = 0.$$

But $Y(f) = bs\, x_3^{s-1}\, Y_3 = 0$ and so $Y_3 \equiv 0$ because $s \neq 0$, contradiction. This proves that $N^2 \equiv 0$ and so $N = N^1\, \partial/\partial x_1$. If we use again that $[N,Y] = 0$ we get

$$N^1 \frac{\partial Y^1}{\partial x_1} = Y^2 \frac{\partial N^1}{\partial x_2} + Y^3 \frac{\partial N^1}{\partial x_3}$$

$$N^1 \frac{\partial Y^2}{\partial x_1} = 0$$

$$N^1 \frac{\partial Y^3}{\partial x_1} = 0$$

If $N^1 \not\equiv 0$, this implies that Y^2 and Y^3 do not depend on x_1. Since N^1 does not depend on x_1, we get from the first equation that $\frac{\partial Y^1}{\partial x_1}$ does not depend on x_1. Hence $\frac{\partial Y^1}{\partial x_1} = h(x_2,x_3)$ and so

$$Y^1 = x_1\, h(x_2,x_3) + g(x_2,x_3).$$

Since the order of Y^1 is greater than 1 we have $Y^1(x_1,0,0) = 0$

and so $Y(x_1,0,0) = 0$, contradicting the fact that 0 is isolated. This ends the proof of Theorem 1.

2.3 - Proof of Theorem 2

Let $S = \sum_{i=1}^{n} \lambda_i x_i \, \partial/\partial x_i$, $Y = \sum_{i=1}^{n} Y^i \, \partial/\partial x_i$ be complex vector fields such that 0 is an isolated zero for Y, $[S,Y] = \alpha Y$ and $\lambda_1,\dots,\lambda_n$, $\alpha \in \mathbb{N}$.

Let $\mu(Y) = \dim_{\mathbb{C}} \mathcal{O}(n)/[Y^i]$. We want to prove that

$$\mu(Y) = \prod_{i=1}^{n} \left(\frac{\alpha+\lambda_i}{\lambda_i}\right)$$

We use the following well known fact (cf. [3]). Let $f\colon (\mathbb{C}^n,0) \to (\mathbb{C}^n,0)$ and $\mu(f) = \dim_{\mathbb{C}} \mathcal{O}(n)/[f^i]$, where $f = (f_1,\dots,f_n)$. There exist $r > 0$ and $\delta > 0$ such that for any regular value ε of f with $\|\varepsilon\| < \delta$ then the number of solutions of $f(x) = \varepsilon$ with $\|x\| < r$ is exactly $\mu(f)$.

Let $f = (Y^1,\dots,Y^n)$. In this case, since $[S,Y] = \alpha Y$, it follows that Y is polynomial and we have a stronger fact: If c is a regular value of f then the number of solutions of $f(x) = c$ is exatly $\mu(f)$.

Proof. Since $[S,Y] = \alpha Y$ it is not difficult to see that

(19) $$f(e^{tS}x) = e^{t(\alpha I+S)} f(x), \qquad t \in \mathbb{C},\quad x \in \mathbb{C}^n.$$

Let x be a solution of $f(x) = c$. Relation (19) implies that

(19') $$f(r^{tS}x) = e^{t(\alpha I+S)}.c.$$

Therefore, if we fix $t \in C$, we see that x is a solution of $f(x) = c$ if and only if $y = e^{tS}.x$ is a solution of $f(y) = c(t) = e^{t(\alpha I+S)}.c$. Moreover, from (19) we get

$$Df(y) = Df(e^{tS}x) = e^{t(\alpha I+S)}.Df(x).e^{-tS}$$

and so c is a regular value of f if and only if $c(t)$ is a regular value of f. Since $\lim_{t\to-\infty} c(t) = 0$, it follows that there exists $t < 0$ such that the equation $f(y) = c(t)$ has at least $\mu(Y) = \mu$ solutions, say $y_1,\dots,y_\mu$. Hence $f(x) = c$ has at least $\mu(Y)$ solutions, namely $x_j = e^{-tS}y_j$, $j = 1,\dots,\mu$. On the other hand, if $f(x) = c$ has more than μ solutions, say $x_1,\dots,x_m$, $m > \mu$, then $y_j = e^{tS}x_j$, $j = 1,\dots,m$ are solutions of $f(y) = c(t)$. But if $t < -M$ where $M > 0$ is big enough, then $\|y_j\| < r$, $j=1,\dots,m$, and $\|c(t)\| < \delta$, contradiction. This proves the assertion.

Now let $F\colon \mathbb{C}^n \to \mathbb{C}^n$ be defined by $F(u_1,\dots,u_n) = (u_1^{\lambda_1},\dots,u_n^{\lambda_n})$ and let $g_i = Y^i \circ F$, $g = (g_1,\dots,g_n)$.

__Assertion__. g_i is a homogeneous polynomial of degree $\lambda_i + \alpha$.

__Proof__. It is sufficient to verify that $R(g_i) = (\lambda_i+\alpha)g_i$, where $R = \sum_{i=1}^{n} u_i\, \partial/\partial u_i$ is the radial vector field. We have

$$R(g_i) = \sum_j u_j \frac{\partial g_i}{\partial u_j} = \sum_j \lambda_j u_j^{\lambda_j} \frac{\partial Y^i}{\partial x_j}(u_1^{\lambda_1},\dots,u_n^{\lambda_n}).$$

Since $[S,Y] = \alpha Y$ we have $S(Y^i) = (\alpha+\lambda_i)Y^i$ and so

$$R(g_i) = (\alpha+\lambda_i)Y^i \circ F = (\alpha+\lambda_i)g_i$$

which proves the assertion.

Now $g(u) = f(u_1^{\lambda_1},\dots,u_n^{\lambda_n})$ and so $g(u) = 0$ implies $u = 0$, hence 0 is the unique solution of $g(u) = 0$. Let c be a regular value of g. Since $g_1,\dots,g_n$ are homogeneous polynomials, it follows from Bézout's theorem that the number of solutions of $g(u) = c$ is

$$\prod_{i=1}^{n} \text{degree}(g_i) = \prod_{i=1}^{n} (\lambda_i+\alpha). \tag{20}$$

Let c be a regular value for both f and g. Let $u^o = (u_1^o,\dots,u_n^o)$ be a solution of $g(u) = c$. Since $g(u) = f(u_1^{\lambda_1},\dots,u_n^{\lambda_n})$, it is clear that if $\lambda_j > 1$ then $u_j^o \neq 0$, $j = 1,\dots,n$. Hence for each solution $x^o = (x_1^o,\dots,x_n^o)$ of $f(x) = c$ there are $\lambda_1\cdots\lambda_n$ solutions of $g(u) = c$, namely the solutions of the system $u_j^{\lambda_j} = x_j^o$, $j = 1,\ldots,n$. Therefore, if k is the number of solutions of $g(u) = c$ we have

$$\mu(Y) = \mu(f) = k/\lambda_1 \ldots \lambda_n .$$

From (20) we get finally $\mu(Y) = \prod_{i=1}^{n} \left(\frac{\lambda_i+\alpha}{\lambda_i}\right)$. Q.E.D.

References

[1] Brjuno - Analytic form of differential equations, Trans. Moscow Math. Soc. 25 (1971) pp. 131-282.

[2] C. Camacho, A. Lins Neto - The topology of integrable differential forms near a singularity. To appear in Publ. Math. de l'IHES.

[3] Griffiths, P. Harris J. - Principles of algebraic geometry, John Wiley, N.Y., 1978.

[4] I. Kupka - The singularities of integrable structurally stable Pfaffian forms, Proc. Nat. Acad. Sci. U.S.A., 52 (1964) pp. 1431-1432.

[5] A. Lins Neto - Structural stability of C^2 integrable forms, Ann. Inst. Fourier, 27 (2) (1977) pp. 197-225.

[6] A. Medeiros - Structural stability of integrable differential forms, Springer Lec. Notes 597 (1977) pp. 395-428.

[7] G. Reeb - Sur certaines propriétés topologiques des variétés feuilletées, Act. Sc. et Ind. Hermann, Paris, 1952.

[8] R. Roussarie - Modêles locaux de champs et de formes, Astérisque #30 (1975).

Instituto de Matemática Pura e Aplicada
Estrada Dona Castorina 110

20460 - Rio de Janeiro - RJ

Brasil

AN EXAMPLE OF POLYNOMIAL INTERPOLATION OF AN HYPERBOLIC ATTRACTOR

by

Arthur Oscar Lopes

§0. We will present here an alternative method to interpolate real polynomial diffeomorphism on R^2. We will show an example in which the interpolation is obtained by means of geometric considerations about the behavior of the polinomial that we want to abtain. This kind of consideration can be important in the case we are interested in the dynamics of the polynomial diffeomorphism. In general, the interpolation formulas that are known give us information about the value of the polynomial in just a finite number of points. In the example above, we will obtain a real polynomial diffeomorphism that interpolates an attractor. Geometric considerations allow us to conjecture that this attractor is hyperbolic.

The polynomial that we'll obtain has a very large degree, however it's the composition of polynomials of small degree. We'll indicate a procedure that will make possible to decrease the degree of the polynomial obtained by means of an extra amount of computer work. It will be shown that 7 is a lower bound for the degree of this polynomial that interpolates Plykin attractor.

The example will be obtained in the following way: in §1 we will make several considerations about the main question that we are interested. In §2 we will define several real polynomial diffeomorphisms and in §3 we will show that the composition of all of them will have the properties that we are looking for.

We will not present here the work made in the computer because it would make this paper very long. We hope the geometric considerations will be enough to convince the reader.

§1

The first example of a non-trivial hyperbolic attractor on R^2 was presented by Plykin [1] and its picture is shown in Figure 1.

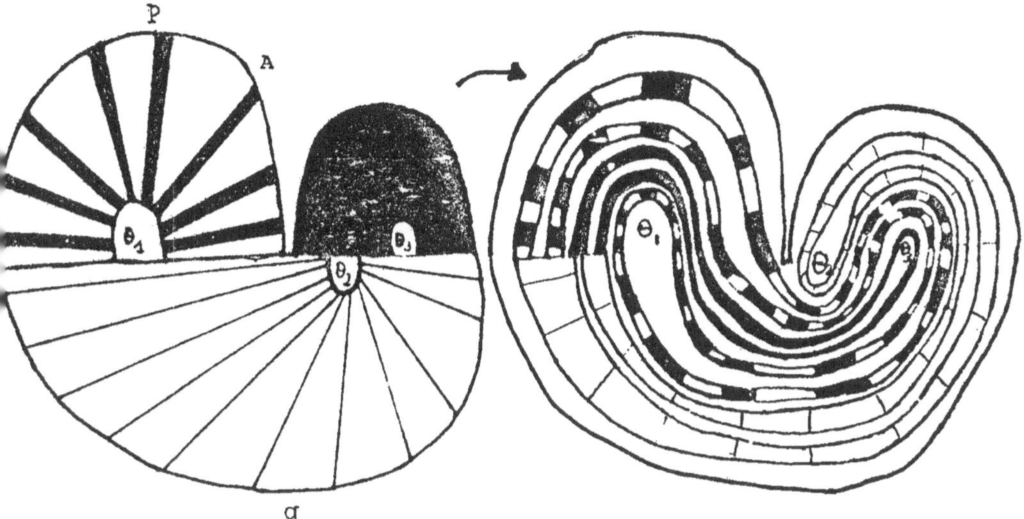

Fig 1

Note that there are three fixed repeling points θ_1, θ_2, θ_3 that are not in the attractor and the attractor is in the interior of Figure A.

For each point of the attractor there exist directions of contraction and expansion. This can be seen geometrically in the radius lines and the semicircles of the picture. More precise analytical considerations are presented in [1].

The following question was proposed to mee by J. Sotomayor: What is the minimal degree of a real polynomial diffeomorphism carrying an hyperbolic attractor? By a real polynomial diffeomorphism we mean a function f on an open subset U of R^2 such that there exist $a_1, a_2, \dots a_r, b_1, b_2, \dots b_s \in R$ $m, n \in N$ such that

$$f(x,y) = (a_1x^m + a_2y^m + a_3x^{m-1}y + \dots + a_{r-2}x + a_{r-1}y + a_r ,$$
$$b_1x^n + b_2y^n + b_3x^{n-1}y + \dots + a_{s-2}x + a_{s-1}y + a_s)$$

We also will suppose that f is one-to-one on U. The degree of the polynomial by definition is the sum of the exponents of x and y for the smallest i such that a_i or b_i is non-zero.

A polynomial with such a minimum degree must exist because the real polynomials are dense and a hyperbolic attractor is structurally stable (see Smale [2] for definitions).

If you take a line conecting the point p with the point q in fig. 1 you'll have that the image of this line have a polynomial expression with the same degree of the polynomial we want to obtain. Therefore this polynomial difeomorphism have degree at least equal 7 because the image of the line p q by the difeomorphism in the y-axis have at least 6 critical points.

My guess is that it can be obtained as a polynomial of degree 16. (see page 9).

Using the usual interpolation formulas in R^n ([3][4]) , we were not able to obtain any good solution to the problem. Even if we try to interpolate a Smale's horse-shoe, the turning arounds that appears on that map, will not allow you to obtain a solution for this simpler case. As far as know the interpolation formulas will not be able to assure you that you will obtain a polynomial diffeomorphism either. I was trying to find a way to interpolate a horse-shoe when Devaney and Nitecky [5] showed that for certain values of the parameter of the Henon map (see [6], [7], [8]) you have a horse-shoe. This real polynomial diffeomorphism has the expression

$$F(x,y) = (A + By - x^2 , x) \qquad A > \frac{(5+2\sqrt{5}) (1+|B|)^2}{4}$$

and therefore has degree 2. The picture of the geometric horse-shoe in the example above is shown in Figure 2.

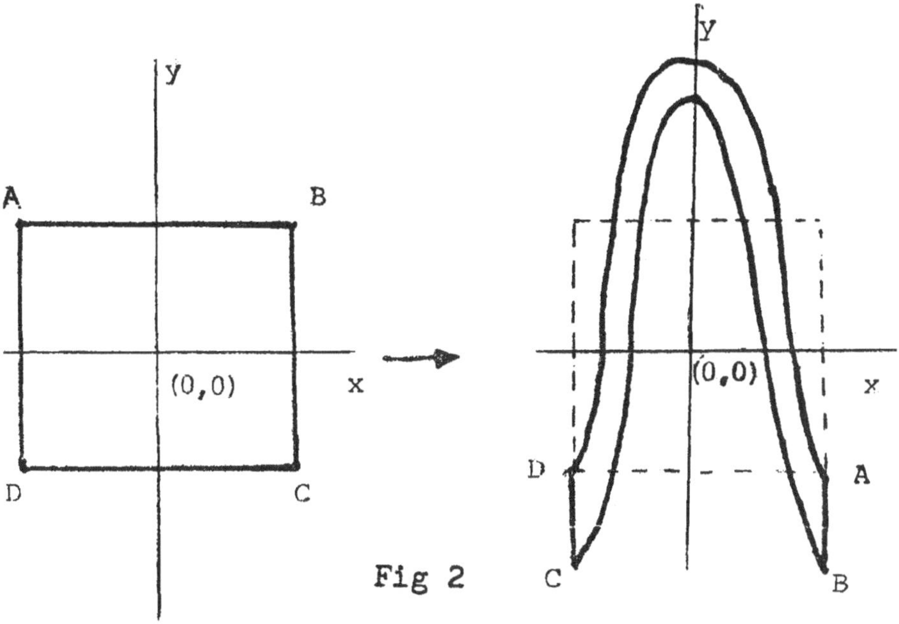

Fig 2

In order to prove that the non-wandering set is hyperbolic they found a field of cones and applied the techniques presented in [9]. This example solved the question of finding the smallest possible degree of a real polynomial diffeomorphism carrying a horse-shoe. The question for hyperbolic attractor remains unsolved. The question is much harder in this case because even if we are able to find a polynomial with small degree, that is candidate to carry an hyperbolic attractor, it would be very difficult to find a field of cones as in the example of Devaney and Nitecky. The reason is that there exist a lot of turn arounds in the stable (unstable) directions of the points of the hyperbolic attractor. Another reason is that is very dificult to prove in general inequalities for polynomials of degree 7. A reasonable approach is to find a real polynomial diffeomorphism that interpolates the attractor, and then make computer experiments that can indicate that we are in fact dealing with an hyperbolic attractor. We will be here concerned with the first part of the approach, that is, we will present a real polynomial diffeomorphism which interpolates Plykin's attractor. As the way we will obtain this polynomial is very geometric, you will be able to see that there exist directions of expansion and contractions as in the hyperbolic attractor we are looking for. The degree of this polynomial that we obtained in APL-GRAPHICS is larger than 1000. I believe that with the same geometric idea and some small variations of the method we can decrease the degree of the polynomial obtained. We will return to this point later.

§2

a) Let $g_1 : R^2 \to R^2$

$$(x,y) \to (\lambda x, \lambda y)$$

with $\lambda \in R$ that we will specify later

b) Let $g_2 : R^2 \to R^2$ $\qquad c=0.11 \quad d= 8.17$

$$(x,y) \to \left(\frac{1}{d-c}(cx+dy)^2 + cx \;,\; \frac{1}{c-d}(cx + dy)^2 + dy\right).$$

g_2 is obtained by a linear change of coordinates and a translation of the example of Devaney-Nitecky for the values $A = 8{,}6$ $B= -0.8987$.

Therefore g_2 is conjugated to the map of that example. The point (0,0) is a fixed point for g_2 and the stable(unstable) direction in (0,0) is the x-axis (y-axis). We can see the position of the stable and unstable mainfolds in Figure 3. This position will be important for us in the picture of fig 7.

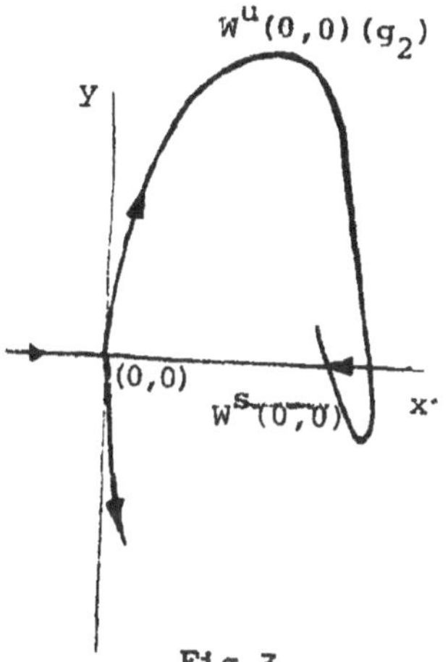

Fig 3

c) Let $g_3 : R^2 \to R^2$

$$(x,y) \to (x,-y)$$

d) Let (c_1,d_1), (c_2,d_2), $(c_3,d_3) \in R^2$ and $p_1,q_1,v_1,p_2,q_2,v_2 \in R$. The exact values of these points will be specified later. Let

$g_4 : R \to R$

be the polynomial that interpolates $0,c_1,p_1,c_2,q_1,c_3,v_1$ with the graphic that is shown in Figure 4. This polynomial has degree 7.

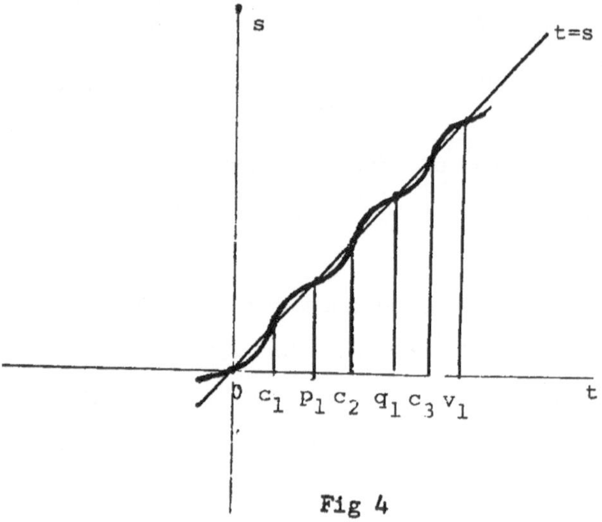

Fig 4

Note that therc exist $\varepsilon>0$ such that g_4 is one-to-one one the set $(-\varepsilon, v_1 + \varepsilon)$. The points c_1, c_2, c_3 are expansive fixed points and $0, p_1$, q_1, v_1 are contracting fixed points.

Let $g_5 : R \to R$ be the polynomial that interpolates p_2, d_2, 0, d_1, q_2, d_3, l_2, s_2, v_2 with the graphic that is shown in Figure 5. This polynomial has degree 9.

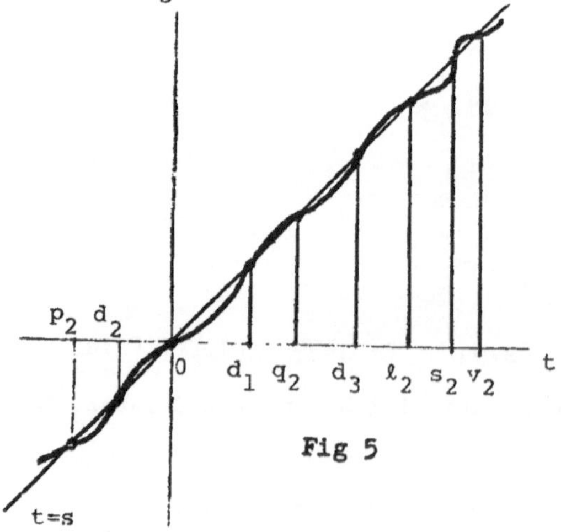

Fig 5

Note that there exist $\varepsilon > 0$ such that g_5 is one-to-one on the set $(p_2-\varepsilon, y_2+\varepsilon)$. The points d_1, d_2, d_3, s_2 are expansive fixed points and p_2, q_2, 0, 1_2, v_2 are contracting fixed points.

Let now g_6 be $g_6 = g_4 \times g_5$, that is, $g_6(x,y)=(g_4(x), g_5(y))$. Therefore g_6 is a real polynomial diffeomorphism on $(-\varepsilon, v_1+\varepsilon) \times (p_2-\varepsilon, v_2+\varepsilon)$. It is not one-to-one in R^2.

§3

Before we begin to show the way we will obtain the attractor, we will find a square-root of Plykin attractor. Let be $g : R^2 \to R^2$ the diffeomorphism shown in Figure 6. It is easy to see that we can obtain such g in order that g^2 is the Plykin attractor shown in Figure 1.

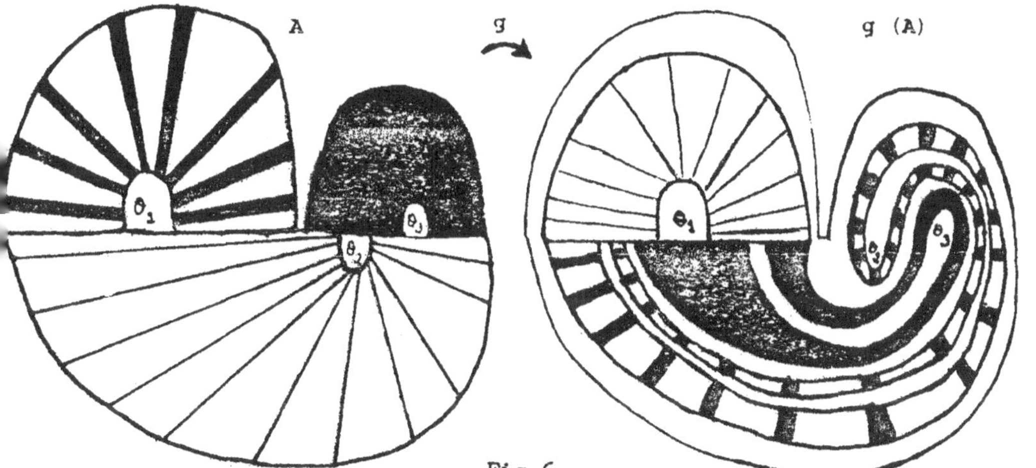

Fig 6

Therefore, to find a polynomial expression for such g will allow us immediately to have a real polynomial expression for the Plykin attractor. My guess is that it can be obtained as a polynomial of degree 4. (see page 9).

We can almost obtain g as the composition of the real polynomial diffeomorphism presented in a), b), c) of §2. In Figure 7 you can see that for certain values λ of the parameters the

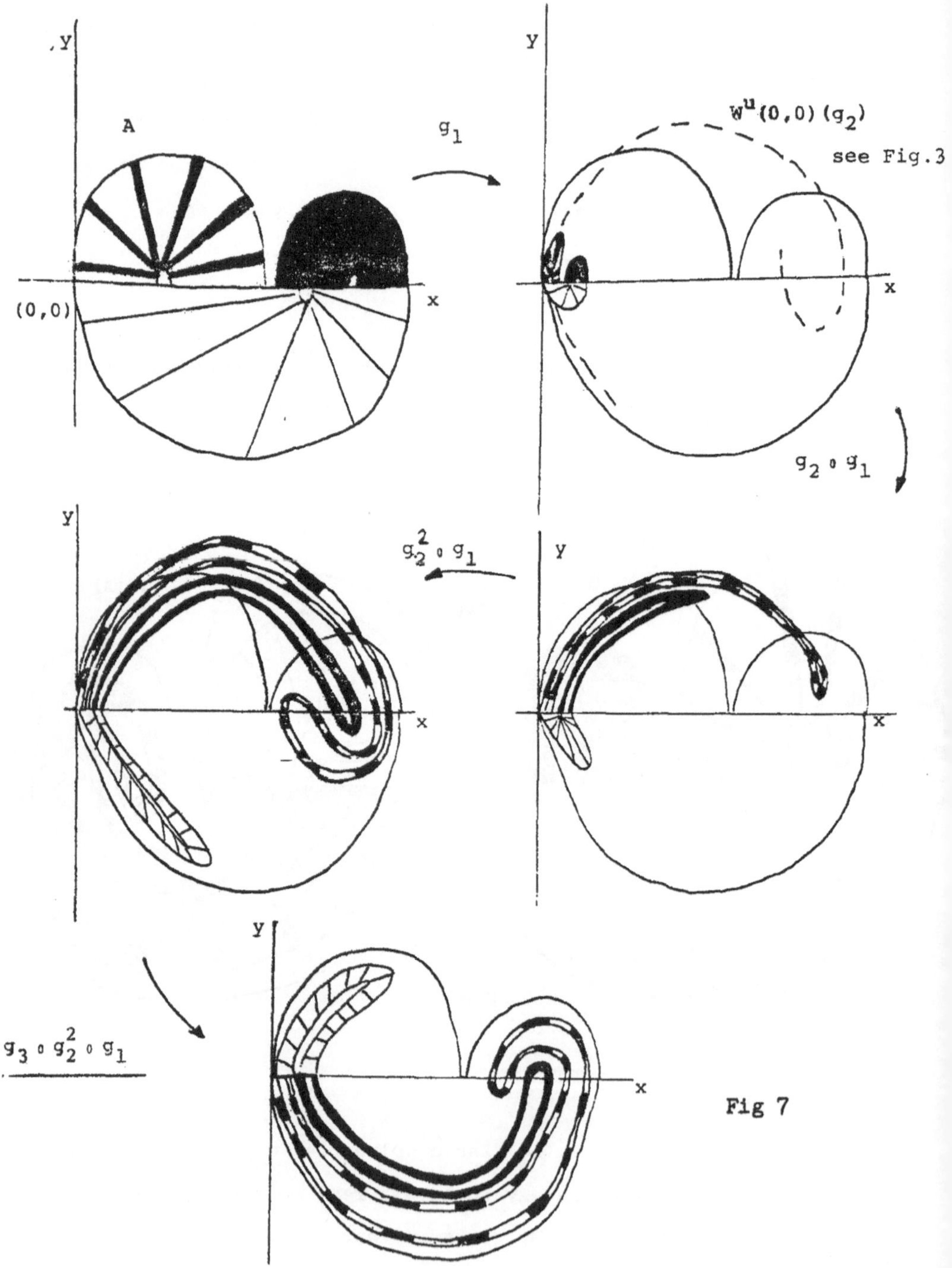
y
A
(0,0)
x
g_1
y
$W^u(0,0)(g_2)$
see Fig.3
x
$g_2 \circ g_1$
y
x
$g_2^2 \circ g_1$
y
x
y
$g_3 \circ g_2^2 \circ g_1$
x

Fig 7

composition of $g_3 o g_2^2 o g_1$ is similar to the action of g in the set A. The polynomial $g_3 o g_2^2 o g_1$ have degree four. This polynomial is not g because, Plykin attractor have three holes. That is, there exist three open sets that a·e contained in its image. This cannot happen for $g_3 o g_2^2 o g_1$ because each one of g_i $i \varepsilon \{1,2,3\}$ have constant jacobian. Let us now define $g_7 = (g_3 o g_2^2 o g_2)^2$. Now we select points (c_1,d_1), (c_2,d_2) and (c_3,d_3) in the position shown in Figure 8.

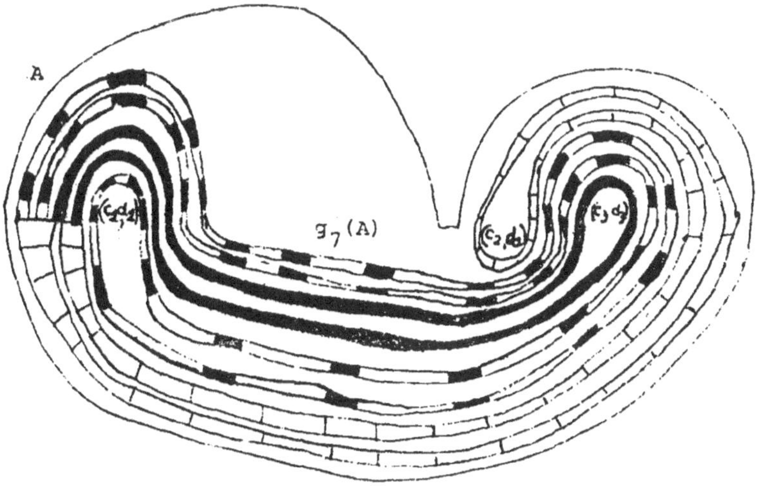

Fig 8

They will be the fixed repelling points θ_1, θ_2, θ_3 of Plykin attractor (see Figure 1). If we select convenient values of $m \varepsilon N$, $0 < c_1 < p_1 < c_2 < q_1 < c_3 < v_1$ and $p_2 < d_2 < 0 < d_1 < q_2 < d_3 < l_2 < s_2 < v_2$ in such way that the image of $g_7(A) \subset (0,v_1) \times (p_2,v_2)$ we can get the geometric picture shown in Figure 9, a), b) for $g_6^m o g_7$. This is a real polynomial diffeomorphism that looks like Plykin attractor. Note that it is not one-to-one on R^2, but just in A. We Changed A in order to fit with the composition with g_6. The geometric behavior and considerations made before remain the same.

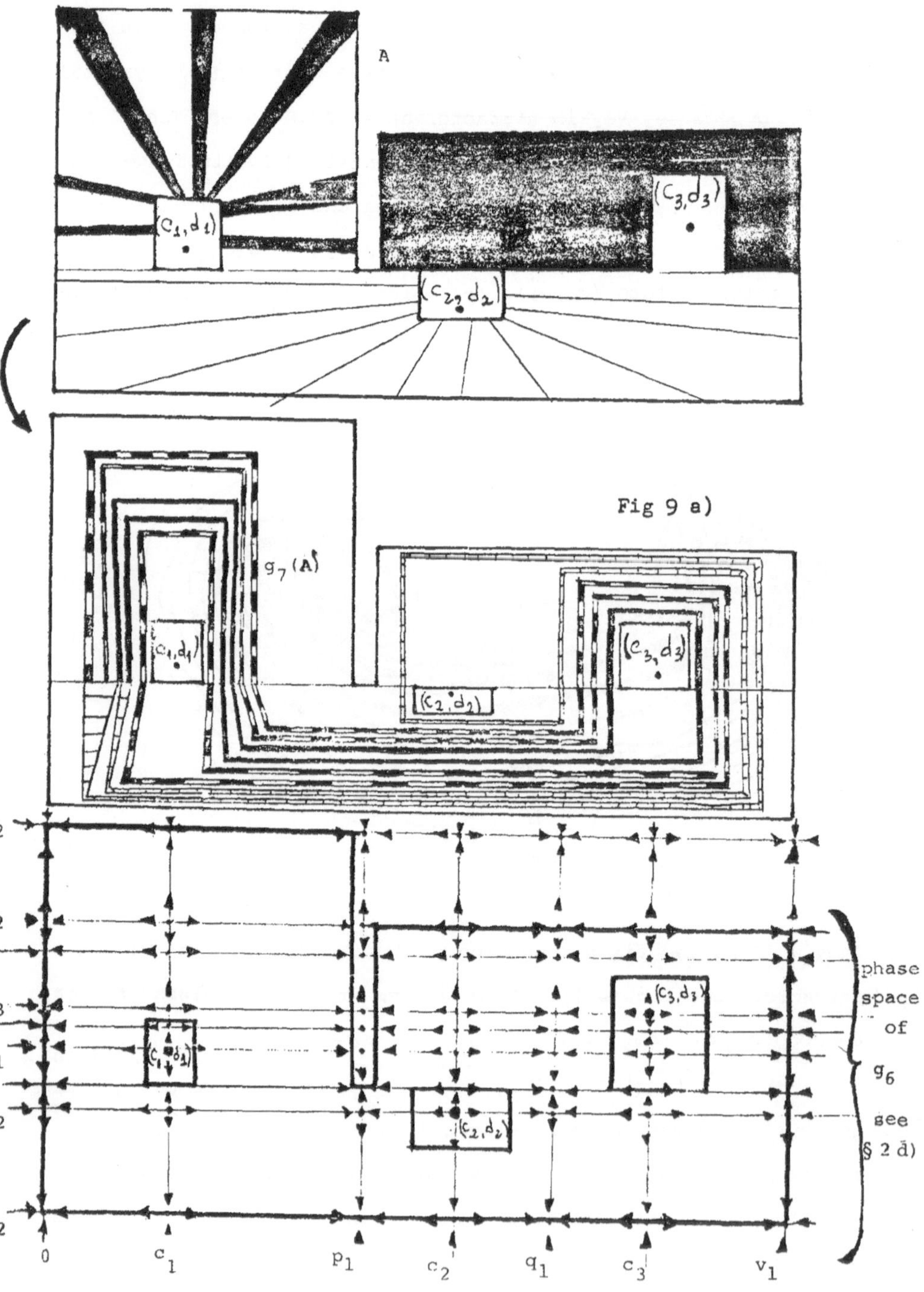

A
g7
(c1, d1)
(c2, d2)
(c3, d3)
g7 (A)
v2
s2
ℓ2
d3
d1
0
d2
p2
0
c1
p1
c2
q1
c3
v1
phase space of g6 see § 2 d)

Fig 9 a)

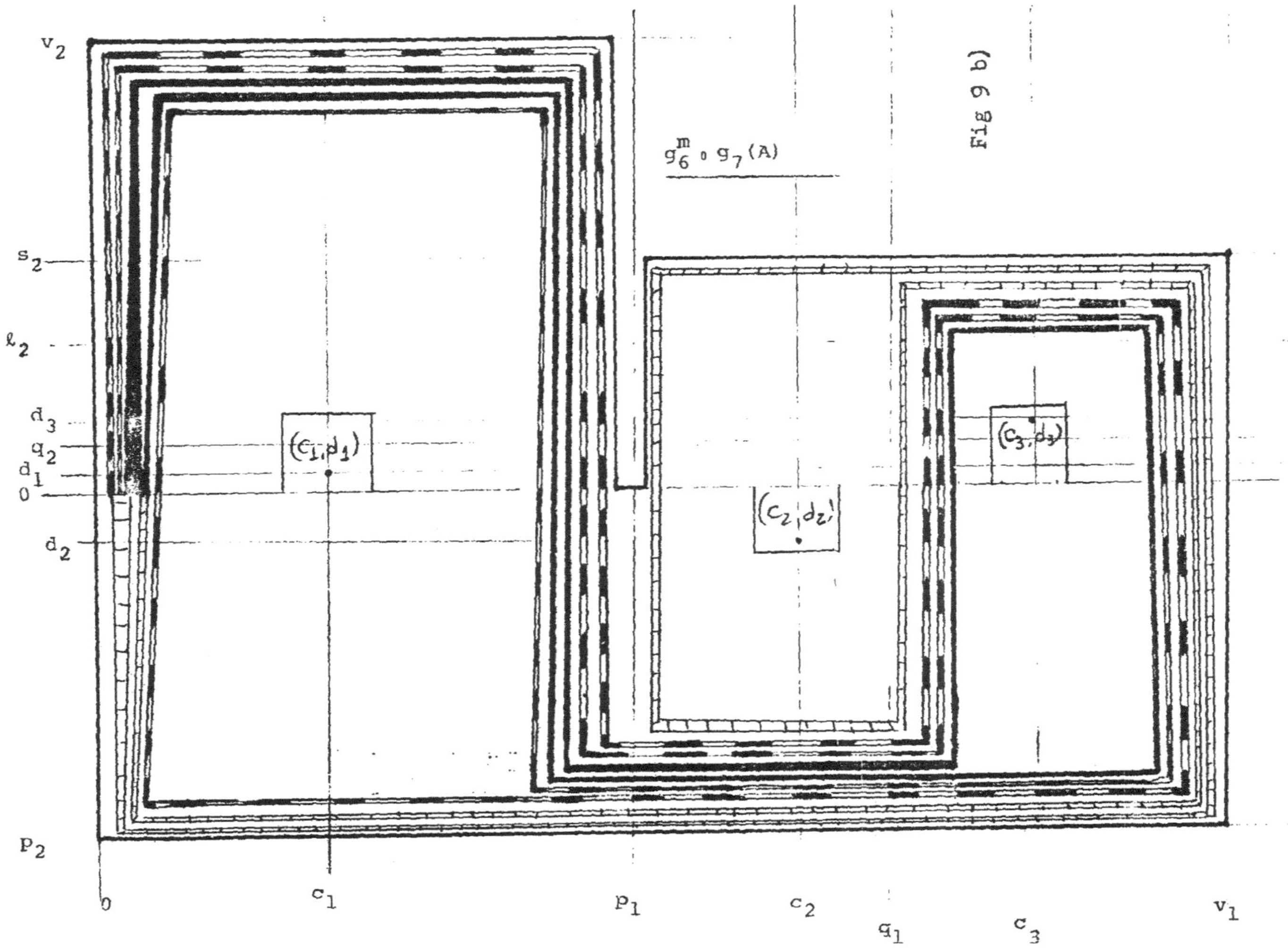

Fig 9 b)

The m above must be very large in order to create fixed expanding points in (c_1,d_1), (c_2,d_2) and (c_3,d_3).

The geometric considerations above were tested in APL-GRAPHICS and we were able to obtain the picture that we wanted to get. In the example that we obtained the value of m was 4. We believe that if we take a perturbation of g_2 with a slightly bigger degree we will be able to decrease the value of the m. Remember that g_2 is structurally stable, therefore the geometric and qualitative behavior of g_2 or a small perturbation of it are the same.

The problem that makes m big is that in the particular case that I worked (see the values a A and b in pag 5) in the computer,the distances between the points of fig 4 (and fig 5) are very small.

Therefore the expansions and contractions that we obtain for g_6 are very weak. If you are able to find optimal positions for these points you can decrease the degree of g_6^m a lot.

Another way to decrease the degree of the polynomial is to take a slight perturbation of g (or g_2) in such way that the jacobian is not constant. This will require some expriments with the computer. In this case g_6 will be not use to find the polynomial expression.

We point out that is transparent from the geometric construction above the existence of a field of directions of expansion and a field of directions of contraction.

This small contribution in the direction of finding a solution to the question stated in §1 leaves us with several open questions:

1) Is the attractor defined above an hyperbolic attractor, or at least have the numerical properties that one expects for an hyperbolic attractor ?
2) Is it possible to obtain any hyperbolic attractor by means of composition of real polynomial diffeomorphism similar to the ones of §2 ?

[1]-R.V.Plykin -Sources and Sinks of A-Diffeomorphisms of Surfaces-Math. Urss Sbornick (1974)

[2]-S.Smale -Differentiable Dynamical Systems-Bull.AM.Math Soc(1967)

[3]-P.Kergin-A Natural Interpolation of C^k Functions-J. of Aproximation Theory (1980)

[4]-C.Micchelli and P.Milman-Aformula for Kergin Interpolation in R^k-J. of Aproximation Theory (1980)

[5]-R.Devaney and Z.Nitecky-Shift Authomorphisms in the Henon Mappin Comm. in Math Physics (1979)

[6]-M.Henon Atwo-dimensional map with a strange attractor.- Comm in Math. Physics.(1976)

[7]-S.Feit -Characteristic exponenets ans strange attractor. Comm in Math Physics (1978)

[8]-J.Curry-On the structure of the Henon attractor-(preprint) National Center for Atmospheric Research- (1977)

[9]-J.Moser -Stable and Random Motions in Dynamical Systems-Princenton University Press. (1973).

ALGEBRAIC PROBLEMS ARISING FROM

MORSE-SMALE DYNAMICAL SYSTEMS *

Michael Maller

Much work has been done to detect when a diffeomorphism is isotopic to one which is Morse-Smale ([SS], [FS], [H], [M], [B2]). Here we report progress in the non-simply connected case. We would like to thank Michael Shub and John Franks for many helpful discussions.

Definition: A square integral matrix is a virtual permutation (v.p.) if it has the form

$$\begin{bmatrix} p_1 & & & * \\ & p_2 & & \\ & & \ddots & \\ 0 & & & p_r \end{bmatrix}$$

where the p_i are signed permutation matrices or zero. Suppose C_* and D_* are chain complexes over a ring R and $E: C_* \to C_*$, $F: D_* \to D_*$ are chain maps. We will say (C_*,E) is chain homotopy equivalent ($\sim$) to (D_*,F) if there exists a chain homotopy equivalence $h: C_* \to D_*$ such that $hE \simeq Fh$; i.e. the diagram commutes up to chain homotopy

$$\begin{array}{ccc} C_* & \xrightarrow{h} & D_* \\ \downarrow E & & \downarrow F \\ C_* & \xrightarrow{h} & D_* \end{array}$$

Problem: When is $(C_*,E) \sim (D_*,F)$ where F is represented by virtual permutations?

Let $C_*(M)$ be the Z-chain complex induced by a handle decomposition of M, $f : M \to M$ a diffeomorphism preserving the decomposition, and $f_* : C_*(M)\circlearrowleft$ the induced map. Shub and Sullivan proved that provided dim M > 5 and $\pi_1(M) = 0$, f is isotopic to a Morse-Smale

*Research supported in part by NSF Grant MCS 79-02647 and PSC-CUNY Research Award 1981-13723.

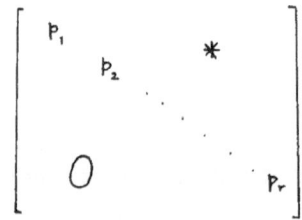

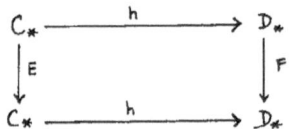

diffeomorphism if and only if $(C_*(M), f_*)$ is chain homotopy equivalent to a virtual permutation of a free Z-complex [SS].

Shub and Franks introduced an obstruction group, now called SSF, which detects when this condition is satisfied. A linear map is called quasi-unipotent if all its eigenvalues are roots of unity, and quasi-idempotent if we allow the zero eigenvalue as well. Let $\underline{\underline{QI}}$ be the category of abelian groups with quasi-idempotent endomorphisms. SSF is the torsion subgroup of $K_0(\underline{\underline{QI}})$. If $E : C_* \to C_*$ is quasi-idempotent they define an Euler characteristic $\chi(E) = \sum_{i=0}^{n} (-1)^i [E_i]$ where $[E_i]$ is the class in SSF. Shub and Franks prove that (C_*, E) is chain homotopy equivalent to a virtual permutation of a free Z-complex if and only if $E_* : H_*(C) \circlearrowleft$ is quasi-unipotent and $\chi(E) = \chi(E_*) = 0$ in SSF. ([FS] Theorem 3.3). The group SSF was subsequently proved non-zero by Lenstra [L].

When M is not simply connected one uses the chain complex in the universal cover $C_*(\tilde{M})$ over the group ring ZG, $G = \pi_1(M)$. A virtual permutation matrix over ZG has the same form above, but the blocks p_i can have non-zero entries $\pm g$, $g \in G$. If f is isotopic to a Morse-Smale diffeomorphism $(C_*(\tilde{M}), f_*)$ is chain homotopy equivalent to a virtual permutation of a free ZG complex. However the converse requires additional algebraic conditions designed to guarantee geometric cancelling in the 1- and 2-handles (see [M]).

In ([M], Theorem 3) we required that an equivalence $h : (C_*, E) \to (D_*, F)$ be a <u>simple</u> chain homotopy equivalence. In this context this is no longer necessary. If $h : (C_*, E) \to (D_*, F)$ is not simple but F is v.p., choose a contractible complex B_* such that $\tau(B_*) = -\tau(h)$ and consider

$$(C_*, E) \xrightarrow{h} (D_*, F) \xrightarrow{inc} (D_* \oplus B_*, F \oplus Id).$$

It follows that (inc o h) is simple and $(F \oplus Id)$ is v.p.

Let $\underline{\underline{P}}$ be the category of ZG-modules M with endomorphisms $e : M \to M$ such that (M, e) has a free finite v.p. resolution. Let

$\underline{\underline{Q}} \supset \underline{\underline{P}}$ be the category of objects (M,e) which are direct summands of objects of $\underline{\underline{P}}$ i.e. where there exists (M',e') such that $(M \oplus M', e + e') \in \underline{\underline{P}}$. If M is free and M' can be chosen to be free we will write $(M,e) \in \underline{\underline{QF}}$. We define a generalization of the group SSF, SSF(G) = $K_0(\underline{\underline{Q}})/K_0(\underline{\underline{P}})$. If (C_*,E) is given with each $(C_i,E_i) \in \underline{\underline{Q}}$ let $\chi(E) = \sum_{i=0}^{n} (-1)^i [E_i]$ where $[E_i]$ is the class of (C_i,E_i) in SSF(G).

Let $\underline{\underline{QC}}$ be the category of complexes with endomorphisms (C_*,E) which are chain homotopy equivalent to (C_*',E') with each $(C_i',E_i') \in \underline{\underline{QF}}$. $\underline{\underline{QC}}$ will play the role of quasi-unipotence on homology. We prove the following

Theorem: (C_*,E) is chain homotopy equivalent to a virtual permutation of a free ZG-complex if and only if $(C_*,E) \in \underline{\underline{QC}}$ and $\chi(E) = 0$ in SSF(G).

In general we are unable to identify the objects of $\underline{\underline{Q}}$ or of $\underline{\underline{QC}}$ directly. When G = 0 it follows from ([FS] Proposition 2.7) that the objects of $\underline{\underline{Q}}$ are precisely the quasi-idempotent endomorphisms. The proof relies on the fact that $Z[\theta]$ is a Dedekind domain, where θ is a primitive n-th root of unity. For $G \neq 0$, ZG is never a Dedekind domain. However when $G = Z^n$, ZG is Noetherian and integrally closed. Using these properties it is proved in [MW] that when $G = Z^n$, $e : M \to M$ is ZG-linear, and M is projective, $(M,e) \in \underline{\underline{Q}}$ if and only if all eigenvalues of e are roots of units of ZG.

Next we consider the definition of SSF(G) in more detail.

Let $\underline{\underline{C}}$ be the category whose objects are pairs (C_*,E) where C_* is a free, finitely generated ZG chain complex and $E : C_* \to C_*$ is a chain map.(†) A morphism in $\underline{\underline{C}}$, $h : (C_*,E) \to (D_*,F)$ is a ZG-linear chain map such that $Fh = hE$. A sequence of morphisms in $\underline{\underline{C}}$ is exact if it is

---

(†) In general E is Z-linear but not ZG-linear. If $f_*: C_*(M) \circlearrowleft$, $A \in C_*(M)$, $g \in G$, and $\phi : G \to G$ is the induced homomorphism of f, then $f_*(gA) = \phi(g) f_*(A)$. Such a map is called an operator homomorphism associated with ϕ. ([W])

exact as a sequence of chain complexes. A sub-category is said to be closed under short exact sequences if whenever two objects of a sequence are in the sub-category the third is as well. Let $\underline{\underline{VP}} \subset \underline{\underline{C}}$ be the full sub-category of objects chain homotopy equivalent to virtual permutations. Our aim is to detect $\underline{\underline{VP}} \subset \underline{\underline{C}}$.

Lemma 1. $\underline{\underline{VP}}$ is closed under short exact sequences in $\underline{\underline{C}}$.

Proof:

$$\begin{array}{ccccccccc} 0 & \longrightarrow & C' & \xrightarrow{h} & C & \xrightarrow{k} & C'' & \longrightarrow & 0 \\ & & \downarrow{\scriptstyle E'} & & \downarrow{\scriptstyle E} & & \downarrow{\scriptstyle E''} & & \\ 0 & \longrightarrow & C' & \xrightarrow{h} & C & \xrightarrow{k} & C'' & \longrightarrow & 0 \end{array}$$

Since all chain groups are free ZG-modules $C_i \cong C_i' \oplus C_i''$ but boundaries and chain maps may have off-diagonal terms.

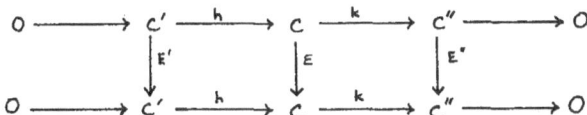

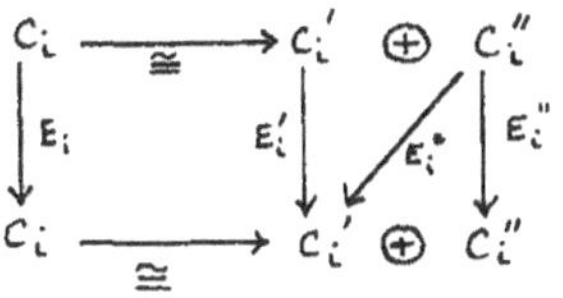

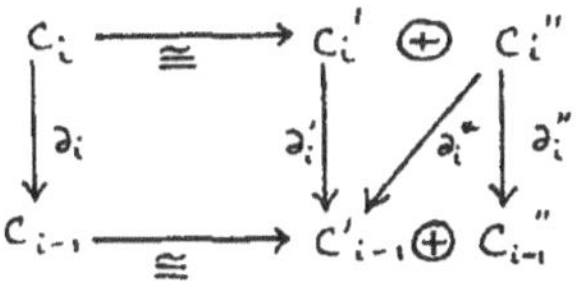

If E' and E" are themselves v.p. so is E. If E' and E" are chain homotopic to virtual permutations, E is chain homotopic to their direct sum with an off-diagonal term. If E' and E" are chain homotopy equivalent to virtual permutations, one constructs a new exact sequence adding contractible complexes to C_*' and C_*'' and their direct sum to C_*; in this new sequence the chain maps on each end are chain homotopic to virtual permutations. It follows that $(C_*, E) \in \underline{\underline{VP}}$.

If the two left hand terms are in $\underline{\underline{VP}}$ we exploit the algebraic mapping cone $M_*(h)$. Since the sequence is exact k induces a chain homotopy equivalence of $(M_*(h), E' \oplus E)$ with (C_*'', E'') so the latter is in $\underline{\underline{VP}}$. Finally, if the right hand terms are in $\underline{\underline{VP}}$ we use duality and the previous step.

□

Let Mod be the category with objects (M,e) where M is a finitely generated ZG-module and $e : M \to M$ is an operator homomorphism. A morphism in Mod, $h : (M,e) \to (N,f)$ is a ZG-linear homomorphism such that $fh = he$. A resolution of (M,e) in Mod is a long exact sequence

$$0 \to (N_K, f_K) \to \ldots (N_0, f_0) \xrightarrow{\varepsilon} (M,e) \to 0.$$

A resolution is free (resp. projective) if all the N_i are free (resp. projective). A resolution can be regarded as a complex $(N_*, f) = (N_i, f_i)$ $i \geq 0$ with chain map $\varepsilon : (N_*, f) \to (M,e)$ inducing homology isomorphisms. If $N_* \xrightarrow{\varepsilon} M$ and $N_*' \xrightarrow{\varepsilon'} M$ are two resolutions of the module M (without an endomorphism) and N_* is projective, then given $h : M \to M$ there exists $H : N_* \to N_*'$ such that $\varepsilon' H = h\varepsilon$, and any two such maps H are chain homotopic. ([CE] Prop. 1.1 page 76. Their argument easily adapts for h an operator homomorphism).

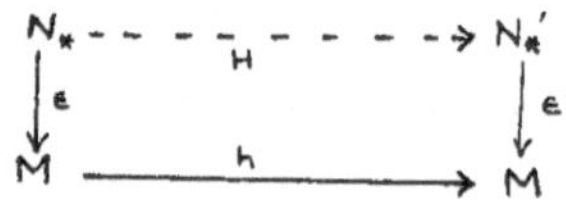

If $(N_*, F) \xrightarrow{\varepsilon} (M,e)$ and $(N_*', F') \xrightarrow{\varepsilon'} (M,e)$ are two projective resolutions we obtain a diagram

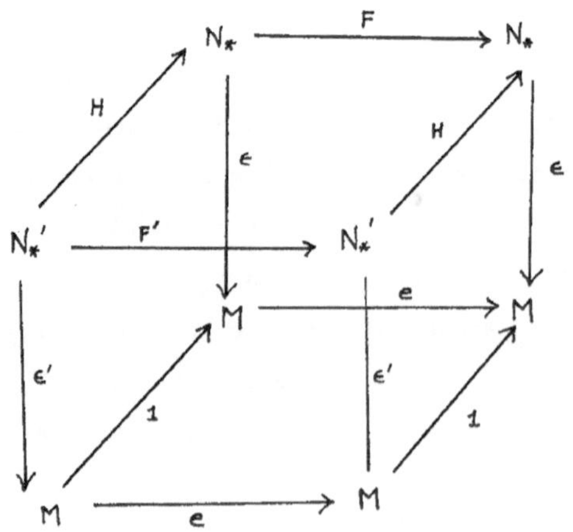

where H is a chain homotopy equivalence and $FH \simeq HF'$. It follows that projective resolutions of objects of $\underline{\underline{Mod}}$ are unique up to chain homotopy equivalence.

Let $\underline{\underline{M}} \subset \underline{\underline{Mod}}$ be the sub-category of objects with finite free resolutions, and let $\underline{\underline{P}} \subset \underline{\underline{M}}$ be the sub-category of objects with finite free v.p. resolutions, i.e. where all the F_i are represented by v.p. matrices.

<u>Lemma 2</u>. $\underline{\underline{P}} \subset \underline{\underline{M}}$ is closed under short exact sequences.

<u>Proof</u>: If $0 \to (M',e') \xrightarrow{h} (M,e) \xrightarrow{k} (M'',e'') \to 0$ is an exact sequence in $\underline{\underline{M}}$, choose free resolutions $(N_*',f') \xrightarrow{\varepsilon'} (M',e')$ and $(N_*'',f'') \xrightarrow{\varepsilon''} (M'',e'')$. By ([CE] Prop. 2.2 and Prop. 2.3 p. 80,81) there exists a free resolution $(N_*,f) \xrightarrow{\varepsilon} (M,e)$ and maps H, K so that

$$\begin{array}{ccccccccc}
0 & \longrightarrow & (N_*',f') & \xrightarrow{H} & (N_*,f) & \xrightarrow{K} & (N_*'',f'') & \longrightarrow & 0 \\
 & & \downarrow{\scriptstyle \varepsilon'} & & \downarrow{\scriptstyle \varepsilon} & & \downarrow{\scriptstyle \varepsilon''} & & \\
0 & \longrightarrow & (M',e') & \xrightarrow{h} & (M,e) & \xrightarrow{k} & (M'',e'') & \longrightarrow & 0
\end{array}$$

the top row is exact and both squares commute. If two of the objects in the bottom row are in $\underline{\underline{P}}$, by uniqueness of resolutions two of the resolutions are in $\underline{\underline{VP}}$, so by Lemma 1 the third resolution is in $\underline{\underline{VP}}$ as well.

□

<u>Lemma 3</u>. If $(C_*,E) \in \underline{\underline{C}}$ and each $(C_i,E_i) \in \underline{\underline{P}}$ then $(C_*,E) \in \underline{\underline{VP}}$.

<u>Proof</u>: We first replace each resolution by a 2-step v.p. resolution

$$0 \to (N_1,f_1) \to (N_0,f_0) \to (C_i,E_i) \to 0$$

by <u>folding</u> as in [SS] Appendix A. This is possible since the C_i are all free. Then we inductively splice in the resolutions starting at the bottom as in ([FS] 3.10). Only projectiveness of the chain modules is used so the proof for Z-complexes adapts immediately to ZG-complexes.

□

Recall that $\underline{\underline{Q}} \subset \underline{\underline{M}}$ is the sub-category of objects which are direct summands of objects of $\underline{\underline{P}}$. We show next that $K_0(\ )$ detects $\underline{\underline{P}}$ inside $\underline{\underline{Q}}$. Let $i : K_0(\underline{\underline{P}}) \to K_0(\underline{\underline{Q}})$ be the inclusion induced homomorphism. It is easy to see that i is injective. We will write $K_0(\underline{\underline{P}}) \subset K_0(\underline{\underline{Q}})$ for the isomorphic image $i(K_0(\underline{\underline{P}}))$.

Definition: $SSF(G) = K_0(\underline{\underline{Q}})/K_0(\underline{\underline{P}})$.

Lemma 4. Suppose $(M,e) \in \underline{\underline{Q}}$. Then $(M,e) \in \underline{\underline{P}}$ if and only if $[M,e] = 0$ in SSF(G).

Proof: It is convenient to work first with a K_0 group defined with direct sum ([B1]). Let $K_0'(\underline{\underline{M}})$ be the abelian group with one generator for each isomorphism class of objects of $\underline{\underline{M}}$, and relations $[M_1,e_1] + [M_2,e_2] = [M_1 \oplus M_2,\ e_1 \oplus e_2]$. It follows that $[M,e] = [N,f]$ in $K_0'(M)$ if and only if there exists (L,h) in $\underline{\underline{M}}$ such that $(M \oplus L,\ e \oplus h) \cong (N \oplus L,\ f \oplus h)$. ([B1] p. 17). We will suppress the endomorphism e when it is clear from the context.

Suppose $[M,e] = [N,f]$ in $K_0'(\underline{\underline{Q}})$ and (N,f) is in $\underline{\underline{P}}$. Choose (L,h) in $\underline{\underline{Q}}$ as above and an inverse (L',h'). Then $M \oplus L \oplus L' \cong N \oplus L \oplus L'$ and the right hand term is in $\underline{\underline{P}}$. The sequence $0 \to M \to M \oplus L \oplus L' \to L \oplus L' \to 0$ is exact so by Lemma 2 $(M,e) \in \underline{\underline{P}}$. If $[M,e] \in K_0'(\underline{\underline{P}})$ then $[M,e] = [N_1,f_1] - [N_2,f_2]$ where $(N_i,f_i) \in \underline{\underline{P}}$. Therefore $[M \oplus N_2,\ e \oplus f_2] = [N_1,f_1]$ so by the previous step $(M \oplus N_2,\ e \oplus f_2) \in \underline{\underline{P}}$. Now the

$$0 \to M \to M \oplus N_2 \to N_2 \to 0$$

is exact so (M,e) is in $\underline{\underline{P}}$.

Consider the quotient group $K_0'(\underline{\underline{Q}})/K_0'(\underline{\underline{P}})$. The class in this group is additive over short exact sequence i.e. the middle term is equal to the sum of the ends. If

$$0 \to (M_1,e_1) \to (M_2,e_2) \to (M_3,e_3) \to 0$$

is exact in $\underline{\underline{Q}}$ choose inverses mod $\underline{\underline{P}}$, M_i' $i = 1,3$ so that $[M_i,e_i] = -[M_i',e_i']$ in the quotient group. The sequence

$$0 \to M_1 \oplus M_1' \to M_2 \oplus M_1' \oplus M_3' \to M_3 \oplus M_3' \to 0$$

is exact. Again by Lemma 2 the middle term is in $\underline{\underline{P}}$ so

$$[M_2 \oplus M_1' \oplus M_3'] = [M_2] - [M_1] - [M_3] = 0$$

in the quotient group. It follows that the natural map

$$\Phi : K_0'(\underline{\underline{Q}})/K_0'/(\underline{\underline{P}}) \to K_0(\underline{\underline{Q}})/K_0(\underline{\underline{P}})$$

is an isomorphism. Therefore if $(M,e) \in \underline{\underline{Q}}$ then $(M,e) \in \underline{\underline{P}}$ if and only if $[M,e] = 0$ in $K_0(\underline{\underline{Q}})/K_0(\underline{\underline{P}}) = SSF(G)$.

□

If (C_*,E) is given with each $(C_i,E_i) \in \underline{\underline{Q}}$ recall that $\chi(E) = \sum_{i=0}^{n} (-1)^i [E_i]$ in $SSF(G)$. If $h : (C_*,E) \to (D_*,F)$ is a chain homotopy equivalence it follows that $\chi(E) = \chi(F)$. If $Fh \underset{\xi}{\sim} hE$ where ξ is a chain homotopy we define an endomorphism K on the mapping cone $M_*(h)$ by

$$\begin{array}{ccccc} M_i & \cong & D_i & \oplus & C_{i-1} \\ \downarrow K_i & & \downarrow F_i & \overset{\xi}{\swarrow} & \downarrow E_{i-1} \\ M_i & \cong & D_i & \oplus & C_{i-1} \end{array}$$

It follows that for each i, $(M_i,K_i) \in \underline{\underline{Q}}$ and the sequence

$$0 \to (D_*,F) \to (M_*(h),K) \to (C_*[-1],E) \to 0$$

is exact. Therefore $\chi(K) = \chi(F) - \chi(E)$. Since h is a chain homotopy equivalence $H_*(M_*(h)) = 0$, so by ([B1], Prop. 6.3, page 21) $\chi(K) = \chi(K_*) = 0$.

Recall that $(M,e) \in \underline{\underline{QF}} \subset \underline{\underline{Q}}$ if M is free and has a free inverse (M',e') mod $\underline{\underline{P}}$. $\underline{\underline{QC}} \subset \underline{\underline{C}}$ is the category of objects (C_*,E) chain homotopy equivalent to (C_*',E') where all $(C_i',E_i') \in \underline{\underline{QF}}$. If $(C_*,E) \in \underline{\underline{QC}}$ then $\chi(E)$ is defined by choosing such a representative (C_*',E') of its chain homotopy class.

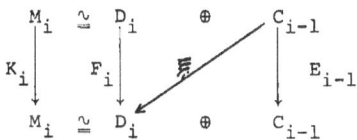

Theorem. $(C_*,E) \in \underline{\underline{VP}}$ if and only if $(C_*,E) \in \underline{\underline{QC}}$ and $\chi(E) = 0$.

Proof: Working in $\underline{\underline{QC}}$ the proof of ([FS] 3.3) adapts easily. Since $(C_0,E_0) \in \underline{\underline{QF}}$ choose a free inverse (C_0',E_0') mod $\underline{\underline{P}}$. Let (B_*,F) be the contractible complex

$$0 \to (C_0',E_0') \xrightarrow{Id} (C_0',E_0') \to 0$$

concentrated in dimensions 1, 0. Therefore

$$inc : (C_*,E) \to (C_* \oplus B_*, E \oplus F)$$

is a chain homotopy equivalence, the right hand term is in $\underline{\underline{QC}}$ and has its 0-dimensional module in $\underline{\underline{P}}$. Continuing inductively we can assume that all but the top dimension (C_n,E_n) are in $\underline{\underline{P}}$. Therefore $0 = \chi(E) = [E_n]$ so by Lemma 4 $(C_n,E_n) \in \underline{\underline{P}}$. By Lemma 3 $(C_*,E) \in \underline{\underline{VP}}$. The converse is immediate.

□

In conclusion this result poses two algebraic problems.

Problem 1: Identify the objects of $\underline{\underline{Q}}$. As noted above, in the special case where G is free abelian, $e : M \to M$ is ZG-linear, and M is projecting, it is shown in [MW] that $(M,e) \in \underline{\underline{Q}}$ if and only if all eigenvalues of e are roots of elements $\pm g$, $g \in G$ i.e. roots of units of ZG.

Problem 2: Identify the objects of $\underline{\underline{QC}}$. In the special case above the natural conjecture is that all eigenvalues on homology should be roots of units of ZG.

References

[B1] H. Bass: Introduction to some methods of algebraic K-Theory. CBMS Regional Conf. Ser. 20(1974).

[B2] H. Bass: The Grothendieck group of the category of abelian group automorphisms of finite order. (pre-print)

[FS] J. Franks and M. Shub: The existence of Morse-Smale diffeomorphisms. Topology 20(1981) pp. 273-290.

[H] B. Halpern: Morse-Smale diffeomorphisms on tori. Topology 18 (1979) pp. 105-111.

[L] H. Lenstra. Letter to H. Bass. 1979.

[M] M. Maller: Fitted diffeomorphisms of non-simply connected manifolds. Topology 19(1980) pp. 395-410.

[MW] M. Maller and J. Whitehead: Virtual permutations of ZG-complexes (to appear).

[SS] M. Shub and D. Sullivan: Homology theory and Dynamical systems. Toplogy 14(1975) pp. 109-132.

[W] J. H. C. Whitehead: Combinatorial homotopy II. B.A.M.S. 55 (1949) pp. 453-496.

[CE] H. Cartan and S. Eilenberg. Homological Algebra. Princeton.

Department of Mathematics
Queens College
Flushing, N.Y. 11367
U.S.A.

LYAPOUNOV EXPONENTS AND STABLE MANIFOLDS FOR COMPACT TRANSFORMATIONS

by

Ricardo Mañé

Instituto de Matemática Pura e Aplicada (IMPA)
Rio de Janeiro, Brazil

Introduction

The aim of this work is the extension of the Lyapounov exponents and stable manifolds theories developed by Oseledec [4], Pesin [5] and Ruelle [6], [7], to compact non linear maps of infinite dimensional Banach spaces. The interest of such extension is its possibility of application to improve the understanding of the dynamics associated to parabolic semilinear equations or retarded functional differential equations, that are described by non linear maps of open sets in Banach spaces with the remarkable property of having compact derivative at every point of its domain. Such feature will be the main hypothesis in all the theorems we shall prove. Unfortunately, for the best and more precise results, hypothesis like the injectivity of these derivatives, density of its ranges and certain injectivity hypothesis of the transformation itself are needed. These kind of hypothesis, quite reasonable for transformations arising from semilinear parabolic equations, are not always satisfied, or at least its validity is difficult to decide, in the case of retarted functional differential equations. The reader will find the basic information about how these equations generate compact transformations in Hale [1] (for retarded functional differential equations) and Henry [2] (for semilinear parabolic equations).

In this section we shall only give a brief description of the main results. Precise complete statements will be given in the next section.

Let E be a Banach space, $U \subset E$ an open set and $f: U \to E$ a C^1 map with compact derivatives. Let $K \subset E$ be a compact set such that $f(K) \subset K$. We want to study the asymptotic behaviour of the linear maps $D_x f^n$ for $x \in K$ and $n \geq 0$. Following Oseledec we introduce the concept of Lyapounov exponents: A Lyapounov exponent of f at x is a number of the form:

$$\limsup_{n\to+\infty} \frac{1}{n} \log\|(D_x f^n)v\| \tag{1}$$

where $0 \neq v \in E$. We shall agree in setting $\log 0 = -\infty$. Now observe that when x is a fixed point there are at most a countable number of Lyapounov exponents of f at x and that they are the limits (and not only the superior limits) of expressions like (1). All these properties follow from the spectral theorem for compact linear transformations (applied to $(D_x f)$) and the Lyapounov exponents at x are precisely the logaritms of the absolute values of the numbers in the spectrum of $D_x f$. Such nice property cannot be expected to hold for any $x \in K$ (as we shal see in a while). The point is that it does hold for a set of points $x \in K$ that is large enough from the probabilistic (or ergodic) view point. The exact statement of this properties require the introduction of the concept of total probability set: We say that a set $K_o \subset K$ has total probability if every f-invariant probability μ on the Borel sets of K (i.e. such that $\mu(f^{-1}(A)) = \mu(A)$ for every Borel set $A \subset K$) satisfies $\mu(K_o) = 1$. The dynamical meaning of this property is the following: If K_o has total probability then:

$$\limsup_{n\to+\infty} \frac{1}{n} \#\{0 \leq i \leq n \mid d(f^i(x),K_o) > \varepsilon\} = 0$$

for every $x \in K$ and $\varepsilon > 0$. In other words if every point of K

spends most of the time nearby K_o. It is well know that the set of recurrent points of K (i.e. points $x \in K$ that satisfy $\liminf_{n\to+\infty} d(f^n(x),x) = 0$) has total probability. Periodic points are contained in every total probability set and Birkhoff's ergodic theorem states that there exists a total probability set such that for any x in it, the orbital average:

$$\lim_{n\to+\infty} \frac{1}{n} \sum_{j=0}^{n} \varphi(f^n(x))$$

exists for any continuous $\varphi: K \to R$.

The aim of this work is to prove that there exists a total probability set $K_o \subset K$ such that for all $x \in K_o$ the limit

$$\lim_{n\to+\infty} \frac{1}{n} \log\|(D_x f^n)v\|$$

exists for every $v \neq 0$ (it can be $-\infty$) and can take only a countable or finite set of values $\lambda_1(x) > \lambda_2(x) > \dots$ in $[-\infty,+\infty)$. Moreover there exist subspaces

$$E = G_1(x) \supset G_2(x) \supset \dots$$

such that $v \neq 0$ satisfies $\lim_{n\to+\infty} n^{-1} \log\|(D_x f^n)v\| = \lambda_j(x)$ if and only if $v \in G_j(x) \setminus G_{j+1}(x)$. When f/K and $D_x f$ are <u>injective</u> for all $x \in K$ this result can be considerably improved as follows: There exist subspaces $E_1(x), E_2(x), \dots, F_1(x) \subset F_2(x) \subset \dots$ such that:

$$E = E_1(x) \oplus \dots \oplus E_j(x) \oplus F_j(x)$$

for all j and

$$\lim_{n\to\pm\infty} \frac{1}{n} \log\|(D_x f^n)v\| = \lambda_j(x)$$

for all j and $0 \neq v \in E_j(x)$. Formally the expression $(D_x f^n)v$ doesn't make sense for $n < 0$ because compact linear maps are never invertible, but we shall agree that to write $(D_x f^n)v$ with $n < 0$ means that there exists w satisfying $(D_{f^n(x)} f^{-n})w = v$ and we define $(D_x f^n)v = w$. The injectivity of $D_x f$ for all x grants the uniqueness of such w.

Once having a Lyapounov exponents theory is natural to develop a stable manifold theory. If $x \in K$ and $\lambda \geq 0$ define $W_\lambda^s(x)$ as the set of points $y \in \bigcap_{n\geq 0} f^{-n}(K)$ (i.e. such that $f^n(y)$ is well defined for all $n \geq 0$) satisfying

$$\limsup_{n\to+\infty} \frac{1}{n} \log\|f^n(y) - f^n(x)\| < -\lambda$$

when f/K is a homeomorphism and $D_x f$ is injective and has dense range for all $x \in K$ we shall prove that if f is $C^{k,\gamma}$ (i.e. class C^k with k-th derivative Hölder continuous with Hölder constant γ) then <u>for a total probability set</u> $K_1 \subset K_0$ <u>of points</u> $x \in K$ <u>the set</u> $W_\lambda^s(x)$ <u>is a</u> $C^{k,\gamma}$ <u>finite codimensional immersed submanifold of</u> E <u>for all</u> $\lambda \geq 0$. This means that $W_\lambda^s(x)$ is an increasing union of $C^{k,\gamma}$ submanifolds of E with the same finite codimension. Moreover it will be proved that if $\lambda' > \lambda'' \geq 0$ then $W_{\lambda'}^s(x) = W_{\lambda''}^s(x)$ if and only if $(-\lambda', \lambda'']$ doesn't contain any Lyapounov exponent of f at x. The codimension of $W_\lambda^s(x)$ can vary with λ and x but we shall prove that for each λ is uniformely bounded on $x \in K_1$.

Similar results hold for the unstable manifolds $W_\lambda^u(x)$ defined, when f is injective and $\lambda \geq 0$, as the set of points $y \in \bigcap_{n\geq 0} f^n(U)$ such that:

$$\limsup_{n\to+\infty} \frac{1}{n} \log\|f^{-n}(y) - f^{-n}(x)\| < -\lambda.$$

If $D_x f$ is injective for all $x \in U$ and f is $C^{k,\gamma}$ then <u>there exists a total probability set</u> $K_2 \subset K_0$ <u>such that if</u> $x \in K_2$ <u>then</u> $W_\lambda^u(x)$ <u>is a finite dimensional</u> $C^{k,\gamma}$ <u>immerse submanifold for all</u> $\lambda \geq 0$ and given $\lambda' > \lambda'' \geq 0$, the equality $W_{\lambda'}^u(x) = W_{\lambda''}^u(x)$ holds if and only if $[-\lambda', -\lambda')$ doesn't contain Lyapounov exponents of f at x.

The methods we shall use to prove the results on Lyapounov exponents are radically different of those used by Oseledec in [3] or Ruelle in [6] and [7]. The stable manifold theory follows more conventional techniques. Essentially we shall use a suitable adaptation of the implicit function method of constructing invariant manifolds (or Perron method). In the works of Pesin [4] and Ruelle [6] the graph transform (or Hadamard) method is used.

I - Basic definitions and statement of the theorems

Let K be a topological space. A Banach vector bundle on K is a triad $(F,\pi,\|\cdot\|)$ where F is a topological space, $\pi: F \to K$ a continuous map (called the canonical projection), $\|\cdot\|: F \to \mathbb{R}$ a continuous function and for every $x \in K$ the set $F_x = \pi^{-1}(x)$ is endowed with a vector space structure such that $\|\cdot\|/F_x$ is a Banach norm on F_x and the topology induced by this norm coincides with the relative topology of F_x. Moreover we require that for all $x \in K$ there exist a neighbourhood U and a homeomorphism $\varphi: F_x \times U \to \pi^{-1}(U)$ such that for all $p \in U$ the map $\varphi(\cdot,p): F_x \to \pi^{-1}(U)$ is an isomorphism between the Banach spaces F_x and $\pi^{-1}(p)$. The set F_x is called the fiber of x. As usual we shall shorten all this just saying that F is a Banach vector bundle on K, the maps π and $\|\cdot\|$ assumed implicit in the definition.
If F_i, $i = 1,2$ are Banach bundles on K_i, $i = 1,2$, we say that a continuous map $L: F_1 \to F_2$ is a homomorphism convering a continuous map $\varphi: K_1 \to K_2$ if $\pi_2 L = \varphi\pi_1$ (where π_i, $i = 1,2$, are the canonical projection associated to F_i, $i = 1,2$) and $L/F_{1,x}$ is bounded linear map of $F_{1,x}$ into $F_{2,\varphi(x)}$ for all $x \in K_1$.
Finally we shall say that a subset $F \subset F_o$ is a subbundle if $(F_o,\|\cdot\|/F_o,\pi/F_o)$ is a Banach vector bundle on K. If K_o is a subset of K we shall denote F/K_o the Banach vector bundle isomorphism given by $(\pi^{-1}(K_o),\ \pi/\pi^{-1}(K_o),\ \|\cdot\|/\pi^{-1}(K_o))$.

In what follows we shall consider a Banach vector bundle F on a compact metric space K and a homomorphism $L: F \circlearrowleft$ covering a homeomorphism $\varphi: K \circlearrowleft$ such that L/F_x is injective for all $x \in K$. If $n > 0$ and $v \in F$ we define $L^{-n}v = w$ if $L^n v = w$. Obviously w doesn't necessarily exist but we shall adopt the convention that the expression $L^{-n}v$ carries the claim that $L^{-n}v$ exists. The injectivity of L/F_x for all x grants the uniqueness of $L^{-n}v$.

Now we are ready to introduce the central definition of this paper:

Definition I.1. We say that $x \in K$ is a regular point of L if satisfies one of the following hypothesis:

I) $\lim_{n\to+\infty} \frac{1}{n} \log\|L^n/F_x\| = -\infty$

II) There exists $k \in \mathbb{Z}^+$ and a splitting

$$F_x = E_1(x) \oplus \ldots \oplus E_k(x) \oplus F_\infty(x)$$

and numbers $\lambda_1(x) > \ldots > \lambda_k(x)$ such that:

a) $\dim E_j(x) < +\infty$ for all $1 \le j \le k$

b) $\lim_{n\to+\infty} \frac{1}{n} \log\|L^n v\| = \lambda_j(x)$ for all $0 \ne v \in E_j(x)$, $1 \le j \le k$.

c) $\lim_{n\to+\infty} \frac{1}{n} \log\|L^n/F_\infty(x)\| = -\infty$.

d) If $0 \ne v \in F_\infty(x)$ and $L^{-n}v$ exists for all $n \ge 0$ then:

$$\lim_{n\to+\infty} \frac{1}{n} \log\|L^{-n}v\| = +\infty.$$

III) There exist subspaces $E_j(x)$, $F_j(x)$, $j = 1,2,\ldots$ and numbers $\lambda_1(x) > \lambda_2(x) > \ldots$ such that:

a) $\dim E_j(x) < +\infty$ for all j

b) $\lim_{j\to+\infty} \lambda_j(x) = -\infty$

c) $E_1(x) \oplus \ldots \oplus E_j(x) \oplus F_j(x) = F_x$ for all j

d) For all j and $0 \neq v \in E_j(x)$:

$$\lim_{n\to\pm\infty} \frac{1}{n} \|L^n v\| = \lambda_j(x)$$

e) For all j:

$$\lim_{n\to+\infty} \frac{1}{n} \log\|L^n/F_j(x)\| = \lambda_{j+1}(x)$$

f) For all j and $0 \neq v \in F_j(x)$ such that $L^{-n}v$ exists for all $n \geq 0$:

$$\liminf_{n\to+\infty} \frac{1}{n} \log\|L^{-n}v\| \geq -\lambda_{j+1}(x).$$

Let us observe some elementary useful properties of regular points that can be deduced directly from the definition. Denote Σ_i, $i = 1,2,3$ the sets of regular points satisfying (I), (II) or (III) respectively. Then these sets are disjoint and $\varphi(\Sigma_i) = \Sigma_i$, $1 = 1,2,3$. Moreover if $x \in \Sigma_2 \cup \Sigma_3$ it is easy to see that the subspaces $E_j(x)$, $F_j(x)$ and the numbers $\lambda_1(x) > \dots$ are unique and satisfy:

$$L(E_j(x)) = E_j(\varphi(x))$$

$$L^{-1}(F_j(x)) = F_j(\varphi^{-1}(x))$$

$$L^{-1}(F_\infty(x)) = F_\infty(\varphi^{-1}(x))$$

$$\lambda_j(\varphi(x)) = \lambda_j(x).$$

If $x \in \Sigma_3$ and we <u>define</u> $F_\infty(x) = \bigcap_j F_j(x)$ (that can be $\{0\}$) then the third equality also holds for x in Σ_3. If x is a regular point we say that the Lyapounov exponents of L at x are the numbers $\lambda_j(x)$ if $x \in \Sigma_2 \cup \Sigma_3$, plus $-\infty$ if either $x \in \Sigma_1 \cup \Sigma_2$ or $x \in \Sigma_3$ and $F_\infty(x) \neq \{0\}$. In other words the Lyapounov exponents of L at a regular point $x \in K$ are the possible values of

$$\limsup_{n\to+\infty} \frac{1}{n} \log\|L^n v\|$$

for $0 \neq v \in F_x$. Denote $\Sigma(L) = \Sigma_1 \cup \Sigma_2 \cup \Sigma_3$.

<u>Theorem A</u>. <u>If</u> L/F_x <u>is compact for all</u> $x \in K$ <u>then</u> $\Sigma(L)$ <u>is a total probability Borel set</u>.

The next theorem examines the dependence of the Lyapounov exponents and the spaces $E_j(x)$, $F_j(x)$ and $F_\infty(x)$ on $x \in \Sigma(L)$. Let $G^+(n)$, $G^-(n)$ denote the Grassmannian bundles of n-dimensional and closed n-codimensional subspaces of the fibers of F i.e. $G^+(n)$ (resp. $G^-(n)$) is the set of n-dimensional subspaces (resp. closed n-codimensional subspaces) of the fibers of F endowed with the unique metrizable topology such that a sequence $E_n \subset F_{x_n}$ converges to $E \subset F_x$ if and only if $x_n \to x$ and there exists a neighborhood U of x, a canonical chart $\varphi: F_x \times U \to \pi^{-1}(U)$ and a splitting $F_x = E \oplus G$ such that for large values of n the subspace $\varphi^{-1}(E_n)$ is the graph of a linear map $T_n: E \to G$ and $\lim_{n\to+\infty} \|T_n\| = 0$. Let $G_2(n_1,\dots,n_k)$ be the set of points in Σ_2 with k finite Lyapounov exponents and $\dim E_j(x) = n_j$ for $1 \le j \le k$. Let $G_3(n_1,\dots,n_k)$ be the set of points in Σ_3 with $\dim E_j(x) = n_j$ for $1 \le j \le k$. When X and Y are topological spaces we say that a map $f: X \to Y$ is <u>strongly measurable</u> if there exist Borel sets $X_n \subset X$, $n = 1,2,3,\dots$ such that $\bigcup_n X_n = X$ and f/X_n is continuous for very n. Obviously strongly measurable maps are measurable but the converse property is false. (The reader can easily verify that the function $f: [0,1] \circlearrowleft$ defined as $f(t) = 0$ when t is rational and $f(t) = 1$ if t is irrational is not strongly measurable).

<u>Theorem B</u>. <u>For every set of positive integers</u> $n_1,\dots,n_k$ <u>the sets</u> $G_i(n_1,\dots,n_k)$ <u>are Borel sets and the functions</u>:

$$G_i(n_1,\dots,n_k) \ni x \mapsto E_j(x) \in G^+(n_j) \qquad i = 2,3, \qquad j = 1,\dots,k$$

$$G_2(n_1,\dots,n_k) \ni x \mapsto F_\infty(x) \in G^-(\sum_j n_j)$$

$$F_3(n_1,\dots,n_k) \ni x \mapsto F_k(x) \in G^-(\sum_j n_j)$$

<u>are strongly measurable. The functions</u>

$$G_i(n_1,\ldots,n_k) \ni x \mapsto \lambda_j(x) \qquad j = 1,\ldots,k$$

<u>are measurable.</u>

As a corollary to this theorem follows that $\Sigma(L)$ is a Borel set. This is proved observing that Σ_1 is Borel (because is the preimage of $-\infty$ under the measurable function $K \ni x \mapsto \lim_{n\to+\infty} n^{-1} \log\|L^n/F\|$) and

$$\Sigma_2 = \bigcup G_2(n_1,\ldots,n_k)$$

$$\Sigma_3 = \bigcup G_3(n_1,\ldots,n_k)$$

where the unions are taken over all k-uples of positive integers and $k \in \mathbb{Z}^+$. Denote $\Sigma_2^{(k)}$ the set of points in Σ_2 with exactly k finite Lyapounov exponents. Since the dependence of the subspaces $E_j(x)$, $F_j(x)$ on x is not continuous, the "angles" between these subspaces can be very small.

To study this problem we consider for a point $x \in \Sigma_2^{(k)} \cup \Sigma_3$ the projection $\pi_k(x)\colon F_x \to F_k(x)$ (if $x \in \Sigma_3$) or $\pi_k(x)\colon F_x \to F_\infty(x)$ (if $x \in \Sigma_3$) associated to the splitting $F_x = E_1(x) \oplus \ldots \oplus E_k(x) \oplus \oplus F_k(x)$ (if $x \in \Sigma_3$) or to the splitting $F_x = E_1(x) \oplus \ldots \oplus E_k(x) \oplus \oplus F_\infty(x)$ (if $x \in \Sigma_2^{(k)}$).

<u>Theorem C. The equality:</u>

$$\lim_{n\to\pm\infty} \frac{1}{n} \log\|\pi_k(\varphi^n(x)\| = 0$$

<u>holds for a total probability set in</u> $\bar{\Sigma}_2^{(k)} \cup \bar{\Sigma}_3$.

Now we shall consider the case of a vector bundle homomorphism $L\colon F\circlearrowleft$ covering a continuous map $\varphi\colon K\circlearrowleft$ <u>without any assumption on the injectivity of</u> φ <u>or the maps</u> L/F_x. <u>We still assume that</u> L/F_x <u>is compact for every</u> $x \in K$. We say that $x \in K$ is <u>almost regular</u> if there exist a countable or finite set $\lambda_1(x)>\lambda_2(x)>\ldots\subset[-\infty,+\infty)$

with $\inf \lambda_j(x) = -\infty$ and a family of subspaces $F_x = F_1(x) \supset F_2(x) \supset \dots$ with finite condimensions and such that

$$\lim_{n\to+\infty} \frac{1}{n} \log\|L^n v\| \le \lambda_j(x)$$

for every j and $0 \ne v \in F_j(x)$ and

$$\lim_{n\to+\infty} \frac{1}{n} \log\|L^n v\| = \lambda_j(x)$$

if $0 \ne v \in F_j(x)\backslash F_{j+1}(x)$. It is easy to see that the numbers $\lambda_1(x) > \lambda_2(x) > \dots$ and the subspaces $F_1(x) \supset F_2(x) \supset \dots$ are unique. The numbers $\lambda_1(x) > \dots$ are called the Lyapounov exponents of L at x and we shall agree in including $-\infty$ as a Lyapounov exponent when $\bigcap_{j\ge1} F_j(x) \ne \{0\}$. As before it follows that the Lyapounov exponents of L at a point $x \in \Lambda$ are possible values of (1). Denote $\Lambda(L)$ the set of almost regular points of L.

<u>Theorem D</u>. $\Lambda(L)$ <u>is a total probability set</u>.

The following result shows that the codimension of the subspace $F_j(x)$ are bounded by a constant determined by $\lambda_j(x)$.

<u>Theorem E</u>. <u>For all</u> $\lambda \ge 0$ <u>there exists</u> $N(\lambda)$ <u>such that if</u> $x \in \Lambda(L)$ <u>and</u> $\lambda_j(x) \ge -\lambda$ <u>then codim</u> $F_j(x) \le N(\lambda)$.

Finally let us give the precise statements of the theorems on stable and unstable manifolds that we sketched in the introduction. Let E be a Banach space $U \subset E$ an open set and $f\colon U \to E$ a C^1 map. If $x \in \bigcap_{n\ge0} f^{-n}(U)$ and $\lambda \ge 0$ we define:

$$W_\lambda^s(s) = \{y \in \bigcap_{n\ge0} f^{-n}(U) \mid \limsup_{n\to+\infty} \frac{1}{n} \log\|f^n(y)-f^n(x)\| < -\lambda\}.$$

If f is injective and $x \in \bigcap_{n\ge0} f^n(U)$ we define:

$$W_\lambda^u(x) = \{y \in \bigcap_{n\ge0} f^n(U) \mid \limsup_{n\to+\infty} \frac{1}{n} \log\|f^{-n}(y)-f^{-n}(x)\| < -\lambda\}.$$

Before stating the results let us briefly recall the definitions of $C^{r,\gamma}$ maps and submanifolds. If E_1, E_2 are Banach spaces and $V \subset E_1$ is an open set we say that $f: V \to E_2$ is a $C^{r,\gamma}$ map (or a Hölder C^r map with Hölder constant γ) if it is C^r and there exists $A > 0$ such that $\|(D_x^r f) - (D_x^r f)\| \leq A\|x-y\|^{\gamma}$ for all x, y in V. We say that a subset $M \subset E_1$ is a $C^{r,\gamma}$ submanifold if it is a C^1 submanifold and every point $x \in M$ has a neighbourhood W such that $W \cap M$ is the image of $\{v \in T_x M \mid \|v\| < 1\}$ by a $C^{r,\gamma}$ homeomorphism. We say that $N \subset E_1$ <u>is an immerse</u> $C^{r,\gamma}$ <u>submanifold if it is the union of a monotone sequence</u> $N_1 \subset N_2 \subset \ldots$ <u>of</u> $C^{r,\gamma}$ <u>submanifolds such that</u> N_j <u>is open in</u> N_{j+1} <u>for all</u> $j \geq 1$. In this case we say that codim $N = n$ (resp. dim $N = n$) if codim $N_j = n$ (resp. dim $N_j = n$) for all $j \geq 1$.

<u>Theorem F</u>. Let $K \subset U$ be a compact set such that $f(K) = K$. Then if f is $C^{r,\gamma}$ and $D_x f$ <u>is injective and compact for all</u> $x \in K$ <u>and has dense range for all</u> $x \in U$ <u>there exists a total probability set of points</u> $\Gamma \subset K$ <u>contained in the set of regular points of</u> $Df: E \times K \circlearrowleft$ (where $Df: E \times K \circlearrowleft$ is the homomorphism of the Banach vector bundle $E \times K$ defined by $(x,v) \to (f(x), (D_x f)v))$ <u>such that the following properties hold for every</u> $x \in \Gamma$:

a) For every $\lambda \geq 0$ $W_\lambda^s(x)$ is a finite codimensional immerse $C^{r,\gamma}$ submanifold.

b) For all $\lambda' > \lambda'' \geq 0$, $W_{\lambda'}^s(x) = W_{\lambda''}^s(x)$ if and only if $(-\lambda', -\lambda'']$ doesn't contain any Lyapounov exponent of Df at x.

c) If $\lambda_1(x) > \lambda_2(x) > \ldots$ are the Lyapounov exponents of Df at x and $F_j(x)$, $j = 1,2,\ldots$ the subspaces given in the definition of regular points, then $T_x W_\lambda^s(x) = F_{j_o}(x)$ where $j_o = \min\{j / \lambda_j(x) < -\lambda\}$.

Finally we state the version for unstable manifolds of Theorem F.

Theorem F. If f is $C^{r,\gamma}$ and $K \subset U$ is a compact set such that $f(K) = K$, then if f is injective and $D_x f$ is compact for all $x \in U$, there exists a total probability set Γ^u contained in the set of regular points of $Df\colon E \times K \circlearrowleft$ such that the following properties hold for every $x \in \Gamma^u$:

a) For every $\lambda \geq 0$, $W^u_\lambda(x)$ is a finite dimensional immerse $C^{r,\gamma}$ submanifold.

b) For all $\lambda' > \lambda'' \geq 0$, $W^u_{\lambda'}(x) = W^u_{\lambda''}(x)$ if and only if $(-\lambda', \lambda'']$ doesn't contain any Lyapounov exponent of Df at x.

c) If $\lambda_1(x) > \lambda_2(x) > \ldots$ are the Lyapounov exponents at x and $E_1(x), E_2(x), \ldots$ are the subspaces given by the definition of regular points, then

$$TW^u_\lambda(x) = \bigoplus_{\lambda_j(x) > \lambda} E_j(x).$$

II - Some technical lemmas on compact homomorphisms.

Let K and H be compact metric spaces, F and G Banach vector bundles on K and H respectively and $L\colon F \to G$ a homomorphism covering a continuous map $\varphi\colon K \to G$. Suppose that L/F_x is compact for every $x \in K$.

Lemma II.1. $L(S)$ is relatively compact if $S \subset F$ is bounded.

Proof. If $v_n \in F_{x_n}$, $n \in \mathbb{Z}^+$ is a sequence in S we can assume that $x_n \to x \in K$. Since all our arguments will involve only fibers at points of the convergent sequences x_n and $\varphi(x_n)$ we can suppose that F and G are trivial bundles, say $F = F_o \times K$, $G = G_o \times H$. Then, since $L_x = L/F_o \times \{x\}$ is compact we can assume that the sequence $L_x V_n$ converges to some $v \in G_o$. Therefore:

$$\|L_{x_n} v_n - v\| \leq \|L_{x_n} v_n - L_x v_n\| + \|L_x v_n - v\| \leq$$

$$\leq \|L_{x_n} - L_x\| \sup_n \|v_n\| + \|L_x v_n - v\|.$$

Since $L_{x_n} \to L_x$ and $L_x v_n \to v$ it follows that $L_{x_n} v_n \to v$.

Lemma II.2. Let $m \in \mathbb{Z}^+$ and $E_n \in G^+(m)$, $n \in \mathbb{Z}^+$ be a sequence such that there exists $c > \gamma$ satisfying $\|Lv\| \geq c\|v\|$ for all $v \in E_n$, $n \in \mathbb{Z}^+$. Then there exists a subsequence E_{n_j}, $j \in \mathbb{Z}^+$ such that $L(E_{n_j})$, $j \in \mathbb{Z}^+$, is convergent in $G^+(m)$.

Proof. Suppose that $E_n \subset F_{x_n}$ and that $x_n \to x$. Let $\{v_1^{(n)},\ldots,v_m^{(n)}\}$ be a basis of $L(E_n)$ with $\|v_j^{(n)}\| = 1$ for all j and n and such that if $S_{n,k}$ is the subspace spanned by $\{v_1^{(n)},\ldots,v_k^{(n)}\}$ then:

$$\|v_{k+1}^{(n)} - w\| \geq 1/2 \qquad (1)$$

for all $n \in \mathbb{Z}^+$, $1 \leq k < m$ and $w \in S_{n,k}$. Set $w_j^{(n)} = L^{-1}v_j^{(n)}$. Then $\sup\{\|w_j^{(n)}\| \mid 1 \leq j \leq m,\ n \in Z^+\} \leq c^{-1}$. By Lemma II.1 we can assume that for every $1 \leq j \leq m$ the sequence $v_j^{(n)} = Lw_j^{(n)}$ converges to some $v_j \in G_x$. We claim that the set $\{v_1,\ldots,v_n\}$ is linearly independent. To see this we shall prove by induction that $\{v_1,\ldots,v_k\}$ is linearly independent and that if S_k is the space spanned by $\{v_1,\ldots,v_k\}$ then $\lim_{n\to+\infty} S_{n,k} = S_k$. Clearly the first property holds for k=1. If it is true for $k = \ell$ then to show that holds for $k = \ell+1$, it is sufficient to observe that $v_{\ell+1} \notin S_{n,\ell}$ would imply by the induction hypothesis that $v_{\ell+1} = \lim w_n$ with $w_n \in S_{n,\ell}$. Then: $0 = \|v_{\ell+1}-v_{\ell+1}\| = \lim_{n\to+\infty} \|v_{\ell+1}^{(n)}-w_n\|$. But since $w_n \in S_{n,\ell}$, property (1) implies that $\|v_{\ell+1}^{(n)} - w_n\| \geq 1/2$ for all n thus contradicting its convergence to 0. This complete the proof of the claim. Then, since $v_j = \lim_{n\to+\infty} v_j^{(n)}$ for all $1 \leq j \leq m$ it follows that $L(E_n)$ converges to the subspace spanned by $\{v_1,\ldots,v_m\}$.

In the statement of the next lemma we shall use the following definition of angle between subspaces of a Banach space. Let S_1, S_2 be closed subspaces of a Banach space F_o such that $S_1 \cap S_2 = \{0\}$.

We define the angle $\alpha(S_1,S_2)$ by:

$$\alpha(S_1,S_2) = \inf\{\|\pi v\|/\|v\| \mid 0 \neq v \in S_2\}$$

where $\pi\colon F_o \to F_o/S_1$ is the canonical homomorphism.

Lemma II.3. Let $m \in \mathbb{Z}^+$, $\{x_n \mid n \in \mathbb{Z}^+\}$ a sequence in K and at each x_n a splitting $F_{x_n} = E_n \oplus T_n$ with $\dim E_n = m$. Suppose that there exists $c > 0$ satisfying

$$\|Lv\| \geq c\|v\|$$

$$\alpha(L(E_n),\overline{L(T_n)}) \geq c$$

for all $n \in \mathbb{Z}^+$, $v \in E_n$. Then the sequence $\{T_n \mid n \in \mathbb{Z}^+\} \subset G^-(m)$ has a convergent subsequence.

Proof. We can assume that $x_n \to x$ and (by Lemma I.2) that $L(E_n) \to S \subset G_{\varphi(x)}$. As in the proof of Lemma I.1 we shall suppose that the vector bundles F and G are trivial, say $F = F_o \times K$, $G = G_o \times H$. Take a closed subspace $S' \subset G_o$ such that $S \oplus S' = G_o$. Then if n is large, say $n \geq N$ there exists a closed subspace $S'_n \subset S'$ and a linear continuous map $A'_n\colon S'_n \to S$ such that:

$$\sup_n \|A'_n\| < +\infty$$

$$\overline{L(T_n)} = \{x + A'_n x \mid x \in S'_n\}.$$

We leave the proof of these properties to the reader just remarking that the subspaces S'_n are given by:

$$S'_n = \pi(L(T_n))$$

where π is the projection of G onto S associated to the splitting $S \oplus S' = G$ and the maps A'_n by:

$$A'_n x = -x + (\pi/\overline{L(T_n)})^{-1} x.$$

By the Hahn-Banach theorem there exist continuous linear maps $A_n\colon S' \to S$ such that

$$A_n/S'_n = A'_n$$

$$\sup\|A_n\| < +\infty .$$

Set $R_n = \{x + A_n x \mid x \in S'\}$. We claim that

$$L_{x_n}^{-1}(R_n) = T_n$$

for large values of n. Obviously $L_{x_n}^{-1}(R_n) \supset T_n$. If $L_{x_n}^{-1}(R_n) \not\subset T_n$ there exists $v \notin T_n$ such that $Lv_{x_n} \in R_n$. Write $v = v_1 + v_2$ with $v_1 \in T_n$ and $v_2 \in E_n$. Since $v \notin T_n$ we must have $v_2 \neq 0$. Then $L_{x_n} v = L_{x_n} v_1 + L_{x_n} + v_2$ and since $L_{x_n} v_1 \in L(T_n) \subset R_n$ and $L_{x_n} v \in R_n$ it follows that $L_{x_n} v_2 \in R_n$. Then $L_{x_n} v_2 \in R_n \cap L(E_n)$. But clearly $R_n \oplus S = G_o$ and since $L_{x_n}(E_n) \to S$ it follows that for large values of n $R_n \cap L_{x_n}(E_n) = \{0\}$. Then $L_{x_n} v_2 = 0$. But L_{x_n}/E_n is injective. Hence $v_2 = 0$ contradicting the property $v_2 \neq 0$ implied by $v \in T_n$. This completes the proof of the claim. Now define B_n: $L_{x_n}^{-1}(S') \to L_{x_n}^{-1}(S)$ by:

$$B_n = L_{x_n}^{-1} A_n(L_{x_n}/L_{x_n}^{-1}(S')).$$

It is easy to see that $L_{x_n}^{-1}(R_n) = \{x + B_n x \mid x \in L_{x_n}^{-1}(S')\}$. Then, for large values of n:

$$T_n = L_{x_n}^{-1}(R_n) = \{x + B_n x \mid x \in L_{x_n}^{-1}(S')\}.$$

But $L_{x_n}^{-1}(S') \to L_x^{-1}(S')$ and $L_{x_n}^{-1}(S) \to L_x^{-1}(S)$. Hence there exist continuous injective linear maps i_n: $L_x^{-1}(S') \to F_o$, j_n: $L_x^{-1}(S) \to F_o$, such that $i_n(L_x^{-1}(S')) = L_{x_n}^{-1}(S')$, $j_n(L_x^{-1}(S)) = L_{x_n}^{-1}(S)$ and if i: $L_x^{-1}(S') \to F_o$, j: $L_x^{-1}(S) \to F_o$ are the inclusion maps, $\lim_{n\to+\infty} \|i_n - i\| = \lim_{n\to+\infty} \|j_n - j\| = 0$. We claim that the sequence $j_n^{-1} B_n i_n$: $L_x^{-1}(S') \to L_x^{-1}(S)$ has a convergent subsequence. If the claim is true, the lemma is proved because the sequence of subspaces

$$T'_n = \{x + (j_n^{-1} B_n i_n)x \mid x \in L_x^{-1}(S')\}$$

converges in $G^-(m)$ and the property $\lim_{n\to+\infty} \|i_n - i\| = \lim_{n\to+\infty} \|j_n - j\| = 0$ implies that the distances between T_n and T'_n converge to zero. Hence the convergence of T'_n implies that of T_n. To prove the clai take the closed linear hull $S'' \subset S'$ of $\bigcup_{n>0} L_{x_n}(L_{x_n}^{-1}(S'))$. It is a Banach separable space because every L_{x_n} is compact. Then the uniformely bounded sequence of linear maps $j_n^{-1} L_{x_n}^{-1} A_n/S'' \to L_x^{-1}(S)$ has a pointwise convergent subsequence and, to simplify the notation, we shall assume that the sequence itself is convergent. But if B is the unitary ball of $L_x^{-1}(S')$ the set $\bigcup_{n>0} L_{x_n} i_n(B) = L(\bigcup_{n>0} i_n(B))$ is relatively compact by Lemma II.1. Then the sequence $j_n^{-1} L_{x_n}^{-1} A_n / \bigcup_{n>0} L_{x_n} i_n(B)$ converges uniformely. Therefore so it does the sequence $j_n^{-1} L_{x_n}^{-1} A_n L_{x_n} i_n/B$ and the claim is proved.

<u>Lemma II.4</u>. For every $c > 0$ there exists $N > 0$ such that if $x \in K$ and $v_i \in F_x$, $i = 1,\dots,n$ are unitary vectors satisfying $\|L_x v_i - L_x v_j\| \geq c$ for all $1 \leq i \leq j \leq n$, then $n \leq N$.

<u>Proof</u>. If this property is false, for some $c > 0$ there exists a sequence of points $x_n \in K$ and in each F_{x_n} a set of unitary vectors $\{v_1^{(n)},\dots,v_n^{(n)}\}$ such that

$$\|L_{x_n} v_j^{(n)} = L_{x_n} v_j^{(n)}\| \geq c$$

if $1 \leq i < j \leq n$. Suppose that $x_n \to x$ and as we did in the proofs of II.3 and II.1 that $F = F_o \times K$, $G = G_o \times K$. By II.1 we can suppose that for each j the sequence $L_{x_n} v_j^{(n)}$ converges to some w_j. Then $\|w_j - w_i\| \geq c$ if $i < j$. For each j take n_j satisfying:

$$\|L_{x_n} v_j^{(n_j)} - w_j\| \leq c/3.$$

By II.1 the sequence $L_{x_n} v_n^{(n_j)}$ must have a convergent subsequence. However this is impossible because:

$$\|Lv_k^{(n_k)} - Lv_\ell^{(n_\ell)}\| \geq \|w_k - w_\ell\| - \|Lv_k^{(n_k)} - w_k\| -$$
$$- \|Lv_\ell^{(n_\ell)} - w_\ell\| \geq c - \frac{1}{3}c - \frac{1}{3}c = \frac{1}{3}c$$

for all k and ℓ.

III - Proof of Theorems A and C

Let K, F and L be as in the statement of Theorem A. By the ergodic decomposition theorem in order to prove Theorems A and B it is sufficient to show that if μ is a φ-invariant ergodic probability on K then $\mu(\Sigma(L)) = 1$ and that if $\mu(\bar{\Sigma}_2^{(k)} \cup \bar{\Sigma}_3) \neq 0$ then the equality of Theorem B holds μ-a.e. (observe that by the ergodicity of μ the property $\mu(\bar{\Sigma}_2^{(k)} \cup \bar{\Sigma}_3) \neq 0$ implies, by the φ-invariance of $\bar{\Sigma}_2^{(k)}$ and $\bar{\Sigma}_3$ and Theorem A, that $\mu(\Sigma_2^{(k)}) = 1$ or $\mu(\Sigma_3) = 1$). Then fix any φ-invariant ergodic probability measure μ and we shall prove these properties. For this purpose we shall need measurable versions of the concepts of Banach vector bundles and homomorphisms. We say that G is a μ-measurable Banach vector bundle on K if there exists a sequence of compact sets K_n, $n \in \mathbb{Z}^+$, called a normal sequence, such that $\mu(\bigcup_n K_n) = 1$, a map $\pi: G \to K$ (the canonical projection) and a function $\|\cdot\|: G \to \mathbb{R}$ such that $(\pi/\pi^{-1}(K_n), \|\cdot\|/K_n)$ is a Banach vector bundle on K for all n. A map $T: G \circlearrowleft$ is a μ-measurable homomorphism covering a continuous map $\varphi: K \circlearrowleft$ if the normal sequence K_n can be taken such that for all $m \in \mathbb{Z}^+$ there exists n satisfying $\varphi(K_m) \subset K_n$ and $T/\pi^{-1}(K_n)$ is a vector bundle homomorphism of $(\pi^{-1}(K_m), \pi/\pi^{-1}(K_m), \|\cdot\|/\pi^{-1}(K_m))$ in $(\pi^{-1}(K_n), \pi/\pi^{-1}(K_n), \|\cdot\|/\pi^{-1}(K_n))$. To simplify the notation we define $G/K_m = (\pi^{-1}(K_m)\cdot\pi/\pi^{-1}(K_m), \|\cdot\|/\pi^{-1}(K_m))$.

Before going into the proof of the property $\mu(\Sigma(L)) = 1$ we shall prove some lemmas, in whose statements K will be a compact metric space, $\varphi: K \circlearrowleft$ a homeomorphism, μ a φ-invariant ergodic probability on the Borel σ-algebra of K, G a μ-measurable Banach vector bundle on K and $T: G \circlearrowleft$ a μ-measurable homomorphism covering φ such that T/G_x is injective and compact for a.e. $x \in K$

and $\sup\|T/G_x\| < \infty$. If $\lambda \in \mathbb{R}$ and $x \in K$ define $E_\lambda(x)$ as the set of vectors $v \in G_x$ such that $T^{-n}v$ exists for all $n \ge 0$ and $\limsup n^{-1} \log\|T^{-n}v\| \le \lambda$.

Lemma III.1. For every $\lambda \in \mathbb{R}$, $\dim E_\lambda(x) < \infty$ for a.e. $x \in K$. Moreover the set $\bigcup_x E_\lambda(x)$ is a μ-measurable Banach vector bundle on K.

Define:

$$k(T) = -\limsup_{n\to+\infty} \frac{1}{n} \log\|T^n/G_x\|.$$

Since μ is ergodic and $\sup\|T/G_x\| < +\infty$ this lim sup is constant a.e. and $> -\infty$.

Lemma III.2. $E_{k(T)}(x) \neq \{0\}$ a.e. if $k(T) < +\infty$,

Lemma III.3. $\lim_{n\to\pm\infty} \frac{1}{n} \log\|T^n v\| = -k(T)$ for a.e. x and every $0 \neq v \in E_{k(T)}(x)$.

Lemma III.4. For every $\lambda \in \mathbb{R}$ there exists a μ-measurable Banach vector bundle $G_\lambda \subset G$ such that if $G_\lambda(x)$ denotes its fiber at x the following properties hold for a.e. x:

a) $T(G_\lambda(x) \subset G_\lambda(\varphi(x))$

b) $E_\lambda(x) \oplus G_\lambda(x) = G_x$

c) $\limsup_{n\to+\infty} \frac{1}{n} \log\|T^n/G_\lambda(x)\| < -\lambda$

d) $\liminf_{n\to+\infty} \frac{1}{n} \log\|T^{-n}v\| \ge \lambda$ for every $0 \neq v \in G_\lambda(x)$ such that $T^{-n}v$ exists for all $n \ge 0$.

e) If $\pi(x)\colon G_x \to G_\lambda(x)$ is the projection associated to the splitting $G_x = E_\lambda(x) \oplus G_\lambda(x)$ then $\lim_{n\to\pm\infty} \frac{1}{n} \log \|\pi(\varphi^n(x)\| = 0$ a.e.

Before proving these lemmas we shall show how they imply the property $\mu(\Sigma(L)) = 1$. If $k(L) = +\infty$ then $\mu(\Sigma_1) = 1$ and we are done. If it is $< +\infty$, putting $\lambda_1 = -k(L)$ we have, by Lemmas III.4 and III.2, a μ-measurable bundle $G_{\lambda_1} \subset G$ such that for a.e. x

satisfies:

$$F_x = E_{\lambda_1}(x) \oplus G_{\lambda_1}(x)$$

$$0 < \dim E_{\lambda_1}(x) < +\infty$$

$$L(G_{\lambda_1}(x)) \subset G_{\lambda_1}(\varphi(x)).$$

By Lemma III.3:

$$\lim_{n\to\pm\infty} \frac{1}{n} \log\|T^n v\| = \lambda_1$$

for a.e. x and every $0 \neq v \in E_{\lambda_1}(x)$. Define $E_1(x) = E_{\lambda_1}(x)$. Moreover, Lemma III.4 implies:

$$k(T/G_{\lambda_1}) > -\lambda_1 \; .$$

If $k(T/G_{\lambda_1}) = +\infty$ we define $F_\infty(x) = G_{\lambda_1}(x)$ and we have proved that $\mu(\Sigma_1^{(2)}) = 1$ and that, by part (e) of III.4, that the equality of Theorem C holds a.e. in $\Sigma_1^{(2)}$. If $k(T/G_{\lambda_1}) < +\infty$, we put $F_1(x) = G_{\lambda_1}(x)$, and we have obtained the first subspaces of the families $E_1(x),\ldots,F_1(x),\ldots$ that will satisfy property II or III of the definition of regular points. We continue this process by applying Lemma III.4 and III.2 again, but this time to the μ-measurable homorphism $T/G_{\lambda_1} = T/F_1$ and putting $\lambda_2 = -k(T/F_1)$ we obtain a μ-measurable bundle $G_{\lambda_2} \subset F_1$ such that for a.e. x satisfies:

$$F_1(x) = G_{\lambda_2}(x) \oplus E_{\lambda_2}(x)$$

$$0 < \dim E_{\lambda_2}(x) < \infty$$

$$L(G_{\lambda_2}(x)) \subset G_{\lambda_2}(\varphi(x))$$

and by Lemma III.3:

$$\lim_{n\to\pm\infty} \frac{1}{n} \log\|T^n v\| = \lambda_2$$

for a.e. x and every $0 \neq v \in E_{\lambda_2}(x)$. Again, if $k(T/G_{\lambda_2}) = +\infty$, we proved $\mu(\Sigma_2^{(2)}) = 1$ and by part (e) of III.4 the equality of Theorem C holds a.e. If it is $< +\infty$ we put $E_2(x) = E_\lambda(x)$,

$F_2(x) = G_{\lambda_2}(x)$ and we have obtained the second elements in the sequences $E_1(x), E_2(x), \ldots, F_1(x), F_2(x), \ldots$ required to satisfy properties II or III of the definition of regular point. Continuing this process either we arrive to some λ_m such that $k(T/G_{\lambda_m}) = +\infty$, and then $\mu(E_2^{(m)}) = 1$ (and again, by (e) of III.4, the equality of Theorem C holds a.e.), or the process continues indefinitely and then we have proved $\mu(\Sigma_3) = 1$ and that Theorem C holds a.e. In this last case we still have to show that $\lim_{n\to+\infty} \lambda_n = -\infty$. If there exists λ satisfying $\lambda < \lambda_n$ for all n, we must have $E_\lambda(x) \supset E_1(x) \oplus \ldots \oplus E_n(x)$ a.e. for all n. But this contradicts the property $\dim E_\lambda(x) < +\infty$ a.e. given by III.1. This proves Theorem A.

Proof of Lemma III.1. Let $K_o \subset K$ be a compact set such that for all $n \geq 0$ $G/K_o \cup \varphi^n(K_o)$ is a Banach vector bundle and $T\colon G/K_o \to G/\varphi^n(K_o)$ is continuous. Let K' be the set of points $x \in K$ where:

$$\mu(K_o) = \lim_{n\to+\infty} \frac{1}{n} \#\{1 \leq j \leq n \mid \varphi^{-j}(x) \in K_o\}.$$

By the ergodicity of μ, $\mu(K') = 1$. We shall prove that $\dim E_\lambda(x) < \infty$ for all $x \in K'$. Suppose that $\dim E_\lambda(x) = +\infty$ for some $x \in K'$. We shall show this contradicts Lemma II.4.

Let $\{v_j^{(n)} \mid j \in \mathbb{Z}^+\} \subset E_\lambda(\varphi^{-n}(x))$, $n \in \mathbb{Z}^+$, be a sequence of unitary vectors such that if $G_j^{(n)}$ is the space spanned by $\{v_1^{(n)}, \ldots, v_j^{(n)}\}$ then:

$$T^{-1}(G_j^{(n)}) = G_j^{(n+1)} \tag{1}$$

and:

$$\|v_j^{(n)} - w\| \geq 1/2 \tag{2}$$

for all $j > 1$, $n \geq 1$, $w \in G_{j-1}^{(n)}$. This sequences can be easily constructed by induction on j as follows: we take $v_1^{(1)} \in E_\lambda(\varphi^{-1}(x))$ with $\|v_1^{(1)}\| = 1$ and then we define $v_1^{(n)} = T^{-(n-1)}v_1^{(1)}/\|T^{-(n-1)}v_1^{(1)}\|$. Suppose that we have constructed $\{v_1^{(n)}, \ldots, v_k^{(n)}\}$ for all $n \in \mathbb{Z}^+$ satisfying (1) and (2) for $1 \leq j \leq k$. Then we take $v_{k+1}^{(1)} \in E_\lambda(\varphi^{-1}(x))$

with norm 1 and such that (2) holds for $n=1$, $j = k+1$ and every $w \in G_k^{(1)}$. Set $G_{k+1}^{(n)} = T^{-(n-1)}(G_{k+1}^{(1)})$ and take $v_{k+1}^{(n)}$ with $\|v_{k+1}^{(n)}\| = 1$ and $\|v_{k+1}^{(n)} - w\| \geq 1/2$ for all $w \in G_k^{(n)}$. Then the set $\{v_1^{(n)},\ldots,v_{k+1}^{(n)}\}$ satisfy (1), (2) for all $1 \leq j \leq k+1$, $n \in Z^+$. Now let $E_j^{(n)}$ be the one dimensional space generated by $v_j^{(n)}$ and let $\pi_j^{(n)}: G_j^{(n)} \to E_j^{(n)}$ be the projection associated to the splitting $G_j^{(n)} = E_j^{(n)} \oplus G_k^{(n)}$. Then the sets $\{v_1^{(n)},\ldots,v_{k+1}^{(n)}\}$ satisfy (1) and (2) for all $1 \leq j \leq k+1$, $n \in Z^+$. Now let $E_j^{(n)}$ be the one dimensional space generated by $v_j^{(n)}$ and let $\pi_j^{(n)}: G_j^{(n)} \to E_j^{(n)}$ be the projection associated to the splitting $G_j^{(n)} = E_j^{(n)} \oplus G_{j-1}^{(n)}$. By (2)

$$\|\pi_j^{(n)}\| \leq 2 \tag{3}$$

for all $j \geq 1$, $n \in \mathbb{Z}^+$. Define $P_j^{(n)} = \|\pi_j^{(n+1)}T^{-1}/E_j^{(n)}\|$. By (3)

$$\frac{1}{N}\log\prod_{n=1}^{N} P_j^{(n)} = \frac{1}{N}\log\|\pi_j^{(n+1)}T^{-N}/E_j^{(1)}\| \leq$$

$$\leq \frac{1}{N}\log 2 + \frac{1}{N}\log\|T^{-N}/E_j^{(1)}\| =$$

$$= \frac{1}{N}\log 2 + \frac{1}{N}\log\|T^{-N}v_j^{(1)}\| .$$

Hence:

$$\limsup_{N\to+\infty}\frac{1}{N}\sum_{n=1}^{N}\log P_j^{(n)} \leq \lambda .$$

Denote $S_N = \{1 \leq n \leq N \mid \varphi^{-n}(x) \in K_o\}$. Since $x \in K'$:

$$\lim \frac{1}{N}\# S_N = \mu(K_o).$$

Take $0 < b < 1$ and N_o such that:

$$\# S_N \geq b\mu(K_o)N$$

if $N \geq N_o$. Take $c > 0$ such that:

$$\log P_j^{(n)} \geq -c$$

for all $j \geq 1$, $n \in \mathbb{Z}^+$. This number exists because the norm of T

is by hypothesis bounded on the fibers of G. Define:

$$S_N^{(j)} = \{n \in S_N \mid \log P_j^{(n)} \le c\}$$

$$\alpha_N^{(j)} = \frac{\# S_N^{(j)}}{\# S_N} .$$

Take N_j such that $N \ge N_j$ implies:

$$\sum_1^N \log P_j^{(n)} \le (\lambda+1)N.$$

Hence, for every $j \ge 1$:

$$(\lambda+1)N \ge c\,\#(S_N \setminus S_N^{(j)}) - c(N - \#(S_N \setminus S_N^{(j)})) \ge$$

$$\ge c(1-\alpha_N^{(j)})\,\#\,(S_N \setminus S_N^{(j)}) - c(N-(1-\alpha_N^{(j)})\# S_N) \ge$$

$$\ge c(1-\alpha_N^{(j)}) b\mu(K_o)N - cN.$$

Then:

$$\alpha_N^{(j)} \ge 1 - \frac{\lambda+1+c}{2cb\mu(K_o)}$$

Then

$$\alpha_N^{(j)} \ge 1/3 \tag{4}$$

for all j, if $\mu(K_o)$ and b are taken near to 1 and c is large enough. Now we shall use the following elementary combinatorial lemma:

<u>Lemma III.5</u>. If A is a finite set and $\mathcal{a} = \{A_1,\ldots,A_n\}$ a family of subsets there exists a point in A that belongs to
$\sum_{j=1}^{n} (\# A_j/\# A)$ sets of $\mathcal{a}$.

<u>Proof</u>. Suppose that k is the maximum number of elements of $\mathcal{a}$ with non empty intersection. If $p \in A$ and $1 \le j \le n$ define $m(p,j) = \#\{1 \le i \le j \mid p \in A_i\}$. Define subsets $\tilde{A}_j$ of $A \times \{1,\ldots,k\}$ as follows:

$$\tilde{A}_j = \{(p,m(p,j)) \mid p \in A_j\}.$$

These sets are disjoint because if $(p,q) \in \tilde{A}_{j_1} \cap \tilde{A}_{j_2}$ with $j_1 > j_2$, then

$p \in A_{j_1}$, $p \in A_{j_2}$ and $q = m(p,j_1) = m(p,j_2)$. But since $p \in A_{j_2}$ we must have $m(p,j_2) \geq 1 + m(p,j_1)$.

Then:

$$k \# A = \#(A\times\{1,\ldots,k\}) \geq \# \bigcup_{j=1}^{n} \tilde{A}_j = \sum_{j=1}^{n} \# \tilde{A}_j =$$

$$= \sum_{j=1}^{n} \# A_j = \left(\sum_{j=1}^{n} (\#A_j/\#A) \right)\#A.$$

Hence:

$$k \geq \left[\sum_{j=1}^{n} (\#A_j/\#A) \right]$$

that is what we wanted to prove.

Now apply this lemma to $A = \{1,\ldots,N\}$ and $G = \{ S_N^{(1)},\ldots, S_N^{(n)} \}$. By (4) and the lemma there exists a point m that belongs to $S_N^{(j_1)},\ldots,S_N^{(j_q)}$, with $q \geq [n/4]$. Let $y = \varphi^{-m}(x)$. Then

$$\log P_{j_i}^{(m)} \leq c$$

for $i \leq i \leq q$. This means:

$$\| \pi_{j_i}^{(r)} T \, v_{j_i}^{(r+1)} \| \exp(-c).$$

If $1 \leq k < i \leq q$ we can write:

$$\| Tv_{j_i}^{(r+1)} - Tv_{j_k}^{(r+1)} \| = \| \pi_{j_i}^{(r)} Tv_{j_k}^{(r+1)} - \pi_{j_k}^{(r+1)} Tv_{j_i}^{(r+1)} - Tv_{j_i}^{(r+1)} \| =$$

$$= \| \pi_{j_i}^{(r)} T v_{j_i}^{(r+1)} \| \cdot \| v_{j_i}^{(r)} + \| \pi_{j_i}^{(r)} Tv_{j_i}^{(r+1)} \|^{-1} (Tv_{j_i}^{(r+1)} - \pi_{j_i}^{(r)} Tv_{j_i}^{(r+1)} - Tv_{j_k}^{(r+1)}) \|$$

But the expression between brackets is a vector in $G_{j-1}^{(r)}$ because $Tv_{j_k}^{(r+1)} \in TG_{j_k}^{(r+1)} \subset G_{j_k}^{(r)} \subset G_{j_i-1}^{(r)}$ (recall that $k < i$, hence $j_k < j_i$) and the range of $(I - \pi_{j_i}^{(r)})$ is $G_{j_i-1}^{(r)}$. Using (4) and (2) we obtain

$$\| Tv_{j_i}^{(r+1)} - Tv_{j_k}^{(r+1)} \| \geq \exp(-c) \cdot \frac{1}{2}$$

for all $1 \leq i < k \leq q$. But Lemma II.4 applied to the homomorphisms

$T: G/\varphi^{-1}(K_o) \to G/K_o$ gives a constant that bounds q as a function of $(1/2)\exp(-c)$. On the other hand $q \geq [n/4]$ and n is arbitrary. This contradiction completes the proof of $\dim E_\lambda(x) < +\infty$ for $x \in K'$.

To complete the proof of III.1 we still have to show that $\bigcup\{E_\lambda(x) \mid x \in K\}$ is a μ-measurable Banach vector bundle on K. Since $E_{\lambda+\epsilon}(x)$ is finite dimensional for all $\epsilon > 0$ and a.e. $x \in K$ it follows that we can take $\epsilon > 0$ so small that $E_{\lambda+\epsilon}(x) = E_\lambda(x)$ a.e. Let $K(C)$ be a compact subset of K_o such that for all $n \in \mathbb{Z}^+$:

$$\|T^{-n}/E_\lambda(x)\| \leq C\exp(\lambda+\epsilon)n.$$

Denote K_1 the set of points where $E_{\lambda+\epsilon}(x) = E_\lambda(x)$ and the dimensions of these subspaces is constant, say m. We claim that the function $K_1 \cap K(C) \to G^+(m)$ is continuous. Let $\{x_n \mid n \in \mathbb{Z}^+\} \subset$ $\subset K_1 \cap K(C)$ be a sequence converging to $x \in K_1 \cap K(C)$. Suppose that it is not true that $E_\lambda(x_n) \to E_\lambda(x)$. Then it is not true that $T^{-1}(E_\lambda(x_n)) \to T^{-1}(E_\lambda(x))$. But $\|Tv\| \geq C^{-1}\exp(-(\lambda+\epsilon)n)$ for all $v \in \bigcup_n T^{-1}(E_\lambda(x_n))$. Then by II.6 there exists a convergent subsequence $T^{-1}(E_\lambda(x_{n_j})) \to E \subset G_{\varphi^{-1}(x)}$, $E \neq T^{-1}(E_\lambda(x))$. Then $E_\lambda(x_{n_j}) \to T(E)$. But then, since $x \in K(C) \subset K_o$,

$$\|T^{-n}/T(E)\| \leq C\exp(\lambda+\epsilon)n$$

for all $n \in \mathbb{Z}^+$. Therefore $T(E) \subset E_{\lambda+\epsilon}(x)$. But, since $x \in K_1$, $E_{\lambda+\epsilon}(x)$ and $E_\lambda(x)$ coincide. Hence $T(E) = E_\lambda(x)$ and $E = T^{-1}(E_\lambda(x))$. This completes the proof of the claim. It is clear that we can choose $K(n)$, $n \in \mathbb{Z}^+$, satisfying:

$$\bigcup_{n\in\mathbb{Z}^+} K(n) \cap K_1 = K_o \cap K_1 \bmod(0).$$

But the definition of μ-measurable vector bundle implies that K_o can be taken with measure arbitrarily near to 1. Since $\mu(K_1) = 1$ the lemma is proved.

<u>Proof of Lemma III.2</u>. We shall prove $\dim E_\lambda(x) \geq 1$ a.e. for every $\lambda > k(T)$. Observing that

$$E_{k(T)}(x) = \bigcap_{\lambda > k(T)} E_\lambda(x)$$

this will complete the proof of the theorem. Fix any $\lambda > k(T)$. Let A_n be the set of points $x \in K$ where there exists $0 \neq v \in G_x$ satisfying $\|T^{-m}v\| \leq \lambda^m \|v\|$ for all $1 \leq m \leq n$. We claim that there exists $c > 0$ such that $\mu(A_n) \geq c$ for all $n \geq 0$. Before proving this claim we shall show how it implies that $\dim E_\lambda(x) \geq 1$ a.e. Take a compact set $K_o \subset K$ such that for all n $G/\varphi^{-n}(K_o)$ is a Banach vector bundle and $T/(G/\varphi^{-n}(K_o))$ is continuous. As we mentioned before, K_o can be chose with measure arbitrarily near to 1. Suppose then that $\mu(K_o) \geq 1 - c/2$. Set $K_n = K_o \cap A_n$. Then $\mu(K_n) = \mu(K_n) = \mu(K_o \cap A_n) \geq \mu(A_n) - \mu(K_o^c) \geq c/2$. This implies $\mu(\bigcap_{m=1}^{\infty} \bigcup_{n=m}^{\infty} A_n) \geq c/2$. If we prove that $\dim E_\lambda(x) > 0$ for $x \in \bigcap_{m=1}^{\infty} \bigcup_{n=m}^{\infty} A_n$ we are done because as we just proved the measureof this set is >0 and if $\dim E_\lambda(x) > 0$ then $\dim E_\lambda(\varphi^n(x)) > 0$ for all n. Then $\dim E_\lambda(x) > 0$ holds for all x in the φ-invariant set $\bigcup_j \varphi^j(\bigcap_{m=1}^{\infty} \bigcup_{n=m}^{\infty} A_n)$. Since this set has positive measure (because it contains $\bigcap_{m=1}^{\infty} \bigcup_{n=m}^{\infty} A_n$), the ergodicity of μ implies that $\dim E_\lambda(x) > 0$ a.e. Now, if $x \in \bigcap_{m=1}^{\infty} \bigcup_{n=m}^{\infty} A_n$ there exists a sequence of positive integers $n_1, n_2, \ldots$ such that $x \in A_{n_j}$ for all j. Let $v_j \in G_x$, with $\|v_j\| = 1$, such that $x \in A_{n_j}$ for all j. Let Let $v_j \in G_x$, with $\|v_j\| = 1$, such that $\|T^{-n}v_j\| \leq \lambda^n$ for all $0 \leq n \leq n_j$. Then the set $\{T^{-1}v_j \mid j \in \mathbb{Z}^+\}$ is bounded. Hence we can assume that $v_j = T(T^{-1}v_j)$ converges to some vector $v \in G_x$. It is clear that $\|v\| = 1$ and $\|T^{-n}v\| \leq \lambda^n$ for all $n \geq 0$ and then $v \in E_\lambda(x)$.

Now let us prove the claim. We shall use the following lemma due to V.I. Pliss.

<u>Lemma III.6</u>. Given $H > \lambda$ and $\varepsilon > 0$ there exist $N_o = N_o(H,\lambda,\varepsilon)$ and $\delta = \delta(H,\lambda,\varepsilon)$ such that if $a_i \in \mathbb{R}$, $i = 0,\dots,N$ satisfy:

$$N \geq N_o$$

$$a_n \geq H \tag{1}$$

$$\sum_{0}^{N} a_n \leq (\lambda-\varepsilon)(N+1) \tag{2}$$

then there exist $0 \leq n_1 < n_2 < \dots < n_\ell \leq N$ such that:

$$\ell \geq \delta N \tag{3}$$

and

$$\sum_{n=n_j+1} a_n \leq (m-n_j)\lambda \tag{4}$$

for all $n_j < m \leq N$, $1 \leq j \leq \ell$.

We shall apply this lemma taking $0 < \varepsilon < \lambda-k(T)$ and N such that:

$$\log \frac{\|Tv\|}{\|v\|} \leq -H$$

for all $0 \neq v \in G$. We shall prove that for all n:

$$\mu(A_n) = \delta = \delta(H,\lambda,\varepsilon).$$

For this define:

$$\tau_N(x,A_n) = \frac{1}{N} \#\{0 < j \leq N \mid \varphi^j(x) \in A_n\}.$$

Chose x such that

$$\lim \tau_N(x,A_n) = \mu(A_n)$$

for all n. Since $-(\lambda-\varepsilon) < -k(T)$, for every large N there exists $0 \neq v \in G_x$ satisfying

$$\log \frac{\|T^{N+1}v\|}{\|v\|} \geq -(\lambda-\varepsilon)(N+1).$$

Hence, defining

$$a_k = \log \frac{\|T^{N-k}v\|}{\|T^{N-k+1}v\|}$$

for $0 \leq k \leq N$, we obtain:

$$\sum_{k=0}^{N} a_k = \log \frac{\|v\|}{\|T^{N+1}v\|} \leq (\lambda-\varepsilon)(N+1)$$

and, by the definition of H,

$$a_n \geq H.$$

Without loss of generality we can suppose:

$$a_n \geq N_o(H,\lambda,\varepsilon).$$

Then Lemma III.6 gives integers $0 < n_1 < n_2 < \ldots < n_\ell \leq N$ satisfying (1) and (2). But (2) means:

$$(m-n_j)\lambda \geq \sum_{n=n_j+1}^{m} a_n = \sum_{m=n_j+1}^{m} \log \frac{\|T^{N-n}v\|}{\|T^{N-n+1}v\|} = \log \frac{\|T^{N-m}v\|}{\|T^{N-n_j}v\|}.$$

Set

$$u_j = \frac{T^{N-n_j}v}{\|T^{N-n_j}v\|}.$$

Then, for all $n_j \leq m \leq N$

$$\|T^{-(m-n_j)}u_j\| \leq \exp \lambda(m-n_j)$$

or:

$$\|T^{-k}u_j\| \leq \exp \lambda k$$

for all $0 \leq k \leq N-n_j$. Since $N - n_j \geq n$ for $j > n$, it follows that $\varphi^{N-n_j}(x) \in A_n$ for $n < j \leq \ell$ (because $u_j \in G_{\varphi^{N-n_j}(x)}$ and satisfies the required conditions). But

$$\ell-n \geq \delta N-n$$

and then:

$$\frac{1}{N}\#\{0 < i \leq N \mid \varphi^i(x) \in A_n\} \geq \frac{1}{N}\#\{0 < i \leq N \mid i = N-n_j,\ n < j \leq \ell\} \frac{1}{N}(\ell-n) \geq \delta - \frac{n}{N}.$$

If for a fixed value of x we take limits for $N \to +\infty$ we conclude

$$\mu(A_n) = \lim_{N\to+\infty} \tau_N(x,A_n) \geq \delta.$$

<u>Proof of Lemma III.6</u>. Define $b_n = a_n-\lambda$ and $S_n = \sum_0^m b_n$. These numbers satisfy:

$$S_N \leq -\varepsilon(N+1) \tag{3}$$

$$b_n \geq H-\lambda \ . \tag{4}$$

Define $0 \leq n_1 < \ldots < n_\ell \leq N$ as follows:

$$n_1 = \min\{m \mid S_m = \max_{0\leq k\leq N} S_k\}$$

$$n_{j+1} = \min\{m \mid S_m = \max_{n_j<k\leq N} S_k\} .$$

Then:

$$S_{n_1} \geq H-\lambda$$

$$S_{n_{j+1}} \leq S_{n_j} \ .$$

To continue this argument we shall need that the set $n_1,\ldots,n_\ell$ contains at least two elements, i.e. $\ell > 1$. But since obviously $n_\ell = N$ we have only to show $n_1 < N$. But $n_1 = N$ would imply $S_N \geq S_o = b_o$, that means by (3) and (4):

$$-\varepsilon(N+1) \geq S_N \geq S_o = b_o \geq H-\lambda$$

and

$$N+1 \leq \frac{1}{\varepsilon}\ (H-\lambda) . \tag{5}$$

Then if we impose on $N_o(H,\lambda,\varepsilon)$ the requirement

$$N_o(H,\lambda,\varepsilon) > \frac{1}{\varepsilon}\ (H-\lambda) - 1 \tag{6}$$

we make (5) impossible and then $\ell > 1$ as we wanted.

Now observe that by the way we choose the n_j, we have:

$$S_m - S_{n_j} \leq 0$$

for $m \geq n_j$. But then:

$$\sum_{n=n_j+1}^{m} b_n = \sum_{0}^{m} b_n - \sum_{0}^{n_j} b_n = S_m - S_{n_j} \leq 0.$$

Recalling that $b_n = a_n-\lambda$ we obtain

$$\sum_{n=n_j+1}^{m} a_n \leq (m-n_j)\lambda$$

for all $1 < j \le \ell$, $n_j < m \le N$. This proves property (2) in III.6. To show (4) observe that hypothesis (1) and (2) of III.6 imply $H < \lambda$ and define

$$\delta(H,\lambda,\varepsilon) = \frac{\varepsilon}{\lambda - H}.$$

Then:

$$S_{n_{j+1}} \ge S_{n_j+1} = S_{n_j} + b_{n_j+1} \ge S_{n_j} + H-\lambda.$$

Applying this inequality for $j = 1,\dots,\ell-1$ and adding:

$$S_{n_\ell} \ge S_{n_1} + (\ell-1)(H-\lambda).$$

But $S_{n_1} \ge H-\lambda$ and $S_{n_\ell} = S_N \le -\varepsilon(N+1)$. Hence:

$$-\varepsilon(N+1) \ge (H-\lambda)(\ell-1).$$

Therefore:

$$\ell \ge 1 + \frac{\varepsilon}{\lambda-H}(N+1) \ge \delta(H,\lambda,\varepsilon)N.$$

Proof of Lemma III.3. We shall prove first that

$$\liminf_{n\to+\infty} \frac{1}{n} \log\|T^n v\| \ge -k(T) \tag{1}$$

for a.e. x and every $0 \ne v \in E_{k(T)}$. Together with the definition of $k(T)$ this proves III.3 for $n \to +\infty$. Instead of (1) we shall prove the following more general property that will be useful later:

Lemma III.7. If $\lambda > 0$, for a.e. $x \in K$ the inequality:

$$\liminf_{n\to+\infty} \frac{1}{n} \log\|T^n v\| \ge -\lambda$$

holds for all $v \in E_\lambda(x)$.

Proof. Take $\varepsilon > 0$ and define:

$$C_\varepsilon(x) = \sup \frac{\|T^{-n}/E_\lambda(x)\|}{\exp(\lambda+\varepsilon)n}.$$

Then if $x \in K$ and $v \in E_\lambda(x)$:

$$\|T^n v\| \ge \|(T/E_\lambda(x))^{-n}\|^{-1}\|v\| \ge C_\varepsilon^{-1}(x)\exp(-(\lambda+\varepsilon)n)\|v\|.$$

Hence:

$$\liminf_{n\to+\infty} \frac{1}{n} \log\|T^n v\| \geq -(\lambda+\varepsilon) - \liminf_{n\to+\infty} \frac{1}{n} \log C_\varepsilon(\varphi^n(x)).$$

Since ε is arbitrary our problem is reduced to show that for all $\varepsilon > 0$:

$$\liminf_{n\to+\infty} \frac{1}{n} C_\varepsilon(\varphi^n(x)) \leq 0 \qquad \text{a.e.}$$

This will follow from the following lemma. In fact it is the easy part of it, the others conclusions in it will be very useful later.

Lemma III.8. Let $(X,\mathcal{G},\mu)$ be a probability space and $\varphi: X \circlearrowleft$ a measure preserving transformation. If $A: X \to \mathbb{R}$ is a measurable function, then:

$$\liminf_{n\to+\infty} \frac{1}{n} A(\varphi^n(x)) = 0 \qquad \text{a.e.} \qquad (1)$$

Moreover, if there exists $F \in \mathcal{L}^1(X,\mathcal{G},\mu)$ such that

$$A(\varphi(x)) - A(x) \leq F(x) \qquad \text{a.e.} \qquad (2)$$

or if there exists $G \in \mathcal{L}^1(X,\mathcal{G},\mu)$ such that

$$A(\varphi(x)) - A(x) \geq G(x) \qquad \text{a.e.} \qquad (3)$$

Then $A\circ\varphi - A \in \mathcal{L}^1(X,\mathcal{G},\mu)$ and

$$\lim_{n\to+\infty} \frac{1}{n} A(\varphi^n(x)) = 0 \qquad \text{a.e.} \qquad (4)$$

Finally, if φ is invertible,

$$\lim_{n\to-\infty} \frac{1}{n} A(\varphi^n(x)) = 0 \qquad \text{a.e.} \qquad (5)$$

Proof. Suppose that (1) is false. Then there exist $S \in \mathcal{G}$ with $\mu(S) > 0$ and $c > 0$ such that

$$\liminf_{n\to+\infty} \frac{1}{n} A(\varphi(x)) > c$$

for all $x \in S$. Since A is measurable, there exists $S_o \in \mathcal{G}$ contained in S with $\mu(S_o) > 0$ and $K > 0$ such that $\varphi(x) \leq K$ for all $x \in S_o$. By Poincaré recurrence theorem we can take $x \in S_o$ and a sequence $0 < n_1 < n_2 < \ldots$ satisfying $\varphi^{n_j}(X) \in S_o$

for all $j \geq 1$. Then $A(\varphi^{n_j}(x)) \geq cn_j$ if j is large enough. But $\varphi^{n_j}(x) \in S_o$. Then $K \geq A(\varphi^{n_j}(x)) \geq cn_j$ for every large value of j. This contradiction proves (1). The integrability of $A\circ\varphi - A$ and (4) will be proved under hypothesis (2). When hypothesis (3) holds, the proof of these properties are reduced to the case of hypothesis (2) because

$$(-A)(\varphi(x)) - (-A(x)) = A(x) - A(\varphi(x)) \leq -G(x).$$

Then $(-A)\circ\varphi - (-A)$ is integrable thus implying the integrability of $A\circ\varphi - A = -((-A)\circ\varphi - (-A))$ and

$$\lim_{n\to+\infty} \frac{1}{n} A(\varphi^n(x)) = -\lim \frac{1}{n} (-A)(\varphi^n(x)) = 0 \qquad \text{a.e.}$$

So let us assume that (2) holds. If $N \in \mathbb{Z}^+$ define $F_N\colon X \to \mathbb{R}$ by

$$F_N(x) = A(\varphi(x)) - A(x)$$

if $A(\varphi(x)) - A(x) \geq -N$, and

$$F_N(x) = -N$$

otherwise. Then the sequence $\{F_N \mid N \in \mathbb{Z}^+\}$ is monotone and converges a.e. to $A\circ\varphi - \varphi$. Hence to prove the integrability of $A\circ\varphi - \varphi$ it suffices to show that

$$\inf \int_X F_N \, d\mu > -\infty.$$

We have:

$$\int_X F_N \, d\mu = \int_X \left(\lim_{n\to+\infty} \frac{1}{n} \sum_{j=0}^{n-1} F_N\circ\varphi^j\right) d\mu \geq$$

$$\geq \int_X \left(\limsup_{n\to+\infty} \frac{1}{n} \sum_{j=0}^{n-1} (A\circ\varphi^{j+1} - A\circ\varphi^j\right) d\mu =$$

$$= \int_X \limsup_{n\to+\infty} \frac{1}{n} (A\circ\varphi^n - A) d\mu =$$

$$= \int_X \limsup_{n\to+\infty} \frac{1}{n} (A\circ\varphi^n) d\mu.$$

But

$$\limsup_{n\to+\infty} \frac{1}{n} A(\varphi^n(x)) = -\limsup_{n\to+\infty} \frac{1}{n} (-A)(\varphi^n(x)) \geq 0. \tag{6}$$

Then:

$$\int_X F_N \, d\mu \geq -\infty$$

completing the proof of the integrability of $A\circ\varphi - A$. Now let us show (4). From (6) and (1) it follow that it suffices to show that the sequence $n^{-1} A(\varphi^n(x))$ converges a.e. Write:

$$\frac{1}{n} A(\varphi^n(x)) = \frac{1}{n} \sum_{j=0}^{n-1} (A\circ\varphi-A)(\varphi^j(x)) + \frac{1}{n} A(x).$$

By Birkhoff's theorem the sequence

$$\frac{1}{n} \sum_{j=0}^{n-1} (A\circ\varphi-A)(\varphi^j(x))$$

converges a.e. Since $n^{-1} A(x)$ converges to 0 the proof of (4) is completed. To prove (5) it suffices to apply (4) to A and φ^{-1} and observe that if $A(\varphi(x)) - A(x) \leq F(x)$ a.e. then

$$A(x) - A(\varphi^{-1}(x)) \leq F(\varphi^{-1}(x)) \quad \text{a.e.}$$

that implies

$$A(\varphi^{-1}(x)) - A(x) \geq -(F\circ\varphi^{-1})(x)$$

so that A and φ^{-1} satisfy hypothesis (3). In the same way it is proved that if A and φ satisfy hypothesis (3), then A and φ^{-1} satisfy hypothesis (2).

It remains to prove III.3 when $n \to -\infty$. By the definition of $E_{k(T)}(x)$ it suffices to show that:

$$\liminf_{n\to+\infty} \frac{1}{n} \|T^{-n}v\| \geq k(T) \tag{4}$$

for a.e. x and every $0 \neq v \in E_{k(T)}(x)$. Given $\varepsilon > 0$ define:

$$C_\varepsilon(x) = \sup_{n\geq 0} \frac{\|T^n/E_{k(T)}(x)\|}{\exp(-k(T)+\varepsilon)^n}.$$

The number $C_\varepsilon(x)$ is finite by the definition of $k(T)$ and obviously C_ε is a measurable function. Then, if $v \in E_\lambda(x)$

$$\|v\| = \|T^n T^{-n} v\| \le C_\varepsilon(\varphi^{-n}(x)) \exp(-k(T)+\varepsilon)n \, \|T^{-n}v\| .$$

Hence:

$$\liminf_{n\to+\infty} \frac{1}{n} \|T^{-n}v\| \ge k(T)-\varepsilon - \limsup_{n\to+\infty} \frac{1}{n} \log C_\varepsilon(\varphi^{-n}(x)).$$

Therefore, since ε is arbitrary, (4) will follow if we show:

$$\lim_{n\to+\infty} \frac{1}{n} \log C_\varepsilon(\varphi^{-n}(x)) \qquad \text{a.e.}$$

for all $\varepsilon > 0$. By III.8 we have only to show that $\log C_\varepsilon \circ \varphi^{-1} - \log C_\varepsilon$ is bounded by an integrable function. But if $n \ge 1$

$$\begin{aligned} \|T^n/E_{k(T)}(\varphi^{-1}(x)\| &\le \|T/E_{k(T)}(\varphi^{-1}(x))\| \|T^{n-1}/E_{k(T)}(x)\| \le \\ &\le \|T/E_{k(T)}(\varphi^{-1}(x))\| C_\varepsilon(x)\exp(-k(T)+\varepsilon)(n-1) = \\ &= \|T/E_{k(T)}(\varphi^{-1}(x))\| \exp(k(T)-\varepsilon)\exp(-k(T)+\varepsilon)n \, C_\varepsilon(x). \end{aligned}$$

Therefore:

$$\begin{aligned} C_\varepsilon(\varphi^{-1}(x)) &\le \sup\{1, \sup_{n\ge1} \frac{\|T^n/E_{k(T)}(\varphi^{-1}(x))\|}{\exp(-k(T)+\varepsilon)n}\} \\ &\le \sup\{1, \|T/E_{k(T)}(\varphi^{-1}(x))\| \exp(k(T)-\varepsilon) \, C_\varepsilon(x)\}. \end{aligned}$$

But $C_\varepsilon(x) \ge 1$ for all x. Then the last expression is less than or equal to:

$$C_\varepsilon(x)\sup\{1, \|T/E_{k(T)}(\varphi^{-1}(x))\| \exp(k(T)-\varepsilon)\}$$

and then

$$\log C_\varepsilon(\varphi^{-1}(x)) - \log C_\varepsilon(x) \le \sup\{0, \log\|T/E_{k(T)}(\varphi^{-1}(x))\| + k(T)-\varepsilon\}$$

that is integrable.

<u>Proof of Lemma III.4</u>. Let $F \subset G$ be a μ-measurable vector bundle such that $F_x \oplus E_\lambda(x) = G_x$ a.e. and such that if $\pi_x\colon G_x \to F_x$, $P_x\colon G_x \to E_\lambda(x)$ are the projections associated with this splitting

there exists $P > 0$ such that $\|p_x\| \le P$, $\|\pi_x\| \le P$ a.e. Define a μ-measurable homomorphism $\tilde{T}\colon F \circlearrowleft$ by $\tilde{T}/F_x = \pi_{\varphi(x)} T/F_x$. We claim that $k(\tilde{T}) > \lambda$. If $k(\tilde{T}) \le \lambda$ the previous lemmas imply that there exists a μ-measurable vector bundle $E = E_{k(\tilde{T})} \subset F$ such that a.e.:

$$\tilde{T}(E_x) = E_{\varphi(x)}$$

$$\lim_{n\to\pm\infty} \frac{1}{n} \log\|\tilde{T}^n/E_x\| = -k(\tilde{T}) \ge -\lambda .$$

Take $\hat{E}_x = \pi_x^{-1}(E_x)$. Then $\hat{E}_x \supset E_\lambda(x)$ and $\hat{E}_x \ne E_\lambda(x)$.

By III.7:

$$\liminf_{n\to+\infty} \frac{1}{n} \log\|T^n v\| \ge -\lambda$$

for a.e. x and every $0 \ne v \in E_\lambda(x)$. Then:

$$\liminf_{n\to+\infty} \frac{1}{n} \log\|T^n v\| \ge -\lambda \tag{1}$$

for a.e. x and every $0 \ne v \in \hat{E}_\lambda$. On the other hand

$$k(T^{-1}/\hat{E}) = -\limsup_{n\to+\infty} \frac{1}{n} \log\|T^{-n}/\hat{E}_x\| < -\lambda$$

because otherwise we should have $\hat{E} \subset E_\lambda$. Let $\tilde{E} \subset \hat{E}$ be the μ-measurable vector bundle $E_{k(T^{-1}/\hat{E})}$ associated to T^{-1}. Then

$$\lim_{n\to\pm\infty} \frac{1}{n} \log\|T^{-n}/\tilde{E}_x\| = -k(T^{-1}/\hat{E}) > \lambda .$$

In particular

$$\lim_{n\to+\infty} \frac{1}{n} \log\|T^n/\tilde{E}_x\| = k(T^{-1}/\hat{E}) < -\lambda$$

contradicting (1) and completing the proof of the claim. Now let M be the μ-measurable vector bundle on K whose fibers M_x are the spaces of continuous linear maps $L\colon F_x \to E_\lambda(x)$. Define a μ-measurable homomorphism $\Phi\colon M \circlearrowleft$ by:

$$\Phi(L) = T \circ L \circ (\tilde{T}^{-1}/F_{\varphi(x)}).$$

Let $\Gamma(M)$ be the space of measurable sections of M, i.e. measurable maps that to a.e. $x \in K$ associate a linear map $L_x \in M_x$.

To each $\Phi \in \Gamma(M)$ we can associate a μ-measurable vector bundle $G(\Phi)$ whose fibers are given by:

$$G_x(\Phi) = \{v + \Phi_x v \mid v \in F_x\}.$$

This μ-measurable vector bundle is T-invariant, i.e.

$$T(G_x(\Phi)) = G_{\varphi(x)}(\Phi) \qquad \text{a.e.}$$

if and only if:

$$(T^{-1} \Phi_{\varphi(x)}(\tilde{T}/F_x)) - \Phi_x = P_{\varphi(x)}(T/F_x) \quad \text{a.e.} \tag{1}$$

Then, to find Φ satisfying this condition, we define

$$A_x = -P_{\varphi(x)}(T/F_x)$$

and

$$\Phi_x = \sum_{n=0}^{\infty} T^{-n} A_{\varphi^n(x)}(\tilde{T}^n/F_x). \tag{2}$$

If we show that a.e. the sequence:

$$\sum_{n=0}^{\infty} \# \; T^{-n}/E_\lambda(\varphi^n(x))\| \cdot \| \; \|\tilde{T}^n/F_n\| \tag{3}$$

converges, then (2) will define an element of $\Gamma(M)$ satisfying (1). Then we shall define $G_\lambda = G(\Phi)$ and show that satisfies the required properties. To prove the convergence a.e. of the series (3) take $\varepsilon > 0$ so small that:

$$2\varepsilon < k(\tilde{T}) - \lambda. \tag{4}$$

Define

$$C(x) = \sup_{n \geq 0} \frac{\|T^{-n}/E_\lambda(x)\|}{\exp(\lambda+\varepsilon)n}$$

$$C_1(x) = \sup \frac{\|\tilde{T}^n/F_x\|}{\exp(-k(\tilde{T})+\varepsilon} \, .$$

Then:

$$\|\Phi_x\| \leq P \sum_{n=0}^{\infty} \|T^{-n}/E_\lambda(\varphi^n(x))\| \cdot \|\tilde{T}(x)\| \cdot \|\tilde{T}^n/F_x\| \leq$$

$$\leq P \sum_{n=0}^{\infty} C(\varphi^n(x)) C_1(x) \exp(\lambda - k(\tilde{T}) + 2\varepsilon)n.$$

But it is not difficult to show, using methods similar to those applied in the proof of Lemma III.3, that $\log C\circ\varphi - \log C$ is bounded by an integrable function. Then Lemma III.8 proves that given $0 < \delta < k(\tilde{T}) - \lambda$, the number:

$$C_2(x) = \sup_{n\geq 0} \frac{C(\varphi^n(x))}{\exp \delta n}$$

is finite. Then:

$$\|\Phi_x\| \leq P \sum_{n=0}^{\infty} C_2(x)C_1(x) \exp(\lambda-k(\tilde{T})+2\varepsilon+\delta)n =$$

$$= P\, C_2(x)C_1(x)(1-\exp(\lambda-k(\tilde{T})+2\varepsilon+\delta)^{-1} = \bar{P}\, C_2(x)C_1(x)$$

where $\bar{P} = P(1-\exp(\lambda-k(\tilde{T})+2\varepsilon+\delta)^{-1})$. Now define $G_\lambda = G(\Phi)$. The previous considerations show that satisfy part (a) of III.4. Part (b) is clear. To prove part (c) set $B_x: F_x \to G_\lambda(x)$ defined by:

$$B_x v = v + \Phi_x v.$$

Then Φ_x is an isomorphism with

$$\|B_x\| \leq 1 + \|\Phi_x\|$$

$$\|B_x^{-1}\| \leq p'$$

a.e., where p' depends only on P. Moreover

$$T^n/G_\lambda(x) = B_{\varphi^n(x)} \tilde{T}^n B_x^{-1} . \tag{5}$$

Then:

$$\|T^n/G_{\lambda,x}\| \leq \|\tilde{T}^n\|\ P'(1 + \|\Phi_{\varphi^n(x)}\|) \leq$$

$$\leq \|\tilde{T}^n\|\ p'(1+\bar{P}C_2(\varphi^n(x))C_1(\varphi^n(x)))$$

and by III.8:

$$\limsup_{n\to+\infty} \frac{1}{n} \log\|T^n/G_\lambda(x)\| \leq \limsup_{n\to+\infty} \frac{1}{n} \log\|\tilde{T}^n\| +$$

$$+ \limsup_{n\to+\infty} \frac{1}{n} \log C_2(\varphi^n(x)) + \limsup_{n\to+\infty} \frac{1}{n} \log C_1(\varphi^n(x)) \leq$$

$$\leq \limsup_{n\to+\infty} \frac{1}{n} \log\|\tilde{T}^n\| = k(\tilde{T}) < -\lambda$$

proving part (c) of III.4. Here we used only, the fact of $\log C_1$,

$\log C_2$ being measurable. But in fact it is not difficult to check that they satisfy also the property that $\log C_i \circ \varphi - \log C_i$ is bounded by an integrable function. Then $\lim_{n\to+\infty} n^{-1} C_i(\varphi^n(x)) = 0$ for $i = 1,2$. From this property together with (5), part (d) of III.4 follows immediately.

To prove part (e) of III.4 we start observing that there exist constants $0 < k_1 < k_2$, depending only on P, such that

$$k_1\|\Phi(x)\| \le \|\pi(x)\| \le k_2\|\Phi(x)\|.$$

We leave to the reader the verification of the existence of k_1 and k_2. This reduces the problem to prove

$$\lim \frac{1}{n} \log\|\Phi(\varphi^n(x))\| = 0 \quad \text{a.e.}$$

But since $\|\Phi(x)\|$ is a measurable function of x we have by Lemma III.8 that:

$$\limsup_{n\to\pm\infty} \frac{1}{|n|} \log\|\Phi(\varphi^n(x))\| = - \liminf_{n\to\pm\infty} \frac{1}{|n|} (-\log\|\Phi(\varphi^n(x))) \ge 0.$$

On the other hand:

$$\liminf_{n\to\pm\infty} \frac{1}{|n|} \log\|\Phi(\varphi^n(x))\| \le \liminf_{n\to\pm\infty} \frac{1}{|n|} \bar{P}\, C_2(\varphi^n(x))C_1(\varphi^n(x)) = 0$$

thus completing the proof.

IV - Proof of Theorem B

Of the statement of Theorem B we shall only prove the properties concerning the sets $G_3(n_1,\dots,n_k)$. The corresponding properties for $G_2(n_1,\dots,n_k)$ follow applying the same methods with minor modifications.

Let $\hat{G}(n_1,\dots,n_k)$ be the set of points $x \in K$ such that there exists a splitting

$$F_x = E_1(x) \oplus \dots \oplus E_k(x) \oplus F_k(x)$$

and numbers $\lambda_1(x) >\ldots> \lambda_{k+1}(x)$ such that $\dim E_j(x) = n_j$ and:

$$\lim_{n\to+\infty} \frac{1}{n} \log\|L^n/F_k(x)\| = \lambda_{k+1}(x)$$

$$\lim_{n\to-\infty} \frac{1}{n} \log\|L^{-n}v\| = -\lambda_{k+1}(x)$$

for all $0 \neq v \in F_k(x)$ such that $L^{-n}v$ exists for all $n \geq 0$, and

$$\lim_{n\to\pm\infty} \frac{1}{n} \log\|L^n/E_j(x)\| = \lambda_j(x)$$

for all $1 \leq j \leq k$, $0 \neq v \in E_j(x)$. Let $\hat{G}_k$ be the union of the sets $\hat{G}(n_1,\ldots,n_k)$ when $n_1,\ldots,n_k$ varies over all the possible k-uples of positive integers. Then:

$$\Sigma_3 = \bigcap_{k\geq 1} \hat{G}_k \ .$$

Moreover if $x \in \Sigma_3 \cap \hat{G}(n_1,\ldots,n_k)$ the spaces $E_j(x)$, $F_k(x)$ and the numbers $\lambda_1(x) >\ldots> \lambda_k(x)$ coincide with those given in the definition of Σ_3 and:

$$\Sigma_3 \cap \hat{G}(n_1,\ldots,n_k) = G_3(n_1,\ldots,n_k). \tag{2}$$

Then if prove that the functions:

$$\hat{G}(n_1,\ldots,n_k) \ni x \mapsto E_j(x) \in G^+(n_j)$$

$$\hat{G}(n_1,\ldots,n_k) \ni x \mapsto F_k(x) \in G^-(\Sigma n_j)$$

are strongly measurable, that the functions:

$$\hat{G}(n_1,\ldots,n_k) \ni x \mapsto \lambda_j(x)$$

are measurable and that $\hat{G}(n_1,\ldots,n_k)$ is a Borel set, the proof of Theorem B will be complete. The measurability of the functions λ_j follows from (1). To prove the other properties we introduce the set $\hat{G}_m(n_1,\ldots,n_k)$ of points $x \in K$ such that there exists a splitting $F_x = E_1(x) \oplus\ldots\oplus E_k(x) \oplus F_k(x)$ and numbers $\lambda_1(x) >\ldots> \lambda_{k+1}(x)$ satisfying:

a) $\lambda_j(x) - \frac{1}{m} > \lambda_{j+1}(x) + \frac{1}{m}$, $0 \le j \le k$

b) $\|v\| \frac{1}{m} \exp(\lambda_j(x) - \frac{1}{m})n \le \|L^n v\| \le \|v\| m \exp(\lambda_j(x) + \frac{1}{m})n$

for all $1 \le j \le k$, $n \ge 0$, $v \in E_j(x)$

c) For all $n \ge 0$:

$$\frac{1}{m} \exp(\lambda_{k+1}(x) - \frac{1}{m})n \le \|L^n/F_k(x)\| \le m \exp(\lambda_{k+1}(x) + \frac{1}{m})n$$

d) For all $n \ge 0$ and every $0 \ne v \in F_k(x)$ such that $L^{-n}v$ exists for all $n \ge 0$:

$$\|L^{-n}v\| \ge \|v\| \frac{1}{m} \exp(\lambda_{k+1}(x) - \frac{1}{m})n$$

e) If $\alpha(\cdot,\cdot)$ is the angle defined as in Section II, then, for all $1 \le j < k$,

$$\alpha(E_1(x) \oplus \ldots \oplus E_j(x), E_{j+1}(x) \oplus \ldots \oplus E_k(x) \oplus F_k(x)) \ge \frac{1}{m}$$

and

$$\alpha(E_1(x) \oplus \ldots \oplus E_k(x), F_k(x)) \ge \frac{1}{m} .$$

It is clear that:

$$\hat{G}(n_1,\ldots,n_k) = \bigcup_{N=1}^{\infty} \bigcap_{m=N}^{\infty} \hat{G}_m(n_1,\ldots,n_k) \qquad (3)$$

and at points $x \in \bigcap_{m=N}^{\infty} \cap \hat{G}_m(n_1,\ldots,n_k)$ the spaces $E_j(x)$, $F_k(x)$ and the numbers $\lambda_1(x) > \ldots > \lambda_{k+1}(x)$ given by the definition of $\hat{G}(n_1,\ldots,n_k)$ and $\hat{G}_m(n_1,\ldots,n_k)$ concide. Therefore (3) reduces our problem to show that each set $\hat{G}_m(n_1,\ldots,n_k)$ is <u>closed</u> and that the functions:

$$\hat{G}_m(n_1,\ldots,n_k) \ni x \mapsto E_j(x) \qquad (4)$$

$$\hat{G}_m(n_1,\ldots,n_k) \ni x \mapsto F_k(x) \qquad (5)$$

are <u>continuous</u>. To show these properties take a sequence $\{x_n\} \subset \hat{G}_m(n_1,\ldots,n_k)$ converging to $x \in K$. By Lemmas II.2 and II.3 we can suppose that the numbers $\lambda_j(x_n)$, $j=1,\ldots,k+1$ and the subspaces $E_j(x_n)$, $j=1,\ldots,k$ and $F_k(x_n)$ converge to numbers λ_j,

$j=1,\ldots,k+1$, and subspaces E_j, $j=1,\ldots,k$, F_k. The uniformity of properties (a)-(e) imply that they are satisfied putting $\lambda_j(x) = \lambda_j$, $E_j(x) = E_j$, $F_k(x) = F_k$. Therefore $x \in \hat{G}_m(n_1,\ldots,n_k)$. To show the continuity of the functions λ_j, E_j, F_k at x we proceed by contradiction. Suppose for instance that $F_k(x_n)$ doesn't converge to $F_k(x)$. Then we can assume (by Lemma II.3) that the sequences $\lambda_j(x_n)$, $E_j(x_n)$, $F_k(x_n)$ converge to λ_j, E_j, F_k and that $F_k \neq F_k(x)$. The same argument used before shows that $\lambda_j = \lambda_j(x)$, $E_j = E_j(x)$ and $F_k = F_k(x)$ contradicting that $F_k \neq F_k(x)$. This concludes the proof of Theorem B.

V - Proof of Theorem D

Let K, F, φ and L as in the statement of Theorem D. First we shall prove the theorem under the assumption that L/F_x is injective for all x. Suppose that μ is a φ-invariant probability on K and define $K_1 = \bigcap_{n\geq 0} \varphi^n(K)$. Then:

$$\varphi(K_1) = K_1. \tag{1}$$

Moreover:

$$1 = \mu(K) = \mu(\varphi^{-n}(\varphi^n(K)) = \mu(\varphi^n(K)).$$

Then $\mu(K_1) = 1$. Let $\tilde{K} = \{\theta\colon \mathbb{Z} \to K_1 \mid \varphi(\theta(n)) = \theta(n+1),\ n \in \mathbb{Z}\}$. Define $\tilde{\varphi}\colon \tilde{K}\circlearrowleft$ by $\tilde{\varphi}(\theta)(n) = \varphi(\theta(n))$. This map is a homeomorphism with inverse given by $\tilde{\varphi}^{-1}(\theta)(n) = \theta(n-1)$. Let $p\colon \tilde{K} \to K_1$ be defined by $p(\theta) = \theta(0)$. By (1), p is surjective and $\varphi p = p\tilde{\varphi}$. Then there exists a $\tilde{\varphi}$-invariant probability $\tilde{\mu}$ on $\tilde{K}$ such that $\tilde{\mu}(p^{-1}(A)) = \mu(A)$ for every Borel set $A \subset K$. It is not clear that in general p maps Borel sets onto Borel sets. Consider the Banach vector bundle $\tilde{F}$ on $\tilde{K}$ with fibers $\tilde{F}_\theta = F_{\theta(0)}$ and the homomorphism $\tilde{L}\colon \tilde{F}\circlearrowleft$ covering $\tilde{\varphi}$ given by $\tilde{L}/\tilde{F}_\theta = L/F_{\theta(0)}$. To $\tilde{L}$ we can apply Theorem A because all the injectivity assumptions are satisfied and it is easy to see that $p(\Sigma(\tilde{L})) \subset \Lambda(L)$ because if $\theta \in \Sigma(\tilde{L})$ and

$\lambda_1(\theta),\dots,E_1(\theta),\dots,F_1(\theta),\dots$ are given by the definition of $\Sigma(\tilde{L})$ then the numbers $\lambda_1(\theta),\dots$ and the spaces $F_{\theta(0)} \supset F_1(\theta) \supset F_2(\theta) \supset \dots$ satisfy the definition of $\Lambda(L)$. Now we are tempted to write $\mu(\Lambda(L)) = \tilde{\mu}(p^{-1}(\Lambda(L))) \geq \tilde{\mu}(\Sigma(\tilde{L})) = 1$. But we dont know whether $\Lambda(L)$ is a Borel set. Anyhow there exists a sequence $U_1 \subset U_2 \subset \dots$ of <u>compact</u> sets contained in $\Sigma(\tilde{L})^c$ such that

$$\lim_{n\to+\infty} \tilde{\mu}(U_n) = 1.$$

Then

$$\Lambda(L)^c \subset p(\Sigma(\tilde{L}))^c \subset p(\bigcup_{n\geq 1} U_n)^c \subset (\bigcup_{n\geq 1} p(U_n))^c = \bigcap_{n\geq 1} p(U_n)^c.$$

Since U_n is compact, $p(U_n)$ is compact and a fortiori a Borel set. Then:

$$\mu(p(U_n)) = \tilde{\mu}(p^{-1}(p(U_n))) \geq \tilde{\mu}(U_n)$$

and

$$\mu(p(U_n)^c) \leq 1 - \tilde{\mu}(U_n).$$

Hence $\Lambda(L)^c$ is a measure zero set with respect to μ.

Now let us consider the general case. Define a new Banach vector bundle $\tilde{F}$ on K whose fiber $\tilde{F}_x$ on x is the space of bounded sequences $\theta: \mathbb{Z}^+ \to F_x$ endowed with the norm $\|\theta\|_\infty = \sup\|\theta(n)\|$. On $\tilde{F}$ define a homomorphism $\tilde{L}: \tilde{F}\circlearrowleft$ covering φ by:

$$(\tilde{L}\theta)(0) = L\theta(0)$$

$$(\tilde{L}\theta)(m) = \frac{1}{a^m}\,\theta(m-1) \qquad m=1,2,\dots$$

where $a > \max\{1, \sup_x \|L/F_x\|\}$. Then:

$$(\tilde{L}^n\theta)(0) = L^n\theta(0)$$

$$(\tilde{L}^n\theta)(m) = \frac{1}{a^{mn}}\,L^{n-m}\,\theta(0) \qquad m=1,2,\dots,n-1$$

$$(\tilde{L}^n\theta)(m) = \frac{1}{a^{mn}}\,\theta(m-n) \qquad m=n,\dots$$

Using this expression it is easy to check that for any $x \in K$ and

$v \in F_x$, the limit $\lim_{n\to+\infty} n^{-1} \log\|\tilde{L}^n\theta\|$ exists if and only if $\lim_{n\to+\infty} n^{-1} \log\|L^n\theta(0)\|$ exists and in that case:

$$\lim_{n\to+\infty} \frac{1}{n} \log\|L^n\theta(0)\| = \lim \frac{1}{n} \log\|\tilde{L}^n\theta\| \tag{2}$$

(where we agree in setting $\log 0 = -\infty$). Now define $\pi_x: \tilde{F}_x \to F_x$ by $\pi_x\theta = \theta(0)$. Since $\tilde{L}/\tilde{F}_x$ is injective for all x we can apply to $\tilde{L}$ the restricted version of Theorem D we proved above. Then $\Lambda(\tilde{L})$ is a total probability set. Therefore (2) implies that $\Lambda(L) \supset \Lambda(\tilde{L})$ because if $x \in \Lambda(\tilde{L})$ and $\lambda_1(x) > \lambda_2(x) > \ldots$, $F_1(x) \supset \supset F_2(x) \supset \ldots$ are given by the definition of $\Lambda(\tilde{L})$, then (2) implies that the numbers $\lambda_1(x) > \lambda_2(x) > \ldots$ and the subspaces $\pi_x F_1(x) \supset \pi_x F_2(x) \supset \ldots$ satisfy all the properties required to grant $x \in \Lambda(L)$.

VI - Proof of Theorem E

We shall consider only the invertible case, i.e. when φ is a homeomorphism and L is injective on each fiber. In this case the statement of the theorem is equivalent to show that for all $\lambda > 0$ there exists $N = N(\lambda)$ such that if $x \in \Sigma(L)$ and $\lambda_j(x) > -\lambda$ then $\dim E_j(x) \le N(\lambda)$. Once this is proved the general case is reduced to invertible case using the method employed in Section V to prove Theorem D. The proof of the statement above for the invertible case is contained in the proof of Lemma III.1. In fact the argument given in the proof of that lemma bounds $\dim E_\lambda(x)$ by a constant depending only on a number c on which the only conditions imposed is that $\log P_j^{(n)} \ge -c$ for all n and j and that b and $\mu(K_o)$ can be chosen near to one in order to have

$$\frac{\lambda + 1 + c}{2cb_\mu(K_o)} \ge \frac{2}{3}.$$

For the first requirement, recalling the definition of $P_j^{(n)}$ and that $\|\pi_j^{(n)}\| \le 2$, it is sufficient that

$$c \geq \exp 2(\sup\{\|L/F_x\| \mid x \in K\})^{-1}.$$

For the second it suffices to take c such that:

$$\frac{\lambda+1+c}{2c} > \frac{2}{3}.$$

So c can be taken depending only on λ.

VII - Proof of Theorems F and H.

The main tool for the proof of Theorem F is the following lemma:

Lemma VII.1. Let E be a Banach space, $U \subset E$ an open set, $f: U \to E$ a $C^{k,\gamma}$ map and $K_o \subset U$ a subset satisfying the following proporties:

a) $f^n(K_o) \subset U$ for all $n \geq 0$

b) For all $n \geq 0$ and $x \in f^n(K_o)$ there exists a splitting $E = H_x^{(n)} \oplus G_x^{(n)}$ such that:

b_1) $H_x^{(0)}$ and $G_x^{(0)}$ depend continuously on x.

b_2) For all $n \geq 0$ and $x \in f^n(K_o)$:

$$(D_x f)H_x^{(n)} = H_{f(x)}^{(n+1)}$$

$$(D_x f)G_x^{(n)} \subset G_{f(x)}^{(n+1)}$$

b_3) $(D_x f)/H_x^{(n)}$ is injective for all $n \geq 0$ and $x \in f^n(K_o)$

b_4) There exist $C > 0$, $\mu > 0$ and $0 < \varepsilon < \gamma\mu/3$ satisfying:

$$\|(D_{f^n(x)} f^m)/G_{f^n(x)}^{(n)}\| \leq C \exp \varepsilon n \exp(-\mu m)$$

for all $x \in K_o$, $n \geq 0$, $m \geq 0$. Moreover, if $n-m \geq 0$:

$$\|((D_{f^{n-m}(x)} f^m)/H_{f^{n-m}(x)}^{(n-m)})^{-1}\| \leq C \exp \varepsilon n \exp(\mu-3\varepsilon)m.$$

b$_5$) If $\pi_x^{(n)}: E \to G_x^{(n)}$ is the projection associated to the splitting $E = H_x^{(n)} \oplus G_x^{(n)}$, then:

$$\|\pi^{(n)}_{f^n(x)}\| \le C \exp \epsilon n$$

for all $x \in K_o$, $n \ge 0$.

c) There exists $A > 0$ such that:

$$d(f^n(K_o), E \setminus U) \ge A \exp(\mu+\epsilon)n$$

for all $n \ge 0$.

Then there exist $r > 0$ and $\delta > 0$ arbitrarily small such that if $\Gamma_r = \{(x,v) \mid x \in K_o,\ v \in G_x^{(0)},\ \|v\| < r\}$ there exists a map $\varphi: \Gamma_r \to E$ satisfying the following properties:

I) $\pi_x^{(0)}\ \varphi(x,v) = v$ for all $(x,v) \in \Gamma_r$

II) For all $(x,v) \in \Gamma_r$, $\varphi(x,v)$ is the unique vector in E with $\pi_x^{(0)}(\varphi(x,v)-x) = v$ and:

$$\|f^n(\varphi(x,v)) - f^n(x)\| \le \delta \exp(-n\mu)$$

for all $n \ge 0$. In particular this implies $\varphi(x,0) = x$.

III) φ is continuous and for every x the map $\varphi(x,\cdot)$ is $C^{k,\gamma}$ and its derivatives depend continuously on (x,v). Moreover its first derivative at $(x,0)$ is the inclusion map.

For the proof of Theorem G we shall need the dual version of this lemma:

<u>Lemma VII.2</u>. Let E, U and f be as in Lemma VII.1. Let $K_o \subset U$ be a subset satisfying the following properties:

a) $f^{-n}(K_o) \subset U$ and $f/f^{-n}(K_o)$ is injective for all $n \ge 0$.

b) For all $n \ge 0$ and $x \in f^{-n}(K_o)$ there exists a splitting $E = H_x^{(n)} \oplus G_x^{(n)}$ such that:

b$_1$) $H_x^{(0)}$, $G_x^{(0)}$ depend continuously on $x \in K_o$.

b$_2$) $(D_x f)H_x^{(n)} = H_{f(x)}^{(n-1)}$

$(D_x f)G_x^{(n)} \subset G_{f(x)}^{(n-1)}$

for all $n \geq 1$ and $x \in f^{-n}(K_o)$.

b$_3$) $(D_x f)/H_x^{(n)}$ is injective for all $n \geq 1$, $x \in f^{-n}(K_o)$.

b$_4$) There exist $C > 0$, $\mu > 0$ and $0 < \epsilon < \mu\gamma/3$ such that:

$$\|(D_x f^m)/G_x^{(n)}\| \leq C \exp \epsilon n \exp(\mu - 3\epsilon)m$$

for all $x \in f^{-n}(K_o)$ and $0 \leq m \leq n$. Moreover:

$$\|((D_{f^{-m}(x)} f^m)/H^{(n+m)}_{f^{-m}(x)})^{-1}\| \leq C \exp \epsilon n \exp(-\mu m)$$

for all $x \in f^{-n}(K_o)$, $n \geq 0$, $m \geq 0$.

b$_5$) If $\pi_x^{(n)}: E \to H_x^{(n)}$ is the projection associated to the splitting $E = G_x^{(n)} \oplus H_x^{(n)}$, then:

$$\|\pi^{(n)}_{f^{-n}(x)}\| \leq C \exp \epsilon n$$

for all $x \in K_o$, $n \geq 0$.

c) There exists $A > 0$ such that:

$$d(f^{-n}(K_o), E \setminus U) \geq A \exp(\mu + \epsilon)n$$

for all $n \geq 0$.

Then there exists $r > 0$ and $\delta > 0$ arbitrarily small such that if $\Gamma_r = \{(x,v) \mid x \in K_o, v \in H_x^{(0)}, \|v\| < r\}$, there exists a map $\varphi: \Gamma_r \to E$ satisfying the following properties:

I) $\pi_x^{(0)} \varphi(x,v) = v$ for all $(x,v) \in \Gamma_r$.

II) For all $(x,v) \in \Gamma_r$, $\varphi(x,v)$ is the unique vector in E with $\pi_x^{(0)}(\varphi(x,v) - x) = v$ and such that there exists a sequence $v_0, v_1, \ldots$ with $v_0 = \varphi(x,v)$, $f(v_j) = v_{j-1}$ for all $j \geq 1$ and $\|v_n\| \leq \delta \exp(-\mu n)$. In particular $\varphi(x,0) = x$.

III) φ is continuous and for every x the map $\varphi(x,\cdot)$ is $C^{k,\gamma}$ and its derivatives depend continuously on (x,v). Moreover its first derivative at $(x,0)$ is the inclusion map.

Remark. In the proof of both lemmas the hypothesis of f being $C^{k,\gamma}$ will be used via the following property: set

$$p(x,v) = f(x+v) - f(x) - (D_x f)v$$

and let $(D_2^j p)(x,v)$ be the j-th derivative of p at (x,v) with respect to the second variable. Then there exist $C_1 > 0$

$$\|p(x,v)\| \le C_1 \|v\|^{1+\gamma} \qquad (*)$$

$$\|p(x,v+w) - p(x,v) - \sum_{j=2}^{k} \frac{1}{j} (D_2^j p)(x,v)w^j\| \le C_1 \|v\|^{k+\gamma} \qquad (**)$$

for all $n \ge 0$, $x \in f^n(K_o)$ in the case of VII.1, $x \in f^{-n}(K_o)$ in the case of VII.2, $\|v\| \le 2^{-1} A \exp(\mu+\epsilon)n$, $\|w\| \le 2^{-1} A \exp(\mu+\epsilon)n$.

Let us see how Theorems F and H follow from this lemmas. To prove Theorem F we have to show that for any φ-invariant ergodic measure ν on K, the properties of Theorem F hold for the points of a set $K' \subset K$ with $\nu(K') = 1$. For this we need the following corollary of VII.1:

Lemma VII.3. Suppose that the Lyapounov exponents of $Df: E \times K \circlearrowleft$ are $\lambda_1 > \lambda_2 > \ldots$ ν-a.e. If $k \ge 1$, $\mu > 0$ and $\lambda_k > -\mu > \lambda_{k+1} \ge -\infty$ there exist a compact set K_μ with measure arbitrarily near to 1 and $\delta(\mu) > 0$, $r(\mu) > 0$ arbitrarily small such that the maps $x \to E_j(x)$, $x \to F_k(x)$, $1 \le j \le k$, are continuous on K_μ and if $\Gamma_{r(\mu)} = \{(x,v) \mid x \in K_\mu, v \in F_k(x), \|v\| < r(\mu)\}$, there exists a map $\varphi_\mu = \Gamma_{r(\mu)} \to E$ such that for all $(x,v) \in \Gamma_{r(\mu)}, \varphi_\mu(x,v)$ is the unique vector in E of the form $\varphi_\mu(x,v) = x + v + w$ where $w \in E_1(x) \oplus \ldots \oplus E_k(x)$, such that:

$$\|f^n(\varphi_\mu(x,v)) - f^n(x)\| \le \delta(\mu)\exp(-\mu n).$$

The map φ_μ is continuous and for every $x \in K_\mu$, the map $\varphi_\mu(x,\cdot)$ is $C^{k,\gamma}$ and its derivatives vary continuously with (x,v).

The set $\{\varphi_\mu(x,v) \mid \|v\| < r(\mu)\}$ will be denoted $W^s_{\mu,loc}(x)$. Lemma VII.3 states that $W^s_{\mu,loc}(x)$ is a $C^{k,\gamma}$ submanifold characterized as the set of vectors $Y \in E$ of the form $y = x + v + w$ with $v \in F_k(x)$, $\|v\| < r(\mu)$, $w \in E_1(x) \oplus \dots \oplus E_k(x)$, and satisfying:

$$\|f^n(y)-f^n(x)\| \le \delta(\mu)\exp(-\mu n)$$

for all $n \ge 0$. This characterization implies that if $x \in K_\mu$ and $f^n(x) \in K_\mu$, then $f^n(W^s_{\mu,loc}(x)) \subset W^s_{\mu,loc}(x)) \subset W^s_{\mu,loc}(f^n(x))$. If $x \in K$ define

$$\tilde{W}^s_\mu(x) = \bigcup \{f^{-k}(W^s_{\mu,loc}(f^k(x))) \mid f^k(x) \in K_\mu,\ k \in \mathbb{Z}^+\}.$$

If f satisfies the hypothesis of Theorem F it follows that $f^{-k}(W^s_{\mu,loc}(x))$ is a $C^{r,\gamma}$ submanifold because the density of the range of D_yf at every $y \in U$ implies that the map f^k is transversal to the finite codimensional $W^s_{\mu,loc}(x)$. Moreover the sets $f^{-k}(W^s_{\mu,loc}(f^k(x)))$ with $f^k(x) \in K_\mu$ increase with k because of the characterization of the sets $W^s_{\mu,loc}(f^k(x)))$ with $f^k(x) \in K_\mu$. increase with k because of the characterization of the sets Then $\tilde{W}^s_\mu(x)$ is an embedded submanifold. Now let $\Lambda \subset K$ be the set of points x such that for every positive rational μ the property $f^n(x) \in K_\mu$ holds for infinitely many positive values of n. We shall prove that all the properties in Theorem F are valid at this points. Since $\nu(\Lambda) = 1$ (because of the ergodicity of ν) this will prove that those properties hold ν-a.e. and then that they hold for a total probability set. Let $\lambda > 0$ and $x \in \Lambda$. Take $k \ge 1$ such that $\lambda_k > -\lambda \ge \lambda_{k+1}$. Such k exists if the sequence $\lambda_1 > \lambda_t > \dots$ contains positive elements, and we can always assume that it does because otherwise we could replace $f\colon U \times \mathbb{R} \to E \to \mathbb{R}$ defined by $f(x,t) = (f(x),2t)$, that contains 2 in the sequence. Now we claim that $W^s_\lambda(x) = \tilde{W}^s_\mu(x)$ for every $\mu > 0$ with $\lambda_k > -\mu > \lambda_{k+1}$. Clearly this claim and the previous remarks on $\tilde{W}^s_\mu(x)$ prove that $W^s_\lambda(x)$ is an embedded submanifold and that $W^s_{\lambda'}(x) = W^s_{\lambda''}(x)$ if the interval $(-\lambda',\lambda'']$ doesn't contain elements of the sequence $\{\lambda_j\}$. To prove the claim assume first that $\lambda > \mu$. To show $\tilde{W}^s_\mu(x) \supset W^s_\lambda(x)$ if $x \in \Lambda$, take $p \in W^s_\lambda(x)$ $A > 0$ such that $\|f^n(p)-f^n(x)\| \leq A\exp(-\mu n)$ for all $n \geq 0$. Such A exists because of the definition of $W^s_\lambda(x)$ and the inequality $-\mu > -\lambda$. Let $\pi_y\colon E \to F_k(y)$ be the projection associated to the splitting $E = F_k(y) \oplus (E_1(y) \oplus \ldots \oplus E_k(y))$ and set $B = \sup\{\|\pi_y\| \mid y \in K_\mu\}$. Fix $n_0 \geq 0$ such that $f^{n_0}(x) \in K_\mu$, $BA\exp(-\mu n_0) < r(\mu)$ and $A\exp(-\mu n_0) \leq A\exp(-\mu n_0) \leq \delta(\mu)$. Then:

$$\|\pi_{f^{n_0}(x)}(f^{n_0}(p) - f^{n_0}(x))\| \leq$$
$$\leq BA\exp(-\mu n_0) < r(\mu)$$

and:

$$\|f^n(f^{n_0}(p)) - f^n(f^{n_0}(x))\| \leq$$
$$\leq A\exp(-\mu(n+n_0)) \leq \delta(\mu(\exp(-\mu n))).$$

By the characterization of the submanifolds $W^s_{\mu,loc}(y)$, these inequalities imply $f^{n_0}(p) \in W^s_{\mu,loc}(f^{n_0}(x))$. Then $p \in f^{-n}(W^s_{\mu,loc}(f^{n_0}(x))) \subset \tilde{W}^s_\mu(x)$. To show $W^s_\lambda(x) \supset \tilde{W}^s_\mu(x)$ take $\epsilon > 0$ and a rational $\mu_1 > 0$ satisfying $\min(\mu,-\lambda+\epsilon) > -\mu_1 > \lambda_{k+1}$. Without lose of generalizty we can assume that the constant $\delta(\mu)$ were taken satisfying $\delta(\mu') < \delta(\mu'')$, if $\mu' > \mu''$. Then $\delta(\mu_1) < \delta(\mu)$. If $p \in \tilde{W}^s_\mu(x)$ and $x \in \Lambda$ there exists $A > 0$ satisfying $\|f^n(p)-f^n(x)\| \leq \exp(-\mu n)$ for all $n \geq 0$. Take $N > 0$ such that:

$$BA\exp(-\mu N) \leq \min(r(\mu),r(\mu_1))$$
$$f^N(x) \in K_{\mu_1} \cap K_\mu$$
$$A\exp(-\mu N) \leq \delta(\mu).$$

Then:

$$\|f^n(f^N(p)) - f^n(f^N(x))\| \le$$
$$\le A\exp(-(n+N)\mu) \le \delta(\mu)\exp(-\mu n) \qquad (1)$$

and:

$$\|\pi_{f^N(x)}(f^N(p) - f^N(x))\| \le$$
$$\le B\|f^N(p) - f^N(x)\| \le BA\exp(-\mu N) \le \min(r(\mu), r(\mu_1)).$$

Since $f^N(x) \in K_\mu$ these inequalitites imply $f^N(p) \in W^s_{\mu,loc}(f^N(x))$. The uniqueness properties of VII.3 grant that if some $q \in E$ satisfies:

$$\pi_{f^N(x)}(q-f^N(x)) = \pi_{f^N(x)}(f^N(p) - f^N(x))$$
$$\|f^n(q) - f^n(f^N(x))\| \le \delta(\mu)\exp(-\mu n)$$

then $q = f^N(p)$. Moreover

$$(f^N(x), \pi_{f^N(x)}(f^N(p)-f^N(x)) \in \Gamma_{r(\mu_1)}$$

because $f^N(x) \in K_{\mu_1}$. Then there exists $\varphi_{\mu_1}(f^N(x), \pi_{f^N(x)}(f^N(p)-f^N(x)))$ satisfying:

$$\pi_{f^N(x)}(\varphi_{\mu_1}(f^N(x), \pi_{f^N(x)}(f^N(p)-f^N(x)))) =$$
$$= \pi_{f^N(x)}(f^N(p) - f^N(x))$$

and

$$\|f^n(\varphi_{\mu_1}(f^N(x), \pi_{f^N(x)}(f^N(p)-f^N(x)))) - f^n(f^N(x))\| \le$$
$$\le \delta(\mu_1)\exp(-n\mu_1) \le \delta(\mu_1)\exp(-n\mu_1) \le \delta(\mu)\exp(-n\mu).$$

Therefore $f^N(x) + \varphi_{\mu_1}(f^N(x), \pi_{f^N(x)}(f^N(p)-f^N(x)))$ satisfies the properties of q and then it coincides with $f^N(p)$. But:

$$\limsup_{n\to+\infty} \frac{1}{n}\log\|f^n(p)-f^n(x)\| =$$
$$= \limsup_{n\to+\infty} \frac{1}{n}\log\|f^n(f^N(p))-f^n(f^N(x))\|$$

and since $f^N(p) = f^N(x) + \varphi_{\mu_1}(f^N(x), \pi_{f^N(x)}(f^N(p)-f^N(x)))$ it follows that the last limit is $\leq \mu_1 \leq -\lambda+\varepsilon$. This inequality holds for all $\varepsilon > 0$. Then $p \in W^s_\lambda(x)$. This completes the proof of $W^s_\lambda(x) = \tilde{W}^s_\mu(x)$ when $\lambda > \mu$. With minor charges the same methods can be applied to show the same property for $\mu \geq \lambda$.

Similar techniques can be used to prove Theorem G. Instead of VII.3 we shall use the following lemma:

Lemma VII.4. Let ν be a ergodic φ-invariant probability on K. Suppose that the Lyapounov exponents of $Df\colon E\times K \circlearrowleft$ are $\lambda_1 > \lambda_2 > \dots$ ν-a.e. If $k \geq 1$, $\mu > 0$ and $\lambda_{k+1} < \mu < \lambda_k$ there exists a compact set K_μ with measure arbitrarily near to 1 and $\delta(\mu) > 0$, $r(\mu) > 0$ arbitrarily small such that the maps $x \mapsto F_k(x)$, $x \mapsto H_k(x) = E_1(x) \oplus \dots \oplus E_k(x)$ are continuous on K_μ and if $\Gamma_{r(\mu)} = \{(x,v) \mid x \in K_\mu,\ v \in H_k(x),\ \|v\| < r(\mu)\}$, there exists $\varphi_\mu\colon \Gamma_{r(\mu)} \to E$ such that for all $(x,v) \in \Gamma_{r(\mu)}$, $\varphi_\mu(x,v)$ is the unique vector of the form $x + v + w$ with $w \in F_k(x)$, such that $f^{-n}(x)$ exists for all $n \geq 0$ and satisfies:

$$\|f^{-n}(\varphi_\mu(x,v)) - f^{-n}(x)\| \leq \delta(\mu)\exp(-\mu n).$$

Moreover φ_μ is continuous and has continuous derivative with respect to the second variable.

To prove Theorem G using this lemma we proceed as we did to deduce Theorem F from Lemma VII.3. We have to show that for any φ-invariant ergodic probability ν the properties of Theorem H hold ν-a.e. For this we take any $\mu > 0$ such that there exists $k \geq 1$ satisfying $\lambda_{k+1} < \mu < \lambda_k$. Then we define $W^u_{\mu,loc}(x) = \{\varphi_\mu(x,v) \mid v \in H_k(x),\ \|v\| < r(\mu)\}$ if $x \in K_\mu$. Define $\tilde{W}^u_\mu(x) = \bigcup \{f^n(W^u_{\mu,loc}(f^{-n}(x))) \mid n \geq 0,\ f^{-n}(x) \in K_\mu\}$. By Lemma VII.4, $W^u_{\mu,loc}(y)$ is a $C^{r,\gamma}$ finite dimensional submanifold. Since f and its first derivative at every point of U are injective then $f^n(W^u_{\mu,loc}(f^{-n}(x)))$ is, for all $n \geq 0$, a $C^{r,\gamma}$ finite dimensional submanifold. As in the proof of Theorem F, we prove that $f^{n_1}(W^u_{\mu,loc}(f^{-n_1}(x))) \supset f^{n_2}(W^u_{\mu,loc}(f^{-n_2}(x)))$ if $n_1 > n_2$ and $f^{n_1}(x) \in K_\mu$, $f^{n_2}(x) \in K_\mu$. Then $\tilde{W}^u_\mu(x)$ is a finite dimensional $C^{r,\gamma}$ immerse submanifold. Now consider the set $\Lambda \subset K$ of points x such that $f^{-n}(x) \in K_\mu$ for infinitely many values of $n \geq 0$. This set satisfies $\nu(\Lambda) = 1$ because of the ergodicity of ν. As in the proof of Theorem F we show that if $\lambda > 0$ and $\mu > 0$ satisfy $\lambda_{k+1} < \lambda \leq \lambda_k$ and $\lambda_{k+1} < \mu < \lambda_k$ then $W^u_\lambda(x) = \tilde{W}^u_\lambda(x)$ for all $x \in \Lambda$. From this follows that the properties of Theorem H hold for every point in Λ. Then they hold ν-a.e. as we wished to prove.

It remains to show how VII.3 and VII.4 follow from VII.1 and VII.2 respectively. In fact we shall only prove VII.3. The proof of VII.4 is basically the same with some obvious modifications.

Take $\varepsilon > 0$ such that $\lambda_k > -\mu + 3\varepsilon > \lambda_{k+1}$ and define:

$$C_1(x) = \sup_{n \geq 0} \|(D_x f^n)/F_k(x)\| \exp \mu n$$

$$C_2(x) = \sup_{n \geq 0} \|(D_{f^{-n}(x)} f^n)/E_1(x) \oplus \ldots \oplus E_k(x)\| \exp(-\mu+3\varepsilon)n.$$

The basic properties of the regular points stated in its definition imply that $C_1(x) < +\infty$ ν-a.e. Let $\pi_y\colon E \to F_k(y)$ be the projection associated to the splitting $E = (E_1(x) \oplus \ldots \oplus E_k(x)) \oplus F_k(x)$. Define

$$C_3(x) = \sup_{n \geq 0} \|\pi_{f^n(x)}\| \exp(-\varepsilon n).$$

By Theorem B we have $C_3(x) < +\infty$ ν-a.e. Moreover it is easy to check that $\log(C_1(\varphi(x))/C_1(x))$ and $\log(C_2(x)/C_2(\varphi(x)))$ are bounded by integrable functions. Then we can apply Lemma III.8 to define the function:

$$C_4(x) = \sup_{n \geq 0} \exp(-n\varepsilon)\max(C_1(\varphi^n(x)), C_2(\varphi^n(x))).$$

By Lemma III.8, $C_4(x) < +\infty$ ν-a.e. Then we can take a compact set K_μ with arbitrarily large measure such that C_4 is bounded on K_μ by a constant $C > 0$. We can assume that also C_3 is bounded by C on K_μ. Then we can apply Lemma VII.1 to the compact set K_μ setting $H_x^{(n)} = E_1(x) \oplus \ldots \oplus E_k(x)$, $G_x^{(n)} = F_k(x)$ for $x \in f^n(K_\mu)$. Obviously $\varepsilon > 0$ can be taken so small that $2\varepsilon < \gamma\mu$ as required by condition (b) of VII.1. Applying VII.1 we obtain $r = r(\mu)$, $\delta = \delta(\mu)$ and $\varphi = \varphi_\mu : \Gamma_{r(\mu)} \to E$ satisfying the properties required by VII.3.

Now let us prove VII.1. Let Σ be the space of sequences $\xi : Z^+ \to E$ such that:

$$\sup_{n\geq 0} \|\xi(n)\| \exp \mu n < +\infty$$

endowed with the norm:

$$\|\xi\| = \sup_{n\geq 0} \|\xi(n)\| \exp \mu n$$

Σ is a Banach space. Set:

$$\Sigma_\rho = \{\xi \in \Sigma \mid \|\xi\| < \rho\}$$

$$\Gamma_r = \{(x,v) \mid x \in K_o,\ v \in G_x^{(0)},\ \|v\| < r\}$$

and $\tilde{\pi}_x^{(n)} = I - \pi_x^{(n)}$ for $x \in f^n(K_o)$. Take $0 < \rho < d(K_o, E\ U)$ and define a map $F : \Gamma_\rho \times \Gamma_\rho \to \Sigma$ by:

$$F(x,v,\xi)(n) =$$

$$= (D_x f^n)v + \sum_{j=0}^{n-1} (D_{f^{j+1}(x)} f^{(n-1)-j})\pi^{(j+1)}_{f^{j+1}(x)} p(f^{j+1}(x),\xi(j)) -$$

$$- \sum_{j=n}^{\infty} (D_{f^n(x)} f^{j-(n-1)})^{-1} \tilde{\pi}^{(j+1)}_{f^{j+1}(x)} p(f^j(x),\xi(j))$$

for $n \geq 1$ and

$$F(x,v,\xi)(0) = v - \sum_{j=0}^{\infty} (D_x f^{j+1})^{-1} \tilde{\pi}^{(j+1)}_{f^{j+1}(x)} p(f^j(x),\xi(j)).$$

The series that appear in this definition and the fact that the

Lemma VI.5. Let $x \in K_o$, $0 < \delta < \rho$ and $v \in G_x^{(0)}$ with $\|v\| \leq \rho$. Then $y \in E$ satisfies:

$$\pi_x^{(0)}(y-x) = v \tag{1}$$

$$\|f^n(y)-f^n(x)\| \leq \rho \exp(-\mu n) \tag{2}$$

for all $n \geq 0$ if and only if there exists $\xi \in \Sigma_\rho$ such that

$$\xi(0) = y-x \tag{3}$$

$$F(x,v,\xi) = \xi . \tag{4}$$

Proof. Suppose y satisfies (1), (2). It is easy to see that if $w_n = f^n(y) - f^n(x)$ then:

$$f^n(y)-f^n(x) = (D_x f^n)w_o + \sum_{j=0}^{n-1} (D_{f^{j+1}(x)} f^{(n-1)-j}) p(f^j(x),w_j).$$

and

$$\tilde{\pi}^{(n)}_{f^n(x)} (f^n(y)-f^n(x)) = (D_x f^n)\tilde{\pi}_x^{(0)} w_o +$$

$$+ \sum_{j=0}^{n-1} (D_{f^{j+1}(x)} f^{(n-1)-j}) \tilde{\pi}^{(j+1)}_{f^{j+1}(x)} p(f^j(x),w_j) =$$

$$= (D_x f^n)(\tilde{\pi}_x^{(0)} w_o + \sum_{j=0}^{n-1} (D_x f^{j+1})^{-1} \tilde{\pi}^{(j+1)}_{f^{j+1}(x)} p(f^j(x),w_j).$$

Hence:

$$\rho \exp(-\mu n) C \exp \epsilon n \geq \|\tilde{\pi}^{(n)}_{f^n(x)} (f^n(y)-f^n(x))\| \geq$$

$$\geq \frac{1}{C \exp(\mu-3\epsilon)n} \|\tilde{\pi}_x^{(0)} w_o + \sum_{j=a}^{n-1} (D_x f^{j+1})^{-1} \tilde{\pi}^{(j+1)}_{f^{j+1}(x)} p(f^j(x),w_j)\| .$$

Finally we obtain that for all $n \geq 0$:

$$C^2 \rho \exp(-\mu n)\exp(-2\epsilon n) \geq \|\tilde{\pi}_x^{(0)} w_o + \sum_{j=0}^{n-1} (D_x f^{j+1})^{-1} \tilde{\pi}^{(j+1)}_{f^{j+1}(x)} p(f^j(x),w_j)\| .$$

This implies:

$$\tilde{\pi}_x^{(0)} w_o + \sum_{j=0}^{\infty} (D_x f^{j+1})^{-1} \tilde{\pi}^{(j+1)}_{f^{j+1}(x)} p(f^j(x),w_j) = 0. \tag{5}$$

Therefore for $n > 0$ we can write:

sequence $n \to F(x,v,\xi)(n)$ belongs to Σ follow from the inequality $\|(D_x f^n)v\| \le C\exp(-\mu n)$ and the inequality (*) in the remark after the statement of VII.1. In fact:

$$\|\sum_{j=0}^{n-1} (D_{f^{j+1}(x)} f^{(n-1)-j})\pi^{(j+1)}_{f^{j+1}(x)}\, p(f^j(x),\xi(j))\| \le$$

$$\le \sum_{j=0}^{n-1} C_1 C^2 \exp 2(j+1)\varepsilon\ \exp(-\mu(n-1-j))\|\xi\|^{1+\gamma} \exp(-(1+\gamma)\mu n) \le$$

$$\le C_1 C^2\|\xi\|^{1+\gamma} \exp(-(2+\gamma)\mu n) \sum_{j=0}^{n-1} \exp(2(j+1)\varepsilon + \mu(j+1)) =$$

$$= C_1 C^2\|\xi\|^{1+\gamma} \exp(-(2+\gamma)\mu n)\exp(2\varepsilon+\mu)\ \frac{1-\exp(2\varepsilon+\mu)n}{1-\exp(2\varepsilon+\mu)} \le$$

$$\le \frac{C_1 C^2\|\xi\|^{1+\gamma} \exp(-(2+\gamma)\mu n)\exp(2\varepsilon+\mu)}{\exp(2\varepsilon+\mu)-1}$$

$$\frac{1}{\exp(2\varepsilon+\mu)-1}\, C_1 C^2\|\xi\|^{1+\gamma} \exp(2\varepsilon+\mu)\exp(-(\mu+\gamma\mu-2\varepsilon)n).$$

Since $2\varepsilon < \gamma\mu$ the last exponential is dominated by $\exp(-\mu n)$.

Finally:

$$\|\sum_{j=n}^{\infty} (D_{f^n(x)} f^{j-(n-1)})^{-1}\, \tilde{\pi}^{(j+1)}_{f^{j+1}(x)}\, p(f^j(x),\xi(j))\| \le$$

$$\le \sum_{j=n}^{\infty} C_1 C^2 \exp 2\varepsilon(j+1)\exp(\mu-3\varepsilon)(j-n+1)\|\xi\|^{1+\gamma} \exp(-(1+\gamma)\mu j) \le$$

$$\le C_1 C^2\|\xi\|^{1+\gamma} \exp(\mu-\varepsilon)\exp(2\varepsilon-(1+\gamma)\mu)n \sum_{j=n}^{\infty} \exp(-\varepsilon-\gamma\mu)(j-n) \le$$

$$\le C_1 C^2\|\xi\|^{1+\gamma} \exp(\mu-\varepsilon)\exp(-\mu n)\ \frac{1}{1-\exp(-\varepsilon-\gamma\mu)}.$$

Similar computations show that F is continuous and, using inequality (**) of the remark after Lemma VII.2, that for each x the function $F(x,\cdot,\ )$ is $C^{r,\gamma}$ and its derivatives depend continuously on (x,v,ξ).

The main property of F is stated in the next lemma:

$$w_n = f^n(y) - f^n(x) =$$

$$= (D_x f^n)v + \sum_{j=0}^{n-1} (D_{f^{j+1}(x)} f^{(n-1)-j})\, \pi^{(j+1)}_{f^{j+1}(x)}\, p(f^{j+1}(x), w_j) +$$

$$+ (D_x f^n)\tilde{\pi}^{(0)}_x w_o + \sum_{j=0}^{n-1} (D_{f^{j+1}(x)} f^{(n-1)-j})\, \tilde{\pi}^{(j+1)}_{f^{j+1}(x)}\, p(f^{j+1}(x), w_j) =$$

$$= (D_x f^n)v + \sum_{j=0}^{n-1} (D_{f^{j+1}(x)} f^{(n-1)-j})\pi^{(j+1)}_{f^{n+1}(x)}\, p(f^{j+1}(x), w_j) +$$

$$+ (D_x f^n)(\tilde{\pi}^{(0)}_x w_o + \sum_{j=0}^{n-1} (D_x f^{j+1})^{-1}\, \tilde{\pi}^{(j+1)}_{f^{j+1}(x)}\, p(f^{j+1}(x), w_j)) =$$

$$= (D_x f^n)v + \sum_{j=0}^{n-1} (D_{f^{j+1}(x)} f^{(n-1)-j})\pi^{(j+1)}_{f^{j+1}(x)}\, p(f^{j+1}(x), w_j) +$$

$$- (D_x f^n)(\sum_{j=0}^{\infty} (D_x f^{j+1})^{-1}\, \tilde{\pi}^{(j+1)}_{f^{j+1}(x)}\, p(f^{j+1}(x), w_j) = F(x,v,w)(n). \quad (6)$$

Moreover, (5) implies:

$$v = \pi^{(0)}_x w_o = w_o - \tilde{\pi}^{(0)}_x w_o =$$

$$= w_o + \sum_{j=0}^{\infty} (D_x f^{j+1})^{-1}\, \tilde{\pi}^{(j+1)}_{f^{j+1}(x)}\, p(f^{j+1}(x), w_j) \quad (7)$$

Then (6) and (7) imply:

$$F(x,v,w) = w.$$

Where $w \in \Sigma_\rho$ is the sequence $w_o, w_1, \ldots$ Clearly this sequence satisfies $y-x = \pi^{(0)}_x w_o$ so that (3) and (4) are satisfied. Conversely suppose that there exists $\xi \in \Sigma_\rho$ satisfying (3) and (4). By induction on n it is easy to prove that:

$$f^n(y) - f^n(x) = F(x,v,\xi)(n).$$

Then $\|f^n(y)-f^n(x)\| = \|F(x,v,\xi)(n)\| = \|\xi(n)\| \le \rho \exp(-\mu n)$. Moreover $\pi^{(0)}_x(y-x) = \pi^{(0)}_x \xi(0) = \pi^{(0)}_x F(x,v,\xi)(0) = v$.

Then $y \in E$ satisfies (1), (2) if and only if $y = x + \xi$ where ξ satisfies

$$\Phi(x,v,\xi) = 0 \quad (8)$$

where $\Phi: \Gamma_{\rho_1} \times \Sigma_\rho \to \Sigma$ is just $\Phi(x,v,\xi) = \xi - F(x,v,\xi)$.

Moreover the derivative of G respect to the third variable at $(a,0,0)$ is the identity and $G(x,0,0) = 0$. Then we can apply the implicit function theorem to solve (8) and we obtain $\rho > r > 0$ and $\rho > \delta > 0$ arbitrarily small such that if $(x,v) \in \Gamma_r$ there exists a unique $\xi \in \Sigma_\delta$ such that $G(x,v,\xi) = 0$ and defining $\xi = \Psi(x,v)$ the function $\Psi: \Gamma_r \to \Sigma_\delta$ satisfies that for all $x \in K_0$ the map $\varphi(x,\cdot)$ is C^r,γ and its derivatives depend continuously on (x,v). By VI.4 it follows that for all $(x,v) \in \Gamma_r$ there exists a unique $\gamma \in E$ satisfying (1), (2) and this y is given by $y = x + \Psi(x,v)$. Then the function $\varphi: \Gamma_r \to \Sigma_\delta$ defined by $\varphi(x,v) = x + \Psi(x,v)$ satisfies all the required properties.

REFERENCES

[1] J. Hale - Theory of functional differential equations. Springer-Verlag, 1977.

[2] D. Henry - Geometric Theory of Semilinear Parabolic Equations. Lecture Notes in Mathematics (1980) 840

[3] V.I. Oseledec - A multiplicative ergodic theorem. Trans. Moscow Math. Soc. 19 (1968) 197-231.

[4] Ya. Pesin - Invariant manifolds which correspond to non vanishing exponents. Math. USSR 10, (1976).

[5] V.A. Pliss - On a conjecture due to Smale, Diff. Uravnenija, 8 (1972).

[6] D. Ruelle - Ergodic theory of differentiable systems. Publications IHES 50 (1979).

[7] D. Ruelle - Characteristic exponents and invariant manifolds in Hilbert space. Annals of Mathematics, 115 (1982).

HYPERBOLIC DYNAMICAL SYSTEMS WITH ISOLATED POINTS

Servet Martínez A.*

ABSTRACT

We prove that the non-wandering set of an hyperbolic topologically transitive dynamical system whose basis space is an infinite non perfect set, is the union of two disjoint orbits.

By a (compact) dynamical system we shall mean a couple (X,T) where the basis space X is a compact metric space and $T: X \to X$ is an homeomorphism. Two dynamical systems (X,T) and (X',T') are said to be topologically conjugate if there exists an homeomorphism, called a topological isomorphism: $\phi: X \to X'$ such that $\phi \circ T = T' \circ \phi$. $P(T)$ will be the set of periodic points, and the set of points of period n will be $P_n(T)$. The positive and negative limit sets of $x \in X$ will be:

$$\omega^{+(-)}(x) = \{z \in X \ / \ \exists\, n_k \to +(-)\infty \text{ such that } T^{n_k}x \to z\}$$

We define the non-wandering set:

$$\Omega(T) = \{x \in X \ / \ \forall\, V \text{ neighbourhood of } x \ \exists\, n > 0 \text{ so that } V \cap T^{-n}V \neq \emptyset\}$$

If ϕ is a topological isomorphism then:

$$\phi(\omega^{+(-)}(x)) = \omega^{+(-)}(\phi(x)), \Omega(\phi\circ T\circ\phi^{-1}) = \phi\Omega(T)$$

A dynamical system will be topologically transitive (respectively topologically mixing) if for every couple V, V' of non empty open sets of X there is an $n \in Z$ such that (respectively there exist some $n_0 \in Z$ such that for all $n \geq n_0$ we have): $V \cap T^{-n}V' \neq \phi$. These properties are preserved by topological isomorphisms. If (X,T) is topologically transitive it is easy to see that $X = \overline{\bigcup_{n\in Z} T^n V}$ for every open non empty set V.

We say that a dynamical system (X,T) is hyperbolic if:

(a) $\forall \varepsilon > 0 \ \exists \ \delta(\varepsilon) > 0$ so that $d(x,y) < \delta(\varepsilon)$ implies

$$W_\varepsilon^+(x) \cap W_\varepsilon^-(y) \neq \phi$$

where $W_\varepsilon^{+(-)}(x) = \{z \in X: d(T^{+(-)n}x, T^{+(-)}z)\} \leq \varepsilon, \quad \forall n \geq 0\}$
(condition for the existence of canonical coordinates);

(b) $\exists \ \varepsilon_0 > 0, \ 0 < \lambda < 1, \ c \geq 1$ such that:

$$W_{\varepsilon_0}^{+(-)}(x) = \{z \in X \mid d(T^{+(-)n}x, T^{+(-)n}y) \leq c\lambda^n d(x,z), \quad \forall n \geq 0\}$$

(hyperbolicity condition).

The following fundamental results are true on these systems:

<u>Anosov's closing lemma</u>: An hyperbolic dynamical system satisfies the closing property $\Omega(T) = \overline{P(T)}$

<u>Proof</u>.: See [1], p. 74-76.

<u>Smale-Bowen spectral decomposition theorem</u>: If (X,T) is an hyperbolic system satisfying $X = \overline{P(T)}$ then there exist $s>0$ and $\{n_k \ / \ k=0,\ldots,s-1\}$ such that;

$$X = \sum_{k=0}^{s-1} X_k, \quad X_k = \sum_{\ell=0}^{n_k-1} \quad , \quad T\, X_k^{\ell} = X_k^{\ell+1} (\mathrm{mod}\ n_k)$$

where $T|X_k$ is topologically transitive and $T^{n_k} \mid X_k^{\ell}$ is topologically mixing.

<u>Proof</u>.: See [1] p. 69-74

We can remark that condition $X = \overline{P(T)}$ may be replaced by the condition $X = \Omega(T)$ by an application of Anosov's closing lemma.

This means we can extend all the results proved on hyperbolic topologically mixing dynamical systems to the hyperbolic dynamical systems that satisfy the closing property $X = \Omega(T)$, using the following commutative diagram (see [4]):

$$(1) \qquad \begin{array}{ccc} X_k^{\ell} & \xrightarrow{T^{n_k}} & X_k^{\ell} \\ T \downarrow & & \downarrow T \\ X_k^{\ell+1} & \xrightarrow{T^{n_k}} & X_k^{\ell+1} \end{array}$$

We know that there are hyperbolic dynamical systems which do not satisfy $X=\Omega(T)$. In the following we will obtain a complete characterization of the hyperbolic topologically transitive dynamical systems satisfying $X\neq\Omega(T)$.

Except for these exceptional cases we shall be able to extend confidently the results obtained for hyperbolic and topologically mixing systems.

Theorem 1. If (X,T) is an hyperbolic and topologically transitive dynamical system then;

(a) $\Omega(T) = X$ if X is finite or perfect.

(b) $\Omega(T)$ is the union of two disjoint orbits if X is an infinite non perfect set.

Proof.: (a) is obvious from the definitions.

(b) By the topological transitivity of (X,T) we have:

$X = \overline{\bigcup_{n\in Z} T^n x_0}$, where x_0 is an isolated point.

We obtain $\Omega(T) = \omega^+(x_0) \cup \omega^-(x_0)$. It is easy to see that there exists $0<\tilde{\varepsilon}_0\leq\varepsilon_0$ such that for $0<\varepsilon\leq\tilde{\varepsilon}_0$: $W^+_\varepsilon(T^{-p}x_0)=\{T^{-p}x_0\}$ and $W^-_\varepsilon(T^p x_0) = \{T^p x_0\}$ when $p\geq 0$. If $x\in\omega^+(x_0)$ there exists $0\leq p<q$ such that $d(T^p x_0, T^q x_0) \leq \delta(\varepsilon)$ so $d(T^{p+n}x_0, T^{q+n}x_0) \leq c\delta(\varepsilon)\lambda^n$ for every $n\geq 0$. Then it is easy to prove that the sequence $\{T^n x_0\ n\geq 0\}$ has at most q-p accumulation points and by the continuity of T they are a finite orbit. This means that $\omega^+(x_0)$ and (analogously) $\omega^-(x_0)$

are finite orbits. If $\omega^+(x_0) \cap \omega^-(x_0) \neq \emptyset$ there will exist p>0, q>0 such that $d(T^p x_0, T^{-q} x_0) < \delta(\varepsilon)$. Then $\emptyset \neq W_\varepsilon^-(T^p x_0) \cap W_\varepsilon^+(T^{-p} x_0) = \{T^p x_0\} \cap \{T^{-q} x_0\}$ so $T^{p+q} x_0$ and the point x_0 is periodic. This proves the theorem.

Q.E.D.

We obtain the following results as corollaries of this theorem:

Corollary 1: If (X,T) is an hyperbolic and topologically transitive dynamical system we have:

(a) $h_{top}(T) = 0$ if X is non perfect

(b) $h_{top}(T) > 0$ if X es perfect and there is only one maximal measure. It is ergodic and its support is X.

Proof.: (a) follows from the fact that $\Omega(T)$ is finite and every invariant measure is concentrated on $\Omega(T)$.

(b) follows from the Bowen result in the topologically mixing case (see [2]) and by applications of the commutative diagram (1). It is easy to see that the maximal measure is also a Bowen measure in the transitive case, that is to say it is the only weak limit of probability measures uniformilly distributed over $P_n(T)$.

Q.E.D.

Corollary 2. Let $(\Lambda,\sigma|\Lambda)$ be a finite type subshift of order 2 (see [2] for definitions) and $L(\Lambda)$ its associated transition matrix. Then $(\Lambda,\sigma|\Lambda)$ is topologically transitive if and only if:

(a) $L(\Lambda)$ is irreducible,or

(b) $L(\Lambda)$ can be transformed, under a permutation of the states, into the following matrix of order $p+q+r-1$, for some $p,q,r\geq 1$:

$C_{p,r-1,q}$ =

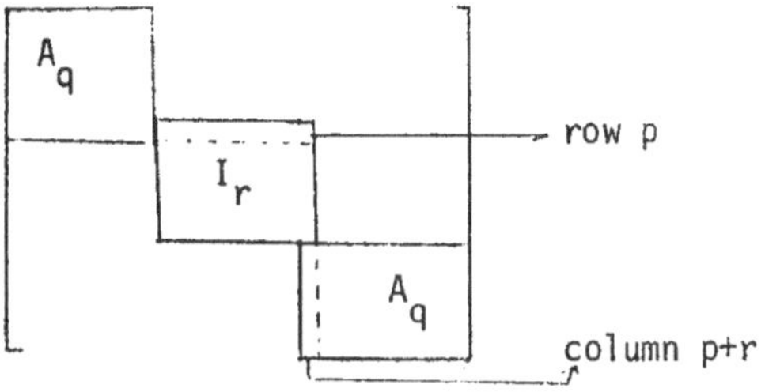

where A_p is the following transition matrix of order p:

$$A_p = \begin{bmatrix} 0 & 1 & 0,\ldots\ldots,0 \\ 0 & 0 & 1.,\ldots\ldots,0 \\ 0 & 0 & 0,\ldots\ldots,1 \\ \vdots & & \\ 1 & 0 & 0,\ldots\ldots,0 \end{bmatrix}$$

if $p\geq 1$; and I_r is the identily matrix of order r.

Proof.: For the detailed proof of (b) see [3].

References

[1] BOWEN, R. "Equilibrium States and the Ergodic Theory of Anosov Diffeomorphisms". Lecture Notes in Mathematics, Springer 470 (1975).

[2] DENKER, M.; GRILLENBERG, C.: SIGMUND, K. "Ergodic Theory on Compact Spaces". Lecture Notes in Mathematics, Sprin ger 527 (1976).

[3] MARTINEZ, S. "Characterization of Reducible Matrices Defining Topologically Transitive Subshifts". Preprint (1981).

[4] RUELLE, D. "A measure associated with Axiom A attractors". Amer. Journal of Math. 98 (1976), 619-654.

Universidade de Chile - Facultad de Ciencias Fisicas y Matematicas
Departamento de Matemáticas y Ciencias de la Computación
Casilla 5272 - Correo 3
Santiago - Chile

ON MODULI OF STABILITY OF VECTOR FIELDS ON OPEN MANIFOLDS

by

Pedro Mendes*

ABSTRACT. In this paper the author shows that the interior of the set of vector fields, with infinite non-denumerable moduli of stability, on open manifolds, are never dense. The same is true in the world of gradients of proper real functions, defined on open manifolds. It is also shown that finite modulus of stability, on two dimensional open manifolds, implies a finite number of saddle connections. As a consequence, moduli of stability of gradients of real proper Morse functions, defined on such manifolds, are either, infinite non-denumerable or zero.

1. INTRODUCTION

Let M be an open manifold, that is, a C^∞, noncompact (finite-dimensional), connected manifold without boundary. Let $\mathfrak{X}^r(M)$ be the space of all C^r-vector fields (generated by flows) on M, with the C^r-Whitney topology, $\mathcal{G}_1^r(M) \subset \mathfrak{X}^r(M)$ be the set of the vector fields with hyperbolic critical elements (singularities and closed orbits), and S(X) be the set of singular points of X. Let $G^r_{pr}(M) = \{\text{grad } f = \nabla f \mid f: M \to \mathbb{R} \text{ is } C^{r+1} \text{ and proper}\}$. and $GM^r_{pr}(M) \subset G^r_{pr}(M)$ be the subset of the gradients of the C^{r+1} proper Morse functions.

Two vector fields $X, Y \in \mathfrak{X}^r(M)$ are topologically equivalent if there is a homeomorphism $h: M \to M$, such that $h|_{S(X)}$ is near the identity in the C^0-Whitney topology (strong topological equivalence) h sending X-orbits onto Y-orbits and preserving the orientation of the orbits. If h is such that $X_t \circ h = h \circ Y_t$, where X_t, Y_t are the

* This paper was partially supported by CNPq, and was partially done when the author was a Visiting Professor at IMPA, Rio de Janeiro, Brazil, in the summer, 1981.

flows of X,Y, then h is called a conjugacy between X and Y. The study of the conjugacy class of same convenient reparametrization of a vector field X is often used in the study of its topological equivalence class.

An important concept in Dynamical Systems is the concept of structural stability. A vector field X is structurally stable,if its topological equivalence class is open. Another important concept is the concept of bifurcation point (point where the structural stability is broken), for a n-parameter family of vector fields.Moduli of stability have been studied as a step in the understanding of bifurcation theory. ([3], [7]).

Now we are going to state this concept, the results and some related problems.

DEFINITION 1.1.

Let A be an index set, $U_a \subset \mathbb{R}^{n_a}$ be an open set, $0 \in U_a$, $X \in \mathfrak{X}^r(M)$, and $\mathcal{N}(X)$ be a neighbourhood of X. Let $\{\xi_a : U_a \to N(X)\}$ be a family of continous functions, such that ξ_a is constant iff $n_a=0$. We say that $\{\xi_a, a \in A\}$ represents $\mathcal{N}(X)$, under topological equivalence, if, for each $Y \in \mathcal{N}(X)$, there are $a \in A$ and $u_a \in U_a$, such that $\xi_a(u_a)$ is topologically equivalent to Y. We say that

i) X has finite modulus of stability if A is at most denumerable, and $n_a \neq 0$ only for a finite number of $a \in A$. The minimum of $\sum_{a \in A} n_a$, when $\mathcal{N}(X)$ changes, is called the modulus of stability of X.

ii) X has infinite denumerable modulus of stability if A is denumerable and i) does not occur.

iii) X has infinite non-denuberable modulus of stability if i) and ii) do not occur.

There are three important questions related with the above definition.

PROBLEM 1. Is finite modulus of stability a generic property?

PROBLEM 2. Characterize finite modulus of stability.

PROBLEM 3. Does infinite denumerable modulus of stability imply finite modulus of stability?

The answer for the first problem is no, on any open manifold M, with dimension greater than one. In order to state this precisely, we need the following definitions.

DEFINITION 1.2.

Let $X \in \mathfrak{X}^r(M)$. A saddle connection for X is an ordered triple (p,q,γ), such that p,q are hyperbolic saddles of X, γ is a regular orbit of X, and γ is simultaneously a connected component of $W^u(p) - \{p\}$ and of $W^s(q) - \{q\}$.

DEFINITION 1.3.

Let $X \in \mathfrak{X}^r(M)$ and $x \in M$. We say that X has the density saddle connection property (DSCP) at x if there are an embedded codimension one disc $\bar{D} \subset M$, $x \in D$, and a denumerable set $\{(p_i, q_i, \gamma_i),\ i=1,2,\ldots\}$ of saddle connections, such that $\{p_i, q_i,\ i=1,2\ldots\}$ is a closed set, and $\{a_i = b_i = \gamma_i \cap D,\ i=1,2,\ldots\}$ is a dense subset of $\bar{D}$, where a_i is the intersection of γ_i with D. Here, a_i is considered as $W^u(p_i) \cap D$, and b_i as $W^s(q_1) \cap D$.

The following theorem and a remark will solve the first problem.

THEOREM A.

If $X \in \mathfrak{X}^r(M)$ has the DSCP at some $x \in M$, then X is in the interior of vector fields with infinite non-denumerable moduli of stability.

M.Peixoto - C.Pugh ([5]) has shown that, in any open manifold with dimension greater than one, there is a vector field satisfying the DSCP. Thus, Theorem A gives the answer to Problem 1. Moreover, we have the following.

REMARK 1.4

Theorem A holds for $G^r_{pr}(M)$, with a similar proof. In the end of this paper we will exhibit an element of $GM^r_{pr}(M)$ with the DSCP.

For the second problem, we have only the following partial result.

THEOREM B.

If $X \in \mathfrak{X}^r_1(M)$, where M is a two dimensional open manifold, has finite modulus of stability, then X has a finite number of saddle connections.

Our next result presents a particular situation where the three problems above are completely solved.

To state it precisely two technical conditions are necessary. These conditions, so called Condition A-a) and Condition A-b) (Condition A means both Conditions, A-a) and A-b), will be described later on. We have that $\nabla f \in GM^r_{pr}(M)$ satisfies those conditions iff the following happens:

"There exist a neighbourhood $\mathcal{N}(\nabla f) \subset GM^r_{pr}(M)$ such that if $\nabla g \in \mathcal{N}(\nabla f)$, then ∇g has no saddle connections".

We have the following

THEOREM C.

Let M be a two-dimensional open manifold and $r \geq 2$. If $\nabla f \in GM^r_{pr}(M)$, then 1) ∇f has infinite nondenumerable modulus of stability iff ∇f does not satisfy the Condition A;

2) ∇f has zero modulus of stability iff ∇f satisfies the Condition A;

3) ∇f is structurally stable iff ∇f satisfies the Condition A-a) and has no saddle connections.

Thus, in the world of gradients of Morse functions on two-dimensional open manifolds, finite modulus of stability is not generic, (this works for any open manifold; see Remark 1.4); it is possible to characterize finite modulus of stability; and infinite denumerable modulus of stability implies finite modulus of stability, (in fact, zero modulus of stability).

Actually we have the following conjectures:

CONJECTURE 1.

If $X \in \mathcal{G}^r_1(M)$; then X has either infinite non-denumerable modulus of stability, or zero modulus of stability.

CONJECTURE 2.

If $X \in GM^r_{pr}(M)$, then X is structurally stable iff ∇f satisfies Condition A-a) and has no saddle connections.

In these conjectures M is any open manifold.

I thank Z.Nitecki, J.Palis, P.Sad and J.Sotomayor for helpfull conversations and suggestions.

2. PROOF OF THEOREM A.

The main Dynamical System result needed in the proof of Theorem A, is essentially contained in ([5]), and can be stated as follows.

LEMMA 2.1.

Let M be an open manifold and $X \in \mathfrak{X}^r(M)$ be a vector field holding the DSCP at some point $x \in M$. If D is the closed (m-1) disc as in (1.3) and $D_1 \subset \bar{D}_1 \subset \text{int } D$ is a given (m-1) - open disc, then there is a neighbourhood $\mathcal{N}(X)$ such that, for each $Y \in \mathcal{N}(X)$, the sets $\{a_i(Y),\ i=1,2,\dots\}$ and $\{b_i(Y),\ i=1,2\dots\}$ are both dense in D_1.

The proof of this Lemma will be omitted.

LEMMA 2.2

Let $r \geq 1$ be an integer, $\varepsilon > 0$ a given real number, and $[-a,a] \subset \mathbb{R}$ a given closed interval. There are a number $\delta > 0$, and a C^∞ function $\varphi : \mathbb{R} \to \mathbb{R}$, such that:

i) $\varphi(x) \in [0,\delta]$ for all $x \in \mathbb{R}$;

ii) $\varphi(x) = 0$ if $x \notin [-a,a]$

iii) $\varphi(x) = \delta$ if $x \in \left[-\frac{a}{2}, \frac{a}{2}\right]$

iv) $\|\varphi\|_r < \varepsilon$.

PROOF.

Starting by a C^{∞} function $\tilde{\varphi}: \mathbb{R} \to \mathbb{R}$, such that $\tilde{\varphi}(x) \in [0,1]$, for all $x \in \mathbb{R}$, $\tilde{\varphi}(x) = 0$ if $x \notin [-a,a]$, $\tilde{\varphi}(x) = 1$ if $x \in \left[-\frac{a}{2}, \frac{a}{2}\right]$, dividing it by the maximum of the norms of its derivatives, up to the order r, and multiplying the result by $\frac{\varepsilon}{2}$, we get the function φ, as stated in the Lemma.

From (2.2) we are going to prove an approximation lemma, which will be usefull in the proof of Theorem A.

LEMMA 2.3.

Let $X(x) = (1,0,\ldots,0)$ be a vector field defined on the set $Q = \{(x_1,\ldots x_n) \in \mathbb{R}^n : |x_i| \leq a_i, \ i=1,\ldots,n\}$. Given $\varepsilon > 0$, there is a $\delta > 0$ such that, for each $y = (a_1, y_2, \ldots y_n)$, with $|y_i| < \delta$, $i=2,\ldots,n$, it is possible to construct a C^{∞} vector field Y, ε - C^r near X in Q, for a given $r \geq 1$, with the following properties:

i) $Y(x) = X(x)$ if $x_i \notin \left[-\frac{a_i}{2}, +\frac{a_i}{2}\right]$ for some $i=1,\ldots,n$;

ii) the orbit of $(-a_1,0,\ldots,0)$ by Y contains y.

PROOF. Let φ be the function built in (2.2), for the given ε, taking $a = \min\{a_i, i=1,\ldots,n\}$. Then, the vector field $Y_{\lambda}(x) = X(x) + \lambda \ . \ \varphi(\max_{1 \leq i \leq n} |x_i|)$. y, will satisfy the required properties, for some $\lambda \in [0,1]$, where $\delta > 0$ is given by (2.2)

Now we are going to construct a topological space, which is the key to the proof of Theorem A. It has not the Cantor property.

We will indicate a sequence $\{a_1, a_2, \ldots\}$, of positive integers, simply by a. Let

$\mathcal{G} = \{ a : a_i = 0$, if i is odd, and $a_i \in \{1,2\}$, if i is even$\}$.

It is clear that $\mathcal{G}$, with the discrete topology, is a non-denumerable collection of non-empty disjoint open sets. That is, $\mathcal{G}$ has not the Cantor property.

We will start now the

PROOF OF THEOREM A.

Let M be a m-dimensional open manifold and $X \in \mathfrak{X}^r(M)$ be a vector field, holding the DSCP at some point $x \in M$.
Let $D \subset M$ be the closed embedded (m-1)-disc, as in (1.3), and $D_1 \subset \bar{D}_1 \subset \text{int } D$ an open embedded (m-1)-disc, with $x \in D_1$. Let $\mathcal{N}(X)$ be the neighbourhood of X, given by (2.1), relative to those discs.

To prove the Theorem it suffices to show that, to each $Y \in \mathcal{N}(X)$ and for any neighbourhood $\eta(Y) \subset \mathcal{N}(X)$, it is possible to construct a vector field $Z \in \eta(Y)$, with infinite non-denumerable modulus of stability.

Let $Y_o \in \eta(Y)$ be a Kupka-Smale vector field. We take a small flow box, centered in $a_1(Y_o)$, whose intersection with D is contained in D_1. By (2.1), we can choose $b_j(U_o)$, in that flow box, arbitrarilly near $a_1(Y_o)$ and using (2.3), we get a vector field $Y_1 \in \eta(Y)$, such that $Y_1 = Y_o$ outside $\tilde{D}_1 = \bigcup_{t \in \mathbb{R}} (Y_o)_t(D_1)$, and $a_1(Y_1) = b_j(Y_1) = c_1(Y_1)$. Proceeding in the same way with $a_2(Y_1)$, and taking care that the flow box does not contain $a_1(Y_1)$, we get a vector field $Y_2 \in \eta(Y)$, such that $Y_2 = Y_o$ outside $\tilde{D}_1$, and $a_i(Y_2) = c_i(Y_2)$, $i = 1,2$. Continuing this process, we obtain, by induction, a vector field $Z \in \eta(Y)$, such that $Z = Y_o$ outside $\tilde{D}_1$, and $a_i(Z) = c_i(Z)$, $i = 1,2,\ldots$. Along that process, we need to take $\|Y_{n+1} - Y_n\|_r < \frac{\varepsilon}{n^2}$, in a compact neighbourhood of D_1, where $\varepsilon > 0$ is small, depending on this neighbourhood and on $\eta(Y)$, to assume that $Y_n \to Z$. But, this is clearly possible. We will prove that Z has infinite non-denumerable modulus of stability.

Let $\tilde{\eta}(Z) \subset \eta(Y)$ be a given neighbourhood of Z

Let $D_i \subset \bar{D}_i \subset D_1$, i=2,3..., be a family of embedded open (m-1)-discs such that $D_i \supset \bar{D}_{i+1}$, $\bigcap_{i=1}^{\infty} \bar{D}_i = \{x\}$ and $\bar{D}_1 = \bigcup_{i=1}^{\infty} \bar{S}_i$, where $S_i = D_i - \bar{D}_{i+1}$, i=1,2,3,... Set $\mathcal{O}_Z(A) = \mathcal{O}(A) = \bigcup_{t \in \mathbb{R}} Z_t(A)$, for each $A \subset M$.

Let $\{(p_i, q_i, \gamma_i), i = 1,2,...\}$ be the set of all saddle connections of Z, $a_i(Z) = W^u(p_i) \cap D_1 = \gamma_i \cap D_1$ and $c_i(Z) = W^s(q_i) \cap D_1 = \gamma_1 \cap D_1$.

For each i = 1,2,..., we can choose two saddle connections $(p_{r_i}, q_{r_i}, \gamma_{r_i})$ of Z, such that:

1) $a_{\ell_i}(Z) = c_{\ell_i}(Z) \in S_{2i}$, if $\ell_i \in \{r_i, r'_i\}$;

2) there are neighbourhoods $V_{r_i} \ni p_{r_i}$, $V_{r'_i} \ni p_{r'_i}$, such that

$\mathcal{O}(D_1) \cap \mathcal{O}(V_{r_i}) \cap \mathcal{O}(V_{r'_i}) = \emptyset$; $\mathcal{O}(D_1) \cap \mathcal{O}(V_{\ell_i}) \cap \mathcal{O}(V_{\ell_j}) = \emptyset$

if $i \neq j$, $\ell_i \in \{r_i, r'_i\}$, $\ell_j \in \{r_j, r'_j\}$; and

3) $V_{\ell_i} \cap V_{\ell_j} = \emptyset$, $\ell_i \in \{r_i, r'_j\}$, $\ell_j \in \{r_j, r'_j\}$, $\ell_i \neq \ell_j$,

We define $\mathcal{B} \subset \tilde{\eta}(Z)$ as follows:

$Z' \in \mathcal{B}$ iff

i) $a_{\ell_i}(Z') \neq c_{\ell_i}(Z') \Rightarrow \ell_i \in \{r_i, r'_i\}$ for some i=1,2,...;

ii) $Z' = Z$ on $M - \left[\bigcup_{\{\ell_i : a_{\ell_i}(Z') \neq c_{\ell_i}(Z')\}} \mathcal{O}_Z(V_{\ell_i}) \right]$;

iii) $a_{\ell_i}(Z') \neq c_{\ell_i}(Z')$, at least for one value of ℓ_i, i=1,2,...

Let $\varphi:\mathcal{B} \to G$ be a defined by $\varphi(Z') = a$, where $a_{2i}=2$ if iii) occurs for both values of ℓ_i, and $a_{2i}=1$ otherwise.

It is easy to see that φ is continous. (The topology on $\mathcal{B}$ is the relative topology.).

Moreover, φ is onto. The proof of this fact is similar to the construction of Z, and in this proof property 2) above is strongly used. The details are left to the reader.

Let Γ and $\tilde{\mathcal{B}} \subset \Gamma$ be the spaces of the topological equivalence classes of $\tilde{\eta}(Z)$ and $\mathcal{B}$, with the quotient topology, and $\pi:\tilde{\eta}(Z) \to \Gamma$ be the natural projection.

Let $Z_1, Z_2 \in \mathcal{B}$ be topologically equivalent.

From the property 3) above the definition of $\mathcal{B}$, taking the topological equivalence between Z_1 and Z_2 sufficiently near the identity in the C^0-Whitney topology on $S(X)$ and using the definition of DSCP, we get a homeomorphism $h:D_1 \to D_1$ such that $h(a_{\ell_i}(Z_1)) = a_{\ell_i}(Z_2)$ and $h(c_{\ell_i}(Z_1)) = c_{\ell_i}(Z_2)$, whenever $\ell_i \in \{r_i, r_i'\}$ is such that $a_{\ell_i}(Z_i) \neq c_{\ell_i}(Z_i)$, $i = 1,2,\ldots$. This implies that there is a continous map $\tilde{\varphi}:\tilde{\mathcal{B}} \to G$, , which is onto, and makes the following diagram

$$\begin{array}{ccc} \mathcal{B} & \xrightarrow{\ \varphi\ } & \\ \downarrow \pi & & G \\ \tilde{\mathcal{B}} & \xrightarrow{\ \tilde{\varphi}\ } & \end{array}$$

commutative.

Now, suppose by contradiction that there exists a denumerable family of continuous functions $\{\xi_i : U_i \subset \mathbb{R}^{n_i} \to \tilde{\eta}(Z)\}$, representing $\tilde{\eta}(Z)$. Let $U = \bigcup_{i=1}^{\infty} U_i$, be endowed with the disjoint union topology, and $\xi : U \to \tilde{\eta}(Z)$ be the natural continuous map induced by the family $\{\xi_i,\ i=1,2,\ldots\}$. Then, from (1.1), we get a continuous map $\xi' : U \to \Gamma$, which is onto. Let $\tilde{U} = (\xi')^{-1}(\tilde{\beta})$ and $\Psi = \tilde{\varphi} \circ \xi'|_{\tilde{U}}$. Thus, $\Psi : \tilde{U} \to G$ is continuous and onto. But $\tilde{U}$ has the Cantor property and G has not. This is a contradiction, and Theorem A is proved.

3. PROOF OF THEOREM B

Using the orientable double covering of a manifold, we can suppose that M is an oriented two-dimensional open manifold.

We indicate by $\Sigma \subset M$ a closed embedded interval.

We say that a saddle connection (p,q,γ) is isolated, if there is Σ, transversal to X, such that γ is the unique saddle separatrix of X crossing Σ.

Suppose, by contradiction, that X has finite modulus of stability, and infinitely many saddle connections.

Then we have the following three possibilities:

1) There is Σ, transversal to X, such that the set of its intersections with separatrices of saddle connections is dense in Σ.

2) There is Σ , transversal to X, such that the set of its intersections with separatrices of saddle connections is a monotone sequence in Σ.

3) All the saddle connections of X are isolated, and 2) does not occur.

The first possibility is impossible by Theorem A.

For the second possibility, we consider a point $x_i \in \Sigma \cap \gamma_i$, where (p_i,q_i,γ_i) is a saddle connection. Let $x = \lim_{i\to\infty} x_i$, and γ be its orbit by X. Using that M is an oriented two-dimensional manifold, it possible to prove that $F = \bar{\gamma} \ (\bigcup_{i=1} \bar{\gamma}_i)$ is closed in M. Let $\varepsilon : M \to \mathbb{R}$ be a continuous arbitrarilly small function, such that $\varepsilon|_F \equiv 0$ and $\varepsilon|_{(M-F)} > 0$. There is a C^r Kupka-Smale vector field Y_1 on M-F, ε-C^r near $X|_{(M-F)}$. Let $Y=Y_1$ on M-F, and Y = X on F. Y is a C^r vector field, arbitrarilly near X, and its saddle connections are exactly (p_i,q_i,γ_i), i=1,2,..., and perhaps more (p,q,γ), if it is a saddle connections for X.

Procceding as in the proof of Theorem A, we get that Y has infinite non-denumerable modulus of stability, which is a contradiction.

Finally , in the third possibility, we say that two saddle connections (p_1,q_1,γ_1) and $(q_1,q_2\gamma_2)$ are adjacent. By a small perturbation, we can suppose that X has not adjacent saddle connections and has infinitely many isolated saddle connections (p_i,q_i,γ_i), i =1,2,... . We notice that 3) was strongly used in

these reductions. From 3) and the orientation of M, it is possible
to build a C^∞ regular curve,$\alpha:[1,\infty)\to M$,with close unbounded image,
such that $\alpha(i) \in \gamma_i$ and $\{\alpha'(i), X(\alpha(i))\}$is a positive   basis   of
$T_{\alpha(i)}M$, $i=1,2,\ldots$.

Using that all saddle connections of X are transversal
to α , as described above, we get, as in the proof of Theorem A,
that X has infinite modulus of stability.

This is a contradicition, and Theorem B is proved.

4. PROOF OF THEOREM C.

We start with some notations.

Let M be an open two-dimensional manifold,
$\nabla f, \nabla g \in GM_{pr}^r$ (M), $r \geq 2$, and $a \in \mathbb{R}$ a regular level of f.

Let $F_a^u(g) = \left[\bigcup_{i=1}^{\infty} W^u(\bar{p}_i)\right] \cap f^{-1}(a)$, $F_a^s(g) = \left[\bigcup_{i=1}^{\infty} W^s(\bar{p}_i)\right] \cap f^{-1}(a)$,

$F_a^u(f) = F_a^u$, $F_a^s(f) = F_a^s$, where $\bar{p}_i$, $i=1,2,\ldots$ (p_i, $i=1,2,\ldots$) are
the saddles of ∇g (∇f). Let a_j , $b_j \in \mathbb{R}$ be regular levels of f,
such that $a_1 < b_1$, $a_{j+1} < a_j$, $b_j > b_{j+1}$, $j=1,2,\ldots$.

Set $F_{a_j}^u(g) = F_{-j}^u(g)$, $F_{a_j}^s(g) = F_{-j}^s(g)$, $F_{b_j}^u(g) = F_j^u(g)$,

$F_{b_j}^s(g) = F_j^w(g)$, $M_{-j}=f^{-1}([\ a_{j+1}, a_j])$, $M_j=f^{-1}([\ b_j,b_{j+1}])$, $j=1,2,\ldots$.

The condition A mentioned in the Introduction can be
stated as folows.

CONDITION A.

We say that $\nabla f \in GM^r_{pr}(M)$ satisfies the condition A if ∇f has only a finite number of saddle connections and

a) there are regular levels $a < b$, $a,b, \in \mathbb{R}$, of f such that the saddle connections of ∇f are contained in $f^{-1}([a,b])$, and $F^u_c \cap F^s_c = \emptyset$ for all regular levesl $c \in \mathbb{R}$ of f such that $c \notin f^{-1}((a,b))$; or b) ∇f does not satisfy a), and there are regular levels $a_j, b_j \in \mathbb{R}$ of f, as above, such that $[\bigcup_{j=1}^{\infty} \partial(F^u_{-j} \cap F^s_{-j}) \cup [\bigcup_{j=1}^{\infty} \partial(F^u_j \cap F^s_j)]$ is contained in the union of a finite number of orbits, and each element of $\partial(F^u_{-j} \cap F^1_{-j})$ or $\partial(F^u_j \cap F^s_j)$ is accumulated from one side by unstable separatrices, and from the other by stable separatrices of ∇f.

Now we start the proof of Theorem C.

Firstly, we will prove that if ∇f does not satisfy Condition A, then ∇f has infinite non-denumerable modulus of stability.

From The Theorem B, if ∇f has infinitely many saddle connections, then ∇f has infinite non-denumerable modulus of stability. Suppose that ∇f does not satisfy the condition A and has finitely many saddles connection. Then $\left[\bigcup_{j=1}^{\infty} \partial(F^u_{-j} \cap F^s_{-j})\right] \cup \cup \left[\bigcup_{j=1}^{\infty} \partial(F^u_j \cap F^s_j)\right]$ contains infinitely many regular orbits, and by an arbitrarily small perturbation of ∇f, we get $\nabla g \in GM^r_{pr}(M)$ with infinitely many saddle connections. This implies again, by Theorem B, that ∇f has infinite non-denumerable modulus of stability

Now we are going to prove that, if ∇f satisfies condition A-a), then ∇f has zero modulus of stability.

This will be done in three steps.

FIRST STEP: The Condition A-a) is an open property

Let $a_j, b_j \in \mathbb{R}$ be regular levels of f, $a_1=a.$ $b_1=b$, $a_{j+1} < a_j$, $b_j < b_{j+1}$; $a_j \to -\infty$, $b_j \to +\infty$, where $a, b \in \mathbb{R}$ are given by A-a), and $j=1,2,\ldots$.

Let $U \subset f^{-1}(b)$ be a neighbourhood of $F_b^s = F_1^s$. We will construct a neighbourhood of ∇f in $GM_{pr}^r(M)$, such that $F_1^s(g) \subset U$ and b_j, $j=1,2,\ldots$, are regular levels of g, whenever ∇g is in that neighbourhood.

Let $\vartheta_{\nabla g}^j(V_i) = \left[\bigcup_{t\in\mathbb{R}} (\nabla g)_t(V_i)\right] \cap f^{-1}(b_j)$, $\vartheta^j(V_i) = \vartheta_{\nabla f}^j(V_i)$, where $V_i \subset f^{-1}(b_i)$ is any subset.

By the Long Flow-Box Theorem, if V_i is open, then $\vartheta^j(V_i)$ is open, $i,j = 1,2,\ldots$.

Let $U' \subset \bar{U}' \subset U$ be an open neighbourhood of F_1^s .

Then there is $\varepsilon_1 > 0$ o such that:

P1 - if $\nabla g \in GM_{pr}^r(M)$ and $\|\nabla g - \nabla f\|_r < \varepsilon_1$ on M_1, then

i) $W^s(\bar{p}_i) \cap f^{-1}(b_1) \in U'$ for all saddles $\bar{p}_i \in M_1$ of ∇g ;

ii) $F_2^s \subset \vartheta_{\nabla g}^2(U') \subset \vartheta_{\nabla g}^2(\bar{U}') \subset \vartheta^2(U)$;

iii) b_1, b_2 are regular levels of g.

The above property follows from the monotonicity of f along the ∇f-orbits, and from the fact that f is proper.

With the same argument, taking $\Theta^2_{\nabla g}(U')$ instead of U, and $U'_2, F^s_2 \subset U'_2 \subset \bar{U}'_2 \subset \Theta^2_{\nabla g}(U')$, as an open neighbourhood of F^s_2 in $f^{-1}(b_2)$, we obtain $\varepsilon_2 > 0$ such that:

P2 - if $\nabla g \in GM^r_{pr}(M)$ and $\|\nabla g - \nabla f\|_r < \varepsilon_2$ on M_2, then

i) $W^s(\bar{p}_i) \cap f^{-1}(b_2) \in U'_2$ for all saddles $\bar{p}_i \in M_2$ of ∇g;

ii) $F^s_3 \subset \Theta^3_{\nabla g}(U'_2) \subset \Theta^3_{\nabla g}(\bar{U}'_g) \subset \Theta^3(\Theta^2_{\nabla g}(U'))$;

iii) b_2, b_3 are regular levels of g.

Continuing this process, we get by induction the desired neighbourhood.

Given an open neighbourhood V of F^u_b in $f^{-1}(b)$, we get, by similar arguments, a neighbourhood of ∇f in $GM^r_{pr}(M)$ such that $F^u_1(g) \subset V$ and a_j, b_1, j=1,2,..., are regular levels of g, whenever ∇g is in that neighbourhood.

Taking $U \cap V = \emptyset$, it follows from the above facts that A-a) is an open property, and the first step is finished.

<u>SECOND STEP</u>.

There is a neighbourhood of ∇f in $GM^r_{pr}(M)$ such that, for each ∇g in this neighbourhood there exists a homeomorphism $h: F^u_a(g) \cup F^s_a(g) \to F^u_a \cup F^s_a$ with $h(F^u_a(g)) = F^u_a$, $h(F^s_z(g)) = F^s_a$, and h is in a given neighbourhood of the identity.

To prove this we will use a Structural Stability Theorem of ([1]), which we will state below.

Following ([1]), let M be a two-dimensional open manifold and $X \in \mathfrak{X}^r(M)$.

A compact X-invariant subset $K \subset M$ is minimal if there is no compact X-invariant proper subset of K. If K is a singularity or a closed orbit, we say that K is a trivial minimal subset.

If $x \in M$ is such that $\omega_X(x)$ ($\alpha_X(x)$) is non compact, then we say that $\mathcal{O}(x)$ is an oscillating orbit.

The first positive (negative) prolongational limit set of $x \in M$ is the set

$$J^{\pm}(x) = \{ y \in M \mid \exists\ x_n \to x,\ t_n \to \pm\infty : X_{t_n}(x) \to y \} .$$

Let $x, y \in M$ be such that $\omega_X(x) = \emptyset$ and $\alpha_X(x) = \emptyset$. We say that $\mathcal{O}_+(x)$ and $\mathcal{O}_-(y)$ constitute a saddle at infinite if $y \in J^+(x)$ (i.é. $x \in J^-(x)$). In this case, $\mathcal{O}(x)$ and $\mathcal{O}(y)$ are the stable and unstable manifold of that saddle. We indicate by $W^+(W^-)$ the union of the stable (unstable) manifolds of the saddles of X (fixed or at infinite).

With these notations we have the following

<u>THEOREM ([1]).</u>

If $x \in \mathfrak{X}^r(M)$ is such that

i) there is neither minimal non trivial subsets and nor oscillating orbits;

ii) the singularities and closed orbits of X are hyperbolic;

iii) $W^+ \cap W^- \subset \mathrm{Per}\ X = \{$ singularities and closed orbits of X$\}$;

then a) $\mathrm{Per}\,X = \Omega(X)$;

and b) X is structurally stable.

If $\nabla f \in GM^r_{pr}(M)$ then ∇f satisfies the hypothesis i) and ii) of this Theorem.

But f is proper and monotone along the ∇f-orbits. Thus, ∇f satisfies also the hypothesis iii) of that theorem, iff $\partial(F^u_c \cap F^s_c) = \emptyset$, for all regular levels $c \in \mathbb{R}$ of f. Since $\partial(F^u_c \cap F^s_c) = \emptyset$,, for all regular levels $c \in \mathbb{R}$ of f, is a necessary condition for structural stability, we have in particular that $\nabla f \in GM^r_{pr}(M)$, $r \geq 1$, is structurally stable, iff $\partial(F^u_c \cap F^s_c) = \emptyset$, for all regular levels $c \in \mathbb{R}$ of f.

Suppose that $\nabla f \in GM^r_{pr}(M)$ satisfies A-a). Looking the connected component of $f^{-1}(a)$ as sinks of ∇f restricted to $f^{-1}((-\infty,a])$, it is easy to prove from the above Theorem, that this restriction is structurally stable. Similarly, the restriction of ∇f to $f^{-1}([b,+\infty))$ is also structurally stable.

Then, for ∇g sufficiently near ∇f in $GM^r_{pr}(M)$, there are homeomorphisms $h_a : F^u_a(g) \to F^u_a$ and $h_b : F^s_b(g) \to F^s_b$. But, by the first step, we can also suppose that $F^u_b(g) \cap F^s_b(g) = \emptyset$, and then

$$[\Theta_{\nabla g}(F^s_b(g)) \cap f^{-1}(a)] \cap F^u_a(g) = \emptyset \ .$$

Moreover, we can suppose that

$$\#([\Theta_{\nabla g}(F_b^s(g)) \cap f^{-1}(a)] \cap f^{-1}(a)] - F_a^s(g)) = \#([\Theta_{\nabla f}(F_b^s) \cap f^{-1}(a)] - F_a^s)$$

is finite.

This allow the construction of h in the natural way, and the second step is finished.

THIRD STEP.

In this step we will use the steps above to prove that, if ∇f satisfies A-a), then ∇f has zero modulus of stability.

Let $\mathcal{N}(\nabla f)$ be a neighbourhood of ∇f in $GM^r_{pr}(M)$, contained in all neighbourhoods, considered in the first two steps, and ∇g_1, $\nabla g_2 \in \mathcal{N}(\nabla f)$.

Let $\bar{p}_{ri}$, $i=1,\dots,\ell$ be the saddles of ∇g_r such that $\bar{p}_{ri} \in M_o = f^{-1}([a,b])$, $r=1,2$. We can suppose that $\bar{p}_{1i}$ and $\bar{p}_{2i}$ are arbitrarily near, for $i=1,\dots \ell$.

Suppose that $(\bar{p}_{1i_1}, \bar{p}_{1i_2}, \gamma_{i_1 i_2})$ is a saddle connection of ∇g_1 iff $(\bar{p}_{2i_1}, \bar{p}_{2i_2}, \tilde{\gamma}_{i_1 i_2})$ is a saddle connection of ∇g_2. Then ∇g_1 and ∇g_2 are topologically equivalent.

In fact, considering the connected component of $F^{-1}(a)$ as sources for ∇g_1, ∇g_2, and the ones of $f^{-1}(b)$ as sinks of ∇g_1, ∇g_2, and using the tubular family theory ([4]), it is possible to prove that $\nabla g_1|M_o$ and $\nabla g_2|M_o$ are topologically equivalent, and the topological equivalence is an extension of h to M_o. Here the first two steps are used. (For more details see also ([2])).

This topological equivalence can be extended to the manifolds M_{-j}, M_j, j=1,2,..., as a topological equivalence between ∇g_1 and ∇g_2, by the same argument. But the behavior of $\nabla g \in GM^r_{pr}(M)$ on M_o, characterize its equivalence class. Then, from ([2]) it follows that $\mathcal{N}(\nabla f)$ intersects finitely many topological equivalence classes, and then ∇f has zero modulus of stability. This ends the third step.

By the second step above, the part 3) of Theorem C is already proved.

Finally we outline the proof that if ∇f satisfies A-b), then ∇f has zero modulus of stability.

We remark that if ∇f satisfies A-b), then there is a neighbourhood $\mathcal{N}(\nabla f)$, such that ∇g satisfies A-a) or A-b) whenever $\nabla g \in \mathcal{N}(\nabla f)$.

Using the above arguments one can prove the following facts:

- The number of topological equivalence classes of $\nabla g \in \mathcal{N}(\nabla f)$ satisfying A-b) is finite

- The collection of the topological equivalences classes of $\nabla g \in \mathcal{N}(\nabla f)$ satisfying A-a) is denumerable.

From these facts it follows easily that ∇f has zero modulus of stability, and Theorem C is proved.

5. CONSTRUCTION OF A C^∞ PROPER MORSE FUNCTION WHOSE GRADIENT SATISFIES THE DSCP. (REMARK 1. 4.)

We will make this construction in two steps. In the first step we will construct a C^∞ - proper Morse function $G: \mathbb{R}^m \to \mathbb{R}$, such that ∇G satisfies the DSCP. In the second step we will use results of ([6]) and ∇G, to get $\nabla f \in GM^\infty_{pr}(M)$ satisfying the DSCP, on any open manifold M.

FIRST STEP.

Let $a \in (0, \frac{3\pi}{16})$ be such that $\mathrm{sen}\,(x+\frac{\pi}{8}) - x < 0$ for all $x > 2a$. We consider the bump functions φ and Ψ, given by the following pictures:

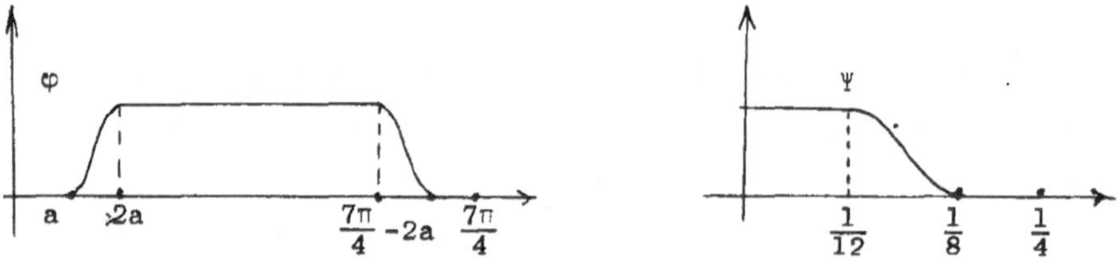

Fig. 1

Now we define the functions $\tilde{G}_o : [0, \frac{7\pi}{4}] \times \mathbb{R}^{m-1} \to \mathbb{R}$ and $G_o : [0,1] \times \mathbb{R}^{m-1} \to \mathbb{R}$ by

$$\tilde{G}_o(x,y) = \Psi(\|y\|)\,[\varphi(x).(\mathrm{sen}(x+\tfrac{\pi}{8}) + \|y\|^2) + (1-\varphi(x)).x] + (1-\Psi(\|y\|))\,x$$

and $$G_o(x,y) = \tilde{G}_o\left(\frac{7\pi}{4}\cdot x,\ \frac{1}{4}\|y\|\right).$$

It is a simple exercise in Differential Calculus (left to the reader) to prove that $\tilde{G}_o$ (and thus G_o) is a C^∞ -proper Morse function, with exactly one source and one saddle as its critical points.

The phase diagram of ∇G_o is given by Fig.2.

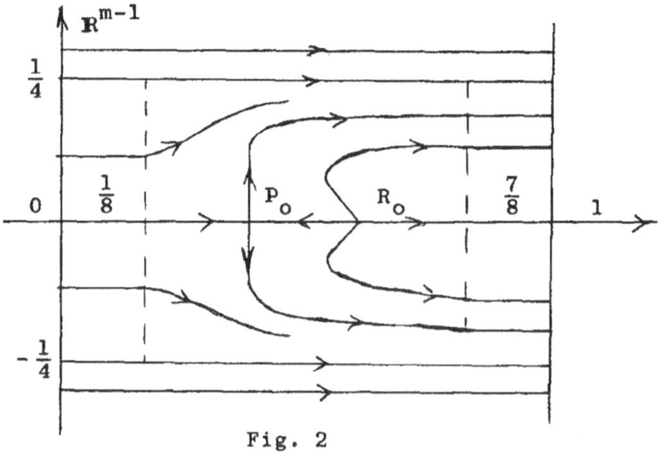

Fig. 2

Using the function G_o and proceeding by induction, we are going to build the function G, as mentioned above.

Let $\mathring{D}^{m-1}(1) = \{y \in \mathbb{R}^{m-1} : \|y\| < 1\}$ and $\{q_o = 0, q_1, q_2, \dots\} \subset \mathring{D}^{m-1}(1)$ be a dense set.

Suppose that we have built a C^∞- proper Morse function $G_{n-1} : [0,n] \times \mathbb{R}^{n-1} \longrightarrow \mathbb{R}$, such that

1) The singularities of ∇G_{n-1} on $[i,i+1] \times \mathbb{R}^{m-1}$ are the source $R_i = (i+\frac{11}{14}, q_i)$ and the saddle $P_i = (i+\frac{3}{14}, q_i)$, i=0,..., n-1;

2) $G_{n-1}(x,y) = x$ if $\|y\| \geq 1$ or $|x-i| \leq \frac{1}{8}$, for $x \in [0,n]$ and i=0,...,n;

3) $W^u(R_i) \cap (\{n\} \times \mathring{D}^{m-1}(1)) = \{n\} \times \mathring{D}^{m-1}_{\bar{q}_i}(\epsilon_i)$ $i=1,\ldots,n-1$,

where $D^{m-1}_{\bar{q}_i}(\epsilon_i) \cap D^{m-1}_{\bar{q}_j}(\epsilon_j) = \emptyset$, whenever $i \neq j$;

4) $W^s(P_i) \cap (\{0\} \times \mathring{D}^{m-1}(1)) = q_i$, for $i=0,\ldots,n-1$.

We are going to extend G_{n-1} to a C^∞-proper Morse function $G_n: [0,n+1] \times \mathbb{R}^{m-1} \to \mathbb{R}$ with the above properties.

Let Y_n be the ∇G_{n-1}-orbit of $(0,q_n)$, and $(n,\bar{q}_n) = Y_n \cap (\{n\} \times \mathring{D}^{m-1}(1))$. It is clear that $\bar{q}_n \notin \bigcup_{i=o}^{n-1} D^{m-1}_{\bar{q}_i}(\epsilon_i)$. Then, there exist $\epsilon_n > 0$ such that $D^{m-1}_{\bar{q}_n}(\epsilon_n) \subset \mathring{D}^{m-1}$ and $D^{m-1}_{\bar{q}_n}(\epsilon_n) \cap (\bigcup_{i=1}^{n-1} D^{m-1}_{\bar{q}_i}(\epsilon_i)) = \emptyset$. We define $G_n: [0,n+1] \times \mathbb{R}^{m-1} \to \mathbb{R}$ by

$$\begin{cases} G_n(x,y) = G_{n-1}(x,y), & \text{if } x \leq n; \\ G_n(x,y) = G_o\left(x-n, \dfrac{y-\bar{q}_n}{\epsilon_n}\right) + n, & \text{if } (x,y) \in [n,n+1] \times D^{m-1}_{\bar{q}_n}(\epsilon_n); \\ \text{and } G_n(x,y) = x, & \text{otherwise} \end{cases}$$

From the induction hypotesis and the fact that G_o satisfies the properties 1), 2), 3) and 4) above, it follows that G_n satisfies these properties.

Let $G_{-n}: [-n,0] \times \mathbb{R}^{m-1} \to \mathbb{R}$, $n \geq 1$, be defined by $G_{-n}(x,y) = -G_n(-x,y)$ and $G: \mathbb{R} \times \mathbb{R}^{m-1} \to \mathbb{R}$ be defined by

$$G(x,y) = \begin{cases} G_n(x,y) & \text{if } x \in [0,n] , \\ G_{-n}(x,y) & \text{if } x \in [-n,0] \end{cases}$$

for $n \geq 1$.

It is easy to verify that G is well defined and satisfies the required properties. This finishes the first step.

SECOND STEP.

In this step we build a C^∞-proper Morse function $F:M \to \mathbb{R}$, $r \geq 3$, on any open manifold M, by carrying G to a C^∞-function F_o, defined on an open set of M, and extending F_o to f carefully in order to have ∇f satisfying the DSCP.

Let $\gamma:\mathbb{R} \to M$ be a C^∞ embedding with closed image $\Gamma \subset M$, and $g:\mathbb{R} \times \mathring{D}^{m-1}(5) \subset \mathbb{R}^m \to U_5 \subset M$ be a C^∞ tubular neighbourhood of Γ, where $\mathring{D}^{m-1}(5)$ is the interior of the disc $D^{m-1}(5) = \{y \in \mathbb{R}^{m-1} : \|y\| \leq 5\}$, and $m = \dim M$.

Let $U_i = g(\mathbb{R} \times \mathring{D}^{m-1}(i))$, $0 < i \leq 5$, $\tilde{F}_1 : M - \bar{U}_2 \to \mathbb{R}$ be a C^∞-proper function, and $\xi:\mathbb{R} \to \mathbb{R}$ be a C^∞ function such that:

$$\begin{cases} \xi(t) = 1, & \text{if } |t| \leq 3, \\ \xi(t) = 0, & \text{if } |t| \geq 4, \\ \xi(t) \in (0,1), & \text{otherwise}. \end{cases}$$

We define $F_2 : \mathbb{R} \times \overset{\circ}{D}{}^{m-1}(5) \to \mathbb{R}$ by

$F_2(x,y) = \xi(\|y\|) \cdot G(x,y) + (1-\xi(\|y\|)).(\tilde{F}_1 \circ g^{-1})(x,y)$, and

$F_o : U_5 \to \mathbb{R}$ by $F_o = F_2 \circ g$. Let $F_1' : M \to \mathbb{R}$ be given by $F_1' | U_5 = F_o$ and $F_1' | M - U_5 = \tilde{F}_1$.

Now we are going to modify F_1' on $W=M-\bar{U}_2$ in order to get a C^2-aproximation F_1, such that F_1 is a C^∞ - proper Morse function, and the invariant manifolds of the singularities of ∇F_1 are in general position. This will be done by using Theorems A and B of ([6], pg. 199-200).

There exist $a_n \in \mathbb{R}$, $n \in \mathbb{Z}$, $a_n < a_{n+1}$, such that a_n is a regular value of F_1' and $F_1'^{-1}([a_n, a_{n+1}] = W_n$ is a compact submanifold of W, and ∇F_1 is transversal $\partial W_n = F_1'^{-1}(a_n) \cup F_1'^{-1}(a_{n+1})$, for each $n \in \mathbb{Z}$, because F_1' is proper.

It is clear that the set of C^∞ proper Morse functions g, defined on M such that ∇g is transversal to ∂W_n and $\nabla g|_{W_n}$ is Morse-Smale, for all $n \in \mathbb{Z}$, is open in $C^\infty_{pr}(W, \mathbb{R})$, with the C^r Whitney topology, $r \geq 2$. From Theorems A and B of ([6]) , it follows that this set is dense in $C^\infty_{pr}(W, \mathbb{R})$ with the C^2 Whitney topology.

Thus, given a continuous function $\epsilon : M \to \mathbb{R}$ such that $\epsilon|_{\bar{U}_2} \equiv 0$ and $\epsilon|_W > 0$, there exists $F_1 \in C^\infty_{pr}(W, \mathbb{R})$, ϵ-C^2 near $F_1'|_W$, with the following properties: a) F_1 is a C^∞ proper Morse function b) the singularities of ∇F_1 have invariant manifolds in general position.

Let $F:M \to \mathbb{R}$ be defined by $F_{|\bar{U}_2} = F_o$ and $F_{|W} = F_1$. We have by the constructions, that F is a C^2 proper Morse Function, which is C^∞ on $M-\partial U_2 = M - \partial M$, and such that the invariant manifolds of the singularities of $(\nabla F)_{|M-U_1}$ are general position

We define $f:M \to \mathbb{R}$ by

$$\begin{cases} f_{|M-\bar{U}_4} = F \\ f \circ g(x,y) = \xi(\|y\|-\delta)\cdot F_1' \circ g(x,y) + (1-\xi(\|y\|-\delta))\cdot(F \circ g)(x,y), \\ \quad \text{for all } (x,y) \in \mathbb{R} \times \mathring{D}^{m-1}(5), \text{ where } o < \delta < 1 \end{cases}$$

We can suppose that $f_{|M-\bar{U}_2}$ is arbitrarily C^2 near $F_{|M-U_2}$, in the C^2 Whitney topology, by taking the function ξ suficiently small, because $f \circ g(x,y) - F \circ g(x,y) = \xi(\|y\|-\delta)(F_1' \circ g(x,y) - F_1 \circ g(x,y))$, if $\|y\| \in [3-\delta, 4-\delta]$, and $f=F$ otherwise.

Thus f is a C^∞ - proper Morse function, $f_{|M-\bar{U}_1}$ has the properties a) and b) above, and ∇f satisfies the DSCP, as required.

REMARK . In the second step we have proved that the Kupka-Smale Theorem works in the world of gradients of proper functions on any open manifold. This fact is essential in the proof or our Theorem A in this context.

REFERENCES

[1] J.Kotus, M.Krych, Z.Nitecki - Global Structural Stability of flows on open surfaces. To appear.

[2] P.Mendes - On moduli of stability of two-dimensional vector fields. To appear in Boletim da SBM.

[3] J.Palis - Moduli of stability and bifurcation theory. Proceedings of the International Simposium of Mathematicians Helsinki - 1978.

[4] ______________ - On Morse-Smale dynamical systems. Topology 4 (1969) 385-404.

[5] M.Peixoto, C.Pugh - Structurally stable systems on open manifolds are never dense. Annals of Math. 87 (1969) 423-430.

[6] S.Smale - On gradient dynamical systems. Annals of Math. 74 (1961) 199-206.

[7] J.Sotomayor - Generic one-parameter families of vector fields on two manifolds. Publ. IHES 43 (1973).

PEDRO MENDES
Departamento de Matemática
ICEX - UFMG
30.000 - Belo Horizonte - MG - Brazil

DYNAMICS OF CERTAIN SKEW PRODUCTS

by

Sheldon E. Newhouse and Lai-sang Young

1. Introduction and statement of results

We present here a structure theory for the dynamics of certain homeomorphisms of compact metric spaces. These mappings arise as non-wandering set restrictions of certain diffeomorphisms including some of those studied by Abraham and Smale [1] , and Hirsch, Pugh, and Shub [10] . The non-wandering sets of these diffeomorphisms are not hyperbolic, but they have enough "partial hyperbolicity" and rich topological structure so that one can give a nearly complete analysis of their dynamics. Our results provide the first detailed orbit structure theory for open sets of non-Axiom A diffeomorphisms.

Precise definitions will be given later, but for now let us say that two diffeomorphisms are almost conjugate if after neglecting small invariant sets they become topologically conjugate to each other. Elements in certain open sets of non-Ω-stable diffeomorphisms (e.g. [1] , [14]) are shown to be equivalent and tractable in this sense as are some other systems which are difficult to describe completely from a purely topological point of view. There are various ways of defining the small sets to be neglected. In this paper we shall do this for mappings with finitely many ergodic measures of maximal entropy. For such a map we shall neglect sets of measure zero for any such ergodic measure (equivalently, we neglect any set of measure zero for some measure of maximal entropy). For a very strong notion of almost conjugacy, see [2] .

The present results were announced in [13]. The hypotheses here, however, are slightly different. With the present hypotheses we are able to greatly simplify the proofs which were sketched in [13]. In particular, the use of branched manifolds has been eliminated. Moreover, since in most examples the present hypotheses are satisfied, it seems desirable to work with them. A significant part of this simplification is due to C. Robinson who showed us how to get a globally defined semi-conjugacy in place of one which we had only obtained between certain dense sets of our systems. It should be remarked, however, that the present assumptions may not cover all the cases considered in [13] and vice-versa.

We now state our results.

Let $\phi : \Lambda \longrightarrow \Lambda$ be a homeomorphism of the compact metric space Λ. Let $\tilde{L} : \mathbb{R}^n \longrightarrow \mathbb{R}^n$ be a linear automorphism such that $\tilde{L}(\mathbb{Z}^n) \subset \mathbb{Z}^n$, $\det \tilde{L} = \pm 1$, $\tilde{L}$ has one real eigenvalue λ of multiplicity one with $|\lambda| < 1$, and the other eigenvalues of $\tilde{L}$ have norm greater than one. Let $T^n = \mathbb{R}^n/\mathbb{Z}^n$ be the n-torus and let $L : T^n \longrightarrow T^n$ be the automorphism induced by $\tilde{L}$. Let $\mathcal{D}^1(T^n)$ be the space of C^1 diffeomorphisms of T^n with the uniform C^1 topology. Let $F : \Lambda \times T^n \longrightarrow \Lambda \times T^n$ be a homeomorphism satisfying the following conditions.

1. $F(x,y) = (\phi(x), f_x(y))$ where $x \longrightarrow f_x$ is a continuous map from Λ into $\mathcal{D}^1(T^n)$

2. F is homotopic to $\phi \times L$ as a bundle map. That is, there is a continuous map $G : \Lambda \times T^n \times [0,1] \longrightarrow T^n$ such that $G(x,y,0) = f_x(y)$ and $G(x,y,1) = Ly$ for $(x,y) \in \Lambda \times T^n$.

3. There is a one-dimensional lamination $\mathcal{F}$ of $\Lambda \times T^n$ which is F-invariant and normally expanded. This means that each leaf $\mathcal{F}_{(x,y)}$ through (x,y) is a smoothly immersed line in $\{x\} \times T^n$ and $F\mathcal{F}_{(x,y)} = \mathcal{F}_{F(x,y)}$. Moreover, there is a splitting $E^u_{(x,y)} \oplus E^s_{(x,y)}$ of the tangent space to $\{x\} \times T^n$ at (x,y) which varies continuously with (x,y) such that

a. $T_y f_x(E^u_{(x,y)}) = E^u_{F(x,y)}$ and

$$T_y f_x(E^s_{(x,y)}) = E^s_{F(x,y)}$$

). $E^s_{(x,y)} = T_{(x,y)} F_{(x,y)}$

:. There is a Riemann metric on $\{x\} \times T^n$ for $x \in \Lambda$ with induced
norm $|\cdot|$ such that

$$\inf_{\substack{|v|=1 \\ v\in E^u_{(x,y)} \\ (x,y)\in\Lambda\times T^n}} |T_y f_x(v)| > \max(1, \sup_{\substack{|v|=1 \\ v\in E^s_{(x,y)} \\ (x,y)\in\Lambda\times T^n}} |T_y f_x(v)|)$$

For a homeomorphism ϕ , let $h(\phi)$ denote its topological entropy, and let $M(\phi)$ denote the set of ϕ-invariant Borel probability measures. If $\mu \in M(\phi)$, let $h_\mu(\phi)$ denote its metric entropy. It is well-known that $h(\phi) = \sup_{\mu\in M(\phi)} h_\mu(\phi)$. If $\mu \in M(\phi)$ satisfies $h_\mu(\phi) = h(\phi)$, we call μ a maximal measure. If ϕ has a unique maximal measure, then ϕ is called intrinsically ergodic.

Theorem 1 : Suppose assumptions 1 through 3 above are satisfied. Then the following results hold

1. $h(F) = h(\phi) + h(L)$
2. If ϕ has maximal measures, then so does F .
3. If $\pi_1 : \Lambda \times T^n \longrightarrow \Lambda$ is the natural projection, then any maximal measure for F projects by π_1 to a maximal measure for ϕ .

Let ν be normalized Haar measure on T^n , and for $(x,y) \in \Lambda \times T^n$, let

$$\chi(x,y) = \liminf_{m\to\infty} \frac{1}{m} \log |T_{(x,y)} F^{-m}| E^s_{(x,y)}| \quad .$$

where we write $T_{(x,y)}F^{-m}$ for $T_y \pi_2 \circ F^{-m}$ with $\pi_2 : \Lambda \times T^n \longrightarrow T^n$ the projecti

That is, $\chi(x,y)$ is the lower Lyapunov exponent of F^{-1} at (x,y) in the $\mathcal{F}$-direction.

<u>Theorem 2</u> : In addition to the assumptions of theorem 1, suppose there is a ϕ-invariant set B such that for all $(x,y) \in B \times T^n$, $\chi(x,y) > 0$. Then, $F \mid B \times T^n$ is topologically conjugate to $\phi \times L \mid B \times T^n$. Moreover, if ϕ is intrinsically ergodic with maximal measure μ_ϕ and $\mu_\phi(B) = 1$, then F is intrinsically ergodic, say with maximal measure μ_F. Further (F, μ_F) and $(\phi \times L, \mu_\phi \times \nu)$ are measure theoretically isomorphic.

<u>Remarks</u> : 1. Some well-known examples of intrinsically ergodic systems are topologically transitive subshifts of finite type and diffeomorphisms of hyperbolic basic sets. Other examples involving commutative skew-products are given in [12]. The arguments there are mainly measure theoretic and make strong use of the commutativity of the fiber maps. Theorem 2 gives new examples. These are non-commutative skew-products in the sense that the fiber maps f_x do not commute. Various topological conditions are used to overcome this lack of commutativity.

2. The maximal measure μ of an intrinsically ergodic measure is indeed an ergodic measure and gives a natural way of describing "most" of the orbit structure of the system. To wit, μ-null sets can be neglected without destroying the essence of the dynamics. More precisely, let $f : X \longrightarrow X$ and $g : Y \longrightarrow Y$ be intrinsically ergodic systems with maximal measures μ and $\bar{\mu}$, respectively. Say that f and g are almost conjugate if there are sets $A \subset X$ and $B \subset Y$ such that $f(A) = A$, $g(B) = B$, $\mu(A) = 0$, $\bar{\mu}(B) = 0$ and $f|X-A$ is topologically conjugate to $g|Y-B$. In our examples in section 4 below, we show that elements in certain open sets of diffeomorphisms are almost conjugate to Axiom A diffeomorphisms. This brings up a natural question : For any compact manifold M, let $B(M)$ denote the set of C^r diffeomorphisms f such that

(i) f has only finitely many ergodic measures of maximal entropy.

(ii) On each support of an ergodic measure of maximal entropy, f is almost conjugate to the restriction of an Axiom A diffeomorphism to a topologically transitive basic set.

Is $\mathcal{B}(M)$ residual in $\mathcal{D}^r(M)$?

3. The additional asumption of theorem 2 states roughly that F^{-m} expands lengths in leaves of $\mathcal{F}$ for large m .

2. Proof of theorem 1 :

The following lemma is due to Clark Robinson. It is a parametrized version of a theorem due to Franks [8] .

Lemma 1 : There is a continuous surjection $\psi : \Lambda \times T^n \longrightarrow \Lambda \times T^n$ of the form $\psi(x,y) = (x,\psi_x(y))$ such that $(\phi \times L)\psi = \psi F$ and each $\psi_x : T^n \longrightarrow T^n$ is homotopic to the identity.

Proof : Consider the natural projection $\pi : \mathbb{R}^n \longrightarrow T^n$. Let $\tilde{L} : \mathbb{R}^n \longrightarrow \mathbb{R}^n$ be the lift of $L : T^n \longrightarrow T^n$ such that $\tilde{L}(y_1+y_2) = \tilde{L}(y_1)+\tilde{L}(y_2)$ for $y_1,y_2 \in \mathbb{R}^n$. We first observe that there is a lift $\tilde{F} : \Lambda \times \mathbb{R}^n \longrightarrow \Lambda \times \mathbb{R}^n$ of F such that $\tilde{F}_x(y+z) = \tilde{F}_x(y)+\tilde{L}(z)$ for $y \in \mathbb{R}^n$ and $z \in \mathbb{Z}^n$. Indeed, we consider the homotopy $G : \Lambda \times T^n \times [0,1] \longrightarrow T^n$ between $(x,y) \longrightarrow f_x(y)$ and $(x,y) \longrightarrow Ly$. Let $\tilde{G} : \Lambda \times \mathbb{R}^n \times [0,1] \longrightarrow \mathbb{R}^n$ be such that

$$\begin{array}{ccc} \Lambda \times \mathbb{R}^n \times [0,1] & \xrightarrow{\tilde{G}} & \mathbb{R}^n \\ {\scriptstyle id\times\pi\times id}\downarrow & & \downarrow{\scriptstyle \pi} \\ \Lambda \times T^n \times [0,1] & \xrightarrow{G} & T^n \end{array}$$

commutes. Thus, $\tilde{G}$ is a lift of G . Let $\tilde{F}_x(y) = \tilde{G}(x,y,0)$, and let $\tilde{F}_{x,t}(y) = \tilde{G}(x,y,t)$. Then, for $z \in \mathbb{Z}^n$, $\tilde{F}_{x,t}(y+z) - \tilde{F}_{x,t}(y)$ is continuous in t and belongs to $\mathbb{Z}^n$ for each t . Since $\tilde{F}_{x,1}(y+z) - \tilde{F}_{x,1}(y) =$

$\widetilde{L}(y+z)-\widetilde{L}(y) = \widetilde{L}(z)$, we have $\widetilde{F}_x(y+z)-\widetilde{F}_x(y) = \widetilde{L}(z)$. Now consider the maps $L = \phi \times \widetilde{L}$ and $\widetilde{F}$ from $\Lambda \times \mathbb{R}^n$ to itself. To find $\psi : \Lambda \times T^n \longrightarrow \Lambda \times T^n$ as required, we shall find $H : \Lambda \times \mathbb{R}^n \longrightarrow \Lambda \times \mathbb{R}^n$ of the form $H(x,y) = (x,H_x(y))$ so that $H_x(y+z) = H_x(y)+z$ for $y \in \mathbb{R}^n$ and $z \in \mathbb{Z}^n$ and $LH = H\widetilde{F}$. Then ψ will be the map induced by H .

We write $H_x(y) = y+K_x(y)$ and $\widetilde{F}_x(y) = \widetilde{L}y + \overline{F}_x(y)$.

Then the equation

$$LH = H\widetilde{F}$$

becomes

$$(\phi \times \widetilde{L})(x,H_x(y)) = H(\phi x,\widetilde{F}_x(y))$$

or

$$(\phi x,\widetilde{L}H_x(y)) = (\phi x,H_{\phi x}(\widetilde{F}_x(y)))$$

or

$$\widetilde{L}(y+K_x(y)) = \widetilde{F}_x(y) + K_{\phi x}(\widetilde{F}_x(y))$$

$$= \widetilde{L}y + \overline{F}_x(y) + K_{\phi x}(\widetilde{L}y+\overline{F}_x(y))$$

or

$$\widetilde{L}K_x(y) = \overline{F}_x(y) + K_{\phi x}(\widetilde{L}y+\overline{F}_x(y))$$

or

(*) $$K_x(y)-\widetilde{L}^{-1}K_{\phi x}(\widetilde{L}y+\overline{F}_x(y)) = \widetilde{L}^{-1}\,\overline{F}_x(y) \quad .$$

Also, rewriting the equation

$$H_x(y+z) = H_x(y)+z \quad , \quad z \in \mathbb{Z}^n \quad ,$$

we have

$$y + z + K_x(y+z) = y + K_x(y) + z$$

or

$$K_x(y+z) = K_x(y) \qquad \text{for all } z \in \mathbb{Z}^n \quad .$$

Consider the Banach space

$$\mathcal{K} = \{\hat{K} : \mathbb{R}^n \longrightarrow \mathbb{R}^n \,|\, \hat{K}(y+z) = \hat{K}(y) \quad \text{for} \quad y \in \mathbb{R}^n, z \in \mathbb{Z}^n\}$$

with sup norm and $\mathcal{C} = C^0(\Lambda,\mathcal{K})$. The equation (*) is the same as

$$(I-\Phi)K = \gamma$$

where $I : \mathcal{C} \longrightarrow \mathcal{C}$ is the identity map $\Phi : \mathcal{C} \longrightarrow \mathcal{C}$ is the linear map

$$\Phi(K)_x(y) = \tilde{L}^{-1} K_{\phi x}(\tilde{L}y + \overline{F}_x(y))$$

and,

$$\gamma : \mathcal{C} \longrightarrow \mathcal{C}$$

is the constant map

$$\gamma(K)_x(y) = \tilde{L}^{-1} \overline{F}_x(y) \qquad .$$

That $K \in \mathcal{C}$ implies $\Phi(K) \in \mathcal{C}$ and γ maps $\mathcal{C}$ to $\mathcal{C}$ follow from the facts that if $z \in \mathbb{Z}^n$, then

$$\tilde{F}_x(y+z) = \tilde{F}_x(y) \qquad \text{mod} \quad \mathbb{Z}^n$$

and

$$\overline{F}_x(y+z) = \overline{F}_x(y) \qquad .$$

Next we claim $I-\Phi$ has a right inverse J . From this it follows that $(I-\Phi)J\gamma = \gamma$, so we may take $K = J\gamma$. To prove the existence of J , consider $\mathbb{R}^n = \mathbb{R}^s + \mathbb{R}^u$ where $\tilde{L}(\mathbb{R}^s) = \mathbb{R}^s$, $\tilde{L}(\mathbb{R}^u) = \mathbb{R}^u$, $|\tilde{L}|\,\mathbb{R}^s| < 1$ and $|\tilde{L}^{-1}|\mathbb{R}^u| < 1$. Then, write $K = (K^s,K^u)$ where $K^s_x(y) \in \mathbb{R}^s$ and $K^u_x(y) \in \mathbb{R}^u$ for $x \in \Lambda$, $y \in \mathbb{R}^n$. Let $L_s = \tilde{L}\,|\,\mathbb{R}^s$, $L_u = \tilde{L}\,|\,\mathbb{R}^u$. Then,

$$\Phi(K^s,K^u)_x(y) = (L_s^{-1}K^s_{\phi x}\tilde{F}_x(y), L_u^{-1}K^u_{\phi x}(\tilde{F}_x(y)))$$

$$\equiv (\Phi_s(K^s), \Phi_u(K^u)) \quad .$$

Since $|\Phi_u| < 1$ and $|\Phi_s^{-1}| < 1$, we may set

$$J = (I-\Phi)^{-1} = (-\sum_{i=1}^{\infty} \Phi_s^{-i}, \sum_{i=0}^{\infty} \Phi_u^{i}) .$$

Note that $(x,y) \longmapsto H_x(y)-y$ is uniformly bounded on $\Lambda \times \mathbb{R}^n$.

Given that assumptions1,2, and 3 of theorem 1 hold, it follows from [10] that there is a strong unstable foliation W^u of $\mathcal{F}$. The leaves of W^u are C^1 injectively immersed copies of $\mathbb{R}^{n-1}$ and are contained in the fibers $\{x\} \times T^n$ for $x \in \Lambda$. Let λ_1 be a real number between the two quantities in 3.c. Then for $(x,y) \in \Lambda \times T^n$, $W^u_{(x,y)}$ is characterized as

$$W^u_{(x,y)} = \{(x,z) : \text{for some } K > 0, d(F^{-m}(x,z),F^{-m}(x,y)) \le K\lambda_1^{-m} \text{ for } m \ge 0\} \quad .$$

Let $\tilde{W}^u$, $\tilde{\mathcal{F}}$ be the lifts of W^u and F to $\Lambda \times \mathbb{R}^n$. Since $\tilde{W}^u|_{\{x\}\times\mathbb{R}^n}$ is a codimension one foliation of $\{x\} \times \mathbb{R}^n$ by planes, it follows that each leaf of $\tilde{W}^u \,|\{x\}\times T^n$ is properly embedded. If the foliation $\tilde{W}^u|\{x\} \times T^n$ were C^2 , this would follow from Sacksteder [15] . In the C^1 case, it follows from the argument of Franks [7;pp 81-85] . In the C^0 case, it also holds (private communication with S. Goodman, J. Plante, and D. Epstein). We have been told that the arguments needed for the C^0 case are expected to appear in a forthcoming book by D. Epstein. It could be noted parenthetically here that if our mapping $x \longmapsto f_x$ is a continuous map from Λ into $\mathcal{D}^2(T^n)$ instead of $\mathcal{D}^1(T^n)$, then the methods of Hirsch and Pugh [9] can be used to prove the foliation $\tilde{W}|\{x\} \times \mathbb{R}^n$ is C^1

Lemma 2 : Let W^u be the strong unstable foliation of F and let $W^{u'}$ be the unstable foliation of $\phi \times L$ on $\Lambda \times T^n$. Let F' be the stable foliation of $\phi \times L$. Let $\tilde{W}^u$, $\tilde{W}^{u'}$, $\tilde{F}$, and $\tilde{F}'$ be the lifts to $\Lambda \times \mathbb{R}^n$. Then, H_x maps $\tilde{W}^u_{(x,y)}$ homeomorphically onto $\tilde{W}^{u'}_{H(x,y)}$ and $H_x \tilde{F}_{(x,y)} = \tilde{F}'_{H(x,y)}$.

Proof : Let d denote the metric on $\Lambda \times \mathbb{R}^n$ induced by the Riemann metric on $\Lambda \times T^n$ described in assumption 3.c. Suppose $(x,y_1) \in \tilde{W}^u_{(x,y)}$ with $x \in \Lambda$, $y,y_1 \in \mathbb{R}^n$. Then, $d(\tilde{F}^{-m}(x,y),\tilde{F}^{-m}(x,y_1)) \to 0$ as $m \to \infty$, so $d(H\tilde{F}^{-m}(x,y),H\tilde{F}^{-m}(x,y_1)) = d(\tilde{L}^{-m}H_x(y),\tilde{L}^{-m}H_x(y_1)) \to 0$, so $H_x(y) \in \tilde{W}^{u'}(H_x(y_1))$. Thus, $H(\tilde{W}^u) \subset \tilde{W}^{u'}$. Also, if $y_1 \neq y$, then $d(\tilde{F}^m(x,y_1), \tilde{F}^m(x,y)) \to \infty$ as $m \to \infty$ in $\tilde{W}^u_{\tilde{F}^m(x,y_1)}$. Since each leaf of $\tilde{W}^u$ is properly embedded, we have that $d(\tilde{F}^m(x,y_1),\tilde{F}^m(x,y)) \to \infty$ in $\Lambda \times \mathbb{R}^n$. Since H is proper, it follows that $d(H\tilde{F}^m(x,y_1),H\tilde{F}^m(x,y)) = d(L^mH(x,y_1),L^mH(x,y)) \to \infty$, so $H(x,y_1) \neq H(x,y)$. Thus H maps $\tilde{W}^u(x,y)$ bijectively into $\tilde{W}^{u'}H(x,y)$. Since H is continuous and proper and $\tilde{W}^u(x,y)$ and $\tilde{W}^{u'}H(x,y)$ are properly embedded, we have that $H|\tilde{W}^u(x,y)$ is a homeomorphism onto $\tilde{W}'^u(H(x,y))$. We now prove that $H\tilde{F}_{(x,y)} = \tilde{F}'_{H(x,y)}$. Let us first prove that $H\tilde{F}_{(x,y)} \subset \tilde{F}'_{H(x,y)}$. Suppose $(x,y) \in F_{(x,y_1)}$ and $H(x,y_1) \notin \tilde{F}'_{H(x,y)}$. Then $d(L^mH(x,y),L^mH(x,y_1)) \to \infty$ as $m \to \infty$ and so will $d(\tilde{F}^m(x,y), \tilde{F}^m(x,y_1))$. Thus we need only to verify that bounded segments of $\tilde{F}$-leaves do not become arbitrarily long under iterates of $\tilde{F}$. For this it suffices in turn, to show that forward orbits of bounded segments of F leaves remain bounded.

Let C be a bounded curve lying in an F-leaf in $\{x\} \times T^n$. Let $D^u(x,y)$ be a neighborhood of (x,y) in $W^u(x,y)$. We call U a "special neighborhood" of C if $U = \bigcup_{(x,y)\in C} D^u(x,y)$ where $D^u(x,y) \cap D^u(x,y_1) = \emptyset$ for (x,y), $(x,y_1) \in C$, $y \neq y_1$. If the unstable discs of radius r about distinct points of C are disjoint, then we say that C has a "special neighborhood of radius r" , written $U(C,r) = \bigcup_{(x,y)\in C} D^u_r(x,y)$. Observe that :

1. If U is a special neighborhood of an $\mathcal{F}$-curve C , then $F^m U$ is a special neighborhood of $F^m C$,

2. every $\mathcal{F}$-curve C has a special neighborhood of radius r for some $r > 0$,

3. given $r > 0$, there is an $r_o = r_o(r) > 0$ such that if $U(C,r)$ is a special neighborhood of C of radius r , and $m > 0$, then $F^m(U(C,r)) \supset U(F^m C, r_o)$.

Now if the iterates $F^m C$ have unbounded length, then for some m and some $y_1 \neq y_2$ with (x,y_1) and (x,y_2) in C we have $F^m(x,y_1) \in D^u_{r_o}(F^m(x,y_2))$. This contradicts the fact that $F^m(U(C,r))$ is a special neighborhood of $F^m C$.

Thus, the iterates $F^m C$ have uniformly bounded length for $m \in \mathbb{Z}^+$. This completes the proof that $H\tilde{\mathcal{F}}_{(x,y)} \subset \tilde{\mathcal{F}}'_{H(x,y)}$.

To prove that $H\tilde{\mathcal{F}}_{(x,y)} = \tilde{\mathcal{F}}'_{H(x,y)}$ first note that each leaf of $\tilde{\mathcal{F}}$ is properly embedded (a non-properly embedded leaf F of $\tilde{\mathcal{F}}$ would meet some leaf of $\tilde{W}^u$ in at least two points forcing the forward orbit of some bounded interval in F to become unbounded). Suppose there is a point (x,y) such that $H\tilde{\mathcal{F}}_{(x,y)} \neq \tilde{\mathcal{F}}'_{H(x,y)}$. Set $V(x,y) = \cup\{\tilde{W}^u_{(x,y')} : (x,y') \in \tilde{\mathcal{F}}_{(x,y)}\}$. Then, $V(x,y)$ is open in $\{x\} \times \mathbb{R}^n$. We claim it is closed also, so $V(x,y) = \{x\} \times \mathbb{R}^n$. To prove this, let $(x,y_m) \in V(x,y)$ converge to $(x,\bar{y}) \in \{x\} \times \mathbb{R}^n$ as $m \to \infty$. We may assume $(x,y_m) \in \tilde{\mathcal{F}}_{(x,\bar{y})}$ for all m . Let $z_m \in \tilde{W}^u_{(x,y_m)} \cap \tilde{\mathcal{F}}_{(x,y)}$. If $(x,\bar{y}) \notin V(x,y)$, then $\mathrm{dist}(z_m,(x,y_m)) \to \infty$ as $m \to \infty$. Properness of $H|\{x\}\times\mathbb{R}^n$ then forces $\mathrm{dist}(H(z_m),H(x,y_m)) \to \infty$ as $m \to \infty$. But $H(z_m) \in \tilde{W}^{u'}_{H(x,y_m)} \cap H\tilde{\mathcal{F}}_{(x,y)}$ and $H(x,y_m) \in \tilde{W}^{u'}_{H(x,y_m)} \cap H\tilde{\mathcal{F}}_{(x,\bar{y})}$ Since the distance between any leaves of $\tilde{\mathcal{F}}'$ along a leaf of $\tilde{W}'$ is bounded, we get a contradiction.

Now, $H|\{x\}\times\mathbb{R}^n = H\,|\,V(x,y)$ is surjective, so there is a point $(x,y_1)\in V(x,y)$ such that $H(x,y_1)\in \tilde{\mathcal{F}}'_{H(x,y)} - H\tilde{F}_{(x,y)}$. This means $H\,\tilde{W}^u_{(x,y_1)} = \tilde{W}^{u'}_{H(x,y_1)}$ meets $\tilde{\mathcal{F}}'_{H(x,y)}$ in at least two points which is a contradiction. This proves lemma 2.

Now it is appropriate to recall some definitions and results about entropy. Let $f : X\circlearrowleft$ be a continuous mapping of a compact metric space into itself and let d denote the metric in X. If $K\subset X$ is a compact subset, $\varepsilon>0$ and $n\in\mathbb{Z}^+$, we say a set $E\subset K$ is (n,ε,K)-separated if for $x\neq y$ in E there is a $j\in[0,n)$ such that $d(f^jx,f^jy)>\varepsilon$. We let $r(n,\varepsilon,K)$ be the maximal cardinality of an (n,ε,K)-separated set and let

$$h(K,f) = \lim_{\varepsilon\to 0}\limsup_{n\to\infty}\frac{1}{n}\log r(n,\varepsilon,K) \quad .$$

Then $h(f) = h(X,f)$ is the topological entropy of f.

Suppose now that $g : Y\circlearrowleft$ is also a continuous mapping of a compact metric space into itself and that there is a continuous surjection $\pi : X \longrightarrow Y$ satisfying $g\pi = \pi f$. In general we have

$$h(g) \le h(f) \quad .$$

Bowen [4] proved that

$$h(f) \le h(g) + \sup_{y\in Y} h(\pi^{-1}y,f) \quad .$$

A sharper measure-theoretic version of this result was proved by Walters and Ledrappier [11]. It states that for $\mu\in M(g)$

$$\sup_{\substack{\bar\mu\in M(f)\\ \pi_*\bar\mu=\mu}} h_{\bar\mu}(f) = h_\mu(g) + \int_Y h(\pi^{-1}y,f)\,d\mu \quad .$$

<u>Lemma 3</u> : Let $\psi : \Lambda\times T^n \longrightarrow \Lambda\times T^n$ be the semi-conjugacy of lemma 1. Then, $h(\psi^{-1}(x,y),F) = 0$ for each $(x,y)\in\Lambda\times T^n$.

<u>Proof</u> : We first claim

(*) for each $(x,y) \in \Lambda \times T^n$, $\psi^{-1}(x,y)$ is contained in an interval $J(x,y)$ in a single leaf of F , and the length of each such interval is bounded by a constant $K > 0$ independent of (x,y).

Let us assume (*) for the moment. Fix $(x,y) \in \Lambda \times T^n$. Let $m > 0$ and $\varepsilon > 0$. We shall estimate $r(m,\varepsilon,\psi^{-1}(x,y))$. For $j \in [0,m)$, we have $F^j\, J(x,y) \subset J(\phi^j x, L^j y)$. Let $K_1 > 0$ be such that for each z, w in the same leaf of F , $d(z,w) \leq K_1\, d_z(z,w)$. Here $d(z,w)$ is the distance in $\Lambda \times T^n$ and $d_z(z,w)$ is the distance in F_z . Since the length of $J(\phi^j x, L^j y)$ is less than K , we may choose a finite subset B_j of the leaf of F containing $J(\phi^j x, L^j y)$ such that B_j contains less than $\frac{3KK_1}{\varepsilon} + 1$ elements, and, for each $z \in J(\phi^j x, L^j y)$, there is a $z' \in B_j$ such that $d_z(z,z') < \frac{\varepsilon}{2K_1}$. Let $B = \bigcup_{j=0}^{m-1} F^{-j} B_j$. Let $E \subset J(x,y)$ be an $(m,\varepsilon,J(x,y))$ separated set. For $z \in E$, there is $\zeta(z) \in B$ such that

$$d_{f^j z}(f^j z, f^j \zeta(z)) < \frac{\varepsilon}{2K_1} \quad \text{for } j \in [0,m) .$$

If $\zeta(z) = \zeta(z')$ with $z,z' \in F$, then

$$\begin{aligned} d(f^j z, f^j z') &\leq K_1\, d_{f^j z}(f^j z, f^j z') \\ &\leq K_1 [d_{f^j z}(f^j z, f^j \zeta(z)) + d_{f^j z}(f^j \zeta(z'), f^j z')] \\ &\leq \varepsilon \end{aligned}$$

for $j \in [0,m)$. Hence $z = z'$, so ζ is injective. Now B contains at most $m[\frac{3KK_1}{\varepsilon} + 1]$ elements, so

$$\limsup_{m\to\infty} \frac{1}{m} \log r(m,\varepsilon,J(x,y)) = 0 \qquad .$$

Thus, it suffices to prove (*) to obtain lemma 3 .

We have already seen that the lift H of ψ maps leaves of $\tilde{W}^u$ onto leaves of $\tilde{W}^{u'}$ and leaves of $\tilde{F}$ onto leaves of $\tilde{F}'$. If we show

(**) H^{-1} (leaf of $\tilde{F}'$) is contained in a single leaf of $\tilde{F}$,

then the properness of H guarantees that for $(x,y) \in \Lambda \times \mathbb{R}^n$, $H^{-1}(x,y)$ is contained in an interval $\tilde{J}(x,y)$ in a leaf of $\tilde{F}$, and the lengths of those intervals are bounded independent of (x,y). This will imply (*) by projecting down to $\Lambda \times T^n$. We now prove (**).

Let $\tilde{F}_1$ be a leaf of $\tilde{F}$ and let $V(\tilde{F}_1) = \bigcup_{(x,y)\in\tilde{F}_1} \tilde{W}^u(x,y)$. Since H maps each leaf of $\tilde{F}$ onto one of $\tilde{F}'$ and maps each leaf of $\tilde{W}^u$ homeomorphically onto one of $\tilde{W}^{u'}$, it follows that $H(V(\tilde{F}_1)) = \{x\} \times \mathbb{R}^n$ and distinct leaves of $\tilde{F}$ in $V(\tilde{F}_1)$ are mapped to distinct leaves of $\tilde{F}'$. Moreover, $V(\tilde{F}_1)$ is an open subset of $\{x\} \times \mathbb{R}^n$. Now the sets $V(\tilde{F}_z)$ $z \in \{x\} \times \mathbb{R}^n$, form a partition of $\{x\} \times \mathbb{R}^n$ into open sets. Thus there is only one of them and (**) is proved.

Proof of Theorem 1 : The proof that $h(F) = h(\phi) + h(L)$ now follows easily. First, it is standard that $h(F) \geq h(\phi \times L) = h(\phi) + h(L)$. On the other hand by lemma 3 and Bowen's result we have

$$h(\phi \times L) \leq h(F) + \sup_{(x,y)\in\Lambda\times T^n} h(\psi^{-1}(x,y),F)$$

$$= h(F)$$

so $h(F) = h(\phi) + h(L)$.

Now suppose ϕ has a maximal measure, say μ_ϕ. Then $\mu_\phi \times \nu$ is a maximal measure for $\phi \times L$. Thus if $\rho \in M(F)$ is such that $\psi_*\rho$ $\psi_*\rho = \mu_\phi \times \nu$ (such a ρ exists by the Hahn-Banach theorem), then ρ is a maximal measure for F. Finally if ρ is an F-maximal measure, then

$$h(F) = h_\rho(F) = \sup_{\substack{\eta\in M(F) \\ \psi_*\eta=\psi_*\rho}} h_\eta(F)$$

$$= h_{\psi_*\rho}(\phi \times L) + \int_{\Lambda\times T^n} h(\psi^{-1}(x,y),F)\,d\psi_*\rho$$

by the Ledrappier-Walters result, and, hence

$$= h_{\psi_*\rho}(\phi \times L) \qquad \text{by lemma 3 .}$$

Thus $\psi_*\rho$ is a maximal measure for $\phi \times L$. Repeating the argument above, we have

$$h(\phi \times L) = h_{\pi_{1*}\psi_*\rho}(\phi) + \int_{\Lambda} h(\pi_1^{-1}x,L)\,d\pi_{1*}\psi_*\rho$$

$$= h_{\pi_{1*}\psi_*\rho}(\phi) + h(L) \qquad .$$

That is, $\pi_{1*}\psi_*\rho$ is a maximal measure for ϕ . Since $\pi_1\psi = \pi_1$, we have $\pi_{1*}\rho$ is a maximal measure for ϕ .

3. Proof of theorem 2

Lemma 4 : Suppose the hypotheses of theorem 2 are satisfied. Let $\psi : \Lambda \times T^n \longrightarrow \Lambda \times T^n$ be the semi-conjugacy of lemma 1. Then $\psi \mid B \times T^n$ is a homeomorphism.

Proof : It suffices to prove that $\psi \mid B \times T^n$ is one-to-one. For if $(x_i,y_i) \longrightarrow B \times T^n$ is such that $\psi(x_i,y_i) \longrightarrow \psi(x,y)$ as $i \to \infty$, then $x_i \to x$ as $i \to \infty$. If y' and y'' are limit points of the sequence $\{y_i\}$, then continuity of ψ requires that $\psi(x,y') = \psi(x,y'') = \psi(x,y)$. So if $\psi \mid B \times T^n$ is one-to-one, we have $y' = y'' = y$. Hence, $(x_i,y_i) \longrightarrow (x,y)$ and $\psi^{-1} \mid B \times T^n$ is continuous.

Let us now prove that $\psi \mid B \times T^n$ is one-to-one. If not, then for some $(x,y) \in B \times T^n$, $\psi^{-1}(x,y)$ contains more than one point. Let $J(x,y)$ be the shortest F-leaf segment containing $\psi^{-1}(x,y)$. Then,

$F^{-m} J(x,y) = J((\phi \times L)^{-m}(x,y))$ and $\sup_{m\in\mathbb{Z}^+}$ length $J((\phi \times L)^{-m}(x,y))$ is finite.
By hypothesis, for each point $(x,z) \in B \times T^n$,

$$\liminf_{m\to\infty} \frac{1}{m} \log |TF^{-m} \mid E^s_{(x,z)}| > 0$$

which implies in particular that

$$|TF^{-m} \mid E^s_{(x,z)}| \to \infty \quad \text{as} \quad m \to \infty \quad .$$

Let $\alpha : [0,s] \longrightarrow \Lambda \times T^n$ be a parametrization of $J(x,y)$ by arclength.
We then have

$$\text{length } J((\phi \times L)^{-m}(x,y)) = \int_0^s |TF^{-m} \mid E^s_{\alpha(t)}| \, dt \to \infty \quad ,$$

a contradiction.

3. Proof of theorem 2

From lemma 4, we have that $\psi \mid B \times T^n$ is a topological conjugacy between $F \mid B \times T^n$ and $\phi \times L \mid B \times T^n$.

Now suppose ϕ is intrinsically ergodic with maximal measure μ_ϕ and $\mu_\phi(B) = 1$. Now L is intrinsically ergodic with measure ν and (L,ν) is a K-automorphism. Hence by a theorem of K. Berg [3] (see [12] for a generalization), $\phi \times L$ is intrinsically ergodic with maximal measure $\mu_\phi \times \nu$. Suppose μ_1 and μ_2 are two maximal measures for F . Then $\pi_{1*}\mu_1$ and $\pi_{1*}\mu_2$ are maximal measures for ϕ by theorem 1.3. Hence, $\pi_{1*}\mu_1 = \pi_{1*}\mu_2 = \mu_\phi$. Since $\mu_\phi(B) = 1$, we have $\mu_1(B \times T^n) = \mu_2(B \times T^n) = 1$. Since $\psi \mid B \times T^n$ is a topological conjugacy between $F \mid B \times T^n$ and $\phi \times L \mid B \times T^n$, $\phi_* \mu_1$ and $\psi_*\mu_2$ are two measures on $\Lambda \times T^n$ invariant by $\phi \times L$ with maximal entropy. Thus, $\psi_*\mu_1 = \psi_*\mu_2 = \mu_\phi \times \nu$. So, $\mu_1 = \psi_*^{-1}(\mu_\phi \times \nu) = \mu_2$ and F is intrinsically ergodic. Also, ψ is the measure theoretic conjugacy between (F,μ_1) and $(\phi \times L, \mu_\phi \times L)$.

4. Examples

In this section we show that the hypotheses of theorems 1 and 2 are satisfied by the non-wandering set restrictions of certain open sets of non-Ω-stable diffeomorphisms. The first such diffeomorphisms were given by Abraham and Smale [1] on $S^2 \times T^2$, the product of the two-sphere and the two-torus. An explicit special case of their construction is in [6, p. 40]. Related examples are studied in [10; § 8]. The results of the present paper grew out of attempts to understand the dynamics of the examples in [6] and [10].

Let us consider some systems to which our results apply. Let $\phi : M \longrightarrow M$ be a C^1 diffeomorphism satisfying Smale's Axiom A and No cycle conditions (see [14] for definitions). Let $\Omega(\phi) = \Lambda$ be the non-wandering set of ϕ, and let $L : T^n \longrightarrow T^n$ be a linear Anosov diffeomorphism. In [10], it is shown that the product $\phi \times L$ can be deformed to a diffeomorphism $F : M \times T^n \longrightarrow M \times T^n$ so that no F' C^1 near F is Ω-stable, but F is topologically Ω-stable. This last statement means that each F' near F has $\Omega(F')$ homeomorphic to $\Omega(F)$. While the topology of $\Omega(F')$ is controllable, there has up to now been no good description of the orbit structure of F' on $\Omega(F')$. Theorems 1 and 2 will enable us to give a reasonable description of this structure under some additional assumptions.

Using techniques as in [10, § 8] we may find a smooth diffeomorphism $F : M \times T^n \longrightarrow M \times T^n$ such that

(a) $F(x,y) = (\phi x, f_x(y))$ satisfies assumptions 1, 2, and 3 of theorem 1.

(b) F is normally hyperbolic to the lamination $\{\{x\} \times T^n\}_{x \in \Omega(\phi)}$.

(c) F is plaque expansive in the sense of [10;p.116].

(d) $\Omega(F) = \Lambda \times T^n$.

(e) No F' C^1 near F is Ω-stable.

From the results of [10] , if $F' : M \times T^n \longrightarrow M \times T^n$ is C^1 near F , then there is a homeomorphism $H : \Omega(F') \longrightarrow \Lambda \times T^n$ such that $HF'H^{-1} : \Lambda \times T^n \longrightarrow \Lambda \times T^n$ is a bundle map covering ϕ , and $HF'H^{-1}$ satisfies conditions (a), (b), and (c) above. Hence, theorem 1 holds for $HF'H^{-1}$. Thus $h(F') = h(HF'H^{-1}) = h(F) = h(\phi) + h(L)$.

Let $\Lambda = \Lambda_1 \cup \dots \cup \Lambda_t$ be the spectral decomposition of $\Omega(\phi) = \Lambda$. Set $\Omega_i = \Lambda_i \times T^n$ and $\Omega_i' = H^{-1}(\Omega_i)$. From Bowen [5] , we know that ϕ is intrinsically ergodic on each basic set Λ_i . Thus, $HF'H^{-1}$ has maximal measures on each Ω_i , and, hence, so does F' on Ω_i' .

Now we impose further conditions on $F \mid \Omega_i$ so that the dynamics of $F' \mid \Omega_i'$ for F' near F may be described by theorem 2. Let μ_i be the unique maximal measure of $\phi \mid \Lambda_i$. Let $\zeta(x) = \inf_{y \in T^n} |T_{(x,y)}F^{-1}|F_{(x,y)}|$ for $x \in \Lambda$. Suppose that

(f) $$\int_{\Lambda_i} \log \zeta(x) \, d\mu_i > 0 \quad .$$

Note that the ergodic theorem and (f) imply that the assumptions of theorem 2 hold for $F \mid \Omega_i$. Moreover, for F' C^1 near F , the homeomorphism $H : \Omega(F') \longrightarrow \Lambda \times T^n$ is C^0 near the inclusion $\Omega(F') \subset M \times T^n$, so the function ζ' for $HF'H^{-1}$ defined analogous to ζ again satisfies an inequality as in (f) . Then, theorem 2 applies to $F' \mid \Omega_i'$. So, $F' \mid \Omega_i'$ is intrinsically ergodic and almost conjugate to $\phi \times L \mid \Omega_i$. Thus, neglecting sets of measure zero for the maximal measures, the dynamics of $F' \mid \Omega(F')$ is the same as that of the known system $\phi \times L$ on $\Omega(\phi) \times T^n$.

<u>Remark</u> : In the theorems presented here, very strong use was made of the fact that the foliation F was one-dimensional. There are certain analogous results which hold when the leaves have larger dimension. In this case, the fiber entropies $h(\psi^{-1}(x,y),F)$ are not necessarily zero. To compensate for this, one has to require that the set U_i of points $x \in \Lambda_i$ at which $F_x : \{x\} \times T^n \longrightarrow \{\phi x\} \times T^n$ is not Anosov (identifying $\{x\} \times T^n$ and $\{\phi x\} \times T^n$ with T^n) is a small set. For instance, one could require that $\mu_i(U_i)$ is small relative to the hyperbolicity of L and

$$\sup_{(x,y)\in\Lambda \times T^n} h(\psi^{-1}(x,y),F) .$$

REFERENCES

[1] R. Abraham and S. Smale, "Non-genericity of Ω-stability", Proc. Symp. in Pure Math. 14 (1970), Amer. Math. Soc., 5-8.

[2] R. Adler and B. Marcus, "Topological entropy and equivalence of dynamical systems", Mem. AMS 219, Amer. Math. Soc., Providence, R.I.

[3] K. Berg, "Convolution and invariant measures, maximal entropy", Math. Sys. Theory 3 (1969), 146-150.

[4] R. Bowen, "Entropy for group endomorphisms and homogeneous spaces", Trans. Amer. Math. Soc. 153, (1974), 401-413.

[5] ________ , "Some systems with unique equilibrium states"; Math. Sys. Theory 8 (1974), 193-202.

[6] ________ ,"On Axiom A diffeomorphism", CBMS Regional Conference Series in Math. 35, Amer. Math. Soc., Providence, R.I., 1978.

[7] J. Franks, "Anosov diffeomorphisms", Proc. Symp. Pure Math. 14 (1970), Amer. Math. Soc., 61-93.

[8] ________ , "Anosov diffeomorphisms on tori", Trans. Amer. Math. Soc. 145 (1969), 117-124.

[9] M. Hirsch and C. Pugh, "Stable manifolds and hyperbolic sets", Proc. Symp. in Pure Math. 14 (1970), Amer. Math. Soc., 133-163.

[10] M. Hirsch, C. Pugh, and M. Shub, "Invariant manifolds", Springer Lecture Notes in Math. 583, 1977.

[11] F. Ledrappier and P. Walters, "A relativized variational principle for continuous transformations", J. London Math. Soc. 16 (1977), 568-576.

[12] B. Marcus and S. Newhouse, "Measures of maximal entropy for a class of skew-products", Lecture Notes in Math. 729.

[13] S. Newhouse, "Dynamical properties of certain non-commutative skew-products", Lecture Notes in Math. 819, Springer-Verlag, N.Y., 353-364.

[14] ————, "Lectures on Dynamical Systems", Progress in Math. 8, Birkhäuser, Boston, 1980, 1-115.

[15] R. Sacksteder, "Foliations and pseudo-groups", Amer. J. Math. 87 (1965), 79-102.

Sheldon E. Newhouse and Lai-sang Young

Institute des Hautes Etudes Scientifiques
35, route de Chartres
91440 - Bures-sur-Yvette (France)

April 1981

IHES/M/81/24

A Note on the Inclination Lemma (λ-Lemma) and Feigenbaum's Rate of Approach

by

J. Palis

In recent years, striking results has been obtained concerning "universal constants" for cascades of bifurcations of dynamical systems such as maps of the interval and area preserving transformations (see [1], [3]). If $\mathfrak{F}_\mu$, $\mu \in \mathbb{R}$, is such a family and $\{\mu_n\}$, $\mu_1 < \ldots < \mu_n < \mu_{n+1} < \ldots$ with $\mu_n \to \mu_\infty$, are successive bifucation values of certain pattern, Feigenbaum suggested that the ratio $\mu_\infty - \mu_n / \mu_\infty - \mu_{n+1}$ should converge to a "universal" number, for a generic family in a (Banach) space E of transformations. Such a constant should be the expanding eigenvalue of the linear part of an operator F in the Banach space at a hyperbolic fixed point with codimension one stable manifold. The one-parameter family is a differentiable curve $L: I \subset \mathbb{R} \to E$ in E crossing the stable manifold transversally at $L(\mu_\infty)$ while Σ should also cross the unstable manifold transversally, where Σ is the codimension one submanifold of transformations exhibiting a (unique) bifurcating orbit of the required pattern.

The purpose of this note is to present an elementary discussion of the dynamical behavior of a transformation in Banach space, near a hyperbolic fixed point having its stable manifold of finite codimension. To relate it to part of Feigenbaum's program, we will be especially concerned with (local) iterates of submanifolds that intersect transversally the stable and unstable manifolds, respectivelly. Roughly speaking, the transformation being of class C^1, their iterates approach the stable and unstable manifolds, respectively, in the C^1 sense. This is the content of the Inclination Lemma (λ-Lemma) in [6, 7], and it was used in several contexts in Dynamics. The key novelty introduced in [1] was to show, in the codimension one case for a transformation of class C^2, that the (negative) iterates of a codimension one submanifold cutting the unstable manifold transversally approach the stable manifold as before but now at a geometric rate (the inverse of the expanding eigenvalue). With a different proof, we present this result in Theorem A for C^1 transformations with (one of the) partial derivatives being Lipschitz. The Lipschitz condition is necessary as shown in the counter-example we provide. Finally, in Theorem B we observe that a similar result is also valid in codimension k, $k \geq 1$ (corresponding to k-parameter families), under mild genericity conditions.

Two final remarks are in order. In the applications to compute Feigenbaum's constants ([1], [3]), the Banach space transformation is C^∞ (not only C^2 or C^1 plus Lipschitz). We think, however, that it makes sense to announce the precise theorem in this setting and, perhaps, to open up some perspective for k-parameter families of systems with such geometric view. We also observe that similar approach appears in the study of saddle-connections (orbits of tangency) generating moduli of stability (see [4], [5], [8]).

We now present Theorems A and B and the counter-example mentioned above.

Let $F: V(0) \subset E \to E$ be of class C^r, $r \geq 1$, with 0 being a hyperbolic fixed point of codimension k; i.e., its unstable manifold is of dimension k. Here, $V(0)$ is a neighborhood of the origin 0 in the (real) Banach space E. If $\rho_1, \rho_2, \ldots, \rho_k$ are the expanding eigenvalues of $DF(0)$, we assume that $|\rho_1| < |\rho_j|$ for $2 \leq j \leq k$ or $|\rho_1| = |\rho_2| < |\rho_j|$ for $2 \leq j \leq k$ in case ρ_1 is not real and ρ_2 is its conjugate. From the Stable Manifold Theorem (see [2]), there are C^r stable manifold W^s, unstable manifold W^u and strong unstable manifold $W^{uu} \subset W^u$, which is of dimension $k-1$ or $k-2$ depending on ρ_1 being real or not. These F-invariant submanifolds are C^r plus Lipschitz if F is C^r plus Lipschitz.

Theorem A. Let $F: V(O) \to E$ as above be of class C^1 and suppose O is a hyperbolic fixed point for F of codimension one. Let Σ be a codimension one submanifold intersecting W^u transversally and let L be a C^1 curve intersecting W^s transversally at z_∞. Then, there exist $n_0 \in \mathbb{N}$ and a sequence of points $z_n \in L$, $n \geq n_0$, such that $F^n(z_n) \in \Sigma$. Moreover, if F is C^1 plus Lipschitz then $d(z_n, z_\infty)/d(z_{n+1}, z_\infty)$ converges to $|\rho|$ (ρ being the expanding eigenvalue).

Proof. The first part is an immediate consequence of the Inclination Lemma: given $\epsilon > 0$, a bounded disk D^u containing O in W^u and a small ("product") neighborhood $\mathcal{U}^u$ of D^u in $V(O)$, then there exist $n_0 \in \mathbb{N}$ such that the connected components of $F^n(L) \cap \mathcal{U}$ through $F^n(z_\infty)$ is ϵ-C^1 close to D^u for all $n \geq n_0$.

To prove the second part, let us analyse somewhat more the inclination of the tangent vectors to the local inverse images by F of the submanifold Σ. Since F is not necessarily injective, we consider the components of $F^{-n}(\Sigma)$, $n \geq 0$, which are diffeomorphic to Σ. They exist, by the Implicit Function Theorem, and they will be denoted simply by Σ^{-n}. By the same Inclination Lemma, we have that given $\epsilon > 0$ and a (bounded) disk $D^s \subset W^s$ containing O and z_∞, there exist $n_0 \in \mathbb{N}$ such that $\Sigma^{-n} \cap \mathcal{U}^s$ is ϵ-C^1 close to D^s for all $n \geq n_0$, where $\mathcal{U}^s$ is a neighborhood of D^s. We can express F as $F(x,y) = (Ax + \Phi^s(x,y), \rho y + \Phi^u(x,y))$, $x \in W^s$ and $y \in W^u$, where $\|A\| \leq a < 1$, $D\Phi^s(0,0) = 0$, $D\Phi^u(0,0) = 0$, $\Phi^s(0,y) = 0$, $\Phi^u(x,0) = 0$. Furthermore, $|\frac{\partial \Phi^u}{\partial x}(x,y)| \leq K|y|$ for some $K > 0$. By taking a finite number of iterates of the curve L by F, we may assume that in $\mathcal{U}^s$ all partial derivatives of Φ^s, Φ^u have norm bounded by a small number $\delta > 0$.

We now claim that, for ϵ and δ small enough and $n \geq n_0$, in every point $(x,y) \in \Sigma^{-n}$ and for every nonzero vector $v = (v^s, v^u)$ tangent to Σ^{-n} at (x,y), we have that $\lambda(v)/|y|$ is uniformly bounded, say, by a constant $C > 0$. Here, $\lambda(v)$ means the inclination of v; i.e., $\lambda(v) = |v^u|/\|v^s\|$. This readily implies the result. In fact, for $\tilde{y}_n = \Sigma^{-n} \cap W^u$ it is clear that $\lim |\tilde{y}_n|/|\tilde{y}_{n+1}| = |\rho|$, since $\frac{\partial \Phi^u}{\partial y}(0,0) = 0$. For $z_n = \Sigma^{-n} \cap L$, $z_n = (x_n, y_n)$, we conclude from our claim that $e^{-C\ell} \le |y_n|/|\tilde{y}_n| \le e^{C\ell}$, where ℓ is the diameter of a disk in W^s centered at O and containing z_∞ in its interior. From this we get

$$e^{-2C\ell} \cdot |\tilde{y}_n|/|\tilde{y}_{n+1}| \le |y_n|/|y_{n+1}| \le e^{2C\ell} \cdot |\tilde{y}_n|/|\tilde{y}_{n+1}|.$$

But, for any convergent subsequence $|y_k|/|y_{k+1}|$ its limit is the same as that of $d(z_k, z_\infty)/d(z_{k+1}, z_\infty)$, which in turn is the same as that of $d(t_k, t_\infty)/d(t_{k+1}, t_\infty)$, where $t_k = \Sigma^{-k} \cap F^j(L)$ and $t_\infty = F^j(t_\infty)$, for any (fixed) $j \in \mathbb{N}$. Thus, by making ℓ small, we conclude that $d(z_n, z_\infty)/d(z_{n+1}, z_0)$ converges to $|\rho|$.

Let us prove the above claim. Formally we will be using the Inclination Lemma, but the proof we present can be easily adapted to also prove this Lemma. Let $(x,y) \in \Sigma^{-n}$ and $v = (v^s, v^u)$ be a nonzero tangent vector to Σ^{-n} at (x,y), and let $(w^s, w^u) = w = DF(v)$ and $(X,Y) = F(x,y)$. From the expressions for F and DF, we have that

$$\lambda(w) \ge [(\rho - \delta)\lambda(v) - K|y|]/[a + \delta + \delta\varepsilon].$$

Let $\rho - \delta = \tilde{\rho} > 1$, $a + \delta + \delta\varepsilon = \tilde{a} < 1$. We then get $1/\tilde{\rho}\ [\tilde{a}/|y| \cdot \lambda(w) + K] \ge \lambda(v)/|y|$. Thus, $1/\tilde{\rho}\ [\tilde{a} \cdot |Y|/|y| \cdot \lambda(W)/|Y| + K] \ge \lambda(v)/|y|$. Again, for δ small, we have that $\tilde{a}/\tilde{\rho} \cdot |Y|/|y| \le a^* < 1$ for some a^*. And so, $\lambda(v)/|y| \le a^*[\lambda(w)/|Y| + K]$. Proceeding by induction, using the fact that $a^* < 1$, we easily conclude our claim and Theorem A is proved.

The following result is a generalization of Theorem A and its proof is quite similar to the previous one.

<u>Theorem B</u>. Let $F: V(O) \to E$ as before be of class C^1. Suppose O is a hyperbolic fixed point of codimension $k \ge 1$. Let Σ be a co-

dimension k submanifold intersecting W^u transversally at w and let P be a k-dimensional surface intersecting W^s transversally at z_∞. Then, there exist $n_0 \in \mathbb{N}$ and a sequence $z_n \in L$, $n \geq n_0$, such that $z_n \to z_\infty$, $F^n(z_n) \in \Sigma$ and $\lim F^n(z_n) = w$. Moreover, if F is C^1 plus Lipschitz and w is off W^{uu}, then $\lim d(z_n,z_\infty)/d(z_{n+1},z_\infty) = |\rho_1|$.

The Counter-Example. Let $V = \{(x,y) \in \mathbb{R}^2,\ y < 1\}$ and consider the diffeomorphism $F\colon V \to F(V) \subset \mathbb{R}^2$, defined as follows:

$$F(x,y) = (ax, a^{-1}y + \Phi^u(x,y)),$$

where $0 < a < 1$, and

i) $\Phi^u(x,y) = 0$ for $x \leq y$;

ii) $\Phi^u(x,y) = 0$ for $y \leq 0$;

iii) $\Phi^u(x,y) = 0$ for $a^{2k+1/2} \leq y/x \leq a^{2k}$, $k \in \mathbb{N}$;

iv) $\Phi^u(x,y) = -y/\log y \cdot \sin^2(2\pi \cdot \log(y/x)/\log a)$ otherwise.

One can check that F is of class C^1 (but not C^1 plus Lipschitz), O is a hyperbolic fixed point, a and a^{-1} are the contracting and expanding eigenvalues, the x-axis is the stable manifold and the y-axis is the unstable manifold. Consider Σ as the horizontal line $y = a^d$, $0 < d < a^{1/4}$, and L the vertical line $x = a$. As before, there exist a sequence $z_n \to z_\infty = (a,0)$ such that $F^n(z_n) \in \Sigma$. We will show that $d(z_n,z_\infty)/d(z_{n+1},z_\infty)$ does not converge to a^{-1}. To see this, observe that for n odd we have $z_n = a^{n+d}$. Let now n to be even. If we write $z_n = a^{n+d+\ell_n}$, then ℓ_n cannot converge zero. This is clear because $\Sigma_{j=1}^{n/2} \cdot 1/n-j > 1/2$ and so

$$-1/\log a^2 \cdot \sin^2(2\pi d) \cdot \Sigma_{j=1}^{n/2} 1/n-j > \delta > 0.$$

Thus, $d(z_n,z_\infty)/d(z_{n+1},z_\infty)$ does not converge to a^{-1}. In fact, this sequence does not converge at all. Also, by varying the line Σ, we can get subsequences of the above sequence converging to uncountably many different numbers.

References

[1] Collet, P. and Eckmann, J.-P., Iterated Maps of the Interval as Dynamical Systems, Progress in Physics, Birkhäuser (1980).

[2] Hirsch, M. and Pugh, C. and Shub, M., Invariant Manifolds, Lecture Notes in Math. 583, Springer-Verlag (1977).

[3] Lanford, O., A computer-assisted proof of the Feigenbaum conjectures, Bull. Am. Math. Soc. 6 (1982), 427-434.

[4] Newhouse, S. and Palis, J. and Takens, F., Bifurcations and stability of families of diffeomorphisms, Publ. Math. IHES, to appear.

[5] Palis, J., A differentiable invariant of topological conjugacies and moduli of stability, Astérisque 51 (1978), 335-346.

[6] Palis, J., On Morse-Smale dynamical systems, Topology 8 (1969), 385-405.

[7] Palis, J. and de Melo, W., Geometric Theory of Dynamical Systems, An Introduction, Springer-Verlag (1982).

[8] Palis, J. and Takens, F., Stability of parametrized families of gradient vector fields, Annals of Math., to appear.

Instituto de Matemática Pura e Aplicada (IMPA)
Estrada Dona Castorina, 110
22.460 - Rio de Janeiro, RJ - Brasil

A Special C^r Closing Lemma

by

Charles C. Pugh*

A general, open question in dynamical systems is:

"When can a small change of a diffeomorphism or differential equation convert recurrence to periodicity?"

More specifically, let M be a smooth, compact manifold, $p \in M$, and let $f: M \circlearrowleft$ be a C^r diffeomorphism, $0 \leq r \leq \infty$. If

$$(1) \qquad f^{n_k}(p_k) \longrightarrow p$$

for some $n_k \rightarrow \infty$ and $p_k \rightarrow p$ as $k \rightarrow \infty$ then p is said to be <u>nonwandering</u> under f. If, in (1), p_k can be chosen to always equal p then p is ω-<u>recurrent</u>. Equivalent to (1) is the property that for each neighborhood U of p there is $n \neq 0$ with

$$(2) \qquad f^n(U) \cap U \neq \emptyset .$$

The set of nonwandering points of f is denoted by $\Omega = \Omega(f)$. A similar discussion can be given for the flow φ_t of a differential equation on M, replacing n by t.

C^r <u>Closing Lemma for</u> $r = 0,1$. <u>If</u> $p \in \Omega(f)$ <u>then</u> f <u>can be</u> C^r-<u>approximated by</u> g <u>such that</u> $p \in$ <u>Per</u>(g); <u>i.e.</u>,

$$p = g^n(p) \qquad \underline{\text{for some}}\ n \neq 0$$

When $r = 0$, this turns out to be easy; when $r = 1$ it is proved in [3,4,5]. For $r > 1$, the C^r Closing Lemma has yet to be verified, even generically. In this article, two conditions

---

*Supported in part by NSF grants MCS77 - 17907, 77-224/A01, and SERC grant No. GR/B 82363

are presented, each of which implies the C^r Closing Lemma, $r > 1$. (Curiously, the cases $0 < r < 1$ and r=Lipschitz have never been studied.)

Definition 1. $p \in \Omega(f)$ almost wanders if there is a neighborhood U of p and a bound N on the number of iterates $f^n(x)$ which lie in U, uniform over all $x \in M$.

Theorem 1. If $p \in \Omega(f)$ almost wanders then f can be C^r-approximated by g with $p \in Per(g)$.

Example. An almost wandering point $p \in \Omega(f)$ has an extremely non-recurrent orbit as is seen in the following diffeomorphism of Jacob Palis. Let $f : S^2 \rightarrow S^2$ have two sources. sinks, and saddles as in figure 1.

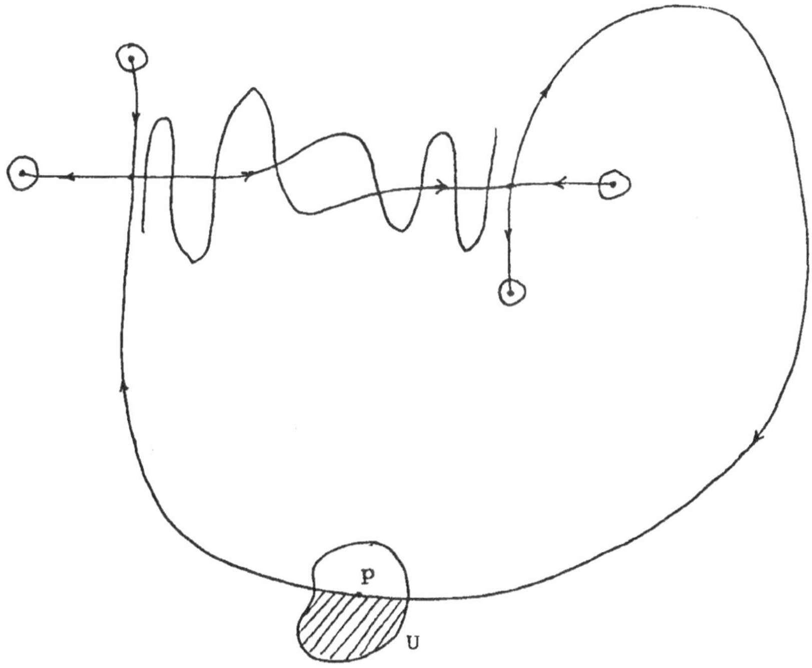

Figure 1. An almost wandering point p

Consider $f^n x$ for $x \in U$. If x lies in the outer (shaded) half of U then $f^n x$ falls directly into the left-most sink as $n \to \infty$, while if x lies in the inner half of U then $f^n x$ emerges directly from the right-most source as $n \to -\infty$. Thus, no point x has more than two iterates $f^{n_1}(x)$, $f^{n_2}(x)$ in U. By Cloud Chasing as in [6], some points x in the inner half of U do return to the outer half of U, so $p \in \Omega(f)$ but p is almost wandering.

Lemma 1. *If $p \in \Omega(f)$ almost wanders then p has a small neighborhood U_0 containing a sequence $p_k \to p$ as $k \to \infty$ such that*

$$f^{n_k}(p_k) \to p \qquad \text{for some } n_k > 0$$

and $f^n(p_k) \notin U_0$, $0 < n < n_k$. See figure 2.

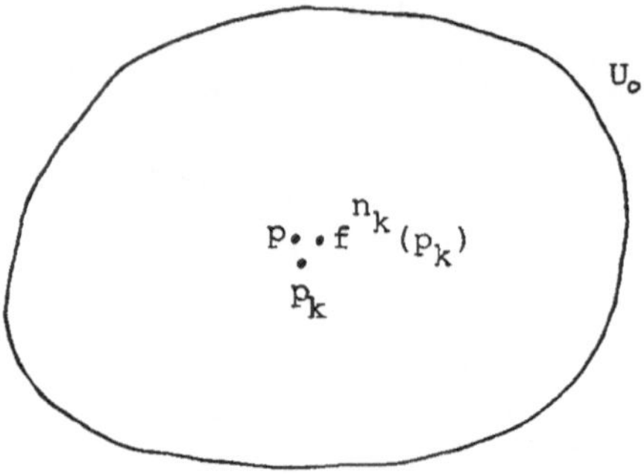

Figure 2. $f^n(p_k)$ first revisits U_0 near p.

Proof. Since p almost wanders, there is a neighborhood U of p such that $f^n(x) \in U$ for no more then N values of n and $2 \le N < \infty$ is fixed.

Let $U_0 \subset U$ be a neighborhood of p and let $p_k \to p$ be a sequence with

$$f^{n_k}(p_k) \to p$$

for some $n_k > 0$. Such sequences exist since $p \in \Omega(f)$. Call N_0 the maximum number of iterates $f^n(p_k) \in U_0$, $0 \le n \le n_k$. Clearly $N_0 \le N$. If $N_0 = 2$ there is nothing to prove. If

$N_0 \geq 3$ here is how to reduce it. That $N_0 \geq 3$ means

$$f^{\ell_k}(p_k) \in U_0$$

for some ℓ_k, $0 < \ell_k < n_k$. If $f^{\ell_k}(p_k)$ accumulate at p then choose a subsequence with

$$f^{\ell_n}(p_k) \to p \qquad \text{as} \quad k \to \infty$$

and replace n_k by ℓ_k. This reduces N_0. If $f^{\ell_k}(p_k)$ does not accumulate at p then replace U_0 by a smaller neighborhood U_1 of p so that $f^{\ell_k}(p_k) \notin U_1$. This also reduces N_0. Inductively reduce N_0 to 2 as required. Q.E.D

Proof of Theorem 1. By lemma 1, we can find U_0 and $p_k \to p$ with $f^{n_k}(p_k) \to p$ but

$$f^n(p_k) \notin U_0 \qquad 0 < n < n_k .$$

There are diffeomorphisms $g_k, h_k : M \leftrightarrows$ such that

$$g_k = h_k = \text{identity outside } U_0$$

$$g_k(p) = p_k$$

$$h_k(f^{n_k}(p_k)) = p$$

Since U_0 is fixed we can choose $g_k, h_k \to$ id in the C^∞-sense as $k \to \infty$. Then

$$g = h_k \circ f \circ g_k$$

C^r-approximates f and has $p \in \text{Per}(g)$. For, under g,

$$p \mapsto g_k(p) = p_k \mapsto f(p_k) \mapsto f^2(p_k) \mapsto \cdots \mapsto f^\ell(p_k)$$

$$p \leftarrow f^{n_k}(p_k) \leftarrow f^{n_k-1}(p_k) \leftarrow \cdots \leftarrow f^{\ell+1}(p_k)$$

Q.E.D

Remark. A condition weaker than almost wandering suffices for this proof - namely the conclusion of Lemma 1: p has a neighborhood U_0 containing a sequence of points $p_k \to p$ with $f^{n_k}(p_k) \to p$ but $f^n(p_k) \notin U_0$ for $0 < k < n_k$.

The second condition that implies the C^r Closing Lemma is more interesting. We formulate it first on the 2-torus T^2.

Definition 2. The diffeomorphism $f : T^2 \to T^2$ is non-reversing if for each $v \in \mathbb{R}^2 - 0$ there is a half-space H_v of $\mathbb{R}^2$ at 0 such that

$$(Df^n)_x(v) \in H_v \qquad \forall n \geq 0 \quad \forall x \in T^2$$

$$\measuredangle(v, \partial H_v) \geq \beta > 0$$

β is some constant independent of v.

Theorem 2. *If $f : T^2 \to T^2$ is nonreversing and $p \in \omega(p)$, i.e. p is recurrent, then f can be C^r-approximated by g such that $p \in \mathrm{Per}(g)$.*

Remarks. (a) If $\measuredangle((Df^n)_x(v), v) \leq \pi/2$ for all $v \in \mathbb{R}^2$, all $n \geq 0$, and all $x \in T^2$ then f is non-reversing because we may take

$$H_v = \{w \in \mathbb{R}^2 : w.v \geq 0\}$$

(b) In general we do not require H_v to depend continuously on v.

(c) A linear Anosov diffeomorphism of T^2, which preserves the orientation of its invariant bundles, E^u and E^s, is non-reversing, e.g.

$$f = \begin{pmatrix} 2 & 1 \\ 1 & 1 \end{pmatrix}$$

Of course, Theorem 2 is already well known for such an f, indeed f itself has Per(f) dense in T^2.

(d) Translations of T^2 are non-reversing.

(e) Non-reversing is already significant in dimension 1. Peixoto's proof of the C^r genericity of Morse-Smale flows [2] is valid (only) on orientable surfaces because there the Poincaré maps are non-reversing. However, see Theorem 3 and the remark after.

(f) "Recurrent" can be replaced by "non-wandering". See Theorem 3.

Proof of Theorem 2. For $v \in \mathbb{R}^2$, let τ_v be a translation of T^2 by $x \mapsto x+v$ and let

$$f_v = \tau_v \circ f$$

When $|v|$ is small, f_v C^∞-approximates f.

Fix any $v \in \mathbb{R}^2 - 0$ and make the following construction along the orbit of the recurrent point p. At f(p) draw a segment σ_1 parallel to and of the same length as v. At the end of $f(\sigma_1)$ draw a second such segment σ_2. At the end of $f(\sigma_2)$ draw a third, σ_3, and so on. See figure 3.

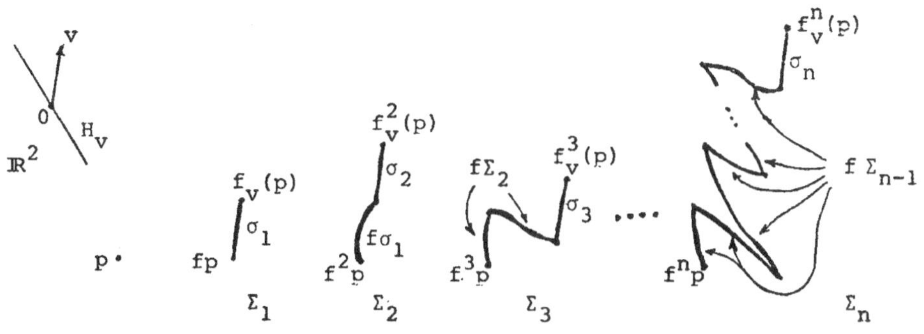

Figure 3. Segments σ_k

Now examine the curve Σ_n at $f^n p$ as drawn:

$$\Sigma_n = f^{n-1}(\sigma_1) \cup f^{n-2}(\sigma_2) \cup \ldots \cup \sigma_n$$

which ends at $f_v^n(p)$. Since $Df^n(v) \in H_v$ we see that $f^\ell(\sigma_{n-\ell})$ is a curve pointing everywhere into the translates of the half space H_v, $0 \le \ell \le n-1$ and that the final segment σ_n of Σ_n is straight, has length $|v|$, and makes an angle $\ge\beta$ with ∂H_v. Thus

(3) the last end point of Σ_n, $f_v^n(p)$, lies inside H_v translated to $f^n p$ and lies no closer than $|v|\sin\beta$ to $f^n p$.

When ε is small and $|v| \le \varepsilon$, f_v C^∞-approximates f. Fix such an ε. Then choose

(4) $\delta < \varepsilon \sin(\beta)$

and fix an n with

(5) $|f^n(p) - p| < \delta$

Call $B_\varepsilon = \{v \in \mathbb{R}^2 : |v| \le \varepsilon\}$ and examine the map

$$F : \begin{array}{l} B_\varepsilon \longrightarrow \mathbb{R}^2 \\ v \longmapsto f_v^n(p) - f^n p \end{array}$$

From (3) we have

(6) $$\begin{array}{l} |F(v)| \ge |v|\sin(\beta) . \\ \measuredangle(F(v),v) > \pi \qquad v \ne 0 . \end{array}$$

From (4), (6) it follows that $F|\partial B_\varepsilon$ has index 1 respecting each point of B_δ. Hence

$F(B_\varepsilon) \supset B_\delta$

and hence, from (5), for some $v \in B_\varepsilon$,

$f_v^n(p) - f^n(p) = p - f^n(p)$

i.e., $f_v^n(p) = p$.

Q.E.D.

To formulate a local non-reversing condition that implies the Closing Lemma seems to require the following concept.

Definition 3. A codimension-one foliation $\mathfrak{F}$ of a manifold M is monotone if no loop everywhere transverse to $\mathfrak{F}$ is homotopic to zero.

Remarks. Monotonicity of $\mathfrak{F}$ requires that the lifted foliation $\tilde{\mathfrak{F}}$ of the universal cover $\tilde{M}$ of M has no closed transversals. Linear 2-foliations of T^3 are monotone, as are all analytic, codimension-one foliations [1]. All 1-foliations of 2-manifolds are monotone. The Reeb foliation of S^3 is not monotone.

Definition 4. $f : M \to M$ is non-reversing at p if there exist a coordinate neighborhood U of p and an open set N, $M \supset N \supset U$, such that

(7) $f(N) \subset N$

(8) for each constant vectorfield v in U there is a monotone oriented, codimension-one C^1 lamination $\mathcal{H}_v$ of N and

(i) $\measuredangle(v, \mathcal{H}_v) \geq \beta > 0$ on U

(ii) if γ is any v-integral curve then $f^n(\gamma)$ points non-negatively across $\mathcal{H}_v$, $n \geq 0$.

The angle is measured in the U-coordinate chart; β is a constant independent of v. Recall that a C^1 lamination is a C^0 foliation whose leaves are C^1 and whose leaf tangent field is continuous.

Remarks. (a) We permit $M = N$ so we include the global non-reversing condition of Definition 2: the lamination is formed by the parallel translations of the planes ∂H_v.

(b) If $M = N$ and $\dim(M) = 2$ then M is either T^2 or K^2 because other surfaces support no laminations.

(c) If f has a source s then the complement to the local basin or repulsion of s serves as a possible N in Definition 3, and this lets us handle diffeomorphisms on S^2.

(d) Non-linear Anosov diffeomorphisms of T^2 which preserve the orientations of their E^u and E^s bundles satisfy Definition 4 as do such DA-diffeomorphisms.

(e) $\mathcal{H}_v$ need neither be invariant nor depend continuously on v.

Theorem 3. If $f : M \to M$ is non-reversing at $p \in \Omega(f)$ then f can be C^r-approximated by g with $p \in \text{Per}(g)$.

Remarks. (a) Flows which have locally non-reversing Poincaré maps can be treated similarly; however, the semi-global nature of the hypothesis (namely N) becomes un-natural. For example, if p is a flow on an orientable M^2 having a recurrent point p then the Poincaré map f defined on a circle-transveral Γ through p preserves orientation ((8) of Definition 4) but existence of N in (7) precludes recurrent separatrices of saddle-points, i.e. Γ must be the domain of f and φ its suspension. Thus, we exclude the interesting (and open!) problem of C^2-closing recurrent orbits of flows on surfaces.

(b) I do not know if Theorem 3 is true without monotonicity of the foliations $\mathcal{H}_v$.

Lemma 2. If S_0 is an embedded (m-1)-sphere in M^m that separates one point, x, from another, y, and if S_t is a homotopy of S_0 which misses x and y then S_1 also separates x and y.

Proof. Draw any arc α from x to y. The intersection number of α and S_0 equals 1 and it is independent of t, $0 \le t \le 1$. Hence S_1 also separates x from y. Q.E.D.

Proof of Theorem 3. Let p, U, N, β be as given in Theorem 3. Take a C^∞ bump function $\psi : M \to [0,1]$ such that $\text{supp}(\psi) \subset U$ and $\psi \equiv 1$ near p; say $p = 0$ in the U-coordinates and

$$|x| \le \nu_0 \Longrightarrow \psi(x) = 1$$

with $\nu_0 > 0$. For each constant vector field v on U let V_v be the vectorfield on M defined by

$$V_v(x) = \begin{cases} \psi(x)v & x \in U \\ 0 & x \notin U \end{cases}$$

and let $\tau_v : M \to M$ be its time-one map. Clearly

$$f_v = \tau_v \circ f$$

C^r-approximates f provided $|v|$ is small, say $|v| \le \varepsilon \le \frac{1}{2}\nu_0$. Fix this ε and determine ν

(9) $\quad \nu < \frac{1}{2}\varepsilon \sin \beta$

Since $p \in \Omega(f)$ we can find $q \in U$ and $n > 0$ such that

(10) $\quad |q - p| < \nu \qquad |f^n q - p| < \nu$

Fix such q and n. To prove Theorem 3 it suffices to find v, $|v| \le \varepsilon$, such that

(11) $\quad f_v^n(q) = q$

For then we set

$$g = h^{-1} \circ f_v \circ h$$

where h C^r-approximates the identity and sends p to q. Clearly g C^r-approximates f and $p \in \mathrm{Per}(g)$ by (11).

As in the proof of Theorem 2, construct a segmented curve $\Sigma_n = \Sigma_n(v)$ from $f^n q$ to $f_v^n(g)$ as follows, $n \ge 1$.

σ_1 is the unit V_v-integral curve starting at f(p)
σ_2 is the unit V_v-integral curve starting where $f(\sigma_1)$ ends
$\vdots$
σ_n is the unit V_v-integral curve starting where $f(\sigma_{n-1})$ ends

$$\Sigma_n = f^{n-1}(\sigma_1) \cup f^{n-2}(\sigma_2) \cup \ldots \cup f(\sigma_{n-1}) \cup \sigma_n = f(\Sigma_{n-1}) \cup \sigma_n$$

By "unit" we mean "time-length-one". Also, we reconsider the map

$$F : \begin{array}{l} B_\varepsilon \longrightarrow M \\ v \longmapsto f_v^n(q) \end{array}$$

By (7), $F(B_\varepsilon) \subset N$.

Fix a constant vectorfield u on U with $|u| = 1$ and regard the foliation $\mathcal{H}_u \cap U$. Think of u as "vertical" and think correspondingly of "above" and "below". Let $H_u^{\pm}$ denote the regions in U above and below the component in U of the $\mathcal{H}_u$-leaf through $f^n q$. See figure 4.

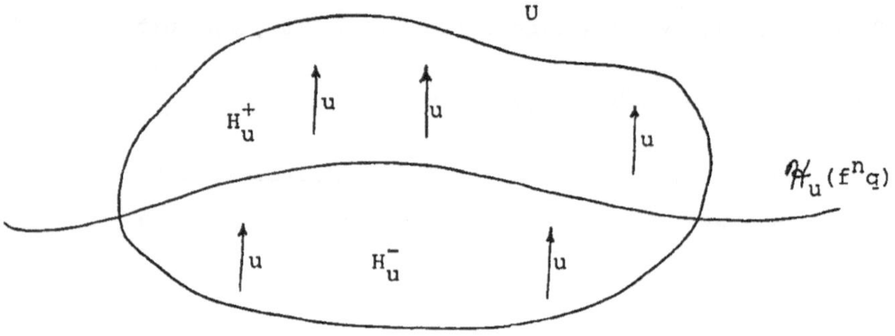

Figure 4. The regions $H_u^{\pm}$ in U.

For μ very small, say $0 \le \mu \le \mu_0$, the entire segmented curve $\Sigma_n(\mu u)$ lies in U; its final segment, $\sigma_n = \sigma_n(\mu u)$, has length μ and lies in H_u^+. Since the leaves of $\mathcal{H}_u$ make an angle of at least β with direction u, σ_n does not touch the sector of angle β around $-u$ at $f^n q$. See figure 5.

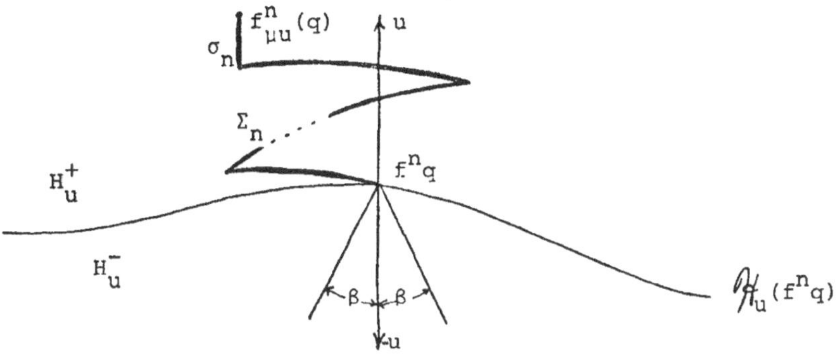

Figure 5. The curve $\Sigma_n(\mu u)$ for small μ

The closest that the upper end point of such a segment σ_n can be to $f^n q$ is $\mu \sin \beta$. Thus

$$\text{(12)} \qquad \begin{aligned} &|f^n_{\mu u}(q) - f^n q| \geq \mu \sin \beta \\ &\measuredangle(u, f^n_{\mu u}(q) - f^n q) < \pi \end{aligned}$$

for $0 < \mu \leq \mu_0$. Therefore

$$\text{(13)} \qquad \text{Index } (F|S_{\mu_0}, f^n q) = 1$$

where S_μ denotes the sphere of radius μ at 0 in B_ε. Now F is C^1, so (12) implies $(DF)_0$ is an isomorphism and $F|S_{\mu_0}$ is an embedded sphere. By (13) it surrounds $f^n q$ and since μ_0 is small, it separates $f^n q$ from q.

The curve $t \mapsto F(tv)$ is not necessarily transverse to $\mathcal{H}_v$, but it is homotopic to $\Sigma_n(v)$ and $\Sigma_n(v)$ is transverse to $\mathcal{H}_v$. To see this, select a continuous parameterization of the family of curves $\Sigma_n(\mu v)$ $\quad 0 \leq \mu \leq 1$:

$$t \mapsto \Sigma_n(\mu v)(t) \qquad 0 \leq t \leq 1$$

and consider the homotopy

$$H(s,t) = \begin{cases} F(tv) & t \geq s \\ \Sigma_n(sv)(t/s) & t \leq s \end{cases}$$

Note that for $t = s$ we have equality:

$$\Sigma_n(sv)(1) = F(sv)$$

As $s,t \to 0$, the curve $\Sigma_n(sv)$ tends uniformly to the point $f^n q$ so continuity of H is assured. At $s = 0$, H is $F(tv)$ and at $s = 1$ it is $\Sigma_n(v)(t)$. By monotonicity of $\mathcal{H}_v$ we conclude

(14) If $F(v) = f^n q$ for some $v \neq 0$ then the loop $t \mapsto F(tv)$ is not homotopic to zero in N.

Let $\tilde{N}$ be the universal covering of N. Lift the segment $[f^n q, q]$ in U to a segment $\alpha = [\alpha_0, \alpha_1]$ in $\tilde{N}$ and lift F to $\tilde{F}$ so that

$$\tilde{F}(0) = \alpha_0 \qquad F(0) = f^n q \qquad \begin{array}{ccc} & & \tilde{N} \\ & \overset{\tilde{F}}{\nearrow} & \downarrow \\ B_\varepsilon & \xrightarrow{F} & N \end{array} \qquad \begin{array}{c} [\alpha_0, \alpha_1] \\ \downarrow \\ [f^n q, q] \end{array}$$

By construction, $\tilde{F}|S_{\mu_0}$ is a sphere separating α_0 from α_1 in $\tilde{N}$ and by Lemma 2 either

(a) $\alpha_0 \in \tilde{F}(S_\mu)$ for some $\mu, \mu_0 \leq \mu \leq \varepsilon$

or (b) $\alpha_1 \in \tilde{F}(S_\mu)$ for some $\mu, \mu_0 \leq \mu \leq \varepsilon$

or (c) $\tilde{F}|S_\varepsilon$ intersects $[\alpha_0, \alpha_1]$.

(a) is impossible. For if $\tilde{F}(\mu u) = \alpha_0, \mu \neq 0$, then $F(\mu u) = f^n q$ and by (14) the loop $t \mapsto F(t\mu u)$ is not homotopic to zero. But $t \mapsto \tilde{F}(t\mu u)$ covers this loop and begins at α_0; since $\tilde{N}$ is simply connected, it cannot end at α_0 also; i.e. $\tilde{F}(\mu u) \neq \alpha_0$.

(c) is impossible. Suppose $\tilde{F}(\varepsilon u) \in [\alpha_0, \alpha_1]$. Then $F(\varepsilon u)$ lies in $B_\nu(p)$ and the curve $\Sigma_n(\varepsilon u)$ has its final segment σ_n of length ε. Our choice of ν in (9), $\nu < \frac{1}{2}\varepsilon \sin \beta$, forces the lower endpoint of σ_n, say y, to lie below $\mathcal{H}_u(f^n q)$; i.e. $y \in H_u^-$. See figure 6.

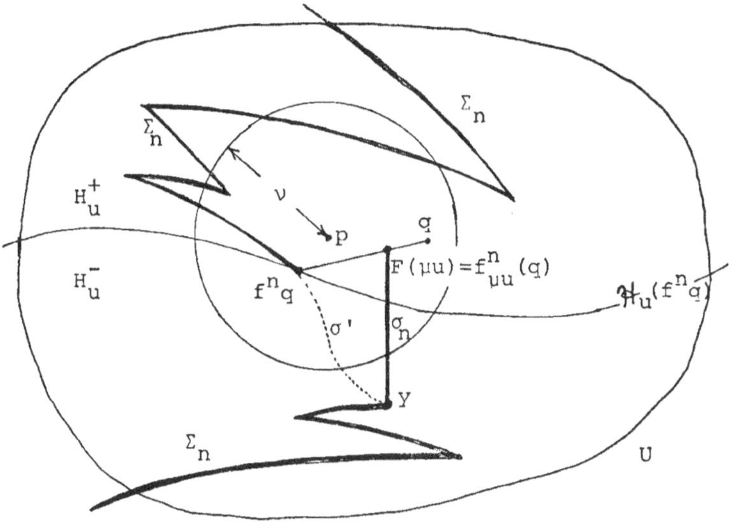

Figure 6. Σ_n approaching $f^n q$ from below

The curve $\sigma_n \cup [F(\mu u), f^n q]$ from y to $f^n q$ is homotopic to a curve σ' from y to $f^n q$ which is transverse to $\mathcal{H}_u$. Let Σ' be the loop formed by adjoining σ' to the part of Σ_n from $f^n q$ to y. Then Σ' is a loop at $f^n q$ transverse to $\mathcal{H}_u$. Monotonicity of $\mathcal{H}_u$ implies Σ' is not homotopic to zero. Lift Σ_n to a curve $\tilde{\Sigma}_n$ in $\tilde{N}$ starting at α_0. The corresponding lift $\tilde{\Sigma}'$ of Σ' does not end at α_0, but at some other α_0' projecting onto $f^n q$. Points near α_0' (such as the second end point of $\tilde{\Sigma}_n$) are far from α_0; α_0 and α_0' are on different sheets of $\tilde{N}$. Since $t \mapsto F(tv)$ is homotopic to $\Sigma_n(v)$, the second end point of $\tilde{\Sigma}_n$ equals $\tilde{F}(\varepsilon u)$, so $\tilde{F}(\varepsilon u) \notin [\alpha_0, \alpha_1]$.

Therefore (b) must hold:

$$\tilde{F}(\mu u) = \alpha_1$$

for some μ, $0 < \mu \leq \varepsilon$. Hence $F(\mu u) = q$, i.e. (11), which completes the proof of Theorem 3. Q.E.D.

References

1. A. Haefliger, Structures feuilletées et cohomologie à valeurs dans un faisceau de groupoïdes. Comm. Math. Helv. 32 (1958) p.317.

2. M. Peixoto, Structural Stability on Two-dimensional Manifolds, Topology, 1 (1962), pp.101-120.

3. C. Pugh, The Closing Lemma, Amer. J. Math. 89 (1967) pp.956-1009.

4. C. Pugh, An Improved Closing Lemma and General Density Theorem, Amer. J. Math., 89 (1967), 1010-1021.

5. C. Pugh and R.C. Robinson, The C^1 Closing Lemma including Hamiltonians, to appear in the Journal of Ergodic Theory and Dynamical Systems.

6. S. Smale, Differentiable Dynamical Systems, Bull AMS, 73 (1967), p.780.

Capture in Resonance: Opening a Homoclinic Orbit through Slowly Varying Coefficients

by Clark Robinson*

Abstract: Capture in sustained resonance is proved to occur for a system of nonlinear ordinary differential equations with slowly varying coefficients which model a rolling and pitching reentry vehicle. Capture is proved to occur by showing that the homoclinic orbit opens up for the small parameter $\varepsilon > 0$. The size of the opening is measured using the Melnikov integral. Since this integral has usually been applied to time periodic perturbations, it is derived for systems with slowly varying coefficients.

§1. Statement of Results

J. Kevorkian introduced the following equations as a model for roll/pitching resonance of a reentry vehicle with $r(u,\sigma) = u/2$, [5]:

$$\begin{aligned} \ddot{q} &= -(p^2+u^2)q \\ \dot{p} &= \varepsilon u^2 q \sin \psi \\ \dot{\psi} &= 2^{1/2} p \\ \dot{u} &= \varepsilon r(u,\sigma) \qquad \sigma = \varepsilon t \end{aligned} \tag{1.1}$$

where q is the pitch angle, ψ is the roll angle, p is the roll rate, u is the natural pitch frequency when roll is not present, and the small dimensionless parameter ε is related to the change of atmospheric density at high altitude and also the distance

---

* This research was partially supported by NSF Grant MSC 81-02177. A.M.S. Classification 34C35, 70K30.

of the center of mass from the axis of the vehicle. W. Kath suggested allowing the more general form of $r(u,\sigma) > 0$, [13].

If $p(0) > u(0)$, then the trajectory for (1.1) enters the resonant band where $p \approx u$ but does not always remain there. A trajectory of equations (1.1) is said to be in <u>sustained resonance</u> if $|u(t)-p(t)| = O(\varepsilon^{1/2})$ for a time interval $t_o \leq t \leq C_1/\varepsilon$, [6, p. 748]. For a trajectory in sustained resonance, the roll rate p increases as fast as u. The following theorem gives the existence of such trajectories.

1.2 <u>Theorem</u>: For $2r(u_o,\sigma) < u_o^2 w_o$ and

$$0 > \frac{d}{d\sigma}\left[r(u,\sigma)u^{-13/8}\right]\Big|_{u=u_o} = \frac{\partial r}{\partial u} r u_o^{-13/8} + \frac{\partial r}{\partial \sigma} u_o^{-13/8} - \frac{13}{8} u_o^{-21/8} r^2,$$

there exist $t_o > 0$, $C_1 > 0$, $C_o > 0$, and $\varepsilon_o(w_o,u_o) > 0$ such that for $0 < \varepsilon \leq \varepsilon_o(w_o,u_o)$ there exist trajectories $(q(t,\varepsilon),p(t,\varepsilon),\psi(t,\varepsilon),u(t,\varepsilon))$ for (1.1) with $u(0,\varepsilon) = u_o$, $p(0,\varepsilon) > u_o$, and $w_o = [q(0,\varepsilon)^2+\dot{q}(0,\varepsilon)^2 u_o^{-2}]^{1/2}$ which are captured in sustained resonance with $|u(t,\varepsilon)-p(t,\varepsilon)| \leq C_o\varepsilon^{1/2}$ for $t_o \leq t \leq C_1/\varepsilon$. For these trajectories the roll rate $p(t,\varepsilon)$ increases on this time interval at the same rate as $u(t,\varepsilon)$.

As remarked by Kath, [13], if u increases too fast, $2r(u,\sigma) > u^2 w$, then the resonance is broken. This rate of increase can be modified by changing either the reentry angle or velocity, see [13] or [5] for related discussion.

The proof of this theorem is given in section 3. The first step is to use the method of higher order averaging to change equations (1.1) into equations of the form

$$(1.3)\qquad \begin{aligned} \phi' &= p_o(\phi,v,w) + \mu p_1(\phi,v,w) + O(\mu^2) \\ v' &= q_o(\phi,v,w) + \mu q_1(\phi,v,w) + O(\mu^2) \\ w' &= \mu r(\phi,v,w) + O(\mu^2) \end{aligned}$$

where ϕ and v are scalars, w is a vector quantity, and $\mu = \varepsilon^{1/2}$. Letting $x = (\phi,v)$ equations (1.3) can be written as

$$(1.3)'\qquad \begin{aligned} x' &= f_o(x,w) + \mu f_1(x,w) + O(\mu^2) \\ w' &= \mu r(x,w) + O(\mu^2). \end{aligned}$$

The derived equations satisfy the following homoclinic assumptions: for $\mu = 0$, the equations $x' = f_o(x,w)$ have a hyperbolic saddle fixed point with a homoclinic orbit (saddle connection). The following theorem is an application of the persistence of a normally hyperbolic invariant compact manifold.

1.4 Theorem: Let K be a compact domain in w-space such that for w in K, equations (1.3) for $\mu = 0$ have a hyperbolic fixed point. Then there exists $\mu_1(K) > 0$ such that for $0 \leq \mu \leq \mu_1(K)$ there is an invariant (see remark below) manifold for (1.3)

$$M_\mu = \{(x,w) : x = g(w,\mu),\ w \in K\}$$

where g is a C^1 function of w and μ and $g(w,0)$ are the hyperbolic saddle points. This manifold is normally hyperbolic with codimension one stable and unstable manifolds $W^s(\mu)$ and $W^u(\mu)$ which depend differentially on μ.

Remark: For $\mu > 0$, $w' \neq 0$ so trajectories can have $w(t)$ leave K. The manifold M_μ is invariant in the weaker sense that the only way a trajectory $(x(t),w(t))$ can leave M_μ is for $w(t)$ to leave the compact domain to which the theorem applies. This takes time $O(1/\mu)$.

Proof:

Since K is not (necessarily) a compact boundaryless manifold, it is necessary to embed K into a compact boundaryless manifold, S, and extend equations (1.3) to a neighborhood of the manifold

$$M_o = \{(x,w) : w \in S,\ x = g(w,0)\}$$

so it is normally hyperbolic when $\mu = 0$. Then for $\mu > 0$, the existence of M_μ, stable manifold $W^s(\mu)$, and unstable manifold $W^u(\mu)$ near all of S follows from the usual theorem, e.g. [3, Theorem 4.1]. Restricting back to a neighborhood of K gives the result.

The separation of the manifolds $W^s(\mu)$ and $W^u(\mu)$ for $\mu > 0$ can be calculated using the Melnikov integral. Fix w_o and let $x_o(t) = x_o(t,w_o)$ be the homoclinic orbit and $f_o(x_o(t)) = f_o(x_o(t),w_o)$. Let L be a line in x-space through $x_o(0)$ and perpendicular to $f_o(x_o(0))$. Let $x^s(t,\mu)$ be the orbit of (1.3) in $W^s(\mu)$ with $x^s(0,\mu)$ in L (with w(t) having $w(0) = w_o$). Similarly let $x^u(t,\mu)$ be an orbit in $W^u(\mu)$. Define

$$\Delta(t,w_o,\mu) = [x^u(t,\mu)-x^s(t,\mu)] \wedge f_o(x_o(t))$$

and

$$\Delta_1(t,w_o) = \frac{\partial}{\partial\mu}\Delta(t,w_o,\mu)\Big|_{\mu=0}$$

$$= \Delta_1^u(t) - \Delta_1^s(t)$$

where

$$\Delta_1^j(t) = \frac{\partial}{\partial\mu} x^j(t,\mu) \wedge f_o(x_o(t)) \qquad \text{for } j = u,s.$$

The function $\Delta_1(0,w_o)$ measures the infinitesimal separation of the stable and unstable manifolds along L as a function of μ. Figure 1 shows the position of the manifolds for different signs of Δ_1.

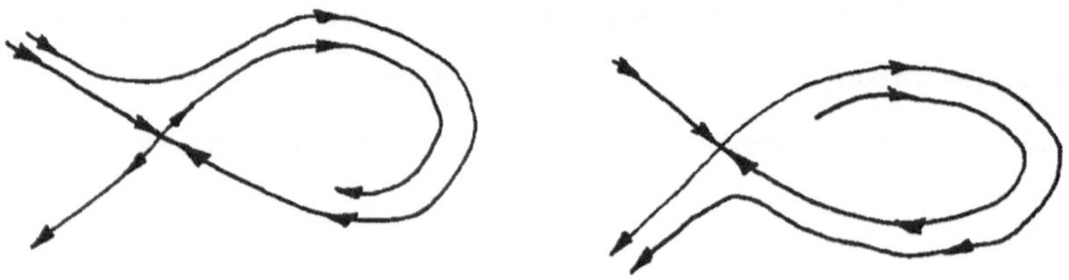

A. $\Delta_1 > 0$, $\mu > 0$ B. $\Delta_1 < 0$, $\mu > 0$

Figure 1

The following result shows how $\Delta_1(0,w_o)$ can be calculated. Compare with [4], [12], and [1].

1.5 <u>Theorem</u>: Assume equations (1.3) satisfy the homoclinic assumption and f_o is divergence free as a function of x. Then

$$\Delta_1(0,w_o) = \int_{-\infty}^{\infty}\left\{f_1(x_o(t),w_o) + \frac{\partial f_o}{\partial w}(x_o(t),w_o)\frac{\partial w}{\partial\mu}\right\}\wedge f_o(x_o(t),w_o)dt$$

$$= \int_{-\infty}^{\infty} q_o p_1 - p_o q_1 + \left(q_o\frac{\partial p_o}{\partial w} - p_o\frac{\partial q_o}{\partial w}\right)\frac{\partial w}{\partial\mu}\,dt$$

where $\partial w/\partial\mu$ satisfies

$$\left(\frac{\partial w}{\partial \mu}\right)' = r(x_o(t),w_o) \quad \text{and} \quad \left(\frac{\partial w}{\partial \mu}\right) = 0 \quad \text{at } t = 0.$$

The proof is given in section 2. We end this section with a simple example which illustrates the calculation of Δ_1. A similar calculation is given for (1.1) in section 3 after it is put in a standard form by the method of higher order averaging.

1.6 Example: Consider the system of equations

$$\dot{\phi} = v$$

$$\dot{v} = f(w) \cos \phi - g(w)$$

$$\dot{w} = \mu r(\phi,v,w)$$

where ϕ and v are scalars and w can be a vector. Letting $f_w = (\partial f/\partial w)$ and $g_w = (\partial g/\partial w)$, assume there is a domain K such that for w in K the following conditions are satisfied:

(i) $-1 < g(w)/f(w) < 1$ and $f(w) > 0$

(ii) $\mu^{-1} \frac{d}{dt} f(w) = f_w \cdot r \geq 0$ (i.e. f is nondecreasing on the homoclinic orbit)

(iii) $-\mu^{-1} d[g(w)/f(w)]/dt = f^{-2}(-fg_w+gf_w)\cdot r \geq 0$ (i.e. f increases at least as fast as g)

(iv) one of the inequalities of (ii) and (iii) is strict at some point on the homoclinic orbit.

Then the system has a saddle point with a homoclinic orbit and $\Delta^1(0,w) > 0$ for all w in K.

Proof:

The first step is to take a new time scale s which solves $ds/dt = f(w)^{1/2}$ and let (') be d/ds. This is equivalent to taking a scalar multiple of the vector field which varies by the point. (This does not change the position of $W^u(\mu)$ and $W^s(\mu)$.) Letting $V = vf(w)^{-1/2}$,

$$V' = \dot{v}f^{-1} - v(1/2)f^{-1}f_w \cdot \mu r f^{-1/2},$$

so the equations become

$$\phi' = V$$

$$V' = \cos\phi - g(w)/f(w) - \mu(1/2)Vf(w)^{-1}f_w \cdot r$$

$$w' = \mu r(\phi, Vf(w)^{1/2}, w)f(w)^{-1/2}.$$

Take the parametrization of the homoclinic orbit so that $V(0) = 0$ and $-sV(s) \geq 0$. The Melnikov integral is

$$\Delta_1(0,w) = -\int_{-\infty}^{\infty} p_o q_1 ds - \int_{-\infty}^{\infty} p_o(\partial q_o/\partial w)\cdot(\partial w/\partial\mu)ds$$

$$= \int_{-\infty}^{\infty} (1/2)V^2 f^{-1} f_w \cdot r ds + \int_{-\infty}^{\infty} (-V)f^{-2}(-fg_w + gf_w)\cdot(\partial w/\partial\mu)ds.$$

The first integral is positive by (ii). In the second integral f, f_w, g, and g_w are constant along the homoclinic orbit (w is constant) so

$$\frac{d}{ds}\left\{(-fg_w+gf_w)\cdot(\partial w/\partial\mu)\right\} = (-fg_w+gf_w)\cdot rf^{-1/2} \geq 0$$

by (iii). Thus $(-fg_w+gf_w)\cdot(\partial w/\partial\mu)$ has the same sign as s. Since $-V$ also has the same sign as s, the second integral is ≥ 0. By (iv) one of these two integrals is strictly positive.

§2. Proof of Theorem 1.5

The fact that $(\partial w/\partial\mu)$ satisfies $(\partial w/\partial\mu)' = r(x_o(t), w_o)$ follows from the theorem on dependence of solutions on a parameter, [2, Theorem V.3.1], using the fact that for $\mu = 0$ $w(t) = w_o$ is a constant.

The fact that the derivative with respect to t of $\Delta_1^s(t,w_o)$ equals the integrand is essentially the same as [4] using the fact that

$$(\partial x^s/\partial\mu)' = Df_o(\partial x^s/\partial\mu) + f_1 + (\partial f_o/\partial w)\cdot(\partial w/\partial\mu)$$

by [2, Theorem V.3.1]. Thus

$$-\Delta_1^s(0) = -\Delta_1^s(T) + \int_0^T \left\{\ \right\} \wedge f_o dt$$

Similarly

$$\Delta_1^u(0) = \Delta_1^u(-T) + \int_{-T}^0 \left\{\ \right\} \wedge f_o dt.$$

Taking the limit as T goes to infinity, it remains to prove the following lemma.

2.1 Lemma: $$\lim_{T \to \infty} \Delta_1^s(T) = 0 = \lim_{T \to \infty} \Delta_1^u(-T).$$

Proof:

The two limits are similar. The term $\Delta_1^s(T) = (\partial x^s/\partial\mu) \wedge f_o(x_o(T))$ goes to zero because $f_o(x_o(T))$ goes to zero exponentially fast (as $x_o(T)$ approaches the fixed point) and $\partial x^s/\partial\mu$ is shown below to be bounded by CT.

To check the growth of $\partial x^s/\partial\mu$ for $\mu = 0$, its time derivative is compared with $d/dT\left[\frac{\partial}{\partial\mu} g(w(T,\mu),\mu)\right]$ for $\mu = 0$ where $x = g(w,\mu)$ defines M_μ and $w(T,\mu)$ is the value of w at time T on the trajectory corresponding to $x^s(T,\mu)$. See Figure 2. At $\mu = 0$

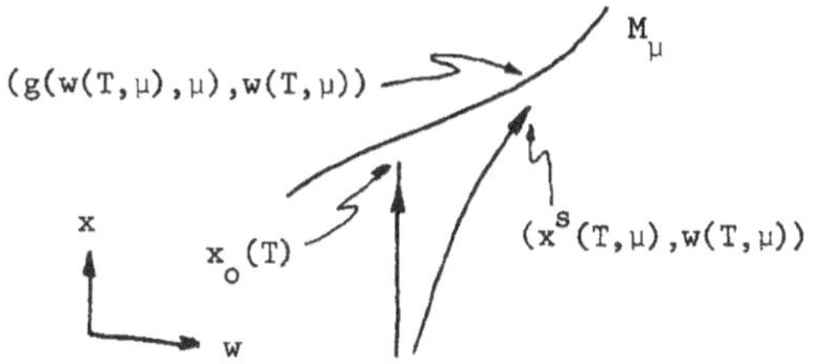

Figure 2

$$\frac{d}{dT}\left[\frac{\partial g}{\partial\mu}(w(T,\mu),\mu)\right] = \frac{d}{dT}\left[\frac{\partial g}{\partial\mu}(w_o,0)\right] + \frac{d}{dT}\left[\frac{\partial g}{\partial w}(w_o,0)\frac{\partial w}{\partial\mu}(T,0)\right]$$
$$= \frac{\partial g}{\partial w}(w_o,0)\, r(g(w_o,0),w_o)$$

is independent of T. As T goes to infinity $x^s(T,\mu)$ approaches $g(w(T,\mu),\mu)$ exponentially fast, so $\frac{d}{dT}[\partial x^s/\partial\mu(T,0)]$ is also a bounded function of T and $\partial x^s/\partial\mu(T,0)$ is bounded by CT for some C.

§3 Proof of Theorem 1.2

The first step of the proof is to put equations (1.1) in a standard form. Compare with [11]. The new variables w, ξ, x are defined by

$$q = w \sin \xi$$
$$(3.1) \qquad \dot{q} = w(p^2+u^2)^{1/2} \cos \xi$$
$$x = p/u .$$

Expanding the resulting differential equations in finite Fourier series yields sines and cosines in $\psi-\xi$, $\psi+\xi$, $3\xi+\psi$, and $3\xi-\psi$. Since $\dot{\psi} = 2^{1/2}xu$ and $\dot{\xi} = u(1+x^2)^{1/2}+0(\varepsilon)$, the only resonance of these angles occurs at x = 1 where $\theta = \psi-\xi$ varies slowly. At x = 1 all the Fourier terms average to zero except those with θ. Then by the method of higher order averaging, [8], there exists a change of variables

$$(3.2) \qquad (\phi,z,y,\eta,u) = (\theta,x,w,\psi,u) + O(\varepsilon)$$

such that the equations eliminate the terms which average to zero and leave error terms of order ε^2. It turns out that η does not appear in the resulting equations except for the terms of order ε^2. Since the terms of order ε determine the behavior, we drop η from further consideration.

In the resonant band about z = 1, letting $\mu = \varepsilon^{1/2}$, $\tau = \mu t$, $v = d\phi/d\tau$, and $h = uy \cos \phi - 2u^{-1}r$, the equations become

$$\frac{d\phi}{d\tau} = v$$
$$\frac{dv}{d\tau} = (8^{-1/2})uh - \mu(3/16)hv + \mu(1/8)u^{-1}rv + O(\mu^2)$$
$$(3.3) \qquad \frac{dy}{d\tau} = -\mu(1/4)yu^{-1}r - \mu(1/16)uy^2 \cos \phi + O(\mu^2)$$
$$\frac{du}{d\tau} = \mu r(u,\sigma) \qquad \sigma = \mu\tau$$

and $z-1 = \mu 2^{1/2}vu^{-1} + O(\mu^2)$. These equations are similar to [11, 8.19]. For $2r < uy$ and $\mu = 0$ equations (3.3) have hyperbolic saddle fixed points so by Theorem 1.4 there are stable and unstable manifolds for the invariant set for $\mu > 0$. The following lemma proves that for $\mu > 0$ these manifolds are as in Figure 1A.

3.4 Lemma: For $2r < u^2 y$ and $0 > \frac{d}{d\sigma}[r(u,\sigma)u^{-13/8}]$ the integral $\Delta_1(0,y,u) > 0$ for equations (3.3).

Proof:

As in Example 1.6, it is helpful to reparametrize the trajectories by introducing a new time scale s by solving $ds/d\tau = uy^{1/2}$. Letting (') be d/ds, $V = \phi'$, and $H = 8^{-1/2}\cos\phi - 8^{-1/2}2y^{-1}u^{-2}r$, the equations become

$$
\begin{aligned}
\phi' &= V\\
V' &= H - \mu 8^{1/2}(5/32)y^{1/2}VH - \mu(11/16)u^{-2}y^{-1/2}Vr + O(\mu^2)\\
y' &= -\mu(3/16)y^{1/2}u^{-2}r - \mu(8^{1/2}/16)y^{3/2}H + O(\mu^2) \qquad (3.4)\\
u' &= \mu y^{1/2}u^{-1}r\\
\sigma' &= \mu y^{-1/2}u^{-1}.
\end{aligned}
$$

Then as in Example 1.6

$$
\begin{aligned}
\Delta_1(0,y,u) &= -\int_{-\infty}^{\infty} p_o\left[q_1 + \frac{\partial q_o}{\partial u}\frac{\partial u}{\partial \mu} + \frac{\partial q_o}{\partial y}\frac{\partial y}{\partial \mu} + \frac{\partial q_o}{\partial \sigma}\frac{\partial \sigma}{\partial \mu}\right]ds\\
&= 8^{1/2}(5/32)y^{1/2}\int_{-\infty}^{\infty} V^2 H\,ds + (13/16)y^{-1/2}u^{-2}r\int_{-\infty}^{\infty} V^2\,ds\\
&\quad + 8^{-1/2}u^{-3}y^{-3/2}\left[-2r\frac{\partial r}{\partial u} - 2\frac{\partial r}{\partial \sigma} + (13/4)u^{-1}r^2\right]\int_{-\infty}^{\infty} s(-V)\,ds.
\end{aligned}
$$

The first integral is different than those in Example 1.6 but it integrates to zero:

$$
\begin{aligned}
\int_{-\infty}^{\infty} V^2 H\,ds &= \int_{-\infty}^{\infty} V^2 V'\,ds\\
&= \frac{1}{3}V^3\Big|_{-\infty}^{\infty}\\
&= 0.
\end{aligned}
$$

The second integral is clearly positive. The last integral is not exactly like Example 1.6 but is positive if

$$
-2r\frac{\partial r}{\partial u} - 2\frac{\partial r}{\partial \sigma} + (13/4)u^{-1}r^2 > 0.
$$

Since $\frac{du}{d\sigma} = r(u,\sigma)$, this is equivalent to

$$0 > \frac{\partial r}{\partial u}\frac{du}{d\sigma} u^{-13/8} + \frac{\partial r}{\partial \sigma} u^{-13/8} - (13/8)u^{-21/8}r^2$$

or

$$0 > \frac{d}{d\sigma}[r(u,\sigma)u^{-13/8}].$$

To complete the proof of Theorem 1.2, take $u_o w_o > 1$. By the change of variables (3.1) and (3.2) this corresponds to $u_o y_o > 1$. Take a compact domain K of $\{(y,u) : uy > 1\}$ containing (y_o,u_o). By Lemma 3.4, $\Delta_1(0,y,u) > 0$ for (y,u) in K. Therefore there is a $\mu_o(K) > 0$ such that for $0 < \mu \leq \mu_o(K)$ the manifolds for equations (3.3) are as in Figure 1A or 3. There is an opening between the stable and the unstable manifolds through which a trajectory starting outside the resonant band can enter and become trapped in the region where v is bounded. (The trajectory can

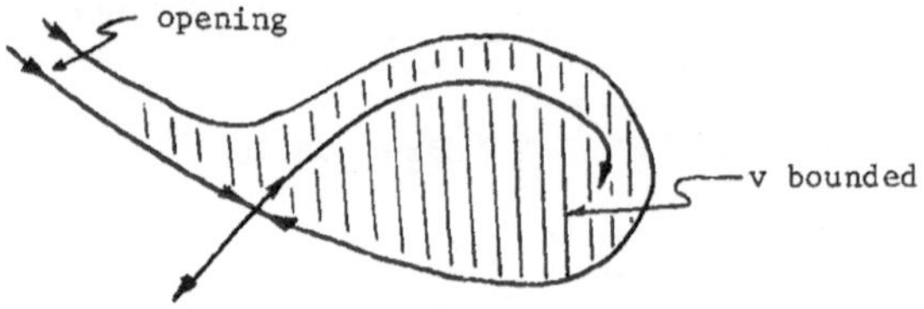

Figure 3

not cross $W^s(\mu)$.) The only way the trajectory can leave this region is for $(y(t,\varepsilon),u(t,\varepsilon))$ to leave K which takes time at least $s > C_1/\mu$ or $t > C_1/\varepsilon$. Therefore $|v(t)| \leq C_2$ for $t_o \leq t \leq C_1/\varepsilon$. For C_1 possibly smaller u is bounded by C_3 for $0 \leq \sigma \leq C_1$ or $0 \leq t \leq C_1/\varepsilon$. Then

$$\begin{aligned} |p(t)-u(t)| &= |u(t)[x(t)-1]| \\ &= |u(t)[z(t)-1+0(\varepsilon)]| \\ &= |\varepsilon^{1/2}v(t)+u(t)0(\varepsilon)| \\ &\leq \varepsilon^{1/2}C_2 + C_3 0(\varepsilon) \\ &\leq \varepsilon^{1/2}C_o. \end{aligned}$$

Therefore for such trajectories the roll rate increases at the same rate as $u(t,\varepsilon)$.

Remark 1: The trajectory can not cross the boundary $u^2y - 2r = 0$ because of the form of (3.3): on $u^2y - 2r = 0$ with $0 > u^{13/8}\frac{d}{d\sigma}[ru^{-13/8}] = \frac{\partial r}{\partial u} r + \frac{\partial r}{\partial \sigma} - \frac{13}{8}u^{-1}r^2$,

$$\frac{d}{d\sigma}(u^2y-2r) = 2uyr + u^2(-\frac{1}{4}yu^{-1}r - \frac{1}{16}uy^2\cos\phi) - 2\frac{\partial r}{\partial u}r - 2\frac{\partial r}{\partial \sigma}$$

$$\geq uyr(2 - \frac{1}{4} - \frac{1}{8}\cos\phi) - \frac{13}{4}u^{-1}r^2$$

$$\geq \frac{13}{8}u^{-1}r(u^2y-2r) = 0.$$

The only reason a compact set K is used is to get a uniform μ_o. If all the estimates could be shown to hold uniformly then there would be a single μ_o for all $u^2y - 2r(u,\sigma) > 0$ and the sustained resonance would hold for all $t \geq t_o$.

Remark 2: Equations (3.3) preserve the energy function $E = \frac{1}{2}V^2 - \int H d\phi$ for $\mu = 0$. Let E_o be the energy on the homoclinic orbit. For $E < E_o$ (inside the homoclinic orbit) and $\mu > 0$ E is not monotonically decreasing (i.e. it is not a Lyapunov function), but the position of the trajectory on $v = 0$ after one loop around the energy surface relative to where it left is given by an integral like Δ_1:
if $V(0,\mu) = 0 = V(\tau(\phi_o),\mu)$ then

$$\frac{d}{d\mu}[\phi(\tau(\phi_o),\mu)-\phi_o]\cdot H(\phi_o)$$

$$= (13/16)y^{-1/2}u^{-2}r\int_0^{\tau(\phi_o)} V^2 ds$$

$$+ 8^{-1/2}u^{-3}y^{-3/2}[-2r\frac{\partial r}{\partial u} - 2\frac{\partial r}{\partial \sigma} + (13/4)u^{-1}r^2]\int_0^{\tau(\phi_o)} s(-V)ds$$

$$> 0.$$

Thus the trajectories spiral in toward an attracting invariant manifold which corresponds to the elliptic fixed points of equations (3.3) when $\mu = 0$.

References

1. S.N. Chow, J. Hale, and J. Mallet Paret, An example of bifurcation to homoclinic orbits, J. Diff. Equat. 37 (1980), pp. 351-373.

2. P. Hartman, Ordinary Differential Equations, John Wiley & Sons, New York/London/Sydney, 1964.

3. M. Hirsch, C. Pugh, M. Shub, Invariant Manifolds, Lecture Notes in Math 583 (1977), Springer Verlag, Berlin/Heidelberg/New York.

4. P. Holmes and J. Marsden, A partial differential equation with infinitely many periodic orbits, Archive for Rational Mechanics and Analysis, 76 (1981), pp. 131-166.

5. J. Kevorkian, On a model for reentry resonance, SIAM J. Appl. Math. 26 (1974), pp. 638-669.

6. L. Lewin and J. Kevorkian, On the problem of sustained resonance, SIAM J. Appl. Math. 35 (1978), pp. 738-754.

7. V.K. Melnikov, On the stability of the center for time periodic perturbations, Trans. Moscow Math. Soc. 12 (1963), pp. 1-57.

8. L.M. Perko, Higher order averaging and related methods for perturbed periodic and quasi-periodic systems, SIAM J. Appl. Math. 17 (1968), pp. 698-723.

9. C. Robinson, Sustained resonance for a nonlinear system with slowly varying coefficients, preprint, Northwestern Univ. 1981.

10. C. Robinson and J. Murdock, Some mathematical aspects of spin/orbit resonance II, Celestial Mech. 24 (1981), pp. 83-107.

11. J. Sanders, On the passage through resonance, SIAM Math. Anal. 10 (1979), pp. 1220-1243.

12. J. Sanders, Melnikov's method and averaging, to appear in Celestial Mechanics.

13. W. Kath, Necessary conditions for sustained reentry roll resonance, preprint, Cal. Tech., 1981.

Northwestern University, Evanston, Illinois

SMALL RANDOM PERTURBATIONS AND THE DEFINITION OF ATTRACTORS

David RUELLE

Introduction.

The computer study of dynamical systems has produced a number of "strange attractors" of dubious mathematical status. For instance the Hénon attractor (see Hénon [3]) has resisted mathematical understanding. It is probably not simply a long attractive periodic orbit. It is also uncertain *) whether it is of the form $\Lambda = \bigcap_{n \geq 0} f^n U$ where f is the Hénon map and U a neighborhood of Λ. We may call <u>attracting set</u> (for a map f) a compact set Λ such that $\Lambda = \bigcap_{k \geq 0} f^k U$ for some compact neighborhood U of Λ such that $f^k U \subset U$ for all sufficiently large n. This notion has been introduced and studied by Thom [1] and Smale [10]. (See also Williams [12]).

Since the "attractors" produced in computer studies are long orbits (or pseudoorbits) they must have some irreducibility property. We are thus led to trying to decompose an attracting set into indecomposable parts. It is however not always possible to decompose an attracting set into irreducible attracting sets (in any reasonable way). A convenient definition of attractors as irreducible objects, which are in general not attracting sets, follows the ideas of Conley [2] (and Hurley [4] : they would be chain-recurrent quasi attractors in the terminology of [4] **)). The main point of the present

*) In particular because of the work of Newhouse [6].

**) Conley [2] calls attractors what we want to call attracting sets.

paper will be a proof of some stability property of attractors (as we define them) under small random perturbations. In view of this it will appear reasonable that the computer produced "attractors" are indeed attractors in our sense. As one would expect, in any individual case a detailed study is necessary to be able to be more definite.

We shall here restrict our attention to discrete time dynamical systems on finite dimensional manifolds. Remarks on more general systems*) are made in another paper [8] which also contains considerations on physical applications.

Attracting sets.

Let M be a finite dimensional manifold, and $f : M \mapsto M$ a continuous map**). As indicated above, we say that a subset $\Lambda \neq \emptyset$ of M is an attracting set if it has a compact neighbourhood U with $\Lambda = \bigcap_{k \geq 0} f^k U$, and $f^k U \subset U$ for all sufficiently large n. In particular Λ is compact. We say that U is a fundamental neighborhood of the attracting set Λ. The open set $W = \bigcup_{n \geq 0} f^{-n} U$ is the basin of attraction of Λ; it is independent of the choice of U and consists of those $x \in M$ such that $f^t x \to \Lambda$ when $n \to \infty$.

If M is compact we may take $U = M$ and $\Lambda = \bigcap_{k \geq 0} f^k M$.

1. Proposition. If V is any neighborhood of Λ we have $f^n U \subset V$ for n large enough.

---

*) Unfortunately, it turns out that little can be said about continuous time dynamical systems (flows and semiflows)

**) The application which we have in mind is to differentiable maps. Our results are however valid for continuous maps of a locally compact metric space.

Taking V open, this is obvious by compactness since $\cap(f^k U \setminus V) = \emptyset$.

2. Proposition. $f\Lambda = \Lambda$

First note that

$$f\Lambda = f \bigcap_{k\geq o} f^k U \subset \bigcap_{k\geq o} f^{k+1} U = \bigcap_{k\geq 1} f^k U$$

Since $f^k U \subset U$ for large k , we have $f\Lambda \subset U$, hence

$$f\Lambda \subset \Lambda$$

Suppose now that $x \notin f\Lambda$, so that $\Lambda \subset f^{-1}(M\setminus\{x\})$ and $f^{-1}(M\setminus\{x\})$ is a neighborhood of Λ . By proposition 1, $f^n U \subset f^{-1}(M\setminus\{x\})$ for sufficiently large n , thus $f^{n+1}U \subset M\setminus\{x\}$, hence $\Lambda \subset M\setminus\{x\}$, hence $x \notin \Lambda$. We have shown that

$$\Lambda \subset f\Lambda$$

concluding the proof of the proposition.

Attractors.

Let again $f : M \mapsto M$ be a differentiable map of the manifold M . We recall that a sequence $(x_j)_{n_1}^{n_2}$ of elements of M is called an ε pseudo-orbit (of length $n_2 - n_1 \geq 0$, going from x_{n_o} to x_{n_2}) if

$$d(fx_j, x_{j+1}) < \varepsilon \quad \text{for } j = n_1, \ldots, n_2-1$$

Some choice of a Riemann metric d has been made*). Putting together an

*) Our study is largely independent of the choice of d , but we do not analyze this here.

ε pseudoorbit of length L going from a to b and an ε pseudoorbit of length L' going from b to c we obtain an ε pseudoorbit of length $L+L'$ going from a to c. Following Conley we say that $a \in M$ is *chain recurrent* if for each $\varepsilon > 0$ there is an arbitrarily long ε pseudoorbit going from a to a. The set of chain recurrent points (chain recurrent set) is closed.

Given $a,b \in M$ we write $a \succ b$ (which may be read "a goes to b") if for each $\varepsilon > 0$ there is an ε pseudoorbit going from a to b. It is easy to see that the relation $\succ$ is closed, and is a preorder. (In particular $a \succ a$ because $(a)_1^1$ is an ε pseudoorbit of length 0). Write $a \sim b$ if $a \succ b$ and $b \succ a$, then $\sim$ is an equivalence relation, and every equivalence class $[a]$ is closed. The quotient of $\succ$ by $\sim$ is an order relation. We write $[a] \geq [b]$ if $a \succ b$.

3. *Proposition. The following conditions on* $x \in M$ *are equivalent*

(a) x *is chain recurrent*

(b) $x \sim fx$

(c) *either* x *is a fixed point or* $\operatorname{card}[x] > 1$.

We always have $x \succ fx$. If x is chain recurrent, a δ pseudoorbit of length n from x to x will give an ε pseudoorbit of length $n-1$ from fx to x when δ is small enough. This proves (a) $\Rightarrow$ (b). It is clear that (b) $\Rightarrow$ (c) and it remains therefore to prove (c) $\Rightarrow$ (a). Clearly a fixed point is chain recurrent (use the ε pseudoorbit $(x,\ldots,x)$). If $\operatorname{card}[x] > 1$ we may choose $y \in [x]$ with $y \neq x$, and there is a long pseudoorbit going from x first to y and then back to x, so that x is chain recurrent.

We say that $[x]$ is a basic class if it satisfies any of the equivalent properties of Proposition 3. In particular $f[x] \subset [x]$.

We say that $[a]$ is an attractor if it is a minimal equivalence class, i.e., $a \succ x$ implies $a \sim x$.

4. Proposition. If $[a]$ is an attractor, then a is a basic class.

Since $a \succ fa$ we have indeed $a \sim fa$.

Basic classes in attracting set.

We shall now assume that an attracting set Λ is given, with a fundamental neighborhood U , and we shall study the basic classes in U . (It will turn out that they are actually in Λ) .

5. Lemma. Given any neighborhood V of Λ there are $\varepsilon > 0$ and N such that all ε pseudoorbits of length $> N$ starting at a point of U end at a point of V .

We may choose $\delta > 0$ so small that there is (by compactness) a neighborhood V' of Λ such that the δ neighborhood of V' is contained in $V \cap U$. In view of Proposition 1 we may then choose N such that $f^n U \subset V'$ for $n \geq N$ by uniform continuity of f on compact sets, we may choose $\varepsilon > 0$, such that all ε pseudoorbits of length N , $N+1,\ldots,2N-1$ starting at a point of U end at a point of V . It then follows that any ε- pseudoorbit of length $\geq N$ starting at a point of U ends at a point of V .

6. Proposition. Let U be a fundamental neighborhood of the attracting set Λ

(a) If $a \in \Lambda$ and $a \succ b$, then $b \in \Lambda$

(b) <u>If</u> $a \in U$, $a \succ b$, <u>and</u> b <u>is chain recurrent, then</u> $b \in \Lambda$

(c) <u>If</u> $a \in \Lambda$ <u>and</u> a <u>is chain recurrent, then</u> $f[a] = [a]$.

(d) <u>If</u> $a \in U$ <u>there is at least one attractor</u> $[b]$ <u>with</u> $a \succ b$

(a) In view of Proposition 2 there is $c \in \Lambda$ with $f^N c = a$ and therefore an ε pseudoorbit of length $\geq N$ from $c \in \Lambda$ to b . The Lemma then yields $b \in U$ for any neighborhood U of Λ , hence $b \in \Lambda$

(b) There are an ε pseudoorbit from a to b and an ε pseudoorbit of length $\geq N$ from b to b , hence an ε pseudoorbit of length $\geq N$ from a to b and the conclusion follows from the lemma.

(c) We have already noted after Proposition 3 that $f[a] \subset [a]$. If x is the last point but one on an ε pseudoorbit of length L going from a to a , let c be a limit of x when $\varepsilon \to 0$, $N \to \infty$. Then $c \in [a]$ and $fc = a$, so that $[a] \subset f[a]$.

(d) If c is an accumulation point of the forward orbit $\{f^k a : k \geq 0\}$, then $a \succ c$ and c is chain recurrent. It suffices thus to find an attractor $[b]$ with $a \succ b$. This is obtained by applying Zorn's Lemma to the basic classes $[x] \leq [b]$. We indeed see that a totally ordered family $([x_\alpha])$ is minorized by $[y]$ where y is a limit of the net (x_α) (such a limit exists by compactness).

7. <u>Corollary</u>. <u>An attracting set</u> Λ <u>is an attractor if and only if it is a basic class</u> (i.e., if it is chain recurrent).

If Λ is an attractor, it is a basic class by Proposition 4. If Λ is chain recurrent, it is an attractor by Proposition 6 (d).

Attractors in an attracting set.

We have seen in the previous section that an attracting set necessarily contains attractors (at least one). Here we characterize the attractors contained in an attracting set according to Conley.

8. Proposition. Let Λ be an attracting set and $a \in \Lambda$.

(a) If the class $[a]$ is an intersection of attracting sets then $[a]$ is an attractor.

(b) Conversely, if $[a]$ is an attractor it is the intersection of the attracting sets containing it.

(a) If $[a] = \cap \Lambda_\alpha$ and $a \succ b$, then $b \in \Lambda_\alpha$ by Proposition 6 (a), so that $b \in [a]$, showing that $[a]$ is an attractor.

(b) Let U_ε be the closure of the set of points x such that there is an arbitrarily long ε pseudoorbit from a to x. Then, the ε neighborhood of fU_ε is contained in U_ε. If U is a (compact) fundamental neighborhood of Λ, we may (by Lemma 5) take ε so small that $U_\varepsilon \subset U$, and therefore (again by Lemma 5)

$$K_\varepsilon = \bigcap_{k\geq 0} f^k U_\varepsilon \subset \bigcap_{k\geq 0} f^k U \subset \Lambda$$

Thus, K_ε is a compact subset of Λ. Since $K_\varepsilon \subset fU$, we see that U_ε is a neighborhood of K_ε. Since $f^n U_\varepsilon \subset U_\varepsilon$ for $n \geq 1$, K_ε is an attracting set with fundamental neighborhood U_ε. Finally, since $[a]$ is an attractor

$$[a] \supset \bigcap_{\varepsilon>0} U_\varepsilon \supset \bigcap_{\varepsilon>0} K_\varepsilon \supset [a]$$

so that $[a] = \bigcap_{\varepsilon > 0} K_\varepsilon$ as announced.

9. *Lemma*[*)]. *Let* Λ *be an attracting set and* K *a compact subset of* Λ .
Then

$$K^* = \bigcup_{z \in K} [z]$$

is compact, and for all $\theta > 0$ *there is* $\varepsilon > 0$ *such that* $d(y,K^*) < \theta$ *whenever* $y \succ x \in K$ *and there is a* 2ε *pseudoorbit from* x *to* y .

K^* is compact because $\sim$ is a closed relation. If, for some $\theta > 0$, ε did not exist as indicated, one could find sequences (x_n) , (y_n) such that $y_n \succ x_n$, there is a $\frac{1}{n}$ pseudoorbit from x_n to y_n , and the distance from y_n to K is $\geq \theta$. If $(\bar{x},\bar{y}) \in K \times \Lambda$ were a limit of (x_n,y_n) , we would have $\bar{x} \sim \bar{y}$ hence $\bar{y} \in K^*$ in contradiction with $d(\bar{y},K^*) \geq \theta$.

10. *Remark*.

According to Proposition 8, which is essentially due to Conley [2], we could define a set to be an attractor if it is chain recurrent and an intersection of attracting sets. (This definition is used by Hurley [4], with another terminology). It seems however that the definition in terms of the relation $\succ$ is applicable in more general situations (see [8]), and should be preferred.

Small random perturbations.

The map f induces a linear map f on Radon measures on M such that

---

*) To be used in the proof of Proposition 12.

$$(f^*\mu)(\varphi) = \mu(\varphi\circ f)$$

More generally we may say that a linear map F on Radon measures is a <u>diffusion</u> if it is positive and preserves total mass. Equivalently, we could define F as an affine map on probability measures. Here we shall be interested in diffusions close to f^* .

Let $\varepsilon > \delta > 0$. We say that an affine map F from the space of probability measures with compact support to itself is an (ε,δ) <u>diffusion</u> associated with f if the following conditions are satisfied*).

(A) <u>Continuity</u> : For every compact set $A \subset M$ there is a compact set $A^* \subset M$ such that F is continuous from the vague topology of measures on A to the vague topology of measures on A^* .

(B) <u>Support</u> : $\text{supp } F\delta_x \subset \bar{B}_{fx}(\varepsilon)$.

(C) "<u>Absolute continuity</u>"**) $\text{supp } F\delta_x \supset f\bar{B}_x(\delta)$.

If φ is a real valued continuous function on M , (a) implies that $x \mapsto (F\delta_x)\varphi$ is continuous on M , and

$$(F\mu)\varphi = \int \mu(dx)[(F\delta_x)\varphi] \quad . \tag{1}$$

---

*) δ_x denotes the unit mass at x, $B_x(\varepsilon)$ the open ball of radius centered at x , and $\bar{B}_x(\varepsilon)$ its closure.

**) This conditions is of course weaker than absolute continuity with respect to Lebesgue measure. The choice of $f\bar{B}_x(\delta)$ rather than $\bar{B}_{fx}(\delta)$ in the statement of the condition simplifies the formulation of

(This is checked by taking a finite sum $\Sigma\, \alpha_i \delta_{x_i}$ tending vaguely to μ).
In particular

$$\text{supp } F\mu = \text{closure} \bigcup_{x \in \text{supp } \mu} \text{supp } F\delta_x \tag{2}$$

11. Proposition. *Let* a *be such that* $[a]$ *is not an attractor for* $f : M \mapsto M$. *For sufficiently small* ε , *if* F *is an* (ε,δ)*-diffusion associated with* f , *and* ν *a probability measure with compact support in* M , *then*

$$\lim_{n\to\infty} (F^n \nu)(B_a(\delta)) = 0$$

We take b such that $a \succ b$ and $b \notin [a]$. For sufficiently small ε there is then no 2ε pseudoorbit from b to a . In particular, there is no ε pseudoorbit from $B_{fb}(\varepsilon)$ to $B_a(\delta)$.

If $x \in \overline{B}_a(\delta)$, property (C) yields $\text{supp } F\delta_x \supset f\overline{B}_x(\delta) \ni fa$. Since $a \succ b$ there is a δ pseudoorbit going from a to b , say $(x_o, x_1, \ldots, x_N)$ with $x_o = a, x_N = b$. We have just proved that $fx_o \in \text{supp } F\delta_x$ Using (C) and (2) we also see that $fx_1 \in \text{supp } F^2\delta_x, \ldots, fx_N \in \text{supp } F^{N+1}\delta_x$. Let now φ be a real continuous function on M with

$$\varphi(y) \begin{cases} \in (0,1] & \text{if } y \in B_{fb}(\varepsilon) \\ \\ = 0 & \text{if } y \notin B_{fb}(\varepsilon) \end{cases}$$

In view of property (A), the function $x \mapsto (F^{N+1}\delta_x)$ is continuous. Furthermore

$$(F^{N+1}\delta_x)\varphi > 0 \quad \text{if } x \in \overline{B}_a(\delta)$$

because we have just shown that $fb \in \operatorname{supp} F^{N+1}\delta_x$. By compactness of $\overline{B}_a(\delta)$, there is $\beta > 0$ such that

$$(F^{N+1}\delta_x)\varphi \geq \beta \quad \text{if} \quad x \in \overline{B}_a(\delta)$$

Therefore, if μ is a positive measure,

$$(F^{N+1}\mu)(B_{fb}(\varepsilon)) \geq (F^{N+1}\mu)(\varphi) \geq \beta\mu(B_a(\delta))$$

where we have used (1) with F replaced by F^{N+1} . This means that mass "leaks" from $B_a(\delta)$ to $B_{fb}(\varepsilon)$ at a rate of at least β in $N+1$ steps.

On the other hand, if $y \in B_{fb}(\delta)$, $\operatorname{supp} F^x\delta_y$ is disjoint from $B_a(\delta)$ because of (B), (2), and the fact that there is no ε pseudoorbit from $B_{fb}(\varepsilon)$ to $B_a(\delta)$. This implies that no mass can go back from $B_{fb}(\varepsilon)$ to $B_a(\delta)$.

Given the probability measure ν suppose now that

$$(F^n\nu)(B_a(\delta)) \geq \gamma > 0 \quad \text{for} \quad n = n_1,\dots,n_k$$

Then, for $L \geq n_1,\dots,n_k$, we find

$$1 \geq (F^{L+N+1}\nu)(M \setminus B_a(\delta))$$
$$\geq k\beta\gamma$$

Thus $k \leq (\beta\gamma)^{-1}$, and from this the proposition immediately results. (The argument which we have used is standard in the theory of Markov processes)

12. Proposition. Let Λ be an attracting set with respect to f , A the union of all attractors in Λ and A^* the union of all classes [z]

for z *in the closure of* A .

Given a fundamental neighborhood U *of* Λ *and a neighborhood* *of* A^* , *then, for sufficiently small* ε ,

$$\lim_{n\to\infty} (F^n\nu)(\Theta) = 1$$

whenever ν *is a probability measure with support in* U *and* F *an* (ε,δ) *diffusion associated with* f .

In view of Lemma 5 we may assume that ε is small enough to ensure $\operatorname{supp} F^n\nu \subset U$ for large n . It is therefore sufficient to prove that

$$\lim_{n\to\infty} (F^n\nu)(U \setminus \Theta) = 0 \qquad (3)$$

We choose $\theta > 0$ such that Θ contains the θ-neighborhood of A^* , and assume that ε is such that Lemma 9 holds with $K = \overline{A}$. Let $a \in U \setminus \Theta$. We may assume (by Proposition 6 (d)) that $a \succ b \in A$ and, (by Lemma 9) that there is no 2ε pseudoorbit from b to a . The proof of Proposition 11 applies again and gives $\lim_{n\to\infty} (F^n\nu)(B_a(\delta)) = 0$. Therefore (3) holds, completing the present proof.

Conclusion.

If A is closed, Proposition 12 shows that the orbits of a dynamical system, with small random independently distributed errors, approximate attractors. In the study of dynamical systems by digital computers, there appear roundoff errors which may, in many cases, be treated as random and independent. It seems therefore reasonable to think that the "strange attractors" obtained are in fact attractors in the present sense.

Suppose that asymptotic measures may be defined as zero noise limits of stationary measures under a diffusion associated with a differentiable map f . Proposition 12 gives information on the support of these measures. It is interesting that, while the study of the asymptotic measures themselves is hard*), their supports are relatively easy to analyze. This is due to the fact that only the topology is involved in our study of supports, while the identification of the asymptotic measures themselves involves the differentiable structure.

*) See Sinai [9], Ruelle [7], Bowen and Ruelle [1], Kifer [5].

References.

[1] R. Bowen and D. Ruelle. The ergodic theory of Axiom A flows. Inventiones math. 29, 181-202 (1975).

[2] Ch. Conley. Isolated invariant sets and the Morse index. CBMS Regional Conference Series N° 38, A.M.S., Providence R.I., 1978.

[3] M. Hénon. A two-dimensional mapping with a strange attractor. Commun. Math. Phys. 50, 69-77 (1976).

[4] M. Hurley. Attractors : persistence and density of their basins. To appear.

[5] Yu. I. Kifer. On small random perturbations of some smooth dynamical systems. Izv. Akad. Nauk SSSR Ser. Mat. 38 N° 5, 1091-1115 (1974). English translation : Math. USSR Izvestija 8, 1083-1107 (1974).

[6] S. Newhouse. Diffeomorphisms with infinitely many sinks. Topology 13, 9-18 (1974).

[7] D. Ruelle. A measure associated with Axiom A attractors. Amer. J. Math. 98, 619-654 (1976).

[8] D. Ruelle. Small random perturbations of dynamical systems and the definition of attractors. Commun. Math. Phys. To appear.

[9] Ya. G. Sinai. Gibbsian measures in ergodic theory. Uspeki Mat. Nauk 27 N°4, 21-64 (1972).

[10] S. Smale. Differentiable dynamical systems. Bull. A.M.S. 73, 747-817 (1967).

[11] R. Thom. Stabilité structurelle et morphogénèse. W.A. Benjamin, Reading, Mass., 1972.

[12] R.F. Williams. Expanding attractors. Publ. Math. I.H.E.S. 43, 169-203 (1974).

ON THE EXISTENCE OF INVARIANT CURVES OF TWIST MAPPINGS OF AN ANNULUS

Helmut Rüssmann

1. Introduction

We consider a mapping of the form

(1) $$(\theta,r) \mapsto G(\theta,r) = (\theta+r,\ r+g(\theta,r))$$

where g is a real function defined in a strip

(2) $$S = \{(\theta,r) \in \mathbb{R}^2 \mid\ a < r < b\}\ ,$$

and of period 2π in θ. For small g we have a perturbed mapping, and the question is if the lines r = const., $a < r < b$ which are invariant under the unperturbed mapping

$$(\theta,r) \mapsto (\theta+r,r)$$

can be continued to invariant curves of the perturbed mapping G.

Identifying points θ mod 2π we have the problem of preserving invariant embedded circles of twist mappings defined in an annulus under small perturbations. But in this paper we prefer working on the universal cover of this annulus.

The first result concerning this problem is due to Moser [1] who proved in 1962 the existence of invariant differentiable curves

(3) $$r = \phi(\theta) = \phi(\theta+2\pi), \qquad a < \phi(\theta) < b$$

under the conditions that g is of class C^{333} and sufficiently small, and that G possesses the intersection property in the sense that each curve of the form (3) with a small derivative ϕ' intersects its image under G.

In 1970 we succeeded in dropping the differentiability condition for g from C^{333} to C^5 using a different method also going back to Moser [2]. Obviously there was C^p, $p > 4$ available in our paper [3]. However, Moser [4], [5] pointed out that our proof leads to invariant curves even for g belonging to class C^p, $p > 3$.

In the meantime Herman [6] has also given a proof in the case C^p, $p > 3$. He surprisingly obtained invariant curves (3) of class C^{p-1} the rotation numbers of which, however, have to be of constant type.

Our efforts in this paper aim into annother direction. We are more interested in weak conditions for g than in high differentiability properties of the constructed invariant curves. Of course, we do not try to get invariant curves in the case that g is of class C^p, $p < 3$ because of Herman's contribution in this volume. Rather we try to obtain a good estimate for the admitted size of the norm of g in the C^p-topology which makes clear the dependence of p for $p \downarrow 3$. Moreover we take into consideration numerical aspects such that all estimates are given explicitly and optimal to a certain extend.

The method of proof goes back to Kolmogorov [7], Arnold [8], [9], [10], Moser [2], [11], and to [3]. However, there is a new aspect. Our iteration process here yields invariant curves which are only continuous, and not necessarily of the form (3). These curves have a parameter representation

$$(4) \qquad \begin{cases} K : \mathbb{R} \to S\,, \\ (\theta,r) = K(\xi) = K(\xi+2\pi) - (2\pi,0) \end{cases}$$

with a continuous function K. Clearly we like K to be injective, but from our construction of a curve (4) being invariant under (1) above all we will only obtain for K the functional equation

$$(5) \qquad G \circ K = K \circ \Omega \quad , \xi \mapsto \Omega(\xi) = \omega + \xi$$

where ω is a real number such that $\omega/2\pi$ is irrational. However, this irrationality implies injectivity of K. We do not know if this fact is new. Anyway we have

Theorem 1: Let a mapping

$$(6) \qquad \begin{cases} G : S \to \mathbb{R}^2 \quad \text{continuous,} \\ (\theta,r) \mapsto G(\theta,r) = G(\theta+2\pi,r) - (2\pi,0) \end{cases}$$

be given (not necessarily of the form (1)) possessing a continuous solution (4) of equation (5) with $\omega/2\pi$ irrational. Then K is injective.

By means of this theorem it follows from equation (4) not only that the curve defined in (4) is invariant under G but also that the mapping induced by G on this invariant curve is a homeomorphism, and has rotation number ω .

One of the conditions necessary for the construction of invariant curves is the intersection property. The following formulation is used in this paper.

Definition 1: We say that a mapping G defined in (6) has the intersection property iff any curve (4) for which K is injective and of class C^1 with $\frac{dK}{d\xi}(\xi) \neq 0$ for all $\xi \in \mathbb{R}$ satisfies the relation

$$G \circ K(\mathbb{R}) \cap K(\mathbb{R}) \neq \emptyset\,.$$

A simple example of a mapping having the intersection property is the exact symplectic diffeomorphism

$$\mathbb{R}^2 \ni (\theta,r) \mapsto (\theta+r,\ r+g(\theta+r))$$

with $g : \mathbb{R} \to \mathbb{R}$ being of class C^1 and satisfying

$$g(\theta+2\pi) = g(\theta), \quad \int_0^{2\pi} g(\theta)d\theta = 0.$$

In the case $g(\theta) = k \sin \theta$ the existence of invariante curves has been studied by means of computers (see e.g. Greene [12]).

For simplicity we assume that the function g in (1) has period 2π in both variables, that is, we assume

$$(7) \qquad \begin{cases} g : \mathbb{R}^2 \to \mathbb{R} \\ (\theta,r) \mapsto g(\theta,r) = g(\theta+2\pi,r) = g(\theta,r+2\pi). \end{cases}$$

Moreover we assume that g is of class C^p, and we define

$$|x| = \max(|\theta|,|r|) \quad \text{for} \quad x = (\theta,r) \in \mathbb{R}^2,$$

$$|g|_{\mathbb{R}^2} = \sup_{x \in \mathbb{R}^2} |g(x)|,$$

$$\|g\|_p = \sup_{\substack{x \in \mathbb{R}^2 \\ |k|=p}} |D^k g(x)|$$

if $p \geq 0$ is an integer, and

$$\|g\|_p = \sup_{\substack{x \neq y \\ |k|=l}} \frac{|D^k g(x) - D^k g(y)|}{|x-y|^\alpha}$$

if $p = l+\alpha$, $l \geq 0$ an integer, $\alpha \in]0,1[$ where we put

$$D^k = \left(\frac{\partial}{\partial\theta}\right)^{k_1} \circ \left(\frac{\partial}{\partial r}\right)^{k_2}, \quad |k| = |k_1|+|k_2|, \quad k = (k_1,k_2).$$

We choose a rotation number ω satisfying the inequalities

$$(8)\qquad \begin{cases} a + 12^{-3}\gamma \leq \omega \leq b - 12^{-3}\gamma \\ |k \frac{\omega}{2\pi} - l| \geq \gamma k^{-\tau}, \quad k = 1,2,\ldots; \; l = 0,\pm 1,\ldots \end{cases}$$

with some constants γ, τ satisfying

$$(9)\qquad 0 < \gamma < \frac{1}{2} \min (1, 12^3(b-a)) \ , \quad 1 \leq \tau \ .$$

It is well known that for fixed τ and sufficiently small γ there exist numbers ω with the properties (8).

Our main result is

Theorem 2: Let G be a mapping of class C^p $(p>2\tau+1)$ defined by (1), (7), and having the intersection property in the strip (2). Then for any number ω satisfying the inequalities (8) with some constants γ, τ satisfying (9) the equation (5) has a continuous solution (4) under the following smallness conditions for g :

$$(10)\qquad |g|_{\mathbb{R}^2} \leq \frac{q}{300c_0} \left(\frac{1}{72}\right)^{\tau} \left(\frac{\gamma}{\Gamma(\tau+1)}\right)^2 ,$$

$$(11)\qquad \|g\|_p \leq \frac{q(1-q)}{3600(3c_1+c_2)} \left(\frac{1}{288}\right)^{\tau} \left(\frac{\gamma}{\Gamma(\tau+1)}\right)^2$$

where Γ is the Gamma function, q is a number satisfying

$$(12)\qquad 0 < q \leq \min \left(\frac{p-2\tau-1}{p+1} \log 2, \ 10^{-2}4^{-\tau}\right) \ ,$$

and c_0, c_1, c_2 are positive constants depending on p.

The constants c_0, c_1, c_2 depend on how well functions of class C^p can be approximated by analytic ones. We have

Lemma 1: Let g be a function of class C^p as defined in (7). Then for any $\delta > 0$ there exists a holomorphic function

$$g_\delta : \mathbb{C}^2 \to \mathbb{C}, \; g_\delta(\mathbb{R}^2) \subseteq \mathbb{R}$$

$$(\theta,r) \mapsto g_\delta(\theta,r) = g_\delta(\theta+2\pi,r) = g_\delta(\theta,r+2\pi)$$

such that the following inequalities hold:

$$|g_\delta|_E \leq c_0 |g|_{\mathbb{R}^2} ,$$

$$|g-g_\delta|_{\mathbb{R}^2} \leq c_1 \|g\|_p \delta^p ,$$

$$|g_\delta - g_{\delta'}|_E \leq c_2 \|g\|_p \delta'^p$$

for $0 < \delta < \delta'$ where

$$E = \{(\theta,r) \in \mathbb{C}^2 \mid |\mathrm{Im}\,\theta| < \delta,\ |\mathrm{Im}\,r| < \delta\} ,$$

$$|\cdot|_E = \sup_{x \in E} |\cdot(x)| ,$$

and c_0, c_1, c_2 are positive constants only depending on p.

There are proofs of lemma 1 available in the literature. See Jacobowitz [13, p. 205 209], Moser [2, p. 528-529], Rüssmann [3, p. 74-78], Zehnder [14, p. 110-113].

It is not so easy to get optimal values for the constants c_0, c_1, c_2. By cumbersome calculations using all facts known in approximation theory about "best constants" we obtained

$$c_0 \leq 11, \quad c_1 \leq 17, \quad c_2 \leq 125 \qquad (p > 0)$$

and

$$c_0 \leq 11, \quad c_1 \leq \frac{1}{8}\pi^p, \quad c_2 \leq \frac{11}{12}\pi^p \qquad (3 < p < 4).$$

We do not prove these results in this paper. In section 2 we prove theorem 1. The remaining sections are reserved for the proof of theorem 2.

2. Proof of Theorem 1

We assume that $K(\xi_1) = K(\xi_2)$ für K defined in (4) and two points $\xi_1, \xi_2 \in \mathbb{R}$ with $\xi_1 < \xi_2$. Repeated application of G to this equation leads by means of (5) to

$$K(\xi_1 + k\omega) = K(\xi_2 + k\omega) \quad , \quad k = 1,2,\ldots \quad .$$

Hence using (4) we get

$$K(\xi_1+k\omega-2\pi l) \;=\; K(\xi_2+k\omega-2\pi l)\;,$$

$$k=1,2,\ldots\;;\; l = 0,\pm1,\pm2,\ldots\;.$$

Since $\omega/2\pi$ is irrational the set $\{\xi_1+k\omega-2\pi l \mid k=1,2,\ldots;\; l\in\mathbb{Z}\}$ is dense in $\mathbb{R}$ as it is well known. So the continuity of K and the equations above give

$$K(\xi) = K\,(\xi+\xi_2-\xi_1)$$

for all $\xi\in\mathbb{R}$, that is, $\xi_2-\xi_1 > 0$ is a period of K. Consequently K is bounded, and we have

$$M := \sup_{\xi\in\mathbb{R}} \|K(\xi)\| \;=\; \sup_{\xi_1\le\xi\le\xi_2} \|K(\xi)\| \;<\; \infty$$

where $\|\cdot\|$ is some norm in $\mathbb{R}^2$. On the other hand we see from (4)

$$(2\pi l,\ 0) = K(\xi+2\pi l) - K(\xi)$$

such that we obtain

$$2\pi l\,\|(1,0)\| \;\le\; 2M\;,\qquad l = 1,2,\ldots\;,$$

that is, a contradiction. Theorem 1 is proved.

3. The Iteration Process

In this section we present an iteration process leading to the proof of theorem 2.

First of all we introduce new variables by the linear transformation

$$(x,y) \mapsto (\theta,r) = (x,\omega+\varepsilon_0 y) \tag{13}$$

where ω is the chosen rotation number satisfying (8), and ε_0 is defined by

$$\varepsilon_0 = \sqrt{3}\;6^{-\tau}\,\Gamma\,(\tau+1)^{-1}\gamma\;. \tag{14}$$

In the new coordinates the given mapping (1), (7) having the intersection property in the strip (2) gets the form

$$(x,y) \mapsto A(x,y) = (\omega+x+\varepsilon_0 y,\ y+\varepsilon_0^{-1} g(x,\omega+\varepsilon_0 y)).$$

Clearly the intersection property is preserved and holds in the strip

(15) $$S^* = \{(x,y) \in \mathbb{R}^2 \mid |y| < 600^{-1}\}$$

where we have used (14) and $\Gamma(\tau+1) \geq 0$ for $\tau \geq 1$.

Since the function g is of class C^p and of period 2π in both variables by assumptioı we may apply lemma 1 in order to obtain a family of holomorphic functions $g_\delta(\delta > 0)$ with which we define the mappings

$$(x,y) \longrightarrow A_\delta(x,y) = (\omega+x+\varepsilon_0 y,\ y+\varepsilon_0^{-1} g_\delta(x,\omega+\varepsilon_0 y)).$$

We define a sequence

(16) $$\delta_k = \left(\frac{1+q}{2}\right)^k \quad , \quad k = 0,1,\ldots$$

where q is a real number satisfying (12), and we put

$$E_k = \{(x,y) \in \mathbb{C}^2 \mid |\operatorname{Im} x| < \delta_k, |\operatorname{Im} y| < \delta_k\}$$

$$A_k = A_{\delta_k} \quad , \quad k = 0,1,\ldots \quad .$$

Then the estimates of lemma 1 can be written in the form

(17) $$\begin{cases} |A_0 - \Omega_0|_{E_0} \leq \varepsilon_0^{-1} c_0 |g|_{\mathbb{R}^2} , & \\ |A - A_k|_{\mathbb{R}^2} \leq \varepsilon_0^{-1} c_1 \|g\|_p \delta_k^p & \\ & k = 0,1,\ldots \\ |A_k - A_{k+1}|_{E_{k+1}} \leq \varepsilon_0^{-1} c_2 \|g\|_p \delta_k^p & \end{cases}$$

where we use the norms

$$|(x,y)| = \max(|x|,|y|) \qquad \text{for } (x,y) \in \mathbb{C}^2 ,$$

$$|f|_B = \sup_{z \in B} |f(z)| \qquad \text{for } f : B \longrightarrow \mathbb{C}^2$$

and introduce the mapping

(18) $$(x,y) \mapsto \Omega_0(x,y) = (\omega+x+\varepsilon_0 y, y).$$

Before we describe the iteration process some more definitions and notations are useful.

Definition 2:

(i) Suppose that we are given subsets $D_1,\dots,D_m$ of $\mathbb{C}^n$ and functions $F_j : D_j \to \mathbb{C}^n$, $j=1,\dots, m-1$. Then we say "the diagram

$$D_1 \xrightarrow{F_1} D_2 \xrightarrow{F_2} D_3 \xrightarrow{F_3} \dots \xrightarrow{F_{m-1}} D_m$$

exists" iff we have

$$F_j(D_j) \subseteq D_{j+1}, \quad j=1,\dots, m-1.$$

In the case $F_j = i = i_{D_j}$ (inclusion) this condition means

$$D_j \subseteq D_{j+1}, \; i(x) = x \text{ for all } x \in D_j.$$

(ii) For positive r,s we define

$$D(r,s) = \{(x,y) \in \mathbb{C}^2 \mid |\mathrm{Im}\, x| < r, \; |y| < s \}.$$

(iii) For $D = D(r,s)$ we denote by $T(D) = T(r,s)$ the set of all holomorphic functions $F : D \to \mathbb{C}^2$ satisfying the identities

$$\Pi \circ F = F \circ \Pi|D, \quad \sigma \circ F = F \circ \sigma|D$$

where Π and σ are defined by

$$(x,y) \mapsto \Pi(x,y) = (x+2\pi, y),$$
$$(x,y) \mapsto \sigma(x,y) = (\bar{x},\bar{y})$$

for all $(x,y) \in \mathbb{C}^2$ with $\bar{x} = a-bi$ for $x = a+bi$; $a,b \in \mathbb{R}$.

(iv) We define the mappings Ω_k, $k=0,1,\ldots$ by

$$(x,y) \mapsto \Omega_k(x,y) = (\omega+x+\varepsilon_k y, y)$$

for all $(x,y) \in \mathbb{C}^2$ with

$$\varepsilon_k = 2^{-k\tau}\varepsilon_o$$

where ε_o is given in (14).

(v) In $\mathbb{C}^n$, $n=1,2,\ldots$, we use the norm

$$|x| = \max_j |x_j| \text{ for } x = (x_1,\ldots, x_n) \in \mathbb{C}^n$$

if nothing else is stated.

(vi) Given $f : D \to \mathbb{C}^n$ with $D = D(r,s)$, we define

$$|f|_{\rho,\sigma} = \sup_{\substack{|\mathrm{Im} x|<\rho \\ |y|<\sigma}} |f(x,y)| \leq \infty$$

for $0 < \rho \leq r$, $0 < \sigma \leq s$, and

$$|f|_D = |f|_{r,s} .$$

This last definition has some formal advantages. Actually we will be concerned only with bounded functions f, and it lies in the nature of the iteration process to be described below that this fact has to be stated in any case by

$$|f|_D \leq M$$

with a positive constant M.

Now we go back to the mappings $A_k : E_k \to \mathbb{C}^2$ defined above. We try to fix domains

$$D_k = D(r_k, s_k), \quad D_k' = D(r_k', s_k')$$

and to find mappings $Z_k \in T(D_k')$, $H_k \in T(D_k)$ such that the diagrams

$$(19)_k \qquad \begin{array}{ccc} D_k & \xrightarrow{\;Z_k | D_k\;} & E_k \\ {\scriptstyle H_k}\downarrow & & \downarrow{\scriptstyle A_k} \\ D_k' & \xrightarrow[\;Z_k\;]{} & \mathbb{C}^2 \end{array}$$

exist and commute for $k = 0,1,\dots$. A proper choice for the constants r_k, s_k, r_k', s_k' is

$$(20) \qquad \begin{cases} r_k = 2^{-k}, \quad s_k = 2^{-k} s_0, \quad s_0 = 300^{-1} 2^{-\tau}, \\ r_k' = \frac{4}{3}(r_k - s_k), \quad s_k' = \frac{4}{3} s_k, \quad k = 0,1,\dots . \end{cases}$$

Then we obviously have $D_k \subsetneq D_k'$, $k = 0,1,\dots$. For the mappings Z_k, H_k the following relations are needed:

$$(21)_k \qquad Z_k(D_k) \subsetneq D_0,$$

$$(22)_k \qquad \begin{cases} b_k |\zeta - \zeta'| \le |Z_k(\zeta) - Z_k(\zeta')| \le p_k |\zeta - \zeta'| \\ \text{for all } \zeta, \zeta' \in D_k', \\ b_k = 2^{-k\tau}(1-q)^k, \quad p_k = (1+q)^k, \end{cases}$$

$$(23)_k \qquad \begin{cases} |Z_{k+1}(\zeta) - Z_k(\zeta)| \le \frac{2}{3} q p_k s_k \\ \text{for all } \zeta = (\xi, 0) \in D_{k+1}, \end{cases}$$

$$(24)_k \qquad \begin{cases} |H_k - \Omega_k|_{D_k} \le M_k, \\ M_k = 2^{-k(\tau+1)} M_0, \quad M_0 = 3^{-1} q \varepsilon_0 s_0 \end{cases}$$

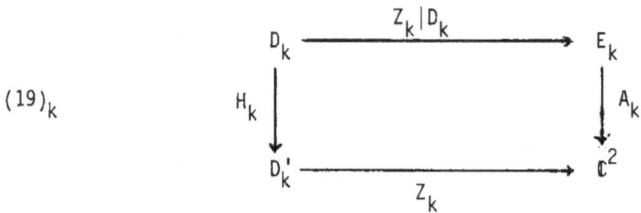

for k = 0,1,... . As in (16) q is a real number satisfying (12), and ε_o is defined in (14).

Finally we define (see definition 2)

$$Z_o = i_{D_o'} \in T(D_o').$$

Then by means of this iteration process - if it exists - the assertion of theorem 2 can easily be proved.

In fact, from (20), $(22)_k$, and $(23)_k$ we see that the sequence $Z_o, Z_1, \ldots$ converges uniformly on $\mathbb{R} \times \{0\}$ such that the limit

$$\xi \mapsto Z_\infty(\xi) = \lim_{k\to\infty} Z_k(\xi,0)$$

is continuous on $\mathbb{R}$. Since $Z_k \in T(D_k')$, k = 0,1,... we have

$$\Pi \circ Z_\infty = Z_\infty \circ \Pi|\mathbb{R}, \quad Z_\infty(\mathbb{R}) \subseteq \mathbb{R}^2.$$

Now the commutativity of $(19)_k$ yields

$$A_k \circ Z_k|D_k = Z_k \circ H_k,$$

hence we get

$$A\circ Z_k(\xi,0) - Z_k(\Omega_k(\xi,0)) = (A-A_k)(Z_k(\xi,0)) + Z_k\circ H_k(\xi,0) - Z_k(\Omega_k(\xi,0))$$

and by virtue of $(21)_k$, $(23)_k$ consequently

$$|A\circ Z_k(\xi,0) - Z_k(\omega+\xi,0)| \leq |A-A_k|_{\mathbb{R}^2} + p_k M_k$$

for all $\xi \in \mathbb{R}$. Passing to the limit we find

$$A\circ Z_\infty(\xi) = Z_\infty(\omega+\xi)$$

for all $\xi \in \mathbb{R}$ in view of (16), (17) and $p_k M_k \to 0$. So going back to the original coordinates θ, r via (13) and defining

$$K(\xi) = (0,\omega) + Z_\infty(\xi)\begin{pmatrix}1 & 0\\ 0 & \varepsilon_o\end{pmatrix}$$

we obtain the solution (4) of (5) asserted in theorem 2. The relation $K(\mathbb{R}) \subseteq S$ is satisfied because $Z_\infty(\mathbb{R})$ is contained in the strip (15). In fact, from $(21)_k$ and $Z_k \in T(D_k')$ we get

$$(25)_k \qquad Z_k(\mathbb{R}^2 \cap D_k) \subseteq \mathbb{R}^2 \cap D_o \subseteq S^*$$

where we pass to the limit $k \to \infty$ and notice $s_o \le 600^{-1}$ in (20).

The proof of the assertion

$$(26)_k \qquad \begin{cases} \text{The diagram } (19)_k \text{ exists and commutes} \\ \text{with some } Z_k \in T(D_k'),\ H_k \in T(D_k) \\ \text{satisfying } (21)_k,\ (22)_k \text{ and } (24)_k \end{cases}$$

and of the estimate $(23)_k$ for $k = 0,1,\ldots$ is done by complete induction.

Let us first consider the case $(26)_o$. As a consequence of the definition of Z_o the relations $(21)_o$ and $(22)_o$ are obvious. Moreover the diagram $(19)_o$ exists and commutes if we define

$$H_o = A_o | D_o$$

and require the relations

$$D_o \subseteq E_o, \quad H_o(D_o) \subseteq D_o'.$$

The first of these relations is satisfied by virtue of (16) and (20). The definition of H_o implies $(24)_o$ by virtue of (10), (14), (17) and (20). The second relation above can be derived from $(24)_o$. In fact, from (14) we obtain $\varepsilon_o < 1$ because of (9) and of $\Gamma(\tau+1) \ge 1$ for $\tau \ge 1$. From (12) we have $q < 1$, thus

$$(27) \qquad M_o = \frac{1}{3} q \varepsilon_o s_o < \frac{1}{3} s_o.$$

Since s_o is small enough according to (20) we get by means of (18), (20) and $(24)_o$

$$H_o(D_o) \subseteq D(r_o + \varepsilon_o s_o + M_o,\ s_o + M_o)$$

$$\subseteq D(1 + s_o + \frac{1}{3} s_o,\ s_o + \frac{1}{3} s_o) \subseteq D_o'.$$

Now let us suppose that $(26)_k$ is true for some $k \geq 0$. We have to show $(26)_{k+1}$ and $(23)_k$. On this way the crucial point is the construction of the commuting diagram

(28)

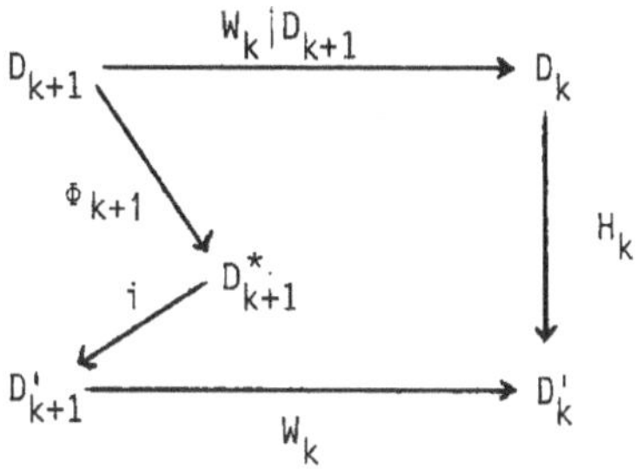

with

(29) $$D^*_{k+1} = D(r'_{k+1} - \frac{1}{7} s_k,\ s'_{k+1} - \frac{1}{7} s_k)$$

and mappings $W_k \in T(D'_{k+1})$, $\Phi_{k+1} \in T(D_{k+1})$.

The existence of such a commuting diagram is guaranteed by the inductive theorem which we prove in section 6. This theorem also delivers the following estimates:

(30) $$\begin{cases} 2^{-\tau}(1-q)|\zeta-\zeta'| \leq |W_k(\zeta) - W_k(\zeta')| \leq (1+q)|\zeta-\zeta'| \\ \text{for all } \zeta,\zeta' \in D'_{k+1}, \end{cases}$$

(31) $$\begin{cases} |W_k(\zeta) - \zeta| \leq \frac{2}{3} q s_k \\ \text{for all } \zeta = (\xi,0) \in D_{k+1} \end{cases}$$

and

(32) $$|\Phi_{k+1} - \Omega_{k+1} - Q_k|_{D_{k+1}} \leq \frac{5}{24} M_{k+1}$$

where Q_k is a polynomial of degree 2 in the second variable only:

(33) $$\begin{cases} Q_k(\eta) = (0,\ a_{ok}+a_{1k}\eta+a_{2k}\eta^2), \\ a_{ok},\ a_{1k},\ a_{2k} \in \mathbb{R}. \end{cases}$$

With these assertions of the inductive theorem we like to show $(26)_{k+1}$ and $(23)_k$.

From the diagrams $(19)_k$ and (28) we see that we may define

$$Z_{k+1} = Z_k \circ W_k$$

and, of course, we have $Z_{k+1} \in T(D'_{k+1})$ and $(21)_{k+1}$. The proof of $(22)_{k+1}$ by means of $(22)_k$ and (30) is obvious if we notice $W_k(D'_{k+1}) \subseteq D'_k$ in (28). The inequality $(23)_k$ follows from $(22)_k$, (31) and $W_k(D_{k+1}) \subseteq D'_k$ as a consequence of $D_{k+1} \subsetneq D'_{k+1}$.

If we combine the commuting diagrams $(19)_k$ and (28) we obtain the commuting diagram

(34)

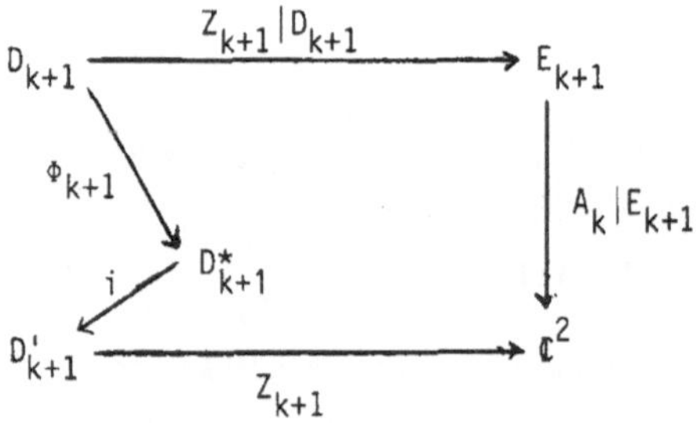

because we are in a position to show $Z_{k+1}(D_{k+1}) \subsetneq E_{k+1}$.

In fact, by definition of E_k, $k = 0,1,\ldots$ we have to form

$$|\mathrm{Im}\, Z_{k+1}(\zeta)| = |\mathrm{Im}\, Z_{k+1}(\mathrm{Re}\,\zeta + i\,\mathrm{Im}\,\zeta)| = |\mathrm{Im}\,\{Z_{k+1}(\mathrm{Re}\,\zeta + i\,\mathrm{Im}\,\zeta) - Z_{k+1}(\mathrm{Re}\,\zeta)\}|$$

for $\zeta \in D_{k+1}$ where we recall that Z_{k+1} is real for real arguments as an element of $T(D'_{k+1})$. Using (16), (20) and $(22)_{k+1}$ we get

$$|\mathrm{Im}\, Z_{k+1}(\zeta)| \leq |Z_{k+1}(\zeta) - Z_{k+1}(\mathrm{Re}\,\zeta)| \leq p_{k+1}|\mathrm{Im}\,\zeta| \leq p_{k+1}\max(r_{k+1}, s_{k+1}) = \delta_{k+1},$$

and the assertion above is proved.

Comparing the diagrams $(19)_{k+1}$ and (34) it remains to prove that we may replace Φ_{k+1} by H_{k+1} if we replace $A_k|E_{k+1}$ by A_{k+1}. Moreover we have to show $(24)_{k+1}$ what is not possible without going back to the original mapping A in order to use the intersection property and to estimate the polynomial (33) well enough such that $(24)_{k+1}$ follows from (32).

All these problems will be solved in the next section. Then the inductive theorem will be proved. Therefore the presented interation process works, and theorem 2 is proved.

4. The Intersection Property

In this section we study the perturbation of Φ_{k+1} in the commuting diagram (34) under the effect of a small perturbation of $A_k|E_{k+1}$. The main result is based on the following lemma in which $\mathbb{K}$ stands for $\mathbb{R}$ or $\mathbb{C}$. We call a function analytic iff it is holomorphic in the case $\mathbb{K} = \mathbb{C}$ and $\mathbb{R}$-analytic in the case $\mathbb{K} = \mathbb{R}$.

By $|\cdot|$ we denote here some norm in $\mathbb{K}^n$, n = 1,2,..., and we define

$$|\cdot|_B = \sup_{x\in B}|\cdot(x)|.$$

Moreover we define for $D \subsetneq \mathbb{K}^n$ and $d > 0$ the set

$$D\text{-}d = \{x \in \mathbb{K}^n \,|\, \{y \,|\, |y-x| \leq d\} \subseteq D\}$$

which may be empty. If D is open so is D-d.

Lemma 2: Let D be an open subset of $\mathbb{K}^n$, and let

$$F : D \to F(D) \subseteq \mathbb{K}^n$$

be an analytic mapping satisfying the estimate

$$(35) \qquad b|z-z'| \leq |F(z)-F(z')|$$

for all $z,z' \in D$ and some $b > 0$. Then F(D) is open, and the inverse mapping $F^{-1} : F(D) \to D$ exists and is analytic. Moreover for any $d > 0$ we have

$$(36) \qquad F(D\text{-}d) \subseteq F(D) - bd.$$

Proof: The existence of F^{-1} follows from (35).
F(D) is open, and F^{-1} is analytic according to the implicit function theorem which can be applied because the Jacobian of F does not vanish in D as a consequence of (35).

For the proof of (36) we assume $z_0 \in D\text{-}d$. Since D is open we have

$$B = \{z \in \mathbb{K}^n \,|\, |z-z_0| \leq d+\delta\} \subseteq D$$

for a sufficiently small $\delta > 0$. So for all $z \in \partial B$ we get from (35) the estimate

$$|F(z) - F(z_0)| \geq b|z-z_0| = b(d+\delta) > bd,$$

that is,

$$F(\partial B) \cap B^* = \emptyset$$

for

$$B^* = \{w \in \mathbb{K}^n \,|\, |w - F(z_0)| \leq bd\} .$$

As a consequence we find the equation

$$B^* = (B^* \cap F(B^0)) \cup (B^* \cap (\mathbb{K}^n \setminus F(B))) \tag{37}$$

where B^0 is the interior of B.
$F(B^0)$ is open because of the first part of the lemma applied to $F|B^0$, and $\mathbb{K}^n \setminus F(B)$ is open because $F(B)$ is compact. Therefore equation (37), the connectedness of B^* and the relation

$$F(z_0) \in B^* \cap F(B^0)$$

lead to

$$B^* \cap (\mathbb{K}^n \setminus F(B)) = \emptyset,$$

hence to

$$B^* \subseteq F(B) \subseteq F(D),$$

and finally to

$$F(z_0) \in F(D) - bd.$$

Since $z_0 \in D\text{-}d$ was arbitrary (36) follows. The lemma is proved.

Now it is useful to introduce an arbitrary bijection

$$\Lambda : \mathbb{K}^n \to \mathbb{K}^n$$

and to denote by

$$\Delta = \Delta(\Lambda, \mathbb{K}^n)$$

the set of all subsets D of $\mathbb{K}^n$ which are invariant under Λ , that is, for which $\Lambda(D) = D$. Furthermore we denote by

$$\Sigma = \Sigma(\Lambda, \mathbb{K}^n)$$

the class of all functions $F : D \longrightarrow \mathbb{K}^n$ such that

$$D \in \Delta \quad , \qquad \Lambda \circ F = F \circ \Lambda|D.$$

Clearly we have $F(D) \in \Delta$ for a function of class Σ with domain D, and if F is injective then also $F^{-1} : F(D) \longrightarrow \mathbb{K}^n$ is of class Σ. Moreover if $F : D \longrightarrow \mathbb{K}^n$, $G : E \longrightarrow \mathbb{K}^n$ are of class Σ, and if $G(E) \subseteq D$ then also $F \circ G$ is of class Σ.

In what follows we also use definition 2(i).

<u>Theorem 3</u>: Let D, D', E be open subsets of $\mathbb{K}^n$ belonging to Δ with $D \subseteq D'$ and let A', Z, Φ' be analytic mappings of class Σ such that the diagram

(38)

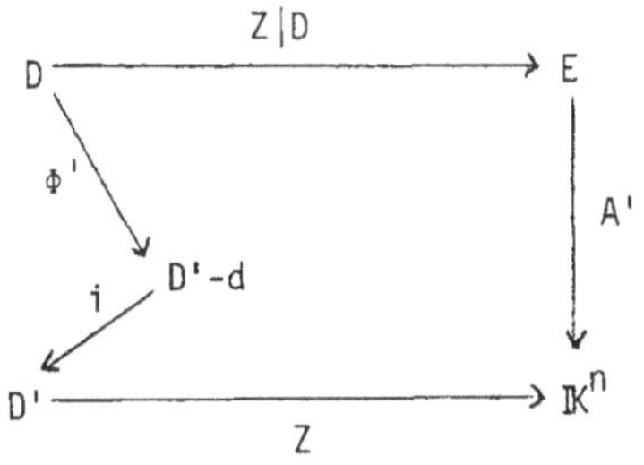

exists and commutes with some $d > 0$, and the estimate

$$b|\zeta - \zeta'| \leq |Z(\zeta) - Z(\zeta')|$$

holds for all $\zeta, \zeta' \in D'$ and some $b > 0$. Then for any continuous mapping $A'' : E \longrightarrow \mathbb{K}^n$ of class Σ satisfying the estimate

(39) $$|A' - A''|_E \leq bd$$

there exists a continuous mapping Φ'' of a class Σ such that diagram

(40)

$$\begin{array}{ccc} D & \xrightarrow{\;Z|D\;} & E \\ \Big\downarrow{\scriptstyle \Phi''} & & \Big\downarrow{\scriptstyle A''} \\ D' & \xrightarrow[\;Z\;]{} & \mathbb{K}^n \end{array}$$

exists and commutes, and the estimate

(41) $$|\Phi' - \Phi''|_D \leq b^{-1}|A' - A''|_E$$

is valid. If A" is analytic so is Φ''.

<u>Proof</u>: The mapping Z satisfies the hypothesis of lemma 2 such that Z(D') is open and

(42) $$Z^{-1} : Z(D') \longrightarrow D'$$

is analytic. In addition we get

$$Z(D'-d) \subseteq Z(D') - bd.$$

Since the diagram (38) commutes we have

$$A'(Z(D)) \subseteq Z(D'-d).$$

These relations lead to

(43) $$A'(Z(D)) \subseteq Z(D') - bd.$$

On the other hand the existence of the diagram (38) includes the relation

(44) $$Z(D) \subseteq E$$

such that the assumption (39) yields

$$|A'(Z(\zeta)) - A''(Z(\zeta))| \leq bd$$

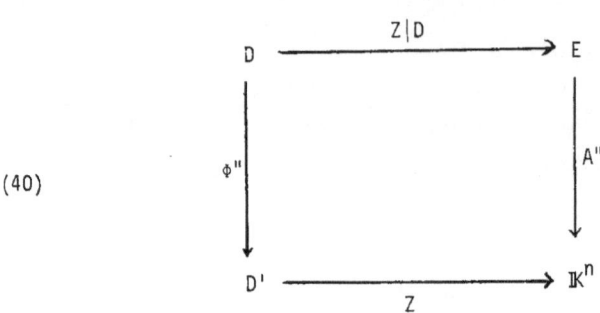

for all $\zeta \in D$. Thus by virtue of (43) we get

$$A''(Z(D)) \subseteq Z(D').$$

This relation and (42) admit the definition

$$\Phi'' = Z^{-1} \circ A'' \circ Z : D \to D',$$

and this mapping is continuous or analytic provided A'' is continuous or analytic because Z and Z^{-1} are analytic. Moreover Φ'' is of class Σ because this is the case for Z, Z^{-1} and A''. Clearly the diagram (40) exists and commutes.

The given estimate for Z leads to the estimate

$$|Z^{-1}(z') - Z^{-1}(z'')| \leq b^{-1}|z'-z''|$$

for all $z', z'' \in Z(D')$. Inserting

$$z' = Z(\Phi'(\zeta)) = A'(Z(\zeta))$$

$$z'' = Z(\Phi''(\zeta)) = A''(Z(\zeta))$$

for all $\zeta \in D$ the estimate (41) follows using (44). The theorem is proved.

Now we apply this theorem to the diagram (34) in order to obtain $(19)_{k+1}$. We have to put

(45) $$A' = A_k|E_{k+1},\quad A'' = A_{k+1},\quad \Phi' = \Phi_{k+1},\quad Z = Z_{k+1}$$

$$D = D_{k+1},\quad D' = D'_{k+1},\quad E = E_{k+1},$$

$$n = 2,\ \mathbb{K} = \mathbb{C},\ b = b_{k+1},\ \text{and}\ d = \frac{1}{7}s_k$$

in view of (29). Moreover Δ has to be the set of all subsets of $\mathbb{C}^2$ which are invariant under $\Lambda = \Pi$ as well as under $\Lambda = \sigma$ (see definition 2 (iii)) such that Σ is the class of all functions $F = D \to \mathbb{C}^2$ with $D \in \Delta$ and $\Pi \circ F = F \circ \Pi|D$, $\sigma \circ F = F \circ \sigma|D$. Then D_{k+1}, D'_{k+1}, E_{k+1} are open sets belonging to Δ, and (45) represents analytic functions of class Σ. For A' and A'' this follows from lemma 1, for Φ' and Z this is true because of $\Phi_{k+1} \in T(D_{k+1})$, $Z_{k+1} \in T(D'_{k+1})$. In addition $(22)_{k+1}$ is valid. Therefore theorem 3 can be applied and gives a function $H_{k+1} = \Phi'' \in T(D_{k+1})$ such that the diagram $(19)_{k+1}$ exists and commutes provided that (39) can be verified. But (39) is not sharp enough to deduce from (41) the inequality

$$(46) \qquad |H_{k+1} - \Phi_{k+1}|_{D_{k+1}} \leq \frac{c_2}{18c_1+6c_2} M_{k+1}$$

which we need for the proof of $(24)_{k+1}$.

The decisive necessary condition which we have to require now is

$$(47) \qquad \frac{(1+q)^p}{1-q} \leq 2^{p-1-2\tau} .$$

For then we get the estimate

$$(48) \qquad \varepsilon_o^{-1} \|g\|_p \delta_k^p \leq \frac{1}{18c_1+6c_2} b_{k+1} M_{k+1}$$

using (11),(14), (16) and the definitions of s_o, b_{k+1}, M_{k+1} in (20), $(22)_{k+1}$ and $(24)_{k+1}$. By means of (17),(41) and (48) the wanted inequality (46) follows as well as (39) where we use $M_{k+1}s_k^{-1} \leq M_o s_o^{-1} \leq 3^{-1}$ by virtue of (27). Therefore we have proved $(26)_{k+1}$ up to the estimate $(24)_{k+1}$ because addition of (32) and (46) gives only

$$(49) \qquad |H_{k+1} - \Omega_{k+1} - Q_k|_{D_{k+1}} \leq \left(\frac{5}{24} + \frac{c_2}{18c_1+6c_2}\right) M_{k+1}$$

with a polynomial Q_k defined in (33).

In order to obtain a proper estimate for Q_k we apply theorem 3 once more to the diagram (34) where this time we restrict D_{k+1} to $D = \mathbb{R}^2 \cap D_{k+1}$ such that we consider (38) with

$$n = 2,\ \mathbb{K} = \mathbb{R},\ b = b_{k+1},\ d = \frac{1}{7} s_k ,$$

$$D = \mathbb{R}^2 \cap D_{k+1},\quad D' = \mathbb{R}^2 \cap D'_{k+1},\ E = \mathbb{R}^2 \cap E_{k+1} = \mathbb{R}^2,$$

$$A' = A_k|\mathbb{R}^2,\ \Phi' = \Phi_{k+1}|\mathbb{R}^2 \cap D_{k+1},\ Z = Z_{k+1}|\mathbb{R}^2 \cap D'_{k+1} .$$

Here we put $\Lambda = \Pi\,|\mathbb{R}^2$ such that D, D', E are open subsets of $\mathbb{R}^2$ belonging to Δ, and A', Φ', Z are analytic functions of class Σ. Also the original mapping A defined at the beginning of section 3 is of class Σ, and it is continuous. Therefore using $(21)_{k+1}$ theorem 3 is again applicable and we obtain a continuous function $\Psi_{k+1} = \Phi''$ of class Σ such that the diagram

(50)
$$\begin{array}{ccc} \mathbb{R}^2 \cap D_{k+1} & \xrightarrow{\ Z_{k+1}|\mathbb{R}^2 \cap D_{k+1}\ } & \mathbb{R}^2 \\ \Big\downarrow{\scriptstyle \Psi_{k+1}} & & \Big\downarrow{\scriptstyle A} \\ \mathbb{R}^2 \cap D'_{k+1} & \xrightarrow[\ Z_{k+1}|\mathbb{R}^2 \cap D'_{k+1}\ ]{} & \mathbb{R}^2 \end{array}$$

exists and commutes provided (39) can be satisfied. But this is possible as above. Furthermore by means of (17), (41) and (48) we get the estimate

$$|\Psi_{k+1} - \Phi_{k+1}|_{\mathbb{R}^2 \cap D_{k+1}} \leq \frac{c_1}{18c_1+6c_2} M_{k+1}$$

which leads with (32) to

(51)
$$|\Psi_{k+1} - \Omega_{k+1} - Q_k|_{\mathbb{R}^2 \cap D_{k+1}} \leq \left(\frac{5}{24} + \frac{c_1}{18c_1+6c_2} \right) M_{k+1} .$$

Now we recall that the mapping A has the intersection property at least in the strip (15). We apply this property to the family of curves

$$\mathbb{R} \ni \xi \mapsto Z_{k+1}(\xi,\eta), \qquad -s_{k+1} < \eta < s_{k+1}$$

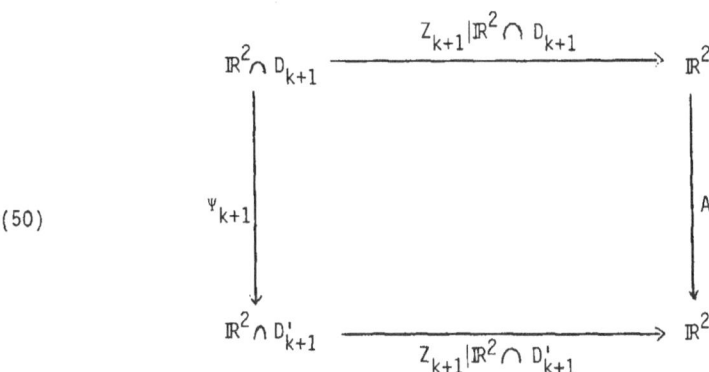

where it is clear from $(25)_{k+1}$ that these curves lie in S*. Moreover these curves satisfy the conditions of definition 1. For the mapping Z_{k+1} is analytic, and as a consequence of $(22)_{k+1}$ it is injective and has non-vanishing Jacobian. So there are for each η with $-s_{k+1} < \eta < s_{k+1}$ real numbers $\xi_0 = \xi_0(\eta)$, $\xi_1 = \xi_1(\eta)$ such that

$$A \circ Z_{k+1}(\xi_0,\eta) = Z_{k+1}(\xi_1,\eta) \ .$$

On the other hand from the commuting diagram (50) we have

$$A \circ Z_{k+1}(\xi_0,\eta) = Z_{k+1}(\Psi_{k+1}(\xi_0,\eta)) \ .$$

Thus the injectivity of Z_{k+1} yields

$$(\xi_1,\eta) = \Psi_{k+1}(\xi_0,\eta),$$

hence

$$\eta = \overset{(2)}{\Psi}_{k+1}(\xi_0,\eta) \qquad , \quad -s_{k+1} < \eta < s_{k+1}$$

where (2) indecates the second component of a vector.

This equation leads to a reasonable estimate for the polynomial (33). Since we use the maximum norm we get for $-s_{k+1} < \eta < s_{k+1}$ with the notation

$$a = a_{ok}, \quad b = a_{1k}, \quad c = a_{2k}$$

the estimate

$$|Q_k(\eta)| = |a + b\eta + c\eta^2| = |\overset{(2)}{\Psi}_{k+1}(\xi_0,\eta) - \eta - \overset{(2)}{Q_k}(\eta)|$$

$$\leq |\Psi_{k+1}(\xi_0,\eta) - \Omega_{k+1}(\xi_0,\eta) - Q_k(\eta)|$$

$$\leq |\Psi_{k+1} - \Omega_{k+1} - Q_k|_{\mathbb{R}^2 \cap D_{k+1}} \leq N$$

where we may put

$$N = \left(\frac{5}{24} + \frac{c_1}{18c_1+6c_2}\right) M_{k+1}$$

by virtue of (51). For $\eta = 0$ we conclude $|a| \leq N$. Therefore we obtain

$$|b\eta + c\eta^2| \leq |a| + |a + b\eta + b\eta^2| \leq 2N.$$

Putting $\eta = \pm\sigma$, $0 < \sigma < s_{k+1}$ and recalling that a, b, c $\in \mathbb{R}$ we get

$$|b|\sigma + |c|\sigma^2 = |\pm b\sigma + c\sigma^2| \leq 2N.$$

The limit $\sigma \uparrow s_{k+1}$ yields

$$|a + b\eta + c\eta^2| \leq |a| + |b||\eta| + |c||\eta^2|$$

$$\leq |a| + |b|s_{k+1} + |c|s_{k+1}^2 \leq 3N$$

for all $\eta \in \mathbb{C}$ with $|\eta| < s_{k+1}$, hence

$$|Q_k|_{D_{k+1}} \leq 3N.$$

This inequality and (49) obviously give the wanted estimate $(24)_{k+1}$.

The proof by induction for justifying the iteration process presented in section 3 has finished.

It remains to find a better form for condition (47). Equivalently we may write

$$f(q) := p \log(1+q) - \log(1-q) \leq (p-2\tau-1) \log 2.$$

For $0 < q \leq \frac{1}{4}$, $p > 2\tau + 1 \geq 3$ we have $\frac{d^2f}{dq^2}(q) \leq 0$, hence

$$f(q) = f(q) - f(0) \leq q \frac{df}{dq}(0) = q(p+1).$$

As a consequence

$$0 < q \leq \frac{p-2\tau-1}{p+1} \log 2$$

is sufficient for (47). This is one of the conditions for q appearing in (12).

5. Linear Difference Equations

In this section we will solve the system of difference equations

$$u(\omega+x,y) - u(x,y) = \varepsilon v(x,y) + f(x,y) \tag{52}$$

$$v(\omega+x,y) - v(x,y) = g(x,y) - \frac{1}{2\pi}\int_0^{2\pi} g(x,y)dx \tag{53}$$

which plays a central role in the proof of the inductive theorem. Here ω is a real number satisfying the diophantic inequalities

$$\left\{\begin{array}{l} |k\frac{\omega}{2\pi} - l| \geq \gamma k^{-\tau},\ k=1,2,\ldots;\ l=0,\pm 1,\ \ldots \\ \text{with some } \gamma,\ \tau,\ 0 < \gamma < \frac{1}{2},\ 1 \leq \tau. \end{array}\right. \tag{54}$$

f,g resp. u, v are given resp. wanted holomorphic functions of the complex variables x,y having period 2π in x. ε is a positive constant to be determined in such a way that the functions u, v will be of the same size.

In order to get estimates for u,v which are good enough for the proof of theorem 2 some technical preparations have to be made.

Lemma 3: For any real number ω satisfying (54) the inequalities

$$\sum_{k=1}^{n} |e^{ik\omega}-1|^{-2} \leq 12^{-1}\gamma^{-2}n^{2\tau} \tag{55}$$

hold for n=1,2,....

Proof: The well known equation

$$\left(\frac{1}{\sin x}\right)^2 = \sum_{l=-\infty}^{\infty} \left(\frac{1}{x-l\pi}\right)^2$$

leads to the estimate

$$\sum_{k=1}^{n} |e^{ik\omega}-1|^{-2} \leq \sum_{\substack{1\leq k\leq n \\ l\in \mathbb{Z}}}^{\infty} (\omega k-2\pi l)^{-2}. \tag{56}$$

If we denumerate the numbers of the set

$$\{\omega k-2\pi l \mid 1 \le k \le n,\ l \in \mathbb{Z}\}$$

according to their natural order

$$\dots < d_{-2} < d_{-1} < 0 < d_1 < d_2 < \dots,$$

$$d_j = \omega k_j - 2\pi l_j,\ j=\pm 1, \pm 2, \dots$$

we obtain from (54) the inequalities $d_1 \ge 2\pi\gamma n^{-\tau}$ and

$$d_{j+1} - d_j = \Big|\omega|k_{j+1}-k_j| \pm 2\pi|l_{j+1}-l_j|\Big| \ge 2\pi\gamma n^{-\tau}$$

for $j \ge 1$, thus

$$d_j \ge 2\pi j\gamma n^{-\tau}, \quad j=1,2,\dots,$$

and finally we have

$$\sum_{j=1}^{\infty} d_j^{-2} \le (2\pi\gamma)^{-2} n^{2\tau} \sum_{j=1}^{\infty} j^{-2} = 24^{-1}\gamma^{-2}n^{2\tau}$$

In the same way we find

$$\sum_{j=-\infty}^{-1} d_j^{-2} \le 24^{-1}\gamma^{-2}n^{2\tau}$$

such that we end with the inequality

$$\sum_{\substack{1 \le k \le n \\ l \in \mathbb{Z}}} (\omega k-2\pi l)^{-2} \le 12^{-1}\gamma^{-2}n^{2\tau}$$

which yields (55) by means of (56). The lemma is proved.

Lemma 4: For some $r > 0$ let

$$f : \{x \in \mathbb{C} \mid |\mathrm{Im}\ x| < r\} \longrightarrow \mathbb{C}$$

be a holomorphic function of period 2π. Then we have the estimate

(57) $$\sum_{k=-\infty}^{\infty} |f_k|^2 e^{2|k|r} \leq 2|f|_r^2$$

where

$$f_k = \frac{1}{2\pi} \int_0^{2\pi} f(x)e^{-ikx}dx, \quad k=0,\pm1,\ldots$$

are the Fourier coefficients of f, and we define

$$|f|_r = \sup_{|\mathrm{Im} x|<r} |f(x)|.$$

Proof: For any $s \in \mathbb{R}$ with $|s|<r$ the function $x \to f(x+is)$ is holomorphic for $|\mathrm{Im}\, x| < r-|s|$ and is of period 2π with Fourier coefficients

$$f_k(s) = \frac{1}{2\pi} \int_0^{2\pi} f(x+is)e^{-ikx}dx, \quad k \in \mathbb{Z}$$

such that Bessel's inequality leads to

(58) $$\sum_{k=-\infty}^{\infty} |f_k(s)|^2 \leq \frac{1}{2\pi} \int_0^{2\pi} |f(x+is)|^2 dx \leq |f|_r^2 .$$

Now we remark that the function

$$s \mapsto f_k(s)e^{ks} = \frac{1}{2\pi} \int_0^{2\pi} f(x+is)e^{-ik(x+is)}dx$$

is independent of s for $-r<s<r$ because its derivative is the mean value of the derivative of a 2π-periodic function and consequently vanishes. Thus we obtain

$$f_k(s)e^{ks} = f_k(0) = f_k$$

and see from (58) that

$$\sum_{k=-\infty}^{\infty} |f_k|^2 e^{-2ks} \leq |f|_r^2.$$

The limit $s \to \pm r$ and addition of the resulting inequalities yield (57). The lemma is proved.

Apart from the notations in definition 2 we use in the sequel

Definition 3:

(i) For $D=D(r,s)$ and $n=1,2,\dots$ we denote by $P^n(r,s)=P^n(D)$ the linear space of all holomorphic functions $f:D \longrightarrow \mathbb{C}^n$ satisfying

$$f\circ\Pi|D = f,\quad f(\mathbb{R}^2)\subseteq\mathbb{R}^n.$$

Clearly $f,\ g\in T(D)$ implies $f-g\in P^2(D)$

(ii) For a function $f\in P^n(D)$ we denote its mean value by $[f]$,

$$[f](y) = \frac{1}{2\pi}\int_0^{2\pi} f(x,y)dx.$$

Theorem 4: Let ω be a real number satisfying (54), and let f be a function belonging to $P^1(r,s)$ for some positive constants r,s. Then the difference equatio

(59) $$u(\omega+x,y) - u(x,y) = f(x,y) - [f](y)$$

has a unique solution $u\in P^1(r,s)$ with $[u]=0$. For this solution the estimate

(60) $$|u|_{r-\rho,s} \le \varepsilon^{-1}|f|_{r,s}\ , \qquad 0<\rho<r$$

holds where ε is defined by

(61) $$\varepsilon = \varepsilon(\rho) = \sqrt{3}\ \Gamma(\tau+1)^{-1}\gamma\rho^{\tau}$$

and Γ is the Gamma function.

Proof: Since the restriction of $f(\cdot,y)$ to $\mathbb{R}$ is a continuously differentiable and 2π-periodic function it can be expanded in its Fourier series

(62) $$f(x,y) = \sum_{k=-\infty}^{\infty} f_k(y)e^{ikx}$$

where

(63) $$f_k(y) = \frac{1}{2\pi}\int_0^{2\pi} f(x,y)e^{-ikx}dx,\quad k=0,\pm1,\dots$$

are the Fourier coefficients of $f(\cdot,y)$, $|y|<s$. An application of lemma 4 to the restriction of $f(\cdot,y)$ to $\{x\in\mathbb{C}\,|\,|\mathrm{Im}x|<\rho\}$ yields

$$(64)\qquad \sum_{k=-\infty}^{\infty} |f_k(y)|^2 e^{2|k|\rho} \le 2|f|^2_{\rho,\sigma}$$

for $0<\rho\le r$, $|y|<\sigma\le s$, and we notice $|f|_{\rho,\sigma}<\infty$ for $\rho<r$, $\sigma<s$. If we insert (62) into the given difference equation we find

$$(65)\qquad u(x,y) = \sum_{k\neq 0} \frac{f_k(y)}{e^{ik\omega}-1} e^{ikx}$$

as the uniquely determined Fourier expansion of the wanted solution u with $[u]=0$.

As a consequence of (54) and lemma 3 we obtain the inequalities (55) from which we conclude

$$|e^{ik\omega}-1| \ge c|k|^{-\tau}, \quad |k| = 1,2,\dots$$

with a positive constant c. On the other hand we get from (64)

$$|f_k(y)| \le 2|f|_{\rho,\sigma} e^{-|k|\rho}, \quad |k| = 1,2,\dots$$

for $0<\rho<r$, $|y|<\sigma<s$ such that the terms of the series (65) which are holomorphic in $D(r,s)$ have the estimate

$$|f_k(y)e^{ikx}(e^{ik\omega}-1)^{-1}| \le 2c^{-1}|f|_{\rho,\sigma}|k|^{\tau}e^{-|k|\delta}$$

for $|k|=1,2,\dots$, $|\mathrm{Im}x|<\rho-\delta$, $0<\delta<\rho$, $|y|<\sigma<r$. Since the series $\sum_{k=-\infty}^{\infty} |k|^{\tau}e^{-|k|\delta}$ converges for $\delta>0$ the series (65) converges absolutely and uniformly in any compact subset of $D(r,s)$. Therefore u is holomorphic in $D(r,s)$. As a consequence of (62) and (65) equation (59) is fullfilled at least for $x\in\mathbb{R}$, but by analytic continuation this is true for all of $D(r,s)$. The reality condition $u(\mathbb{R}^2\cap D(r,s))\subseteq\mathbb{R}$ follows from (63), (65) and $f\in P^1(r,s)$. So u belongs to $P^1(r,s)$, and it remains to prove inequality (60). For this purpose we estimate the sum

$$g_n(y) = \sum_{1\le|k|\le n} |f_k(y)|\,|e^{ik\omega}-1|^{-1}e^{|k|r}$$

for $|y|<s$, $n=1,2,\dots$. Using (55), (64) and the Cauchy-Schwarz-inequality we get

$$g_n(y) \le \sqrt{\sum_{k\in\mathbb{Z}} |f_k(y)|^2 e^{2|k|r}}\ \sqrt{\sum_{0<|k|\le n} |e^{ik\omega}-1|^{-2}} \le \frac{n^\tau}{\sqrt{3}\gamma}|f|_{r,s}.$$

Defining $g_0(y)=0$ we obtain by means of Abel's partial summation

$$\sum_{0<|k|\le N} |f_k(y)|\,|e^{ik\omega}-1|^{-1} e^{|k|(r-\rho)}$$

$$= (1-e^{-\rho})\sum_{n=1}^{N} g_n(y)e^{-n\rho} + g_N(y)e^{-(N+1)\rho}$$

$$\le \frac{1}{\sqrt{3}\gamma}\,|f|_{r,s}\sum_{n=1}^{\infty} n^\tau(e^{-n\rho}-e^{-(n+1)\rho})$$

for $|y| < s$, $0 < \rho < r$ where we have noticed that $g_N(y)e^{-N\rho} \to 0$ for $N \to \infty$. For the last series we get the estimate

$$\sum_{n=1}^{\infty} n^\tau(e^{-n\rho}-e^{-(n+1)\rho}) = \rho\sum_{n=1}^{\infty} n^\tau\int_n^{n+1} e^{-t\rho}dt$$

$$\le \rho\sum_{n=1}^{\infty}\int_n^{n+1} t^\tau e^{-t\rho}dt = \rho\int_1^{\infty} t^\tau e^{-t\rho}dt$$

$$\le \rho^{-\tau}\int_0^{\infty} t^\tau e^{-t}dt = \rho^{-\tau}\Gamma(\tau+1).$$

Therefore we finally arrive for $|\mathrm{Im}x| < r-\rho$, $|y| < s$ at

$$|u(x,y)| \le \sum_{k\neq 0}\left|\frac{f_k(y)}{e^{ik\omega}-1}\right| e^{|k|(r-\rho)} \le \frac{\Gamma(\tau+1)}{\sqrt{3}\gamma}\,\rho^{-\tau}|f|_{r,s},$$

and (60) follows. The theorem is proved.

Now we are ready to solve the equations (52), (53).

<u>Theorem 5:</u> Let ω be a real number satisfying (54), and let f,g be functions belonging to $P^1(r,s)$ and satisfying the estimates

(66) $$|f|_{r,s} \le M, \quad |g|_{r,s} \le M$$

with some positive constants r,s,M. Then the difference equations (52), (53) with ε defined in (61) have a unique solution $u,v \in P^1(r,s)$ with $[u]=0$. For this solutioı the estimates

$$|u|_{r-2\rho,s} \leq 2\varepsilon^{-1}M, \tag{67}$$

$$|v|_{r-\rho,s} \leq 2\varepsilon^{-1}M \tag{68}$$

are valid for $0 < 2\rho < r$.

Proof: In equation (52) the mean value must vanish on both sides. So we get the condition

$$[v] = -\varepsilon^{-1}[f] \tag{69}$$

for the mean value of v. As a consequence we have $[v] \in P^1(r,s)$ and

$$|[v]|_{r,s} \leq \varepsilon^{-1}M \tag{70}$$

in view of (66). Theorem 4 gives a unique solution $v=\tilde{v} \in P^1(r,s)$ of (53) with $[\tilde{v}]=0$. This solution has the estimate

$$|\tilde{v}|_{r-\rho,s} \leq \varepsilon^{-1}M \tag{71}$$

because of (66). Defining $v=\tilde{v}+[v]$ we obtain the uniquely determined solution $v \in P^1(r,s)$ of (53) satisfying (69). This solution has the estimate (68) as a consequence of (70) and (71).

By virtue of (69) equation (52) can be written in the form

$$u(\omega+x,y) - u(x,y) = h(x,y) - [h](y) \tag{72}$$

with $h=\varepsilon\tilde{v}+f$. Since $f,\tilde{v} \in P^1(r,s)$ we have $h \in P^1(r,s)$, and

$$|h|_{r-\rho,s} \leq 2M \tag{73}$$

using (66) and (71). Hence theorem 4 gives a uniquely determined solution $u \in P^1(r,s)$ of (72) with $[u]=0$. For an estimate of u we apply theorem 4 to (72) restricted to $D(r-\rho,s)$ such that in (60) we have to replace f by h and r by $r-\rho$. Then (67) follows by means of (73). The theorem is proved.

6. The Inductive Theorem

First of all we put together constants, domains, and mappings appearing in the formulation of the inductive theorem.

(I) Constants and their Relations

We introduce the constants

(74) $\quad \omega,\ \gamma,\ \tau,\ M,\ q,\ \varepsilon,\ \varepsilon_+,\ r,\ r_+,\ s,\ s_+,\ r',\ r'_+,\ s',\ s'_+$

and the auxiliary constants θ, ρ satisfying the relations

$$|k\frac{\omega}{2\pi} - l| \geq \gamma k^{-\tau}, \quad k = 1,2,\ldots; \quad l = 0, \pm 1,\ldots$$

$$0 < \gamma < \frac{1}{2}, \qquad 1 \leq \tau, \qquad 0 < r \leq 1,$$

$$0 < q \leq (\frac{\theta}{10})^2, \qquad s = \frac{\theta\rho}{50}, \quad \theta = 2^{-\tau}, \qquad \rho = \frac{r}{6}$$

$$\varepsilon = \frac{\sqrt{3}}{\Gamma(\tau+1)}\gamma\rho^\tau, \qquad M = \frac{1}{3}q\varepsilon s, \qquad r' = \frac{4}{3}(r-s), \qquad s' = \frac{4}{3}s,$$

$$\frac{r_+}{r} = \frac{s_+}{s} = \frac{r'_+}{r'} = \frac{s'_+}{s'} = \frac{1}{2}, \qquad \frac{\varepsilon_+}{\varepsilon} = \theta.$$

(II) Domains and Mappings

Using definition 2 we put

$$D = D(r,s), \quad D_+ = D(r_+,s_+), \quad D' = D(r',s'), \quad D'_+ = D(r'_+,s'_+),$$

$$D^*_+ = D(r'_+ - \frac{1}{7}s,\ s'_+ - \frac{1}{7}s) = D'_+ - \frac{1}{7}s,$$

and we introduce the mappings

$$(x,y) \mapsto \Omega(x,y) = (\omega + x + \varepsilon y, y),$$

$$(x,y) \mapsto \Omega_+(x,y) = (\omega + x + \varepsilon_+ y, y),$$

$$(x,y) \mapsto \Theta(x,y) = (x, \theta y)$$

for all $(x,y) \in \mathbb{C}^2$, where we use the same symbol for the mappings Ω, Ω_+, Θ as we as for their restrictions to subsets of $\mathbb{C}^2$.

<u>Theorem 6 (Inductive Theorem)</u>

Let constants (74) and auxiliary constants θ, ρ be given such that the relations in (I) are satisfied, and let domains D, D_+, D', D'_+, D^*_+ and mappings Ω, Ω_+, Θ be given as in (II). Then for any mapping

$$H : D \rightarrow D', \qquad H \in T(D)$$

satisfying the estimate

(75) $$|H - \Omega|_D \leq M$$

there are mappings $W \in T(D'_+)$ and $\Phi_+ \in T(D_+)$ such that the diagram

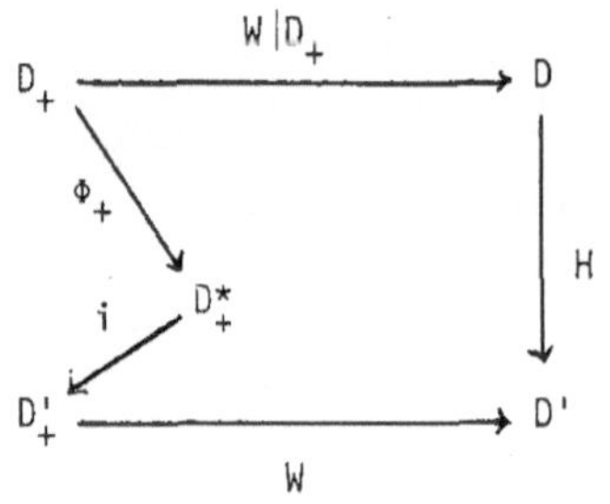

exists and commutes. Moreover the following estimates are valied:

(76) $$|W - \Theta|_{D'_+} \leq \frac{2}{3} qs,$$

(77) $$\begin{cases} \theta(1-q)|\zeta-\zeta'| \leq |W(\zeta) - W(\zeta')| \leq (1+q)|\zeta-\zeta'| \\ \text{for all } \zeta,\zeta' \in D'_+ , \end{cases}$$

(78) $$|\Phi_+ - \Omega_+ - Q|_{D_+} \leq \frac{5}{48} \theta M$$

where $Q : D_+ \rightarrow \mathbb{C}^2$ is defined by

$$(\xi,\eta) \mapsto Q(\eta) = (0,\ a_0+a_1\eta+a_2\eta^2)$$

with some constants a_0, a_1, $a_2 \in \mathbb{R}$.

Remark 1: Putting

$$\varepsilon = \varepsilon_k\ ,\ r = r_k\ ,\ r' = r'_k\ ,\ s = s_k\ ,\ s' = s'_k\ ,\ M = M_k\ ,$$

$$D = D_k\ ,\ D' = D'_k\ ,\ W = W_k\ ,\ H = H_k\ ,$$

and replacing the index + by k + 1 theorem 6 confirms what we have asserted in section 3 concerning the construction of the commuting diagram (28) observing (20), $(24)_k$ (29), (30), (31), (32), and (33).

The conditions (8) and (9) imply the inequalities (54) which correspond to the inequalities for ω,γ,τ in (I). Furthermore the condition (12) implies the inequality for q in (I).

Proof of Theorem 6

a) Heuristic considerations. First of all we define

$$h = (f,g) := H - \Omega\quad ,\quad w = (u,v) := W - \Theta\ ,\quad \phi := \Phi_+ - \Omega_+$$

and remark that by assumption we have $h \in P^2(D)$, and we ought to have $w \in P^2(D'_+)$, $\phi \in P^2(D_+)$.

The results of section 5 enable us to solve the linear difference equations

$$(79)\qquad \begin{cases} u(\omega+x,y) - u(x,y) = \varepsilon v(x,y) + f(x,\theta y) \\ v(\omega+x,y) - v(x,y) = \qquad g(x,\theta y) - \frac{1}{2\pi}\int_0^{2\pi} g(x,\theta y)dx \end{cases}$$

which can be written in the more compact form

$$(80)\qquad w\circ\Omega^* = (d\Omega)\,w + h\circ\Theta - h^*\circ\Theta$$

where we define Ω^* and h by

$$(x,y)\mapsto \Omega^*(x,y) = (\omega+x,y)\quad ,\quad (x,y)\mapsto h^*(y) = (0,[g](y))\ ,$$

and $d\Omega$ is the differential of Ω.

After having obtained a solution $w \in P^2(D'_+)$ we try to determine ϕ from the equation

$$H \circ W|D_+ = W \circ \Phi$$

which holds because the diagram above is to commute, and which gets by means of (80) the form

$$(81)\quad \begin{cases} Z = F(Z) := F_1 + F_2 + F_3 , \\ F_1 = w \circ \Omega_+ - w \circ (\Omega_+ + \phi) , \quad \phi = \Theta^{-1} \circ (Z + h^* \circ \Theta), \\ F_2 = w \circ \Omega^* - w \circ \Omega_+ , \quad F_3 = h \circ (\Theta + w) - h \circ \Theta . \end{cases}$$

Careful estimates will lead to a solution $Z \in P^2(D_+)$ such that $\phi \in P^2(D_+)$ can be determined.

b) Properties of W. Comparing (79) with (52), (53), using theorem 5 with $\rho = r/6$, and respecting (75) we get a solution of (80) with

$$(82)\qquad w \in P^2(r,t), \quad |w|_{4\rho,t} \leq 2\varepsilon^{-1}M, \quad t := \frac{s}{\theta} .$$

Since $D'_+ \subsetneq D(4\rho,t)$ we define $W = \Theta + w|D'_+$ and obtain

$$W \in T(D'_+), \quad |W - \Theta|_{D'_+} \leq 2\varepsilon^{-1}M,$$

hence (76) by (I). From (82) we also have

$$w \circ \Theta^{-1} \in P^2(r,s), \quad |w \circ \Theta^{-1}|_{4\rho,s} \leq \frac{2}{3}\, qs.$$

An application of lemma 5 (see section 7) with $d = \frac{2}{3}\, s$ gives

$$|w \circ \Theta^{-1}(z) - w \circ \Theta^{-1}(z')| \leq q|z-z'| \qquad \text{for all } z,z' \in D(r'_+, \tfrac{s}{3})$$

and consequently

$$(1-q)\,|z-z'| \leq |(Id + w \circ \Theta^{-1})(z) - (Id + w \circ \Theta^{-1})(z')| \leq (1+q)\,|z-z'|$$

for all $z,z' \in D(r'_+, \frac{s}{3})$. Inserting $z = \Theta\zeta$, $z' = \Theta\zeta'$ for $\zeta,\zeta' \in D'_+$ we get (77) because of

$$\theta|\zeta-\zeta'| \leq |z-z'| \leq |\zeta-\zeta'| \text{ and } \Theta(D'_+) \subsetneq D(r'_+, \tfrac{s}{3}).$$

Now we look for the range of W. From (I) we deduce the inequalities

$$qc \leq (1+q)c \leq \frac{3}{4} s \leq \frac{r}{4} \leq \frac{r'}{2}, \quad c := \frac{2}{3} s$$

such that with (76) we get

$$W(D'_+) \subseteq D(r'_+ + qc, \theta s'_+ + qc) \subseteq D(\frac{r'}{2} + \frac{r'}{2}, \frac{s'}{4} + \frac{3}{4} s) \subseteq D',$$

$$W(D_+) \subseteq D(r_+ + qc, \theta s_+ + qc) = D\big(r_+ + (1+q)c, \theta s_+ + (1+q)c\big) - c$$

$$\subseteq D(\frac{r}{2} + \frac{r}{2}, \frac{s}{4} + \frac{3}{4} s) - c = D - \frac{2}{3} s \subseteq D$$

as indicated in the diagram. For later purposes we note

(83) $$W(D_+) \subseteq D - \frac{2}{3} s, \quad \Theta(D_+) \subseteq D - \frac{2}{3} s$$

where the second relation follows as above from

$$\Theta(D_+) = D(r_+, \theta s_+) \subseteq D(r_+ + qc, \theta s_+ + qc) \subseteq D - \frac{2}{3} s.$$

c) Estimate for $F_2 = w \circ \Omega^* - w \circ \Omega_+$

A consequence of (I) is $\varepsilon < 1$ because of $\Gamma(\tau+1) \geq 1$ for $\tau \geq 1$. Hence we get

$$\Omega_+(D_+) \subseteq D(r_+ + \theta\varepsilon s_+, s_+) \subseteq D(r_+ + \frac{\theta}{2} s, s_+).$$

Since $\Omega^*(D_+) = D_+$ and $s_+ < t = \theta^{-1} s$ we have

(84) $$\Omega^*(D_+), \Omega_+(D_+) \subseteq D(4\rho - R, t), \quad R := \rho - \frac{\theta}{2} s.$$

This relation and (82) show that F_2 is well defined as an element of $P^2(D_+)$. Moreover we may apply lemma 5 to $\xi \mapsto w(\xi,\eta)$ with $D = \{\xi \in \mathbb{C} \mid |\mathrm{Im}\,\xi| < 4\rho\}$, $d = R$ such that we obtain

$$|w(\xi,\eta) - w(\xi',\eta)| \leq \frac{2M}{\varepsilon R} |\xi - \xi'| \quad \text{for all } (\xi,\eta),(\xi'\eta) \in D(4\rho - R, t).$$

Using (84) and the definition of Ω, Ω_+ we may write

$$|F_2|_{D_+} \leq \frac{2M}{R} \varepsilon_+ s_+ = \frac{s}{R} \theta M.$$

By (I) and (84) we have

$$R = \rho - \frac{\theta}{2} s = \frac{50}{\theta} s - \frac{\theta}{2} s > 49 \frac{s}{\theta},$$

thus altogether

(85) $$F_2 \in P^2(D_+), \quad |F_2|_{D_+} \leq \frac{1}{49} \theta^2 M.$$

d) Estimate for $F_3 = h \circ W|D_+ - h \circ \Theta$. We have $F_3 \in P^2(D_+)$ by virtue of $h \in P^2(D)$, $W|D_+ \in T(D_+)$, $\Theta = \Theta|D_+ \in T(D_+)$, and (83). From (75) it follows $|h|_D \leq M$ We apply lemma 5 with $d = 2s/3$ in order to obtain

$$|h(z) - h(z')| \leq \frac{3M}{2s} |z-z'| \quad \text{for all } z,z' \in D - \frac{2}{3} s.$$

So respecting (83) and $D_+ \subseteq D_+'$ we get

$$|F_3|_{D_+} \leq \frac{3M}{2s} |W - \Theta|_{D_+'} \leq qM.$$

Since $q \leq 10^{-2}\theta^2$ by (I) we have

(86) $$F_3 \in P^2(D_+), \quad |F_3|_{D_+} \leq \frac{1}{100} \theta^2 M.$$

e) Estimate for $F_1 = w \circ \Omega_+ - w \circ (\Omega_+ + \phi)$.

We assume

(87) $$Z \in P^2(D_+), \quad |Z|_{D_+} \leq \frac{1}{24} \theta^2 M.$$

By definition of h^* and of ϕ in (81) we get $|h^* \circ \Theta|_{D_+} \leq M$, and therefore

(88) $$|\phi|_{D_+} \leq \theta^{-1}(|Z|_{D_+} + |h^* \circ \Theta|_{D_+}) \leq M(\frac{\theta}{24} + \theta^{-1}) \leq \frac{25}{24} \theta^{-1} M \leq \frac{1}{100} s$$

where we use (I) and $\varepsilon < 1$ as in the sequel. Since $s < r/9$ we have

$$\frac{r}{2} + \frac{s}{2} < \frac{4}{7} r = r_+' + \frac{2}{3} s - \frac{2}{21} r < r_+' - \frac{s}{7}$$

hence

$$(\Omega_+ + \phi)(D_+) \subseteq D(r_+ + \varepsilon_+ s_+ + \frac{s}{100}, s_+ + \frac{s}{100})$$

$$\subseteq D(\frac{r}{2} + \frac{s}{2}, \frac{11}{21} s) \subseteq D(\frac{4}{7} r, \frac{11}{21} s) \subseteq D_+^*$$

From (87) we also have $\phi \in P^2(D_+)$. Thus we obtain

$$\text{(89)} \qquad \Phi = \Omega_+ + \phi \in T(D_+) \quad , \quad \Phi(D_+) \subseteq D(\frac{4}{7} r, \frac{11}{21} s) \subseteq D_+^*.$$

Clearly the estimate above gives

$$\Omega_+(D_+) \subseteq D(r_+ + \varepsilon_+ s_+, s_+) \subseteq D(\frac{4}{7} r, \frac{11}{21} s).$$

Moreover it is easy to see that

$$D(\frac{4}{7} r, \frac{11}{21} s) \subseteq D(4\rho - \frac{2s}{3\theta}, \frac{s}{3\theta}) = D(4\rho, t) - \frac{2s}{3\theta}$$

if one remembers the definitions of ρ, t, and takes from (I) the inequalities

$$\frac{s}{\theta} = \frac{r}{300} < \frac{r}{7}, \qquad \frac{11}{21} s \leq \frac{2}{3} s \leq \frac{1}{3}\frac{s}{\theta} .$$

So F_1 is well defined on D_+ by virtue of (82), and we have $F_1 \in P^2(D_+)$. Furthermore an application of lemma 5 is possible with $d = 2s/3\theta$. The result is

$$|w(z) - w(z')| \leq q\theta|z - z'| \quad \text{for all } z,z' \in D(\frac{4}{7} r, \frac{11}{21} s) .$$

As a cosequence we have

$$|F_1|_{D_+} \leq q\theta|\phi|_{D_+} \leq \frac{25}{24} qM$$

using (88). Since $q \leq 10^{-2}\theta^2$ we get

$$\text{(90)} \qquad F_1 \in P^2(D_+) , \qquad |F_1|_{D_+} \leq \frac{1}{96} \theta^2 M .$$

f) Determination of Z. The set of all Z satisfying (87) is a complet metric space. Collecting the results in (85), (86), and (90) we see that in view of (81)

$$Z \to F(Z) = F_1 + F_2 + F_3 \in P^2(D_+)$$

is a mapping of this metric space into itself because the estimate

$$|F(Z)|_{D_+} \leq \left(\frac{1}{49} + \frac{1}{100} + \frac{1}{96}\right) \theta^2 M \leq \frac{1}{24} \theta^2 M$$

is valid. Furthermore putting $\phi' = \Theta^{-1} \circ (Z' + h^* \circ \Theta)$ we have as above

$$|F(Z) - F(Z')|_{D_+} \leq q\theta |\phi - \phi'|_{D_+} = q\theta |\Theta^{-1} \circ (Z - Z')|_{D_+} \leq q|Z - Z'|_{D_+}$$

for all Z, Z' satisfying (87) such that F is a contraction. Hence there is a fixed point $Z = F(Z)$ which leads via (80), (81), and (89) to the existence of a mapping Φ such that the diagram above exists and commutes. As a consequence of $|\Theta^{-1} \circ Z|_{D_+} \leq \theta^{-1} |Z|_{D_+}$ we have the estimate

(91) $$|\Phi - \Omega_+ - \Theta^{-1} \circ h^* \circ \Theta|_{D_+} \leq \frac{1}{24} \theta M .$$

g) Proof of inequality (78). We apply lemma 6 (see section 7) to the function

$$\eta \to \theta^{-1}[g](\theta\eta) = g_0 + g_1 \eta + \dots , \qquad g_j \in \mathbb{R}$$

which is holomorphic for $|\eta| < \theta^{-1} s$ and has the estimate $|\theta^{-1}[g](\theta\eta)| \leq \theta^{-1} M$ because g is the second component of h, and we have $|h(\theta\eta)| \leq M$ for $|\eta| < \theta^{-1} s$ by virtue of (75). We put $q = \theta/2$ and $n=3$ such that for the polynomial

$$Q_2(\eta) = a_0 + a_1 \eta + a_2 \eta^2 , \qquad a_j = \left(1 - \left(\frac{\theta}{2}\right)^{6-2j}\right) g_j$$

we get the estimate

$$|\theta^{-1}[g](\theta\eta) - Q_2(\eta)| \leq \left(\frac{\theta}{2}\right)^3 \frac{M}{\theta} = \frac{1}{8} \theta^2 M \leq \frac{1}{16} \theta M$$

for $|\eta| \leq \frac{\theta}{2} \frac{s}{\theta} = \frac{s}{2} = s_+$. So for

$$Q(\eta) = (0, a_0 + a_1 \eta + a_2 \eta^2) , \qquad \Theta^{-1} \circ h^* \circ \Theta(\eta) = (0, \theta^{-1}[g](\theta\eta))$$

we obtain

$$|\Theta^{-1} \circ h^* \circ \Theta - Q|_{D_+} \leq \frac{1}{16} \theta M .$$

Together with (91) the inequality (78) follows. Theorem 6 is proved.

7. Appendix

For the proof of the inductive theorem we need two technical lemmas, the first of which is standard.

Lemma 5: For positiv integers n,m let D be an open and convex set in $\mathbb{C}^m$, and let $F:D \to \mathbb{C}^n$ be a holomorphic function. Then for any d>0 we have

$$|F(x) - F(y)| \leq d^{-1}|x-y|\sup_{x\in D}|F(x)|$$

for all $x,y \in D-d$.

Remark 2: Here $|\cdot|$ denotes some norm in $\mathbb{C}^m$ resp. $\mathbb{C}^n$, and as in section 4 we define

$$D-d = \{x \in \mathbb{C}^m | \{y||y-x| \leq d\} \subseteq D\}.$$

Proof: Since D is open and convex so is D-d, and for $x,y \in D-d$ we have

$$F(x) - F(y) = \int_0^1 F'(x-t(y-x))(y-x)dt$$

where F' is the derivative of F, hence

(92) $$|F(x) - F(y)| \leq \int_0^1 |F'(x+t(y-x))(y-x)|dt.$$

The function $z \to F(w+zv)$ is holomorphic in the disc $\{z\in\mathbb{C}||z| < d+\delta\}$ for sufficiently small $\delta>0$ provided that $w \in D-d$, $v \in \mathbb{C}^m$, $|v|=1$. So Cauchy's formula gives

$$F(w+zv) = \frac{1}{2\pi i}\int_{|\zeta|=d} \frac{F(w+\zeta v)}{\zeta - z}\,d\zeta.$$

Differentiating with respect to z we get $F'(w+zv)v = \frac{1}{2\pi i}\int_{|\zeta|=d}\frac{F(w+\zeta v)}{(\zeta-z)^2}\,d\zeta$

for all z with $|z| < d$. For z=0 this formula admits the estimate

$$|F'(w+zv)v| \leq d^{-1}\sup_{x\in D}|F(x)|.$$

Putting $w=x+t(y-x)$ and $v=(y-x)/|y-x|$ we obtain by means of (92) the wanted estimate for $x\neq y$.

Lemma 6: For some $r>0$ let $f:\{z\in\mathbb{C}\,|\,|z|<r\}\to\mathbb{C}$ be a holomorphic function with power series expansion

$$f(z) = \sum_{k=0}^{\infty} f_k z^k.$$

Then for the polynomial

$$f_{n-1,q}(z) = \sum_{k=0}^{n-1} (1-q^{2(n-k)}) f_k z^k$$

of degree $n-1$ depending on q, $n=1,2,\ldots,0<q<1$, we have the estimate

(93) $$\sup_{|z|\le qr} |f(z) - f_{n-1,q}(z)| \le q^n \sup_{|z|<r} |f(z)|.$$

Remark 3: For $f_0=f_1=\ldots=f_{n-1}=0$ this estimate is well known. However, the general case seems only to be known to some specialists in approximation theory (see [15]) It is due to Babenko [16].

Proof: Without loss of generality we may assume that f is defined and analytic for $|z|\le r$. For otherwise we could prove the estimate for r replaced by $r'<r$ and then pass to the limit $r'\uparrow r$. By Cauchy's formulae we have

$$\frac{1}{2\pi i}\int_{|\zeta|=r} f(\zeta)\zeta^{-k-1}d\zeta = \begin{matrix} f_k, & k = 0,1,\ldots \\ 0, & k = -1,-2,\ldots . \end{matrix}$$

Multiplying the k-th equation by

$$z^k \text{ for } k \ge n,\ q^{2(n-k)}z^k \text{ for } k<n$$

and summing up we obtain

$$f(z) - f_{n-1,q}(z) = \frac{1}{2\pi i}\int_{|\zeta|=r} \frac{f(\zeta)}{\zeta}\left(\frac{z}{\zeta}\right)^n \left(\frac{\zeta}{\zeta-z} + \frac{q^2\zeta}{z-q^2\zeta}\right) d\zeta$$

for $|z|\le qr$. Putting $z=qre^{i\theta}$, $\zeta=re^{it}$, $0\le\theta,t\le 2\pi$ this equation yields the estimate

$$|f(z) - f_{n-1,q}(z)| \le q^n \sup_{|z|<r}|f(z)|\,\frac{1}{2\pi}\int_0^{2\pi} K(t)dt$$

where

$$K(t) = \left| \frac{1}{1-qe^{i(\theta-t)}} + \frac{q}{e^{i(\theta-t)}-q} \right|$$

$$= \frac{1-q^2}{1+q^2-2q\cos(\theta-t)} = 1 + 2\sum_{k=1}^{\infty} q^k \cos k\,(\theta-t).$$

Thus

$$\frac{1}{2\pi}\int_0^{2\pi} K(t)dt = 1,$$

and (93) follows by means of the maximum principle. The lemma is proved.

References

[1] J. MOSER: On Invariant Curves of Area-Preserving Mappings of an Annulus. Nachr. Akad. Wiss. Göttingen, Math.-Phys. Kl. IIa, Nr. 1 (1962) 1-20.

[2] J. MOSER: A Rapidly Convergent Iteration Method and Nonlinear Differential Equations II. Ann.Scuola Norm. Sup. Pisa (1966) 499-535.

[3] H. RÜSSMANN: Über invariante Kurven differenzierbarer Abbildungen eines Kreisringes. Nachr. Akad. Wiss. Göttingen, Math.-Phys. Kl. II, Nr. 5 (1970) 67-105.

[4] J. MOSER: MR 42 # 8037.

[5] J. MOSER: Stable and Random Motions in Dynamical Systems. Princeton University Press. Princeton, N.J. (1973).

[6] M.R. HERMAN: Demonstration du théorême des courbes translatées de nombre de rotation de type constant. Manuscript 1981.

[7] A.N. KOLMOGOROV: On Quasi Periodic Motions Under Small Perturbations of the Hamiltonian. Doklady Akad. Nauk. SSSR 98 (1954) Nr. 4 527-530.

[8] V.I.ARNOLD: Small Denominators I. On the Mapping of a Circle into Itself. Izv. Akad. Nauk. SSSR Ser. Math. 25 Nr. 1 (1961) 21-86; Am. Mat. Soc. Transl., Ser. 2, 46, 213-284.

[9] V.I. ARNOLD: Proof of A.N. Kolmogorov's Theorem on the Preservation of Quasi-Periodic Motions Under Small Perturbations of the Hamiltonian. Usp. Mat. Nauk. SSSR 18 (1963) 13-40; Russian Math. Surveys 18, 9-36.

[10] V.I. ARNOLD: Small Divisor Problems in Classical and Celestial Mechanics Usp. Mat. Nauk. SSSR 18 (1963) 91-192. Russian Math. Surveys 18, 85-191.

[11] J. MOSER: Convergent Series Expansions for Quasi-Periodic Motions. Math. Ann. 169 (1967) 136-176.

[12] J.M. GREENE: The Calculation of KAM Surfaces. Annals New York Acad. Sci. 357 (1980) 80-89.

[13] H. JACOBOWITZ: Implicit Function Theorem and Isometric Embeddings. Ann. of Math. 95 (1972) 191-225.

[14] E. ZEHNDER: Generalized Implicit Function Theorems with Applications to Some Small Divisor Problems, I. Comm. Pure Appl. Math. 28 (1975) 91-140.

[15] G.G. LORENTZ: Approximation of Functions. Athena Series 1966. Holt, Rinehart and Winston, Inc.

[16] K.I. BABENKO: On the Best Approximation of a Class of Analytic Functions. Izv 22 (1958) 631-640.

Johannes Gutenberg-Universität in Mainz
Fachbereich Mathematik, Saarstr. 21,
D-6500 Mainz, West Germany

This work was partly prepared during the authors visit to the Institut des Hautes Etudes Scientifiques in Bures-sur-Yvette, April-May, 1981.

On the Average Cost of Solving Polynomial Equations

by Mike Shub and Steve Smale*

Given a complex polynomial $f(z) = z^d + a_{d-1} z^{d-1} + \ldots + a_0$ and a complex number z_0 we can attempt to find a root of f by locally inverting f at $f(z_0)$, to $f_{z_0}^{-1}$ taking $f(z_0)$ to z_0, and analytically continuing $f_{z_0}^{-1}$ along the ray from $f(z_0)$ to 0. This process usually works (see Smale for a discussion of this and the history) and defines a curve leading from z_0 to a root ζ of f. This curve is given by

$$f_{z_0}^{-1}((1-h)f(z_0))$$

where $0 \leq h \leq 1$. We may think of h as a complex variable and then $f_{z_0}^{-1}((1-h)f(z_0))$ is an analytic function. Let $h_1 = h_1(f, z_0)$ be the radius of convergence of this function considered as a power series around h=0. If $h_1 > 1$ then substituting h=1 in the power series expansion of $f_{z_0}^{-1}((1-h)f(z_0))$ gives a power series expansion for ζ. But h_1 may not be bigger than one, and even if it is it is not usually computationally practical to evaluate an infinite series, so we truncate this series at some finite power of h. Let

$$E_k(z_0) = E_{k(h,f)}(z_0) = t_k\, f_{z_0}^{-1}((1-h)f(z_0))$$

where t_k stands for truncation of a power series at degree k

$$(t_k(\sum_{i=0}^{\infty} a_i \cdot z^i) = \sum_{i=0}^{k} a_i \cdot z^i).$$ Thus $z_0 \to E_{k(h,f)}(z_0)$ for $0 \leq h$ is a curve which has k^{th} order contact with $f_{z_0}^{-1}((1-h)f(z_0))$ at z_0. Our plan is to iterate E_k for small values of h obtaining $z_n = E_k(z_{n-1})$ which stay close to the curve $f_{z_0}^{-1}((1-h)f(z_0))$ and approach the root ζ. As in Smale we rely heavily on the theory of shlicht functions. A full detailed account of the work described here may be found in Shub-Smale which is in the process of being written.

*Both authors have been partially supported by NSF research grants.

The power series expansion of ζ was given by Euler and he iterated E_k with h=1 and $1 \leq k \leq 5$ to solve equations see Euler. Thus we call our iterative algorithms, $k^{\underline{th}}$ incremental Euler.

It is not difficult to compute that

$$E_{4(h,f)}(z) = z - \frac{f(z)}{f'(z)}\left(h-\sigma_2 h^2 + (2\sigma_2{}^2 - \sigma_3)h^3 - (5\sigma_2{}^3 - 5\sigma_2\sigma_3 + \sigma_4)h^4\right).$$

where $\sigma_i = (-1)^{i-1} \dfrac{f^{(i)}(z)\, f^{i-1}(z)}{i!(f'(z))^i}$. By keeping only the first k powers of h, k = 1,2,3 we obtain E_k for those values. In particular E_1 is just incremental Newton which was studied in Smale.

The first main result is the following:

Theorem 1 There is a universal constant B, $1 \leq B \leq 1.07$ such that for any polynomial f, complex number z with $f'(z) \neq 0$, $f(z) \neq 0$, $z' = E_{k(h,f)}(z)$

$$(*) \qquad \frac{f(z')}{f(z)} = 1 - h + Q(h,f,z)\,\frac{h^{k+1}}{h_1{}^k} \quad \text{and}$$

$|Q| \leq \beta_k(\gamma)$ where

$$\beta_k(\gamma) = \frac{B(k+1)(1-\gamma)^2}{\left[(1-\gamma)^2-4\gamma\right]\left[(1-\gamma)^2-4\gamma(1+B(k+1)\gamma^k\right]}$$

Here $\gamma = \frac{h}{h_1}$ and is assumed to satisfy $0 < \gamma < \gamma_k$ where γ_k is the first positive number for which the denominator of $\beta_k(\gamma)$ vanishes.

Computation shows that γ_1 is approximately $\frac{1}{7}$ and that γ_k increases rapidly to $3 - \sqrt{8}$ which is approximately $\frac{1}{6}$. Thus the range of applicability of the estimate is around $\frac{h_1}{7}$.

We may rewrite (*) $\dfrac{f(z')}{f(z)} = 1 - h + T(h,f,z)h$

where $|T| \leq \beta_k(\gamma)\gamma^k \equiv \alpha_k(\gamma)$.

Thus to make sure that $|f(z')| < |f(z)|$ we want $\alpha_k(\gamma) < 1$. Computation of $\alpha_k'(\gamma)$ shows that $\alpha_k(\gamma)$ is increasing. Thus there is a unique $\bar{\gamma}_k$, $0 < \bar{\gamma}_k < \gamma_k$ such that $\alpha_k(\bar{\gamma}_k) = 1$ and $\alpha_k(\gamma) < 1$ for $0 < \gamma < \bar{\gamma}_k$.

$\bar{\gamma}_k$ also increases with k. $\bar{\gamma}_1$ is approximately $\frac{1}{12}$, $\bar{\gamma}_k > \frac{1}{7}$ for $k > 5$ and $\bar{\gamma}_k$ also tends to $3 - \sqrt{8}$ or approximately $\frac{1}{6}$ as k tends to ∞.

Theorem 1 admits the following Corollary. Let $\rho_f = \min_{\substack{\Theta \\ f'(\Theta)=0}} |f(\Theta)|$.

Corollary: Let $k > 0$. Suppose a polynomial f and a complex number z satisfy

$$|f(z)| = b\left(\frac{\bar{\gamma}_k}{1+\bar{\gamma}_k}\right)\rho_f \text{ for some } b < 1.$$

Then with $h = 1$, $(E_k)^{\ell}(z) = z_{\ell}$ is defined for all ℓ and $z_{\ell} \to z^*$ as $\ell \to \infty$ with $f(z^*) = 0$. Moreover, $|f(z_{\ell})| \leq c|f(z_{\ell-1})|^{k+1}$ all $\ell > 0$ ("$(k+1)^{st}$ order convergence") with $c=\left(\frac{b}{|f(z_0)|}\right)^k$.

Finally

$$|f(z_{\ell})| \leq b^{((k+1)^{\ell})}\frac{\bar{\gamma}_k}{1+\bar{\gamma}_k}\rho_f.$$

We will call z an <u>approximate zero</u> of f <u>relative</u> to <u>k</u> if $|f(z)| < \frac{\bar{\gamma}_k}{1+\bar{\gamma}_k}\rho_f$. An approximate zero of f relative to all $k > 0$ will simply be called an <u>approximate zero</u> of f

$$\frac{\bar{\gamma}_k}{1+\bar{\gamma}_k} \text{ increases with } k.$$

$$\frac{1}{12} > \frac{\bar{\gamma}_1}{1+\bar{\gamma}_1},\quad \frac{\bar{\gamma}_k}{1+\bar{\gamma}_k} > \frac{1}{7} \text{ for } k \geq 5 \text{ and } \frac{\bar{\gamma}_k}{1+\bar{\gamma}_k} < \frac{1}{6} \text{ for all } k.$$

Thus if $|f(z)| < \frac{\rho_f}{12}$, z is an approximate zero of f.

<u>Our goal is to find approximate zeros of **f**!</u>

For $0 < \alpha \leq \frac{\pi}{2}$ let $w_{f,z,\alpha} = \{w \in C \,|\, |w| \leq 2|f(z)|,\ |\arg \frac{w}{f(z)}| < \alpha\}$

Define $w_{f,z}$ be the largest of the $w_{f,z,\alpha}$ to which f_z^{-1} may be analytically continued and $\Theta_{f,z}$ to be the corresponding α. Since the $w_{f,z,\alpha}$ are simply connected, this definition makes sense.

$w_{f,z}$

We navigate in this wedge using Theorem 1 and its corollary.

Let $K(k) = \dfrac{(k+1)^{\frac{k+1}{k}}}{k\,\bar{\gamma}_k\,(1-\bar{\gamma}_k)^{\frac{1}{k}}}$

Theorem 2: Suppose given a polynomial f and a complex number z_0 such that $|f(z_0)| > \rho_f > L > 0$ and $\Theta_{f,z_0} > 0$. Let $C = \dfrac{1}{\Theta_{f,z_0}} \log \dfrac{|f(z_0)|}{L}$.

Then there is an $h_0 \geq \dfrac{\sin \Theta_{f,z_0}}{K(k)(c+1)^{\frac{1}{k}}}$ with this property. For each $0 < h \leq h_0$

and any $s \geq \dfrac{1}{h}\left(\log \dfrac{|f(z_0)|}{L} + \dfrac{\Theta_{f,z_0}}{k+1}\right)$

$|f(z_n)| < L$

K(k) decrease to $\dfrac{1}{3-\sqrt{8}}$ which is less than 6.

k =	1	2	3	4	5	6	7	8	9	10
K(k) <	47.1	21.1	14.7	12.1	10.7	9.9	9.3	8.9	8.6	8.4

Smale gives a better result for Newton's method than the K(1) achieved here

Thus the problem of finding approximate zeros becomes the problem of controlling ρ_f and Θ_{f,z_0}.

This is done in the next two propositions

Let $P_d(1) = \{f \mid f(z) = z^d + a_{d-1} z^{d-1} + \ldots + a_0 \text{ where } |a_i| \leq 1,\ i=0,\ldots, d-1\}$

Proposition 1 (Smale)

$\text{Vol}\ \{f \in P_d(1) \mid \rho_f < \alpha\} \leq (d-1)\alpha^2$

where Vol means normalized Lebesgue measure.

Let S^1_r be the circle of radius r around 0 in C.

Proposition 2 For $\alpha < \frac{\Pi}{2}$, $f \in P_d(1)$

$\text{Vol}\ \{z \in S^1_r \mid \Theta_{f,z} < \alpha\} \leq \frac{2}{\Pi}\left(\frac{d-1}{d}\right)\left(\alpha + 2 \arcsin \frac{1}{r-1}\right)$. In fact $\{z \in S^1_r \mid \Theta_{f,z} < \alpha\}$ is contained in at most 2(d-1) arcs of S^1_r of angle $\frac{2}{d}\left(\alpha + 2 \arcsin \frac{1}{r-1}\right)$.

The last ingedient in the theory is a checking procedure. Let $R(f,\frac{f'}{d})$ be the resultant of f and $\frac{f'}{d}$ see <u>Lang.</u> $R(f,\frac{f'}{d})$ is computable from the coefficients of f alone.

Let $L(f) = \frac{1}{24} \min (1,\rho_f^{\frac{7}{6}})$

<u>Lemma</u> Let $f \in P_d(1)$, then

a) $\rho_f \geq \dfrac{R(f,\frac{f'}{d})}{(d+3)^{2d-1}}$

b) If $|f(z)| < L(f)$ then

$$|f(E^{\ell}_{k(1,f)}(z))| < \frac{\bar{\gamma}}{1+\bar{\gamma}} \frac{R(f,\frac{f'}{d})}{(d+3)^{2d-1}}$$

for $\ell = \lceil \log_{k+1}(3d \log d) \rceil \equiv \ell(k,d)$.

a) is a fairly naive estimate and b) follows from the Corollary.

The use of the lemma is the following.

If we suspect that

$|f(z)| < L(f)$ then

$|f(E^{\ell}_k(z))| < \dfrac{\frac{\bar{\gamma}}{1+\bar{\gamma}} R(f,\frac{f'}{d})}{(d+3)^{2d-1}}$ which is checkable by direct calculation. If this last inequality holds then $E^{\ell}_r(z)$ is an approximate zero of f.

We use the results above to prove the following theorems. Let μ be normalized Lebesgue measure on $S^1_r \times P_d(1)$.

<u>Theorem 3</u> Given $0 < \mu < 1$, $d > 1$ there is an iterative algorithm $E = E_{(k(\mu,d),h(\mu,d))}$ and an $r = r_{(\mu,d)}$ such that:

for (z_0,f) in $S^1_r \times P_d(1)$ and $s = L_1 d(\frac{|\log \mu|}{\mu})^{1+\frac{1}{\lceil \log d \rceil}} + L_2$;

$z = E^s(z_0)$ is an approximate zero for f with probability $1 - \mu$.

Here L_1 and L_2 are small constants, L_1 is less than 20 and L_2 is smaller yet.

The exponents in Theorem 3 improve considerably on those of the Main Theorem of Smale. Yet it seems paradoxical that part of the improvement is achieved by making $r_{(\mu,d)}$ increase like $\frac{1}{\mu}$, which contributes the factor d in the theorem. It would seem more sensible to pick r small, close to 1. It is an excellent problem to carry out the analysis in this case; it becomes more difficult.

By attempting to find a "good" starting point z_0 for each given f we prove "average" theorems for iterative algorithms based on the $k^{\underline{th}}$ incremental Euler algorithms. By "average" we mean the integral with respect to normalized Lebesgue measure on $P_d(1)$.

Theorem 4: There are probabilistic and deterministic iterative algorithms for finding approximate zeros for f $P_d(1)$, with the average number of steps required $O(d)$ and $O(d^2 \log d)$ respectively.

The algorithms in Theorem 3 and 4 may be executed in $O(d \log d)$ arithmetic operations for each step in exact arithmetic with log and real $k^{\underline{th}}$ roots. They are robust.

References

L. Euler, Institutiones Calculi Differentialis, exp IX, Opera Omnia, série I, vol. X, pp. 422-455.

S. Lang, Albegra, Addison Wesley, Reading, Mass.

M. Shub and S. Smale, Computational Complexity: On the geometry of polynomials and a theory of cost: Part I and II, to appear.

S. Smale, 1981, The Fundamental Theorem of Algebra and Complexity Theory, Bull. Amer. Math. Soc.,Vol. 4, No. 1 pp. 36.

CONFORMAL DYNAMICAL SYSTEMS*

by

Dennis Sullivan

1. Iteration

In many parts of geometry and analysis the situation arises in which one considers a set of differentiable transformations of an underlying manifold obtained by iterated composition of a given set of initial or generating transformations. One is then interested in the possible positions and shapes of the images of a neighborhood of a general point by the total set of transformations. Questions as to whether the images of an initial neighborhood return infinitely often to intersect the initial neighborhood or whether the iterated images wander off to accumulate elsewhere -- perhaps even at ∞ if the underlying space is non-compact -- are the basic questions of topological dynamics.

The distribution of the images of a generic point relative to a given measure on the space is the subject of measurable dynamics or ergodic theory (a name especially used for a singly generated set of transformations -- because of the Von Neumann-Birkhoff ergodic theorem). The features relative to topology and measure theory can be quite different when the iterated transformations squeeze of distort the neighborhoods in an unbounded manner.

For example, let C denote a Cantor set of positive measure on the circle S^1 bounding the unit disk D. Let Γ denote the group of transformations of S^1 in $PSL(2,R)$ generated by the set of non-Euclidean reflections of the disk D which interchanges each

---

* This paper is an expanded version of the Colloquium Lectures of the 1982 American Mathematical Society Winter Meeting and of the lecture given at this Conference August 1981.

complementary interval I of the Cantor set with (S^1-I). Then every Γ orbit of a point on S^1 is dense in S^1. But the orbit of the Cantor set partitions a part of the circle having positive linear measure zero into disjoint copies of the Cantor set.

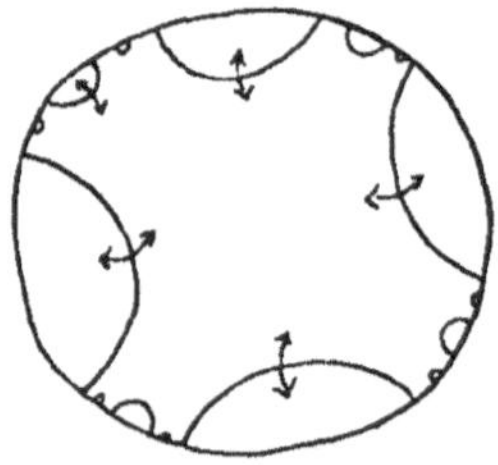

Γ is the group generated by reflections in the sides of non-Euclidean convex hull of the Cantor set.

On the other hand it is easy to show that a group of transformations of a manifold which do not distort distance at all and which has dense orbits acts ergodically -- given two sets A and B of positive measure some image of A intersects B in a set of positive measure.

In the above example the transformations have an unbounded amount of geometric distortion. Thus we are faced with the problem of understanding the distortion produced by large iterated compositions of transformations from a given set. If the set of transformations under study form a continuous group it is only the distortion transversal to the orbit which needs to be considered. To make the problem tractable the generating set will usually be finite or at least compact. For example the phenomenon of the example cannot happen for a finitely generated subgroup of projective transformations, PSL(2,R), acting on the circle. (See Theorem 1.)

Of course, for any composition of differentiable transformations the tangent map is the product of the tangent maps of the successive factors. Thus for the linear part of the distortion we will have a large, random product of matrices taken from a bounded set. Understanding such random products is a field in itself. The main tools are the boundary theory of Lie groups of Furstenberg as used in work of Margulis, the multiplicative ergodic theorem of Osilidec, and the Smale, Anosov, Pesin approach.

Actually, there is already a rich supply of dynamical examples when all the derivatives encountered are scalar multiples of orthogonal transformations (≡ conformal transformations). These include all differentiable examples in a space of one real dimension, all complex analytic examples on a one complex dimensional manifold, and any group of Moebius transformations of the n-sphere. For these <u>conformal dynamical systems</u> the understanding of the linear distortion simplifies to the problem of determining the scalar multiples or conformal factors. This is of course an abelian computation as for dimension one, and we shall see common features of 1-dimensional systems and conformal systems in higher dimensions.

After one has come to terms with the linear part of the distortion of an iterated composition (the tangent map) there remains the difficult problem of the <u>non-linearity</u>, the unbounded deviation of the iterated composition on a neighborhood from any linear approximation.

For conformal transformations the non-linearity problem is similar to the non-linearity problem in dimension-one. In dimension one a natural measure of non-linearity of a transformation f is $L(f) = (\ln f')' = f''/f'$ where prime denotes differentiation. The composition law $L(f\circ g) = (L(f)\circ g)\cdot g' + L(g)$ leads to a general principle of linearity for certain situations in dimension one. (See 2)). This principle is also valid for the volume distortion of an iterated composition of expanding maps.

Again in dimension one a natural measure of non-projectivity of a transformation f is the Schwarzian derivative $S(f) = L(f)' - 1/2(L(f))^2 = f'''/f' - 3/2(f''/f')^2$. The composition law $S(f\circ g) = (S(f)\circ g)(g')^2 + S(g)$ leads to certain results (David Singer, Misuriewicz, Guckenheimer).

2. <u>Distortion lemmas</u>, C^2 <u>Denjoy theory</u>.

We study the non-linear geometric distortion in an iterated composition $g_n = f_n\circ f_{n-1}\circ\dots\circ f_1$. Let f' denote a numerical measure of distortion which multiplies under composition. For example, volume distortion or linear distortion for conformal maps. Suppose the non-linearity measurement, grad(log f'), is bounded by N for each of the generating transformations f_i and that $\sum_{i=0}^{n-1}$ distance $(x_i,y_i) \le L$ where x_0 and y_0 are two initial points and $(x_i,y_i) = (f_i(x_{i-1}),f_i(y_{i-1})) = (g_i(x_0),g_i(y_0))$.

<u>Lemma 1</u>. <u>The ratio of</u> g_n' <u>at the two points</u> x_0,y_0 <u>is bounded on both sides by</u> $\exp \pm(L\cdot N)$.

<u>Proof</u>. (classical)

$$|\log g_n'(x_0)/g_n'(y_0)| = |\log \prod_{i=1}^{n} f_i'(x_{i-1}) - \log \prod_{i=1}^{n} f_i'(y_{i-1})|$$

$$\le \sum_{i=1}^{n} |\log f_i'(x_{i-1}) - \log f_i'(y_{i-1})|$$

$$\le \sum_{i=1}^{n} N \cdot \text{distance } (x_{i-1},y_{i-1}) \le N\cdot L.$$

Q.E.D.

<u>Remark</u>. In particular for an infinite sequence $f_1,f_2,\ldots$ so that $\sum_{i=0}^{\infty}$ distance $(x_i,y_i) = L < \infty$ we have for all n,

$|\log g_n'(x_0)/g_n'(y_0)|$ is bounded independently of n (by $L\cdot N$).

This simple computation has a second corollary for one-dimensional transformations. Let C be a collection of compositions (or words) in the generating transformations which satisfies

$g_n = f_n \circ f_{n-1} \circ \ldots \circ f_1$ in C implies $g_{n-1} = f_{n-1} \circ \ldots \circ f_1$ in C. Let g_n' denote the linear distortion.

<u>Lemma 2</u>. $\{x \mid \sum_{g_n \in C} g_n'(x) < \infty\}$ <u>is an open set</u>.

<u>Proof</u>. (Schwartz). If $\lambda > 1$, one shows by induction on the word length n in C that there is $\delta > 0$ so that $g_n'(y_0) \le \lambda\, g_n'(x_0)$ for distance $(x_0,y_0) < \delta$. Namely,

$$
\begin{aligned}
|\log g_n'(y_0)/g_n'(x_0)| &\leq \sum_{i=1}^{n} N \ \text{distance} \ (x_{i-1},y_{i-1}) \\
&= \sum_{i=1}^{n} N \ \text{distance} \ (g_{i-1}(x_0),g_{i-1}(y_0)) \\
&\leq \sum_{i=1}^{n} N \ g_{i-1}'(x_i^*) \ \text{distance} \ (x_0,y_0) \\
&\leq N\,\lambda\delta \sum_{i=1}^{n-1} g_{i-1}'(x_0) \ \text{ by induction.}
\end{aligned}
$$

If $\sum_{g_i \in C} g_i'(x_0) < \infty$, then we can choose δ small enough to complete the induction. But then distance $(x_0,y_0) < \delta$ implies $\sum_{g_n \in C} g_n'(y_0) \leq \lambda \sum_{g_n \in C} g_n'(x_0) < \infty$.

From these general remarks it is easy to derive much of the C^2 Denjoy theory. For example,

(i) (Denjoy) A C^2 diffeomorphism of the circle without a periodic point has only dense orbits (and thus by Poincaré is topologically conjugate to a rotation).

(ii) (Rosenberg and Sullivan) A complex analytic homeomorphism of a neighborhood of an invariant rectifiable closed curve in $\mathbb{C}$ with no periodic points on the curve has only dense orbits on the curve.

(iii) (Sacksteder) In a C^2-codimension one foliation of a compact manifold by simply connected non-compact leaves all leaves must be topologically dense.

These are all results of topological dynamics. There are also direct corollaries in these three cases relative to the natural one-dimensional measure and measurable dynamics.

Theorem 1. *In each of the examples above there is no set of positive measure which intersects each orbit (or leaf) in at most one point. Namely, these dynamical systems are recurrent or conservative -- for every set of positive measure A infinitely many iterated transformations bring part of A back to A.*

Problem. Is case (iii) ergodic? (i) and (ii) are known to be (Katok, Herman,...)

Proofs and generalizations.

(i) (Denjoy) If a C^2 diffeomorphism f of the circle has an invariant Cantor set, the complementary intervals have finite total length, and are wandering. By Lemma 1 the derivatives of n-fold compositions at various points in one interval are commensurable. Thus the total sum of derivatives along one orbit is comparable to the total length of intervals and so is finite. This is valid even at the endpoints of one interval. By Lemma 2 the sum of derivatives along one orbit is finite on a neighborhood of one of these endpoints x_0. In fact from the proof for an interval I about x_0 $|(f^n)'y|$ is bounded by a constant times $|(f^n)'x_0|$, and so it tends uniformly to zero. By recurrence at x_0 some $f^n(x_0)$ lies close to x_0. Thus we can assume $f^n I \subset I$ and we have a periodic point. This contradicts the assumption, proving f is topologically transitive.

(ii) (Rosenberg-Sullivan) The same argument as in (i) works here. One point of some subtlety perhaps is that the one-dimensional calculations are valid for measuring lengths of images of rectifiable arcs by f^n because f is a conformal map.

Problem. (a) Is such a statement as (ii) true for a complex analytic homeomorphism defined near any invariant topological curve?

(b) Is there a real analytic homeomorphism of the circle giving a Denjoy type example (no periodic points but not topologically transitive)?

(iii) The proof is similar to that of (i) using the holonomy pseudogroup on one transversal which generates a collection C of transformations with (by compactness) a finite number of generators. If there is a non-trivial minimal set of leaves it must intersect the transversal in an invariant Cantor set. The complementary intervals must wander because the Cantor set is a minimal closed invariant set. Now we are in position to do the argument in (i) line by line using the C-orbit here in place of the f^n-orbit there. One constructs a contracting periodic point and thus holonomy in one leaf. This contradicts simply connectivity.

<u>Remark</u>. Lest the reader think this type of argument (invented by Schwarz and Saksteder for the above case) goes forever we mention the following example. If D denotes the endomorphism $\theta \to 2\theta$ on the circle for each irrational α there is (Douady, Sullivan, Thurston) a minimal D-invariant Cantor set $K(\alpha)$ where D identifies exactly 2 endpoints and the order structure of $x, Dx, D^2x, \ldots$ for a general point of $K(\alpha)$ is that of the orbit of a point under the rotation by α.

However, one can easily prove

(iii)′ There is no proper infinite closed invariant set for an expanding endomorphism D of S^1 on which D is a homeomorphism.

<u>Proof</u>. (Douady) An infinite compact metric space cannot have a self homeomorphism which uniformly decreases the distance between sufficiently near pairs of points.

<u>Theorem 1 Proof</u>. A dynamical system is conservative if and only if for almost all points x the sum of volume distortions evaluated at x of all transformations defined at x is a divergent series. (Exercise).

In real dimension one we have by Lemma 2 this convergence set is open. The set is invariant by definition. But we have already proven above these three dynamical systems (i), (ii), (iii) are minimal (all orbits are dense). Thus if this set is non-void it is everything. Then again we construct periodic points which contradict minimality.

3. <u>Analytic functions and mappings</u>.

In the 1880's Poincaré introduced the subject of discrete subgroups Γ of complex linear fractional transformations $w \to aw + b/cw + d$ acting conformally on the plane or Riemann sphere. These were the monodromy groups of 2nd order differential equations $a(z)w'' + b(z)w' + c(z)w = 0$ where the inverse function of the multivalued (ratio) solution defined in $C - \{\text{zeroes of } a(z)\}$ was single valued in some domain. Thus Poincaré investigated complex analytic functions $F(w)$ invariant or automorphic under the action of Γ.

The natural domains of definition of the automorphic functions $F(w)$ were the connected components of the complement of the Poincaré limit set to which every orbit of Γ in C clustered.

In the first part of this century Fatou and independently Julia studied the topological dynamics associated to the iterated compositions of a complex analytic self mapping of the sphere, $z \to R(z)$. Fatou was interested in analytic functions satisfying functional equations associated to rational substitutions of the form $z \to R(z)$. For example, $F(R(z)) = \lambda \cdot F(z)$, studied for $|\lambda| \neq 1$ by Koenig, Shroder, and by Poincaré in the nineteenth century, and for $|\lambda| = 1$ by C.L. Siegel in the 1940's.

Again the natural domains of definition for these functions satisfying functional equations were the connected components of the complement of a limit set associated to the dynamics of $z \to R(z)$. This limit set -- now called the Julia set -- is defined to be the complement of those points which have a neighborhood where the conformal factors (in the spherical metric) of the iterates R, $R \cdot R$, ... are uniformly bounded (the stable points).

We will mention some common features of these two situations related to the conformal properties of the mappings involved.

First the topological dynamics. For the Poincaré limit set $\Lambda(\Gamma)$ of any discrete group Γ one knows the Γ orbit of any point is dense in $\Lambda(\Gamma)$. In complete analogy for the Julia set $J(R)$ of the rational map $z \to R(z)$ one knows (Fatou-Julia) the backward orbit of any point in $J(R)$ is dense in $J(R)$.

On the complement $\Omega(\Gamma)$ of the Poincaré limit set $\Lambda(\Gamma)$ the action of Γ is properly discontinuous (for each compact $K \subset \Omega(\Gamma)$ there are only finitely many $\gamma \in \Gamma$ so that $\gamma K \cap K \neq 0$). One can then form a quotient Riemann surface $\Omega(\Gamma)/\Gamma$. A famous theorem of Ahlfors (1965) is that this Riemann surface for a group Γ of d-generators has finite type (it is obtained from a compact Riemann surface by removing finitely many points.) In particular, the components of the complement of the limit set $\Lambda(\Gamma)$ fall into finitely many Γ

orbits ($\leq 2d-2$), and it is impossible for one of these orbits to be a wandering disk.

The topological picture of the action of $z \to R(z)$ on the complement of Julia limit set $J(R)$ was less well know until recently. For example, could one component be a disk D so that all the images D, $R(D)$, $R \cdot R(D)$, ... are disjoint? If such a wandering disk exists, the total area on the sphere is finite. Thus $\sum_n |R^{n\prime}(z)|^2 < \infty$ for almost all z in D.

One is tempted to use the ideas of 2) to arrive at a contradiction in the manner of Denjoy's theorem for C^2 diffeomorphisms of S^1 (which analogously asserts there is no wandering interval on S^1.)

Such an elementary proof is not available at the present. One proves the following theorem by using the measurable Riemann mapping theorem (Ahlfors-Bers 1960) to construct, if the conclusion is false, an infinite dimensional space of complex analytic self-mappings of $\mathbb{C} \cup \infty$ with a given degree, and thus a contradiction.

Theorem 2. (Sullivan [1982]). Under the forward iteration of a rational map of degree d, $z \to R(z)$ the connected domains of the complement of the Julia limit set $J(R)$ map into finitely many cyclic orbits of domains.

These cyclic stable regions can be classified into five types. The first two types, attractive basins and parabolic basins have fundamental domains for the equivalence relation: $x \sim y$ if and only if $f^n x = f^m y$ some $n,m \geq 0$. The third type, superattractive basins do not, but they are foliated by the closures of the classes of the equivalence relation, $x \approx y$ if and only if $f^n x = f^n y$, $n \geq 0$. The last two types are rotation domains, Siegel disks or Herman rings, which are foliated by the closures of forward orbits. (See figure 1.)

(i) An attractive basin D arises from an attractive periodic cycle γ with nonzero derivative of modulus less than one, $\gamma = z \cup fz \cup \ldots \cup f^{n-1}z$, $f^n z = z$, $0 < |(f^n)'(z)| < 1$, and D consists of the components of $W_s(\gamma) = \bigcup_{x \in \gamma} \{y \mid \lim_{n\to\infty} \text{distance}\,(f^n y, f^n x) = 0\}$

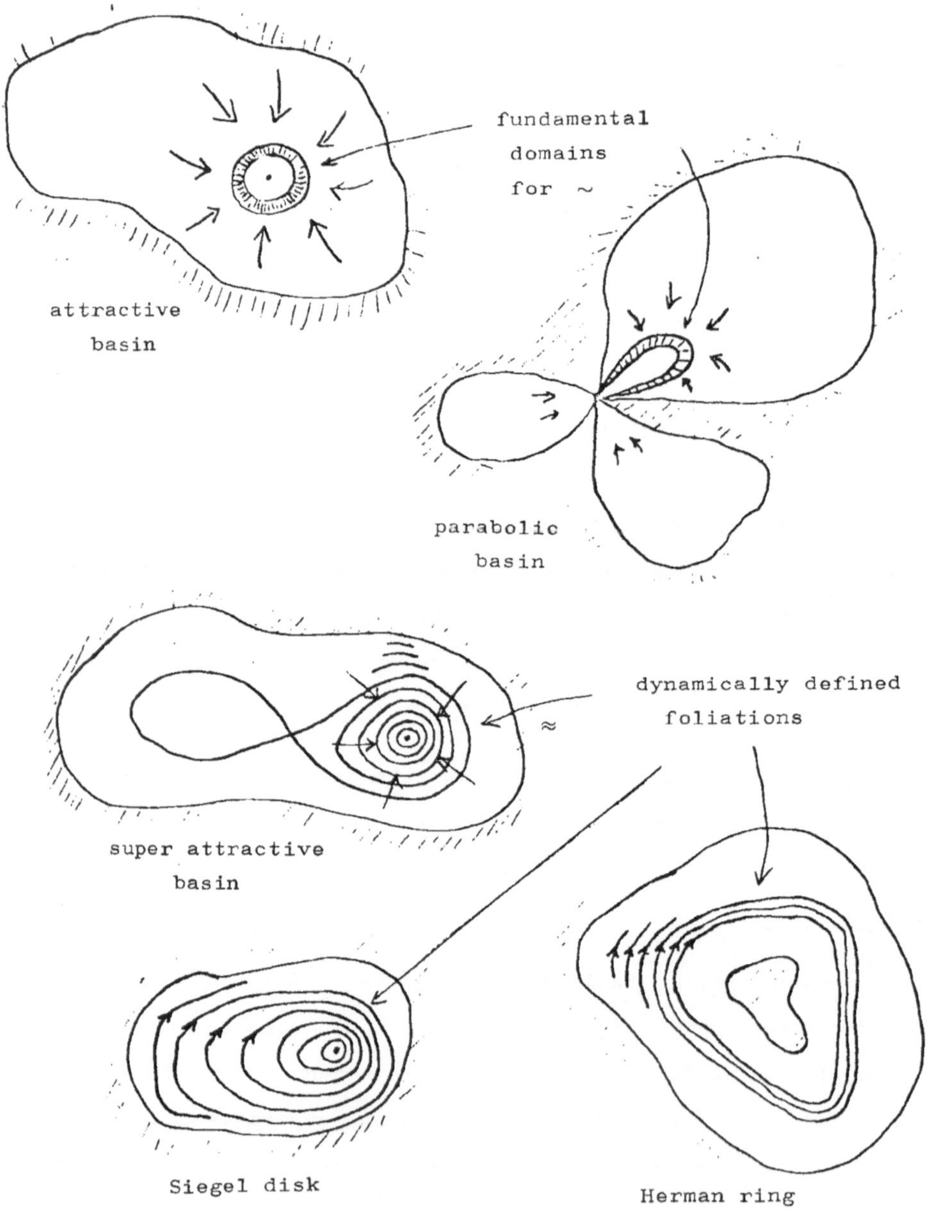
fundamental
domains
for ~
attractive
basin
parabolic
basin
dynamically defined
foliations
≈
super attractive
basin
Siegel disk
Herman ring

figure 1

containing points of γ. Fatou [1918] showed that such a D must contain a critical point of f. Thus there are no more than 2d-2 attractive basins for an endomorphism of degree d.

If we remove from D the inverse orbit of γ, $\{\bigcup_{n\geq 0} f^{-n}\gamma\}$, the set of $\sim$ equivalence classes ($x \sim y$ if and only if $f^n x = f^m y$) defines a torus with branch points corresponding to the critical points of f. This follows easily from the local model of f near γ, where near a fixed point of a power of f we have $z \to \lambda z$, $|\lambda| < 1$.

(ii) A <u>parabolic basin</u> D arises from a non-hyperbolic periodic cycle γ with derivative a root of unity, $\gamma = z \cup f(z) \cup \ldots \cup f^{n-1}z$, $z = f^n(z)$, $((f^n)'(z))^m = 1$, γ is contained in the frontier of D, and each compact in D converges to γ under forward iteration of f. (Fatou [1918].) The local picture of the dynamics consists of parabolic sectors arranged around the fixed point of a power of f which in local coordinates is $z \to z + z^\ell + \ldots$. Fatou [1918], Camacho [1979].

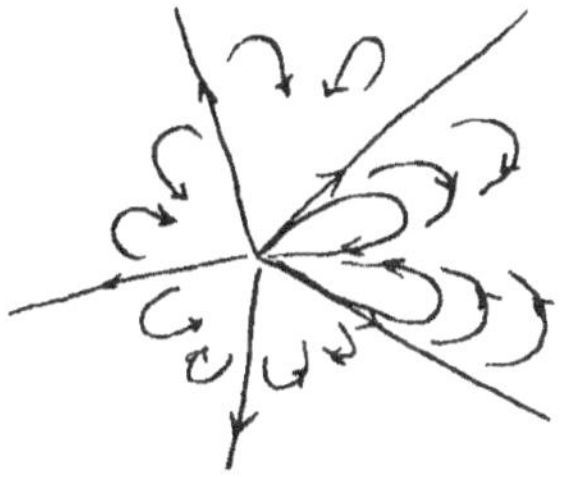

local model
$z \to z + z^\ell + \ldots$

The local model produces a fundamental domain for the global dynamics on D because all orbits in D tend to γ. Looking at the local picture then shows the quotient of D by the $x \sim y$ equivalence classes is a union of twice punctured sphere with branched points coming the critical points of f lying in D (there must be at least one critical point in D, Fatou [1918]).

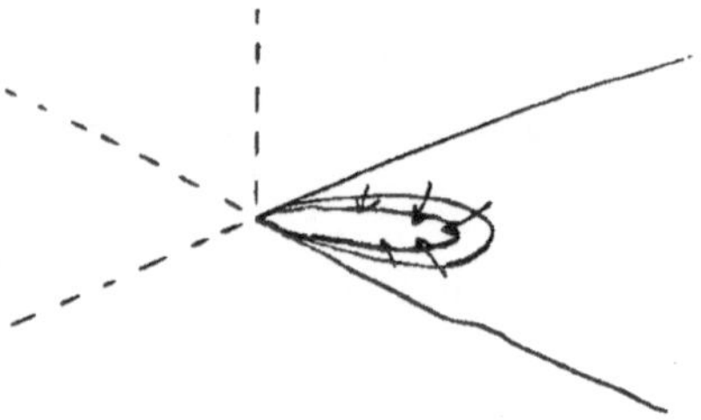

fundamental domain

(iii) A <u>superattractive basin</u> D is defined just like an attractive basin but now the derivative of the power of f having a fixed point is zero. Now points arbitrarily near the attracting cycle are identified by f and there is no true fundamental domain for the $\sim$ equivalence classes. The more precise relation $x \approx y$ if and only if $f^n x = f^n y$, $n \geqq 0$,defines a foliation of $D' = D$ - inverse orbit of γ by the closures of the $\approx$ equivalence classes. The leaves are compact 1-manifolds which are not necessarily connected and which have n-prong singularities at the inverse orbit of other critical points in D. The local linearization near a superattractive fixed point shows the leaves near γ are nearly concentric closed curves around the points of γ. The rest of the foliation of D' is obtained by applying f^{-1} to this concentric foliation near γ.

(iv) A <u>Siegel disk</u> is a stable regions which is cyclic and on which the appropriate power of f is analytically conjugate to a rotation of the standard unit disk. Siegel [1942] proved these occur near a non-hyperbolic periodic point if $1/\pi$-argument of the derivative is far from the rationals. Far from the rationals means $|\vartheta - p/q| > c/q^{\nu}$ for some $c > 0$, $\nu > 0$, and all p/q reduced fractions.

Fatou and Julia showed that if such regions existed their frontiers were contained in the union of the ω-limit sets of critical points.

Siegel disks around the origin occur already in the family $z \to \lambda z + z^2$, $|\lambda| = 1$. However, they do not occur when $1/\pi \arg \lambda$ is sufficiently Liouville because then there are periodic points tending to zero in this case (an easy calculation).

(v) A <u>Herman ring</u> is a stable region similar to a Siegel disk. Now we have a periodic cycle of annuli and a power of f is analytically equivalent to an irrational rotation of the standard annulus. Again the frontier is contained in the ω-limit sets of critical points. Such examples were found by Michel Herman in the family

$$z \to \frac{e^{i\theta}}{z} \cdot \left(\frac{z-a}{1-\bar{a}z}\right)^2$$

appropriate θ, a small. Herman uses Arnold's theorem [1960] about real analytic conjugations of real analytic diffeomorphisms of the circle to rigid rotations when the rotation number is like a Siegel number. Note that both Siegel disks and Herman rings are foliated by the closures of orbits and the leaves are closed curves.

<u>More dynamical properties</u>.

(i) One knows there are only finitely many cyclic stable regions described in (4) Sullivan [1982]. But it is <u>a problem</u> to find the sharp upper bound in terms of the degree. Is it $2d-2$?

(ii) Also for polynomials one knows each <u>bounded stable region is simply connected</u> (apply the maximum principle to $f, f^2, \ldots$). <u>Thus polynomials do not have Herman rings</u>.

(iii) An amusing corollary of the classification of stable regions in (4) is the following: <u>if all critical points of f are eventually periodic but none are periodic then the Julia set of f is all of $\bar{\mathbb{C}}$</u>. (Because each type of cyclic region besides the superattractive basin requires a critical point with an infinite forward orbit.) Examples of this type are $z \to \left(\frac{z-2}{z}\right)^2$ and the quotient of some higher degree endomorphism of a one-dimensional torus by the equivalence relation $x \sim -x$.

(iv) Fatou and Julia showed f on $J(f)$ is topologically transitive. (In fact, for any z in $J(f)$ the inverse orbit $\bigcup_{n\geq 0} f^{-n}(z)$ is dense in $J(f)$.) If no critical points tend to $J(f)$ or touch it Fatou showed some power of f is expanding on $J(f)$. He surmised the dynamical structure was continuous in the coefficients for such examples (now called Axiom A or expanding systems -- see below) and guessed that this property should be true except for special values of the parameters.

Even when $J(f)$ is contaminated by critical points one may think of $J(f)$ as the "hyperbolic" part* of the dynamics. The Siegel disks and Herman rings are in the "elliptic" part of the dynamics. The attractive basins and the parabolic basins are the properly discontinuous part of the dynamics. The superattractive basins are both wandering and of elliptic character.

Newton's method. We recall that it is still a difficult problem to find the zeroes of a complex polynomial $f(z)$. Newton's iterative method $z \to z - f/f'$ provides a natural example where the dynamics above is encountered.

For a general polynomial f of degree $d, N(z) = z - f/f'$ is a rational map of degree d. The zeroes of f determine fixed points of N where $|N'z| < 1$. Thus they determine attracting domains for the dynamics of N. In practice one also finds other periodic attracting domains for $N(z)$.

It is hoped one could account for these in general. Then if one could in addition show the Julia limit set has 2-dimension measure zero, a general understanding of Newton's iteration for almost all points would ensure, Figure 2.

We note that Fatou [1919] proved each contracting periodic domain for a rational map contains a branch point. In curious analogy David Singer proved the analogous result for a smooth endomorphism of the circle with a negative Schwarzian derivative (1975).

Problem: Find a common explanation of Singer's and Fatou's theorem.

---

*The words "hyperbolic" and "elliptic" are meant to suggest chaotic and rigid structure respectively in the dynamics.

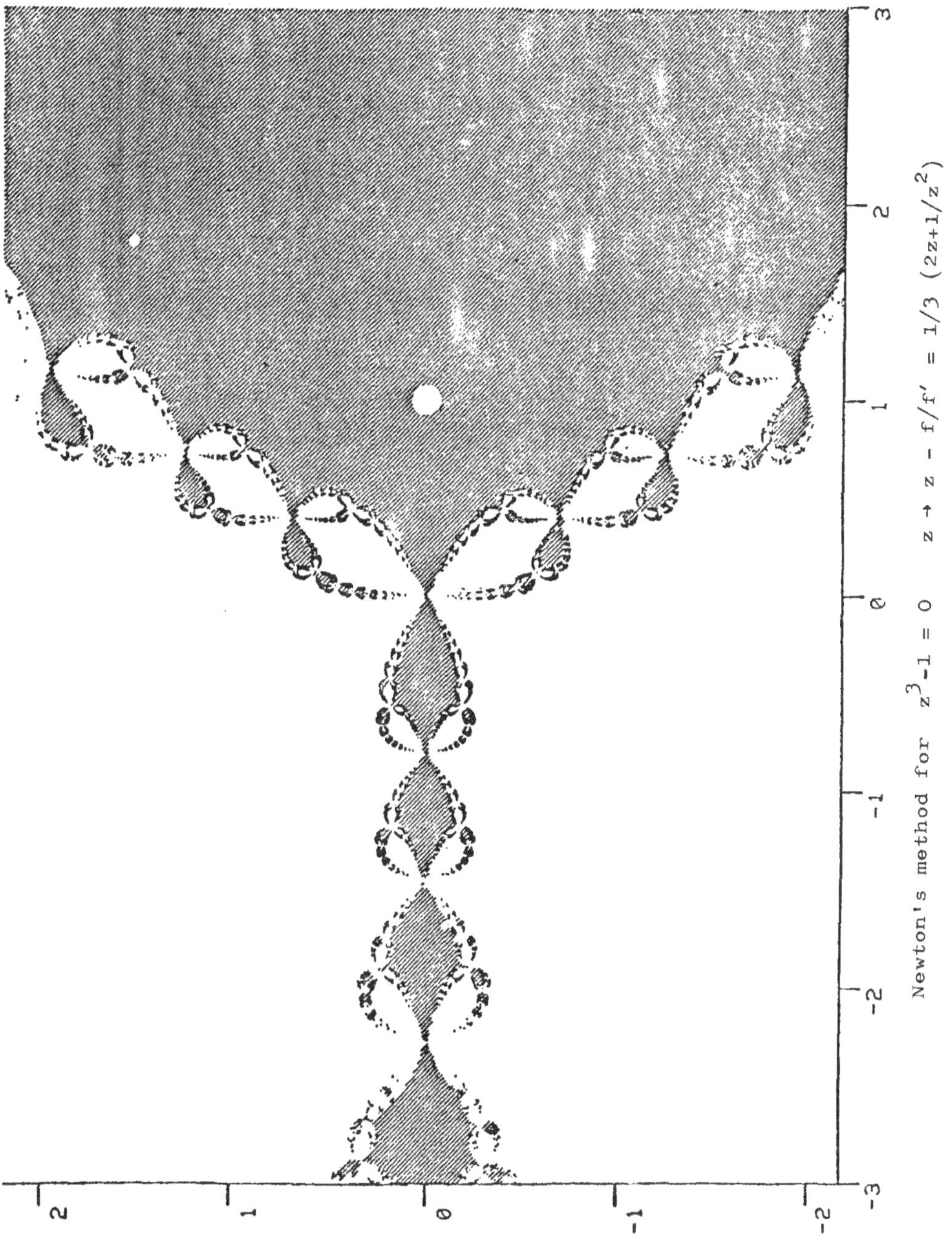

Figure 2
(constructed by J.P. Eckman)

Fractal geometry of limit sets. We construct conformal measures of exponent δ in the Julia set of any rational map and use these to discuss the Hausdorff geometry in the expanding case. This discussion in the analogous case of Kleinian groups was carried out in Sullivan [1980] motivated by papers of Bowen [1980] and Patterson [1976].

Theorem 3. There is a positive, finite measure μ on the Julia set $J(R)$ of a rational map $z \to R(z)$ satisfying for some real number $\delta = \delta(R)$

$$\mu(R(A)) = \int_A |R'(z)|^\delta \, d\mu(z) \qquad (*)$$

for any Borel set $A \subset J(R)$ where R is injective. Moreover, $0 < \delta(R) \leq 2$ and if any δ satisfies (*) for some measure μ then $\delta \geq \delta(R)$.

Proof. If $J(R) = \bar{\mathbb{C}}$ take Lebesgue measure. If $J(R) \neq \bar{\mathbb{C}}$ by Fatou there is an open set U in the complement of $J(R)$ so that the inverse branches of R^{-n} are defined and the inverse images $R_i^{-1}U$, $R_i^{-1}(R_j^{-1}U)$,.. are all disjoint (with the exception possibly of one sequence of choices of inverse branches when U belongs to a disk or annulus on which R is equivalent to an irrational rotation) and converge towards $J(R)$.

For each x in U let $I(x)$ denote all the $R_i^{-n}(x)$ (except for the exceptional sequence if present). If $y \in I(x)$ satisfies $R^n(y) = x$ then define $d(y) = |(R^n)'(y)|^{-1}$ (in the spherical metric). Since all the inverse images of U are disjoint and the area of the sphere is finite $\sum_{y \in I(x)} d(y)^2 < \infty$ a.e. x in U by Lebesgue monotone convergence (solution of exercise above).

Let x_0 in U be any point where the series converges and define $\delta = \inf\{s \mid \sum_{y \in I(x_0)} d(y)^s < \infty\}$. Note $\delta > 0$ because there $\sim d^n$ points in the nth level and the factors $d(y)$ are decreasing no faster than $\sim K^{-n}$ where $|R'(z)| \leq K$ on $J(R)$. Suppose for now $\sum_{y \in I(x_0)} d(y)^\delta = \infty$.

Define measures μ_s by putting atomic masses of weight $d(y)^s$ at y and normalizing to total mass 1. Let μ be any weak limit of the μ_s as $s \searrow \delta$. By our divergence assumption μ is supported at $J(R)$.

If U is a neighborhood of z_0 in $J(R)$ where R is injective, then R is bijective between $I(x_0) \cap U$ and $I(x_0) \cap R(U)$. (If z_0 is a critical point then U can be arranged so that R is k to one between $I(x_0) \cap U$ and $I(x_0) \cap R(U)$.) In the first case further assume U is chosen so that $|R'z|$ is a constant λ up to a factor near 1. (In the second case assume U is chosen so that $|R'z|$ is $< \varepsilon$.)

Thus in the first case we have $\mu_s(R(U))$ is $\lambda^\delta \mu_s(U)$ up to a factor near one. (In the second case $\mu_s(R(U)) < k\cdot\varepsilon^\delta$.) ($s$ is near δ.)

Letting $s \searrow \delta$ and then shrinking U we deduce for any atomic parts of μ we have

$$\mu(\{Rx\}) = |R'(x)|^\delta \, \mu(\{x\}). \qquad (**)$$

We may remove the critical points from consideration and consider (*) in the locally injective part. Then letting $s \searrow \delta$ and shrinking U we deduce (where R is locally injective) $dR^*\mu/d\mu = |R'|^\delta$ a.e. μ and this proves (*).

If $\sum_{y\in I(x_0)} d(y)^\delta$ does not diverge, introduce new weighting factors $h(d(y))d(y)^s$ for masses placed along $I(x_0)$ to define μ_s where

(i) $\sum_{y\in I(x_0)} h(d(y))d(y)^\delta = \infty$, (recall $\Sigma\, d(y)^\delta = \infty$ for $s < \delta$)

(ii) $h(x)$ is a positive function of x increasing to $+\infty$ as $x \searrow 0$ in such a way that for all $\varepsilon > 0$ and $0 < \lambda < \infty$, $|h(\lambda x)/h(x)| \in [1-\varepsilon, 1+\varepsilon]$ for $0 < x \leqq x_0(\varepsilon,\lambda)$. Thus $\alpha > 0$ implies $h(x) \leq x^{-\alpha}$ for x sufficiently small.

Now carry through the argument as above. By (ii) the new factor $h(d(y))$ introduces only a factor near 1 in the computations. (In the second case one has to divide a neighborhood of the critical point into countably many nice annuli and calculate using ii) and $\delta > 0$.) This completes the proof of the existence of conformal measures on any Julia set.

Since the conformal measures form a closed set in the weak topology on the probability measures we can go to the minimal dimension, by definition $\delta(R)$. If $\delta(R)$ were zero there could be no atoms by (**) above (since μ is a finite measure and full orbits are infinite). Then working where R is d to 1 if d = degree R we deduce $R_*\mu = d\mu$ contradicting total mass μ = total mass $R_*\mu$. This proves Theorem 3.

Say that R is expanding on the Julia set if for each x in J(R) there is an n so that $|(R^n)'(x)| > 1$. Then it is easy to see that some fixed m $|(R^m)'(x)| > 1$ for all x in J(R). We now work with R^m and denote it by R.

Let B(x,r) denote any ball of small radius r centered at a point x in J(R). By the distortion lemma 1 there is an n so that if B = B(x,r) and $B' = R^n(B)$, then B' has a definite size and $R^n: B \to B'$ is a "quasi-similarity" (≡ the ratio of derivatives at various point of B are comparable) because in the notation of Lemma 1 the {distance (x_i, y_i)} form a geometric series. One deduces $\mu(B)$ is comparable with fixed bounds to r^δ. (This follows since all B' of a definite size have a definitely positive μ mass since μ is positive on open sets of J(R) by topological transitivity.)

By a relatively simple general proposition (see Federer "Geometric Measure Theory" and §2 of Sullivan [1980]) such a measure μ is boundedly equivalent to the Hausdorff δ-measure. Thus the measure class of any conformal measure μ is determined by the geometric properties of the set J(R).

We collect this information in

Theorem 4. In the expanding case there is one and only one conformal measure μ on the Julia set J(R). The exponent $\delta = \delta(R)$ of μ is the Hausdorff dimension and μ is a constant times the Hausdorff δ-measure H_δ on J(R). Moreover, $0 < \delta(R) < 2$.

Corollary. The Hausdorff δ-measure is a finite and positive measure, the Hausdorff δ-measure of a ball of radius r in J(R) is comparable to r^δ, and the Hausdorff δ-measure is ergodic relatively to R acting on J(R). (In the expanding case).

Proof of Theorem and Corollary. If μ_1 and μ_2 are conformal measures of exponent δ and total mass one so is $1/2\ (\mu_1 + \mu_2) = m$. The ratio of μ_1 to m is defined and is an R-invariant function. Since m being a conformal measure is ergodic (its measure class is determined by the geometry of $J(R)$) this Radon ratio is constant. Thus $m = \mu_1 = \mu_2$ and conformal measures of dimension are unique. By the same reasoning the exponent δ is unique being the Hausdorff dimension of $J(R)$. (Since Lebesgue measure $J(R) = 0$, see proposition below, $\delta < 2$.)

Now we know the Hausdorff measure H_δ by definition satisfies the defining equation (*) to be a conformal measure. In this expanding case we know it is also a finite measure. Thus by the above uniqueness H_δ is a constant times the unique normalized conformal measure. This completes the proof of the theorem.

The statements of the corollary follow directly from the above.

Remark. One can prove Theorem 3 and Theorem 4 for expanding rational maps using Markov partitions and Gibbs measures as Bowen did for quasi-fuchsian surface groups. We have chosen this way because Theorem 3 is more general and Theorem 4 is obvious once Theorem 3 is known. The Bowen proof, however, produces the μ by a method which evidently converges geometrically fast. Lucy Garnett [1983] used finite approximations to calculate the Hausdorff dimension $\delta(R)$ for the family $z \to z^2 + p$ for p real and small. A quadratic curve with minimum at $p = 0$ was found. Based on this calculation by Garnett I asked at this conference whether $\delta(R)$ varied smoothly or even real analytically as R varies in an analytic family of expanding examples. This was answered by Ruelle using Bowen's infinite procedure.

Theorem 5. (Ruelle, 1982) The Hausdorff dimension of $J(R)$ is a real analytic function of the coefficients of $z \to R(z)$ in any open connected set where each such map is expanding.

Problem. Is Theorem 5 true for analytic families of expanding Kleinian groups (for each x in the Poincaré limit set there is

$\gamma \in \Gamma$ so that $|\gamma' x| > 1$) ?

We close with a remark about classifying these expanding groups. They determine compact 3-manifolds (with boundary) and Thurston has characterized which topological 3-manifolds arise in this way. The abstract group structure of the fundamental group Γ determines the topology of the limit set $\Lambda(\Gamma)$ and the topological action of Γ on $\Lambda(\Gamma)$. The simplest example is: Γ is a free group and $\Lambda(\Gamma)$ is the Cantor set of infinite words in Γ. In another class of examples (the acylindrical 3-manifolds) $\Lambda(\Gamma)$ is always homeomorphic to a Sierpinski curve obtained by removing from a 2-disk a dense collection of smaller disks.

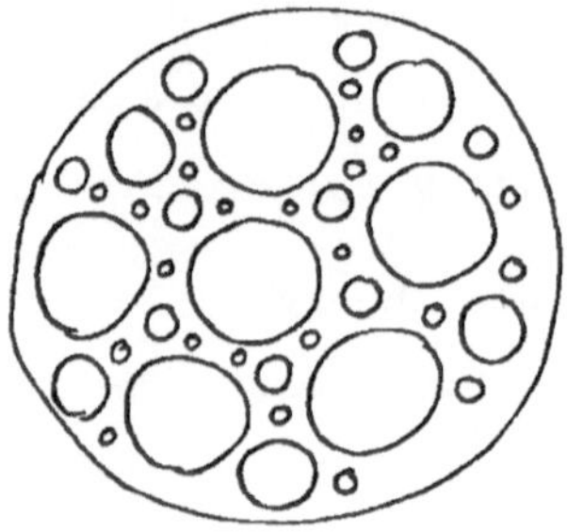

Sierpinski Curve

This is the one dimensional analogue of the 0-dimensional Cantor set.

We ask now the question -- what invariants besides this topology determines the geometric realization of Γ as a discrete group?

Theorem 6. (Sullivan [1981]) The geometric realization of Γ as a discrete group in $PSL(2,\mathbb{C})$ with the expanding property is determined up to isomorphism by the Hausdorff dimension δ of $\Lambda(\Gamma)$ and the abstract measurable dynamics isomorphism class of Γ acting on $\Lambda(\Gamma)$ relative to Hausdorff δ-measure.

The proof uses the ergodicity of the action of Γ on $\Lambda(\Gamma) \times \Lambda(\Gamma)$ and a characterization of Moebius transformations as measurable transformations preserving the cross ratio of almost all 4 tuples of points. The analogue of this theorem for rational maps is not known -- maybe its proof will use the Schwarzian derivative.

Remark. Theorem 3 for all Kleinian groups and Theorem 4 for expanding groups go through just as above, as we mentioned before, using the "Poincaré series method" -- putting mass along the orbit. Now, however, Markov partitions are not always obviously available, and even when they are certain discontinuities arise which don't occur in the rational map case.

Quadratic maps. In the family $z \xrightarrow{R} \lambda z + z^2$ where $|\lambda| < 1$, one knows R is expanding on $J(R)$. (For these the critical point $-\lambda/2$ tends to the attractive fixed point 0.) Thus by Ruelle $D(\lambda)$ = Hausdorff dimension of $J(R)$ is a real analytic function for $|\lambda| < 1$.

Theorem 7. $D(\lambda)$ is strictly greater than one and strictly less than two for $0 < |\lambda| < 1$. (See Bowen [1980], for the analogous theorem on quasi fuchsian groups.) Moreover, $J(k)$ is an Ahlfors quasi-circle.

Proof. The Julia set moves continuously in $0 \le |\lambda| < 1$ and so is always a Jordan curve. See Mañé, Sad, Sullivan [1982] where more is proved.

We need now show that $J(R)$ is not rectifiable and the Lebesgue measure is zero. The second case is ruled out by the following.

Proposition. For an expanding $z \to R(z)$ the Lebesgue measure of $J(R)$ is always zero. Thus $\delta(R) < 2$.

Proof. Take a density point x and radii $r_i \to 0$ so

$$m(J(R) \cap B(x,r_i))/m(B(x,r_i)) \to 1$$

where m is Lebesgue measure. Expand $B(x,r_i)$ up to a definite size B_i' using R^{n_i}. By the quasi-similarity lemma 1 we still have $m(J(R) \cap B_i')/m(B_i') \to 1$. A limit B' of the balls B_i' will satisfy $m(J(R) \cap B')/m(B') = 1$. Thus $B' \subset J(R)$. Then $J(R) = \bar{\mathbb{C}}$ which contradicts the expanding property.

Remark. Because the curve is quasi-self similar it is a quasi-circle. (see Ahlfors book "Quasi-conformal Mappings.")

For the first possiblity using the Riemann mapping theorem on each component of the complement of $J(R)$, we obtain $D_2 \cup D_1 \xrightarrow{\varphi_1 \cup \varphi_2} \bar{\mathbb{C}} - J(R)$ analytic. Conjugating the dynamics back to $D_2 \cup D_1$ we obtain two $2 \to 1$ maps of the standard disk. The one for the component of $C - J(R)$ containg ∞ is $z \to z^2$. The one for the finite component is $z \to z \cdot \left(\frac{z-\lambda}{1-\lambda z}\right)$ (if the Riemann map sends the fixed point to the origin).

The Riemann maps φ_1 and φ_2 extend continuously to the boundary by Caratheodory's work. Moreover, if $J(R)$ is rectifiable these Caratheodory maps are non-singular with respect to arc length measure (by a harmonic measure argument). Thus we obtain a continuous and absolutely continuous conjugacy between $z \to z^2$ and $z \to z \cdot \left(\frac{z-\lambda}{1-\lambda z}\right)$ restricted to $|z| = 1$. But this is impossible. For both are essentially expanding and locally eventually onto and thus each one is ergodic with respect to the Lebesgue measure class on $|z| = 1$. Also each one preserves Lebesgue measure $d\theta$ (which is the harmonic measure of $|z| = 1$ relative to $z = 0$ which is fixed by each map.)

Thus the conjugacy sends $d\theta$ to $d\theta$ and must be a rigid rotation. This contradicts the fact that $z \to z \cdot \frac{z-\lambda}{1-\lambda z}$ has a varying derivative when $\lambda \neq 0$.

Remark. Since for $\lambda \neq 0$ we obtain for the simple map $z \to \lambda z + z^2$ (or $z \to z^2 + c$, $c \neq 0$) Julia sets which are non-rectifiable quasi-self similar fractal curves of Hausdorff dimension > 1 one is tempted to plot these on a computer. Here are some examples, Figure 3.

Bifurcations of conformal dynamical systems. It is very interesting (Figure 3, 4) to study the bifurcations of $J(R)$ for $z \xrightarrow{R} \lambda z + z^2$ as λ varies (see Douady-Hubbard, Mandelbrot,...). We have seen for $|\lambda| < 1$ $J(R)$ is a moving Jordan curve whose Hausdorff dimension is really varying. As λ hits the unit circle $J(R)$ even changes topologically.

If λ hits at a root of unity the curve pinches together at

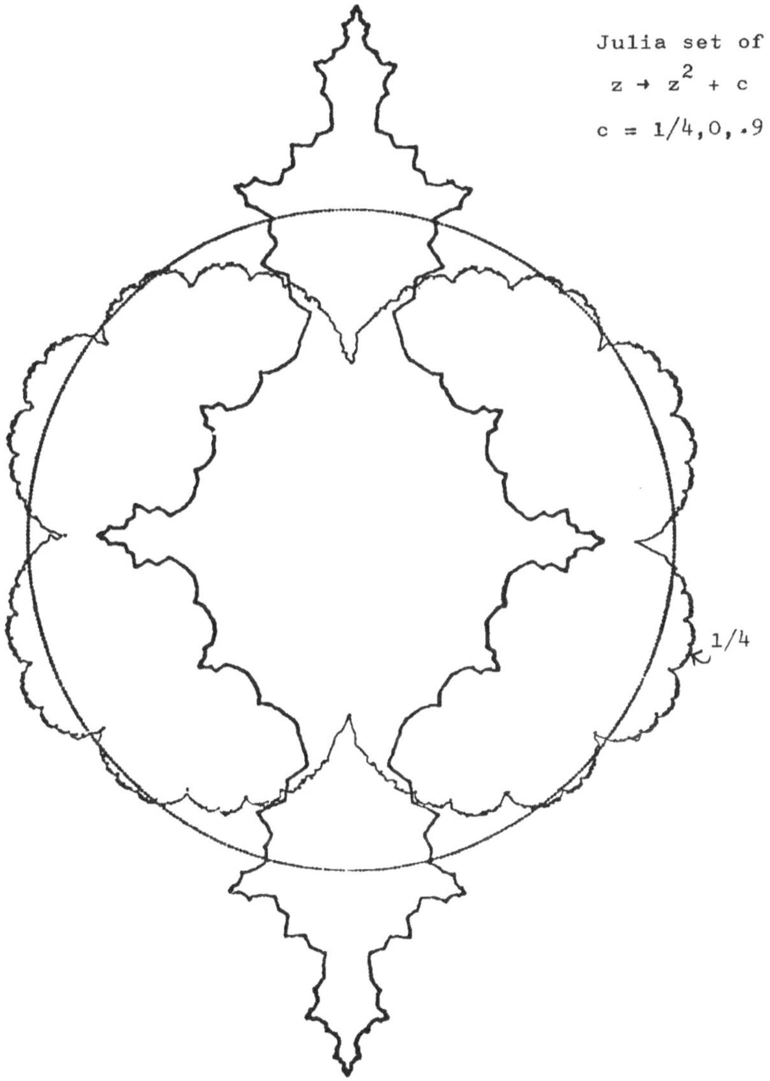

figure 3

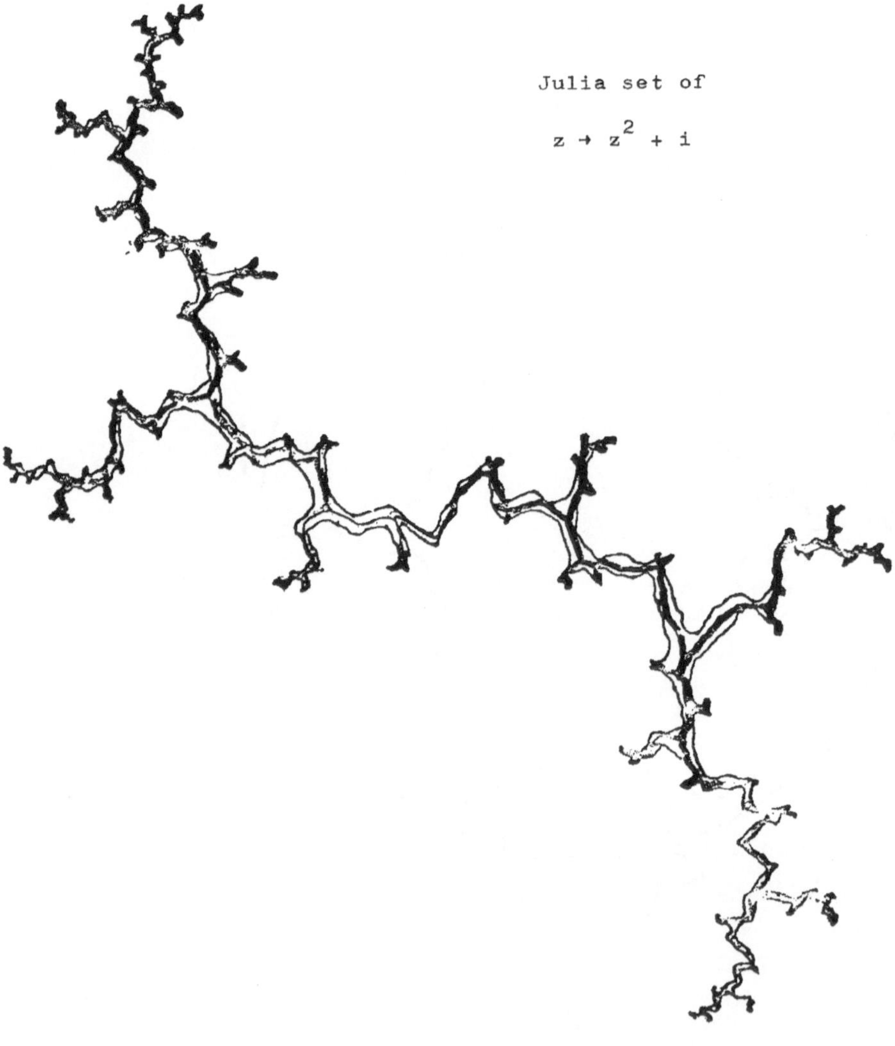

figure 4

the origin and all the preimages. The collapsed circle is still locally connected (Hubbard and Douady). If λ hits at an irrational angle which is far from the rationals $J(R)$ does not reach the origin which is contained in a Siegel rotation disk.

Problem. Is the boundary of the Siegel disk a Jordan curve, and is $J(R)$ locally connected? If so, the topological picture of $J(R)$ can probably be described using the invariant Cantor set for $\theta \to 2\theta$ mentioned above.

The case where λ hits at a Liouville angle leads to a non-locally connected $J(R)$. More precisely,

Theorem 8. (Douady-Sullivan) *If* $z \xrightarrow{R} \lambda z + z^2$ *where* $\lambda = e^{2\pi i \vartheta}$ *with* ϑ *irrational but* R *is not linearizable near* $z = 0$, *then* $J(R)$ *is non-locally connected.*

Proof. The Riemann map of the exterior of the unit disk to the exterior of $J(R)$ extends continuously to the boundary if $J(R)$ is locally connected (Caratheodory). Let C contained in S^1 be those angles (of exterior rays) which land at the fixed point $z = 0$. The Riemann map conjugates $z \to z^2$ to $z \to R(z)$.

If C were finite, some power of R^k would have an invariant line tending to zero. Since the derivative is an irrational rotation this line spirals in to the origin. It is then possible to find a region near zero which is invariant by R^{-k}. But R^{-k} is non-linearizable.

Invariant region (the snail enters its shell)

This is a contradiction. So C must be infinite.

Clearly C is closed. Also $\theta \to 2\theta$ (the action of $z \to z^2$ on rays) restricted to C must be a homeomorphism because R is a bijection on those rays which land at the irrational fixed point. But this contradicts the proposition in the first section (because an infinite compact metric space cannot have a homeomorphism which uniformly expands distances between sufficiently near points).

5. Characterization of complex analytic dynamical systems.

One says that a diffeomorphism of part of the plane is K-quasiconformal if infinitesimal circles are mapped to infinitesimal ellipses of eccentricity $\leq K$. The definition can be extended to homeomorphisms or even branched coverings of any Riemannian surface.

Consider a collection C of transformations of a Riemannian surface S which is closed under composition and which are all K-quasi-conformal in the sense of the extended definition.

Theorem 9. There is a complex analytic structure on S (compatible with its original quasi-conformal structure) so that all the transformations of the dynamical system C become complex analytic.

Note. The condition of the theorem, K-quasi conformality of all the transformations in C is clearly necessary for the statement of the theorem (which asserts it is sufficient).

Corollary. The dynamical systems determined by

(i) complex analytic self-mappings of $\mathbb{C} \cup \infty$

(ii) entire functions $f: \mathbb{C} \to \mathbb{C}$

(iii) a collection of Moebius transformations of S^2, are characterized by the condition of uniform quasi-conformality of all the iterated compositions.

Proof. First one forms an invariant measurable conformal structure by a barycenter construction in the set of similarity structures on the tangent space at each point. Then one introduces complex analytic coordinates using the measurable Riemann mapping theorem.

Closing problem. In either of the conformal dynamical contexts, rational maps or finitely generated Kleinian groups are the expanding systems dense?

Bibliography

R. Bowen, "Dimension of quasi circles", Publ. IHES, Vol. 50 (1980).

A. Camacho, Dijon Conference Asteriks, 1979.

A. Douady, J. Hubbard, "Iteration of complex quadratic polynomials", CRAS, Paris 1982. Longer manuscript in preparation.

M. Fatou, (i) "Functions Satisfying Functional Equations," Bull. of French Math. Soc., 1919-1920.

(ii) "Iterating Entire Functions," Act. Math., Vol. 47.

L. Garnett, "Calculation of Hausdorff dimension," in preparation.

J. Guckenheimer, "Dynamics of rational maps," AMS Conference on Global Analysis, 1968.

G. Julia, "Iteration of rational maps," Journal of Math. Pure et Appliques, 1919.

B. Mandelbrot, "Form, Chance, Dimension," Free an, 1976 (New version, 1982).

R. Mañé, P. Sad, D. Sullivan, "On the dynamics of rational maps," submitted to Journal of the Ecole Normal Superieur.

M. Misuriewicz, "Mapping of the Interval," Publ. IHES, Vol. 48.

S.J. Patterson, "Limit Sets of Fuchian groups," Acta Math., 1976.

H. Poincaré, Collected Works, Vol. 1.

D. Ruelle, "Analytic Repellers," Journal of Ergodic Theory and Dynamical Systems, 1982.

R. Sacksteder, "Invariant measures for pseudogroups," Amer. Journal Math., 1965.

C.L. Siegel, "Iteration of complex analytic transformations," Annals of Math., 1942.

D. Sullivan, (i) "Discrete Conformal Groups and Measurable Dynamics," Proceedings of Poincaré Conference, Bloomington, Ind., 1980. BAMS 1982.

(ii) "Density at infinity..." Pub. IHES, 1980, Vol. 50.

(iii) "Quasi conformal homeomorphisms and dynamics," in preparation 1982; also C.R.A.S. (Paris), 1982.

M.V. Yakobson, "Question of topological classification of rational mappings of the Riemann sphere," Usp. Math. Nauk. 28, No 2, 247-248 (1973), also Math. Sb. (N.S.) 77, 105-124 (1968), and Math. Sb. (N.S.) 80, 365-387 (1967).

Institut des Hautes Etudes Scientifiques
35, route de Chartres
91440 Bures-sur-Yvette (France)

QUADRATIC VECTOR FIELDS WITH FINITELY MANY PERIODIC ORBITS

by

J. SOTOMAYOR
Instituto de Matemática Pura e Aplicada

and

R. PATERLINI
Universidade Federal de São Carlos

Abstract - We prove that vector fields outside an algebraic hypersurface in the space of coefficients of quadratic vector fields in the plane have a finite number of periodic orbits.

1. Introduction

The problem of determining the number and position of periodic orbits of differential equations received great attention in the work of H. Poincaré [P]. He established that in the "general" case for the coefficients of a polynomial differential equation in the plane there appear at most a finite number of isolated periodic orbits (limit cycles).

The "general" conditions of Poincaré were removed in the work of H.Dulac [D]. This work however contains a serious gap in the proof of the claim that the return map associated to a graph of singularities and separatrices (where infinite limit cycles are ultimately likely to accumulate), although not analytic, has the property of isolated fixed points.

Based on the above mentioned work of Poincaré and motivated by a classical result of Harnack which gives a bound of compact components of algebraic plane curves in terms of the degree, D. Hilbert proposed to obtain a bound for the number of limit cycles in terms of the degree of the polynomials which define the differential equation [H].

The relation between the dynamical and algebraic features of the differential equations, the first represented by the number of limit cycles and the second by the degree, as proposed by Hilbert, is unknown even for degree two [Sn].

In this paper we glimpse into the question of the algebraic nature of the finiteness of periodic orbits of polynomial differential equations in the plane. This question focalizes a cruder type of relation between the dynamical and algebraic aspects of differential equations, which nevertheless does not seem to have been previously discussed.

Let $\mathfrak{X}_2$ be the set of polynomial vector fields of degree two, in the plane, endowed with the topology of the coefficients. The main result of this paper is the following:

1.1 Theorem. There exists a non trivial polynomial function $\mathcal{R}:\mathfrak{X}_2 \to \mathbb{R}$ such that if $X \in \mathfrak{X}_2$ satisfies $\mathcal{R}(X) \neq 0$ then X has finitely many periodic orbits.

This paper is organized as follows. In section 2 are reviewed some results that will be used in the proof of Theorem 1.1, given in section 3. Section 4 contains some comments on the meaning of Poincaré's Theorem 17 [P] and on the possibility of extension of Theorem 1.1 to polynomial vector fields of higher degree.

2. Preliminaries

2.1 Let $X:\mathbb{R}^2 \to \mathbb{R}^2$ be a polynomial vector field of degree n. We put $X(x,y) = (P(x,y), Q(x,y))$ and let P_r and Q_r, $0 \le r \le n$ be the homogeneous part of degree r of P and Q respectively. The Poincaré compactification $\mathcal{P}(X)$ of X on $S^2 = \{(x,y,z);\ x^2+y^2+z^2 = 1\}$ is defined by the central projections of $\pi = \{(x,y,z);\ z = 1\}$ on the hemispheres

$H_+ = \{(x,y,z) \in S^2;\ z > o\}$ and $H_- = \{(x,y,z) \in S^2;\ z < o\}$ (see [G] for more details). $S^1 = \{(x,y,z);\ z = o\}$ is invariant by $\mathcal{P}(X)$ and the singular points of $\mathcal{P}(X)$ on S^1 appear symmetrically and are given by the roots of the polynomials $F(s) = Q_n(1,s) - s\,P_n(1,s)$ (if the singularity is in $S^1 \cap \{x > o\}$) and $G(s) = P_n(s,1) - s\,Q_n(s,1)$ (if the singularity is in $S^1 \cap \{y > o\}$). The Jacobian matrix of $\mathcal{P}(X)$ in a singular point given by $F(s_o) = o$ is

$$\frac{1}{\sqrt{s_o^2+1}^{\,n-1}} \begin{pmatrix} F'(s_o) & Q_{n-1}(1,s_o) - s_o\,P_{n-1}(1,s_o) \\ 0 & -P_n(1,s_o) \end{pmatrix}$$

For $G(s_o) = 0$, we have

$$\frac{1}{\sqrt{s_o^2+1}^{\,n-1}} \begin{pmatrix} G'(s_o) & P_{n-1}(s_o,1) - s_o\,Q_{n-1}(s_o,1) \\ 0 & -Q_n(s_o,1) \end{pmatrix}.$$

Multiplying the above matrices by $(-1)^{n-1}$ we obtain the Jacobian matrices for the symmetric singular points.

2.2 A <u>graph</u> of a C^1 vector field X is a compact and connected set of saddle points and separatrices of X satisfying the following conditions:

i) The ω- and α-limit sets of every separatrix of the graph is a saddle point;

ii) Every saddle point of the graph has at least one stable and at least one unstable separatrix in the graph.

Let G be a graph of a C^1 vector field X. Let $p_1,\dots,p_k$ be the saddle points of the graph. Consider the number $\rho(G) = \prod_{i=1}^{k} \frac{|\mu_i|}{\lambda_i}$, where $\mu_i < o < \lambda_i$ are the eigenvalues of $DX(p_i)$, $1 \le i \le k$. We say that the graph G is <u>simple</u> if $\rho(G) \neq 1$.

The proof of the following theorem appears in [S] pg 60, it is based on the fact that simple graphs of C^2 vector fields are isolated from periodic orbits.

2.3 Theorem. Let $X:\mathbb{R}^2 \to \mathbb{R}^2$ be a polynomial vector field satisfying the following conditions:

1. Every singular point of $\mathcal{P}(X)$ is hyperbolic;
2. Every graph of $\mathcal{P}(X)$ is simple.

Then X has finitely many periodic orbits.

The result below follows from the theory of resultants [W]; it is proved in [G]. See also [S], pg 63.

2.4 Proposition. Let $\mathfrak{X}_n$ be the set of polynomial vector fields of degree n in $\mathbb{R}^2$. There exists a polynomial function $\mathcal{R}_o:\mathfrak{X}_n \to \mathbb{R}$ satisfying $\mathcal{R}_o(X) = o$ for all $X \in \mathfrak{X}_n$ such that $\mathcal{P}(X)$ has either a non hyperbolic singular point or a singular point at which the trace of the Jacobian of X vanishes.

3. Proof of Theorem 1.1

The proof of Theorem 1.1 will follow from the next result:

3.1 Propostion. There exists a non trivial polynomial function $\mathcal{R}_1:\mathfrak{X}_2 \to \mathbb{R}$ such that $\mathcal{R}_1(X) = o$ provided that $X \in \mathfrak{X}_2 - \mathcal{R}_o^{-1}(o)$ and that $\mathcal{P}(X)$ has a non simple graph.

In fact, by theorem 2.3 and proposition 2.4 the polynomial function $\mathcal{R} = \mathcal{R}_o \mathcal{R}_1$ is non trivial and if $\mathcal{R}(X) \neq o$ then X has finetely many periodic orbits.

For the proof of proposition 3.1, we need several lemmas proved below.

Firstly, we must assert that only the following a priori possibilities can exist as parts of graphs of $\mathcal{P}(X)$, if $X \in \mathfrak{X}_2 - \mathcal{R}_o^{-1}(o)$:

Type I

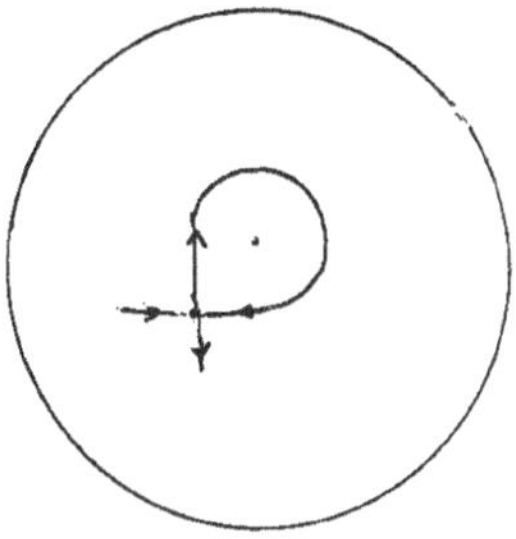

X has a saddle self connection

Type II

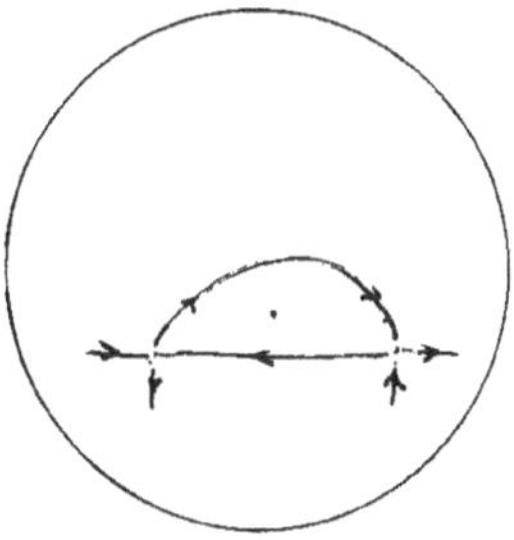

X has a graph with two saddles

Type III

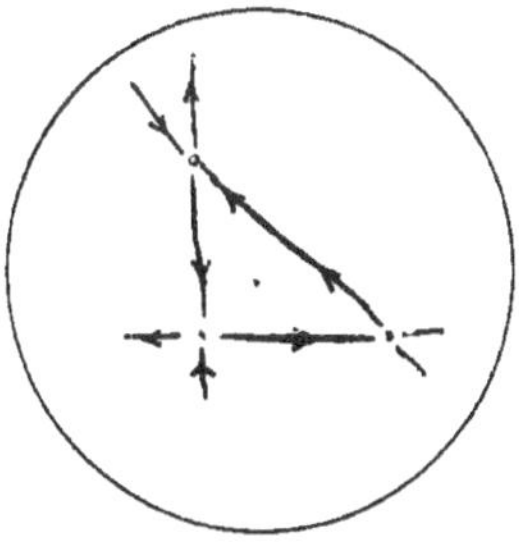

X has a graph with three saddles

(this type does not occur; see Lemma 3.5)

Type IV

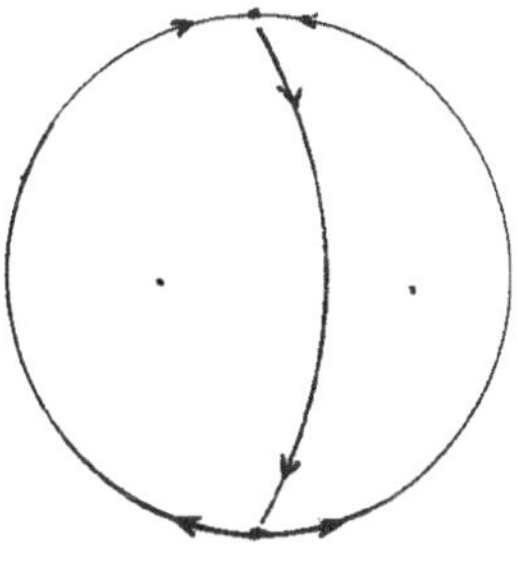

$\mathcal{P}(X)$ has a graph with only two symmetric saddles on S^1.

Type V

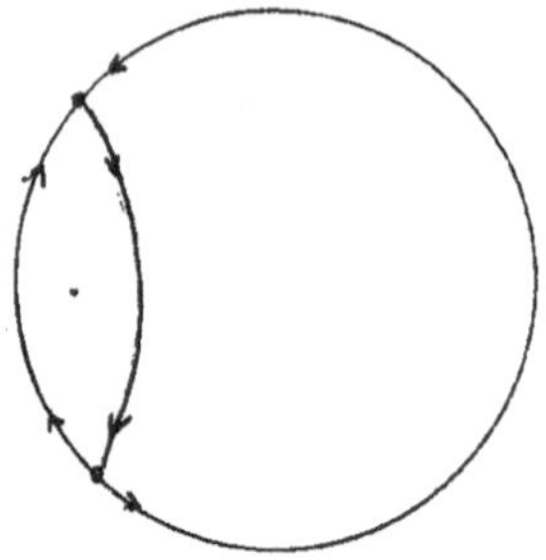

$\mathcal{P}(X)$ has a graph with only two non symmetric saddles on S^1.

Type VI

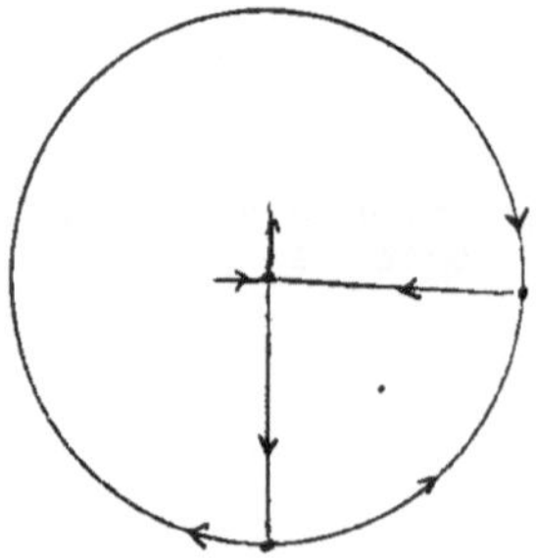

$\mathcal{P}(X)$ has a graph with two saddles on S^1 and one saddle on H_+.

In fact, X has at most four singularities, therefore a graph of X a priori must contain a graph of one of the types I, II or III. Furthermore, $\mathcal{P}(X)$ has at most six singularities in S^1 and one of the two following possibilities occur for the phase portrait in a neighborhood of S^1.

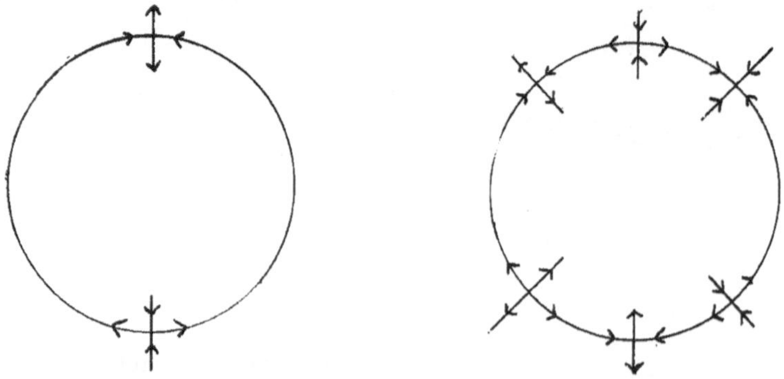

If a graph G of $\mathcal{P}(X)$ has singular points in S^1, the number of such singular points must be even. Then G has exactly two saddle points in S^1. Moreover, the number S of saddle points of G is ≤ 3. Otherwise, $S \geq 4$ would imply that $\mathcal{P}(X)$ has at least two saddle points in each hemisphere and four saddle points in S^1, and therefore that $\mathcal{P}(X)$ would have at least 8 saddle points. Then the number of nodes or foci of $\mathcal{P}(X)$ must be greater then or equal to 10. Since there are exactly two nodes in S^1, we have that there must exist at least four nodes or foci in each hemisphere. Hence X must have at least 4 nodes or foci and 2 saddles, which contradicts the fact that X is quadratic. Then $S \leq 3$, and we must have graphs of one of the types IV, V or VI.

3.2. Lemma. If $X \in \mathfrak{X}_2 - \mathcal{R}_o^{-1}(o)$ has a graph of type I, then the graph is simple.

Proof. Let $\mu < o < \lambda$ be the eigenvalues of the DX at the saddle of the graph. Then the graph is simple if and only if $\lambda + \mu \neq o$. But $\mathcal{R}_o(X) \neq o$ implies that trace $DX \neq o$ on every singularity. Therefore the lemma is proved.

The result below is relevant for the analysis of types II and IV.

3.3 Lemma. There exists a non trivial polynomial function $\rho_1 : \mathfrak{X}_2 \to \mathbb{R}$ such that if $X \in \mathfrak{X}_2$ has an invariant straight line then $\rho_1(X) = o$.

Proof. Let $\bar{X} = (P(x,y),\ Q(x,y))$ where $P(x,y) = \alpha_o + \alpha_1 x + \alpha_2 y + \alpha_3 x^2 + \alpha_4 xy + \alpha_5 y^2$ and $Q(x,y) = \beta_o + \beta_1 x + \beta_2 y + \beta_3 x^2 + \beta_4 xy + \beta_5 y^2$. Suppose that the straight line $\{(x,y);\ y = \bar{a}x + \bar{b}\}$ is invariant by $\bar{X}$. Then $Q(x, \bar{a}x + \bar{b}) = \bar{a}\, P(x, \bar{a}x + \bar{b})$ for all $x \in \mathbb{R}$.

This implies that $(\alpha_o+\alpha_2\bar{b}+\alpha_5\bar{b}^2)\bar{a} -(\beta_o+\beta_2\bar{b}+\beta_5\bar{b}^2) = o$,
$(\alpha_2+2\alpha_5\bar{b})\bar{a}^2+(\alpha_1-\beta_2+\alpha_4\bar{b}-2\beta_5\bar{b})\bar{a} - (\beta_1+\beta_4\bar{b}) = 0$ and
$\alpha_5\,\bar{a}^3+(\alpha_4-\beta_5)\bar{a}^2+(\alpha_3-\beta_4)\bar{a} - \beta_3 = o$.
These relations define three functions $f_i:\mathfrak{X}_2\times\mathbb{R}^2\to\mathbb{R}$, $i=1,2,3$, such that $f_i(\bar{X},\bar{a},\bar{b})=0$. We take now the resultants (see [W]) $R_1(X,b) = R(f_1(X,a,b),\ f_3(X,a,b))$ and $R_2(X,b) = = R(f_2(X,a,b),\ f_3(X,a,b))$. Hence if $R_3(X) = R(R_1(X,b),\ R_2(X,b))$ we have $R_3(\bar{X})=0$. Moreover let $R_4:\mathfrak{X}_2\to\mathbb{R}$ be the projection $R_4(X) = \alpha_5$.
Then if X has an invariant vertical straight line $x=c$ we have $R_4(X) = 0$. Obviously $R_4 \not\equiv 0$. We will verify that $R_3\not\equiv 0$. Let $Y(x,y) = (x^2+2xy+y^2,\ 1-x^2+y^2)$.
We have $f_1(Y,a,b)=b^2a-(1+b^2)$, $f_2(Y,a,b) = 2ba^2$ and $f_3(Y,a,b) = a^3+a^2+a+1$. Since the degree of f_i with respect to a is i, we have the resultants

$$R_1(Y,b) = \begin{vmatrix} b^2 & -(1+b^2) & 0 & 0 \\ 0 & b^2 & -(1+b^2) & 0 \\ 0 & 0 & b^2 & -(1+b^2) \\ 1 & 1 & 1 & 1 \end{vmatrix} = 4b^6+6b^4+4b^2+1$$

and

$$R_2(Y,b) = \begin{vmatrix} 2b & 0 & 0 & 0 & 0 \\ 0 & 2b & 0 & 0 & 0 \\ 0 & 0 & 2b & 0 & 0 \\ 1 & 1 & 1 & 1 & 0 \\ 0 & 1 & 1 & 1 & 1 \end{vmatrix} = 8b^3,$$

which have no common zeros in the complex field. Then $R_3(Y)\neq 0$. Taking $\rho_1=R_3R_4$ we finish the proof of the lemma.

3.4 Lemma. If X has a graph of type II or IV then X has an invariant straight line.

Proof. Suppose that $X \in \mathfrak{X}_2$ has a graph with two saddle points p and q. Let L be the straight line through p and q. Since X is of degree two, we have that X is transversal to L in L-{p,q}. Let S be the separatrix of the graph such that $S \not\subset L$. It is easy to

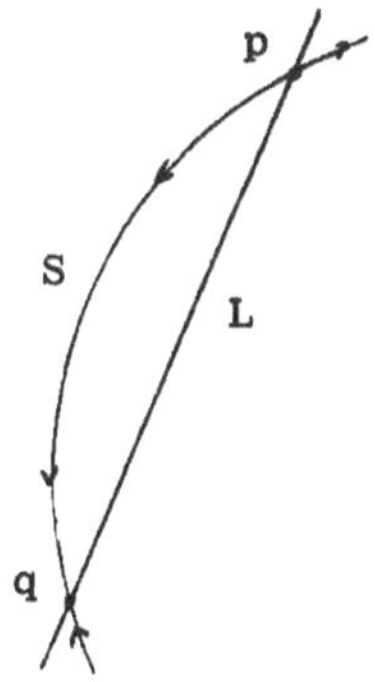

see that the curve $S \cup [p,q]$ is the boundary of an open disc D and the second separatrix S' of the graph satisfies one of the following three possibilities: $S' \subset D$, $S' \subset \mathbb{R}^2 - \bar{D}$ or $S' \subset L$. The first two possibilities imply that X has a third point of tangency with L. Then $S' \subset L$ and L is an invariant line of the vector field X.

Now suppose that $X \in \mathfrak{X}_2$ has a graph of type IV. By means of a rotation we can suppose that p=(1,0,0) and q=(-1,0,0) are saddle points of $\mathcal{P}(X)$ in S^1. Take the maximal circle C of S^2 through p and q and tangent at these points to the saddle connection S (see figure) If $S \not\subset C$, take $C' \neq C$ a

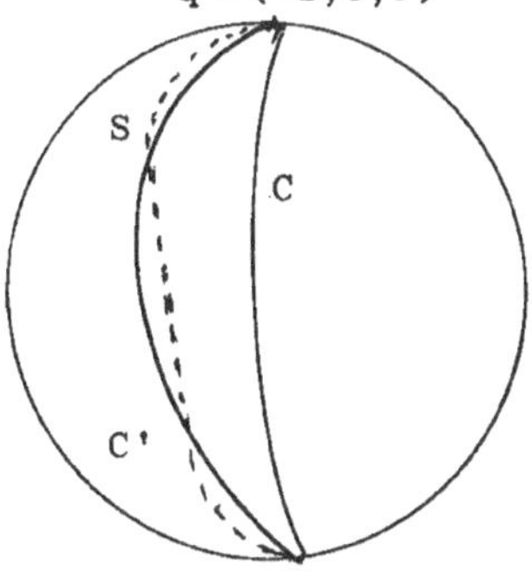

maximal circle cutting S. The vector field $\mathcal{P}(X)$ must have three points of tangency with C' in H_+. This implies that X has three points of tangency with a straight line in R^2. This is impossible since X is quadratic. Terefore we must have $S \subset C$. This finishes the proof of the lemma.

3.5 Lemma. A vector field $X \in \mathfrak{X}_2 - \mathbb{R}_o^{-1}(o)$ cannot have a graph of type III.

Proof. Suppose that $X \in \mathfrak{X}_2$ has a graph with three saddle points p, q and u. We prove that every separatrix of the graph is contained in a straight line. Let S be the separatrix joining p and q and let L be the straight line through p and q. If $S \not\subset L$, $S \cup [p,q]$ is the boundary of an open disc D. Since $u \notin L$, we have $u \in D$ or $u \in \mathbb{R}^2 - \bar{D}$. If $u \in D$, the other two separatrices of the graph are in D, and this implies that L has a third point of tangency (see figure).

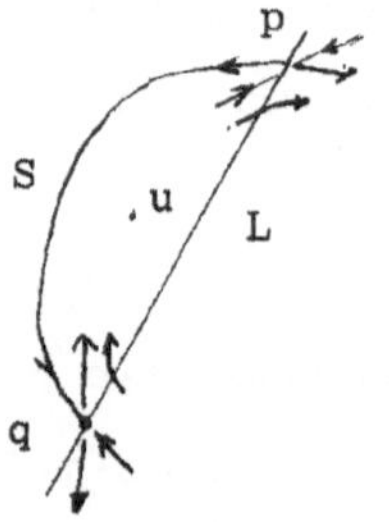

Similarly for $u \in \mathbb{R}^2 - \bar{D}$. Then $S \subset L$ and every separatrix of the graph is contained in a straight line. Now by means of a linear coordinate transformation we can suppose that $p=(0,0)$, $q=(1,0)$ and $u=(0,1)$, then $X(x,y) = (\alpha(x-x^2) + \beta xy,\ \gamma(y-y^2) + \delta xy)$, where $\alpha + \beta + \gamma + \delta = 0$. Moreover the singular point in the interior of the graph has coordinates

$$x_o = \frac{\gamma(\alpha+\beta)}{\alpha\gamma - \beta\delta}, \qquad y_o = \frac{\alpha(\gamma+\delta)}{\alpha\gamma - \beta\delta}.$$

At this point we have trace $DX(x_o,y_o) = \dfrac{-\alpha\gamma(\alpha+\beta+\gamma+\delta)}{\alpha\gamma-\beta\delta} = 0$. This implies that (x_o,y_o) cannot be hyperbolic. This proves the lemma.

<u>3.6 Lemma</u>. There exists a non trivial polynomial function $\rho_2 : \mathfrak{X}_2 \to \mathbb{R}$ such that if $X \in \mathfrak{X}_2$ has a non simple graph of type V then $\rho_2(X) = 0$.

Proof. Let $X(x,y) = (P(x,y), Q(x,y))$, where $P(x,y) = \alpha_o + \alpha_1 x + \alpha_2 y + \alpha_3 x^2 + \alpha_4 xy + \alpha_5 y^2$ and $Q(x,y) = \beta_o + \beta_1 x + \beta_2 y + \beta_3 x^2 + \beta_4 xy + \beta_5 y^2$. Let $f(X,s,t) = F'(s)\, P_2(1,t) - F'(t)\, P_2(1,s)$. Then

$$f(X,s,t) = (t-s)\,[(\beta_4 + 2\alpha_3)\alpha_5(s+t) + (\alpha_4 + 2\beta_5)\alpha_5\, st + \alpha_4\beta_4 - 2\beta_5\alpha_3 + \alpha_4\alpha_3]$$

Let $\bar{f}(X,s,t) = f(X,s,t)\,(t-s)^{-1}$ and let $R_1 : \mathfrak{X}_2 \to \mathbb{R}$ the polynomial function defined by the resultant $R(F(t), R(F(s), \bar{f}(X,s,t)))$. It

follows that if X has $\alpha_5 \neq 0$ and has a non simple graph of type V than $R_1(X) = 0$. Let us prove that $R_1 \neq 0$.
For $Y(x,y) = (xy-y^2, -x^2+y^2)$, $\bar{f}(Y,s,t) = -3ts$, $F(s) = s^3-1$ and the resultant is

$$R(F(s), F(Y,s,t)) = \begin{vmatrix} -3t & 0 & 0 & 0 \\ 0 & -3t & 0 & 0 \\ 0 & 0 & -3t & 0 \\ 1 & 0 & 0 & -1 \end{vmatrix} = 27t^3 .$$

Since $R(t) = 27t^3$ has no common root with $F(t) = t^3-1$, then $R_1(Y) \neq 0$. Finally let $R_2 : \mathfrak{X}_2 \to \mathbb{R}$ be the projection $R_2(X) = \alpha_5$. The polynomial funtion $\rho_2 = R_1 R_2$ satisfies the condition required in the statement of the lemma.

3.7 Lemma. Suppose that $X \in \mathfrak{X}_2$ has a graph of type VI. Then the singular point in the interior of the graph is hyperbolic if and only if the graph is simple.

Proof. Let $X \in \mathfrak{X}_2$ have a graph of type VI. Let $X(x,y) = (P(x,y), Q(x,y))$ where $P(x,y) = \alpha_o + \alpha_1 x + \alpha_2 y + \alpha_3 x^2 + \alpha_4 xy + \alpha_5 y^2$ and $Q(x,y) = \beta_o + \beta_1 x + \beta_2 y + \beta_3 x^2 + \beta_4 xy + \beta_5 y^2$. Firstly we note that the saddle points of the graph which are in S^1 are not symmetric. Then by means of a linear transformation we can assume that (0,0,1), (1,0,0) and (0,1,0) are saddle points of the graph. Then $X(0,0) = (0,0)$, $F(0) = 0$ and $G(0) = 0$ (see 2.1) which implies $\alpha_o = \beta_o = 0$, $\beta_3 = o$ and $\alpha_5 = o$ respectively. Under these conditions either the straight line $y = 0$ is invariant by X or the origin (o,o) is the unique point of tangency between X and $y = 0$. In fact, $Q(x,0) = 0 \Leftrightarrow \beta_1 x = o$. We will prove that the first possiblity actually occurs. Let C be the maximal semi-circle of H_+ through (o,o,1) and (1,o,o). Let S be the saddle connection between (o,o,1) and (1,o,o). If $S \not\subset C$, we can find a point of tangency between $\rho(X)$ and C in H_+ which is different of the origin. Then $S \subset C$ and this proves that the separatrices of the graph are contained in maximal circles. Hence the straight lines $x=o$ and $y=o$ are invariant by X and, taking $-X$ if necessary, we can write

$$X(x,y)=(\alpha x+\beta x^2-\gamma xy, \quad -\lambda y+\eta xy-\tau y^2),$$

where $\alpha>o$, $\lambda>o$, $\eta>\beta>o$ and $\gamma>\tau>o$. The eigenvalues of the jacobian of $P(X)$ in $(0,0,1)$ are $\alpha>o$ and $-\lambda<o$, in $(1,0,0)$ are $\eta-\beta>o$ and $-\beta<o$ and in $(0,1,0)$ are $\tau-\gamma<o$ and $\tau>o$. Hence the graph is simple if and only if $\beta\lambda(\gamma-\tau)\neq\alpha\tau(\eta-\beta)$. Moreover the unique singular point of X in the interior of the graph is the solution (x_o,y_o) of the system $\begin{cases}-\lambda+\eta x-\tau y=o\\ \alpha+\beta x-\gamma y=o,\end{cases}$

and we have that $\text{trace}DX(x_o,y_o)=\frac{1}{\gamma\eta-\beta\tau}[\alpha\tau(\beta-\eta)+\beta\lambda(\gamma-\tau)]$

and $\det DX(x_o,y_o)=\frac{1}{(\gamma\eta-\beta\tau)}(\beta\lambda+\alpha\eta)(\alpha\tau+\lambda\gamma)$.

Hence $\det DX(x_o,y_o)>o$ and $\text{trace } DX(x_o,y_o)\neq o$ if and only if the graph is simple. So the lemma is proved.

3.8 Proof of Proposition 3.1. Follows taking $\mathcal{R}_1=\rho_1\rho_2$, where ρ_1 is defined in 3.3 and ρ_2 in 3.6.

4. Final Remarks

The sense in which the "general" polynomial vector fields have finitely many periodic orbits was not made very precise in [P], Theorem 17. As an essay of interpretation it was proved in [S], Chap. 2, that such finiteness occurs outside the union of an algebraic hypersurface A_n and an immersed (neither proper nor one-to-one) analytic hypersurface B_n in the space $\mathfrak{X}_n$ of coefficients of polynomials of degree less than or equal to n.

It seems to us that the conclusion of Theorem 1.1 may hopefully be also extended to the case of $\mathfrak{X}_n$, with n odd. For n even, greater then 2, we have been led to the belief that B_n may actually be taken to be contained in the union of an algebraic set and an embedded analytic submanifold which contains all the fields of $\mathfrak{X}_n-\mathfrak{R}_o^{-1}(o)$ which have graphs with two symmetric saddles at infinity, like those of type IV.

Meanwhile, the computations involved in the analysis of the assertions above seem to be by no means short, if one wants to extend the methods of this paper to $\mathfrak{X}_n$.

5.Acknowledgement.

We are grateful to F. Dumortier and C. Simó for his comments to a previous version of this paper.

References

[D] Dulac, H. Sur les cycles limits. Bull. Soc. Math. de France Vol. 51, 1923.

[G] Gonzalez Velasco, E., Generic properties of polynomial vector fields at infinity. Transactions A.M.S. 143, 1969.

[H] Hilbert, D. Proceedings International Math. Congress 1900.

[P] Poincaré,H., Mémoire Sur les courbes définies par une équation différentielle, Jour. de Math, 3e série, t.7, p.375-422 (1881) and 5.8 p.251-296 (1882), also in Complete works, Vol.I.

[Sn] Songling,S., Scientia Sinica, Vol. 8, 1980, p.153-158.

[S] Sotomayor, J., Curvas definidas per equações diferenciais no plano, IMPA : 13º Colóquio Brasileiro de Matemática, 1981.

[W] Walker, R.J., Algebraic curves. Dover, 1950.

Instituto de Matemática
Pura e Aplicada, CNPq
Estrada Dona Castorina,110
CEP 22.460, Rio de Janeiro,RJ
Brazil

Universidade Federal de São
Carlos
Rodovia Washington Luiz, KM 235
C.P. 676
CEP 13.560, São Carlos, SP
Brazil

Absolutely continuous invariant measures for rational mappings of the sphere S^2.

W. Szlenk * /Warsaw/

1. The aim of the paper is to present a sufficient condition for existence of a measure absolutely continuous with respect to Lebesgue measure, invariant with respect to a rational mapping of the sphere S^2.

The paper is a continuation of the papers [2] and [3].

The presented condition is rather complicated and it is difficult to check if a given mapping satisfy it.

2. Let $f : S^2 \longrightarrow S^2$ be a rational mapping, i.e. replacing S^2 by the closed complex plane $\bar{C}$ the map f is a rational function: $f(z) = \frac{P(z)}{Q(z)}$, where P and Q are two polynoms. Denote by $z_i = f^i(z) = \underbrace{f \circ \ldots \circ f}_{i\text{-times}}(z)$, $A_i = f^i(A) = \underbrace{f \circ \ldots \circ f}_{i\text{-times}}(A)$, where $z \in C$, $C_1 = \{ c : f'(c) = 0 \}$.

We equip $\bar{C}$ with the Riemaniann metric induced from S^2 by the stereographic projection with respect to the equator (i.e. the plane pass through the equator of the sphere, and the projecting lines pass through the northe pole N). The metric is of the form:

---

* The author gratefully acknowledges the financial support of the Stiftung Volkswagenwerk for a visit to the IHES in 1980, during which this paper was written.

$$ds^2 = \frac{4(dx^2 + dy^2)}{(1 + |z|^2)^2} = \frac{4|dz|^2}{(1 + |z|^2)^2}$$

where $z = x + iy$. Let $\rho(\cdot\ ,\ \cdot)$ be the metric in $\bar{C}$ generated by the Riemaniann metric above. The derivative $df(z)$ maps $T_z(\bar{C})$ onto $T_{f(z)}(\bar{C})$ if $f'(z) \neq 0$ and then $df(z)$ is a conformal mapping at a neighbourhood of z. This implies that $df(z)$ is proportional to an isometry of $T_z(\bar{C})$ and $T_{f(z)}(\bar{C})$. Denote by $\varkappa(z)$ the coefficient of proportionality, which is also the norm of $df(z)$:

$$(1) \qquad \| df(z) \| = \frac{1 + |z|^2}{1 + |f(z)|^2} |f'(z)| \overset{df}{=} \varkappa(z) \ .$$

At a critical point c we have $\varkappa(c) = 0$. The function $\varkappa(z)$ satisfies the Lipschitz condition with a constant ϑ :

$$|\varkappa(z) - \varkappa(w)| \le \vartheta \rho(z,w) \ .$$

Denote by $\varkappa^j(z) = \| df^j(z) \|$. By the Chain Rule $\varkappa^j(z) = \prod_{s=0}^{j-1} \varkappa(z_s)$. The Jacobian of $df(z) = |\det df(z)|$ is equal to $\varkappa^2(z)$.

Let $D_\tau^0 = \{ z : \varkappa(z) < \tau \}$, where τ is a given number. We say that f satisfies the condition (E) if

(E) there exists a number $\tau > 1$ such that if $f^n(z) \in D_\tau$ for $n=1,2,\ldots$, then $\varkappa^n(z) \geq \tau$.

The condition (E) implies that the trajectory of every critical point c stays for ever in the region where $\varkappa(z) \geq \tau$: i.e. $f^n(C_1) \subset \{z : \varkappa(z) \geq \tau\}$ for $n=1,2,\ldots$ (see [2]). We say that a mapping f satisfies condition (F) if for any critical point c (i.e. $f'(c) = 0$) holds $f''(c) \neq 0$.

<u>Theorem</u> If $f : \overline{C} \longrightarrow \overline{C}$ is a rational mapping which satisfies the conditions (E) and (F), then there exists an f-invariant measure absolutely continuous with respect to Lebesgue measure.

For the proof we need some lemmas. Let λ_0 be a number such that $1 < \lambda_0 < \tau$. For given $b > 0$ we define

$$B = B(c) = \{z : \rho(z,c) \leq b\} , \qquad c \in C_1$$

For b small enough we have $B(c) \subset (D^0_{\lambda_0})^c$ (= complement of $D^0_{\lambda_0}$) for all $c \in C_1$.

<u>Lemma 1</u> There exists a number $\delta_1 > 0$ such that for every z with the property: $f^j(z) \in B(c_j)$ for $j \leq k$ (k is an arbitrary integer), we have

$$\delta_1^{-1} \leq \frac{\varkappa^j(z)}{\varkappa^j(c)} \leq \delta_1 \qquad \text{for } 1 \leq j \leq k .$$

<u>Proof</u> By standard estimates we have

$$\frac{\varkappa^j(z)}{\varkappa^j(c)} \leq \prod_{s=0}^{j-1} \left[1 + \frac{|\varkappa(z_s) - \varkappa(c_s)|}{\varkappa(c_s)}\right] \leq$$

$$\leq \prod_{s=0}^{j-1} \left[1 + \frac{\vartheta \rho(z_s, c_s)}{\lambda_0}\right] \leq \exp\left\{\frac{\vartheta}{\lambda_0} \sum_{s=0}^{j-1} \rho(z_s, c_s)\right\} .$$

Since $\rho(z_s, c_s) \leq \frac{\rho(z_{s+1}, z_{s+1})}{\lambda_0}$, we have

$$\sum_{s=0}^{k-1} \rho(z_s, c_s) \leq \sum_{s=0}^{k-1} \frac{\rho(z_k, c_k)}{\lambda_0^{k-s}} \leq \rho(z_k, c_k) \frac{1}{\lambda_0 - 1} \leq \frac{b}{\lambda_0 - 1}$$

Hence

$$\frac{\varkappa^j(z)}{\varkappa^j(c)} \leq \exp\left\{\frac{b\vartheta}{\lambda_0(\lambda_0 - 1)}\right\} \overset{df}{=} \gamma_1$$

Replacing z by c and c by z in the last formula we obtain the other inequality in the statement of Lemma 1.

<u>Lemma 2</u> There exists a neighbourhood U of the set C_1 and a number $d_1 > 0$ such that for every $z \in U - C_1$ there exists an integer $k = k(z)$ such that

$$\varkappa^k(z) \geq \frac{d_1}{\varkappa(z)} ,$$

and $f^j(z) \in (D^o_{\lambda_0})^c$ for $j \leq k$.

<u>Proof</u> Fixing the coordinates in a suitable way we may assume that $\varkappa(\infty) \neq 0$ and $f(c) \neq \infty$ for any $c \in C_1$. Let $c \in C_1$. Assume $f''(c) \neq 0$. Let U be a small disc such that

$$\inf_{z \in U} |f''(z)| \overset{df}{=} \vartheta_1 > 0$$

and set $\vartheta_2 \overset{df}{=} \sup_{z \in U} |f''(z)|$. By Mean Value Theorem we have

$$(2) \qquad \rho(z_1,c_1) = \rho(f(z),f(c)) \leq 2|f(z)-f(c)| \leq 2|f'(\xi)|\cdot|z-c|$$

where $z \in U$ and ξ is a point belonging to the segment joining z and c.

By Taylor's Formula we have

$$|f'(\xi)| = \frac{|f'(\xi)|}{|f'(z)|}|f'(z)| = \frac{|f''(\zeta)|\ |\xi-c|}{|f''(\eta)|\ |z-c|}|f'(z)| \leq \frac{\vartheta_2}{\vartheta_1}|f'(z)|$$

where ζ belongs to the segment joining ξ and c, η - to the segment joining z and c ($|\xi-c| < |z-c|$). On the other hand

$$\varkappa(z) = \frac{1+|z|^2}{1+|f(z)|^2}|f'(z)| = \frac{1+|z|^2}{1+|f(z)|^2}|f'(z)-f'(c)| =$$

$$= \frac{1}{2}\,\frac{1+|z|^2}{1+|f(z)|^2}|f''(\eta)|\cdot|z-c|$$

which implies

$$(3) \qquad |z-c| = \frac{1+|f(z)|^2}{1+|z|^2}\cdot\frac{1}{|f''(\eta)|}\cdot\varkappa(z) \leq L_1\cdot\varkappa(z)$$

where $L_1 = \sup\limits_{z\in U}\left[\frac{1+|f(z)|^2}{1+|z|^2}\right]\cdot\frac{1}{\vartheta_2}$. In the same way we get

$$(4) \qquad |f'(z)| = \frac{1+|f(z)|^2}{1+|z|^2}\varkappa(z) \leq L_2\cdot\varkappa(z)$$

where $L_2 = \sup\limits_{z\in U}\frac{1+|f(z)|^2}{1+|z|^2}$. By (2), (3) and (4) we have

$$(5) \qquad \rho(z_1, c_1) \leq L \cdot \varkappa(z)^2$$

where $\quad L = 2 \dfrac{\vartheta_2}{\vartheta_1} L_1 \cdot L_2$.

Let k be an integer such that

$$\rho(z_k, c_k) \leq b , \qquad \rho(z_{k+1}, c_{k+1}) > b$$

By Mean Value Theorem and by Lemma 1 we get

$$b < \rho(f^k(z_1), f^k(c_1)) \leq \left\| \frac{d}{dz} f^k(\xi_1) \right\| \rho(z_1, c_1) =$$

$$= \prod_{i=1}^{k} \frac{\varkappa(\xi_i)}{\varkappa(c_i)} \cdot \prod_{i=1}^{k} \varkappa(c_i) \rho(z_1, c_1) \leq \rho(z_1, c_1) \cdot \delta_1 \cdot \prod_{i=1}^{k} \varkappa(c_i)$$

The point ξ_1 is a point on the geodesic curve joining z_1 and c_1. Hence by (5)

$$\prod_{i=1}^{k} \varkappa(c_i) \geq \frac{b}{\delta_1} \frac{1}{\rho(z_1, c_1)} \geq \frac{b}{\delta_1} \frac{1}{L} \cdot \frac{1}{\varkappa(z)^2}$$

Therefore

$$\varkappa^{k+1}(z) = \varkappa(z) \prod_{i=1}^{k} \varkappa(z_i) \geq \varkappa(z) \prod_{i=1}^{k} \frac{\varkappa(z_i)}{\varkappa(c_i)} \prod_{i=1}^{k} \varkappa(c_i) \geq$$

$$\geq \varkappa(z) \frac{1}{\delta_1} \frac{b}{\delta_1 L} \frac{1}{\varkappa(z)^2} = \frac{b}{\delta_1^2 L} \cdot \frac{1}{\varkappa(z)} .$$

Replacing $k+1$ by k, we get our estimate:

$$\varkappa^k(z) \geq \frac{d_1}{(z)} , \quad \text{where} \quad d_1 = \frac{b}{\delta_1^2 L} .$$

Replacing U in Lemma 2 by a smaller neighbourhood $c \in U_1 \subset U$ ($\sup_{z \in U_1} \varkappa(z) < \frac{1}{2} d_1$) we obtain the following

<u>Corollary 1</u> There exists a neighbourhood U_1 of C_1, an integer n_1 and two numbers $\beta_1 > 0$ and $d_2 > 1$ such that if $k = k(z)$ is as in Lemma 2, $k \geqslant n_1$, then

$$\varkappa^k(z) \geqslant \beta_1 \cdot d_2^k \quad .$$

Let $\{K_i\}_{i=1}^N$ be a finite cover of C, where K_i are open discs. Assume that: 1) either $K_i \subset D_\tau^0$, or $K_i \cap D_{\lambda_0}^0 = \emptyset$ for every i; 2) each critical point c and each critical value $f(c)$ are centers of some discs; 3) every point $f(c)$, $, c \in C_1$, are covered exactly by one disc K_i; 4) the diameters of K_i are small enough, so that if $C_1 \cap K_i = \emptyset$, then $f^{-1}(z)$ has at most one point in K_i; if a point $c \in K_i$, then for every point $z \in f(K_i)$ the set $f^{-1}(z)$ has at most 2 points in K_i; note that $f''(c) \neq 0$; 5) for every two points $z, w \in K_i$ there exists a geodesic curve, contained in K_i, such that the length of the curve is equal to the distance of z and w; 6) Also we change a little bit the sets $D_\tau^0, D_{\lambda_0}^0$: we replace them respectively by some sets D_τ and D_{λ_0} being unions of K_i and such that $D_{\lambda_0} \subset D_{\lambda_0}^0 \subset D_\tau \subset D_\tau^0$; 7) every disc K_i such that $K_i \cap C_1 \neq \emptyset$ is contained in the set U from Lemma 1. Suppose $c \in K_i$. We shall denote these K_i by K_i^0. Set $M_i = K_i^0 - \bigcup_{K_i \neq K^0} K_i$. Let δ_1 be the Lebesgue number of the cover $\{K_i\}$, let δ_2 be the following number: if $z \in M_i$, then for every w such that $\rho(z,w) < \delta_2$ we have w K_i^0.

We say that two points $u \in f^{-1}(z)$, $v \in f^{-1}(w)$ are in the same preimages of z,w if for every curve l joing z,w the points u,v belong to the same component of $f^{-1}(l)$. Let δ_3 be a number such that if a point $z \in D_{\tau}^{c}$, then $K(z,\delta_3) \cap D_{\lambda_0} = \emptyset$. We note also that there exists a number $\beta_2 > 0$ such that

$$(6) \qquad \rho(z_1,w_1) \geq \beta_2 \rho(z,w)$$

if z,w belong to the same component of the preimages of z_1,w_1. ($f(z) = z_1$, $f(w) = w_1$).

Let δ_4 be equal to the $\frac{1}{2} \min_i \operatorname{dist}(F_r M_i, C_1)$.

<u>Lemma 3</u> Let u,v be two points belonging to the same component of $f^{-n}(z)$, $f^{-n}(w)$: $u \in f^{-n}(z)$, $v \in f^{-n}(w)$. There exists two constant numbers $\beta_3 > 0$ and $0 < \mu < 1$, an increasing function $h(n)$, $h(n) \longrightarrow +\infty$ if $n \longrightarrow +\infty$, such that if $\rho(u_i,v_i) < \min(\delta_1,\delta_2,\delta_3)$ for $i=0,1,\ldots,n$, then

$$(6\ ) \qquad \rho(u,v) \leq \beta_3 \mu^{h(n)} \rho(w,z)^{\frac{1}{r}} .$$

For instance we may set $h(n) = \sqrt{n}$;

<u>Proof</u> We divide the indices $i = 0,\ldots,n$ in some blocks S_j:

$$S_j = [i_j,\ldots,i_{j+1}-1] , \qquad j=0,1,\ldots,k_n .$$

where $u_{i_j}, v_{i_j} \in D_\pi$, and for $i_j < i < i_{j+1}$ $z_i, w_i \notin D_{\lambda_0}$.

The block S_o may not contain any index i_j; the last block S_{k_n} deserves for a special treatment because it does nor precede any other block. Denote by $|S_j| = i_{j+1} - i_j$. In view of definition of D_{λ_0} we have $\varkappa(z) \geqslant \lambda_0$ for $z \in D^c_{\lambda_0}$, therefore

$$\rho(u_i, v_i) \leq \frac{\rho(u_{i_{j+1}}, v_{i_{j+1}})}{\lambda_0^{i_{j+1} - i}}$$

By assumption (E) we have

$$\rho(u_{i_j}, v_{i_j}) \leq \frac{\rho(u_{i_{j+1}}, v_{i_{j+1}})}{\pi}$$

Thus, in view of (6)

$$\text{(7)} \quad \rho(u,v) \leq \frac{1}{\lambda_0^{|S_o|}} \rho(u_{i_1}, v_{i_1}) \leq \frac{1}{\lambda_0^{|S_o|}} \frac{1}{\pi^{k_n - 1}} \cdot$$

$$\cdot \rho(u_{i_{k_n-1}}, v_{i_{k_n-1}}) \leq \frac{1}{\lambda_0^{|S_o|}} \frac{1}{\pi^{k_n-1}} \frac{1}{\beta_2} \rho(u_{i_{k_n-1}+1}, v_{i_{k_n-1}+1})$$

$$\leq \frac{1}{\lambda_0^{|S_o|}} \frac{1}{\pi^{k_n-1}} \frac{1}{\beta_2} \frac{1}{\lambda_0^{|S_{k_n}|}} \rho(z,w)$$

If $z, w \in D_\pi$, then we do not need use (6) and we obtain

$$\text{(8)} \qquad \rho(u,v) \leq \frac{1}{\lambda_0^{|S_o|}} \frac{1}{\pi^{k_n}} \rho(z,w)$$

Now we have two possibilities: (1) either $k_n \geq \sqrt{n}$, and then by (7) we have

$$\rho(u,v) \leq \frac{1}{\tau^{\sqrt{n}}} \rho(z,w) \ ;$$

(2) or $k_n < \sqrt{n}$. Then either there exists a block S_{j_0} such that $|S_{j_0}| \geq \frac{1}{2}\sqrt{n}$. By Corollary 1

$$\rho(u_{i_{j_0}}, v_{i_{j_0}}) \leq \frac{1}{\beta_1}\left(\frac{1}{d_1}\right)^k \rho(u_{i_{j_0+k}}, v_{i_{j_0+k}}), \ i_{j_0+k} < i_{j_0+1} .$$

Hence by (7)

$$\rho(u,v) \leq \frac{1}{\lambda_0^{|S_0|}} \frac{1}{\tau^{j_0-1}} \rho(u_{i_{j_0}}, v_{i_{j_0}}) \leq$$

$$\leq \frac{1}{\lambda_0^{|S_0|}} \frac{1}{\tau^{j_0-1}} \frac{1}{\beta_1} \left(\frac{1}{d_2}\right)^k \rho(u_{i_{j_0+k}}, v_{i_{j_0+k}}) \leq$$

$$\leq \frac{1}{\lambda_0^{|S_0|}} \frac{1}{\tau^{j_0-1}} \frac{1}{\beta_1} \left(\frac{1}{d_2}\right)^k \frac{1}{\lambda_0^{|S_{j_0}|-k}} \rho(u_{i_{j_0+1}}, v_{i_{j_0+1}}) \leq$$

$$\leq \frac{1}{\beta_2 \beta_1} \frac{1}{\lambda_0^{|S_0|}} \frac{1}{\tau^{j_0-1}} \left(\frac{1}{d_2}\right)^k \frac{1}{\lambda_0^{|S_{j_0}|-k}} \frac{1}{\tau^{k_n-1-(j_0+1)}} \frac{1}{\lambda_0^{|S_{k_n}|}} \cdot$$

$$\cdot \rho(z,w)$$

Either $k \geq \frac{1}{4}\sqrt{n}$, or $|S_{j_0}| - k \geq \frac{1}{4}\sqrt{n}$. Denoting by $\gamma_1 = \min(\lambda_0, \tau, \frac{1}{d_2}) > 1$ we get

$$\rho(u,v) \le \text{const}\,(\sqrt[4]{\mu_1})^{\sqrt{n}}\,\rho(z,w) \ .$$

If we do not have any block S_{j_0} with $|S_{j_0}| > \frac{1}{2}\sqrt{n}$, then either S_0 or S_{k_n} are larger than $\frac{n}{4}$, and we obtain even a stronger estimate. Therefore setting $h(n) = \sqrt{n}$ and choosing $\mu = \min(\sqrt[4]{\mu_1}\ , \frac{1}{\tau})$ we get (6).

Let δ_4 be equal to $\frac{1}{2}\min_i \text{dist}(F_r M_i, C_1)$, and let $\delta = \min(\delta_1, \delta_2, \delta_3, \delta_4)$.

Corollary 2 There exists a number $r_1 > 0$ such that if $\rho(z,w) < r_1$, and $(u_n),(v_n)$ are two sequences of points such that u_{n+1}, v_{n+1} belong to the same preimages of $f^{-1}(u_n)$, $f^{-1}(v_n)$, $u_1 \in f^{-1}(z)$, $v_1 \in f^{-1}(w)$, then $\rho(u_n,v_n) < \delta$ for any $n = 0,1,\dots$.

The proof follows from the proof of Lemma 3.

Corollary 3 There exists a constant number $\beta_3 > 0$ such that if u,v belong respectively to the same preimage at $f^{-j}(z)$, $f^{-j}(w)$ for $j = 0,\dots,n$, and $\rho(u_j,v_j) \le$ $\le \min(\delta_1, \delta_2, \delta_3)$, then

$$\sum_{j=0}^{n-1} \rho(u_j,v_j) \le \beta_3\, \rho(z,w) \ .$$

In view of Lemma 3 it is enough to set $\beta_3 = \beta_2 \sum_{n=1}^{\infty} \mu^{\sqrt{n}}$.

<u>Corollary 4</u> There exists a number $\delta_2 > 0$ such that if two points u, v belong respectively to the same preimages of $f^{-j}(z)$, $f^{-j}(w)$, $j = 0, \ldots, n$, and for any j, $u_j, v_j \notin \bigcup_i M_i$, then

$$\delta_2^{-1} \le \frac{\varkappa^j(u)}{\varkappa^j(v)} \le \delta_2$$

for any $j = 0, \ldots, n-1$.

<u>Proof</u> We proceede as in the proof of Lemma 1, using the fact that since $v_j \notin \bigcup_i M_i$, there is $\varkappa(v_j) \geq$ $\geq \inf\{\varkappa(z) : z \in (\bigcup_i M_i)^c\} \overset{\text{df}}{=} \omega > 0$.

Now let $T = [0, 1, \ldots, j]$ be a sequence of indices such that $u, v \in K_i^o$ for some i, $u_i, v_i \notin \overline{\bigcup_i M_i}$ for $0 < i \le j$, and $u_{j+1}, v_{j+1} \in K_s^o$ for some s. We assume u, v belong respectively to the same preimage of $f^{-(j+1)}(z)$, $f^{-(j+1)}(w)$, $z, w \in \overline{C}$.

<u>Lemma 4</u> There exists a number $L_3 > 0$ such that if $\rho(z, w) < r_1$ then there exists an integer $p \in T$ such that

$$\frac{\rho(u, v)}{\varkappa(v)} \le L_3 \rho(u_p, v_p) .$$

<u>Proof</u> Assume $\rho(z, w) < r_1$, where r_1 is from Corollary 2. Since $u, v \in K_i^o \subset U$, there exists an integer $p \in T$ such that

$$\varkappa^p(v_p) \geq \frac{d_1}{\varkappa(v)} \tag{9}$$

We join u_p and v_p by a geodesic curve Γ, contained in a disc K_i, such that the length of Γ is equal to $\rho(u_p,v_p)$:

$$\rho(u_p,v_p) = \int_{\Gamma} ds = \int_{f^{-p}(\Gamma)} \varkappa^{p}(t)dt \quad .$$

In view of Corollary 2 diam $f^{-p}(\Gamma) < \delta$, then by Lemma 2 and 3

$$\rho(u_p,v_p) = \int_{f^{-p}(\Gamma)} \varkappa^{p}(t)dt = \int_{f^{-p}(\Gamma)} \frac{\varkappa^{p}(t)}{\varkappa^{p}(v)}\varkappa^{p}(v)dt \geq$$

$$\geq \quad \gamma_2^{-1}\varkappa^{p}(v)\cdot\text{length } f^{-p}(\Gamma) \geq \gamma_2^{-1}\frac{d_1}{\varkappa(v)}\rho(u,v) \ .$$

We set $L_3 = \dfrac{\gamma_2}{d_1}$.

<u>Lemma 5</u> There exists a number $\gamma_3 > 0$ such that for every disc $P \subset K_i^o$ of radius $r \leq \frac{1}{2}r_1$ for any two points u,v belonging to the same component of $f^{-n}(P)$ there is

$$\gamma_3^{-1} \leq \frac{\varkappa^{n}(u)}{\varkappa^{n}(v)} \leq \gamma_3 \ .$$

<u>Proof</u> We proceede as in Lemma 1. Let $f^n(u) = z$, $f^n(v) = w$. Since $\rho(z,w) < r_1$, we have $\rho(u_i,v_i) < \delta$ for all $i=0,\ldots,n$. Then

$$\frac{\varkappa^{n}(u)}{\varkappa^{n}(v)} \leq \prod_{i=0}^{n-1}\left(1 + \frac{|\varkappa(u_i)-\varkappa(v_i)|}{\varkappa(v_i)}\right) \leq \exp\left\{\vartheta\sum_{i=0}^{n-1}\frac{\rho(u_i,v_i)}{\varkappa(v_i)}\right\}$$

The indices $[0,1,\dots,n-1]$ we split in some blocks T_j, like in Lemma 4. Let i' be an index such that no one of the points u_i, v_i, belongs to M_i; then $\varkappa(v_i) \geq \omega > 0$. Let i'' be an index such that $u_i'', v_i'' \in K_i^o$. Then

$$\sum_{i=0}^{n-1} \frac{\rho(u_i,v_i)}{\varkappa(v_i)} = \sum_{i=i'} \frac{\rho(u_i,v_i)}{\varkappa(v_i)} + \sum_{i=i''} \frac{\rho(u_i,v_i)}{\varkappa(v_i)}$$

The first sum is bounded by Corollary 3:

$$\sum_{i=i'} \frac{\rho(u_i,v_i)}{\varkappa(v_i)} \leq \frac{1}{\omega} \sum_{i=i'} \rho(u_i,v_i) \leq \frac{\beta_3 \rho(z,w)}{\omega} \leq \frac{\beta_3 r_1}{\omega}$$

The second sum is bounded by Lemma 4:

$$\sum_{i=i''} \frac{\rho(u_i,v_i)}{\varkappa(v_i)} \leq \sum_{i=i''} L_3 \rho(u_{p(i)}, v_{p(i)}) \leq L_3 \beta_3 r_1 ,$$

where $p(i)$ is an integer corresponding to $u_i, v_i \in K_s^o$.

We set $\gamma_3 = \exp\left\{ \vartheta \left(\frac{\beta_3 r_1}{\omega} + L_3 \beta_3 r_1 \right) \right\}$.

Let ν_o denotes Lebesgue measure, let $\nu_n = \nu_o \circ f^n$, i.e. $\nu_n(A) = \nu_o(f^{-n}(A))$. We shall show that the measures ν_n are uniformly integrable:

<u>Lemma 6</u> For every $\varepsilon > 0$ there exists an $\eta > 0$ such that for any set F : $\nu_o(F) < \eta$ we have

$$\nu_n(F) < \varepsilon .$$

<u>Proof</u> Without loss of generality we may assume that diam $F < r_1$. Let P be a disc of radius r_1 such that $P \subset F$. Let G be a component of $f^{-n}(F)$, let $Q \supset G$ be a component

of $f^{-n}(P)$. For every $0 \leq j \leq n$ the set $f^j(P)$ is contained in a disc K_{i_j}. We divide the indices $[0,1,\ldots,n]$ in some blocks T_j, as in Lemma 4. Assume that there are k_n+1 blocks:

$$[0,1,\ldots,n] = \bigcup_{j=0}^{k_n} [i_j,\ldots,i_{j+1}-1] \quad ,$$

where $f^i(Q) \subset K_{s_i}$ for $i_j < i < i_{j+1}$, $\overline{K_{s_i}} \cap C_1 = \emptyset$, $f^{i_j}(Q) \subset K^e_{s_j}$. For $0 \leqslant i \leq i_{k_n}$ we have distortion property, i.e.

$$\frac{\varkappa^i(u)}{\varkappa^i(v)} \leq \delta_3 \qquad \text{for any } u,v \in Q \ .$$

Therefore there exists a constant number $L_4 > 0$ such that

$$(10) \qquad \frac{\nu_o(G)}{\nu_o(Q)} \leq L_4 \frac{\nu_o(f^{i_{k_n}}(G))}{\nu_o(f^{i_{k_n}}(Q))}$$

(see for instance [2]). In view of (6) we have

$$(11) \qquad \frac{\nu_o(f^{i_{k_n}+1}(G))}{\nu_o(f^{i_{k_n}+1}(Q))} \geqslant \beta_2^2 \left[\frac{\nu_o(f^{i_{k_n}}(G))}{\nu_o(f^{i_{k_n}}(Q))}\right]^2$$

For $i > i_{k_n} + 1$ we have again distortion property, i.e.

$$(12)\quad \frac{\nu_0\left[f^{|S_{k_n}|-1}(f^{i_{k_n}}(G))\right]}{\nu_0\left[f^{|S_{k_n}|-1}(f^{i_{k_n}+1}(Q))\right]} \leq L_4 \frac{\nu_0(f^n(G))}{\nu_0(f^n(Q))} = L_4 \frac{\nu_0(F)}{\nu_0(P)}$$

The inequalities (10), (11), (12) imply

$$\frac{\nu_0(G)}{\nu_0(Q)} \leq \beta_2\, L_4^{\frac{3}{2}} \sqrt{\frac{\nu_0(F)}{\nu_0(P)}}$$

Hence

$$\nu_n(F) = \nu_0(f^{-n}(F)) = \sum_s \frac{\nu_0(G_s)}{\nu_0(Q_s)}\, \nu_0(Q_s)$$

where $\{Q_s\}$ are all components of $f^{-n}(P)$, $\{G_s\}$ - all components of $f^{-n}(F)$. Therefore if $\nu_0(F) < \varepsilon\,\nu_0(P)\frac{1}{\beta_2^2 L_4^3}$,

then

$$\nu_n(F) < \varepsilon$$

which completes the proof.

Proof of Theorem. We proceede as in [2]. This is a standard way of construction of an invariant measure absolutely continuous with respect to the Lebesgue measure.

<u>Remark 1.</u> Probably the assumption that $f''(c) \neq 0$ is not necessary. Theorem should be true in the case of any rational mapping satisfying the condition (E).

Remark 2. Unfortunately we cannot present any simple example which satisfies the assumption (E). We can only prove that the mappings $f_c(z) = -\frac{1}{4}z + \frac{c}{z} - i\sqrt{c}$ for small $c > 0$ satisfy the condition (E). The proof is rather complicated.

Remark 3. The author of this paper has been informal that D.Fried proved that the mapping $f(z) = \frac{1}{2i}(z - \frac{1}{z})$ preserves an invariant measure absolutely continuous with respect to the Lebesgue measure. It seems that the result has not been published.

References

[1] Misiurewicz M.: Absolutely continuous measures for certain maps of interval, Publications Mathématique No 53, (1981), 17-51.

[2] Szlenk W.: Some dynamical properties of certain differentiable mappings of an interval, Bol. Soc. Mat. Mex. vol. 24, No 2 (1979), 57-82.

[3] Szlenk W.: Some dynamical properties of certain differentiable mappings of an interval, Part II, IHES/M/80/41.

THE ROLE OF QUALITATIVE DYNAMICS IN APPLIED SCIENCES

René THOM

This meeting, devoted to the inauguration of the new IMPA building, had also, in the mind of its organizers, another aim : namely to celebrate the centennial anniversary of the publication of H. Poincaré's fundamental memoir (1881) "Sur les courbes définies par les équations différentielles", in which many see the birth of Qualitative Dynamics. Now it is perhaps time to evaluate the impact of this new discipline on mathematics itself, and also on the applications. The first part of my talk will be devoted to historical considerations, as I intend to describe how the main ideas of qualitative dynamics did evolve. In the second, I shall try to give a survey of their applications in other Sciences (Mechanics, Physics, Chemistry, Biology and Social Sciences) and to judge their possible value for the future of these Sciences.

I. History of Qualitative Dynamics.

Here I intend to speak, not only on Qualitative Dynamics proper, but also on nearby notions, as the notion of structural stability, of general position and transversality which play an important role in many proofs of this discipline as fundamental local tools.

Quite certainly, the idea of "transversality" is a very old one, and for its origin one has to look at the first formulations of the Inverse Function Theorem $F(x,y) = 0$ is solvable with respect to y in $(0,0)$ if $F(0,0) = 0$, $\frac{\partial F}{\partial y}(0,0) \neq 0$, possibly by Cauchy. But a modern formulation had to wait for the clarification of the notion of tangent vector, and tangent bundle, which occurred only in 1940-50 especially with C. Ehresmann's theory of jets. But it is fair to say that the idea of "general position" was one of the basic tools of the Italian Algebraic Geometry of the years 1880-1900. The method of small variation "La piccola variazione" was systematically used by the Italian Geometers to put the projection of a variety in its "generic" form - even if the uncontrolled use of the method led sometimes to mistakes, as for instance the claim by Severi that any variety can be deformed into a smooth one... But it is quite clear that these ideas greatly influenced Lefschetz, who later on played in the years 1955-65 a very important role in emphasing the importance of structural stability - a fact which had a great impact in the orientation of IMPA towards Qualitative Dynamics. The transfer of these techniques from Algebraic Geometry to the C^{∞}-category arose through the fundamental work of Ehresmann (on the theoretical viewpoint), and H. Whitney (Mappings of the plane into the plane), M. Morse(M.Morse's lemma on critical points), followed by the later work of (Thom-Mather) on the structure and stability of smooth maps. Here again, the extension of the preparation theorem (Weierstrass-Rückert) to the C^∞ category by Malgrange was a decisive progress. All these results culminate in the density theorem of topologically stable maps (in the C^m topology), thus establishing a firm basis for genericity. Unfortunately the same result does not hold for flows (even on a local basis except for gradient flows... (cf.Takens) and this to some extent explains the rather difficult state of structural stability questions in general.

Now the global theory of topological stability of flows, originated by Poincaré, and developed by him for the study of the 3-body problem (discovery of homoclinic, heteroclinic points) found its first major development with G.D. Birkhoff (1920), who introduced the fundamental notions of wandering, and non-wandering points. The second decisive progress came from the Sovietic School, which, with Andronov-Pontrjagin, introduced the notion of structural stability of flows (1930). The third decisive progress came with results of S. Smale and M.M. Peixoto, e.g. the density of stable flows on surfaces. Moreover the fine study of diffeomorphisms was greatly enhanced by the interest devoted to several examples (shifts like in the horseshoe diffeomorphism , Anosov diffeomorphism) - interest which is now extended to the study of iteration of maps (an old subject, originated by Fatou-Julia in 1910-20, which is now flourishing). In the West, this was undertaken by many American mathematicians (most of whom being present here, I apologize for not quoting them) and also by the IMPA School. The progresses of the local theory also helped to develop the theory of bifurcation. But it became rapidly obvious that the geometric theory by itself could not hope for a general theorem (like in the situations of smooth maps). Hence the need to come back to the measure theoretic view point and to ergodicity problems. This was to a large extent the work of the Russian School (Arnol'd, Sinai), which devoted also a great interest to hamiltonian flows. (The KAM theorem for instance). The notion of "strange attractor" is new under intensive study - although it is difficult to guess in which direction such a study may develop in the near future.

II. Applications of Qualitative Dynamics.

Strangely enough, it is difficult to pinpoint a specific application of qualitative dynamics (or more generally of stability property) before the last half of the XIXth century There is possibly a reason for that : the conception of a physical law as due to statistics apparently did not appear before the formulation of Statistical Mechanics - with Boltzmann, Maxwell, and Gibbs. This is why the first typical use of genericity arguments in Science is apparently Gibbs phase rule, which states that local phases of a system depending on n parameters may be in equilibrium at a point of R^n if their number does not exceed $(n+1)$. Of course people had earlier the idea that if a system depends on m parameters, if we impose

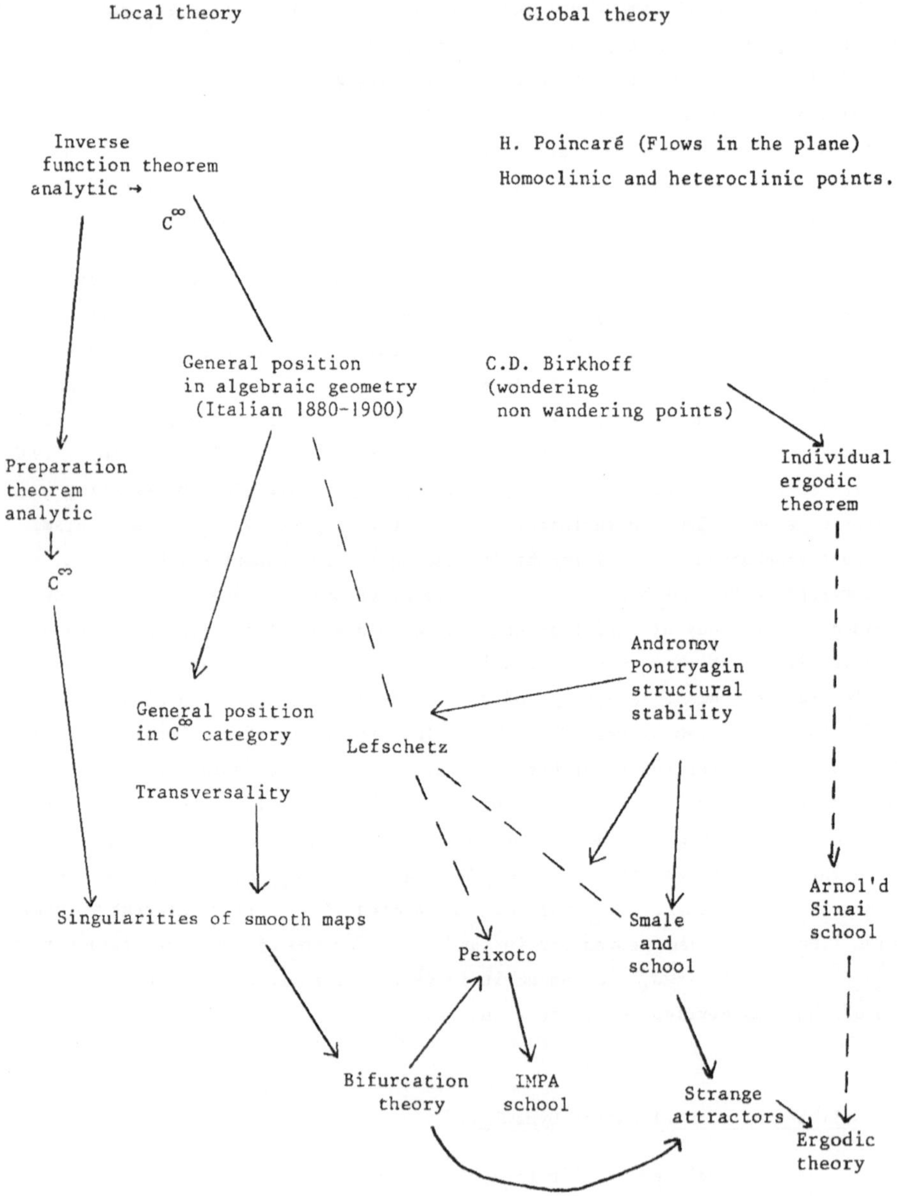
Local theory
Global theory
Inverse function theorem analytic → C^∞
H. Poincaré (Flows in the plane)
Homoclinic and heteroclinic points.
General position in algebraic geometry (Italian 1880-1900)
C.D. Birkhoff (wondering non wandering points)
Preparation theorem analytic ↓ C^∞
Individual ergodic theorem
Andronov Pontryagin structural stability
General position in C^∞ category
Lefschetz
Transversality
Arnol'd Sinai school
Singularities of smooth maps
Smale and school
Peixoto
Bifurcation theory
IMPA school
Strange attractors
Ergodic theory

k-constraints $k < m$ to the system, then the constrained system depends in general on $(m-k)$ parameters. But a precise formulation of this theorem - suggested by the algebraic elimination of variables between equations - could not be obtained before a clear formulation of the inverse function theorem. With the beginning of the approximation methods in applied mathematics, people had a better knowledge of the effect of noise - random perturbations - on some algebraic or analytic system. Hence the notion of structural stability, which tends to preserve at least the qualitative structure (phase portrait) of a differential system under a sufficiently small perturbation. This was certainly the hope of Andronov-Pontrjagin when they introduced their notion of "système grossier". But there again, it is difficult to specify applications of the notion before the development of Catastrophe Theory - inspired by the progress of bifurcation theory.

The relative pragmatic sterility of such an apparently important notion as structural stability should call for an explanation. Physicists are always looking for "quantitative" laws. These laws describe the temporal evolution of phenomena, generally through differential systems (or Partial Differential equations). But - as pointed out to me by D. Berlinski - such systems are in general not structurally stable! (e.g. : hamiltonian systems are not structurally stable). On the other hand, quantitative exact modelization is fundamental for prediction (hence for experimental verification). Hence it is quite easy to understand why physicists are not interested prima facie in structural stability, but mainly in quantitative accuracy of extrapolations, which alone can ensure some control, some mastery on the phenomena. It is true that in some domains of Physics, like Optics, singularity theory may be of some help - at least in the rapid enumeration and classification of stable singularities of the solution (think of the singularities of the solution of the wave equation along a caustic). But this occurs only because the system depends on explicit parameters (as the spatial position of apparatuses), for which genericity has to be proved in each specific case. The development of Catastrophe theoretic models around 1973-6 has led to a controversy, which has profound methodological implications. It was hoped - at first - that such qualitative modeling could lead to some useful quantitative prediction. This amounts to taking seriously Rutherford's maxim : "qualitative is nothing but poor quantitative"... But as a model is defined only up to diffeomorphism (or even homeomorphism), it is difficult to extract out of it some numerical prediction - unless one can give estimates for the effect of perturbation. Purely qualitative models are unfit for prediction, but they may nevertheless be quite useful in understanding the global structure of a system : this is the case of models aimed at providing intelligibility - and not mastery of the phenomena. Obviously the use of such qualitative methods leads to very delicate questions of methods. If we want a system to be generic, what is the topology we have to consider to define genericity ? Obviously no general answer can be given to such a question : only a case study can eventually provide an answer. For instance, the apparent lack of structural stability of the time evolution flows defined by the laws of Physics and/or Mechanics may often be attributed to the existence of unknown - or implicit - symmetry constraints. The generic singularities for symmetric systems are of a more restricted type than those of a general system. For instance, many systems of fundamental Physics owe their lack of structural stability to the requirement of being invariant under a change of observational frame. Up to now, it is probably fair to say that the notions of genericity and structural stability have not been at the source of major progresses in the modelization of phenomena. These notions cannot provide "laws" for phenomena where such laws do not exist. But they may help to a large extent the understanding of the underlying mechanisms of processes; of course, the methods will not be standard, and will require for each case some specific intuition. Here are some examples among the most recent ones :

1) The Debreu theory of economic equilibrium.
2) The notion of weak turbulence - and of strange attractors (Ruelle-Takens theory of turbulence) which is nothing but a first instance of a notion derived from stability theory into physics, and it has already led to some experimental verification. But defining the spatial morphology of attractors in configuration space is an exciting problem. The same can be said about symmetry breakings, defects of ordered media, where stability criteria (Kléman-Toulouse principle) already have been used with success. As a conclusion, purely qualitative models are more difficult to define and to use than quantitative ones;hence it is no wonder that such methods have not been, up to now, as fruitful as it was sometimes expected.

In statistics, I believe also that qualitative notions may be important : for statistics can be defined as the attempt to generate a given cloud of points by a deterministic mechanism as simple as possible. The interpretation of the morphology of the cloud, of its stable accidents requires obviously genericity assumptions. This is a large domain which awaits exploration, and provided one does not expect from qualitative methods results they cannot give, the future of such methods still looks very promising. In some sense, one should perhaps look not so much at possible (precise) applications of qualitative methods, but on the contrary, look for extending and strenghening these methods in specific domains, Biology, Social Sciences... A major test of success will then be if these methods will lead to interesting mathematical developments. This is - to my opinion - the ultimate proof of the validity of a theory.

RATE OF APPROACH TO MINIMA AND SINKS THE C^2 AXIOM A NO CYCLE CASE

by

Helena S. Wisniewski

INTRODUCTION

The dynamical system herein are C^2 Axiom diffeomorphisms and flows on C^∞ compact manifolds. Determining the asymptotic rate of approach to sinks amounts to comparing the Riemannian measure of the entire manifold, which we take normalized to be one, to the measure of the set of points whose orbits remain outside a neighborhood of the sinks after N iterations for diffeomorphisms, or time T for flows. For Axion A systems this rate is bounded by expression of the form: K exp (-DN), where the exponential constants are related to the topological pressure of f.

For the Morse Smale case, [11], we found the exponential constant, D, to be any number smaller than:

$$C = \min_P \{ 1/m \log \operatorname{Jac} D_P f^m \mid W^u (P) \} ,$$

where the minimum is taken over all non-sink P in the non-wandering set of f, and m is the least integer such that $f^m P = P$. However, without the assumption of transversality we found the exponential constant to be more complicated and yield a slower overall rate. Such is the case for general C^2 Axiom A systems with no cycles.

We begin by presenting two examples which illustrate these facts. The first example is with transversality in which we

---

*This paper is part of the author's Ph.D. dissertation completed at the Graduate School & University of C.U.N.Y., under the direction of Prof. Michael Shub.

find the parameter C is actually equal to the Minimum Jacobian, (the Morse Smale case). The other is without the transversality condition in which the parameter C is less than the Minimum Jacobian.

To make the notion of rate precise we cast the problem in terms of volumes.

Theorem: Let M be a compact C^∞ Riemannian manifold. Let μ be the measure induced by the Riemannian metric on M. Let f be a C^2 Axiom A diffeomorphism with no cycles. Let V be an open set in M which:

a) contains all sinks and attractors for f

b) satisfies $f(V) \subset V$,

Then defining: $U_N = \{x \in M\text{-}ClV: f^k(x) \notin V \text{ for } k \leq N\}$

we get for all $\delta > 0$ there exists $K = K(\delta)$, and $C > 0$ such that $\mu(U)_N \leq K\ (1+\delta)^N \exp\ (-CN)$

The precise form of the constant C will be given later in this paper.

Our theorems and their extentions apply as long as we know a volume estimate for the basic sets. For Morse-Smale we know such an estimate when f is C^1; however, for the general Axiom A case we need f to be C^2. These points will be discussed further. Also a discussion of related cases and some open questions will be presented. In a separate paper, [11], we prove the theorems for the Morse Smale case, which yield the results stated earlier.

I wish to acknowledge Prof. M. Shub for presenting me with this topic and for his valuable suggestions . Also I wish to thank Prof. E. Feldman for providing funding through his N.S.F. Grant, and Prof. R. Sacksteder for his many helpful conversations.

§ 1 An Example With Transversal Intersection

Consider the torus $T^2 \subset R^3$ which is tilted back with respect to the horizontal plane. Let the gradient field on it be of the form:

$$\dot{X} = - \text{grad } (h)$$

where h is the height function of points in relation to the horizontal plane. So the flow is downward. Let P_4 be the source, P_3 and P_2 the saddles and P_1 the sink. Next consider the Dehn twist, see Fig. 1 . The effect of it is a transversal intersection between $W^u(P_3)$ and $W^s(P_2)$, which is illustrated in Fig. 1.

Fig. 1

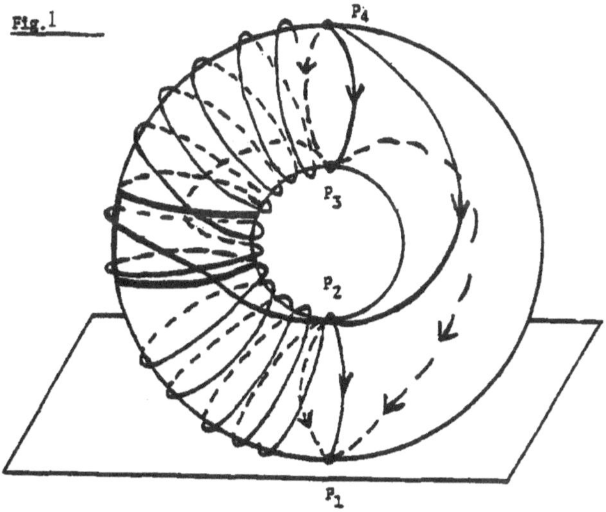

The local intersection between P_3 and P_2 is seen in Fig. 2.

Fig. 2

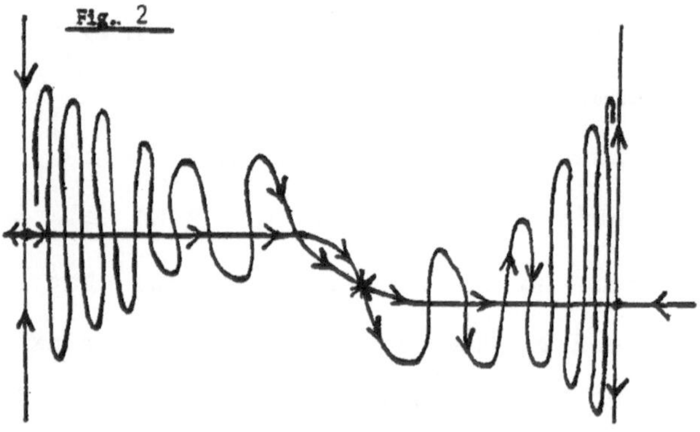

In Fig. 2 there are an infinite number of such intersections tending forward to P_2.

The λ Lemma of Palis [6] tells us that $W^u(P_3)$ in a neighborhood of P_2 becomes close to $W^u(P_2)$ in both distance and slope. In fact it contains $W^u(P_2)$ in its closure.

Next in linearized neighborhoods of P_3 and P_2 the local diffeomorphisms can be given by :

$$(**)\quad \begin{array}{ll} X = X_0 \exp(\gamma) & X = X_0 \exp(-\gamma) \\ Y = Y_0 \exp(-\alpha) & Y = Y_0 \exp(\beta) \end{array}$$

respectively. The logarithms of the eigenvalues are $\gamma, -\alpha, -\gamma, \beta$ For sets moving from P_3 to P_2, the transversality property causes the two unstable directions to become alligned. This is shown in Fig. 3 , where we only consider one point of transversal intersection outside the neighborhoods for clarity.

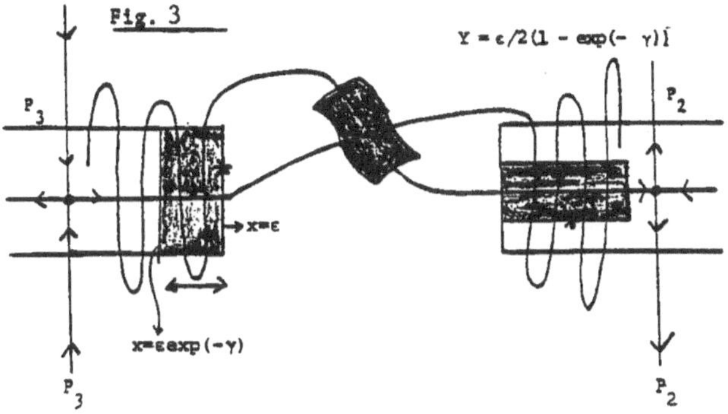

Fig. 3

By the transversality property the height of the set which comes into the neighborhood of P_2 is independent of the number of iterations for which it stays in a neighborhood of P_3 .

We will show that the area of the set whose orbits remain in the neighborhood of P_3 for exactly n iterations, and in the neighborhood of P_2 for at least m iterations is :

$$4\varepsilon^2 \exp(-\beta m)\exp(-\gamma n)$$
$$\leq 4\varepsilon^2 \exp(C)\exp(-CN) \quad \text{where } C = \min\{\gamma,\beta\}$$

and $N = n + 1 + m$.

To see how this estimate was obtained consider the following illustrations. First consider the neighborhood of P_3 .

A Linearized Neighborhood of P_3

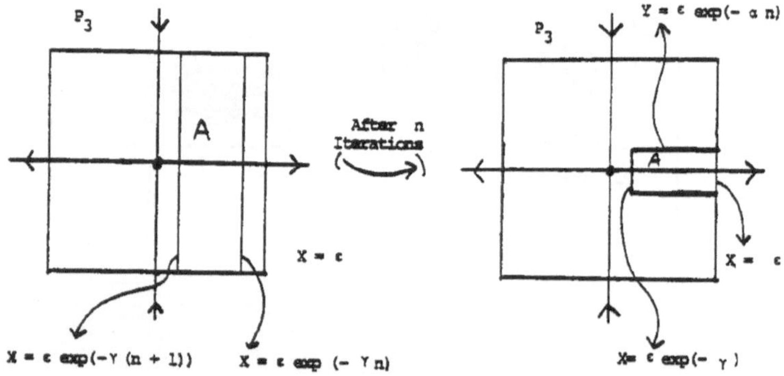

Note the width of A $= (\varepsilon - \varepsilon \exp(-\gamma))$. Let the number of iterations to get from P_3 to P_2 be 1. Consider the neighborhood of P_2 :

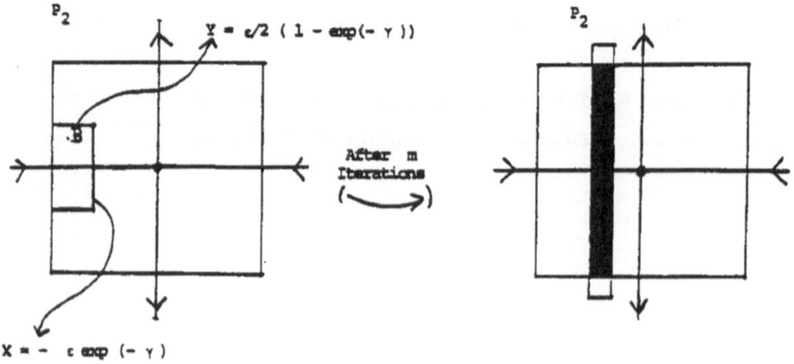

Note the height of B = width of A = $\varepsilon(1-\exp(-\gamma))$, and is independent of the number of iterations that the set was at P_3, due to transversality. Thus B hits the top, or $y = \varepsilon$ in a fixed number of iterations. Next consider the set which stays in the neighborhood of P_2 through m, the shaded strip.

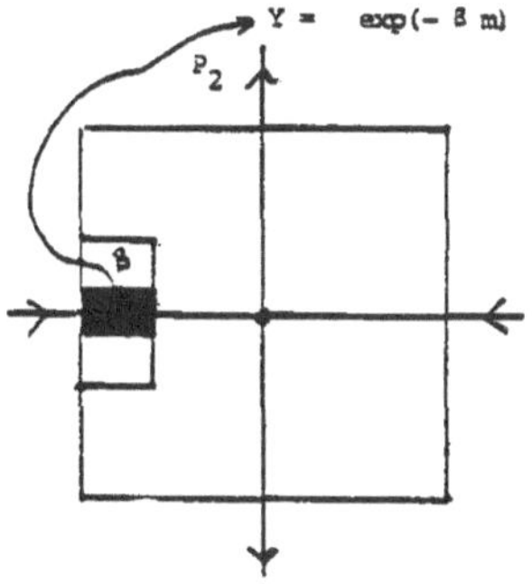

Next translate the shaded set back to P_3. Then 'go back' n iterations at P_3 :

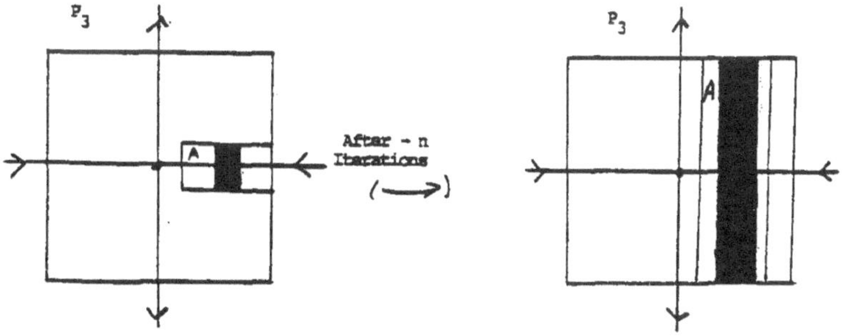

We get the width (shaded strip) = $2\varepsilon \exp(-\beta m) \exp(-\gamma n)$.

So the area (shaded strip) = $4\varepsilon^2 \exp(-\beta m) \exp(-\gamma n)$.

Choose $C = \text{Min}\{\beta,\gamma\}$, to get: Area $\leq 4\varepsilon^2 \exp C \exp(-CN)$, for $N = m + n + 1$. Considering the number of ways to add $n + m$ we have $A \leq \sum_{n=0}^{N} 4\varepsilon^2 \exp C \exp(-CN)$,

so $A \leq 4\varepsilon^2 \exp C\,(N + 1) \exp(-CN)$. We note that this is an example of the Morse Smale case.

§ 2. An Example Without Transversal Intersection

Consider again the gradient system on the torus $T^2 \subset R^3$, but suppose T^2 is not tilted back. Thus we have a nontransversal saddle connection, $W^u(P_3) = W^s(P_2)$, see Fig. 4.

FIG. 4

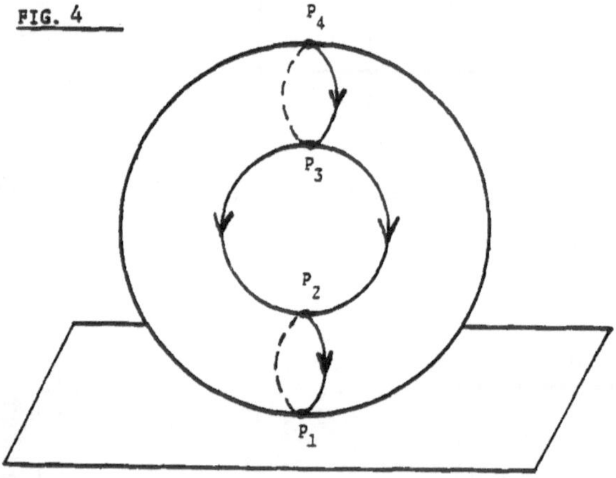

We no longer have the " turning over " property of transversal intersection. Thus the number of iterations for a set to leave the neighborhood of P_2 depends on the number of iterations during which the set remained in the neighborhood of P_3. This is illustrated in Fig. 5.

FIG.5

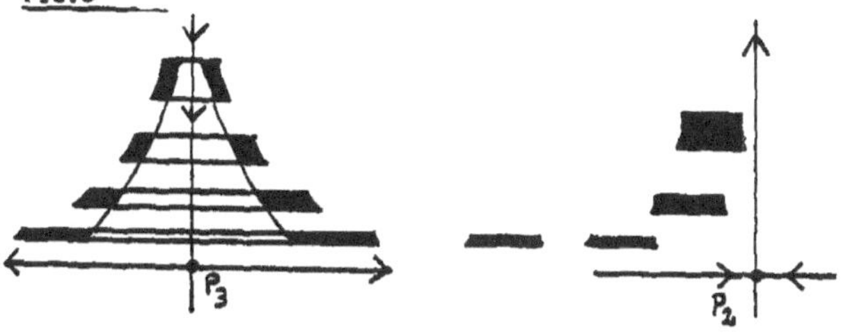

Consider again linearized neighborhoods of P_3 and P_2 and the same diffeomorphisms as in the transversality example. The following sequence of pictures are analogous to the previous example.

In the Neighbornood of P_3 we have the same sequence of illustrations as in the previous example. So consider the linearized neighborhood of P_2 :

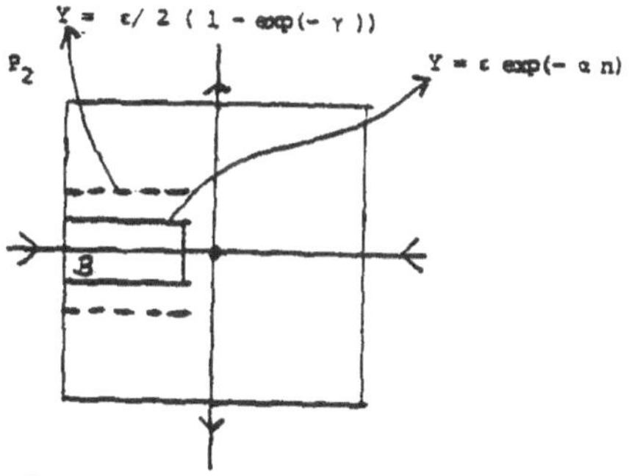

Here the height of B = height of A . Now the height of B is dependent on how long the set stays in the nieghborhood of P_3 .

So except for small values of n , B is not as high as in the transversality case. Thus it takes a number K for B to reach the same height of the transversal intersection case. We refer to K as the 'hung up' parameter . This ' hung up ' parameter causes a slower overall rate . Consider the shaded area, the set which stays in a neighborhood of P_2 through m + k iterations.

<u>Neighborhood of P_2</u> :

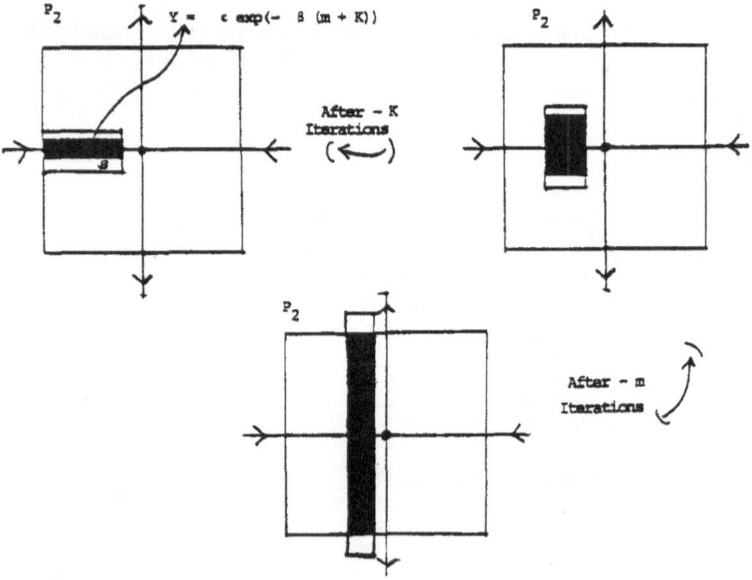

Next transfer the set back to P_3, and 'go back' n iterations.

Neighborhood of P_3 :

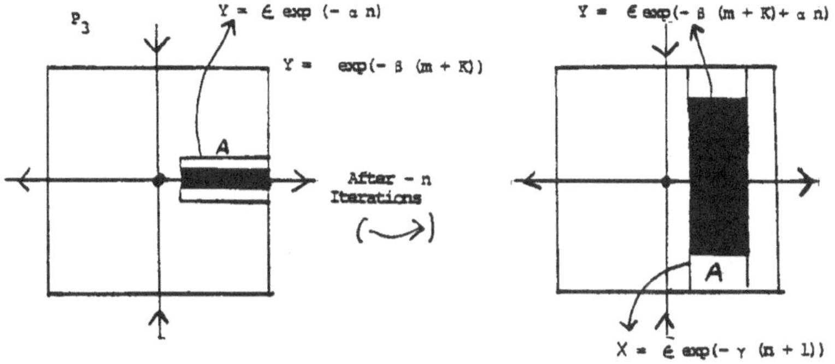

We get the area of the shaded set to be :

Area = $\varepsilon^2 \exp(-\beta(m+K)) \exp(\alpha m) \exp(-\gamma m)\,(1-\exp(-\gamma))$,

for $N = 1 + n + m + K$. By solving the following equation we get an approximation for K : $\exp(-\alpha n)\exp(\beta K) = 1/2(1-\exp(-\gamma)) =$ which yields : $K = \varepsilon'(\alpha/\beta)\,n$, or $K \quad (\alpha/\beta)\,n$.

So : $\exp(-\beta K) = \exp(-\alpha n)$. With $n' = n(1+\alpha/\beta)$,

Area $< \exp(-\beta m)\exp(-\gamma/(1+\alpha/\beta)\,n')$ for $n' + m = n + K + m = N - 1$.

Considering the number of ways to add $n + K + m$ we have :

$$A = \sum_{n'=0}^{N} C \exp(-\gamma/(1+\alpha/\beta)\,n')\exp(-\beta m).$$

Choosing $\gamma/(\alpha+\beta) < 1$, yields $C < \min\{\gamma, \beta\}$. Thus the rate of Volume decay is slower than in the transversal intersection case.

§ 3 The Results for C^2 Axiom A Diffeomorphisms With No Cycles

We begin with some definitions :

Definition: Let $f \in \mathrm{Diff}(M)$. A filtration for f is a sequence of compact manifolds with boundary such that

$M = M_k \supset M_{k-1} \supset \ldots \supset M_1 \supset M_o = \emptyset$, $\dim M_i = \dim M$, $i = 1, \ldots, k$, and $f(M_k) \subset \mathrm{int}\, M_k$.

Given a filtration, $K_i = \bigcap_{n \varepsilon Z} f^n (M_i - \text{int } M_{i-1})$ is the maximal invariant set contained in $(M_i - M_{i-1})$. If $K_i = \Omega \bigcap (M_i - M_{i-1})$, for all i, we say that the filtration is a fine filtration for f. Finally, if we are given closed invariant sets which are disjoint Λ_1 , ..., Λ_k, we say $M = M_k \supset M_{k-1} \supset \ldots M_1 \supset M_o = \emptyset$ is a filtration for Λ_1 ,..., Λ_k if $\Lambda_i = K_i$.

Note: We will denote $(M_i - \text{int } M_j)$ by M [i,j].

<u>Definition</u>: Let $(M_i)_{i=0}^{r}$ be a filtration adapted to f and let Λ_k be the largest invariant set in M_k - int. M_{k-1}. Then $B_{\Lambda_k} (\varepsilon,N) = \{x \varepsilon M : d(f^k x, \Lambda_k) \leq \varepsilon \text{ for } K = 0,1,\ldots,N\}$

<u>Definition</u>: $S(i,j,N) = f^{-N} (f^N (M_i - \text{int } M_k) - \text{int } M_j)$ or $S(i,j,N) = M_i - f^{-N} (\text{int } M_j)$.

To Illustrate S(i,j,N) , consider S(3,1,N) in Fig. 6.

<u>Fig. 6</u>

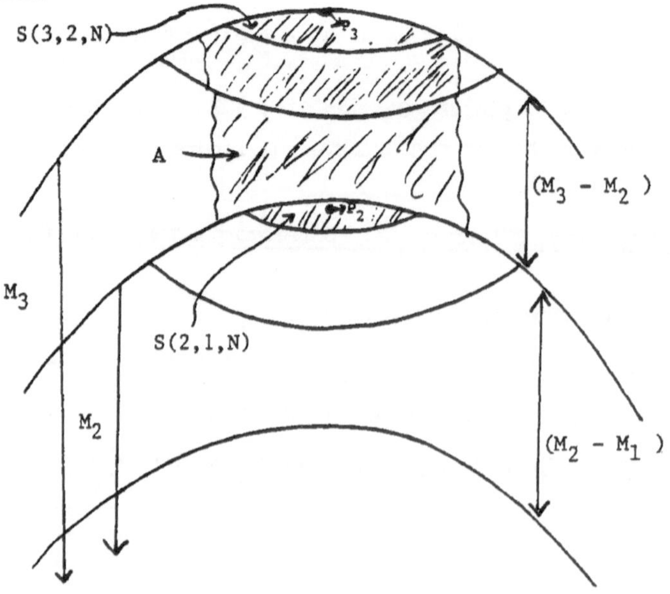

Proposition 3.1 : for small ε there exist L_i, n_i such that

$$S(i,i-1, N+L_i) \subset f^{-n_i} (B_{\Lambda_k} (\varepsilon,N)).$$

where $S(i,i-1,n) = M_i - f^{-n} (\text{int } M_{i-1})$.

Proof: Because $B_{\Lambda_k} (\varepsilon)$ is a neighborhood of Λ_k we know by a proposition from M. Shub's paper [9] that there exist L_i, n_i such that if $x \in (M_i - \text{int } M_{i-1})$ and $f^{Li} x \in \text{int } M_{i-1}$, then $x \in f^{-n_i} (B_{\Lambda_k} (\varepsilon)$. The proof follows.

Theorem 3.2 : Let f be a diffeomorphism of M with $\mathcal{M} = \{M_i\}_{i=0}^r$ a filtration adapted to f with Λ_k the largest invariant set in M_k - int M_k

Assume that $\mu(B_{\Lambda_k} (\varepsilon,N)) \leq D_k \exp(-C_k N)$,

with D_k and C_k positive constants for $k=2,3,\ldots,r$.

Then for $i > j$, $i \leq r$, $j \geq 1$, there exist positive constants $w_{i,j}$, $D_{i,j}$ and $C_{i,j}$ such that $N \geq w_{i,j}$ implies $\mu (S(i,j,N)) \leq D_{i,j} \exp (-C_{i,j}N)$.

Proof: Give $i > 1$ we will first prove the result for $j = i-1$ and then extend the result by induction. By Proposition 2.1 we know that $S(i,i-1, n + L_i) \subset f^{-ni} (B_{\Lambda_i} (\varepsilon,n))$. But

$$\mu(f^{-n_i} (B_{\Lambda_i}(\varepsilon ,n)) = \int_{B_{\Lambda_i}(\varepsilon,n)} \text{Jac } f^{-n_i} du \leq J_i \, \mu(B_{\Lambda_i}(\varepsilon , n)).$$

where $J_i = \max_{M[i,i-1]} (\text{Jac} \mid f^{-ni} \mid)$. Thus we have:

$$\mu(S (i,i-1,n + L_i)) \leq J_i (B_{\Lambda_i} (\varepsilon,n)) \leq J_i D_i \exp (-C_i n),$$

Take $N = n + L_i$ and require $N \geq L_i$, in which case we have:

$\mu(S\ (i,i-1,N)) \leq J_i\ D_i \exp(C_i L_i) \exp(-C_i N)$.

Let $D_{i,i-1} = J_i\ D_i \exp(C_i\ L_i)$ and $C_{i,i-1} = C_i$, and $w_{i,i-1} = L_i$;

We have the desired result. Assume, by way of induction, that the statement holds for $\mu(S\ (i-1,j,N))$, for $i \geq j$, $j \geq 1$, and $N \geq w_{i-1,j}$. Now consider: $S\ (i,j,N)$.

$$S\ (i,j,N) = S(i-1,j,N) \cup S(i,i-1,N) \cup A_N \text{ ,where}$$

$$A_N = \{ x \in M\ [i,i-1] \ : f^N (x) \in M\ [i-1,j]\}.$$

Furthermore decompose $A_N = F \cup E$, where :

$$F = \{x \in A_N : f^{L_i} (x) \notin M[i,i-1] \} \quad \text{and}$$

$$E = \{x \in A_N : f^{L_i} (x) \in M[i-1,j] \}.$$ Thus for $N \geq L_i$

(1) $S(i,j,N) = S(i-1,j,N) \cup S(i,i-1,N) \cup F \cup E$.

From the induction hypothesis:

(2) $\mu S(i-1,j,N)) \leq D_{i-1,j} \exp(-C_{i-1,j}\ N)$ for $N \geq w_{i-1,j}$

A similar result holds for $\mu(S(\ i,i-1,N))$.

For E, observe that $f^{L_i}\ (\ E\ ) \subset S(i-1,j,\ N-\ L\ )$,

so that: $\mu\ (\ f^{L_i}\ (\ E\ ))$ can be estimated.

$$\mu(f^{L_i} (E)) \geq J_i^{L_i}\ \mu(\ E\ ) \text{ , where } J_i^* = \min_{M[i,i-1]} \{|\text{Jac } f|\}.$$

Thus we have:

(3) $\mu(E) \leq J_i^{*-L_i}\ \mu(S(i-1,j,N-L_i) \leq J^{*-L_i}\ (D_{i-1,j} \exp(-C_{i-1,j}N))$,

for $N \quad L_i + w_{i-1,j}$.

Next I claim the following :

<u>Lemma 3.3</u> : For $\delta > 0$ small, and for all

$N \geq \max\{2(L_i + w_{i-1,j}), qw_{i,i-1}\}$ we find that :

$$\mu(F) \leq K(\exp -C_i^{\#}(1-\delta)N)$$

$$C_i^{\#} = \min\{C_i/q, C_{i-1,j}(1-\delta)\}, \text{ and}$$

$q \in Z^*$ is chosen so that it satisfies :

$q \geq 1/\delta\,(1 + (C_i^* / C_{i-1,j}))$, with $C_i^* = \log(\min_{M[i,i-1]} |\mathrm{Jac}\, f|)$.

<u>Proof</u> : Recall that $F = \{x \in A_N : f^{L_i}(x) \in M[i,i-1]\}$.

Let n = (greatest integer $\leq N/q$)

We now decompose F into $F_1 \cup F_2$ where :

$$F_1 = \{x \in F : f^{n/q+L_i}(x) \notin M[i,i-1]\} \text{ and}$$

$$F_2 = \{x \in F : f^{n/q+L_i}(x) \in M[i,i-1]\},$$

with q defined as in the hypotheses.

We have : $F_2 \subset f^{-n_i}(B_{\Lambda_i}(\varepsilon, n/q))$, which yields

an estimate for $\mu(F_2)$.

$$\mu(F_2) \leq J_i(B_{\Lambda_i}(\varepsilon,n/q) \leq J_iD_i \exp(-C_in)$$

$$\leq J_iD_i \exp(-C_i[w/q-1])$$

$$\leq J_iD_i \exp[(-C_i/q)N] \exp C_i$$

$$\leq J_iD_i^* \exp(-C_i/q)N.$$

For $n \geq w_{i,i-1}$ which holds if $N/q \geq w_{i,i-1}$.

Let $F_1(k)$ be such that $f^{n+L_i-k}F_1(k) \subset M[i-1,j]$ and $f^{n+L_i-k}F_1(k) \subset M[i-1,j]$ and $f^{n+L_i-k-1}F_1(k) \subset M[i,i-1]$.

then $F_1 = \bigcup_{k=0}^{n} F(k)$ and $\mu(F) = \sum_{k=0}^{n} \mu(F_1(k))$.

Now notice that $f^{n+L_i-k}\, F_i\,(k) \subset S(i-1,j,N-n-L_i+k)$, thus $\mu(f^{n+L_i-k}\, F_1(k)) \leq D_{i-1,j}\exp(-C_{i-1,j}(N-n-L_i+k))$.

Also

$$\mu[f^{n+L_i-k}\, F_1(k)] = \int_{f^{n+L_i-k} F_1(k)} d\mu \geq (J_i^*)^{n+L_i-k}\,\mu(F_1(k))$$

where $J_i^* = \min_{M[i,i-1]} \{|\mathrm{Jac}\, f|\}$. Thus

$$\mu(F_1(k)) \leq (J_i^*)^{k-n-L_i}\, D_{i-1,j}\exp(-C_{i-1,j}[N-n-L_i+k],$$

$$= ([J_i^*\exp(-C_{i-1,j})]^{-L_i} D_{i-1,j})\,(J_i^*)^{k-n}\exp(-C_{i-1,j}[N-n+k]).$$

This estimate holds for $N-n-L_i+k \geq w_{i-1,j}$ which is true if $N \geq q/q-1\,(L_i+w_{i-1,j})$ which is insured by taking $N \geq 2(L_i+w_{i-1,j})$. Now let $C_i^* = -\mathrm{Log}\, J_i^*$ and $\bar{D} = [\,J_i^*\exp(-C_{i-1,j})]^{-L_i} D_{i-1,j}$.

Then: $\mu(F_1\,(k)) = \bar{D}\exp[-C_{i-1,j}(N-n+k)]\exp[-C_i^*\,(k-n)]$

$= \bar{D}\exp[\,-C_{i-1,j}(N-1/q)[1+C_i^*/C_{i-1,j}]\,(n-k)q\,]$.

Let $1/q\,[\,1+C_i^*/C_{i-1,j}] = \delta$, then $\mu(F_1(k)) \leq \bar{D}\exp[-C_{i-1,j}(N-\delta N)]$.

Hence $\mu(F_1) \leq \bar{D}\,(n+1)\exp[-C_{i-1,j}(1-\delta)N]$ so

$(F) \leq \bar{D}\,(n+1)\exp[C_{i-1,j}(1-\delta)N] + J_i D_i^*\,\exp[-C_{i/q}N]$.

Next let $C_i^{\#} = \min\{C_i/q,\, C_{i-1,j}(1-\delta)\}$ and using the fact that algebraic growth is dominated by exponential decay, then there exists a constant K^* such that: $\mu(F) \leq K^*(\exp(-C_i^{\#}(1-\delta)N)$ for $N \geq \max\{2(L_i+w_{i-1,j})\,,\, qw_{i,i-1}\}$. This concludes the proof. Resuming the proof of Theorem we are ready to state the "total " estimate. The estimate is obtained by using Lemma and the estimates (2), and (3).

$$\mu(S(i,j,N)) \leq K_{i,j} \exp(-C_{i,j}(1-\delta)N),$$

for $N \geq \max\{qw_{i,i-1}, 2(L_i + w_{i-1,j})\}$, with the $C_{i,j}$ defined recursively by $C_{i,j} = C_i^{\#}$, where $C_i^{\#}$ has been previously defined. This completes the proof of the theorem.

<u>Definition:</u> Λ_j is an attractor if for some U a neighborhood of Λ_j, $f(U) \subset U$.

We see from Theorem 2.2.1 that whenever the diffeomorphism f has a filtration $\mathcal{M}$ such that the sets Λ_k have local volume estimates, Λ_k not an attractor, then a global estimate holds. We now show that this is the case for C^2 Axiom A diffeomorphisms with no cycles.

<u>Definition</u> : Let $f \in \mathrm{Diff}(M)$. Then f satisfies Axiom A if and only if :

(a) $\Omega(f)$ is hyperbolic

(b) $\Omega(f) = \mathrm{cl}\,\mathrm{Per}\, f$

<u>Definition</u> : We say that a diffeomorphism f has an Ω decomposition if $\Omega(f)$ may be written as the finite disjoint union of closed invariant sets for f, $\Omega(f) = \Omega_1 \cup \ldots \cup \Omega_k$. Moreover, if $f|\Omega_i$ is topologically transitive for all i, i.e. $f|\Omega_i$ has a dense orbit for all i, then we say that f has a spectral decomposition.

Now given $f \in \mathrm{Diff}^r(M)$ with an Ω - decomposition, we may define a relation on the Ω_i as follows : $\Omega_i > \Omega_j$ if $(W^u(\Omega_i) - \Omega_i) \cap (W^s(\Omega_j) - \Omega_j) \neq \emptyset$, i.e., there is an x which comes from Ω_i and goes to Ω_j.

Definition : Let $f \in \mathrm{Diff}^r(M)$ have an Ω - decomposition $\Omega(f) = \Omega_1 \cup \ldots \cup \Omega_k$. We say that f has no cycles if $\Omega_{i_0} > \Omega_{i_1} > \ldots > \Omega_{i_j} = \Omega_{i_0}$ is impossible for any $j \geq 1$.

If f has an Ω - decomposition and no cycles we may reorder the Ω_i by defining $\Omega_i > \Omega_j$ iff there exists a sequence $\Omega_i > \Omega_{k_1} > \ldots > \Omega_{k_i} > \Omega_j$ and finally we may reindex the Ω_i such that $\Omega = \Omega_1 \cup \ldots \cup \Omega_k$ and $i < j$ implies $\Omega_i \not> \Omega_j$. Henceforth, we will assume that the Ω_i are indexed as above for any f with an Ω - decomposition and the no cycle property.

Spectral Decomposition Theorem (Smale) : If f satisfies Axiom A, f has a spectral decomposition.

Definition: We say that f is an Axiom A diffeomorphism with no cycles if: (a) f satisfies Axiom A and (b) the spectral decomposition of $\Omega(f)$ has no cycles.

Now in [9] we find the following:

Theorem: Let $f \in \mathrm{Diff}^r(M)$. Then f has a fine filtration if and only if f has an Ω decomposition with no cycles.

Therefore, we know that if f is an Axiom A, no cycle diffeomorphism, then f has a fine filtration $\mathcal{M} = \{M_i\}_{i=0}^r$ such that the largest invariant set Λ_k in $M_k - \mathrm{int}\, M_{k-1}$ is the corresponding set in the spectral decomposition of $\Omega(f)$. Λ_k is called a basic set for f, as stated in Bowen, [1].

The following is a simple modification of a result found in Bowen and Ruelle [2]. The difference is that they prove an estimate for $\bigcup_{x\varepsilon\Lambda_j} B_x\ (\varepsilon,N)$ which is not the same as $B_{\Lambda_j}\ (\varepsilon,N)$.

<u>Theorem 3.4 :</u> Let f be a C^2 Axiom A diffeomorphism, with no cycles. Let Λ_j be a C^2 basic hyperbolic set for f, Λ_j not an attractor. Then for ε small and K_j, C_j both positive constants we find: $\mu(B_{\Lambda_j}\ (\varepsilon,N)) \leq K_j \exp\ (-C_jN)$.

Before proving this Theorem, we will prove the following proposition.

<u>Proposition 3.5 :</u> For all $\gamma > 0$, there exists $\varepsilon > 0$ such that $B_{\Lambda_j}\ (\varepsilon,N) \subset \bigcup_{x\varepsilon\Lambda_j} (B_x\ (\varepsilon+\gamma,N))$.

<u>Proof :</u> Let $\gamma > 0$ be given. Take $\beta = \gamma/2$. By Bowen's Shadow Lemma [1] , the local product structure of Λ_j yields such that every α pseudo orbit in Λ_j is β shadowed by some y^* in Λ_j. Furthermore take $\varepsilon = \alpha/\ (K_f + 1)$, where : $K_f = \underset{\substack{x,y\ \varepsilon\ M \\ x \neq y}}{\text{Max}} \{\ d(f(x))\ ,\ f(y)/\ d(x,y)\}$. If $x\ \varepsilon\ B_{\Lambda_j}(\varepsilon,N)$ there exists $y_o\ ,\ y_1\ ,\ \dots\ ,\ y_N\ \varepsilon\ \Lambda j$, such that :

$d(f^k\ (x)\ ,\ y_k) \leq \varepsilon$, for $0 \leq k \leq N$.

<u>Claim</u>: $y_o, y_1, \dots, y_k$, is an α pseudo orbit.

<u>Proof</u>:

$$d(y_{k+1}\ ,\ f(y_k)) \leq d(y_{k+1}\ ,\ f^{k+1}\ (x)) + d(f^{k+1}(x), f(y_k)) \leq (K_f+1)\varepsilon$$

Therefore: $d(y_{k+1}\ ,\ f(y_k)) \leq \alpha$.

So there exists $y^*\ \varepsilon\ \Lambda_j$ which shadows $y_o, \dots, y_k$

i.e. $d(f^k\ (y^*), y_k) \leq \gamma$ for $0 \leq k \leq N$. Then, for $0 \leq k \leq N$,

$$d(f^k(y^*), f^k(x)) \leq d(f^k(y^*), y_k) + d(y_k, f^k(x)) \leq (\gamma+\varepsilon)$$

Therefore: $x \in B_{y^*}(\gamma+\varepsilon, N)$. Hence:

$B\Lambda_j(\varepsilon, N) \subset \bigcup_{x\varepsilon\Lambda_j} B_x(\gamma+\varepsilon, N)$ as desired.

We now return to Theorem 3.4 and begin its proof .

<u>Proof</u> : Let γ be small enough so that we can apply a local estimate $B_x(2\gamma, N]$, i.e., the one obtained from the Local Volume Lemma of [1] , mentioned in Chapter 1. Now let $\varepsilon \leq \gamma$

be small enough so that by Proposition 2.2.2 we have :

$$B_{\Lambda_j}(\varepsilon, N) \subset \bigcup_{x\,\varepsilon\Lambda_j} B_x(\gamma+\varepsilon, N).$$

Since $\gamma+\varepsilon \leq 2\gamma$, and Λ_j is not an attractor, and is a C^2 basic hyperbolic set, then for $(\varepsilon+\gamma)$ sufficiently small, we get by application of Proposition 4.8 and Theorem 4.11 of Bowen [1], that the following is true :

$$\limsup(1/N \log \mu(\bigcup_{x\,\varepsilon\Lambda_j} B_x(\gamma+\varepsilon, N)) = P(f|\Lambda_j, \phi^{(u)}) < 0 .$$

$P(f|\Lambda_j, \phi^{(u)}) \geq 0$, here is the topological pressure of f defined in Bowen [1]. By Theorem 4.11 of [1] we know that $P(f|\Lambda_j, \phi^{(u)}) = 0$ is equivalent to Λ_j being an attractor.

Now let the above lim sup = δ , then for all $\alpha > 0$ we get: $1/N \log \mu(\bigcup_{x\varepsilon\Lambda_j} B_x(\gamma+\varepsilon, N)) \leq (\delta+\alpha)$ for all but finitely many N. Now if one chooses α such that $(\delta+\alpha) < 0$, we get: $\log \mu(\bigcup_{x\varepsilon\Lambda_j} B_x(\gamma+\varepsilon, N)) \leq N(\delta+\alpha) < 0$, or

$\mu(B_{\Lambda_j}(\varepsilon, N)) \leq \mu(\bigcup_{x\varepsilon\Lambda_j} B_x(\varepsilon+\gamma, N)) \leq \exp(N(\delta+\alpha))$ for all but a finitely many N.

In this case we have that $\delta = P(f|\Lambda_j, \phi^{(u)})$ and $(\delta+\alpha) < 0$.

Remark: Thus the Theorem for f a C^2 Axiom A, diffeomorphism with no cycles is proven. This can be stated as:

<u>Theorem 3.6</u> Let M be a compact Riemannian manifold. Let μ be the measure induced by the Riemannian metric on M. Let f be a C^2 Axiom A diffeomorphism with no cycles, defined on M. Then f has a filtration $\mathcal{M} = \{M_i\}_{i=1}^{r}$ for which all attractors f are in M_1. Let $S(i,j,N)$ be as defined before. Then for all $\delta > 0$ there exists $K = K(\delta)$ such that :

$$\mu(S(i,j,N)) \leq K \exp(-C_{i,j}(1-\delta)N),$$

for large N, where $C_{i,j}$ are defined recursively by :

$$C_{i,j} = \min\{C_i/q \;, C_{i-1,j}(1-\delta)\},$$

where C_i can be taken to be any number smaller than

$$-P(f|_{\Lambda_i}, \phi^{(u)}), \text{ where } P(f|_{\Lambda_i}, \phi^{(u)})$$

was defined earlier, and $q \in Z^+$ is chosen so that q satisfies :

$$q \geq (1 + C_i^* / C_{i-1,j}) / \delta \quad , \text{ where}$$

$$C_i^* = \log (\min_{M[i,i-1]} \{|\mathrm{Jac}_{D_{P_i}} f|\}).$$

To obtain our constants $C_{i,j}$ we begin with $C_{i,1} = C_1$ and proceed to calculate $C_{2,1}$, $C_{3,1}$ etc., up to $C_{r,1}$ using $C_i < -P(f,\phi^{(u)})$, and the procedure outlined in the proof.

The best way to follow this procedure is: Take

$$C_{i,j} = \max_{0<\delta<1} \min\{C_i/q,\ C_{i-1,j}\ (1-\delta)\}$$

where q is the least integer $\geq 1/\delta(1+C_i^* / C_{i-1,j})$.

Using Lemma 3.3 we can state: for all $\delta' > 0$ there exists a $K = K(\delta')$ such that:

$$\mu(F) = K \exp(-C_{i,j}(1-\delta')N) \quad , \text{ for all } N =$$

The estimate obtained is of the form : $C(n+2)\exp(-C_{i,j}N)$. Using δ' to control the algebraic growth yields :

$C(n+2) \exp(-C_{i,j}N) = K \exp(-C_{i,j}(1-\delta')N)$, as desired.

§ 4 The Results For C^2 Axiom A Flows With No Cycles

An analogous result for Axiom A flows , to Theorem 3.6 is as follows :

Theorem 4.1 : Let ϕ_t be a C^2 Axiom A flow on M with no cycles. Then :

$\mu(S(i,j,T)) \leq D_{i,j} \exp(-C_{i,j}T)$ for all $i > j$, $i \leq r$, $j \geq 1$, $D_{i,j}$ and $C_{i,j}$ positive constants, and $T \geq w_{i,j}$, with the $C_{i,j}$ defined recursively as in Theorem 3.6 .

Proof : We know from Shub, [9] that ϕ_t has a filtration which we again assume to be $\mathcal{M} = \{M_i\}_{i=1}^{r}$ with all the attractors in M_1 . Here $S(i,j,T)$ is defined analogously to $S(i,j,N)$.

Fix $t = 1$ and consider the diffeomorphism ϕ_1. This is strictly speaking not an Axiom A Diffeomorphism since ϕ_1 may not be hyperbolic on all of $\Omega(\phi_1) = (\phi_t)$. However it does have a spectral decomposition into basic sets Λ_k. Furthermore the basic sets Λ_k have local product structure for flows (see Pugh & Shub [8], & Bowen [3]), and thus we have a shadowing lemma for ϕ_1.

By Proposition 4.4 and Theorem 5.6 of Bowen & Ruelle [2] the basic sets Λ_k for ϕ_1 have a volume estimate similarly related to topological pressure, $P(F|_{\Lambda i}, \phi^{(u)})$.

Thus by Theorem we have : $\mu(S(i,j,N)) = D_{i,j}\exp(-C_{i,j}N)$. For T between n and $N+1$ we get : $S(i,j,T) \subset S(i,j,N)$. $\mu(S(i,j,T)) \leq D_{i,j}\exp(C_{i,j}(T-N))\exp(-C_{i,j}T)$. Define $D^*_{i,j}$ to be : $D_{i,j}\exp(C_{i,j})$ to yield : $\mu(S(i,j,T)) \leq D^* \exp(-C_{i,j}T)$, where the $C_{i,j}$ are defined recursively as in Theorem and C_i can be taken to be any positive number smaller than $-P(F, \phi^{(u)})$.

§ 5 A Discussion Concerning C^1 vs C^2, and Some Open Questions

Since we have proved a Theorem in [11] for a class of C^1 diffeomorphisms, the question arises as to under what conditions Theorem 3.4 holds for C^1 diffeomorphisms of M ? We know from Fried and Shub [5] , that for each $x \in \Lambda$ a basic set for a diffeomorphism f we have a local volume lemma. However [5] does not provide a local estimate for $B_\Lambda(\varepsilon, N)$ only for $B_x(\varepsilon, N)$. While this suffices to yield their results on entropy, it does not yield the uniform $B_\Lambda(\varepsilon, N)$ estimate required herein. In fact Bowen [4] has given an example of a C^1 horseshoe with $\mu(W^s_\Lambda) > 0$ for Λ a basic set which is not an attractor.

For a C^1 diffeomorphism , not Morse-Smale, which has finite Ω Theorem 3.4 holds. In general, however what additional hypothesis beyond Axiom A and no- cycles is needed to insure that Theorem 3.4 holds for a diffeomorphism of M is an open question.

A related open question concerns extending Theorem 3.1 of [1] to a larger class then Morse-Smale diffeomorphisms and flows, and still getting the exponential constant to be as in this Theorem. It may be that Axiom A, and strong transversality suffices.

BIBLIOGRAPHY

1 R . Bowen, Equilibrium States and the Ergodic Theory of Anosov Diffeomorphisms , Springer Lecture Notes in Math. 470 (1975).

2 R. Bowen, and D. Ruelle , The Ergodic Theory of Axiom A Flows, Invent. Math. 29 (1975), pp. 181-202.

3 R. Bowen, Periodic Orbits of Hyperbolic Flows, Amer. J. Math. Vol. 94, (1972), pp. 1 - 37.

4 R. Bowen, A Horseshoe With Positive Measure, Invent. Math. 29 (1975), pp.203-204.

5 D. Fried and Michael Shub , Entropy, Linearity, and Chain Recurrence, Extrait des Publications Mathematiques, No. 50 (1978) .

6 J. Palis, On Morse-Smale Synamical Systems, Topology, Vol. 8, pp. 385-405. Pergamon Press, (1969).

7 J. Palis and S. Smale, Structural Stability Theorems, Proc. Sympos. Pure Math., Vol. 14, Amer. Math. Soc., Providence, R.I., (1970), pp. 223-232.

8 C. Pugh and M. Shub , The Ω - Stability Theorem for Flows, Invent. Math. 11 (1970), 150-158.

9 M. Shub , Stability and Genericity for Diffeomorphisms, Dynamical Systems (Peixoto, ed.), Academic Press (1973), pp. 493.

10 S. Smale, Differentiable Dynamical Systems ,Bull, Amer. Math. Soc. 71 (1967),pp.747.

11 H. Wisniewski, Rate of Approach to Minima and Sinks the Morse-Smale case, Transactions of the Am.Math. Society,(to appear)

Helena S. Wisniewski

Math.Department
Rochester Institute of Technology
Rochester, New York

C^1-conjugaison des difféomorphismes du cercle

J.C. YOCCOZ

0. Introduction

0.1. Le but de cet article est de donner une démonstration simple d'une conjecture de V.I. Arnold ([A]), démontrée par M.R. Herman ([H]): un difféomorphisme du cercle lisse et préservant l'orientation est, pour presque tout nombre de rotation, différentiablement conjugué à une rotation. Plus précisément, notre résultat est le suivant:

THÉORÈME. Soit f un difféomorphisme du cercle de classe C^3, préservant l'orientation. On suppose que le nombre de rotation α de f est irrationnel, et vérifie la condition arithmétique:

$$(C) \qquad \exists\, \beta < 1, \quad \exists\, K > 0, \quad \forall\, p/q \in \mathbb{Q}/\mathbb{Z}, \quad |\alpha - p/q| \geq \frac{K}{q^{2+\beta}} \;;$$

alors, il existe un difféomorphisme du cercle h de classe C^1, préservant l'orientation, qui conjugue f à la rotation R_α d'angle α: $f = h R_\alpha h^{-1}$.

Corollaire. Si, de plus, f est un difféomorphisme de classe C^∞, et α vérifie la condition arithmétique plus forte:

$$(C') \qquad \exists\, \beta < \sqrt{5} - 2, \quad \exists\, K > 0, \quad \forall\, p/q \in \mathbb{Q}/\mathbb{Z}, \quad |\alpha - p/q| \geq \frac{K}{q^{2+\beta}} \;;$$

alors h est aussi un difféomorphisme de classe C^∞.

Remarques: a. Presque tout nombre (pour la mesure de Lebesgue) vérifie la condition (C'); le corollaire implique donc la conjecture d'Arnold. Herman obtient la même conclusion sous une condition arithmétique, notée (A), d'énoncé plus technique et qui est plus restrictive que (C'). En particulier, les nombres algébriques irrationnels et le nombre e vérifient la condition (C'), mais ne vérifient pas (nécessairement) la condition (A).

b. Il n'est pas difficile d'améliorer la borne $\sqrt{5}-2$ du corollaire; mais la longueur de la démonstration s'en ressent, sans que la portée du résultat en soit notablement augmentée.

0.2. Notations et rappels.

Notre cercle est $\mathbb{T}^1 = \mathbb{R}/\mathbb{Z}$. On note $\mathrm{Diff}_+^r(\mathbb{T}^1)$ le groupe des difféomorphismes du cercle de classe C^r qui préservent l'orientation. La rotation d'angle α est notée R_α.

Pour les propriétés élémentaires et moins élémentaires des groupes $\mathrm{Diff}_+^r(\mathbb{T}^1)$ et de l'application "nombre de rotation" $\rho: \mathrm{Diff}_+^0(\mathbb{T}^1) \to \mathbb{T}^1$, nous renvoyons le lecteur à [H].

L'orientation naturelle du cercle le munit d'un ordre cyclique, ce qui donne un sens à la notion d'intervalle sur $\mathbb{T}^1$.

Pour un réel x, $\|x\|$ designe la distance de x au plus proche entier; pour $\alpha \in \mathbb{T}^1$, on pose $\|\alpha\| = \|\tilde{\alpha}\|$, où $\tilde{\alpha} \in \mathbb{R}$ est une préimage quelconque de α. Donc $\|\alpha\|$ est la distance de α à 0 dans la métrique de $\mathbb{T}^1$ héritée de $\mathbb{R}$.

Soit α un nombre réel irrationnel. On peut définir la suite des réduites de α, notée généralement p_n/q_n $(p_n \wedge q_n = 1,\ q_n > 0)$, qui sont les quotients successifs de son développement en fraction continue. Nous utiliserons les propriétés suivantes des réduites:

- $0 < (-1)^n \left(\alpha - \frac{p_n}{q_n}\right) < \frac{1}{q_n q_{n+1}} \leq \frac{1}{q_n^2}$ pour $n \geq 0$;
- si $j \in \mathbb{Z}$, $|j| < q_{n+1}$, alors $\|j\alpha\| \geq \|q_n\alpha\|$;
- pour $n \geq 0$, $\|q_n\alpha\| = a_{n+2}\|q_{n+1}\alpha\| + \|q_{n+2}\alpha\|$, avec $a_{n+2} \in \mathbb{N}^*$.

Soit $\beta \geq 0$; on dira que le nombre α satisfait à une condition diophantienne d'ordre β (C_β), s'il existe $K > 0$ tel que pour tout rationnel p/q, on ait:

$$|\alpha - p/q| \geq \frac{K}{q^{2+\beta}} .$$

Pour les propriétés des réduites et une discussion des conditions arithmétiques, voir [L] et [H, ch V]. Les nombres qui vérifient toutes les conditions (C_β), $\beta > 0$, forment un ensemble de mesure de Lebesgue pleine; les nombres qui ne vérifient aucune des conditions C_β (nombres de Liouville) forment un G_δ-dense de $\mathbb{R}$.

Soit (p_n/q_n) la suite des réduites d'un nombre irrationnel α; la suite des réduites de $\alpha + r$, $r \in \mathbb{Z}$, est $(p_n/q_n + r)$. Donc, lorsque $\alpha \in \mathbb{T}^1 - \mathbb{Q}/\mathbb{Z}$, on peut définir ses réduites $p_n/q_n \in \mathbb{T}^1$. En particulier la suite d'entiers naturels $(q_n)_{n\in\mathbb{N}}$ est déterminée par α.

Soient $f \in \mathrm{Diff}_+^o(\mathbb{T}^1)$, α le nombre de rotation de f. Lorsque α est irrationnel et q est le dénominateur d'une de ses réduites, on désigne par f^q-id l'application de $\mathbb{T}^1$ dans $\mathbb{R}$ définie comme suit: soient $\tilde{f}$ un difféomorphisme de $\mathbb{R}$ qui relève f, p/q la réduite du nombre de rotation (réel) de $\tilde{f}$ qui a pour dénominateur q; alors $\tilde{f}^q$-id-p est une application de $\mathbb{R}$ dans $\mathbb{R}$, $\mathbb{Z}$-périodique; l'application de $\mathbb{T}^1$ dans $\mathbb{R}$, obtenue par passage au quotient, est notée f^q-id et ne dépend pas du relevé $\tilde{f}$ de f.

Pour une application φ de $\mathbb{T}^1$ dans $\mathbb{R}$, on note $|\varphi|_o = \sup_{x\in\mathbb{T}^1} |\varphi(x)|$; si φ vérifie de plus une condition de Hölder d'exposant $\varepsilon \in]0,1[$, on pose:

$$|\varphi|_\varepsilon = \sup_{x\neq y} \left(\frac{|\varphi(x)-\varphi(y)|}{\|x-y\|^\varepsilon} \right) .$$

Si φ est différentiable, on note $D\varphi$ sa dérivée et $D^n\varphi$ sa dérivée d'ordre n.

0.2. Plan de la démonstration.

La méthode de démonstration et beaucoup d'idées se trouvent déjà dans [H].

Pour démontrer la C^1-conjugaison, il suffit (cf [H, ch IV]) de voir que les itérés de f forment une suite bornée dans la C^1-topologie. Soit $(q_n)_{n\in\mathbb{N}}$ la suite des dénominateurs des réduites du nombre de rotation de f. L'estimation de $|\mathrm{Log}\, Df^i|_o$, $i \in \mathbb{N}$, repose sur celle de $|\mathrm{Log}\, Df^{q_n}|_o$, $n \in \mathbb{N}$, par un procédé déjà employé par Denjoy et Herman [H, ch VIII].

La partie nouvelle est l'obtention d'une estimation "a priori" pour $|\mathrm{Log}\, Df^{q_n}|_o$, qui conduit rapidement au résultat désiré. C'est le contenu de la proposition 1:

Proposition 1. Soit $f \in \mathrm{Diff}^3_+(\mathbf{T}^1)$. Il existe une constante $C = C(f) > 0$ telle que, si le nombre de rotation de f est irrationnel et q est le dénominateur d'une de ses réduites, on ait:

$$|\mathrm{Log}\, Df^q|_o \leq C|f^q - id|_o^{1/2}.$$

Dans la démonstration de la proposition 1, les dérivées schwarziennes $Sf = \dfrac{D^3fDf - 3/2(D^2f)^2}{(Df)^2}$, déjà utilisées par Herman ([H, ch IX]), jouent un rôle crucial.

La technique de passage du théorème au corollaire est celle de Herman; elle repose sur trois ingrédients principaux: les inégalités de convexité de Hadamard, le théorème de conjugaison locale de Arnold et Moser, (cf [A], [H, Appendice]) et la possibilité, lorsque f est $C^{1+\varepsilon}$ conjugué à R_α $(0 < \varepsilon < 1,\ \alpha \in \mathbf{T}^1 - \mathbb{Q}/\mathbf{Z})$, de conjuguer différentiablement f à un difféomorphisme $C^{2+\varepsilon'}$ proche de R_α, pour tout $0 < \varepsilon' < \varepsilon$: cf [H, ch VII].

1. Démonstration de la proposition 1.

1.1. Dans cette section, on se donne un difféomorphisme du cercle f, de classe C^3, préservant l'orientation, dont le nombre de rotation α est irrationnel; et deux réduites consécutives de α, p/q et P/Q ($p \wedge q = 1$, $P \wedge Q = 1$, $Q > q \geq 1$). On suppose par exemple que $p/q < \alpha < P/Q$, de sorte que $x \in]f^Q(x), f^q(x)[$ pour $x \in \mathbb{T}^1$ (les modifications à apporter dans l'autre cas étant évidentes).

Les intervalles ordonnés $]x, f^q(x)[$, $]f^{-q}(x), f^q(x)[$ sont respectivement notés I_x et J_x.

<u>Lemme 1</u>: <u>Soit</u> $x \in \mathbb{T}^1$; <u>les intervalles</u> $f^i(I_x)$, $0 \leq i < Q$ <u>sont disjoints</u>; <u>les intervalles</u> $f^j(J_x)$, $0 \leq j < Q$, <u>forment un recouvrement de</u> $\mathbb{T}^1$.

<u>Démonstration</u>: Par le théorème de Denjoy ([H, ch VI]), f est C^0-conjugué à R_α. Les conclusions du lemme étant topologiques, on peut supposer que $f = R_\alpha$, $x = 0$.

Soit K la partie de $\mathbb{T}^1$ constituée par les points $x_i = i\alpha$, $0 \leq i < Q$. Pour démontrer le lemme, il suffit de vérifier que les distances entre points consécutifs de K sont au moins égales à $\|q\alpha\|$ et inférieures à $2\|q\alpha\|$.

Comme p/q, P/Q sont des réduites consécutives de α, on a, en utilisant les propriétés des réduites, pour $0 \leq i < Q$:

$$d(x_i, K-\{x_i\}) = \inf\{\|(j-i)\alpha\| \mid 0 \leq j \leq Q,\ j \neq i\} \geq \|q\alpha\|.$$

D'autre part, pour $0 \leq i < Q$ l'un des points x_{i+q}, x_{i+q-Q} appartient à K; ils vérifient $0 < x_{i+q} - x_i = \|q\alpha\|$, $0 < x_{i+q-Q} - x_i = \|(q-Q)\alpha\| \leq \|q\alpha\| + \|Q\alpha\| < 2\|q\alpha\|$. ■

1.2. On rappelle l'inégalité de Denjoy, valide sous nos hypothèses, ainsi qu'une inégalité similaire ([H, ch VI]).

Lemme 2. *On pose* $\operatorname{Log} C_1 = \operatorname{Var}(\operatorname{Log} Df)$. *Pour* $x \in \mathbb{T}^1$, $y \in I_x$, $0 \leq i < Q$, *on a*:

$$|\operatorname{Log} Df^q(x)| \leq \operatorname{Log} C_1;$$

$$|\operatorname{Log} Df^i(x) - \operatorname{Log} Df^i(y)| \leq \operatorname{Log} C_1.$$

Démonstration: Voir [H, ch VI] pour l'inégalité de Denjoy. On a:

$$\operatorname{Log} Df^i(x) - \operatorname{Log} Df^i(y) = \sum_{j=0}^{i-1} (\operatorname{Log} Df[f^j(x)] - \operatorname{Log} Df[f^j(y)]);$$

les intervalles $f^j(I_x)$ étant disjoints, on peut, si $y \in I_x$, majorer le second membre par $\operatorname{Var}(\operatorname{Log} Df)$. ■

1.3. On pose $M = \max_{x \in \mathbb{T}^1} [f^q(x)-x]$, $m = \min_{x \in \mathbb{T}^1} [f^q(x)-x]$.

Lemme 3. *Tout* $x \in \mathbb{T}^1$ *vérifie*:

$$\sum_{i=0}^{Q-1} [Df^i(x)]^2 \leq C_1^2 \frac{M}{m^2}.$$

Démonstration: On note $|A|$ la longueur d'un intervalle A. Par le théorème des accroissements finis, il existe, pour tout $0 \leq j < Q$, un point $y \in I_x$ tel que $|f^j(I_x)| = Df^j(y)|I_x|$.

Par le lemme 2, on obtient:

$$Df^j(x) \leq C_1 \, Df^j(y) = C_1 \frac{|f^j(I_x)|}{|I_x|}.$$

Les intervalles $f^j(I_x)$, $0 \leq j < Q$, étant disjoints, on a $\sum_{j=0}^{Q-1} |f^j(I_x)| \leq 1$; comme $|I_x| \geq m$, $|f^j(I_x)| \leq M$, on en déduit:

$$\sum_{j=0}^{Q-1} [Df^j(x)]^2 \leq C_1^2 \frac{\Sigma |f^j(I_x)|^2}{|I_x|^2} \leq C_1^2 \frac{M}{m^2}. \quad ■$$

1.4. La dérivée schwarzienne Sg d'une application g de classe C^3 est définie par la formule:

$$Sg = \frac{2D^3 g Dg - 3(D^2 g)^2}{2(Dg)^2} = D^2 \operatorname{Log} Dg - \frac{1}{2} (D \operatorname{Log} Dg)^2 \quad \text{si} \quad Dg > 0.$$

Les formules de composition sont:

$$S(g\circ h) = Sg\circ h(Dh)^2 + Sh,$$

$$Sg^i = \sum_{j=0}^{i-1} Sg\circ g^j(Dg^j)^2 \quad \text{pour} \quad i \geq 1.$$

Lemme 4. Pour tout $0 \leq i < Q$, on a:

$$|D \operatorname{Log} Df^i|_o \leq C_3 \frac{M^{1/2}}{m},$$

où $C_3 = C_1 [2|Sf|_o]^{1/2}$.

Démonstration: Soit $0 \leq i < Q$. En un point où $|D \operatorname{Log} Df^i|$ est maximal, $D^2\operatorname{Log} Df^i$ s'annule, donc $|D \operatorname{Log} Df^i|_o^2 \leq 2|Sf^i|_o$. Mais, par le lemme 3, on a:

$$|Sf^i|_o \leq |\sum_{j=0}^{i-1} Sf\circ f^j(Df^j)^2|_o \leq |Sf|_o\, C_1^2 \frac{M}{m^2}. \quad \blacksquare$$

1.5. Démonstration de la proposition

Soit x un point où $m = f^q(x) - x$, donc $Df^q(x) = 1$. Soit $z \in \mathbf{T}^1$, par le lemme 1, z s'écrit $f^j(y)$, avec $0 \leq j < Q$ et $y \in J_x$. On a alors:

$$\begin{aligned}\operatorname{Log} Df^q(z) &= \operatorname{Log} Df^{q+j}(f^{-j}(z)) + \operatorname{Log} Df^{-j}(z)\\ &= \operatorname{Log} Df^{q+j}(y) - \operatorname{Log} Df^j(y)\\ &= \operatorname{Log} Df^j(f^q(y)) - \operatorname{Log} Df^j(y) + \operatorname{Log} Df^q(y).\end{aligned}$$

Par l'inégalité de Denjoy (Lemme 2), on a $0 \leq f^{2q}(x) - f^q(x) \leq$ $\leq C_1 m$, $0 \leq x - f^{-q}(x) \leq C_1 m$; il en résulte $|y-x| \leq C_1 m$ et $0 \leq f^q(y) - y \leq \max(f^{2q}(x)-x,\ f^q(x)-f^{-q}(x)) \leq (C_1+1)m$.

Le lemme 4 donne alors:

$$|\operatorname{Log} Df^q(y)| = |\operatorname{Log} Df^q(y) - \operatorname{Log} Df^q(x)| \leq C_1 C_3 M^{1/2},$$

$$|\operatorname{Log} Df^j(f^q(y)) - \operatorname{Log} Df^j(y)| \leq (C_1+1)\, C_3 M^{1/2};$$

on obtient donc $|\operatorname{Log} Df^q(z)| \leq C_4 M^{1/2} = C_4 |f^q - id|_o^{1/2}$, avec

$C_4 = \sqrt{2}\ |Sf|_o\ (2C_1^2+C_1)$. ■

On suppose que $q > 1$, ce qui implique $\|q\alpha\| = |q\alpha-p|$ et $\|Q\alpha\| = |Q\alpha-P|$.

On utilise dans la suite le résultat suivant:

Lemme 5. Soit $x \in \mathbb{T}^1$; il existe $y \in I_x$, $z \in \mathbb{T}^1$, tels que

$$\frac{f^{-Q}(y) - y}{\|Q\alpha\|} = \frac{f^{q}(z) - z}{\|q\alpha\|} .$$

Démonstration: On note μ l'unique mesure de probabilité sur T^1 invariante par f. Soient $x \in \mathbb{T}^1$, φ une application continue de $\mathbb{T}^1$ dans $\mathbb{R}$ qui relève l'identité de $\mathbb{T}^1$ sur l'intervalle $[x, f^{q-Q}(x)]$. On a:

$$\int_x^{f^q(x)} (f^{-Q}-id)(t)d\mu(t) = \int_x^{f^q(x)} [\varphi(f^{-Q}(t)) - \varphi(t)]\ d\mu(t)$$

$$= \int_{f^{-Q}(x)}^{f^{q-Q}(x)} \varphi(t)d\mu(t) - \int_x^{f^q(x)} \varphi(t)d\mu(t)$$

$$= \int_{f^{q}(x)}^{f^{q-Q}(x)} \varphi(t)d\mu(t) - \int_x^{f^{-Q}(x)} \varphi(t)d\mu(t)$$

$$= \int_x^{f^{-Q}(x)} [\varphi(f^q(t))-\varphi(t)]\ d\mu(t)$$

$$= \int_x^{f^{-Q}(x)} (f^q-id)(t)d\mu(t).$$

D'autre part, des propriétés du nombre de rotation il résulte que $\int_x^{f^q(x)} d\mu(t) = \|q\alpha\|$, $\int_x^{f^{-Q}(x)} d\mu(t) = \|Q\alpha\|$ (cf [H, ch II]).

Le lemme résulte donc du théorème de la moyenne. ■

2. Démonstration du théorème.

2.1. Soit f un difféomorphisme du cercle de classe C^3, préservant l'orientation. On suppose que le nombre de rotation α de f est irrationnel et on note (p_k/q_k) la suite de ses réduites. On note C une constante qui vérifie, pour tout $k \geq 0$, $|\mathrm{Log}\, Df^{q_k}|_o \leq$ $\leq C|f^{q_k} - id|_o^{1/2}$ et $|Df^{\pm q_k} - 1|_o \leq C|f^{q_k} - id|_o^{1/2}$.

<u>Lemme 6</u>. <u>La suite</u> M_k, <u>définie par</u> $M_k = |f^{q_k} - id|_o^{1/2}$, <u>satisfait la relation</u>:

$$M_k^2 - CM_{k-1}^2\, M_k - M_{k-1}^2 \frac{\|q_k\alpha\|}{\|q_{k-1}\,\alpha\|} \leq 0.$$

<u>Démonstration</u>: On suppose par exemple que $p_k/q_k > \alpha$. Soit $x \in \mathbf{T}^1$ tel que $f^{-q_k}(x) - x = M_k^2$; par le lemme 5, il existe $y \in [x, f^{q_{k-1}}(x)]$, $z \in \mathbf{T}^1$ tels que:

$$f^{-q_k}(y) - y = \frac{\|q_k\alpha\|}{\|q_{k-1}\alpha\|}(f^{q_{k-1}}(z) - z) \leq \frac{\|q_k\alpha\|}{\|q_{k-1}\alpha\|} M_{k-1}^2 .$$

D'autre part, vu le choix du point x, on a:

$$f^{-q_k}(y) - y \geq M_k^2 - |y-x|\ |Df^{-q_k} - 1|_o \geq M_k^2 - CM_{k-1}^2\, M_k.$$

Joindre ces deux inégalités donne le lemme 6. ∎

2.2. On pose $\alpha_k = \|q_k\alpha\|^{1/2}$.

<u>Proposition 2</u>. <u>La suite</u> M_k <u>satisfait la relation de récurrence</u>:

$$M_k \leq M_{k-1}\frac{\alpha_k}{\alpha_{k-1}} + C\, M_{k-1}^2.$$

<u>Démonstration</u>: Pour les réels $a,b,x \geq 0$, l'inégalité $x^2 - ax - b \leq 0$ implique $x \leq a + \sqrt{b}$. Cette observation, appliquée au lemme 6, prouve la proposition. ∎

<u>Corollaire</u>. Soit $\lambda > 2^{-1/4}$; il existe $B = B(f,\lambda)$ tel que $M_k \leq B\lambda^k$.

<u>Démonstration</u>: Des propriétés des réduites, on tire, pour $k \geq 0$:

$$\alpha_k^2 = \|q_k\alpha\| \geq \|q_{k+1}\alpha\| + \|q_{k+2}\alpha\| > 2\|q_{k+2}\alpha\| = 2\alpha_{k+2}^2 .$$

En itérant la relation de la proposition 2, on obtient:

$$M_{k+2} \le \frac{\alpha_{k+2}}{\alpha_k} M_k + C\, M_{k+1}^2 + C \frac{\alpha_{k+2}}{\alpha_{k+1}} M_k^2 .$$

Soit $\lambda > 2^{-1/4}$. Comme M_k décroit vers 0 par le théorème de Denjoy, on a $M_{k+2} \le \lambda^2 M_k$ pour k assez grand, ce qui démontre le lemme. ∎

Remarque: En raffinant, on pourrait choisir $\lambda \in \left]\left(\frac{\sqrt{5}-1}{2}\right)^{1/2}, 1\right[$.

2.3. On suppose maintenant que α satisfait une condition diophantienne (C_β), $\beta < 1$. On a donc:

$$\frac{1}{q_{k+1}} > \|q_k \alpha\| \ge \frac{K}{q_k^{1+\beta}} ,$$

d'où $\|q_{k+1}\alpha\| \ge \frac{K}{q_{k+1}^{1+\beta}} > K\|q_k\alpha\|^{1+\beta}$, et donc $\alpha_{k+1} > K^{1/2}\, \alpha_k^{1+\beta}$.

Proposition 3. La suite $\frac{M_k}{\alpha_k}$ est bornée.

Démonstration: Remarquons tout d'abord que $M_k \ge \alpha_k$: c'est une conséquence de $\int_{\mathbb{T}^1} (f^{q_k} - id)\, d\mu = \|q_k \alpha\|$.

Il réstulte de la condition diophantienne qu'on peut choisir $\varepsilon, \gamma \in]0,1[$, fixés dans la suite, tels que $\alpha_{k+1}^{1+\gamma} > \alpha_k^{2-\varepsilon}$ pour k assez grand. Le corollaire implique que le produit $\prod_k (1+M_k^\varepsilon)$ est convergent, et permet d'en calculer un majorant A (>1); on suppose aussi $A^{1-\varepsilon} > 2$.

On choisit k_o assez grand pour que $M_{k_o} < A^{-2}$, $\alpha_{k+1}^{1+\gamma} > \alpha_k^{2-\varepsilon}$ pour $k \ge k_o$. On pose ensuite $a = a_{k_o} = A^{-2}$, $M_{k_o} = a\, \alpha_{k_o}^{\rho_{k_o}}$; comme $\alpha_{k_o} \le M_{k_o} < a < 1$, on a $0 < \rho_{k_o} \le 1$. Nous allons définir a_k, ρ_k, pour $k > k_o$, inductivement par l'une des deux relations de récurrence suivantes:

(R_k) $\quad a_k = a_{k-1}, \quad \rho_k = \inf(1, \rho_{k-1}(1+\gamma))$;

(R'_k) $\quad a_k = a_{k-1}(1 + M_{k-1}^\varepsilon), \quad \rho_k = \rho_{k-1}$.

Remarquons qu'alors les suites (a_k) et (ρ_k) vérifient toujours $A^{-2} \leq a_k < \prod_j (1 + M_j^{\varepsilon})a \leq A^{-1} < 1$, $0 < \rho_k \leq 1$.

Montrons qu'il est possible d'obtenir ainsi $M_k \leq a_k \alpha_k^{\rho_k}$.

Par définition, $M_{k_o} = \alpha_{k_o} \alpha_{k_o}^{\rho_{k_o}}$; supposons que $M_{k-1} \leq a_{k-1} \alpha_{k-1}^{\rho_{k-1}}$:

1) Si $\frac{\alpha_k}{\alpha_{k-1}} < C\, M_{k-1}^{1-\varepsilon}$, en utilisant $a_{k-1} < A^{-1}$, $2C\, A^{\varepsilon-1} < 1$, et la proposition 2, on obtient:

$$M_k \leq \frac{\alpha_k}{\alpha_{k-1}} M_{k-1} + CM_{k-1}^2 \leq 2CM_{k-1}^{2-\varepsilon} \leq 2C(a_{k-1}\,\alpha_{k-1}^{\rho_{k-1}})^{2-\varepsilon}$$

$$\leq a_{k-1}(2Ca_{k-1}^{1-\varepsilon})(\alpha_{k-1}^{2-\varepsilon})^{\rho_{k-1}} \leq a_{k-1}\,\alpha_k^{(1+\gamma)\rho_{k-1}}$$

De plus $M_k \geq \alpha_k \geq a_{k-1}\,\alpha_k$, donc $(1+\gamma)\rho_{k-1} \leq 1$; on définit donc a_k, ρ_k par la relation (R_k).

2) Dans le cas contraire, on obtient l'estimation:

$$M_k \leq \frac{\alpha_k}{\alpha_{k-1}} M_{k-1}(1 + M_{k-1}^{\varepsilon}) \leq a_{k-1}(1 + M_{k-1}^{\varepsilon})\,\alpha_k^{\rho_{k-1}} ,$$

car $\alpha_k \leq \alpha_{k-1}$, $\rho_{k-1} \leq 1$ implique $\frac{\alpha_k}{\alpha_{k-1}}\,\alpha_{k-1}^{\rho_{k-1}} \leq \alpha_k^{\rho_{k-1}}$.

On définit alors a_k, ρ_k par (R_k'), ce qui nous permet encore d'obtenir $M_k \leq a_k\,\alpha_k^{\rho_k}$.

Le cas 1) ne peut se produire qu'un nombre fini de fois, car on a vu qu'il implique:

$$0 < \rho_k = (1+\gamma)\rho_{k-1} \leq 1.$$

On a donc pour $k \geq k_1$, $M_k \leq \frac{\alpha_k}{\alpha_{k-1}} M_{k-1}(1 + M_{k-1}^{\varepsilon})$; ceci implique $M_k \leq (A \frac{M_{k_1}}{\alpha_{k_1}})\,\alpha_k$ et démontre la proposition. ■

2.4. Il est maintenant facile de démontrer le théorème: en joignant les propositions 1 et 3, on obtient l'estimation $|\mathrm{Log}\, Df^{q_j}|_o \leq$ $\leq \bar{C}\, \|q_j\alpha\|^{1/2}$, où la constante $\bar{C}$ dépend de f mais pas de j.

Soit $n \in N$; par division euclidienne, on écrit (cf [H, ch VIII]) $n = \sum_{j=0}^{k} b_j q_j$, $b_j \in N$, $0 \le b_j < \frac{q_{j+1}}{q_j}$. On a alors:

$$|\text{Log } Df^n|_o \le \Sigma\, b_j |\text{Log } Df^{q_j}|_o \le \bar{C}\, \frac{q_{j+1}}{q_j} \|q_j\alpha\|^{1/2}.$$

La relation $q_{j+1}^{-1} \ge \|q_j\alpha\| \ge K q_j^{-(1+\beta)}$, issue de la condition diophantienne, implique:

$$|\text{Log } Df^n|_o \le \frac{\bar{C}}{K^{1/2}} \sum_{j=0}^{k} q_j^{\frac{\beta-1}{2}} \le \frac{\bar{C}}{K^{1/2}} \sum_{j=0}^{+\infty} q_j^{\frac{\beta-1}{2}} ,$$

la convergence de la série étant assurée par $\beta < 1$ et la croissance au moins géométrique de la suite (q_k) $(q_{k+2} \ge 2q_k$ pour $k \ge 0)$.

Par conséquent, la suite des itérés $(f_i)_{i\in\mathbb{N}}$ est bornée dans la C^1-topologie. Il en résulte ([H], ch IV) que f est C^1-conjugué à R_α. ■

3. Différentiabilité Supérieure.

3.1. Les hypothèses sont les mêmes que précédemment: f est un difféomorphisme du cercle de classe C^3, préservant l'orientation, dont le nombre de rotation α est irrationnel et vérifie une condition diophantienne (C_β), $\beta < 1$; f est donc C^1-conjugué à R_α. On note encore (p_k/q_k) la suite des réduites de α.

3.2. On pose $\Delta = \sup_{i\in\mathbb{N}} |Df^i|_o$. On rappelle (cf 1.4) que les dérivées schwarziennes Sf^{q_k} vérifient:

$$Sf^{q_k} = D^2\text{Log } Df^{q_k} - \frac{1}{2}\,(D \text{ Log } Df^{q_k})^2 = \sum_{i=0}^{q_k-1} Sf\circ f^i (Df^i)^2.$$

On en tire:

$$|Sf^{q_k}|_o \le \Delta^2 |Sf|_o\, q_k;$$

$$|D^2\text{Log } Df^{q_k}|_o \le |Sf^{q_k}|_o + \frac{1}{2}\,|D \text{ Log } f^{q_k}|_o^2 \le 2|Sf^{q_k}|_o ,$$

car $D^2\text{Log } Df^{q_k}$ s'annule lorque $|D \text{ Log } Df^{q_k}|$ est maximal (voir aussi [H, ch IX]). On a donc: $|D^2 \text{Log } Df^{q_k}|_0 \le (2\Delta^2 |Sf|_0) q_k$.

3.3. L'inégalité de convexité de Hadamard que nous utilisons s'énonce comme suit ([H, ch VIII]): pour $0 < \varepsilon < 1$, il existe une constante $C_\varepsilon > 0$ telle que tout h de classe C^2 sur $\mathbf{T}^1$ vérifie:

$$|h|_\varepsilon \le C_\varepsilon \; |h|_0^{1-\varepsilon/2} \; |D^2 h|_0^{\varepsilon/2}.$$

On a obtenu en 2.4 $|\text{Log } Df^{q_k}|_0 \le \bar{C} \|q_k \alpha\|^{1/2}$; donc, en prenant $h = \text{Log } Df^{q_k}$ et en utilisant 3.2, on obtient:

$$|\text{Log } Df^{q_k}|_\varepsilon \le \tilde{C} \; q_k^{\varepsilon/2} \; \|q_k \alpha\|^{\frac{2-\varepsilon}{4}},$$

avec $\tilde{C} = C_\varepsilon \; \bar{C}^{1-\varepsilon/2} (2\Delta^2 |Sf|_0)^{\varepsilon/2}$.

3.4. Pour tout $i \ge 0$, f^i est un difféomorphisme du cercle dont le rapport de Lipschitz est majoré par Δ. Par conséquent, pour toute application φ de $\mathbf{T}^1$ dans $\mathbb{R}$, vérifiant une condition de Hölder d'exposant ε, on a:

$$|\varphi \circ f^i|_\varepsilon \le \Delta^\varepsilon |\varphi|_\varepsilon .$$

Soit $n \in \mathbb{N}$; on écrit à nouveau la décomposition $n = \sum_{j=0}^{k} b_j q_j$ $(b_j \in \mathbb{N}, \; 0 \le b_j < \frac{q_{j+1}}{q_j})$. En utilisant la relation précédente avec $\varphi = \text{Log } Df^{q_j}$, ainsi que la propriété arithmétique $q_{j+1}^{-1} \ge \|q_j \alpha\| \ge$ $\ge K \, q_j^{-(1+\beta)}$, on obtient:

$$|\text{Log } Df^n|_\varepsilon \le \tilde{C} \; \Delta^\varepsilon \sum_{j=0}^{k} \frac{q_{j+1}}{q_j} \|q_j \alpha\|^{\frac{2-\varepsilon}{4}} \; q_j^{\varepsilon/2}$$

$$\le \tilde{\tilde{C}} \sum_{j=0}^{k} q_j^{-\delta} ,$$

avec $\tilde{\tilde{C}} = \tilde{C} \; \Delta^\varepsilon \; K^{-\frac{2+\varepsilon}{4}}$, $\delta = \frac{1-\beta}{2} - \frac{3+\beta}{4} \varepsilon$.

Donc, pour tout $\varepsilon < \frac{2-2\beta}{3+\beta}$, δ est strictement positif, et la suite des itérés $(f^n)_{n \in \mathbb{N}}$ est bornée dans la $C^{1+\varepsilon}$-topologie. Il en résulte alors (cf [H, ch IV]) que f est $C^{1+\varepsilon}$-conjugué à R_α.

Soit $g_n = \frac{1}{n} \sum_{i=0}^{n-1} (f^i - i\alpha)$; on utilise [H, ch VII. 2.7.5] pour conclure que la suite $g_n \circ f \circ g_n^{-1} = f_n$ converge vers R_α dans la $C^{2+\varepsilon'}$-topologie, pour tout $\varepsilon' < \frac{2-2\beta}{3+\beta}$.

3.5. On suppose maintenant que f est de classe C^∞, et que $\beta > 0$ vérifie de plus $\beta^2 + 4\beta - 1 < 0 \Leftrightarrow \beta < \sqrt{5} - 2$. Sous cette hypothèse, $2\beta < \frac{2-2\beta}{3+\beta}$; donc f est C^∞-conjugué à des difféomorphismes f_n arbitrairement proches de R_α dans la $C^{2+2\beta}$-topologie; par le théorème local de Arnold et Moser ([H, Appendice]), f_n est C^∞-conjugué à R_α pour n assez grand. Il en est donc de même pour f. ∎

BIBLIOGRAPHIE

[A] V.I. ARNOLD, Small denominators I; on the mappings of a circle into itself, Izvestijia Akad. Nauk., serie Math., 25, 1 (1961), p. 21-86 = Translations Amer. Math. Soc., 2nd series, 46, p. 213-284.

[H] M.R. HERMAN, Sur la conjugaison différentiable des difféomorphismes du cercle à des rotations, Publ. Math. I.H.E.S., 49 (1979), p. 5-234.

[L] S. LANG, Introduction to diophantine approximation, New York, Addison-Wesley (1966).

J. C. YOCCOZ

Instituto de Matemática Pura e Aplicada - IMPA